Richard C. Larock

Solvomercuration / Demercuration Reactions in Organic Synthesis

Springer-Verlag
Berlin Heidelberg New York
London Paris Tokyo

Professor Richard C. Larock
Department of Chemistry
Iowa State University
Ames, IA 50011 USA

ISBN 3-540-15094-3 Springer-Verlag Berlin Heidelberg New York
ISBN 0-387-15094-3 Springer-Verlag New York Heidelberg Berlin

Library of Congress Cataloging in Publication Data
Larock, R. C. (Richard C.), 1944. Solvomercuration/demercuration reactions in organic
synthesis. Includes bibliographies and index. 1. Chemistry, Organic—Synthesis. 2. Mer-
curation. I. Title. II. Series.
QD262.L364   1985   547'.2   85-12601
ISBN 0-387-15094-3 (U.S.)

Printing: Mercedes-Druck, Berlin. Bookbinding: Lüderitz & Bauer, Berlin.

2152/3020-543210

To the Significant Others in my Life

Preface

The widespread utilization of organometallic reagents in organic synthesis has been one of the major developments in organic chemistry in recent years. The earlier Springer monograph entitled "Organic Synthesis by Means of Transition Metal Complexes" covered many of the more recent advances in this area. The present volume covers applications of the very valuable solvomercuration-demercuration sequence in organic synthesis.

Organomercurials have played an important role in organic chemistry for over a century. They have been known since 1852 and major monographs covering their chemistry have appeared in 1921 ("Organic Compounds of Mercury" by F. C. Whitmore), 1967 ("The Organic Compounds of Mercury", volume 4 in the series "Methods of Elemento-Organic Chemistry", edited by L. G. Makarova and A. N. Nesmeyanov) and 1974 ("Metallorganische Verbindungen-Hg", Houben-Weyl: Methoden der Organischen Chemie, 4th ed., Vol. 13/2b, edited by H. Staub, K. P. Zeller and H. Leditsche).

My own personal interest in the application of organomercurials in organic synthesis goes back to the late 1960's when I first became involved in the development of new synthetic routes to organomercurials. In 1976 I wrote a chapter entitled "Organomercurials as Reagents and Intermediates in Organic Synthesis" for a book entitled "New Applications of Organometallic Reagents in Organic Synthesis". That chapter was subsequently pared down to a brief review article in Angewandte Chemie which appeared in 1978 and a further article in Tetrahedron Reports in 1982. The response to those articles encouraged me to undertake the writing of the recent Springer Verlag monograph "Organomercury Compounds in Organic Synthesis". Unfortunately, that volume did not provide sufficient space to cover the solvomercuration-demercuration sequence in the detail that I believe organometallic and synthetic organic chemists would find desirable. For that reason, this separate volume devoted entirely to synthetic applications of the solvomercuration-demercuration reaction has now been written. The references cited in this monograph cover the chemical literature through mid 1983.

I wish to express my appreciation to my own graduate students for both their patience and their scientific and technical assistance in the writing of this manuscript. In particular, I wish to thank Drs. Douglas Leach and Constance Fellows for assistance in translating the French and Russian literature, respectively. Finally, this volume would not have been possible without the outstanding professional assistance of several very fine secretaries, including Mrs. Kathie Hawbaker, Mrs. Elaine Wedeking, Ms. Colleen Rahfeldt and most particularly Mrs. Denise Junod whose help was invaluable. I would finally like to thank those closest to me for having put up with this project for far longer than I ever imagined it would take.

Ames, Iowa U.S.A.   Richard C. Larock
January 1986

Table of Contents

List of Tables

Abbreviations

The following abbreviations have been used in this book.

Ac	acetyl
n-Am	*n*-amyl
Ar	aryl
n-Bu	*n*-butyl
i-Bu	isobutyl
sec-Bu	*sec*-butyl
t-Bu	*tert*-butyl
Bz	benzoyl
CIDNP	chemically induced dynamic nuclear polarization
d	day(s)
DBU	1,5-diazabicyclo[5.4.0]undecene-5
DMAP	4-dimethylaminopyridine
DME	1,2-dimethoxyethane
DMF	*N,N*-dimethylformamide
DMSO	dimethylsulfoxide
2,4-DNP	2,4-dinitrophenylhydrazone
ee	enantiomeric excess
Et	ethyl
c-Hex	cyclohexyl
n-Hex	*n*-hexyl
HMPA	hexamethylphosphoramide
hr	hour(s)
Me	methyl
min	minute(s)
NBS	*N*-bromosuccinimide
NMR	nuclear magnetic resonance
Ph	phenyl
PNB	*p*-nitrobenzoyl
n-Pr	*n*-propyl
i-Pr	isopropyl
R	alkyl
RT	room temperature
sec	second(s)
SLS	sodium lauryl sulfate
THF	tetrahydrofuran
THP	tetrahydropyranyl
Ts	*p*-toluenesulfonyl

I. Introduction

The reaction of an olefin or acetylene with an electrophilic mercury salt and appropriate solvent or other nucleophile affords adducts in which the mercury moiety and solvent or nucleophile have added across the carbon—carbon unsaturation. This reaction, most commonly called solvomercuration today, has been known since the turn of the century. With the relatively recent development of convenient methods for replacing the mercury moiety by hydrogen or other substituents (demercuration), this reaction has become an extremely valuable tool for the functionalization of olefins and acetylenes. While several short reviews of this subject have appeared [1–6], the emphasis has been on the mechanism of this reaction due to its controversial nature. With the increasing utility of this reaction in organic synthesis, it seems appropriate that this subject be reviewed in depth from a synthetic stand-point. This book will attempt to do that, starting first with the hydroxymercuration (oxymercuration) of alkenes and alkynes. Subsequent chapters will discuss the alkoxy-, peroxy- and acyloxymercuration reactions. The introduction of nitrogen functionality through amino-, amido-, azido- and nitro-mercuration approaches follows. The book concludes with chapters on carbo-, halo- and several miscellaneous mercuration reactions. Each chapter also contains a brief discussion of the various demercuration procedures and other subsequent reactions which have proven important for each type of organo-mercurial.

A number of related mercuration reactions of alkenes and alkynes have been discussed in the book "Organomercury Compounds in Organic Synthesis". For example, Sections I and J in Chapter II of that book cover the preparation of allyl-, vinyl- and certain carbonyl-containing mercurials by mercuration of alkenes and alkynes. In Chapter IV, the oxidation of alkenes by mercury(II) salts was discussed. The use of solvomercuration products for free radical addition to alkenes is another major development of late which was discussed in Chapter VII. Finally, Chapter VIII indicates how solvomercuration can be combined with carbonylation to prepare carbonyl compounds. Let us now turn to new ways in which the mercuration of alkenes and alkynes can be used to advantage in organic synthesis.

References

1. Wright, G. F.: Ann. N.Y. Acad. Sci., *65*, 436 (1957).
2. Chatt, J.: Chem. Rev., *48*, 7 (1951).
3. Zefirov, N. S.: Usp. Khim., *34*, 1272 (1965); Russ. Chem. Rev., *34*, 527 (1965).
4. Kitching, W.: Organometal. Chem. Rev. A, *3*, 61 (1968).
5. Fahey, R. C.: Top. Stereochem., *3*, 237 (1968).
'6. Chandra, G., Muthana, M. S. and Devaprabhakara, D.: J. Sci. Ind. Res., *30*, 333 (1971).

II. Hydroxymercuration

A. Alkenes

The reaction of a wide variety of alkenes and electrophilic mercury salts in water affords a very general approach to β-hydroxyalkylmercurials (Eq. 1). Although this reaction is commonly referred to as oxymercuration, we shall henceforth call it hydroxymercuration to help distinguish it from other reactions which also introduce oxygen functionality on to an alkene. This reaction has been briefly reviewed several times [1–6].

Kucherov in 1892 appears to have been the first to observe the reaction of aqueous solutions of mercury(II) salts and olefins [7]. Deniges also studied these reactions, before Hofmann and Sand in 1900 [8–10] correctly identified the products and subsequently developed the reaction. This reaction has been the most widely studied of all the solvomercuration reactions and today hydroxymercuration — demercuration is one of the most important methods for hydration of the carbon—carbon double bond.

$$\text{>}C=C\text{<} \; + \; HgX_2 \; + \; H_2O \; \longrightarrow \; \underset{\underset{HgX}{|}}{\overset{\overset{HO}{|}}{-}}C\text{-}C\text{-} \; + \; HX \tag{1}$$

The numerous examples of the hydroxymercuration of simple alkenes known to date have been summarized in Tables 2.1–2.4. Table 2.1 covers examples of the hydroxymercuration of alkenes containing only carbon and hydrogen in which the carbon atoms involved in the double bond are not also involved in a ring. The entries are listed according to increasing carbon and, subsequently, hydrogen content of the alkene in question. Where the hydroxymercuration of isomeric alkenes has been reported, the following priorities have been employed: $H_2C=CHR > H_2C=CR_2 > cis$-$RCH=CHR$ $> trans$-$RCH=CHR > RCH=CR_2 > R_2C=CR_2$. Beyond that, the entries are ordered according to the mercury salt utilized, with reactions employing D_2O entered last. Subsequent reactions carried out on the organomercurial (such as reduction or halogenation), the products resulting, and the yields are also given for each entry. Where the structures reported in the literature seem doubtful or incomplete information is available regarding the structure, a question mark has been placed after the structure in question. Actual yields are reported in parentheses, while ratios are designated by a colon.

Table 2.2 is identical to the first table except that the alkenes covered here contain atoms other than just carbon and hydrogen. They are organized

Table 2.1. Hydroxymercuration of Simple Acyclic Alkenes

Alkene	Mercuric salt	Reaction conditions	Organomercurial(s) (% Yield)[a]	Subsequent reactants	Product(s) (% Yield)[a]	Ref.
$H_2C{=}CH_2$	$Hg(OAc)_2$	—	$AcOHgCH_2CH_2OH$	—	—	11
	$Hg(OAc)_2$	KBr	$BrHgCH_2CH_2OH + (BrHgCH_2CH_2)_2O$	—	—	11
	$Hg(O_2CR)_2$	—	$RCO_2HgCH_2CH_2OH$ R = Et, n-Pr, $ClCH_2$, Cl_2CH $+$ $(RCO_2HgCH_2CH_2)_2O$ R = $ClCH_2$, Cl_2CH	—	—	11, 12
	HgX_2 X = NO_3, OAc	H_2O, X^-	$XHgCH_2CH_2OH$ X = Br, I, NO_3	I_2 or I_2/KI (on RHgI)	ICH_2CH_2OH (60)	13–16
	$HgSO_4$ or HgX_2 (X = OAc, NO_3, Cl)	H_2O, X^-	$XHgCH_2CH_2OH + (XHgCH_2CH_2)_2O$ X = Cl, Br X = Cl, Br	—	—	17
	$Hg(NO_3)_2$	H_2O/HOAc, Cl^-	$ClHgCH_2CH_2OH$	—	—	18
	$Hg(NO_3)_2$	KOH pH 1–4	$NO_3HgCH_2CH_2OH$	—	—	19
	$Hg(NO_3)_2$	D_2O	$NO_3HgCH_2CH_2OD$	—	—	20
	$HgSO_4$	H_2O	—	I_2	$(ICH_2CH_2)_2O$ (100)	13
	$HgSO_4$	H_2O, KI	$(IHgCH_2CH_2)_2O$	I_2/KI	$(ICH_2CH_2)_2O$ (90)	16
	$HgSO_4$	H_2O/H_2SO_4, KBr	$BrHgCH_2CH_2OH$	—	—	21
	$Hg(ClO_4)_2$	1 NaOH / 0.01 N $HClO_4$, NaCl	$ClHgCH_2CH_2OH$ (80)	—	—	22
	$Hg(ClO_4)_2$	D_2O	$ClO_4HgCH_2CH_2OD$	—	—	20
	HgO	RCO_2H R = Me, Et, n-Pr	$RCO_2HgCH_2CH_2OH$	—	—	19

Table 2.1. (continued)

Alkene	Mercuric salt	Reaction conditions	Organomercurial(s) (% Yield)[a]	Subsequent reactants	Product(s) (% Yield)[a]	Ref.
	HgO	RCO_2H $R = ClCH_2, Cl_2CH, Cl_3C$	$RCO_2HgCH_2CH_2OH$ $+ (RCO_2HgCH_2CH_2)_2O$	—	—	19
	HgO	$H_2O / HOAc / P_2O_5$	$PO_3HgCH_2CH_2OH$ (?)	—	—	23
	—	H_2O, Br^-	$BrHgCH_2CH_2OH$	$Na(Hg) / H_2O$	CH_3CH_2OH	12, 13
	—	—	$(BrHgCH_2CH_2)_2O$	—	—	12
$H_2C{=}CHCH_3$	$Hg(OAc)_2$	pH 4	$AcOHgCH_2CHOHCH_3$	—	—	24
	$Hg(OAc)_2$	H_2O, NaX	$XHgCH_2CHOHCH_3$ X = Cl (68) Br (75) I (50)	—	—	25
	$Hg(O_2CC_2H_5)_2$	H_2O, KBr	$BrHgCH_2CHOHCH_3$	—	—	26
	$HgSO_4$, $Hg(OAc)_2$, $Hg(NO_3)_2$	H_2O, X^-	$XHgCH_2CHOHCH_3$ X = Cl, Br, I (90)	—	—	9
	$Hg(NO_3)_2$	H_2O	$NO_3HgCH_2CHOHCH_3$	$HNO_3 / NaNO_3$ 50°C	CH_3COCH_3	27
	$HgSO_4$	H_2O 1hr 60–65 °C	—	KOH 100 °C	$CH_3COCH_3 + CH_3CH{-}CH_2$ (51 total) 94 : 6	25
	$Hg(ClO_4)_2$	H_2O 0 °C	$ClO_4HgCH_2CHOHCH_3$ (100)	—	—	28, 29
	$Hg(ClO_4)_2$	D_2O	$ClO_4HgCH_2CHODCH_3$	—	—	20
	—	—	$BrHgCH_2CHOHCH_3$	—	—	12
	—	H_2O	$XHgCH_2CHOHCH_3$ X = ? $+ [BrHgCH_2CH(CH_3)]_2O$	—	—	30
	—	—	$IHgCH_2CHOHCH_3$	—	—	31

Table 2.1. (continued)

Alkene	Mercuric salt	Reaction conditions	Organomercurial(s) (% Yield)[a]	Subsequent reactants	Product(s) (% Yield)[a]	Ref.
$H_2C=CHCH_2CH_3$	$Hg(NO_3)_2$	H_2O	$NO_3HgCH_2CHOHCH_2CH_3$	$HNO_3/NaNO_3$ 50 °C	$CH_3COCH_2CH_3$	27
	$HgSO_4$	H_2O 1hr 60–65 °C	—	KOH 100 °C	$CH_3COCH_2CH_3$ + $CH_3CH_2CH\overset{O}{\frown}CH_2$ (30 total) 93 : 7	25
	—	H_2O, I^-	$IHgCH_2CHOHCH_2CH_3$	—	—	32
	—	—	$XHgCH_2CHOHCH_2CH_3$ X = Cl, OAc, $O_2CC_2H_5$, $O_2C(CH_2)_2CH_3$, $O_2C(CH_2)_{16}CH_3$	—	—	33
$H_2C=C(CH_3)_2$	$HgCl_2$, $Hg(OAc)_2$, $Hg(NO_3)_2$	H_2O, X^-	$XHgCH_2COH(CH_3)_2$ X = Cl, Br	—	—	9
	$Hg(NO_3)_2$	H_2O	$NO_3HgCH_2COH(CH_3)_2$	$HNO_3/NaNO_3$ 50 °C	$(CH_3)_3COH$	27
	$HgSO_4$	H_2O 1hr 60–65 °C	—	KOH 100 °C	$(CH_3)_3COH$ + $(CH_3)_2C\overset{O}{\frown}CH_2$ (63 total) 68 : 32	25
	—	—	$XHgCH_2COH(CH_3)_2$ X = Cl, OAc, $O_2CC_2H_5$, $O_2C(CH_2)_2CH_3$	—	—	33
cis-$CH_3CH=CHCH_3$	$Hg(OAc)_2$	H_2O 1.5hr 25 °C	threo-$CH_3CH(HgOAc)CHOHCH_3$	$NaBD_4$	$CH_3CHDCHOHCH_3$ 1:1 erythro/threo	34
	$Hg(OAc)_2$	H_2O, NaCl	threo-$CH_3CH(HgCl)CHOHCH_3$	—	—	35
	$Hg(NO_3)_2$	H_2O	$CH_3CH(HgNO_3)CHOHCH_3$	$HNO_3/NaNO_3$ 50 °C	$CH_3COCH_2CH_3$	27
	$Hg(NO_3)_2$	H_2O/HNO_3 0 °C	$CH_3CH(HgNO_3)CHOHCH_3$	—	$CH_3COCH_2CH_3$	36
	$Hg(NO_3)_2$	D_2O/DNO_3 0 °C	$CH_3CH(HgNO_3)CHODCH_3$	—	$CH_3COCHDCH_3$	36
	$HgSO_4$	H_2O 1hr 60–65 °C	—	KOH 100 °C	$CH_3COCH_2CH_3$ + $CH_3CH\overset{O}{\frown}CHCH_3$ (62 total) 56 : 44	25

Table 2.1. (continued)

Alkene	Mercuric salt	Reaction conditions	Organomercurial(s) (% Yield)[a]	Subsequent reactants	Product(s) (% Yield)[a]	Ref.
	$Hg(ClO_4)_2$	H_2O 0°C	threo-$CH_3CH(HgClO_4)CHOHCH_3$	—	$CH_3COCH_2CH_3$	28, 37
	$Hg(ClO_4)_2$	1 NaOH / 0.01 N $HClO_4$, NaCl	threo-$CH_3CH(HgCl)CHOHCH_3$ (50)	—	—	22
	—	—	threo-$CH_3CH(HgI)CHOHCH_3$ (53)	—	—	38
trans-$CH_3CH{=}CHCH_3$	$Hg(OAc)_2$	H_2O 1.5 hr 25°C	erythro-$CH_3CH(HgOAc)CHOHCH_3$	$NaBD_4$	$CH_3CHDCHOHCH_3$ 1:1 erythro/threo	34
	$Hg(OAc)_2$	H_2O, NaCl	erythro-$CH_3CH(HgCl)CHOHCH_3$	—	—	35
	$Hg(NO_3)_2$	H_2O	$CH_3CH(HgNO_3)CHOHCH_3$	HNO_3 / $NaNO_3$ 50°C	$CH_3COCH_2CH_3$	27
	$HgSO_4$	H_2O 1hr 60–65°C	—	KOH 100°C	$CH_3CH{-}CHCH_3$ (epoxide) + $CH_3COCH_2CH_3$ 64 : 36 (72 total)	25
	$Hg(ClO_4)_2$	H_2O 0°C	erythro-$CH_3CH(HgClO_4)CHOHCH_3$	—	$CH_3COCH_2CH_3$	28, 37
	$Hg(ClO_4)_2$	1 NaOH / 0.01 N $HClO_4$, NaCl	erythro-$CH_3CH(HgCl)CHOHCH_3$ (60)	—	—	22
	—	—	erythro-$CH_3CH(HgI)CHOHCH_3$ (57)	—	—	38
$H_2C{=}CH(CH_2)_2CH_3$	$Hg(OAc)_2$	1:1 H_2O/THF RT 1hr	—	$NaBH_4$ / NaOH	$CH_3CHOH(CH_2)_2CH_3$ (97)	39
$H_2C{=}C(CH_3)CH_2CH_3$	$Hg(OAc)_2$	1:1 H_2O/THF 5–60 min RT	—	$NaBH_4$ / NaOH	$(CH_3)_2COHCH_2CH_3$ (90, 92)	39, 40
	$Hg(O_2CCF_3)_2$	1:1 H_2O/THF 15 min	—	$NaBH_4$ / NaOH	$(CH_3)_2COHCH_2CH_3$ (88)	41
	$Hg(NO_3)_2$	1:1 H_2O/THF 15 min	—	$NaBH_4$ / NaOH	$(CH_3)_2COHCH_2CH_3$ (48)	41
	$Hg(O_3SCH_3)_2$	1:1 H_2O/THF 15 min	—	$NaBH_4$ / NaOH	$(CH_3)_2COHCH_2CH_3$ (31)	41

Table 2.1. (continued)

Alkene	Mercuric salt	Reaction conditions	Organomercurial(s) (% Yield)[a]	Subsequent reactants	Product(s) (% Yield)[a]	Ref.
cis-$CH_3CH{=}CHCH_2CH_3$	$Hg(OAc)_2$	1:1 H_2O/THF 10–15 min RT	—	$NaBH_4$/$NaOH$	$CH_3CHOH(CH_2)_2CH_3$ + $CH_3CH_2CHOHCH_2CH_3$ 64 : 36 (93–98 total)	39, 40
$trans$-$CH_3CH{=}CHCH_2CH_3$	$Hg(OAc)_2$	1:1 H_2O/THF 15 min	—	$NaBH_4$/$NaOH$	$CH_3CHOH(CH_2)_2CH_3$ + $CH_3CH_2CHOHCH_2CH_3$ 54–56 : 44–46 (91–95 total)	39, 40
$CH_3CH{=}C(CH_3)_2$	$Hg(OAc)_2$	1:1 H_2O/THF	—	$NaBH_4$/$NaOH$	$CH_3CH_2COH(CH_3)_2$ (95–98)	42
	$Hg(OAc)_2$	1:1 H_2O/THF 10–15 min RT	—	$NaBH_4$/$NaOH$	$CH_3CH_2COH(CH_3)_2$ (94, 95)	39, 40
$H_2C{=}CH$ ⊲ CH_3	$Hg(OAc)_2$	1:1 H_2O/THF 30 min	—	$NaBH_4$/$NaOH$	CH_3CHOH ⊲ CH_3 (83)	43
$H_2C{=}CH(CH_2)_3CH_3$	$Hg(OAc)_2$	1:1 H_2O/THF RT 0.1–1 hr	—	$NaBH_4$/$NaOH$	$CH_3CHOH(CH_2)_3CH_3$ (95, 96) + $HO(CH_2)_5CH_3$ (0.5)	39, 40
	$Hg(OAc)_2$	H_2O/THF	—	cat. Li_2PdCl_4/$3\ CuCl_2$	$CH_3CO(CH_2)_3CH_3$ (84)	44
	$Hg(O_2CCF_3)_2$	1:1 H_2O/THF 60 min	—	$NaBH_4$/$NaOH$	$CH_3CHOH(CH_2)_3CH_3$ (93)	41
	$Hg(II)$-(+)-lactate	H_2O/THF 0.5 hr	—	$NaBH_4$/$NaOH$	$(-)$-$CH_3CHOH(CH_2)_3CH_3$ (44) 2.5 % enantiomeric excess	45
	$Hg(NO_3)_2$	1:1 H_2O/THF 60 min	—	$NaBH_4$/$NaOH$	$CH_3CHOH(CH_2)_3CH_3$ (93)	41
	$Hg(O_3SCH_3)_2$	1:1 H_2O/THF 15 min	—	$NaBH_4$/$NaOH$	$CH_3CHOH(CH_2)_3CH_3$ (90)	41
$H_2C{=}CHC(CH_3)_3$	$Hg(OAc)_2$	1:1 H_2O/THF RT 20–60 min	—	$NaBH_4$/$NaOH$	$CH_3CHOHC(CH_3)_3$ (94) + $HOCH_2CH_2C(CH_3)_3$ (3)	39, 40
	0.2 $Hg(O_2CC_2H_5)_2$	H_2O/CH_3COCH_3 H_2SO_4 CrO_3	—	—	$CH_3COC(CH_3)_3$ (86)	46
	$Hg(O_2CCF_3)_2$	1:1 H_2O/THF 15 min	—	$NaBH_4$/$NaOH$	$CH_3CHOHC(CH_3)_3$ (97)	41

Table 2.1. (continued)

Alkene	Mercuric salt	Reaction conditions	Organomercurial(s) (% Yield)[a]	Subsequent reactants	Product(s) (% Yield)[a]	Ref.
	$Hg(NO_3)_2$	1:1 H_2O/THF 60 min	—	$NaBH_4$/NaOH	$CH_3CHOHC(CH_3)_3$ (97)	41
	$Hg(O_3SCH_3)_2$	1:1 H_2O/THF 15 min	—	$NaBH_4$/NaOH	$CH_3CHOHC(CH_3)_3$ (100)	41
$H_2C=C(CH_3)CH_2CH_2CH_3$	$Hg(O_2CCF_3)_2$	1:1 H_2O/THF 15 min	—	$NaBH_4$/NaOH	$(CH_3)_2COH(CH_2)_2CH_3$ (84)	41
	$Hg(NO_3)_2$	1:1 H_2O/THF 15 min	—	$NaBH_4$/NaOH	$(CH_3)_2COH(CH_2)_2CH_3$ (55)	41
	$Hg(O_3SCH_3)_2$	1:1 H_2O/THF 15 min	—	$NaBH_4$/NaOH	$(CH_3)_2COH(CH_2)_2CH_3$ (60)	41
cis-$CH_3CH=CHCH(CH_3)_2$	$Hg(OAc)_2$	1:1 H_2O/THF 1 hr RT	—	$NaBH_4$/NaOH	$CH_3CHOHCH_2CH(CH_3)_2$ 91 : + $CH_3CH_2CHOHCH(CH_3)_2$: 9 (97 total)	39
trans-$CH_3CH=CHCH(CH_3)_2$	$Hg(OAc)_2$	1:1 H_2O/THF 3.5 hr RT	—	$NaBH_4$/NaOH	$CH_3CHOHCH_2CH(CH_3)_2$ 82 : + $CH_3CH_2CHOHCH(CH_3)_2$: 18 (99 total)	39
$(CH_3)_2C=C(CH_3)_2$	$Hg(OAc)_2$	1:1 H_2O/THF 15–35 min RT	—	$NaBH_4$/NaOH	$(CH_3)_2COHCH(CH_3)_2$ (85, 86)	39, 40
	$Hg(O_2CCF_3)_2$	1:1 H_2O/THF 15 min	—	$NaBH_4$/NaOH	$(CH_3)_2COHCH(CH_3)_2$ (9)	41
	$HgSO_4$	H_2O 1 hr 60–65 °C	—	KOH 100 °C	$CH_3COC(CH_3)_3$	25
$H_2C=CH(CH_2)_4CH_3$	$Hg(OAc)_2$	H_2O/THF	—	$NaBH_4$/NaOH	$CH_3CHOH(CH_2)_4CH_3$ (~95–97)	47
	Hg(II)-(+)-tartrate	3:2:3 H_2O/THF/CH_3CN 42 hr	—	$NaBH_4$/NaOH	(−)-$CH_3CHOH(CH_2)_4CH_3$ (43) 13.3 % enantiomeric excess	45
$H_2C=C(CH_3)(CH_2)_3CH_3$	$Hg(OAc)_2$	H_2O/THF	—	—	$(CH_3)_2COH(CH_2)_3CH_3$ (~97–100)	47
$H_2C=C(CH_3)C(CH_3)_3$	$Hg(OAc)_2$	H_2O 1.5 hr	$AcOHgCH_2COH(CH_3)C(CH_3)_3$	—	—	48

Table 2.1. (continued)

Alkene	Mercuric salt	Reaction conditions	Organomercurial(s) (% Yield)[a]	Subsequent reactants	Product(s) (% Yield)[a]	Ref.
cis-$CH_3CH{=}CHC(CH_3)_3$	$Hg(OAc)_2$	1:1 H_2O/THF 3.5 hr RT	—	$NaBH_4$/NaOH	$CH_3CHOHCH_2C(CH_3)_3$ 98 : + $CH_3CH_2CHOHC(CH_3)_3$: 2 (95 total)	39
$trans$-$CH_3CH{=}CHC(CH_3)_3$	$Hg(OAc)_2$	1:1 H_2O/THF 16 hr RT	—	$NaBH_4$/NaOH	$CH_3CHOHCH_2C(CH_3)_3$ 95 : + $CH_3CH_2CHOHC(CH_3)_3$: 5 (91 total)	39
$H_2C{=}CHC_6H_5$	$Hg(OAc)_2$	H_2O/THF 0.5 hr	$AcOHgCH_2CHOHC_6H_5$ (71)	—	—	49
	$Hg(OAc)_2$	1:1 H_2O/THF RT 5–15 min	—	$NaBH_4$/NaOH	$CH_3CHOHC_6H_5$ (91, 96)	39, 40
	$Hg(OAc)_2$	H_2O 24 hr	$AcOHgCH_2CHOHC_6H_5$ (75)	2% Na(Hg)/H_2O, C_6H_5NCO	$C_6H_5CH(CH_3)O_2CNHC_6H_5$ (40)	50
	$Hg(OAc)_2$	H_2O/THF	—	cat. Li_2PdCl_4/3 $CuCl_2$	$CH_3COC_6H_5$ (40)	44
	0.2 $Hg(O_2CC_2H_5)_2$	H_2O/CH_3COCH_3 H_2SO_4 CrO_3	—	—	$CH_3COC_6H_5$	46
	$Hg(O_2CCF_3)_2$	1:1 H_2O/THF	—	$NaBH_4$/NaOH	$CH_3CHOHC_6H_5$ (67)	41
	$Hg(O_2CCHOHCHOHCO_2)(+)$	1:1 H_2O/THF 25°C 4–28d	—	$NaBH_4$/NaOH	$(+)$-$CH_3CHOHC_6H_5$ (4–29) (25.6–32% opt purity)	51
	$Hg(NO_3)_2$	1:1 H_2O/THF 15 min	—	$NaBH_4$/NaOH	$CH_3CHOHC_6H_5$ (82)	41
	$Hg(O_3SCH_3)_2$	1:1 H_2O/THF 15 min	—	$NaBH_4$/NaOH	$CH_3CHOHC_6H_5$ (83)	41
	—	—	$XHgCH_2CHOHC_6H_5$ X = OAc, Cl, Br, I	I_2/KI (on RHgI)	$ICH_2CHOHC_6H_5$	52
	—	—	$XHgCH_2CHOHC_6H_5$ X = CN, CNO	—	—	33

Table 2.1. (continued)

Alkene	Mercuric salt	Reaction conditions	Organomercurial(s) (% Yield)[a]	Subsequent reactants	Product(s) (% Yield)[a]	Ref.
$H_2C=C(\triangleright)_2$	$Hg(OAc)_2$	1:1 H_2O/THF 10 min	—	$NaBH_4$/NaOH	$CH_3COH(\triangleright)_2$ (83)	43
$H_2C=CHCH_2\text{-cyclopentyl}$	$Hg(OAc)_2$	H_2O, Cl^-	$ClHgCH_2CHOHCH_2\text{-cyclopentyl}$	—	—	53
$H_2C=CH(CH_2)_5CH_3$	0.2 $Hg(O_2CC_2H_5)_2$	H_2O/CH_3COCH_3 H_2SO_4 CrO_3	—	—	$CH_3CO(CH_2)_5CH_3$ (82)	46
	Hg(II)-(+)-lactate	H_2O/THF 1 hr	—	$NaBH_4$/NaOH	$(-)-CH_3CHOH(CH_2)_5CH_3$ (63) 4.5% enantiomeric excess	45
	Hg(II)-(+)-tartrate	3:2:3 H_2O/THF/CH_3CN 23 hr	—	$NaBH_4$/NaOH	$(-)-CH_3CHOH(CH_2)_5CH_3$ (58) 15.9% enantiomeric excess	45
$H_2C=C(CH_3)CH_2C(CH_3)_3$	$Hg(OAc)_2$	1:1 H_2O/THF 18–30 min RT	—	$NaBH_4$/NaOH	$(CH_3)_2COHCH_2C(CH_3)_3$ (87, 96)	39, 40
cis-$CH_3CH=CH(CH_2)_4CH_3$	0.2 $Hg(O_2CC_2H_5)_2$	H_2O/dioxane/CF_3CO_2H $Na_2Cr_2O_7$ 18 hr	—	—	$CH_3CO(CH_2)_5CH_3$ + $CH_3CH_2CO(CH_2)_4CH_3$ 64 : 36 (56 total)	46
$trans$-$CH_3CH=CH(CH_2)_4CH_3$	0.2 $Hg(O_2CC_2H_5)_2$	H_2O/dioxane/CF_3CO_2H $Na_2Cr_2O_7$ 18 hr	—	—	$CH_3CO(CH_2)_5CH_3$ + $CH_3CH_2CO(CH_2)_4CH_3$ 63 : 37 (54 total)	46
$(CH_3)_2C=CHC(CH_3)_3$	$Hg(OAc)_2$	1:1 H_2O/THF 1.5 hr 0°C	—	$NaBH_4$/NaOH	$(CH_3)_2COHCH_2C(CH_3)_3$ (86)	39
$H_2C=CHCH_2C_6H_5$	Hg(II)-(+)-lactate	H_2O/THF 0.5 hr	—	$NaBH_4$/NaOH	$(-)-CH_3CHOHCH_2C_6H_5$ (68) 8.2% enantiomeric excess	45
	Hg(II)-(+)-tartrate	3:2:3 H_2O/THF/CH_3CN 28 hr	—	$NaBH_4$/NaOH	$(-)-CH_3CHOHCH_2C_6H_5$ (76) 6.5% enantiomeric excess	45
$H_2C=C(CH_3)C_6H_5$	$Hg(OAc)_2$	1:1 H_2O/THF 10 min RT	—	$NaBH_4$/NaOH	$(CH_3)_2COHC_6H_5$ (95)	39, 40
	$Hg(OAc)_2$	H_2O 12 hr	$AcOHgCH_2COH(CH_3)C_6H_5$ (76)	2% Na(Hg)/H_2O, C_6H_5NCO	$C_6H_5C(CH_3)_2O_2CNHC_6H_5$ (33)	50

Table 2.1. (continued)

Alkene	Mercuric salt	Reaction conditions	Organomercurial(s) (% Yield)[a]	Subsequent reactants	Product(s) (% Yield)[a]	Ref.
cis-$CH_3CH=CHC_6H_5$	$Hg(OAc)_2$	1:1 H_2O/THF 24 hr RT	—	$NaBH_4$/NaOH	$CH_3CHOHCH_2C_6H_5$ + $CH_3CH_2CHOHC_6H_5$ 88 : 12 (77 total)	39
$trans$-$CH_3CH=CHC_6H_5$	$Hg(OAc)_2$	1:1 H_2O/THF 24 hr RT	—	$NaBH_4$/NaOH	$CH_3CH_2CHOHC_6H_5$ + $CH_3CHOHCH_2C_6H_5$ 70 : 30 (55 total) (ratio at 5 min, yield ~5%; ratio gradually approaches that of cis isomer)	39
$H_2C=CH$—(bicyclic structure)	$Hg(OAc)_2$	H_2O	—	$NaBH_4$/NaOH	CH_3CHOH—(bicyclic structure)	54
$H_2C=C(C_6H_5)CH_2CH_3$	$Hg(OAc)_2$	H_2O, KCl	$ClHgCH_2COH(C_6H_5)CH_2CH_3$ (74)	2% Na(Hg)/H_2O	$CH_3COH(C_6H_5)CH_2CH_3$	55
$C_6H_5CH=C(CH_3)_2$	$HgSO_4$	H_2O 2d	—	—	$C_6H_5CHOHCOH(CH_3)_2$ (70)	56
$H_2C=CH(CH_2)_7CH_3$	Hg(II)-(+)-lactate	H_2O/THF 0.7 hr	—	$NaBH_4$/NaOH	(–)-$CH_3CHOH(CH_2)_7CH_3$ (70) 2.3% enantiomeric excess	45
	Hg(II)-(+)-tartrate	3:2:3 H_2O/THF/CH_3CN 47 hr	—	$NaBH_4$/NaOH	(–)-$CH_3CHOH(CH_2)_7CH_3$ (78) 17.4% enantiomeric excess	45
$H_2C=C(C_6H_5)$—(cyclopropyl)	$Hg(OAc)_2$	1:2 H_2O/THF 60 min	—	$NaBH_4$/NaOH	$CH_3COH(C_6H_5)$—(cyclopropyl) (88)	43
$H_2C=CH(CH_2)_8CH_3$	$Hg(OAc)_2$	1:3 H_2O/THF 30 min	—	$NaBH_4$/NaOH	$CH_3CHOH(CH_2)_8CH_3$ (~95–96)	47
	$Hg(OAc)_2$	H_2O/THF	—	cat. Li_2PdCl_4/3 $CuCl_2$	$CH_3CO(CH_2)_8CH_3$ (89)	44
p-$H_2C=C(CH_3)C_6H_4CH(CH_3)_2$	$Hg(OAc)_2$	H_2O	p-$AcOHgCH_2COH(CH_3)C_6H_4CH(CH_3)_2$	—	—	57
$H_2C=CH(CH_2)_9CH_3$	$Hg(OAc)_2$	1:2 H_2O/THF RT 15–70 min	—	$NaBH_4$/NaOH	$CH_3CHOH(CH_2)_9CH_3$ (91)	39, 40
$E+Z$-$C_6H_5C(CH_3)=C(CH_3)$—(cyclopropyl)	$Hg(OAc)_2$	1:1 H_2O/THF	—	$NaBH_4$/NaOH	$C_6H_5CH(CH_3)COH(CH_3)$—(cyclopropyl)	58

Table 2.1. (continued)

Alkene	Mercuric salt	Reaction conditions	Organomercurial(s) (% Yield)[a]	Subsequent reactants	Product(s) (% Yield)[a]	Ref.
$H_2C{=}CHCH_2{-}$ (decalinyl)	$Hg(OAc)_2$	H_2O	$AcOHgCH_2CHOHCH_2{-}$ (decalinyl)	—	—	53
$H_2C{=}C(C_6H_5)_2$	$Hg(OAc)_2$	1:1 H_2O/THF 20°C 1d	$AcOHgCH_2COH(C_6H_5)_2$ (79)	$NaBH_4$/NaOH	$CH_3COH(C_6H_5)_2$	59
	$Hg(OAc)_2$	H_2O	$AcOHgCH_2COH(C_6H_5)_2$ (70)	Na(Hg) [on RHgCl]	$CH_3COH(C_6H_5)_2$	53
	$Hg(OAc)_2$	H_2O 5hr, NaX	$XHgCH_2COH(C_6H_5)_2$ X = OAc (58.5), Cl (95), I (94)	—	—	60
$C_6H_5CH{=}C(CH_3)C_6H_5$	$Hg(OAc)_2$	H_2O/DME ($PhCO_2)_2$ 20-25°C 6d, NaCl	$ClHgCH(C_6H_5)COH(CH_3)C_6H_5$ (28)	N_2H_4/NaOH Δ 7hr	$C_6H_5CH_2COH(CH_3)C_6H_5$ (60)	61
$H_2C{=}CH{-}$ cyclopropyl (C_6H_5, C_6H_5)	$Hg(OAc)_2$	1:1 H_2O/THF RT	—	$NaBH_4$/NaOH	$CH_3CHOH{-}$ + tetrahydrofuranyl 83 : 17 (65 total)	43
	$Hg(OAc)_2$	1:1 H_2O/THF 65°C	—	$NaBH_4$/NaOH	tetrahydrofuranyl + $CH_3CHOH{-}$ 63 : 37 (95 total)	43
$H_2C{=}CH(CH_2)_{15}CH_3$	$Hg(OAc)_2$	1:3 H_2O/THF RT 1hr	—	$NaBH_4$/NaOH	$CH_3CHOH(CH_2)_{15}CH_3$ (93)	39

[a] If there is no entry in the organomercurial column, the yield is based on starting alkene. If there is an organomercurial entry, the yield is based on the organomercurial unless otherwise stated.

Table 2.2. Hydroxymercuration of Heteroatom — Containing Acyclic Alkenes

Alkene	Mercuric salt	Reaction conditions	Organomercurial(s) (% Yield)[a]	Subsequent reactants	Product(s) (% Yield)[a]	Ref.
$HCF=CF_2$	—	H_2O, Cl^-	$ClHgCHFCO_2H$ (87)	—	—	62
trans-$CHD=CHD$	$Hg(NO_3)_2$	$KOH/HOAc/H_2O$, NaCl	*threo*-$ClHgCHDCHDOH$ (74)	—	—	63
$H_2C=C=O$	$HgCl_2$	1:1 H_2O/Et_2O 0°C	$ClHgCH_2CO_2H$ (85–88)	—	—	64
$H_2C=CHCO_2H$	HgO	—	$HOCH_2\overset{\displaystyle Hg-O}{\underset{\displaystyle \quad}{CH-CO}}$?	—	—	65
$H_2C=CHCH_2Cl$	$Hg(OAc)_2$	1:1 H_2O/THF 60 min	—	$NaBH_4/EtOH$	$CH_3CHOHCH_2Cl$ (70) + $H_2C-CHCH_3$ (epoxide) (21) + $H_2C=CHCH_2OH$ (3)	66
$H_2C=CClCH_3$	HgX_2 $X=ClO_4, SO_4, NO_3, BF_4$	H_2O 60°C	—	—	CH_3COCH_3 (89–96)	67
$HDC=CHCH_3$	$Hg(ClO_4)_2$	H_2O 0°C	—	—	DCH_2COCH_3	37
$H_2C=CDCH_3$	$Hg(ClO_4)_2$	H_2O 0°C	—	—	DCH_2COCH_3	37
$H_2C=CHCH_2OH$	$Hg(OAc)_2$	1:1 H_2O/THF 30 min	—	$NaBH_4/NaOH$	$CH_3CHOHCH_2OH$ (97)	68
	$Hg(NO_3)_2$	H_2O, X^-	$XHgCH_2CHOHCH_2OH$? $X=Br$, I	—	—	17
	$Hg(ClO_4)_2$	1 $NaOH/0.01$ N $HClO_4$, NaI	$IHgCH_2CHOHCH_2OH$ (70)	—	—	22
	—	—	$IHgCH_2CHOHCH_2OH$	C_6H_5COCl	$IHgCH_2CH(O_2CC_6H_5)CH_2O_2CC_6H_5$	16
	—	—	$XHgCH_2CHOHCH_2OH$ $X=?$	—	—	13
cis-$HO_2CCH=CHCO_2H$	$Hg(OAc)_2$	H_2O RT	$\underset{\displaystyle AcOHgCHCO_2}{\overset{\displaystyle HOCHCO_2}{\diagdown}}Hg$	$NaOH/H_2S$	$HO_2CCH_2CHOHCO_2H$ (82)	65, 69

Table 2.2. (continued)

Alkene	Mercuric salt	Reaction conditions	Organomercurial(s) (% Yield)^a	Subsequent reactants	Product(s) (% Yield)^a	Ref.
$H_2C=C(CN)OCH_3$	$Hg(OAc)_2$	H_2O	$AcOHgCH_2CO_2CH_3$	—	—	70
$H_2C=CHCH\overset{\displaystyle O}{\diagup\!\!\diagdown}CH_2$	$Hg(OAc)_2$	1:1 H_2O/THF 30 min	—	$NaBH_4$/$NaOH$	$CH_3CHOHCH\overset{\displaystyle O}{\diagup\!\!\diagdown}CH_2$ (56) + $CH_3CH\overset{\displaystyle O}{\diagup\!\!\diagdown}CHCH_2OH$ (5)	66
$H_2C=CHO_2CCH_3$	$Hg(OAc)_2$	H_2O RT, KCl	$ClHgCH_2CHO$ (87, 100)	—	—	71, 72
	$HgCl_2$/HgO	H_2O pH 4–5	$ClHgCH_2CHO$ (95.7)	—	—	73
$trans$-$CH_3CH=CHCO_2H$	$Hg(OAc)_2$	H_2O/HOAc	$CH_3CHOHCH\overset{Hg-O}{\underset{\mid\quad\mid}{-}}CO$?	$NaOH$/H_2S	$CH_3CHOHCH_2CO_2H$	69
	HgO/$Hg(OAc)_2$	H_2O warm	$CH_3CHOHCH\overset{Hg-O}{\underset{\mid\quad\mid}{-}}CO$?	—	—	65
$CH_3CBr=CHCH_3$	HgX_2 X = ClO_4, SO_4, NO_3, BF_4	H_2O 60 °C	—	—	$CH_3COCH_2CH_3$ (50–59)	67
$H_2C=CHCHClCH_3$	$Hg(OAc)_2$	1:1 H_2O/THF 30 min	—	$NaBH_4$/$NaOH$	$CH_3CHOHCHOHCH_3$ (35) + $HOCH_2CH_2CHOHCH_3$ (22) + $H_2C=CHCHOHCH_3$ (14) + $CH_3CH=CHCH_2OH$ (5)	66
$H_2C=CHCH_2CH_2Cl$	$Hg(OAc)_2$	1:1 H_2O/THF 30 min	—	$NaBH_4$/$NaOH$	$CH_3CHOHCH_2CH_2Cl$ (100)	66
$H_2C=C(CH_3)CH_2Cl$	$Hg(OAc)_2$	1:1 H_2O/THF	—	$NaBH_4$/$EtOH$	$(CH_3)_2COHCH_2Cl$ (80)	66
$trans$-$ClCH_2CH=CHCH_3$	$Hg(OAc)_2$	1:1 H_2O/THF 30 min	—	$NaBH_4$/$NaOH$	$CH_3CHOHCHOHCH_3$ (31) + $CH_3CHOHCH_2CH_2OH$ (21) + $H_2C=CHCHOHCH_3$ (9) + $CH_3CH=CHCH_2OH$ (7)	66
$CH_3CCl=CHCH_3$	HgX_2 X = ClO_4, SO_4, NO_3, BF_4	H_2O 60 °C	—	—	$CH_3COCH_2CH_3$ (67–74)	67
cis-$CH_3CD=CHCH_3$	$Hg(ClO_4)_2$	H_2O 0 °C	—	—	$CH_3COCHDCH_3$	37

Table 2.2. (continued)

Alkene	Mercuric salt	Reaction conditions	Organomercurial(s) (% Yield)[a]	Subsequent reactants	Product(s) (% Yield)[a]	Ref.
$trans$-$CH_3CD{=}CHCH_3$	$Hg(ClO_4)_2$	H_2O 0°C	—	—	$CH_3COCHDCH_3$	37
$H_2C{=}CHCH_2OCONH_2$	$Hg(II)$ tartrate	H_2O	$HOCH_2CH(HgO_2CR)CH_2OCONH_2$ (wrong regioisomer ?)	—	—	74
$H_2C{=}CHCH_2NHCONH_2$	$Hg^{197}(OAc)_2$	—	$AcOHg^{197}CH_2CHOHCH_2NHCONH_2$	—	—	75
	$HgCl_2$	H_2O	$ClHgCH_2CHOHCH_2NHCONH_2$	—	—	76
	—	—	$ClHgCH_2CHOHCH_2NHCONH_2$	—	—	77
$H_2C{=}CHOCH_2CH_3$	$Hg(OAc)_2$	H_2O or EtOH, KCl	$ClHgCH_2CHO$ (80 from H_2O)	—	—	78
	$HgO/Hg(OAc)_2$	$H_2O/EtOH$ Δ	$Hg(CH_2CHO)_2$ (87)	—	—	79
	$Hg[C(NO_2)_3]_2$	H_2O	$(NO_2)_3CHgCH_2CHO$ (66)	—	—	80
$H_2C{=}CHCH_2OCH_3$	$Hg(NO_3)_2 \cdot H_2O$	$CH_3OH/KOCH_3$, NaI	$IHgCH_2CHOHCH_2OCH_3$	—	—	32
$H_2C{=}CHCHOHCH_3$	$Hg(OAc)_2$	1:1 H_2O/THF 30 min	—	$NaBH_4/NaOH$	$CH_3CHOHCHOHCH_3$ + $HOCH_2CH_2CHOHCH_3$ 97 : 3 (92 total)	68
$H_2C{=}CHCH_2CH_2OH$	$Hg(OAc)_2$	1:1 H_2O/THF 30 min	—	$NaBH_4/NaOH$	$CH_3CHOHCH_2CH_2OH$ (93)	68
$trans$-$CH_3CH{=}CHCH_2OH$	$Hg(OAc)_2$	1:1 H_2O/THF 30 min	—	$NaBH_4/NaOH$	$CH_3CHOHCH_2CH_2OH$: $CH_3CH_2CHOHCH_2OH$ 94 : 6 (93 total)	68
$H_2C{=}CHCH_2SCH_3$	$Hg(O_2CCF_3)_2$	1:1 H_2O/THF 2 hr	—	$NaBH_4/NaOH$	$CH_3CHOHCH_2SCH_3$ (78)	66
$H_2C{=}CHCH_2NHSO_2CH_3$	1:1 $Hg(OAc)_2/HgCl_2$	H_2O 60°C then 4d RT	$ClHgCH_2CHOHCH_2NHSO_2CH_3$ (66)	—	—	81
$H_2C{=}CH{-}$cyclopropane(Cl,Cl)	$Hg(OAc)_2$	1:1 H_2O/THF	—	$NaBH_4/NaOH$	$CH_3CHOH{-}$cyclopropane(Cl,Cl) (85)	82

Table 2.2. (continued)

Alkene	Mercuric salt	Reaction conditions	Organomercurial(s) (% Yield)[a]	Subsequent reactants	Product(s) (% Yield)[a]	Ref.
$HO_2CCH=CHCH_2CO_2H$	$Hg(OAc)_2$	H_2O 70 °C	$HO_2CCH=CHCH-C=O$? (with Hg—O bridge); + $O=C-CH-CH-CH-C=O$? (with O—Hg, OH, Hg—O bridges)	—	—	83
$Z-HO_2CC(CH_3)=CHCO_2H$	$Hg(OAc)_2$	H_2O 60 °C	structure with CH_3, $HOC-CO_2$, $AcOHgCHCO_2$, Hg ?	—	—	65
$H_2C=C(CN)OCH_2CH_3$	$Hg(OAc)_2$	H_2O	$AcOHgCH_2CO_2CH_2CH_3$ (89)	—	—	70
$H_2C=CHCH_2HgOAc$	$Hg(OAc)_2$	H_2O/THF	—	$NaBH_4/NaOH$	$CH_3CHOHCH_3$ (90)	84
	$Hg(OAc)_2$	$D_2O/THF-d_8$	$AcOHgCH_2CHODCH_2HgOAc$	—	—	84
$cis-CH_3CH=CHHgOAc$	$Hg(OAc)_2$	H_2O/THF 5 min	—	$NaBH_4/NaOH$	$CH_3CHOHCH_3$ (30) + $CH_3CH=CH_2$ (6)	84
	$Hg(OAc)_2$	$D_2O/THF-d_8$	$CH_3CHODCH(HgOAc)_2$	—	—	84
$H_2C=CHCH_2CH\overset{O}{-}CH_2$	$Hg(OAc)_2$	1:1 H_2O/THF	—	$NaBH_4/NaOH$	$CH_3CHOHCH_2CH\overset{O}{-}CH_2$ (96)	66
$CH_3CH=CHCH\overset{O}{-}CH_2$	$Hg(OAc)_2$	1:1 H_2O/THF 30 min	—	$NaBH_4/NaOH$	triols (major) + $CH_3CH=CHCHOHCH_2OH$ (12) + epoxypentanols (<3)	66
$CH_3CH=CHOAc$	$Hg(OAc)_2$	H_2O, KCl	$ClHgCH(CH_3)CHO$ (71)	—	—	72
$H_2C=C(OAc)CH_3$	$Hg(OAc)_2$	H_2O, KCl	$ClHgCH_2COCH_3$ (71.5)	—	—	71
	$Hg(ClO_4)_2$	H_2O	$(ClO_4)HgCH_2COCH_3$	—	—	85
	$Hg(ClO_4)_2$	dil. $HClO_4$, Cl^-	$ClHgCH_2COCH_3$	—	—	86

Table 2.2. (continued)

Alkene	Mercuric salt	Reaction conditions	Organomercurial(s) (% Yield)[a]	Subsequent reactants	Product(s) (% Yield)[a]	Ref.
E-$CH_3CH=C(CH_3)CO_2H$	$Hg(OAc)_2$	H_2O RT 3d	—	$NaBH_4$/$NaOH$	$CH_3CHOHCH(CH_3)CO_2H$ (56) ~1:1 *erythro/threo*	87
	$Hg(OAc)_2$	H_2O RT 3d	—	H_2S/$NaOH$	$CH_3CHOHCH(CH_3)CO_2H$ 95% *erythro*	87
$H_2C=CHCH_2CHBrCH_3$	$Hg(OAc)_2$	1:1 H_2O/THF 15 min	—	$NaBH_4$/$NaOH$	$CH_3CHOHCH_2CHBrCH_3$ (82) + $CH_3CH(OAc)CH_2CHBrCH_3$ (5)	66
$H_2C=CH(CH_2)_3Cl$	$Hg(OAc)_2$	1:1 H_2O/THF 30 min	—	$NaBH_4$/$NaOH$	$CH_3CHOH(CH_2)_3Cl$ (~81) + (tetrahydrofuranyl)—CH_3 (~9) + $H_2C=CH(CH_2)_3OH$ (2) + $CH_3CH(OAc)(CH_2)_3Cl$ (1–2)	66
$H_2C=CHCH_2NHCO_2CH_3$	Hg(II) tartrate	H_2O	$HOCH_2CH(HgO_2CR)CH_2NHCO_2CH_3$ (wrong regioisomer ?)	—	—	74
$H_2C=CHCH_2NHCONHCONH_2$	—	—	$HOHgCH_2CHOHCH_2NHCONHCONH_2$	—	—	88
$H_2C=CHO$-i-C_3H_7	HgO/$Hg(OAc)_2$	H_2O/EtOH Δ	$Hg(CH_2CHO)_2$ (90)	—	—	79
$H_2C=CHCHOHCH_2CH_3$	$Hg(ClO_4)_2$	1 NaOH/0.01 N $HClO_4$, NaI	$IHgCH_2CHOHCHOHCH_2CH_3$ (95)	—	—	22
$H_2C=C(OC_2H_5)CH_3$	HgO/$Hg(OAc)_2$	H_2O/EtOH	$Hg(CH_2COCH_3)_2$ (98)	—	—	79
$H_2C=CHCH_2CH_2SCH_3$	$Hg(OAc)_2$	1:1 H_2O/THF 24 hr	—	$NaBH_4$/$NaOH$	$CH_3CHOH(CH_2)_2SCH_3$ (76) + $HO(CH_2)_4SCH_3$ (9)	66
trans-$CH_3CH=CHCH_2SCH_3$	$Hg(OAc)_2$	1:1 H_2O/THF 24 hr	—	$NaBH_4$/$NaOH$	$H_2C=CHCHOHCH_3$ (25)	66
$H_2C=CHCH_2NHSO_2CH_2CH_3$	1:1 $Hg(OAc)_2$/$HgCl_2$	H_2O 60°C then 4d RT	$ClHgCH_2CHOHCH_2NHSO_2CH_2CH_3$ (90)	—	—	81
	1:1 $Hg(OAc)_2$/$HgBr_2$	H_2O 60°C then 4d RT	$BrHgCH_2CHOHCH_2NHSO_2CH_2CH_3$ (61)	—	—	81

Table 2.2. (continued)

Alkene	Mercuric salt	Reaction conditions	Organomercurial(s) (% Yield)[a]	Subsequent reactants	Product(s) (% Yield)[a]	Ref.
$H_2C{=}CHSi(CH_3)_3$	$Hg(OAc)_2$	1:1 H_2O/THF 5min	—	$NaBH_4$/NaOH	$HOCH_2CH_2Si(CH_3)_3$ (90) + $AcOCH_2CH_2Si(CH_3)_3$ (1) + $(CH_3)_3SiOH$ (6) + CH_3CH_2OH (3)	84
	$Hg(OAc)_2$	H_2O/NaOH, NcCl	$HOCH_2CH(HgCl)Si(CH_3)_3$ (88 crude)	3% Na(Hg)/H_2O	$HOCH_2CH_2Si(CH_3)_3$ (26.1)	89
	$Hg(OAc)_2$	H_2O/NaOH, KI	$HOCH_2CH(HgI)Si(CH_3)_3$ (60, 62)	I_2	$HOCH_2CHISi(CH_3)_3$ (25)	89
$H_2C{=}CH(CH_2)_3CN$	$Hg(NO_3)_2$	THF RT 1hr	—	$NaBH_4$/NaOH	$CH_3CHOH(CH_2)_3CN$ (65) + $H_2C{=}CH(CH_2)_3CONH_2$ (5)	90
$H_2C{=}C(CH_3)(CH_2)_2CN$	$Hg(NO_3)_2$	1:1 H_2O/THF RT 1hr	—	$NaBH_4$/NaOH	[lactone] =O (50) + $H_2C{=}C(CH_3)(CH_2)_2CONH_2$ (50)	90
$H_2C{=}CHCH_2NH{-}$[triazine ring, NH_2]	—	—	$XHgCH_2CHOHCH_2NH{-}$[triazine ring, NH_2] X = OAc, Cl	—	—	91
$H_2C{=}C(CN)OCH_2CH_2CH_3$	$Hg(OAc)_2$	H_2O	$AcOHgCH_2CO_2CH_2CH_2CH_3$ (95)	—	—	70
$H_2C{=}CH(CH_2)_2CH{-}CH$[epoxide]	$Hg(OAc)_2$	1:1 H_2O/THF 25 °C	—	$NaBH_4$/NaOH	[tetrahydrofuran, CH_3, CH_2OH] + [tetrahydropyran, HO, CH_3] 60 : 40 (65 total)	92
	$Hg(OAc)_2$	1:1 H_2O/THF	—	$NaBH_4$/NaOH	[tetrahydropyran, OH, CH_3] ? (~27) + [tetrahydrofuran, CH_3, CH_2OH] (23) + [tetrahydropyran, OH, CH_3] ? (21) + [tetrahydrofuran, CH_3, CH_2OH] (~10) + $CH_3CHOH(CH_2)_2CH{-}CH_2$[epoxide] (5)	66

Table 2.2. (continued)

Alkene	Mercuric salt	Reaction conditions	Organomercurial(s) (% Yield)[a]	Subsequent reactants	Product(s) (% Yield)[a]	Ref.
$CH_3CH_2CH=CHOAc$	$Hg(OAc)_2$	H_2O, KCl	$ClHgCH(C_2H_5)CHO$ (92)	—	—	72
$BrCH=C(OC_2H_5)_2$	$Hg(OAc)_2$	H_2O/EtOH	$Hg(CHBrCO_2C_2H_5)_2$ (17)	—	—	93
	$Hg(OAc)_2$	Et_2O 24 hr	$(AcOHg)_2CBrCO_2C_2H_5$	—	—	93
$H_2C=CH(CH_2)_4Cl$	$Hg(OAc)_2$	1:1 H_2O/THF 30 min	—	$NaBH_4$/NaOH	$CH_3CHOH(CH_2)_4Cl$ (91) + $CH_3CHOAc(CH_2)_4Cl$ (3)	66
$ClCH=C(OC_2H_5)_2$	$Hg(OAc)_2$	H_2O	$Hg(CHClCO_2C_2H_5)_2$ (7.5)	—	—	93
	$Hg(OAc)_2$	Et_2O 24 hr	$(AcOHg)_2CClCO_2C_2H_5$ (90)	—	—	93
$H_2C=CHCH_2NHCO_2C_2H_5$	$Hg(OAc)_2$	H_2O	$AcOHgCH_2CHOHCH_2NHCO_2C_2H_5$	—	—	94
	HgX_2	H_2O	$HOCH_2CH(HgX)CH_2NHCO_2C_2H_5$ (wrong regioisomer ?) X = NO_3, propionate, benzoate, tartrate, nicotinate	—	—	74
$H_2C=CHCH_2NHCOCH_2NHCONH_2$	$Hg(OAc)_2$	H_2O	$AcOHgCH_2CHOHCH_2NHCOCH_2NHCONH_2$	—	—	94
	1:1 $Hg(OAc)_2$/$HgCl_2$	H_2O 12 hr RT	$ClHgCH_2CHOHCH_2NHCOCH_2NHCONH_2$ (42)	—	—	81
$H_2C=CHO(CH_2)_3CH_3$	HgO/$Hg(OAc)_2$	H_2O/EtOH Δ	$Hg(CH_2CHO)_2$ (86)	—	—	79
	$Hg(OAc)_2$	H_2O, KCl	$ClHgCH_2CHO$ (85)	—	—	78
	$Hg(OAc)_2$	H_2O, KBr	$BrHgCH_2CHO$ (83)	Br_2	$BrCH_2CHO$ (45)	78
$H_2C=C(OC_2H_5)CH_2CH_3$	HgO/$Hg(OAc)_2$	H_2O/EtOH	$Hg(CH_2COCH_2CH_3)_2$ (100)	—	—	79
$H_2C=C(OC_2H_5)_2$	$Hg(OAc)_2$	acetone, KCl	$(ClHg)_3CCO_2C_2H_5$ (60)	—	—	93

Table 2.2. (continued)

Alkene	Mercuric salt	Reaction conditions	Organomercurial(s) (% Yield)[a]	Subsequent reactants	Product(s) (% Yield)[a]	Ref.
$H_2C{=}CH(CH_2)_3SCH_3$	$Hg(OAc)_2$	1:1 H_2O/THF 24 hr	—	$NaBH_4$/NaOH	$CH_3CHOH(CH_2)_3SCH_3$ + $CH_3CHOAc(CH_2)_3SCH_3$ 60 : 40 (78 total)	66
$H_2C{=}CHCH_2Si(CH_3)_3$	$Hg(OAc)_2$	1:1 H_2O/THF 5 min	—	$NaBH_4$/NaOH	$(CH_3)_3SiOH$ (44) + $CH_3CHOHCH_3$ (25) + $CH_3CH_2CH_2OH$ (10) + $[(CH_3)_3Si]_2O$ (7)	84
	$Hg(OAc)_2$	D_2O/THF–d_8 33 °C 10 min	$AcOHgCH_2CHODCH_2HgOAc$ (40) + $H_2C{=}CHCH_2HgOAc$ (15)	—	—	84
$H_2C{=}C(CH_3)Si(CH_3)_3$	$Hg(OAc)_2$	H_2O/THF 5 min	—	$NaBH_4$/NaOH	$CH_3CHOHCH_3$ (39) + $(CH_3)_3SiOH$ (38) + $CH_3CH_2CH_2OH$ (9) + CH_3COCH_3 (5)	84
cis-$CH_3CH{=}CHSi(CH_3)_3$	$Hg(OAc)_2$	1:1 H_2O/THF 5 min	—	$NaBH_4$/NaOH	$CH_3CHOHCH_2Si(CH_3)_3$ (44) + $(CH_3)_3SiOH$ (25) + $CH_3CHOHCH_3$ (19) + $CH_3CH_2CH_2OH$ (7)	84
$H_2C{=}CHCH_2N$(succinimidyl)	$Hg(OAc)_2$/HgO	Δ 1 hr	$AcOHgCH_2CHOHCH_2N$(succinimidyl)	—	—	95
$H_2C{=}CHCH_2NHCO(CH_2)_2CO_2H$	$Hg(OAc)_2$	—	$HgCH_2CHOHCH_2NHCO(CH_2)_2C{=}O$ (cyclic)	—	—	96
$H_2C{=}CHCH_2NHCO(CHOH)_2CO_2H$ (D-galacto)	$Hg(OAc)_2$	H_2O/NaOH	$HOHgCH_2CHOHCH_2NHCO(CHOH)_2CO_2Na$	—	—	94
$H_2C{=}CHCH_2NH{-}$(4-amino-6-methyl-1,3,5-triazin-2-yl)	—	—	$XHgCH_2CHOHCH_2NH{-}$(4-amino-6-methyl-1,3,5-triazin-2-yl) X = OAc, Cl, Br	—	—	91
$H_2C{=}C(CN)O(CH_2)_3CH_3$	$Hg(OAc)_2$	H_2O	$AcOHgCH_2CO_2(CH_2)_3CH_3$ (90)	—	—	70
$H_2C{=}CHCH_2NHCONHCOC_2H_5$	$Hg(OAc)_2$	H_2O/NaOH	$HOHgCH_2CHOHCH_2NHCONHCOC_2H_5$	—	—	94
$H_2C{=}CH(CH_2)_3OAc$	$Hg(OAc)_2$	1:1 H_2O/THF 30 min	—	$NaBH_4$/NaOH	$CH_3CHOH(CH_2)_3OH$ + $H_2C{=}CH(CH_2)_3OH$ 82 : 18 (98 total)	68

Table 2.2. (continued)

Alkene	Mercuric salt	Reaction conditions	Organomercurial(s) (% Yield)[a]	Subsequent reactants	Product(s) (% Yield)[a]	Ref.
$H_2C=CHCH_2NHCO_2(CH_2)_2CH_3$	$Hg(OAc)_2/HgO$	—	$HOCH_2CH(HgOAc)CH_2NHCO_2(CH_2)_2CH_3$ — (wrong regioisomer ?)	—	—	74
$H_2C=CHCH_2NHCO_2CH(CH_3)_2$	$Hg(OAc)_2$	H_2O	$HOCH_2CH(HgOAc)CH_2NHCO_2CH(CH_3)_2$ — (wrong regioisomer ?)	—	—	74
$H_2C=CHO(CH_2)_2CH(CH_3)_2$	$Hg(OAc)_2$	H_2O, KCl	$ClHgCH_2CHO$ (38)	—	—	78
$H_2C=CH(CH_2)_5OH$	$Hg(OAc)_2$	1:1 H_2O/THF 30 min	—	$NaBH_4$/NaOH	$CH_3CHOH(CH_2)_5OH$ (77)	68
$H_2C=C(O-n-C_4H_9)CH_3$	$Hg(OAc)_2$	H_2O, KCl	$ClHgCH_2COCH_3$ (58)	—	—	78
$H_2C=CH(CH_2)_2Si(CH_3)_3$	$Hg(OAc)_2$	1:1 H_2O/THF 5 min	—	$NaBH_4$/NaOH	$CH_3CHOH(CH_2)_2Si(CH_3)_3$ (99)	84
$H_2C=CHOC_6H_5$	$HgO/Hg(OAc)_2$	H_2O/EtOH Δ	$Hg(CH_2CHO)_2$ (81)	—	—	79
$H_2C=CH(CH_2)_4OAc$	$Hg(OAc)_2$	1:1 H_2O/THF 30 min	—	$NaBH_4$/NaOH	$CH_3CHOH(CH_2)_4OH$ + $H_2C=CH(CH_2)_4OH$ 95 : 5 (92 total)	68
$H_2C=CHCH_2NHCO_2(CH_2)_3CH_3$	$Hg(OAc)_2$	H_2O	$HOCH_2CH(HgOAc)CH_2NHCO_2(CH_2)_3CH_3$ — (wrong regioisomer ?)	—	—	74
$H_2C=C(OC_2H_5)(CH_2)_3CH_3$	$HgO/Hg(OAc)_2$	H_2O/EtOH	$Hg[CH_2CO(CH_2)_3CH_3]_2$ (100)	—	—	79
$H_2C=C[OSi(CH_3)_3]\!-\!\triangleleft$	$HgO/Hg(OAc)_2$	H_2O/EtOH	$Hg[CH_2CO\!-\!\triangleleft]_2$	$COCl_2$	$H_2C=C(OCOCl)\!-\!\triangleleft$	255e
trans-$C_6H_5CH=CHCO_2H$	$Hg(OAc)_2$	H_2O/HOAc 1 hr 85 °C	$C_9H_8O_2\cdot HgO$ + $C_9H_8O_2HgOAc$	Br_2	$C_6H_5CHOHCHBrCO_2H$ + $C_6H_5CH=CHCO_2H$	97
cis-$C_6H_5CH=CHCO_2H$	$Hg(OAc)_2$	H_2O Δ	$C_6H_5CHCH\!-\!CO$? (HO Hg—O)	$NaOH/H_2S$	$C_6H_5CHOHCH_2CO_2H$ (82)	69
$H_2C=CHCH_2NHCO\!-\!(pyridyl\;Cl)$	$Hg(OAc)_2$	—	$HOCH_2CH(HgOAc)CH_2NHCO\!-\!(pyridyl\;Cl)$ — (wrong regioisomer ?)	—	—	98

Table 2.2. (continued)

Alkene	Mercuric salt	Reaction conditions	Organomercurial(s) (% Yield)[a]	Subsequent reactants	Product(s) (% Yield)[a]	Ref.
$H_2C=CHCH_2NHCO-$(3-pyridyl)	$Hg(OAc)_2$	H_2O	$AcOHgCH_2CHOHCH_2NHCO-$(3-pyridyl)	—	—	94
	$Hg(O_2CR)_2$	—	$HOCH_2CH(HgO_2CR)CH_2NHCO-$(3-pyridyl) $R=CH_3, C_2H_5, C_6H_5$, tartaroxy (wrong regioisomer ?)	—	—	98
$H_2C=CHCH_2NHCO-$(2-hydroxy-5-pyridyl)	$Hg(OAc)_2$	—	$HOCH_2CH(HgOAc)CH_2NHCO-$(2-hydroxy-5-pyridyl) (wrong regioisomer ?)	—	—	98
$H_2C=CHCH_2NHCO(CH_2)_4CO_2H$	$Hg(OAc)_2$	—	$HgCH_2CHOHCH_2NHCO(CH_2)_4C=O$ (cyclic)	—	—	96
$H_2C=CHCH_2NHCO(CHOH)_4CO_2H$	$Hg(OAc)_2$	$H_2O/NaOH$	$HOHgCH_2CHOHCH_2NHCO(CHOH)_4CO_2Na$	—	—	94
$H_2C=CHCH_2-$(1-hydroxycyclohexyl)	$Hg(OAc)_2$	H_2O/THF 15 min	—	$NaBH_4/NaOH$	$CH_3CHOHCH_2-$(1-hydroxycyclohexyl)	162
$H_2C=CHCH_2OCH_2CHCH_2$ (isopropylidenedioxy)	$Hg(OAc)_2$	$H_2O/$dioxane	$AcOHgCH_2CHOHCH_2OCH_2CHCH_2$ (isopropylidenedioxy)	—	—	99
$H_2C=CHCH_2O-$(glucopyranosyl)	$Hg(OAc)_2$	H_2O	$AcOHgCH_2CHOHCH_2O-$(glucopyranosyl)	—	—	100
$H_2C=CHCH_2NHCO_2(CH_2)_2CH(CH_3)_2$	$Hg(OAc)_2$	H_2O	$HOCH_2CH(HgOAc)NHCO_2(CH_2)_2CH(CH_3)_2$ (wrong regioisomer ?)	—	—	74
$H_2C=CHCH_2NHCO(CHOH)_4CH_2OH$ (D-gluco)	$Hg(OAc)_2$	H_2O	$AcOHgCH_2CHOHCH_2NHCO(CHOH)_4CH_2OH$	—	—	94

Table 2.2. (continued)

Alkene	Mercuric salt	Reaction conditions	Organomercurial(s) (% Yield)[a]	Subsequent reactants	Product(s) (% Yield)[a]	Ref.
$(CH_3)_2C=CH(CH_2)_2COH(CH_3)_2$	$Hg(OAc)_2$	H_2O, X^-	$(CH_3)_2COHCH(HgX)(CH_2)_2COH(CH_3)_2$ X = Cl (17), I (25) 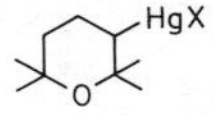 + [tetrahydropyran-HgX structure] X = Cl (25), I (21)	N_2H_4/KOH (on diol RHgI)	$(CH_3)_2COHCH_2CH_2COH(CH_3)_2$ (65)	13, 101
$H_2C=CHCH_2OCH(CHOHCH_2OH)CHOHCHOHCH_2OH$ $Hg(OAc)_2$		H_2O	$AcOHgCH_2CHOHCH_2OCH(CHOHCH_2OH)CHOHCHOHCH_2OH$	—	—	100
$H_2C=C[OSi(CH_3)_3]C(CH_3)_3$	$HgO/Hg(OAc)_2$	$H_2O/EtOH$	$Hg[CH_2COC(CH_3)_3]_2$ (94)	$COCl_2$	$H_2C=C(O_2CCl)C(CH_3)_3$	255e
$H_2C=CHCH_2NHCO-$[3-(pyridine)-2-CO_2H] 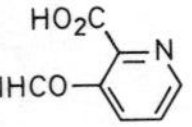	$Hg(OAc)_2$	—	$HOCH_2CH(HgOAc)CH_2NHCO-$[3-(pyridine)-2-CO_2H] (wrong regioisomer ?) 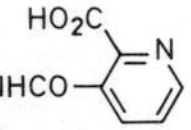	—	—	98
$H_2C=CHCH_2-$[benzodioxole] 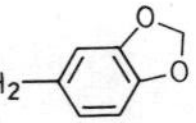	$Hg(OAc)_2$	H_2O RT 4 months	$AcOHgCH_2CHOHCH_2-$[benzodioxole] 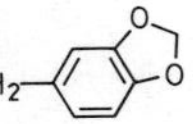	—	—	102,103
	$Hg(OAc)_2$	H_2O 50°C 2.5 hr, NaX	$XHgCH_2CHOHCH_2-$[benzodioxole] X = Cl, Br, I 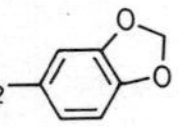	—	—	104, 105
$H_2C=C(C_6H_5)O_2CCH_3$	$Hg(OAc)_2$	H_2O	$AcOHgCH_2COC_6H_5$ (88)	—	—	106
trans-$CH_3CH=CH-$[benzodioxole] 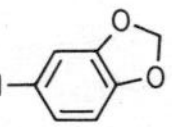	$Hg(OAc)_2$	H_2O	—	—	$CH_3CHOHCHOH-$[benzodioxole] 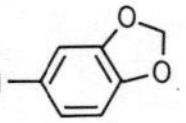	102–104
$H_2C=CHCH_2NHCO-$[2-CH$_3$-6-Cl-pyridin-4-yl] 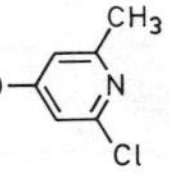	$Hg(OAc)_2$	—	$HOCH_2CH(HgOAc)CH_2NHCO-$[2-CH$_3$-6-Cl-pyridin-4-yl ?] (wrong regioisomer ?) 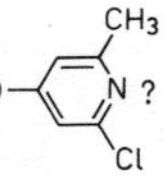	—	—	98
$H_2C=CHCH_2NHCONHCO-$[3-pyridyl] 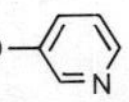	$Hg(OAc)_2$	H_2O	$AcOHgCH_2CHOHCH_2NHCONHCO-$[3-pyridyl] 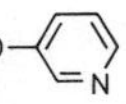	—	—	94

Table 2.2. (continued)

Alkene	Mercuric salt	Reaction conditions	Organomercurial(s) (% Yield)[a]	Subsequent reactants	Product(s) (% Yield)[a]	Ref.
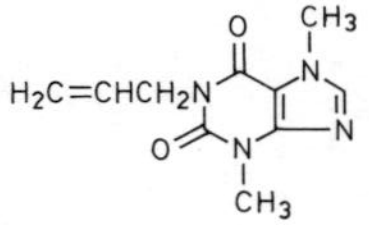 $H_2C=CHCH_2$-(1,3-dimethylxanthine)	$Hg(OAc)_2$	H_2O Δ 30 min, NaX (X = Cl, Br)	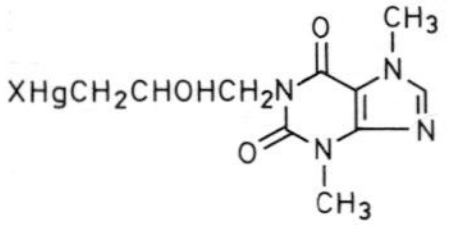 $XHgCH_2CHOHCH_2$-(1,3-dimethylxanthine)	—	—	107
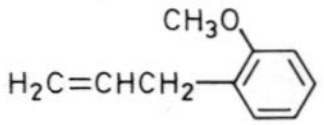 $H_2C=CHCH_2$-(2-CH_3O-phenyl)	$Hg(OAc)_2$	H_2O several hr (NaCl)	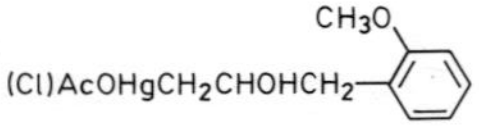 $(Cl)AcOHgCH_2CHOHCH_2$-(2-CH_3O-phenyl)	—	—	108
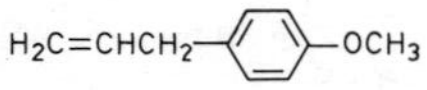 $H_2C=CHCH_2$-(4-OCH_3-phenyl)	$Hg(OAc)_2$	H_2O	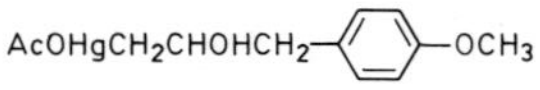 $AcOHgCH_2CHOHCH_2$-(4-OCH_3-phenyl)	—	—	103
	$Hg(OAc)_2$	H_2O, NaCl	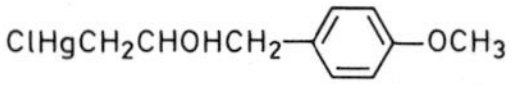 $ClHgCH_2CHOHCH_2$-(4-OCH_3-phenyl)	—	—	104
$H_2C=C(OC_2H_5)C_6H_5$	$Hg(OAc)_2$	H_2O	$AcOHgCH_2COC_6H_5$ (86)	—	—	106
	$HgO/Hg(OAc)_2$	$H_2O/EtOH$ RT	$Hg(CH_2COC_6H_5)_2$ (50)	—	—	79
$C_6H_5CH=CHOCH_2CH_3$	HgX_2/HgO	$H_2O/acetone$ Δ	$XHgCH(C_6H_5)CHO$ X = Cl (88), Br (55)	—	—	106
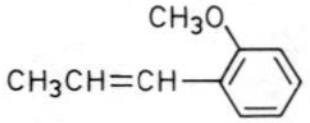 $CH_3CH=CH$-(2-CH_3O-phenyl)	$Hg(OAc)_2$	H_2O 2 weeks	—	—	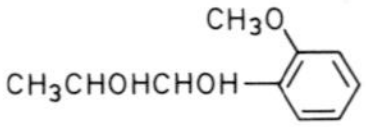 $CH_3CHOHCHOH$-(2-CH_3O-phenyl)	109
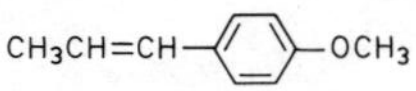 $CH_3CH=CH$-(4-OCH_3-phenyl)	$Hg(OAc)_2$	H_2O 10–12 d	—	—	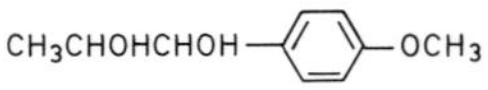 $CH_3CHOHCHOH$-(4-OCH_3-phenyl) (66)	102, 103, 109
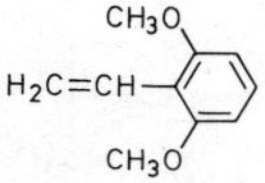 $H_2C=CH$-(2,6-di-CH_3O-phenyl)	$Hg(OAc)_2$	H_2O, KCl	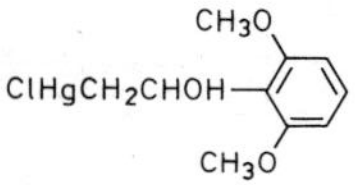 $ClHgCH_2CHOH$-(2,6-di-CH_3O-phenyl)	2% Na(Hg)/H_2O	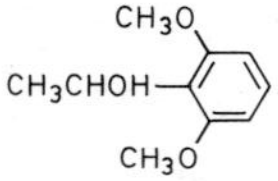 CH_3CHOH-(2,6-di-CH_3O-phenyl)	110

Table 2.2. (continued)

Alkene	Mercuric salt	Reaction conditions	Organomercurial(s) (% Yield)[a]	Subsequent reactants	Product(s) (% Yield)[a]	Ref.
$H_2C{=}CHCH_2NHCO$–[3,4-dimethyluracil]	1:1 $Hg(OAc)_2/HgCl_2$	H_2O Δ	$ClHgCH_2CHOHCH_2NHCO$–[3,4-dimethyluracil]	—	—	81
[2-methyl-5-isopropenylcyclohexanone]	$Hg(OAc)_2$	H_2O/THF	—	$NaBH_4/NaOH$	[2-methyl-5-(2-hydroxyisopropyl)cyclohexanone]	111
$H_2C{=}CHCH_2NHCOCH_2N$[piperidine]	$Hg(OAc)_2$	H_2O	$HOHgCH_2CHOHCH_2NHCOCH_2N$[piperidine]·$^1/_2\,H_2SO_4$ (?)	—	—	94
[tetrahydrofuran-isopropenyl structure] 1:1 *cis/trans*	$Hg(OAc)_2$	1:1 H_2O/THF RT	—	$NaBH_4/NaOH$	[diol] (50) + [bicyclic acetal] (40)	112
$(CH_3)_2C{=}CH(CH_2)_2\overset{\,}{C}CH_3$[dioxolane]	$Hg(OAc)_2$	1:1 H_2O/THF RT 10 min	—	$NaBH_4/NaOH$	$(CH_3)_2COH(CH_2)_3\overset{\,}{C}CH_3$[dioxolane] (85)	113
$CH_3CH{=}C[OSi(CH_3)_3]C(CH_3)_3$	$HgO/Hg(OAc)_2$	$H_2O/EtOH$	$Hg[CH(CH_3)COC(CH_3)_3]_2$	—	—	255a, 255c
$H_2C{=}CHCH_2$–[3,4-dimethoxyphenyl]	$Hg(OAc)_2$	H_2O	$AcOHgCH_2CHOHCH_2$–[3,4-dimethoxyphenyl]	—	—	103
	$Hg(OAc)_2$	H_2O RT 3hr or 50 °C 0.5 hr, NaCl	$ClHgCH_2CHOHCH_2$–[3,4-dimethoxyphenyl]	—	—	104, 105
trans-$CH_3CH{=}CH$–[3,4-dimethoxyphenyl]	$Hg(OAc)_2$	H_2O 25d	—	—	$CH_3CHOHCHOH$–[3,4-dimethoxyphenyl]	103, 104
$H_2C{=}C[OSi(CH_3)_3]C_6H_5$	$HgO/Hg(OAc)_2$	$H_2O/EtOH$	$Hg(CH_2COC_6H_5)_2$ (89)	—	—	255b

Table 2.2. (continued)

Alkene	Mercuric salt	Reaction conditions	Organomercurial(s) (% Yield)[a]	Subsequent reactants	Product(s) (% Yield)[a]	Ref.
$H_2C=CHCH_2NHCOCH_2NHCO-(CHOH)_4CH_2OH$ (D–gluco)	$Hg(OAc)_2$	H_2O	$AcOHgCH_2CHOHCH_2NHCOCH_2NHCO-(CHOH)_4CH_2OH$	—	—	94
$H_2C=CH(CH_2)_8CO_2H$	$Hg(OAc)_2$	$H_2O/HOAc$	$AcOHgCH_2CHOH(CH_2)_8CO_2H$ (99)	KBr/Br_2	$BrCH_2CO(CH_2)_8CO_2H$? (30)	114
	$0.2\ Hg(O_2CC_2H_5)_2$	H_2O/CH_3COCH_3 H_2SO_4 CrO_3	—	—	$CH_3CO(CH_2)_8CO_2H$ (83)	46
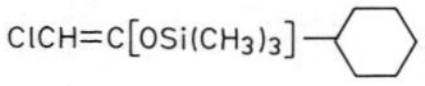 $ClCH=C[OSi(CH_3)_3]\!-\!\bigcirc$	$HgO/Hg(OAc)_2$	$H_2O/EtOH$	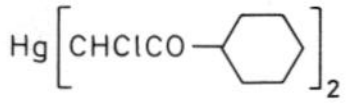 $Hg\!\left[CHClCO\!-\!\bigcirc\right]_2$	CH_3COCl	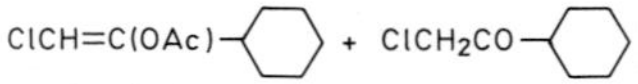 $ClCH=C(OAc)\!-\!\bigcirc$ (major) $+\ ClCH_2CO\!-\!\bigcirc$ (minor)	255g
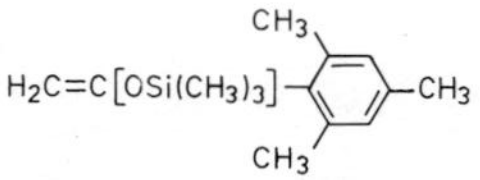 $H_2C=C[OSi(CH_3)_3]$ (mesityl)	$HgO/Hg(OAc)_2$	$H_2O/EtOH$	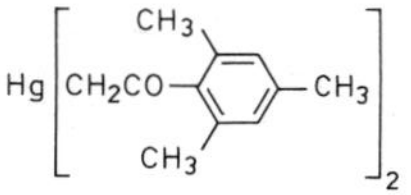 $Hg\!\left[CH_2CO\text{-(mesityl)}\right]_2$	—	—	255c
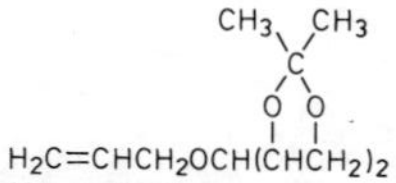 $H_2C=CHCH_2OCH(CHCH_2)_2$ (D–xylitol)	$Hg(OAc)_2$	$H_2O/dioxane$	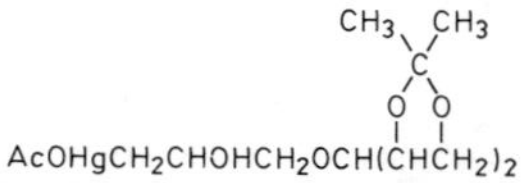 $AcOHgCH_2CHOHCH_2OCH(CHCH_2)_2$	—	—	99
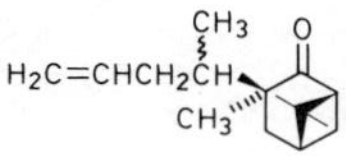 $H_2C=CHCH_2CH$ (pinanone)	$Hg(OAc)_2$	RT	—	cat. $Li_2PdCl_4/CuCl_2$	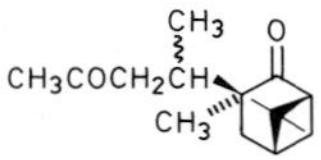 CH_3COCH_2CH (pinanone) (72)	119
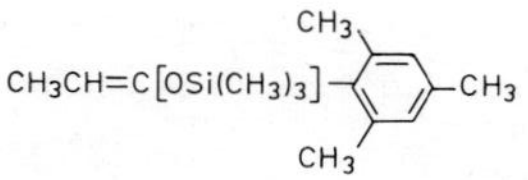 $CH_3CH=C[OSi(CH_3)_3]$ (mesityl)	$HgO/Hg(OAc)_2$	$H_2O/EtOH$	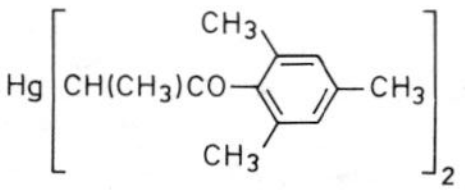 $Hg\!\left[CH(CH_3)CO\text{-(mesityl)}\right]_2$	—	—	255c
$H_2C=CHCH_2CHCO(CH_2)_{10}$	$0.2\ Hg(O_2CC_2H_5)_2$	H_2O/CH_3COCH_3 H_2SO_4 CrO_3	—	—	$CH_3COCH_2CHCO(CH_2)_{10}$ (70)	46

Table 2.2. (continued)

28

Alkene	Mercuric salt	Reaction conditions	Organomercurial(s) (% Yield)[a]	Subsequent reactants	Product(s) (% Yield)[a]	Ref.
$H_2C{=}CHCH_2NHCOCH_2SCH_2-$ $(CHOH)_4CH_2OH(D{-}gluco)$	$Hg(OAc)_2$	H_2O	$AcOHgCH_2CHOHCH_2NHCOCH_2SCH_2-$ $(CHOH)_4CH_2OH$	—	—	94
$H_2C{=}CHCH_2NHCOCH_2NHCOC_6H_5$	1:1 $Hg(OAc)_2/HgCl_2$	H_2O RT 3–5d	$ClHgCH_2CHOHCH_2NHCOCH_2NHCOC_6H_5$ (85)	—	—	81
$H_2C{=}CHCH_2-$ (benzodioxole, CH_3O, OCH_3)	$Hg(OAc)_2$	H_2O	$AcOHgCH_2CHOHCH_2-$ (benzodioxole, CH_3O, OCH_3)	Zn/OH^-	$CH_3CHOHCH_2-$ (benzodioxole, CH_3O, OCH_3)	104
$H_2C{=}CHCH_2-$ (benzodioxole, OCH_3, CH_3O)	$Hg(OAc)_2$	H_2O	$AcOHgCH_2CHOHCH_2-$ (benzodioxole, OCH_3, CH_3O)	—	—	103
$trans{-}CH_3CH{=}CH-$ (benzodioxole, CH_3O, OCH_3)	$Hg(OAc)_2$	H_2O	—	—	$CH_3CHOHCHOH-$ (benzodioxole, CH_3O, OCH_3)	104
$CH_3CH{=}CH-$ (benzodioxole, OCH_3, CH_3O)	$Hg(OAc)_2$	H_2O	—	—	$CH_3CHOHCHOH-$ (benzodioxole, OCH_3, CH_3O)	103
$trans{-}CH_3CH{=}CH-$ (aryl, OCH_3, OCH_3, CH_3O)	$Hg(OAc)_2$	H_2O	—	—	$CH_3CHOHCHOH-$ (aryl, OCH_3, OCH_3, CH_3O)	103
$CH_3C[OSi(CH_3)_3]{=}CHC_6H_5$	$HgO/Hg(OAc)_2$	$H_2O/EtOH$	$Hg[CH(C_6H_5)COCH_3]_2$ (63)	—	—	255b

Table 2.2. (continued)

Alkene	Mercuric salt	Reaction conditions	Organomercurial(s) (% Yield)[a]	Subsequent reactants	Product(s) (% Yield)[a]	Ref.
(structure: OAc-substituted cyclopentane with isopropenyl)	$Hg(OAc)_2$	H_2O	—	—	(structure: dihydroxy cyclopentane)	115
(structure: $H_2C=C$, CH_3, CO_2CH_3, dioxolane-spiro cyclopentane)	$Hg(OAc)_2$	H_2O	—	$NaBH_4/NaOH$	(structure: $(CH_3)_2COH$, CO_2CH_3, dioxolane-spiro cyclopentane)	116
$CH_3(CH_2)_3C[OSi(CH_3)_3]={}$ $=CH(CH_2)_2CH_3$	$HgO/Hg(OAc)_2$	$H_2O/EtOH$	$Hg[CH(n-C_3H_7)CO(CH_2)_3CH_3]_2$ (67)	RCOCl	$CH_3(CH_2)_3C(O_2CR)=CH(CH_2)_2CH_3$ $R = t-Bu$ (39), Ph (62)	255d
$H_2C=CHCH_2NHCO$—(aryl, OCH_2CO_2H)—$NHCOCH_3$	$Hg(OAc)_2$	H_2O	$HOHgCH_2CHOHCH_2NHCO$—(aryl, OCH_2CO_2H)—$NHCOCH_3$	—	—	117
m-$H_2C=CHCH_2NHCOC_6H_4$— $OCH_2CO_2CH_2CH_3$	$Hg(OAc)_2$	H_2O	m-$AcOHgCH_2CHOHCH_2NHCOC_6H_4$— $OCH_2CO_2CH_2CH_3$ (60)	—	—	118
(structure: $H_2C=CHCH_2OCH$ with two acetonide rings, ?)	$Hg(OAc)_2$	$AcOHgCH_2CHOHCH_2OCH$	(structure: organomercurial with two acetonide rings, ?)	—	—	99
$H_2C=C(CH_3)CH_2PO(C_6H_5)_2$	$Hg(OAc)_2$	H_2O/THF	—	$NaBH_4/NaOH$	$(CH_3)_2COHCH_2PO(C_6H_5)_2$	120
(structure: $H_2C=CH$, $C_6H_5CH_2O$, furanose acetonide)	$Hg(O_2CCF_3)_2$	H_2O	—	$NaBH_4/NaOH$	(structures: two furanose products) 68 : 32	121

30 **Table 2.2.** (continued)

Alkene	Mercuric salt	Reaction conditions	Organomercurial(s) (% Yield)[a]	Subsequent reactants	Product(s) (% Yield)[a]	Ref.
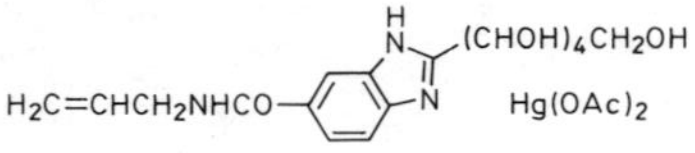 $H_2C=CHCH_2NHCO-$ (benzimidazole with $(CHOH)_4CH_2OH$ substituent)	$Hg(OAc)_2$	H_2O	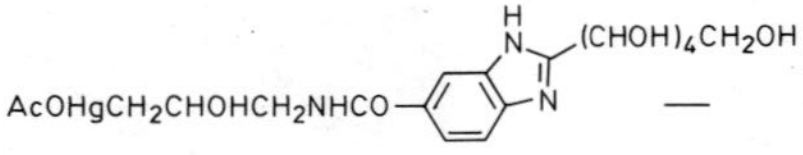 $AcOHgCH_2CHOHCH_2NHCO-$ (benzimidazole with $(CHOH)_4CH_2OH$ substituent)	—	—	94
$R\text{-}cis\text{-}CH_3(CH_2)_5\text{-}CHOHCH_2CH=CH(CH_2)_7CO_2H$	$Hg(OAc)_2$?	H_2O	—	$NaBH_4$?	$CH_3(CH_2)_5CHOH(CH_2)_2CHOH(CH_2)_7CO_2H$ (only product ?)	122
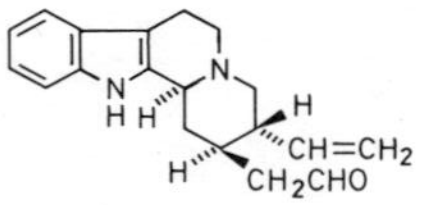 (yohimbane-type structure with $CH=CH_2$, CH_2CHO)	$Hg(NO_3)_2$	2:3 H_2O/THF 0°C 3hr	—	$NaBH_4$/NaOH	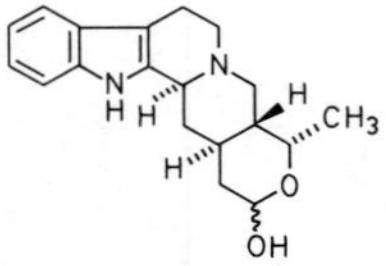 (yohimbane-type structure with CH_3, O, OH)	123
$cis\text{-}CH_3(CH_2)_7CH=CH\text{-}(CH_2)_7CO_2CH_3$	$Hg(OAc)_2$	1:1 H_2O/THF or H_2O/DMF	—	$NaBH_4$	$CH_3(CH_2)_7CHOH(CH_2)_8CO_2CH_3$ + $CH_3(CH_2)_8CHOH(CH_2)_7CO_2CH_3$ (77,71 total)	124
$cis\text{-}CH_3(CH_2)_5CHOHCH_2\text{-}CH=CH(CH_2)_7CO_2CH_3$	$Hg(OAc)_2$	H_2O	—	$NaBH_4$	$CH_3(CH_2)_5CHOHCH_2CHOH(CH_2)_8CO_2CH_3$ (9) + $CH_3(CH_2)_5CHOH(CH_2)_2CHOH(CH_2)_7CO_2CH_3$ (70) 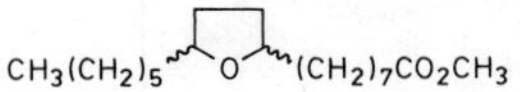 + $CH_3(CH_2)_5$—(tetrahydrofuran)—$(CH_2)_7CO_2CH_3$ (17)	125
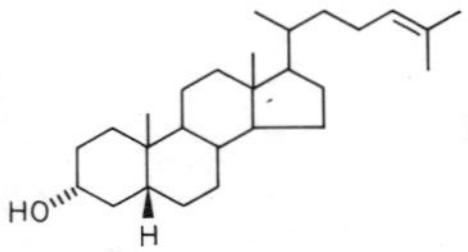 (cholesterol-type structure with HO)	$Hg(O_2CCF_3)_2$	H_2O/THF/DMF 7 hr	—	$NaBH_4$/NaOH	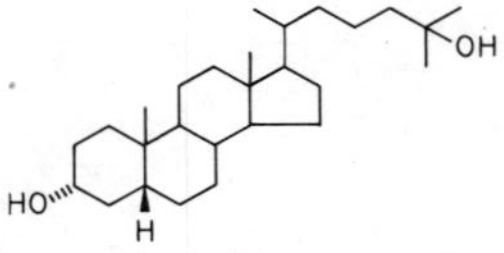 (steroid structure with HO and OH)	126

[a] If there is no entry in the organomercurial column, the yield is based on starting alkene. If there is an organomercurial entry, the yield is based on the organomercurial unless otherwise stated.

Table 2.3. Hydroxymercuration of Cyclic Alkenes

Alkene	Mercuric salt	Reaction conditions	Organomercurial(s) (% Yield)[a]	Subsequent reactants	Product(s) (% Yield)[a]	Ref.
(cyclobutene)	$Hg(OAc)_2$	H_2O	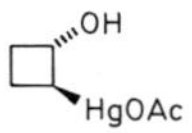 (cyclobutane with OH and HgOAc)	—	—	127
	$Hg(OAc)_2$	3:1 H_2O/CH_3COCH_3 RT, NaCl	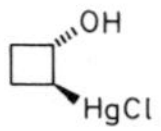 (cyclobutane with OH and HgCl) (33 crude)	$NaBH_4$	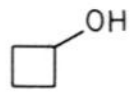 (cyclobutanol)	128
	$Hg(ClO_4)_2$	$H_2O/HClO_4$	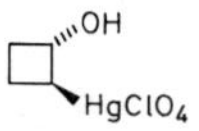 (cyclobutane with OH and $HgClO_4$)	—	—	129
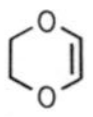 (1,4-dioxene)	$Hg(OAc)_2$	H_2O RT 30 min	—	—	$HOCH_2CH_2OH$ (~100) + HCOCHO (~100) + Hg° (96)	109
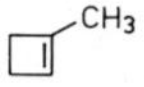 (methylcyclobutene)	$Hg(ClO_4)_2$	$H_2O/HClO_4$	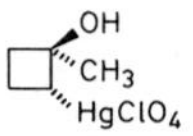 (cyclobutane with OH, CH_3 and $HgClO_4$)	—	CH_3CO—(cyclopropyl) (100)	129, 130
(cyclopentene)	$Hg(OAc)_2$	1:1 H_2O/CH_3COCH_3 10 min, NaCl	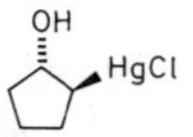 (cyclopentane with OH and HgCl) (63)	—	—	131
	$Hg(OAc)_2$	1:1 H_2O/THF 1 hr RT	—	$NaBH_4/NaOH$	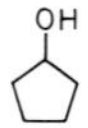 (cyclopentanol) (91)	39, 40
	$Hg(OAc)_2$	H_2O/THF	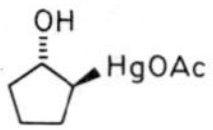 (cyclopentane with OH and HgOAc)	$NaBD_4$	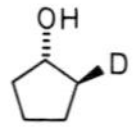 (cyclopentanol with D) > 95% *trans*	34
	—	—	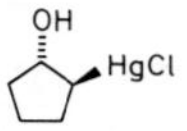 (cyclopentane with OH and HgCl)	$NaBH_4$	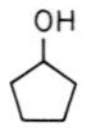 (cyclopentanol)	128
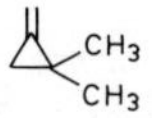 (isopropylidenecyclopropane, CH_3, CH_3)	$Hg(OAc)_2$	H_2O/THF	—	—	$H_2C{=}C(CH_3)COH(CH_3)_2$	132

Table 2.3. (continued)

Alkene	Mercuric salt	Reaction conditions	Organomercurial(s) (% Yield)[a]	Subsequent reactants	Product(s) (% Yield)[a]	Ref.
1-methyl-cyclopentene	Hg(OAc)$_2$	1:1 H$_2$O/THF	—	NaBH$_4$/NaOH	69 : 18 : 11 : 2	133
4-methyl-cyclopentene	Hg(OAc)$_2$	1:1 H$_2$O/THF	—	NaBH$_4$/NaOH	77 : 23	133
1-methyl-cyclopentene	Hg(OAc)$_2$	1:1 H$_2$O/THF 0.1 hr RT	—	NaBH$_4$/NaOH	(93)	39, 40
cyclohexene	Hg(OAc)$_2$	H$_2$O	(63, 100)	—	—	131, 134, 135
	Hg(OAc)$_2$	H$_2$O, NaCl		—	—	136, 137
	Hg(OAc)$_2$	H$_2$O 15 min, KBr		—	—	138
	Hg(OAc)$_2$	1:1 H$_2$O/THF 11–15 min RT	—	NaBH$_4$/NaOH	(96, 99)	39, 40
	Hg(OAc)$_2$	1:1 H$_2$O/THF		I$_2$/HOAc/Ac$_2$O		139

Table 2.3. (continued)

Alkene	Mercuric salt	Reaction conditions	Organomercurial(s) (% Yield)[a]	Subsequent reactants	Product(s) (% Yield)[a]	Ref.
	$Hg(OAc)_2$	$H_2O/HOAc$	—	I_2/Δ	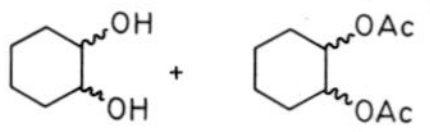 (cyclohexane-1,2-diol, OH/OH) cis and trans + (cyclohexane-1,2-diyl diacetate, OAc/OAc) cis and trans	139
	$Hg(OAc)_2$	D_2O 2–3 hr	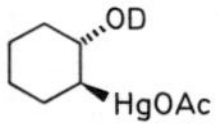 (cyclohexane, OD / HgOAc)	—	—	140
	$Hg(O_2CCF_3)_2$	1:1 H_2O/THF 15 min	—	$NaBH_4/NaOH$	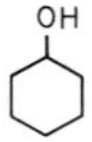 (cyclohexanol, OH) (94)	41
	0.5 $Hg(O_2CC_2H_5)_2$	H_2O/dioxane/CF_3CO_2H $Na_2Cr_2O_7$ 18 hr	—	—	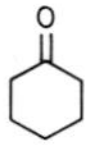 (cyclohexanone) (41)	46
	$Hg(NO_3)_2$	1:1 H_2O/THF 15 min	—	$NaBH_4/NaOH$	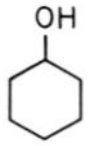 (cyclohexanol, OH) (99)	41
	$Hg(NO_3)_2$	H_2O	—	—	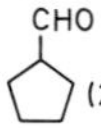 (cyclopentane, CHO) (23) 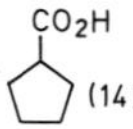 + (cyclopentane, CO_2H) (14)	56
	$Hg(NO_3)_2$	H_2O/HNO_3 10 min, NaCl	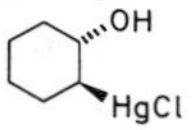 (cyclohexane, OH / HgCl) (79)	—	—	56
	$HgSO_4$	H_2SO_4	—	—	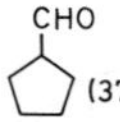 (cyclopentane, CHO) (37) 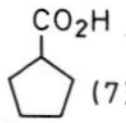 + (cyclopentane, CO_2H) (7)	56
	$HgSO_4$	H_2O 1 hr 60–65 °C	—	KOH 100 °C	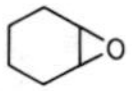 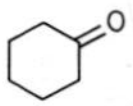 (cyclohexene oxide) + (cyclohexanone) (48 total) 84 : 16	25

34 **Table 2.3.** (continued)

Alkene	Mercuric salt	Reaction conditions	Organomercurial(s) (% Yield)[a]	Subsequent reactants	Product(s) (% Yield)[a]	Ref.
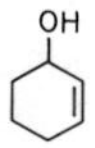	Hg(O$_3$SCH$_3$)$_2$	1:1 H$_2$O/THF 60 min	—	NaBH$_4$/NaOH	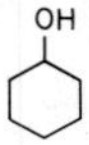 (91)	41
	Hg(ClO$_4$)$_2$	1 NaOH/0.01 N HClO$_4$, NaCl	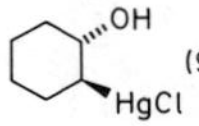 (90)	—	—	22
	Hg(OAc)$_2$	1:1 H$_2$O/THF	—	NaBH$_4$/NaOH	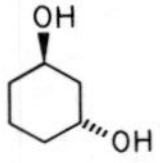 (50)	141
	Hg(OAc)$_2$	1:1 H$_2$O/THF 30 min	—	NaBH$_4$/NaOH	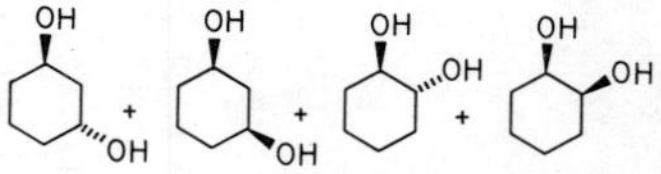 80 : 12 : 1 : 1 (95 total)	68
	Hg(OAc)$_2$	1:1 H$_2$O/THF 25 °C ~0.5 hr	—	NaBH$_4$/H$_2$O	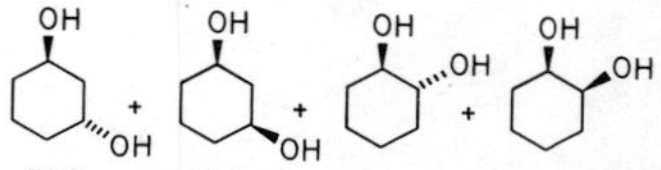 78.2 : 19.6 : 0.9 : 0.8 (86 total)	142
	Hg(OAc)$_2$	1:19 H$_2$O/CH$_3$CN 34 min 25 °C	—	NaBH$_4$/NaOH	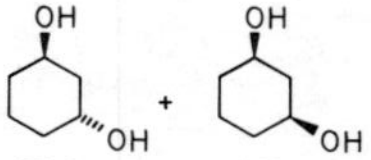 96.3 : 3.7 (83 total)	142
	Hg(OAc)$_2$	1:1 H$_2$O/THF	—	NaBH$_4$/NaOH, Ac$_2$O	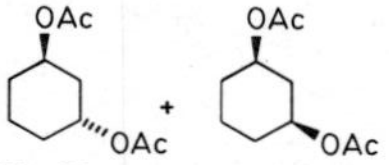 98–99 : 1–2	143
	Hg(OAc)$_2$	1:1 H$_2$O/THF	—	NaBH$_4$/NaOH, Ac$_2$O	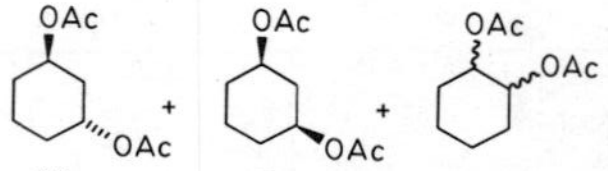 85 : 15 : 1 cis and trans	143

Table 2.3. (continued)

Alkene	Mercuric salt	Reaction conditions	Organomercurial(s) (% Yield)[a]	Subsequent reactants	Product(s) (% Yield)[a]	Ref.
[HO-cyclohexene]	Hg(OAc)$_2$	1:1 H$_2$O/THF	—	NaBH$_4$/NaOH, Ac$_2$O	[OAc-cyclohexane isomers] 79.6 : 18.7 : 0.8 : 0.9 (78 total)	144
	Hg(OAc)$_2$	H$_2$O 10d RT, NaCl	[HO-cyclohexane-HgCl, OH] (94)	N$_2$H$_4$/NaOH	[HO-cyclohexane-OH] (90)	145
	Hg(OAc)$_2$	1:1 H$_2$O/THF 30 min	—	NaBH$_4$/NaOH	[cyclohexanediol isomers] 70 : 14 : 13 : 3 (92 total)	68
[pyranose: OH, OH, O, CH$_2$OH]	cat. HgSO$_4$	H$_2$SO$_4$ RT 2 hr	—	—	[furyl]–CHOHCH$_2$OH (81)	146
[OAc-cyclopentene]	Hg(OAc)$_2$	H$_2$O, KCl	[cyclopentanone-HgCl] (71)	—	—	71
[methylenecyclopentane, CH$_3$]	Hg(OAc)$_2$	1:1 H$_2$O/THF 0°C 15 min	—	NaBH$_4$/NaOH	[CH$_3$, OH cyclopentane isomers] 90 : 10	147
[dimethylcyclopentene, CH$_3$, CH$_3$]	Hg(OAc)$_2$	1:1 H$_2$O/THF	—	NaBH$_4$/NaOH	[CH$_3$-cyclopentanol isomers] 51 : 39 : 6 : 4	133
[dimethylcyclopentene, CH$_3$, CH$_3$]	Hg(OAc)$_2$	1:1 H$_2$O/THF 5 min	—	NaBH$_4$/NaOH	[CH$_3$, OH cyclopentane isomers] 73 : 27 (~100 total)	133, 148

Table 2.3. (continued)

Alkene	Mercuric salt	Reaction conditions	Organomercurial(s) (% Yield)[a]	Subsequent reactants	Product(s) (% Yield)[a]	Ref.
	$Hg(OAc)_2$	1:1 H_2O / THF RT 5 min, NaBr	(86)	$NaBH_4$ / NaOH		149
	$Hg(OAc)_2$	1:1 H_2O / THF RT 5 min, NaBr		Br_2	(92)	149
	$Hg(OAc)_2$	1:1 H_2O / THF 0.08 hr RT	—	$NaBH_4$ / NaOH	(99)	39, 40
	$Hg(OAc)_2$	H_2O, NaCl	? (44) (should be trans isomer)	N_2H_4		150
	$Hg(OAc)_2$	1:1 H_2O / THF	—	$NaBH_4$ / NaOH	79 : 12 : 5 : 4	133
	$Hg(OAc)_2$	H_2O, NaCl	(1.7) + (0.6) (all should be trans isomers) (assumed) (assumed)	$Na(Hg)$ / H_2O	(trans-80) + (trans-85) 5 : 1	150

Table 2.3. (continued)

Alkene	Mercuric salt	Reaction conditions	Organomercurial(s) (% Yield)[a]	Subsequent reactants	Product(s) (% Yield)[a]	Ref.
1-methylcyclohexene (CH₃)	Hg(OAc)₂	1:1 H₂O/THF 30 min	—	NaBH₄/NaOH	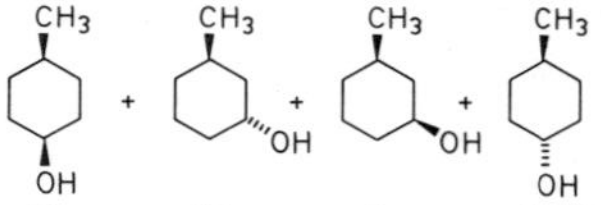 [4-methylcyclohexanol] + [3-methylcyclohexanol] + [3-methylcyclohexanol] + [4-methylcyclohexanol] 51 : 47 : 1 : 1 (90 total)	133
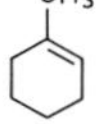	Hg(OAc)₂	H₂O, NaCl	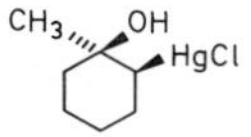 [1-methyl-2-(HgCl)cyclohexanol] (62) (should be trans isomer)	Na(Hg)/H₂O	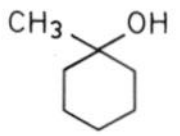 [1-methylcyclohexanol] (51)	150
	Hg(OAc)₂	1:1 H₂O/THF 0.08 hr RT	—	NaBH₄/NaOH	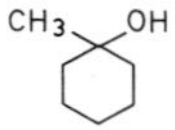 [1-methylcyclohexanol] (100)	39, 40
	Hg(OAc)₂	1:1 H₂O/Et₂O 30 min RT	—	NaBH₄/NaOH	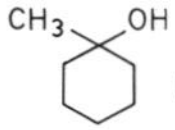 [1-methylcyclohexanol] (70.5–75.4)	151
 cycloheptene	Hg(OAc)₂	3:1 H₂O/CH₃COCH₃ RT, NaCl	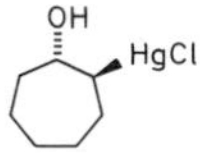 [2-(HgCl)cycloheptanol] (64)	NaBH₄	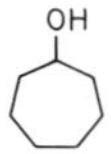 [cycloheptanol]	128
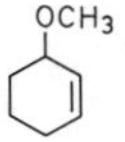 3-methoxycyclohexene (OCH₃)	Hg(OAc)₂	1:1 H₂O/THF	—	NaBH₄/NaOH	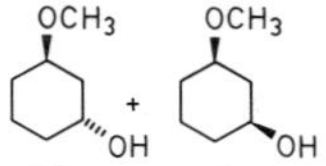 [OCH₃ cyclohexanol] + [OCH₃ cyclohexanol] 85 : 15	144
	Hg(OAc)₂	1:1 H₂O/THF 25 °C ~30 min	—	NaBH₄/NaOH	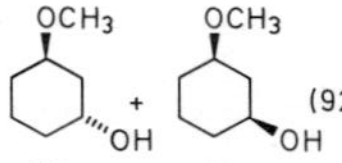 [OCH₃ cyclohexanol] + [OCH₃ cyclohexanol] (92 total) 89 : 11	142

Table 2.3. (continued)

Alkene	Mercuric salt	Reaction conditions	Organomercurial(s) (% Yield)[a]	Subsequent reactants	Product(s) (% Yield)[a]	Ref.
	$Hg(OAc)_2$	1:19 H_2O/CH_3CN 25°C ~30 min	—	$NaBH_4/NaOH$	(93 total) 94 : 6	142
	$Hg(OAc)_2$	H_2O, NaCl		$N_2H_4/NaOH$	(72) + (10)	145
	$Hg(OAc)_2$	H_2O, NaCl		$N_2H_4/NaOH$	(65)	145
	$Hg(OAc)_2$	1:1 H_2O/THF	—	$NaBH_4/NaOH$	(65 total) 71 : 27 : 2	141
	$Hg(OAc)_2$	1:1 H_2O/THF 12 hr	—	$NaBH_4/NaOH$	+ other diols 50 : 20 : 5 : 5	152
	$Hg(OAc)_2$	1:1 H_2O/THF 12 hr	—	$NaBH_4/NaOH$	80 : 10 : 10	152
	$Hg(OAc)_2$	1:1 H_2O/THF	—	$NaBH_4/NaOH$	65 : 15 : 10 : 10	152

Table 2.3. (continued)

Alkene	Mercuric salt	Reaction conditions	Organomercurial(s) (% Yield)[a]	Subsequent reactants	Product(s) (% Yield)[a]	Ref.
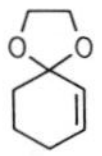	$Hg(OAc)_2$	H_2O, NaCl	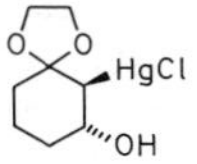 (62)	—	—	153
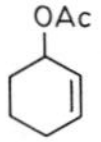	$Hg(OAc)_2$	1:1 H_2O/THF	—	$NaBH_4$/NaOH	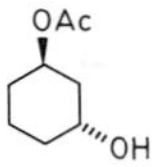 (65–75)	144
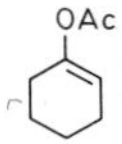	$Hg(OAc)_2$	1:19 H_2O/CH_3CN 25 °C ~30 min	—	$NaBH_4$/NaOH	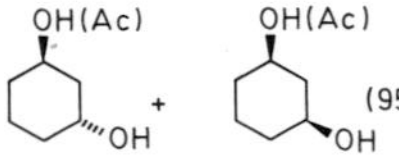 96 : 4 (95 total)	142
	$Hg(OAc)_2$	H_2O, KCl	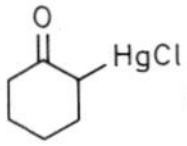 (86)	—	—	71
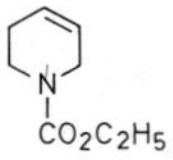	$Hg(NO_3)_2$	1:1 H_2O/THF 5 h	—	$NaBH_4$/NaOH, CrO_3	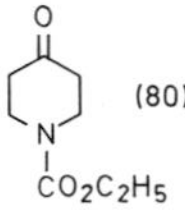 (80)	154
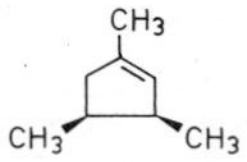	$Hg(OAc)_2$	1:1 H_2O/THF 15 min	—	$NaBH_4$/NaOH	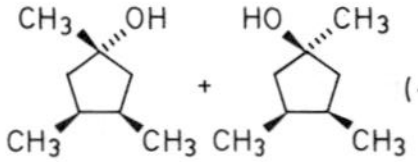 76 : 24 (~100 total)	148
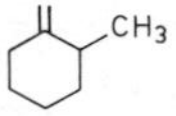	$Hg(OAc)_2$	1:1 H_2O/THF	—	$NaBH_4$/NaOH	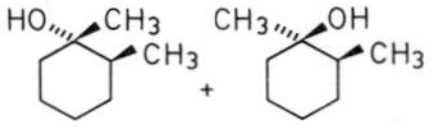 66 : 34 (~100 total)	147, 155, 156
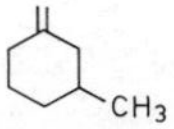	$Hg(OAc)_2$	1:1 H_2O/THF	—	$NaBH_4$/NaOH	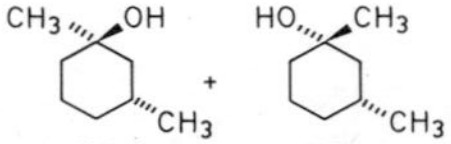 58 : 42 (~100 total)	155, 156

Table 2.3. (continued)

Alkene	Mercuric salt	Reaction conditions	Organomercurial(s) (% Yield)[a]	Subsequent reactants	Product(s) (% Yield)[a]	Ref.
	Hg(OAc)$_2$	1:1 H$_2$O / THF 0 °C 15 min	—	NaBH$_4$ / NaOH	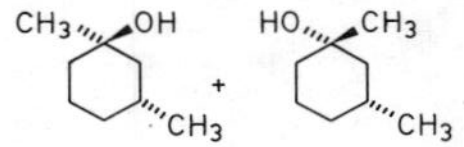 CH$_3$–OH / HO–CH$_3$ 54 : 46	147
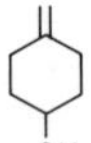 (methylenecyclohexane, CH$_3$)	Hg(OAc)$_2$	1:1 H$_2$O / THF 0 °C 15 min	—	NaBH$_4$ / NaOH	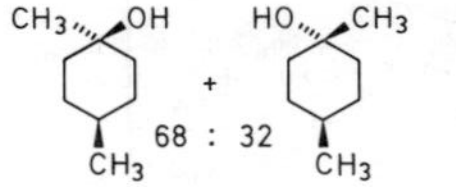 CH$_3$–OH / HO–CH$_3$ 68 : 32	147
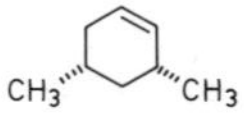 (CH$_3$, CH$_3$-cyclohexene)	Hg(OAc)$_2$	1:1 H$_2$O / THF 25 °C	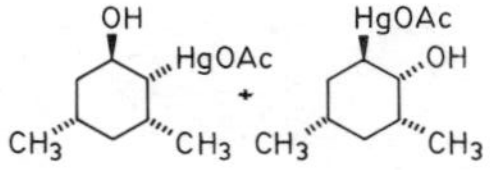 OH, HgOAc, CH$_3$, CH$_3$ + HgOAc, OH, CH$_3$, CH$_3$	NaBH$_4$ / NaOH	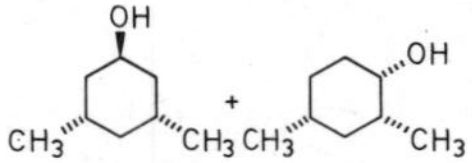 OH, CH$_3$, CH$_3$ + OH, CH$_3$, CH$_3$ 83 : 17	157, 158
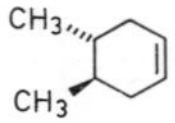 (CH$_3$, CH$_3$-cyclohexene)	Hg(OAc)$_2$	1:1 H$_2$O / THF	—	NaBH$_4$ / NaOH	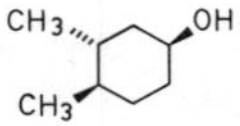 CH$_3$, OH, CH$_3$	155
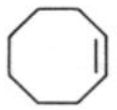 (cyclooctene)	Hg(OAc)$_2$	1:1 H$_2$O / CH$_3$COCH$_3$ 3d, NaCl	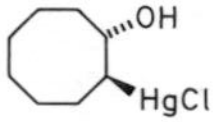 OH, HgCl (50)	—	—	159
	Hg(OAc)$_2$	3:1 H$_2$O / CH$_3$COCH$_3$ RT, NaCl	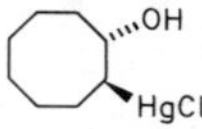 OH, HgCl (59)	NaBH$_4$	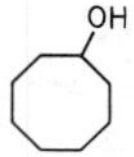 OH	128
	Hg(OAc)$_2$	1:1 H$_2$O / THF 3 hr RT	—	NaBH$_4$ / NaOH	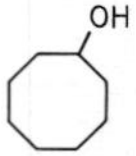 OH (88)	39, 40
 (bicyclic alkene)	Hg(OAc)$_2$	1:1 H$_2$O / CH$_3$COCH$_3$ 5 min, NaCl	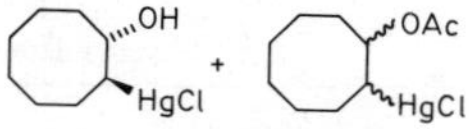 OH, HgCl + OAc, HgCl	—	—	159
	Hg(OAc)$_2$	1:1 H$_2$O / CH$_3$COCH$_3$ several hr, NaCl	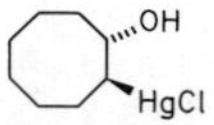 OH, HgCl (70)	—	—	159

Table 2.3. (continued)

Alkene	Mercuric salt	Reaction conditions	Organomercurial(s) (% Yield)^a	Subsequent reactants	Product(s) (% Yield)^a	Ref.
	Hg(OAc)$_2$	3:1 H$_2$O/CH$_3$COCH$_3$ RT, NaCl	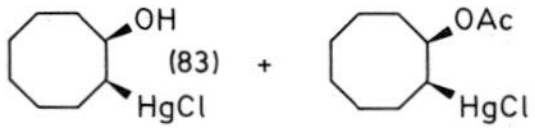 [cyclooctyl OH, HgCl (83)] + [cyclooctyl OAc, HgCl]	NaBH$_4$ (on alcohol)	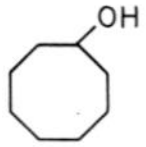 [cyclooctanol OH]	128
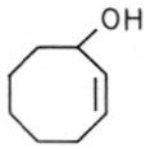 [cyclooctenol OH]	Hg(OAc)$_2$	1:1 H$_2$O/THF	—	NaBH$_4$/NaOH	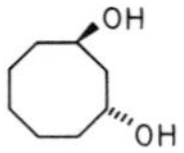 [cyclooctanediol OH, OH] (70)	141
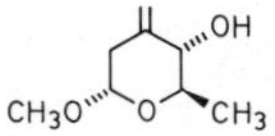 [pyran CH$_3$O, CH$_3$]	Hg(OAc)$_2$	1:1 H$_2$O/THF RT 1 hr	—	NaBH$_4$/NaOH	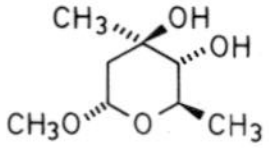 [CH$_3$, OH, OH pyran CH$_3$O, CH$_3$] (25)	160
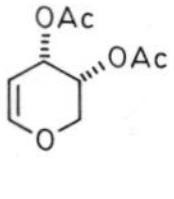 [OAc, OAc dihydropyran]	cat. HgSO$_4$	H$_2$O/H$_2$SO$_4$/dioxane 3 hr	—	—	*trans*-HOCH$_2$CHOAcCH=CHCHO (88)	146
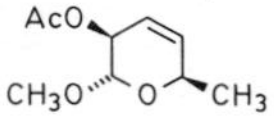 [AcO, CH$_3$O, O, CH$_3$ pyran]	Hg(OAc)$_2$	H$_2$O RT 5 hr	—	NaBH$_4$, Ac$_2$O	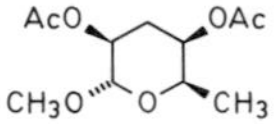 [AcO, OAc, CH$_3$O, O, CH$_3$ pyran] (70)	161
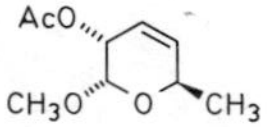 [AcO, CH$_3$O, O, CH$_3$ pyran]	Hg(OAc)$_2$	H$_2$O RT 4 hr	—	NaBH$_4$, Ac$_2$O	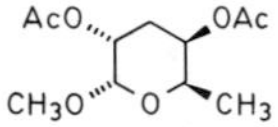 [AcO, OAc, CH$_3$O, O, CH$_3$ pyran] (69)	161
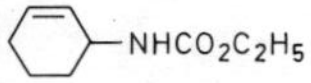 [cyclohexene—NHCO$_2$C$_2$H$_5$]	Hg(NO$_3$)$_2$	1:1 H$_2$O/THF 24 hr	—	NaBH$_4$/NaOH, CrO$_3$	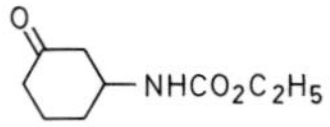 [cyclohexanone—NHCO$_2$C$_2$H$_5$] (57)	154
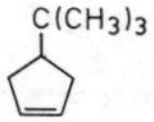 [cyclopentene C(CH$_3$)$_3$]	Hg(OAc)$_2$	1:1 H$_2$O/THF	—	NaBH$_4$/NaOH	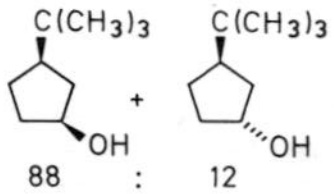 [C(CH$_3$)$_3$ cyclopentanol OH] + [C(CH$_3$)$_3$ cyclopentanol OH] 88 : 12	133

Table 2.3. (continued)

Alkene	Mercuric salt	Reaction conditions	Organomercurial(s) (% Yield)[a]	Subsequent reactants	Product(s) (% Yield)[a]	Ref.
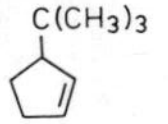	Hg(OAc)$_2$	1:1 H$_2$O/THF 6 hr	—	NaBH$_4$/NaOH	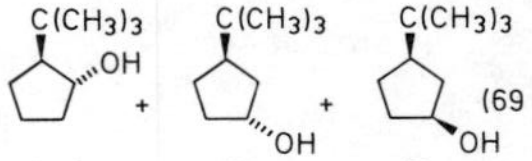 71 : 17 : 12 (69 total)	133
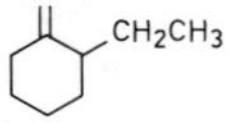	Hg(OAc)$_2$	1:1 H$_2$O/THF	—	NaBH$_4$/NaOH	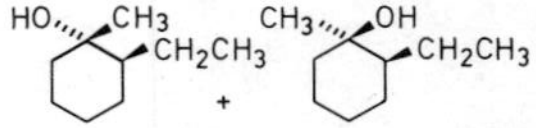 74 : 26 (~100 total)	155, 156
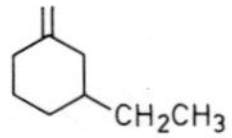	Hg(OAc)$_2$	1:1 H$_2$O/THF	—	NaBH$_4$/NaOH	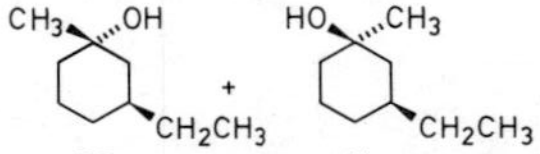 56 : 44	155
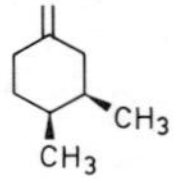	Hg(OAc)$_2$	1:1 H$_2$O/THF	—	NaBH$_4$/NaOH	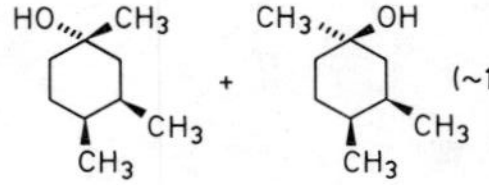 59 : 41 (~100 total)	156
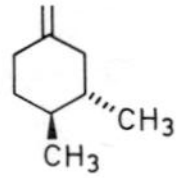	Hg(OAc)$_2$	1:1 H$_2$O/THF	—	NaBH$_4$/NaOH	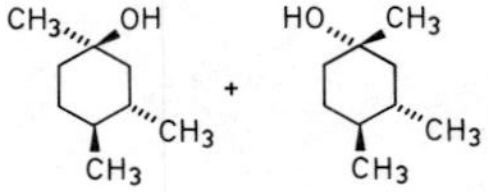 62 : 38 (~100 total)	155, 156
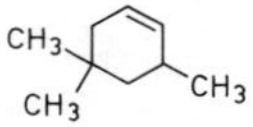	Hg(OAc)$_2$	1:1 H$_2$O/THF 25 °C	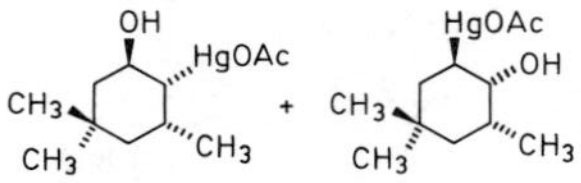	NaBH$_4$/NaOH	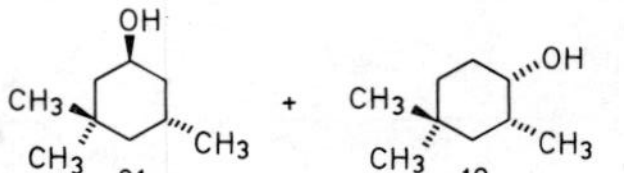 81 : 19	157, 158
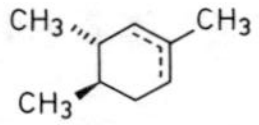	Hg(OAc)$_2$	1:1 H$_2$O/THF	—	NaBH$_4$/NaOH	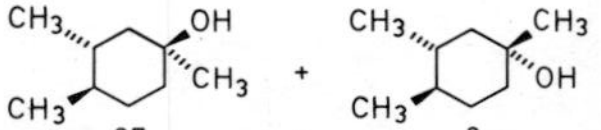 97 : 3	155

Table 2.3. (continued)

Alkene	Mercuric salt	Reaction conditions	Organomercurial(s) (% Yield)[a]	Subsequent reactants	Product(s) (% Yield)[a]	Ref.
 (cyclooctene)	Hg(OAc)₂	3:1 H₂O/CH₃COCH₃ RT, NaCl	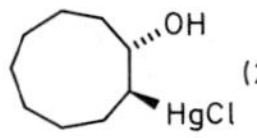 (OH, HgCl structure) (25)	NaBH₄	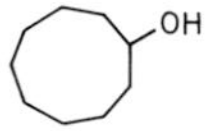 (cyclooctanol, OH)	128
 (cyclooctene)	Hg(OAc)₂	3:1 H₂O/CH₃COCH₃ RT, NaCl	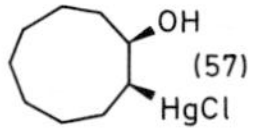 (OH, HgCl) (57) 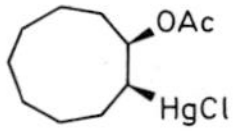 + (OAc, HgCl)	NaBH₄	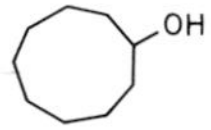 (cyclooctanol, OH)	128
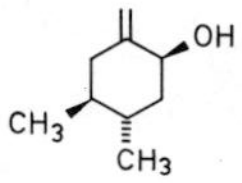 (OH, CH₃, CH₃ cyclohexane structure)	Hg(OAc)₂	1:1 H₂O/THF	—	NaBH₄/NaOH	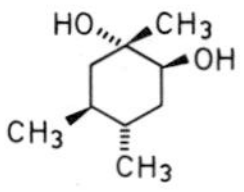 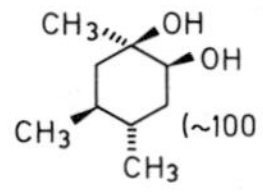 (CH₃, OH, CH₃ structure) + (CH₃, OH, CH₃ structure) (~100 total)	156
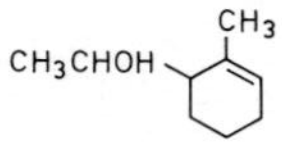 CH₃CHOH (CH₃ cyclohexene structure)	Hg(OAc)₂	1:1 H₂O/THF	—	NaBH₄/NaOH	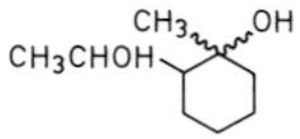 (CH₃CHOH, CH₃, OH structure) (> 80)	163
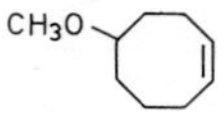 CH₃O (cyclooctene structure)	Hg(OAc)₂	1:1 H₂O/THF 25°C	—	NaBH₄/NaOH	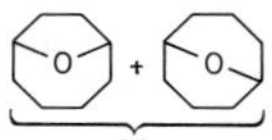 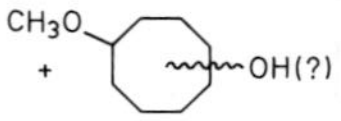 CH₃O (OH(?) structure) + (bicyclic O ethers) 70 : 30 (70 total)	92
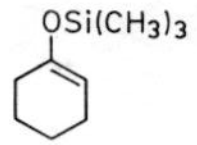 OSi(CH₃)₃ (cyclohexene structure)	HgO/Hg(OAc)₂	H₂O/EtOH	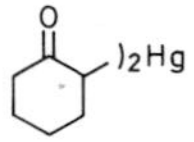 ( )₂Hg (O structure) (66, 85)	—	—	255b, 255d, 255f
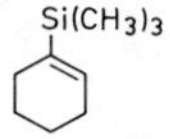 Si(CH₃)₃ (cyclohexene structure)	Hg(OAc)₂	1:1 H₂O/THF	—	NaBH₄/NaOH	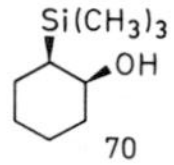 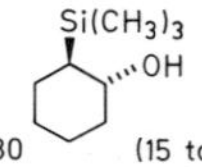 Si(CH₃)₃ (OH structure) 30 + Si(CH₃)₃ (OH structure) 70 (15 total)	164
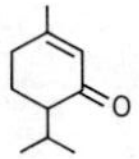 (carvone-type O structure)	Hg(OAc)₂	1:1 H₂O/THF 12 hr RT	—	NaBH₄/NaOH	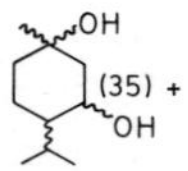 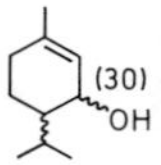 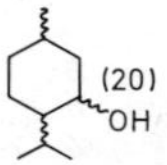 (OH structure) (20) + (OH structure) (30) + (OH structure) (35)	165

43

Table 2.3. (continued)

Alkene	Mercuric salt	Reaction conditions	Organomercurial(s) (% Yield)[a]	Subsequent reactants	Product(s) (% Yield)[a]	Ref.
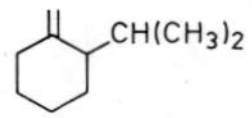	$Hg(OAc)_2$	1:1 H_2O/THF 0 °C 15 min	—	$NaBH_4$/NaOH	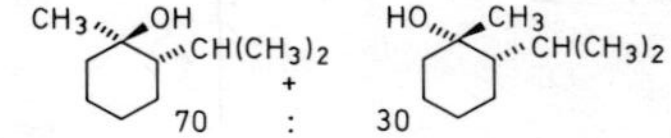 70 : 30	147
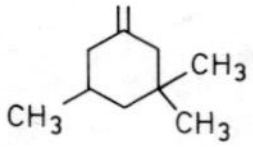	$Hg(OAc)_2$	1:1 H_2O/THF	—	$NaBH_4$/NaOH	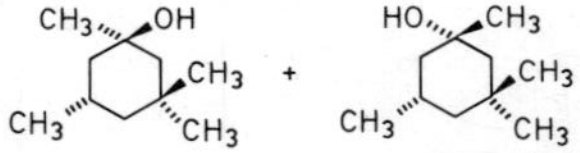 70 : 30 (~100 total)	156
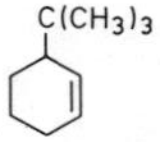	$Hg(OAc)_2$	1:1 H_2O/THF 5 hr	—	$NaBH_4$/NaOH	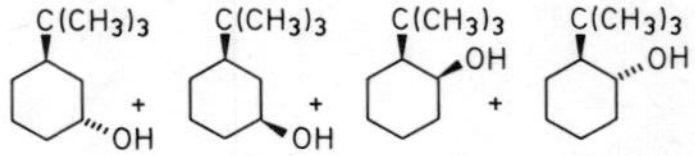 91 : 4 : 4 : 1 (78 total)	133
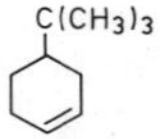	$Hg(OAc)_2$	1:1 H_2O/THF 25 °C 20 min	—	$NaBH_4$/NaOH	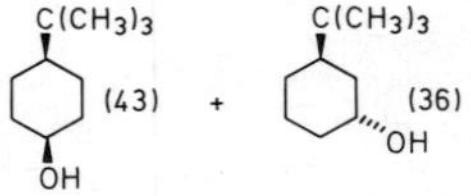 (43) + (36)	166
	$Hg(OAc)_2$	1:1 H_2O/THF 20–25 °C 1 hr	—	$NaBH_4$/NaOH	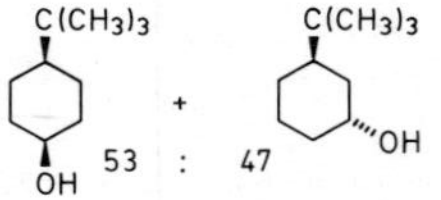 53 : 47	158, 167, 168
	$Hg(OAc)_2$	1:1 H_2O/THF 25 °C	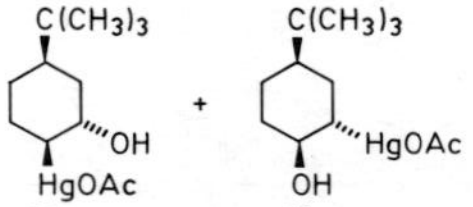 48 : 41	$NaBH_4$	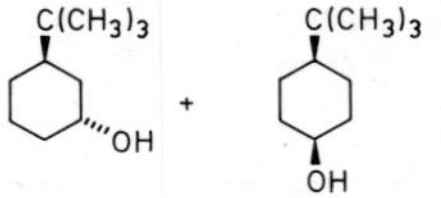	157
	$Hg(OAc)_2$	1:1 H_2O/THF	—	$NaBH_4$/NaOH	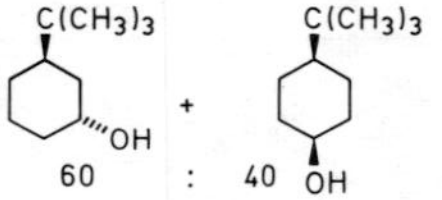 60 : 40	155

Table 2.3. (continued)

Alkene	Mercuric salt	Reaction conditions	Organomercurial(s) (% Yield)[a]	Subsequent reactants	Product(s) (% Yield)[a]	Ref.
	Hg(OAc)$_2$	1:1 H$_2$O/THF 3% HClO$_4$ 1hr	—	NaBH$_4$/NaOH	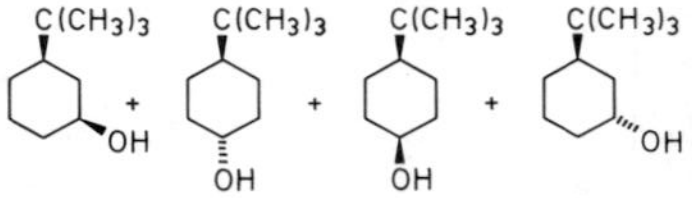 C(CH$_3$)$_3$ ... 46 : 43 : 6 : 5	168
	Hg(OAc)$_2$	244 hr	—	NaBH$_4$	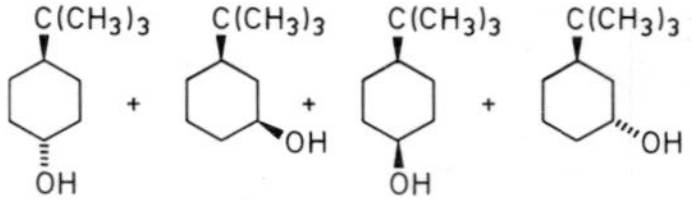 C(CH$_3$)$_3$ ... 41.1 : 37.0 : 10.8 : 10.1	166
	Hg(OAc)$_2$	1:1 D$_2$O/CD$_3$COCD$_3$ 25 °C 10 min	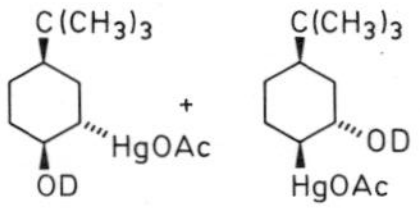 C(CH$_3$)$_3$, OD, HgOAc	—	—	166
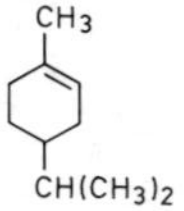 CH$_3$ / CH(CH$_3$)$_2$	Hg(OAc)$_2$	H$_2$O/dioxane/NaOH, NaCl	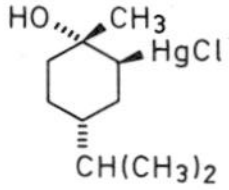 HO, CH$_3$, HgCl, CH(CH$_3$)$_2$ (74)	N$_2$H$_4$/NaOH	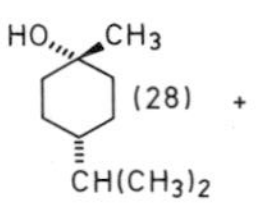 (28) 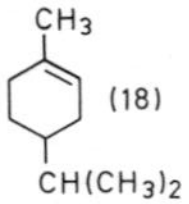 + (18)	169
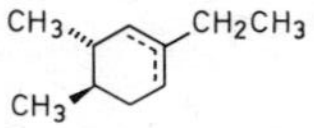 CH$_3$ / CH$_2$CH$_3$ / CH$_3$	Hg(OAc)$_2$	1:1 H$_2$O/THF	—	NaBH$_4$/NaOH	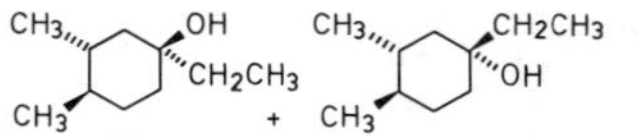 98 : 2	155
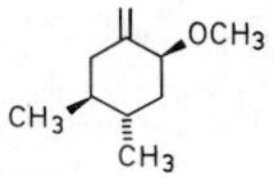 OCH$_3$ / CH$_3$ / CH$_3$	Hg(OAc)$_2$	1:1 H$_2$O/THF	—	NaBH$_4$/NaOH	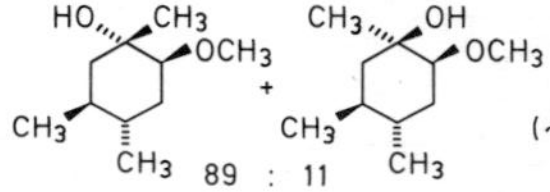 89 : 11 (~100 total)	156
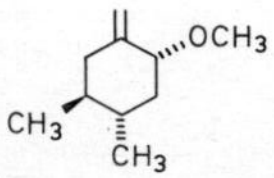 OCH$_3$ / CH$_3$ / CH$_3$	Hg(OAc)$_2$	1:1 H$_2$O/THF	—	NaBH$_4$/NaOH	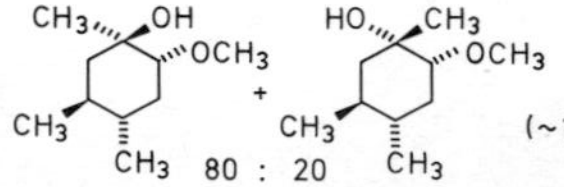 80 : 20 (~100 total)	156

Table 2.3. (continued)

Alkene	Mercuric salt	Reaction conditions	Organomercurial(s) (% Yield)[a]	Subsequent reactants	Product(s) (% Yield)[a]	Ref.
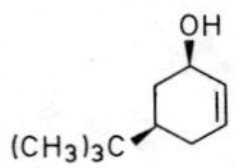 (cyclohexene with OH, (CH₃)₃C substituents)	Hg(OAc)₂	1:1 H₂O/THF	—	NaBH₄/NaOH	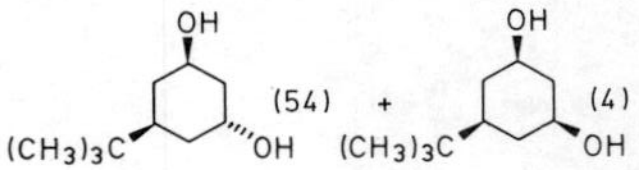 (diol, 54) + (diol, 4)	170
	Hg(OAc)₂	1:1 H₂O/THF	—	NaBH₄/NaOH	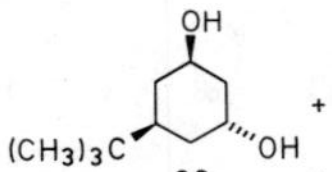 (diol) 96 : 4 + other glycols	167
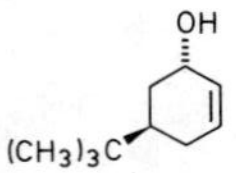 (cyclohexene with OH, (CH₃)₃C substituents)	Hg(OAc)₂	1:1 H₂O/THF	—	NaBH₄/NaOH	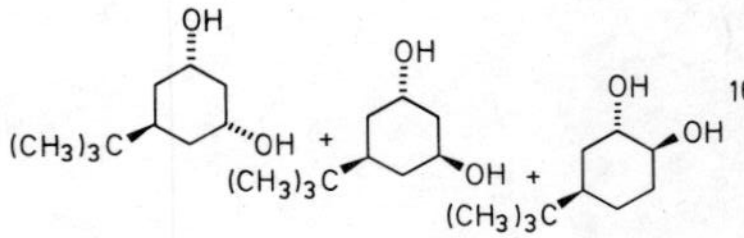 75 : 17 : 8 (diols)	167
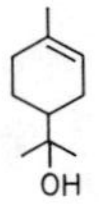 (methylcyclohexene carbinol)	Hg(OAc)₂	H₂O/THF	—	NaBH₄/NaOH	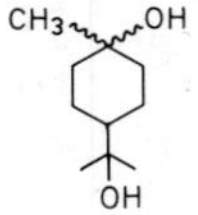 (diol)	171
	Hg(OAc)₂	1:1 H₂O/THF 2 min RT	—	NaBH₄/NaOH	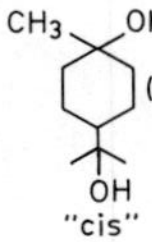 (68.6)  "cis" + (29.2)	172
	Hg(NO₃)₂	H₂O/Et₂O, KOH/KI	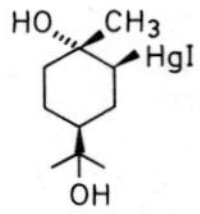 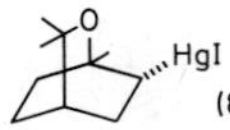 (organomercurials) + (8)	Na(Hg)/H₂O (on diol)	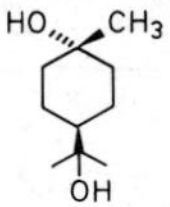 (diol)	173

Table 2.3. (continued)

Alkene	Mercuric salt	Reaction conditions	Organomercurial(s) (% Yield)[a]	Subsequent reactants	Product(s) (% Yield)[a]	Ref.
	HgO	—	—	$NaBH_4$ / NaOH	"trans" (48.8) + "cis" (20.2) + (8.6)	172
	—	—	(organomercurial: HO–CH_3, HgCl, OH)	—	(HO–CH_3, OH)	145
	—	—	—	$Na(Hg)$ / H_2O	(CH_3, OH, OH)	13
$OSi(CH_3)_3$ / CH_3	HgO / $Hg(OAc)_2$	H_2O / EtOH	($)_2$Hg, CH_3, trans > cis	—	—	255f
Br, CH_3, NCH_3, phenyl	$Hg(OAc)_2$	—	Br, HO, HgOAc, CH_3, NCH_3, $(HgOAc)_3$	—	—	174
AcO, CH_3O, CH_2OAc	$Hg(OAc)_2$	H_2O RT 30 hr	—	$NaBH_4$, Ac_2O	AcO, OAc, CH_3O, CH_2OAc (75)	161
AcO, CH_3O, CH_2OAc	$Hg(OAc)_2$	H_2O RT 25 hr	—	$NaBH_4$, Ac_2O	AcO, OAc, CH_3O, CH_2OAc (80)	161

48

Table 2.3. (continued)

Alkene	Mercuric salt	Reaction conditions	Organomercurial(s) (% Yield)[a]	Subsequent reactants	Product(s) (% Yield)[a]	Ref.
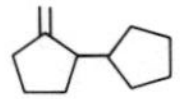	Hg(OAc)$_2$	1:1 H$_2$O/THF 0°C 15 min	—	NaBH$_4$/NaOH	 72 : 28	147
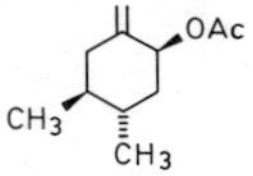	Hg(OAc)$_2$	1:1 H$_2$O/THF	—	NaBH$_4$/NaOH	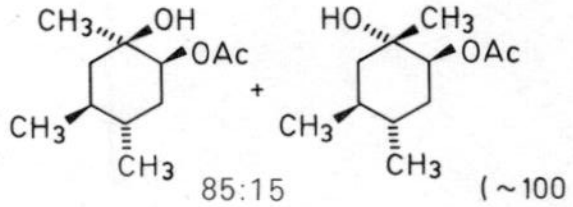 85:15 (~100 total)	156
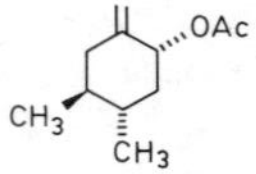	Hg(OAc)$_2$	1:1 H$_2$O/THF	—	NaBH$_4$/NaOH	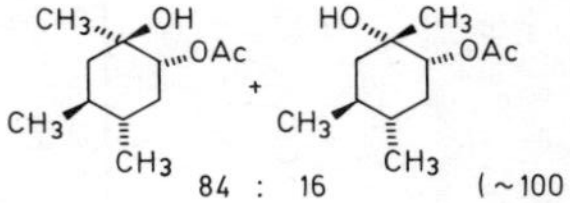 84 : 16 (~100 total)	156
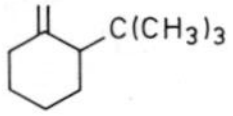	Hg(OAc)$_2$	1:1 H$_2$O/THF	—	NaBH$_4$/NaOH	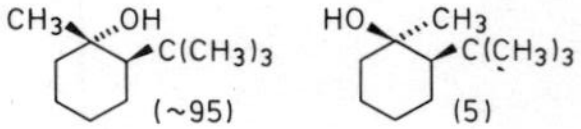 (~95) (5)	175
	Hg(OAc)$_2$	1:1 H$_2$O/THF 0°C 15 min	—	NaBH$_4$/NaOH	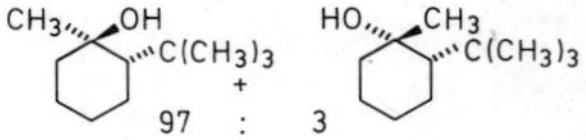 97 : 3	147
	Hg(OAc)$_2$	1:1 H$_2$O/THF 0°C 15 min	—	NaBH$_4$/NaOH	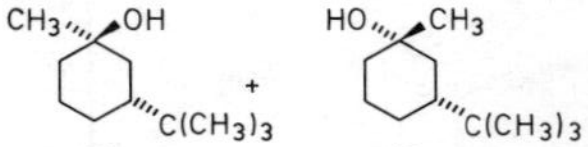 58 : 42	147
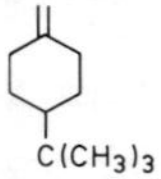	Hg(OAc)$_2$	1:1 H$_2$O/THF 0°C 15 min	—	NaBH$_4$/NaOH	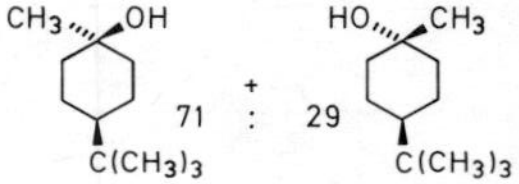 71 : 29	147

Table 2.3. (continued)

Alkene	Mercuric salt	Reaction conditions	Organomercurial(s) (% Yield)[a]	Subsequent reactants	Product(s) (% Yield)[a]	Ref.
	Hg(OAc)₂	1:1 H₂O/THF	—	NaBH₄/NaOH	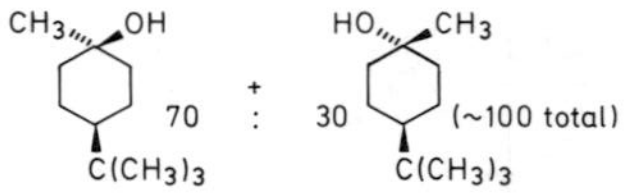 70 : 30 (~100 total)	156
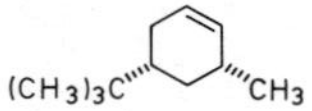 (CH₃)₃C—[cyclohexene]—CH₃	Hg(OAc)₂	1:1 H₂O/THF 25 °C	—	NaBH₄/NaOH	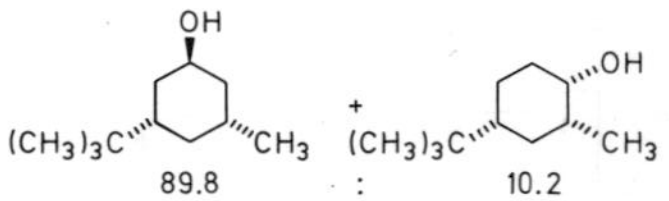 89.8 : 10.2	157, 158
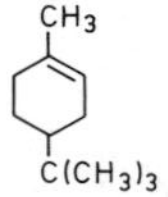 [1-methyl-4-tert-butylcyclohexene]	Hg(OAc)₂	—	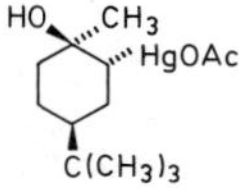 [organomercurial]	NaBH₄	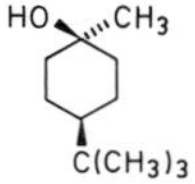 (93.7)	166
	Hg(OAc)₂	1:1 H₂O/THF 30 min	—	NaBH₄/NaOH	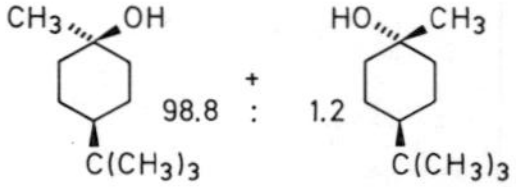 98.8 : 1.2	168
	Hg(OAc)₂	1:1 H₂O/THF 0.2% H₂SO₄ 2C hr	—	NaBH₄/NaOH	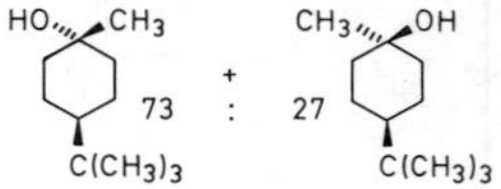 73 : 27	168
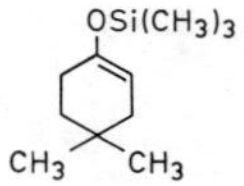 [OSi(CH₃)₃ cyclohexene, 4,4-diCH₃]	HgO/Hg(OAc)₂	H₂O/EtOH	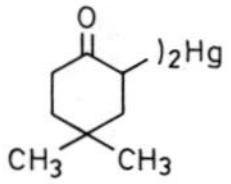 [organomercurial] ₂Hg	—	—	255f
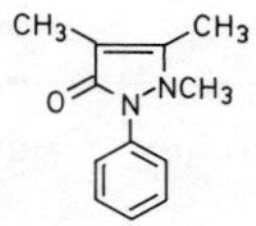 [pyrazolone, NCH₃, phenyl]	Hg(OAc)₂	—	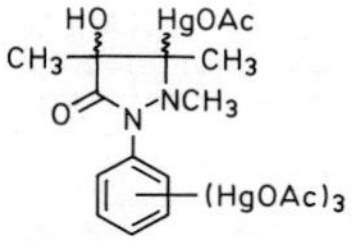 [organomercurial] (HgOAc)₃	—	—	174

Table 2.3. (continued)

Alkene	Mercuric salt	Reaction conditions	Organomercurial(s) (% Yield)[a]	Subsequent reactants	Product(s) (% Yield)[a]	Ref.
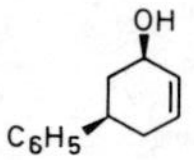 (cyclohexene, C_6H_5, OH)	$Hg(OAc)_2$	1:1 H_2O/THF	—	$NaBH_4$/NaOH	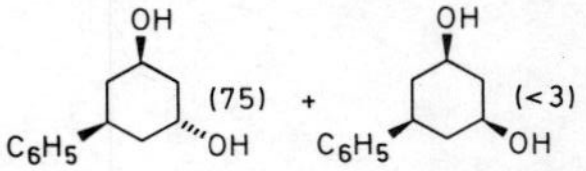 (75) + (<3)	170
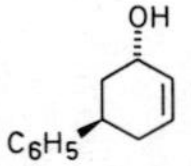 (cyclohexene, C_6H_5, OH)	$Hg(OAc)_2$	1:1 H_2O/THF	—	$NaBH_4$/NaOH	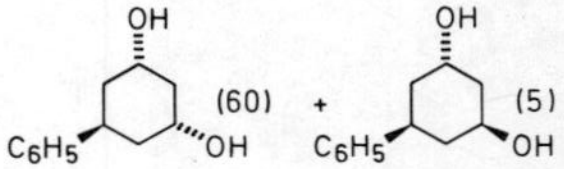 (60) + (5)	170
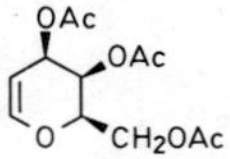 (OAc, OAc, CH_2OAc pyran)	cat. $HgSO_4$	H_2SO_4	—	—	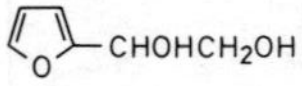 furan–$CHOHCH_2OH$	146
	cat. $HgSO_4$	H_2O/H_2SO_4/dioxane 3 hr	—	—	trans-$AcOCH_2CHOHCHOAcCH=CHCHO$ + trans-$AcOCH_2CHOAcCHOAcCH=CHCHO$ (82 total)	146
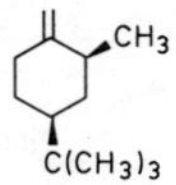 (CH_3, $C(CH_3)_3$ methylenecyclohexane)	$Hg(OAc)_2$	1:1 H_2O/THF 0 °C 15 min	—	$NaBH_4$/NaOH	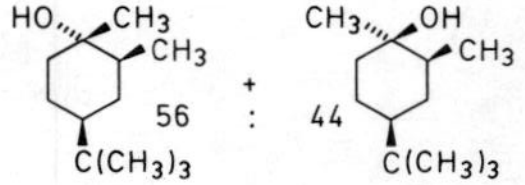 56 : 44	147
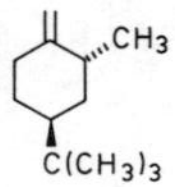 (CH_3, $C(CH_3)_3$ methylenecyclohexane)	$Hg(OAc)_2$	1:1 H_2O/THF 0 °C 15 min	—	$NaBH_4$/NaOH	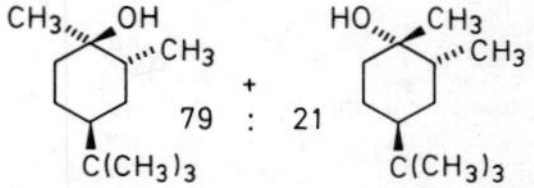 79 : 21	147
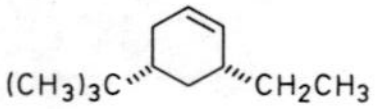 ($(CH_3)_3C$, CH_2CH_3 cyclohexene)	$Hg(OAc)_2$	1:1 H_2O/THF 25 °C	—	$NaBH_4$/NaOH	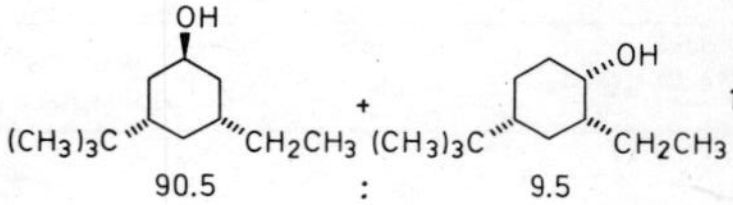 90.5 : 9.5	157, 158

Table 2.3. (continued)

Alkene	Mercuric salt	Reaction conditions	Organomercurial(s) (% Yield)[a]	Subsequent reactants	Product(s) (% Yield)[a]	Ref.
	0.5 Hg(O₂CC₂H₅)₂	H₂O/dioxane/CF₃CO₂H Na₂Cr₂O₇ 18 hr	—	—	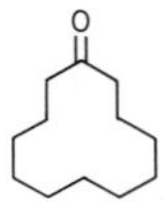 (36)	46
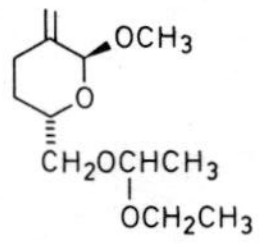	Hg(OAc)₂	H₂O/THF 1.5 hr	—	NaBH₄/NaOH	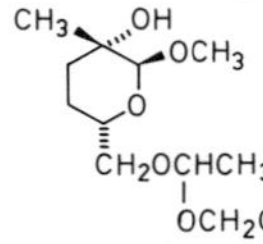 (85)	176
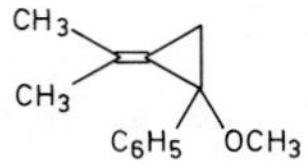	Hg(OAc)₂	—	—	H₂S	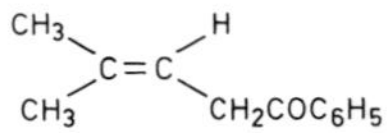 (76)	177
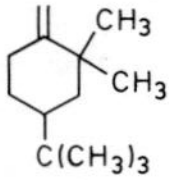	Hg(OAc)₂	1:1 H₂O/THF 0 °C 15 min	—	NaBH₄/NaOH	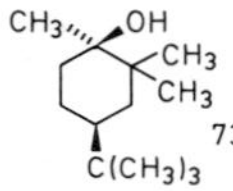 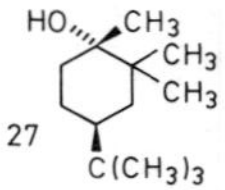 73 : 27	147
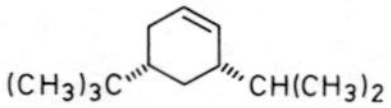	Hg(OAc)₂	1:1 H₂O/THF 25 °C	—	NaBH₄/NaOH	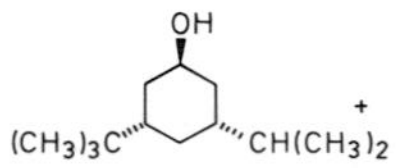 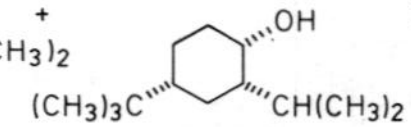 94.2 : 5.8	157, 158
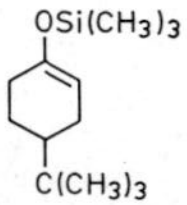	HgO/Hg(OAc)₂	H₂O/EtOH	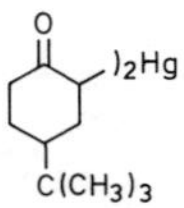 trans > cis	—	—	255f

Table 2.3. (continued)

Alkene	Mercuric salt	Reaction conditions	Organomercurial(s) (% Yield)[a]	Subsequent reactants	Product(s) (% Yield)[a]	Ref.
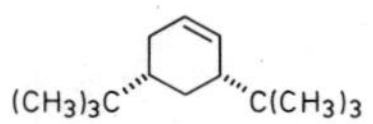 (structure: $(CH_3)_3C$…$C(CH_3)_3$ cyclohexene)	$Hg(OAc)_2$	1:1 H_2O/THF 25 °C	—	$NaBH_4$/NaOH	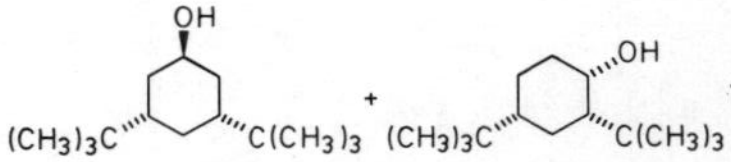 (structures: $(CH_3)_3C$…$C(CH_3)_3$ cyclohexanol, 99.06 : 0.94)	157, 158
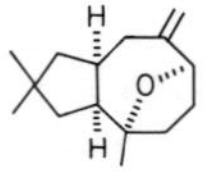 (tricyclic ether structure)	$Hg(NO_3)_2$	H_2O	—	BH_4^-	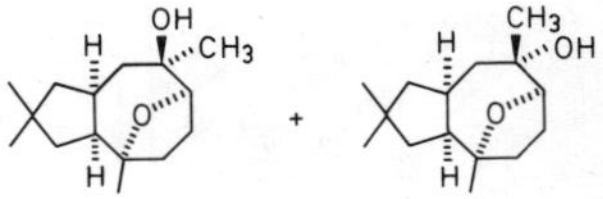 (two tricyclic CH_3/OH ether structures)	178
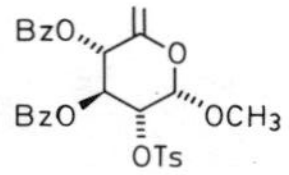 (BzO, BzO, OCH_3, OTs pyran structure)	$HgCl_2$	H_2O/CH_3COCH_3 Δ 4.5 hr	—	—	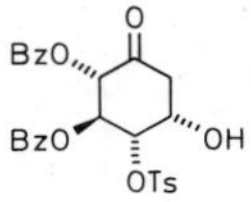 (BzO, BzO, OH, OTs cyclohexanone structure) (83)	179, 180

[a] If there is no entry in the organomercurial column, the yield is based on starting alkene. If there is an organomercurial entry, the yield is based on the organomercurial unless otherwise stated.

Table 2.4. Hydroxymercuration of Bicyclic and Polycyclic Alkenes

Alkene	Mercuric salt	Reaction conditions	Organomercurial(s) (% Yield)[a]	Subsequent reactants	Product(s) (% Yield)[a]	Ref.
(norbornene)	$Hg(OAc)_2$	H_2O, Cl^-	OH / HgCl	Ac_2O, $NaBD_4$	OAc / D	181
(norbornene)	$Hg(OAc)_2$	H_2O	OH (96) / HgOAc + OAc / HgOAc	—	—	182
	$Hg(OAc)_2$	H_2O / THF	OH / HgOAc + OAc / HgOAc 81 : 19 (95 total)	—	—	183
	$Hg(OAc)_2$	H_2O / THF	—	$NaBH_4$ / NaOH	OH + OAc 83 : 17	184
	$Hg(OAc)_2$	1:1 H_2O / THF 30 sec	—	$NaBH_4$ / NaOH	OH (100) >99.8% exo	185, 186
	$Hg(OAc)_2$	D_2O 2–3 hr	OD / HgOAc	—	—	140
	$Hg(OAc)_2$	H_2O / HNO_3 1d, NaCl	OH / HgCl (88)	—	—	187, 188
	1:1 $Hg(OAc)_2$ / HgO	—	OH / HgOAc (58)	—	—	189
	1:1 $Hg(OAc)_2$ / HgO	H_2O 30 hr	OH / HgOAc (61)	—	—	131

Table 2.4. (continued)

Alkene	Mercuric salt	Reaction conditions	Organomercurial(s) (% Yield)[a]	Subsequent reactants	Product(s) (% Yield)[a]	Ref.
	$Hg(O_2CCF_3)_2$	1:1 H_2O/THF 15 min	—	$NaBH_4$/NaOH	(95)	41
	$Hg(NO_3)_2$	3:1 H_2O/CH_3CN, NaCl	(92)	—	—	190
	$Hg(NO_3)_2$	1:1 H_2O/THF 15 min	—	$NaBH_4$/NaOH	(100)	41
	$Hg(O_3SCH_3)_2$	1:1 H_2O/THF 60 min	—	$NaBH_4$/NaOH	(94)	41
	$Hg(OH)ClO_4$	H_2O, NaCl	(86)	Na(Hg)/H_2O		191
	1:1 HgO/$HClO_4$	H_2O 50 °C, NaCl	(86)	—	—	131
	$Hg[C(NO_2)_3]_2$	H_2O overnight		—	—	188
	$Hg(OAc)_2$	1:1 H_2O/THF 10 min RT	—	$NaBH_4$/NaOH	(83)	192
	$Hg(OAc)_2$	1:1 H_2O/THF 10 min RT	—	$NaBH_4$/NaOH	(96)	192
	$Hg(OAc)_2$	H_2O/THF	—	$NaBH_4$/NaOH		193

Table 2.4. (continued)

Alkene	Mercuric salt	Reaction conditions	Organomercurial(s) (% Yield)[a]	Subsequent reactants	Product(s) (% Yield)[a]	Ref.
(bicyclic alkene, CN)	$Hg(OAc)_2$	H_2O/THF	—	$NaBH_4/NaOH$	(bicyclic, OH, CN)	193
	$HgCl_2$	3:1 H_2O/CH_3COCH_3	(OH, HgCl, CN) 97 : 3 (HgCl, NH, O)	—	—	190
(bicyclic diketone, N–N)	$Hg(OAc)_2$	H_2O	(fused ring, OH, HgOAc)	—	—	194
(bicyclic alkene, CO_2H)	$Hg(OAc)_2$	H_2O	—	$NaBH_4$	(bicyclic, OH, CO_2H)	190
(methylenenorbornane)	$Hg(OAc)_2$	1:1 H_2O/THF 5 min	—	$NaBH_4/NaOH$	(OH, CH_3) + (CH_3, OH) (99.5) : (0.5) (~100 total)	148, 186, 195
(1-methylnorbornene, CH_3)	$Hg(OAc)_2$	H_2O/THF 5 min	—	$NaBH_4/NaOH$	(CH_3, OH) + (OH, CH_3) + (CH_3, OH) 48 : 48 : 4 (100 total)	148
	$Hg(OAc)_2$	$H_2O/NaOAc$ 1 hr	—	$NaBH_4/NaOH$	(OH, CH_3) + (CH_3, OH) + (OH, CH_3) 53.5 : 46 : 0.5 (93 total)	185

Table 2.4. (continued)

Alkene	Mercuric salt	Reaction conditions	Organomercurial(s) (% Yield)[a]	Subsequent reactants	Product(s) (% Yield)[a]	Ref.
 (norbornene with CH₃) 90:10 endo:exo	Hg(OAc)₂	1:1 H₂O/THF	—	NaBH₄/NaOH, Ac₂O	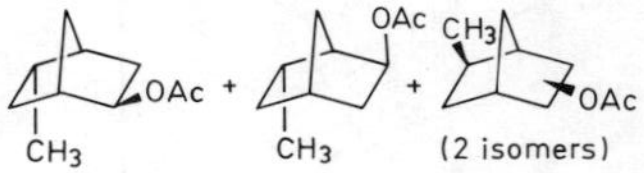 (OAc, CH₃ products) ~61 : ~26 : ~13 (2 isomers)	196
 (methyl norbornene) CH₃	Hg(OAc)₂	H₂O/THF	—	NaBH₄/NaOH	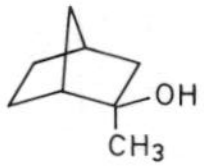 (OH, CH₃) (86) >99.8% exo	148
 (bicyclopentane alkene)	Hg(OAc)₂	1:1 H₂O/THF	—	NaBH₄/NaOH	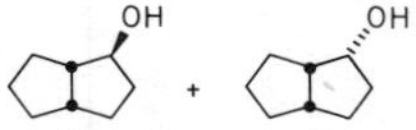 (OH products) 90 : 10	186
	Hg(OAc)₂	1:1 H₂O/THF 1 hr	—	NaBH₄/NaOH	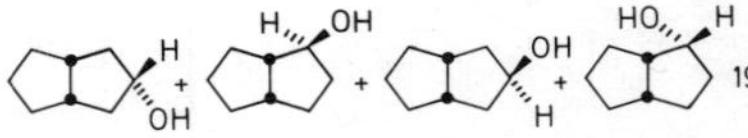 (OH products) 33 : 33 : 31 : 3 (90 total)	195
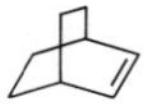 (bicyclo[2.2.2]octene)	Hg(OAc)₂	1:1 H₂O/CH₃COCH₃ 24 hr, NaCl	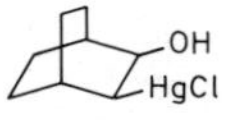 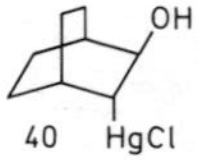 (OH, HgCl products) 60 : 40	—	—	197
	1:1 HgO/TsOH	H₂O 15 hr, HCl/NaCl	(OH, HgCl products) ~50 : ~50	NaBH₄	(OH) (25)	197
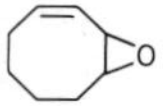 (cyclooctene oxide)	Hg(OAc)₂	1:1 H₂O/THF 8d	—	NaBH₄/NaOH	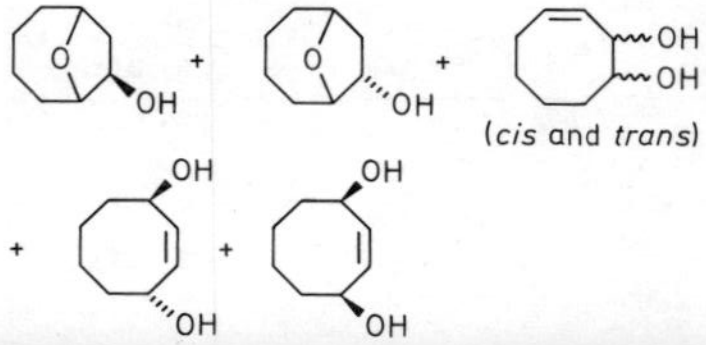 (O, OH products) (cis and trans)	152

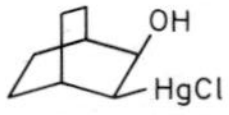

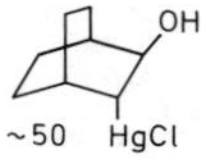

Table 2.4. (continued)

Alkene	Mercuric salt	Reaction conditions	Organomercurial(s) (% Yield)[a]	Subsequent reactants	Product(s) (% Yield)[a]	Ref.
(structure)	Hg(OAc)$_2$	1:1 H$_2$O/THF 15 min	—	NaBH$_4$/NaOH	(72) + (18) + (10) (cis and trans)	152
(structure)	Hg(OAc)$_2$	1:1 H$_2$O/THF 15 min	—	NaBH$_4$/NaOH	4 : 1 (85)	152
	Hg(OAc)$_2$	1:1 H$_2$O/THF 25°C	—	NaBH$_4$/NaOH	76 : 24 (92 total)	92
	Hg(OAc)$_2$	H$_2$O 50 hr RT, KI	HgI + IHg (94 total)	I$_2$	(85 total) major product	198
HO$_2$C CO$_2$H (structure)	Hg(OAc)$_2$	H$_2$O Δ 2 hr	HO$_2$C CO$_2$H HgOH or Hg (lactone) ? (100)	—	—	199
CD$_3$ (structure)	Hg(OAc)$_2$	1:1 H$_2$O/THF	—	NaBH$_4$/NaOH	CD$_3$ CH$_3$ OH (structure)	185, 200
NC (structure)	Hg(OAc)$_2$	H$_2$O/THF	—	NaBH$_4$/NaOH	NC CH$_3$ OH (structure)	193

Table 2.4. (continued)

Alkene	Mercuric salt	Reaction conditions	Organomercurial(s) (% Yield)[a]	Subsequent reactants	Product(s) (% Yield)[a]	Ref.
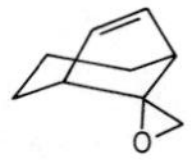	$Hg(OAc)_2$	1:1 H_2O/THF 29 hr RT	—	$NaBH_4$/NaOH	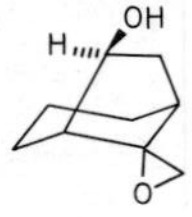 (80) + 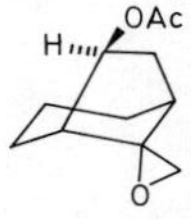 (<10)	201
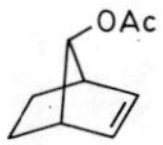	$Hg(OAc)_2$	1:1 H_2O/THF 5 min	—	$NaBH_4$/NaOH	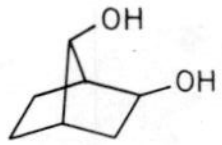 (90)	192
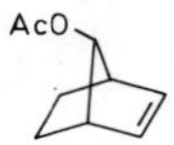	$Hg(OAc)_2$	1:1 H_2O/THF 5 min	—	$NaBH_4$/NaOH	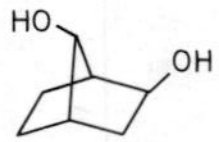 (92)	192
 (opt. active)	$Hg(OAc)_2$	1:1 H_2O/THF	—	$NaBH_4$/NaOH	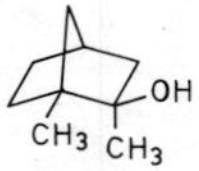 (opt. active)	200
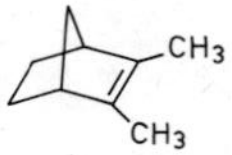	$Hg(OAc)_2$	H_2O	$C_9H_{14}(OH)HgOAc$ (no structure given)	—	—	202
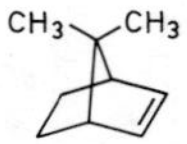	$Hg(OAc)_2$	H_2O/THF	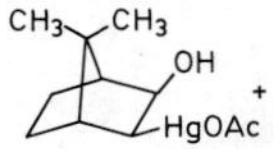 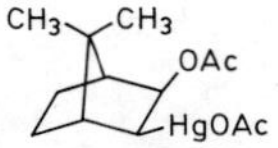 75 : 25 (80 total)	—	—	183
	$Hg(OAc)_2$	H_2O/THF	—	$NaBH_4$/NaOH	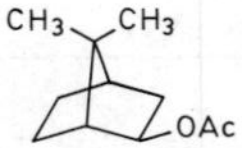 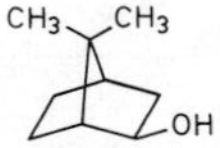 63 : 37	184

Table 2.4. (continued)

Alkene	Mercuric salt	Reaction conditions	Organomercurial(s) ($\%$ Yield)[a]	Subsequent reactants	Product(s) ($\%$ Yield)[a]	Ref.
	Hg(OAc)$_2$	H$_2$O/THF 73 min	—	NaBH$_4$/NaOH	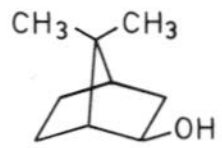 CH$_3$, CH$_3$ ... OH (84) > 99.8 % exo	148
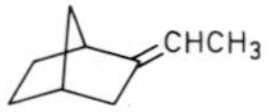 =CHCH$_3$	Hg(OAc)$_2$	1:1 H$_2$O/THF	—	NaBH$_4$/NaOH	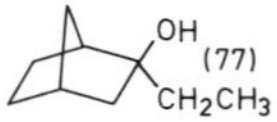 OH (77) 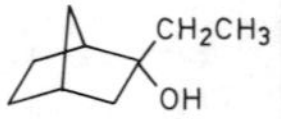 + CH$_2$CH$_3$ ... CH$_2$CH$_3$, OH (5)	196
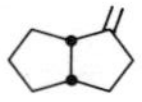 (=CH$_2$)	Hg(OAc)$_2$	1:1 H$_2$O/THF 5 min	—	NaBH$_4$/NaOH	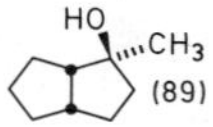 HO ...CH$_3$ (89) 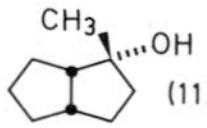 + CH$_3$ ...OH (11)	148, 186, 195
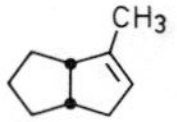 CH$_3$	Hg(OAc)$_2$	1:1 H$_2$O/THF 5 min	—	NaBH$_4$/NaOH	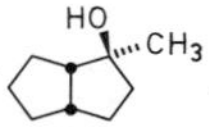 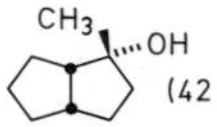 HO ...CH$_3$ + CH$_3$ ...OH (42 total) 84 : 16	148, 195
	Hg(OAc)$_2$	1:1 H$_2$O/THF 180 min	—	NaBH$_4$/NaOH	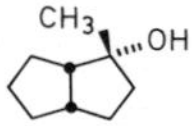 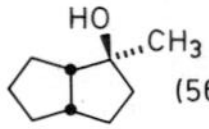 CH$_3$ ...OH + HO ...CH$_3$ (56 total) 74 : 26	195
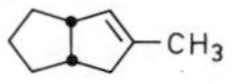 —CH$_3$	Hg(OAc)$_2$	1:1 H$_2$O/THF 5 min	—	NaBH$_4$/NaOH	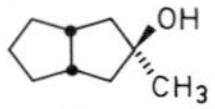 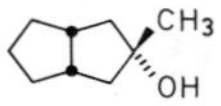 OH, CH$_3$ + CH$_3$, OH 78 : 22 (40–50 total)	148, 195
	Hg(OAc)$_2$	1:1 H$_2$O/THF 180 min	—	NaBH$_4$/NaOH	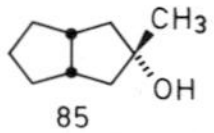 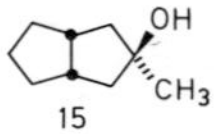 CH$_3$, OH + OH, CH$_3$ 85 : 15	195
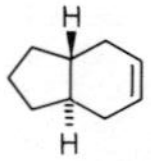 (H, H)	Hg(OAc)$_2$	1:1 H$_2$O/THF	—	NaBH$_4$/NaOH	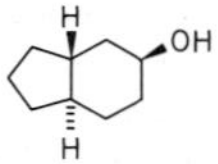 H ...OH H	155

Table 2.4. (continued)

Alkene	Mercuric salt	Reaction conditions	Organomercurial(s) (% Yield)[a]	Subsequent reactants	Product(s) (% Yield)[a]	Ref.
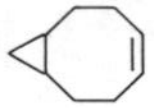	Hg(OAc)$_2$	1:1 H$_2$O/THF 25 °C	—	NaBH$_4$/NaOH	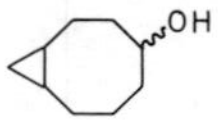 (60) cis and trans	92
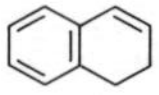	Hg(OAc)$_2$	—	—	—	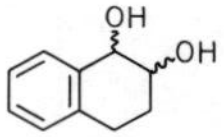 ?	203
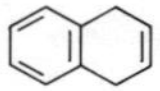	Hg(OAc)$_2$	H$_2$O 24 hr	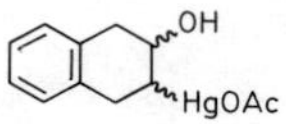	—	—	30
	—	—	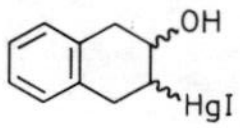 (good)	—	—	13
	Hg(OAc)$_2$	—	Cl$_2$ and Br$_2$ on 1 + Cl$_2$ on 2		X = Cl or Br or	204
	Hg[C(NO$_2$)$_3$]$_2$	H$_2$O overnight	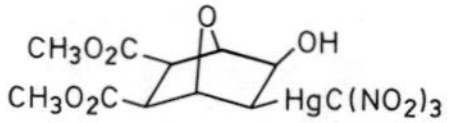	—	—	188
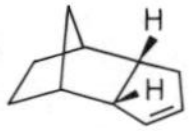	Hg(OAc)$_2$	1:1 H$_2$O/THF 7.5 d	—	NaBH$_4$/NaOH	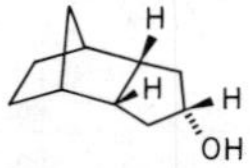 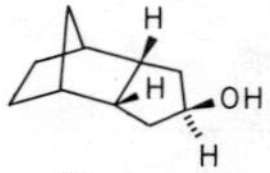 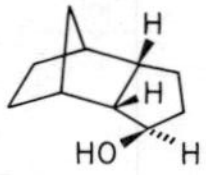 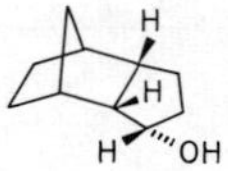 39 : 36 : 17 : 8 (36 total)	195

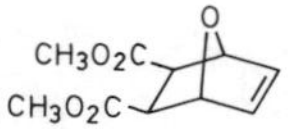

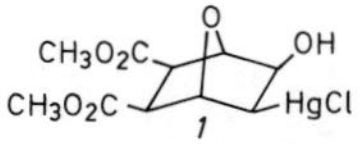

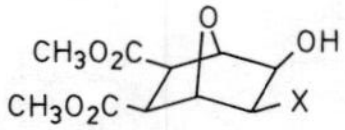

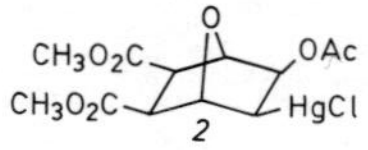

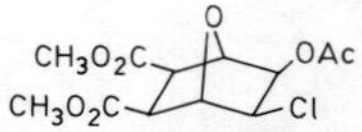

Table 2.4. (continued)

Alkene	Mercuric salt	Reaction conditions	Organomercurial(s) (% Yield)[a]	Subsequent reactants	Product(s) (% Yield)[a]	Ref.
	$Hg(O_2CCF_3)_2$	1:1 H_2O/THF	—	$NaBH_4$/NaOH	(structure) 50 : 31 : 10 : 9	195
(structure)	$Hg(OAc)_2$	H_2O/THF	—	$NaBH_4$/NaOH	(structure) (89)	205
(structure)	1:1 HgO/$HClO_4$	1:1 H_2O/CH_3COCH_3 5 hr, NaCl	(structure) (99)	$NaBH_4$	(structure) (98)	206
(structure)	$Hg(OAc)_2$	H_2O	(structure)	—	—	194
(structure)	$Hg(OAc)_2$	1:1 H_2O/THF 27.5 hr RT	—	$NaBH_4$/NaOH	(structure) + (structure) (95 total)	201
(structure)	$Hg(OAc)_2$	1:1 H_2O/THF 24 hr RT	—	$NaBH_4$/NaOH	(structure) (44) + (structure) (44)	207

Table 2.4. (continued)

Alkene	Mercuric salt	Reaction conditions	Organomercurial(s) (% Yield)[a]	Subsequent reactants	Product(s) (% Yield)[a]	Ref.
$C_2H_5O_2CN$-bicycloalkene; β-pinene	$Hg(NO_3)_2$	1:1 H_2O/THF 0.2 hr	—	$NaBH_4$/NaOH, CrO_3	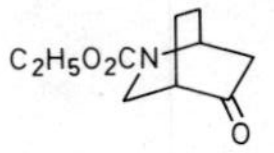 $C_2H_5O_2CN$ ketone (87)	154
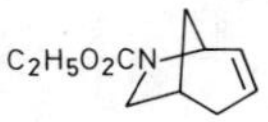	$Hg(NO_3)_2$	1:1 H_2O/THF 48 hr	—	$NaBH_4$/NaOH, CrO_3	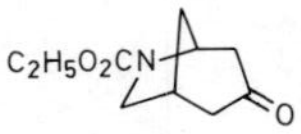 $C_2H_5O_2CN$ ketone (79)	154
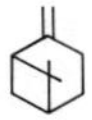	$Hg(OAc)_2$	1:1 H_2O/THF 20 °C 10 min	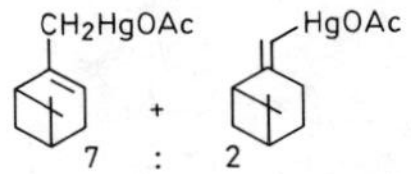 CH_2HgOAc + $HgOAc$ (7 : 2)	$NaBH_4$/NaOH	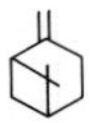	59
	$Hg(OAc)_2$	—	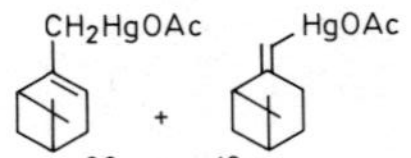 CH_2HgOAc + $HgOAc$ (90 : 10)	$NaBH_4$	+	208
	$HgCl_2$	H_2O	—	$NaBH_4$	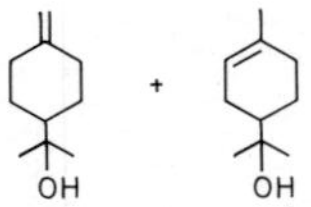 OH + OH	208
	$HgCl_2$	D_2O/CD_3COCD_3	—	—	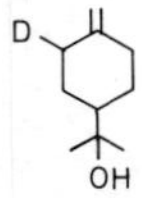 D ... OH	208
	HgX_2 X = NO_3, ClO_4, SO_4	H_2O	—	—	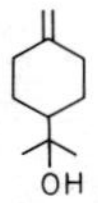 OH (up to 95)	208

Table 2.4. (continued)

Alkene	Mercuric salt	Reaction conditions	Organomercurial(s) (% Yield)[a]	Subsequent reactants	Product(s) (% Yield)[a]	Ref.
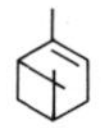	Hg(OAc)$_2$	H$_2$O RT several days	—	—	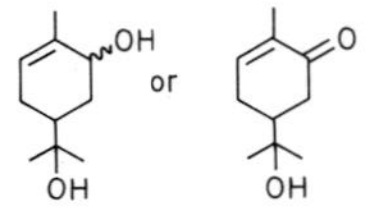	209, 210
	Hg(OAc)$_2$	H$_2$O 3 hr Δ	—	—	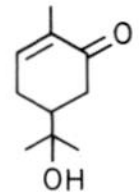 (48)	211
	HgX$_2$ X = ClO$_4$, NO$_3$, SO$_4$	H$_2$O/THF	—	NaBH$_4$	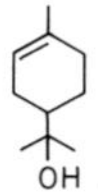 (70)	208
	Hg(OAc)$_2$	1:1 H$_2$O/THF	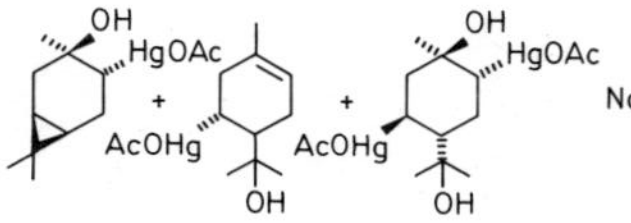	NaBH$_4$/NaOH	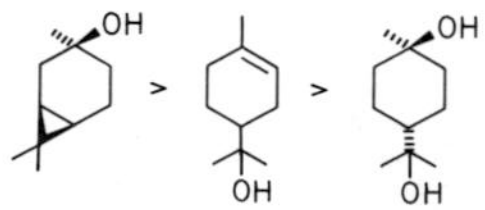	212
	Hg(OAc)$_2$	1:1 H$_2$O/THF 7 hr	—	NaBH$_4$/NaOH	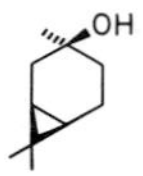 (40)	213
	Hg(OAc)$_2$	1:1 H$_2$O/THF RT	—	NaBH$_4$/NaOH	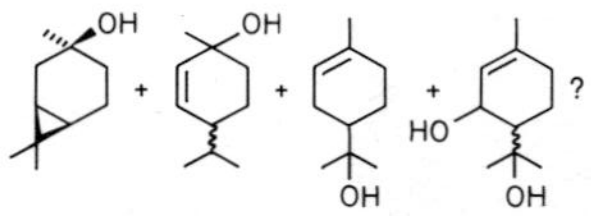	214

Table 2.4. (continued)

Alkene	Mercuric salt	Reaction conditions	Organomercurial(s) (% Yield)[a]	Subsequent reactants	Product(s) (% Yield)[a]	Ref.
	Hg(OAc)$_2$	1:1 H$_2$O/THF 0.25–30 hr	—	NaBH$_4$/NaOH	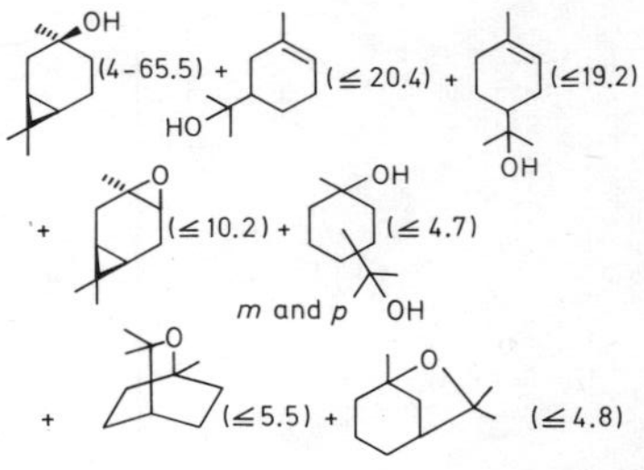 (4–65.5) + (≤20.4) + (≤19.2); (≤10.2) + (≤4.7) m and p; + (≤5.5) + (≤4.8)	215
	Hg(OAc)$_2$	1:1 H$_2$O/THF 36 hr RT, NaCl	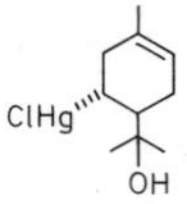	NaBH$_4$/NaOH	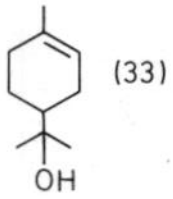 (33)	212
	Hg(OAc)$_2$	1:1 H$_2$O/THF 20°C 5 hr	—	NaBH$_4$/NaOH	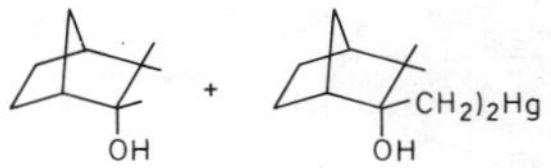	59
	Hg(OAc)$_2$	H$_2$O/THF	—	NaBH$_4$/NaOH	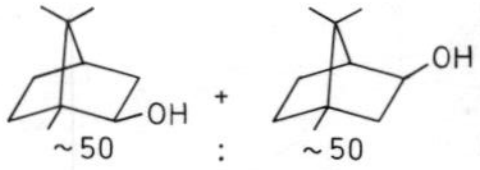 ~50 : ~50	148
	Hg(OAc)$_2$	1:1 H$_2$O/THF 20°C 2.5 hr	—	NaBH$_4$/NaOH	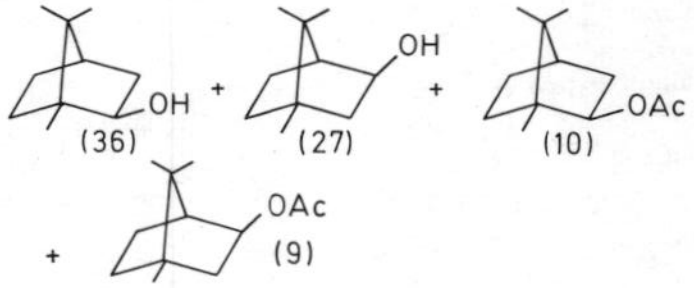 (36) + (27) + (10); + (9)	59

Table 2.4. (continued)

Alkene	Mercuric salt	Reaction conditions	Organomercurial(s) (% Yield)[a]	Subsequent reactants	Product(s) (% Yield)[a]	Ref.
	$HgO/HClO_4$	NaCl	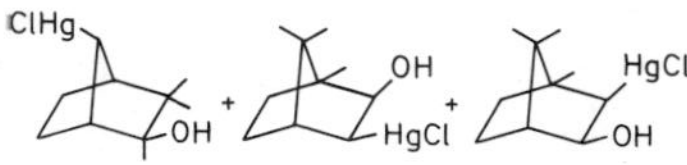 [structures]	$Na(Hg)/H_2O$	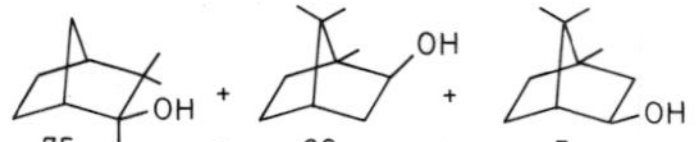 [structures] 75 : 20 : 5	216
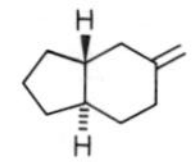 [structure]	$Hg(OAc)_2$	1:1 H_2O/THF	—	$NaBH_4$/NaOH	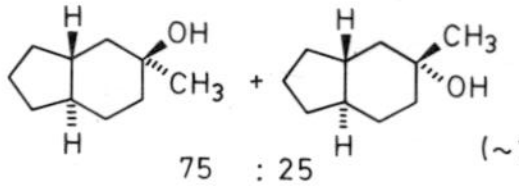 [structures] 75 : 25 (~100 total)	155, 156
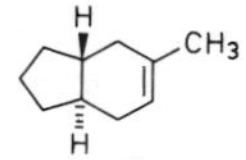 [structure] CH_3	$Hg(OAc)_2$	1:1 H_2O/THF	—	$NaBH_4$/NaOH	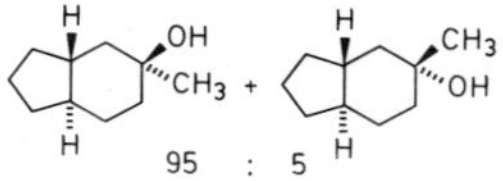 [structures] 95 : 5	155
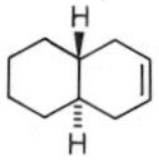 [structure]	$Hg(OAc)_2$	1:1 H_2O/THF	—	$NaBH_4$/NaOH	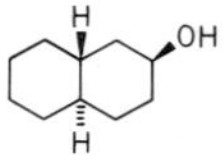 [structure]	155
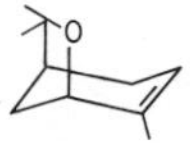 [structure]	$Hg(OAc)_2$	1:1 H_2O/THF 14d	—	$NaBH_4$/NaOH	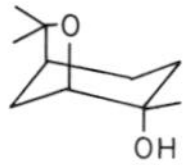 [structure] (53)	217
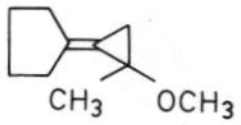 [structure]	$Hg(OAc)_2$	—	—	H_2S	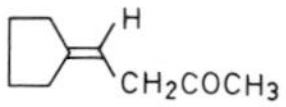 [structure] (94)	177
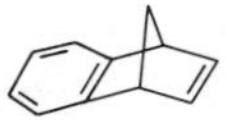 [structure]	1:1 $HgO/Hg(OAc)_2$	H_2O 24 hr, NaCl	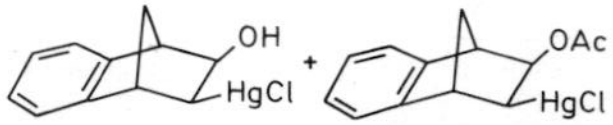 [structures] (50 total)	2% Na(Hg)/NaOH	 [structure]	131
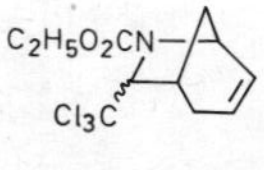 [structure]	$Hg(NO_3)_2$	1:1 H_2O/THF 1 hr	—	$NaBH_4$/NaOH, CrO_3	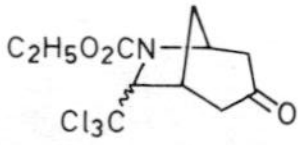 [structure] (75)	154

Table 2.4. (continued)

Alkene	Mercuric salt	Reaction conditions	Organomercurial(s) (% Yield)[a]	Subsequent reactants	Product(s) (% Yield)[a]	Ref.
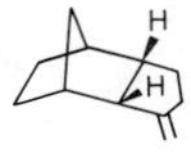	Hg(OAc)$_2$	1:1 H$_2$O/THF 1–5 min	—	NaBH$_4$/NaOH	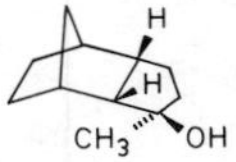 (100) CH$_3$ OH > 99.9% exo	148, 186, 195
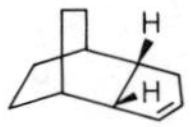	Hg(OAc)$_2$	1:1 H$_2$O/THF 1 week	—	NaBH$_4$/NaOH	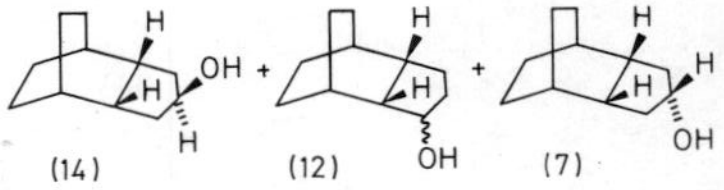 OH (14) + (12) OH + (7) OH	218
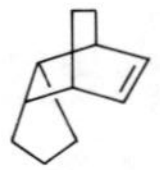	Hg(OAc)$_2$	1:1 H$_2$O/THF 24 hr RT	—	NaBH$_4$/NaOH	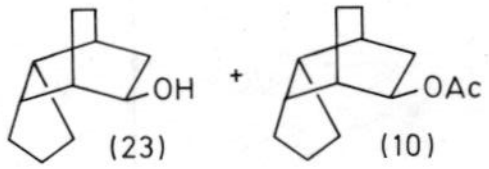 OH (23) + OAc (10)	218
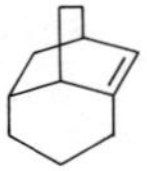	Hg(OAc)$_2$	H$_2$O/THF 4 min	—	NaBD$_4$/NaOH	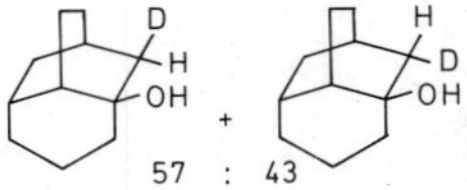 D H OH + H D OH (86 total) 57 : 43	219
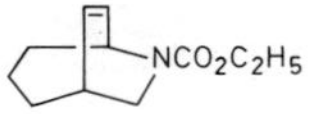 NCO$_2$C$_2$H$_5$	Hg(NO$_3$)$_2$	1:1 H$_2$O/THF 24 hr	—	NaBH$_4$/NaOH, CrO$_3$	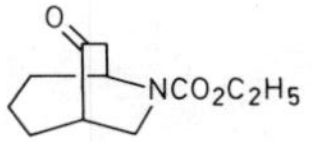 NCO$_2$C$_2$H$_5$ (57)	154
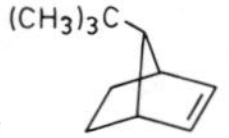 (CH$_3$)$_3$C	Hg(OAc)$_2$	2:3 H$_2$O/THF 5 min	—	NaBH$_4$/NaOH, NaOCH$_3$/CH$_3$OH	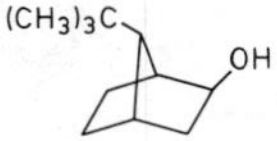 (CH$_3$)$_3$C OH (62)	220
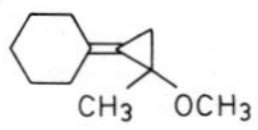 CH$_3$ OCH$_3$	Hg(OAc)$_2$	H$_2$O/EtOH 5 min	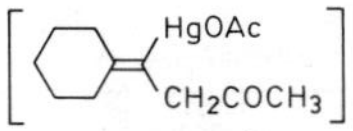 [HgOAc CH$_2$COCH$_3$] presumed	H$_2$S	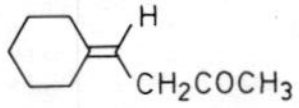 H CH$_2$COCH$_3$ (85)	177
	Hg(OAc)$_2$	H$_2$O/CH$_3$OH 10 min 20 °C	—	I$_2$/KI	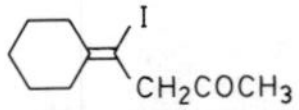 I CH$_2$COCH$_3$ (90)	177

Table 2.4. (continued)

Alkene	Mercuric salt	Reaction conditions	Organomercurial(s) (% Yield)[a]	Subsequent reactants	Product(s) (% Yield)[a]	Ref.
CH_3O_2C, CO_2CH_3 bicyclic alkene	$Hg(OAc)_2$	H_2O/Cl^-	CH_3O_2C, OH, HgCl, CO_2CH_3 (5) + CH_3O_2C, HgCl, lactone (27)	—	—	199
norbornene–C_6H_5	$Hg(OAc)_2$	1:1 H_2O/THF 25°C 70 min	—	$NaBH_4$/NaOH	OH, C_6H_5 (90)	221
CH_3 benz-fused bicyclic	$Hg(OAc)_2$	1:1 H_2O/THF 1 hr	—	$NaBH_4$/NaOH	HO, CH_3 (37) + CH_3, OH (6.6)	222
$(CH_3)_3C$ epoxy bicyclic alkene	$Hg(OAc)_2$	1:1 H_2O/CH_3CN 32 hr RT	—	$NaBH_4$/NaOH	H, OH, $(CH_3)_3C$, epoxide (~80)	201
cyclohexylidene, C_2H_5O, OC_2H_5	$Hg(OAc)_2$	—	—	H_2S	H, $CH_2CO_2C_2H_5$ (74)	177
$(CH_3)_3C$ dioxolane bicyclic alkene	$Hg(OAc)_2$	1:1 H_2O/CH_3CN 30.5 hr RT	—	$NaBH_4$/NaOH	H, OH, $(CH_3)_3C$ dioxolane > H, OAc, $(CH_3)_3C$ dioxolane	201
CH_3O, CH_3, CH_3O bicyclic alkene	$Hg(OAc)_2$	1:1 H_2O/THF 1 hr	—	$NaBH_4$/NaOH	CH_3O, HO, CH_3, CH_3O (29) + CH_3O, CH_3, OH, CH_3O (5.5)	222

Table 2.4. (continued)

Alkene	Mercuric salt	Reaction conditions	Organomercurial(s) (% Yield)[a]	Subsequent reactants	Product(s) (% Yield)[a]	Ref.
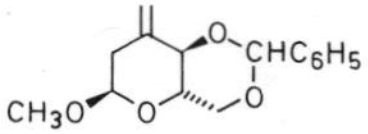 CH$_3$O, CHC$_6$H$_5$ (alkene)	Hg(OAc)$_2$	H$_2$O/THF RT 30 min	—	NaBH$_4$/NaOH	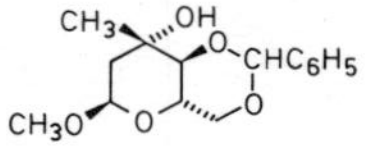 CH$_3$, OH, CH$_3$O, CHC$_6$H$_5$ (72)	223
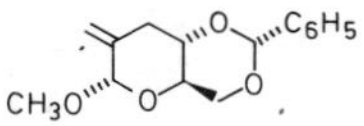 CH$_3$O, C$_6$H$_5$ (alkene)	Hg(OAc)$_2$	H$_2$O/THF RT 24 hr	—	NaBH$_4$/NaOH	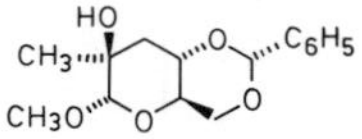 HO, CH$_3$, CH$_3$O, C$_6$H$_5$	176
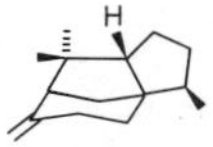 (alkene, H)	Hg(OAc)$_2$	1:1 H$_2$O/THF 10 hr RT	—	NaBH$_4$/NaOH	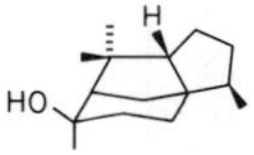 HO (66)	224
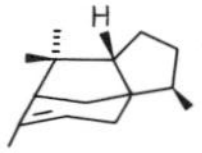 (alkene, H)	Hg(OAc)$_2$	1:1 H$_2$O/THF 18 hr RT	—	NaBH$_4$/NaOH	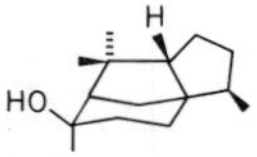 HO (54)	224
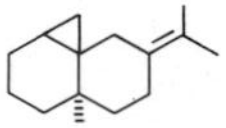 (alkene)	Hg(OAc)$_2$	H$_2$O/THF 2 hr RT	—	NaBH$_4$/NaOH	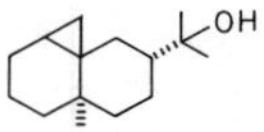 OH (90)	225
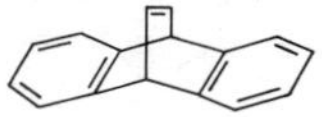 (alkene)	Hg(OAc)$_2$	H$_2$O/CH$_3$COCH$_3$ several days, NaCl	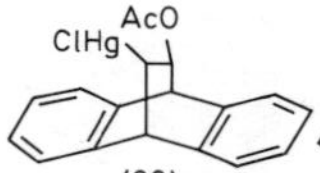 ClHg, AcO (30) 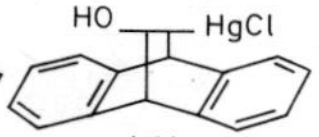 + HO, HgCl (12)	—	—	226
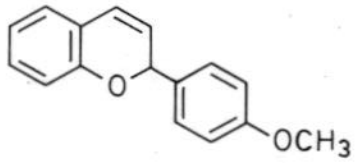 OCH$_3$ (alkene)	Hg(OAc)$_2$	1:1 H$_2$O/THF 18 hr RT	—	NaBH$_4$/NaOH	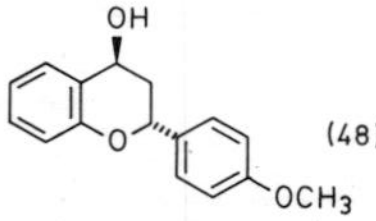 OH, OCH$_3$ (48)	227
	Hg(OAc)$_2$	1:1 H$_2$O/THF 18 hr RT	—	NaBD$_4$/NaOH	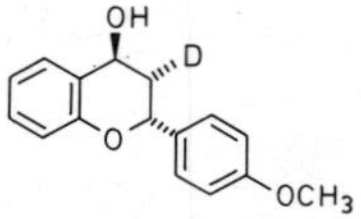 OH, D, OCH$_3$ (47)	227

Table 2.4. (continued)

Alkene	Mercuric salt	Reaction conditions	Organomercurial(s) (% Yield)[a]	Subsequent reactants	Product(s) (% Yield)[a]	Ref.
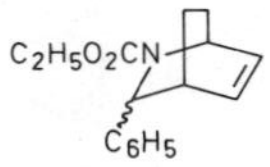 $C_2H_5O_2CN$ ··· C_6H_5 (norbornene)	$Hg(NO_3)_2$	1:1 H_2O/THF 72 hr	—	$NaBH_4$/NaOH, CrO_3	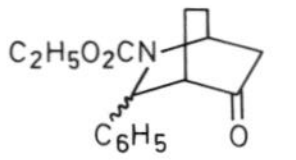 $C_2H_5O_2CN$ ··· C_6H_5, =O (86)	154
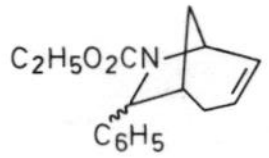 $C_2H_5O_2CN$ ··· C_6H_5 (norbornene)	$Hg(NO_3)_2$	1:1 H_2O/THF 72 hr	—	$NaBH_4$/NaOH, CrO_3	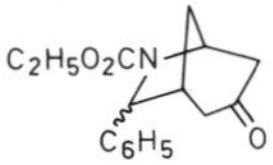 $C_2H_5O_2CN$ ··· C_6H_5, =O (85)	154
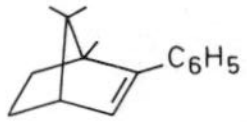 C_6H_5 (pinene)	$Hg(OAc)_2$	H_2O/THF 20°C 7d	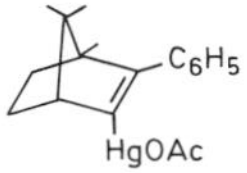 C_6H_5, HgOAc	$NaBH_4$/NaOH	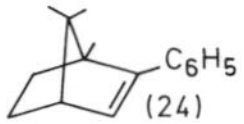 C_6H_5 (24) 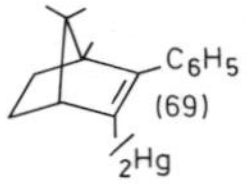 + C_6H_5 , $_2$Hg (69)	59
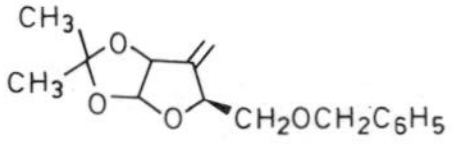 CH_3, CH_3 (dioxolane)··· $CH_2OCH_2C_6H_5$	$Hg(OAc)_2$	H_2O/THF 60°C 30 min	no reaction	—	—	228
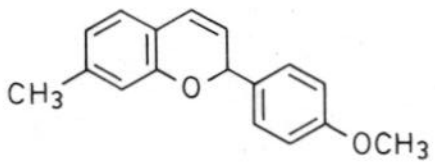 CH_3 —(chromene)— C_6H_4–OCH_3	$Hg(OAc)_2$	1:1 H_2O/THF 18 hr RT	—	$NaBH_4$/NaOH	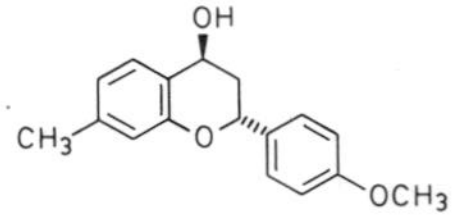 OH, CH_3 —(chroman)— C_6H_4–OCH_3 (42)	227
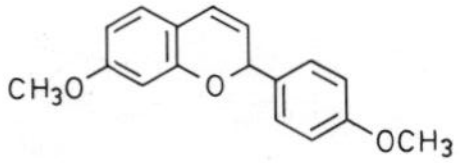 CH_3O —(chromene)— C_6H_4–OCH_3	$Hg(OAc)_2$	1:1 H_2O/THF 18 hr RT	—	$NaBH_4$/NaOH	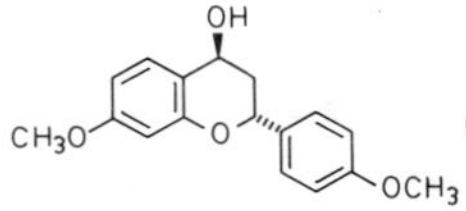 OH, CH_3O —(chroman)— C_6H_4–OCH_3 (21)	227
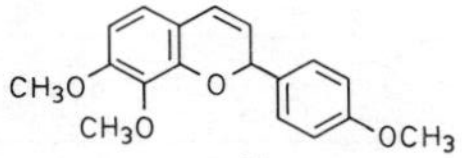 CH_3O, CH_3O —(chromene)— C_6H_4–OCH_3	$Hg(OAc)_2$	1:1 H_2O/THF 18 hr RT	—	$NaBH_4$/NaOH	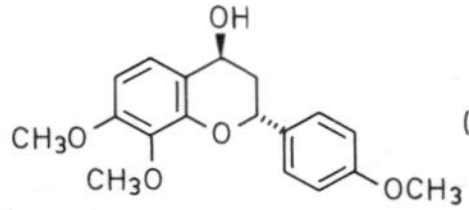 OH, CH_3O, CH_3O —(chroman)— C_6H_4–OCH_3 (22)	227

Table 2.4. (continued)

Alkene	Mercuric salt	Reaction conditions	Organomercurial(s) (% Yield)[a]	Subsequent reactants	Product(s) (% Yield)[a]	Ref.
(structure)	Hg(OAc)$_2$	1:1 H$_2$O/THF 18 hr RT	—	NaBH$_4$/NaOH	(structure) (35) + (structure) (6)	227
(structure)	Hg(O$_2$CCF$_3$)$_2$	2:3 H$_2$O/THF 6d RT	—	—	(structure) (15) + hydroxymercurial	229
(structure)	Hg(O$_2$CCF$_3$)$_2$?	H$_2$O/THF 2d pH ?	—	NaBH$_4$	(structure)	230
(structure)	Hg(OAc)$_2$	1:5 H$_2$O/THF 24 hr RT	—	NaBH$_4$/NaOH	(structure) + (structure) 70 : 30	231
(structure)	Hg(O$_2$CCF$_3$)$_2$	H$_2$O/THF pH 2.1 2d	—	NaBH$_4$	(structure) (40) + (structure) (26) + (structure) (17) + (structure) (2)	230

Table 2.4. (continued)

Alkene	Mercuric salt	Reaction conditions	Organomercurial(s) (% Yield)[a]	Subsequent reactants	Product(s) (% Yield)[a]	Ref.
(steroid, HO)	$Hg(O_2CCF_3)_2$	H_2O/THF pH 3.5	—	$NaBH_4$	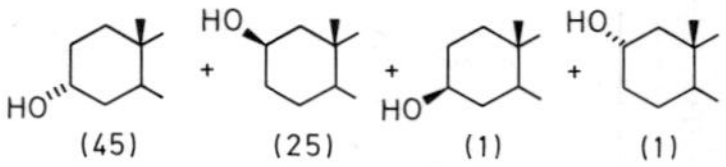 (45) + (25) + (1) + (1)	230
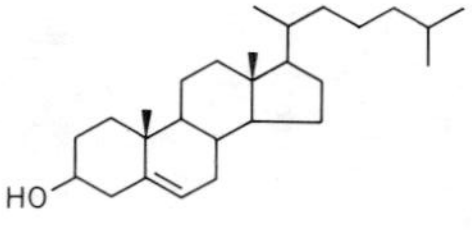 (steroid)	$Hg(O_2CCF_3)_2$	2:3 H_2O/THF 6d RT	—	—	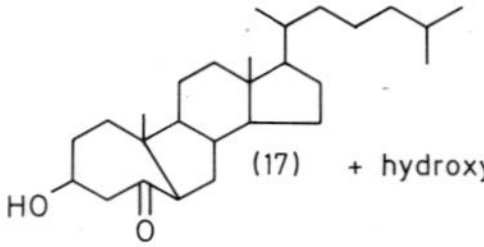 (17) + hydroxymercurial	229
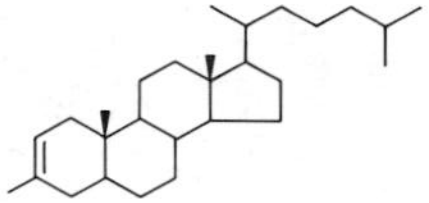 (steroid)	$Hg(O_2CCF_3)_2$	H_2O/THF 2d 40°C	—	$NaBH_4$	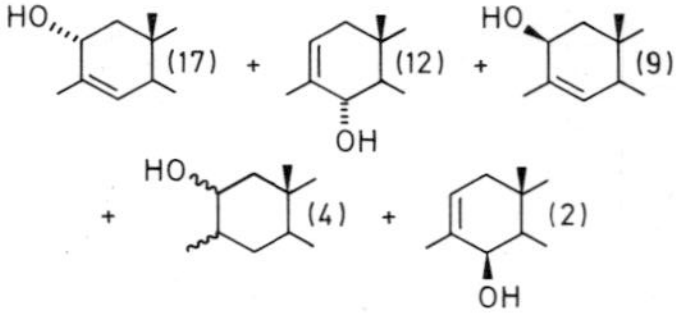 (17) + (12) + (9) + (4) + (2)	232
(steroid)	$Hg(O_2CCF_3)_2$	H_2O/THF pH 3.5	—	$NaBH_4$	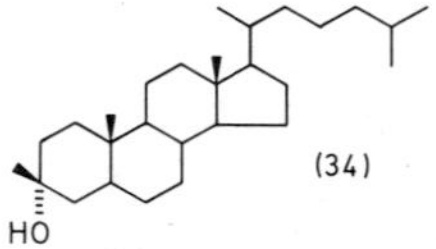 (34)	232
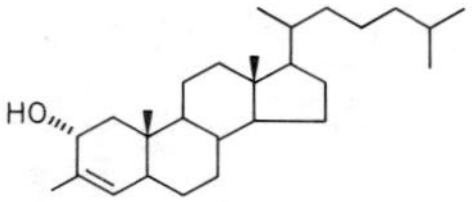 (steroid, HO)	$Hg(O_2CCF_3)_2$	H_2O/THF	—	—	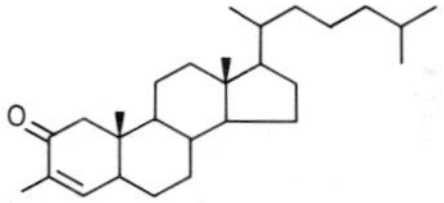	232

[a] If there is no entry in the organomercurial column, the yield is based on the organomercurial unless otherwise stated.

71

according to their carbon and hydrogen content and then alphabetically according to the other atoms. Isomers are again listed according to their substitution pattern about the carbon—carbon double bond as described earlier. The number of carbon atoms in the alkene chain is then taken into account, starting with the lower alkenes first.

Table 2.3 summarizes hydroxymercuration results obtained on monocyclic alkenes where one or both carbons of the double bond are involved in the ring. Table 2.4 covers bicyclic and polycyclic olefins. Isomeric olefins are tabulated first according to ring size (smallest first) and then substitution pattern as described earlier.

The reader is encouraged to examine the tables to determine whether the hydroxymercuration of a particular olefin has previously been examined. The reaction conditions reported, however, may not necessarily be the best for that olefin, since a substantial amount of the work on hydroxymercuration was carried out before a careful search for the best reaction conditions was completed.

The present widespread use of the hydroxymercuration reaction is due in large part to the valuable work of H. C. Brown, who in 1967 reported a very convenient method for hydroxymercuration-demercuration which consisted of adding x mmoles of olefin to a flask containing x mmoles of mercuric acetate, x ml of water and x ml of THF [39, 40]. The reaction is stirred approximately 10–15 minutes or 5–10 times the amount of time it takes for the initial yellow color to disappear. The color change seems to correspond to about 60% reaction. One then adds x ml of 3 M sodium hydroxide, followed by x ml of 0.5 M sodium borohydride in 3.0 M sodium hydroxide. Demercuration is almost instantaneous. The mercury metal is allowed to settle. Sodium chloride is added to saturate the water layer and the THF solution of the alcohol is separated and dried, and the product isolated. This is the procedure in common use today.

A number of variations of this basic procedure have been employed since the initial discovery of the reaction. Many different mercury salts have been utilized as seen in Tables 2.1–2.4. Although mercuric chloride is reported not to react with pure ethylene, impurities apparently can promote reaction [233]. Mercuric chloride has been employed, however, in a number of intramolecular solvomercuration reactions to be discussed later. More electrophilic mercury salts such as the sulfate, perchlorate, nitrate, methanesulfonate and trifluoroacetate salts have been utilized. They appear to hold no advantage over mercuric acetate however. In a comparison of mercuric acetate, mercuric trifluoroacetate, mercuric nitrate and mercuric methanesulfonate, Brown found that all four salts are effective with olefins of the type $RCH=CH_2$ and $RCH=CHR$ [41]. Yields of alcohols of 90–100% are generally obtained with a regioselectivity in excess of 99.5% when terminal olefins are employed. With more highly substituted olefins, only mercuric acetate proved suitable. The yields of alcohol tend to rapidly decrease with time using the other salts. With 1-phenylcyclopentene, these other salts gave substantial amounts of the corresponding allylic alcohol, 2-phenylcyclopenten-3-ol, while mercuric acetate reacted only sluggishly (Eq. 2). Isotopically labelled mercuric acetate has also been employed in the preparation of certain diuretic organomercurials [75].

$$(2)$$

While various solvents have been employed over the years in the hydroxymercuration reaction, THF appears to be the solvent of choice. It affords reasonable miscibility of the olefin and the aqueous layer containing the mercury salt. With olefins of high carbon content, however, one may need to increase the ratio of THF to water to increase the rate of reaction [39]. On occasions, D_2O has been employed for hydroxymercuration, particularly where NMR analysis of the product was desired.

The reaction times for hydroxymercuration can vary appreciably. In general, the less substituted or hindered the carbon—carbon double bond is, the more readily the olefin will react. The actual relative rates will be discussed in more detail later in this section. With unreactive olefins, one occasionally observes a change in the regio- and/or stereochemistry of the products with time due to reversible hydroxymercuration, apparently promoted by the acid generated during the reaction [195]. Several studies have shown that the products can be highly dependent on the pH of the reaction medium or the mercury salt employed [168, 173, 195, 230]. In cyclohexyl systems, this means that one generally obtains axial alcohols under the usual hydroxymercuration-demercuration conditions, but predominantly equatorial alcohols upon long reaction times in the presence of a strong acid, as nicely illustrated by the following example (Eq. 3) [168, 230].

$$(3)$$

| | 30 min | 98.8 | : | 1.2 |
| 0.2% H_2SO_4 | 20 hr | 27 | : | 73 |

A number of different reagents have been employed to effect demercuration. However, only two have really proven synthetically useful. Sodium borohydride in aqueous sodium hydroxide is most commonly utilized. It is convenient, rapid and affords high yields of products. It has the one major disadvantage, however, that the reaction proceeds via free radicals and rearrangements can occur. Should one desire a stereospecific demercuration, sodium-mercury amalgam is the preferable reagent. Other reagents, such as hydrazine or hydrogen sulfide, have been used on occasions, but they are less general and often give low yields. The demercuration step and other subsequent reactions of the organomercurials prepared by solvomercuration will be discussed in more detail at the end of each of the appropriate chapters.

II. Hydroxymercuration

The actual structure of the hydroxymercuration products was originally the center of considerable controversy. In Kucherov's initial work in 1892 on the hydroxymercuration of ethylene, 1,5-hexadiene and allyl alcohol, a number of incorrect structures were presented [7]. In 1900 Hoffmann and Sand prepared what they claimed to be a number of different organomercurial products from ethylene, among which $XHgCH_2CH_2OH$ and $(XHgCH_2CH_2)_2O$ were correctly identified [8, 17]. Several of these products were eventually shown to be identical with those of Biilman [21]. A number of the hypothesized compounds were characterized solely by elemental analysis and have never been correctly identified. The observation that many of these products reverted to ethylene upon reaction with HCl, KSCN or CH_3I caused Manchot to argue for some twenty years that these products should be written as olefin complexes, such as $C_2H_4 \cdot ClHgOH$, and not covalently bound addition compounds [105, 234, 235]. Not until 1958 was convincing physical evidence provided to establish the true nature of these compounds. At that time, Cotton and Leto [18] reported that the NMR spectra for $ClHgCH_2CH_2OH$ and $CH_3OCH_2CH_2HgOAc$ were only consistent with the structures first proposed by Hoffmann and Sand. Subsequent NMR work on related hydroxymercurials supported this conclusion and provided a proof of structure for the related compounds $(XHgCH_2CH_2)_2O$ [20, 236, 237], although the latter spectra were originally incorrectly claimed to provide evidence for an intermediate mercurinium ion [15]. Procedures are now available for the preparation of a number of these mono- and dimercury compounds [11–13, 16, 17, 19].

For some time, a controversy also existed over the structure of the organomercurials derived from hydroxymercuration of allyl alcohol. It now appears that either the hydroxymercurial or a dimercurated dioxane can be obtained depending on the experimental procedure employed (Eq. 4) [13, 16, 17, 21]. The structures of hydroxymercurials derived from unsaturated

$$H_2C{=}CHCH_2OH \xrightarrow{HgX_2} XHgCH_2\overset{\overset{\displaystyle OH}{|}}{C}HCH_2OH \quad \text{or} \quad \text{(dimercurated dioxane)} \tag{4}$$

carboxylic acids have never been fully established either, although a number of cyclic derivatives have been claimed in the literature, including even four-membered ring structures (Eq. 5) [65, 69, 83, 96, 97].

$$C{=}C-C_n-CO_2H \longrightarrow \overset{\overset{\displaystyle HO\ \ Hg}{|\ \ \ |}}{C}-\overset{}{C}-C_n-\overset{\overset{\displaystyle O}{|}}{C}{=}O \quad \text{or} \quad \text{(alt. structure)} \tag{5}$$

Let's take a closer look at the regiochemistry of the hydroxymercuration reaction. The majority of results have been determined by analysis of the resulting demercurated alcohols. Hydroxymercuration of a simple terminal monosubstituted olefin, such as 1-hexene, with mercuric acetate, trifluoroacetate, nitrate or methanesulfonate gives $\geq 99.5\%$ of the secondary alcohol

74

[39, 41]. Even with more sterically demanding olefins such as 3,3-dimethyl-1-butene, only 0.2–0.6% of the primary alcohol is observed [41]. Simple terminal disubstituted olefins and trisubstituted olefins give >99.9% of the Markovnikov hydration product [39].

Steric hindrance can effect dramatically the regiochemistry of addition to acyclic olefins. For example, on changing from *cis*-2-pentene to *cis*-4-methyl-2-pentene to *cis*-4,4-dimethyl-2-pentene, the percentage of 2-pentanol increases dramatically (Eq. 6) [39]. Many other examples of the effect of steric

$$R \overset{CH_3}{\underset{H}{\underset{|}{C=C}}} H \longrightarrow R\overset{OH}{\underset{|}{CH}}CH_2CH_3 \ + \ RCH_2\overset{OH}{\underset{|}{CH}}CH_3 \tag{6}$$

$$\begin{array}{ccccc} R= & Et & 36 & : & 64 \\ & i\text{-Pr} & 9 & : & 91 \\ & t\text{-Bu} & 2 & : & 98 \end{array}$$

hindrance can be seen in the tables. It appears that the major product is determined by the ease of attack of the water on the mercury-olefin intermediate, whatever that may be.

The presence of certain functional groups can provide a powerful directing effect in hydroxymercuration. For example, a mixture of E- and Z-2-cyclopropyl-3-phenyl-2-butene affords only the cyclopropyl carbinol (Eq. 7) [58]. The hydroxymercuration of unsaturated alcohols introduces the new

$$C_6H_5C(CH_3){=}C(CH_3){-}\triangleleft \ \longrightarrow \ C_6H_5CH(CH_3)\overset{OH}{\underset{|}{C}}(CH_3){-}\triangleleft \tag{7}$$

alcohol group on the carbon furthest removed from the original alcohol group, except where intramolecular alkoxymercuration can result in five- or six-membered ring ethers (Eq. 8) [68]. Ether functionality can even reverse the

$$CH_3CH{=}CHCH_2OH \ \xrightarrow{93\%} \ CH_3\overset{OH}{\underset{|}{CH}}CH_2CH_2OH \ + \ CH_3CH_2\overset{OH}{\underset{|}{CH}}CH_2OH \tag{8}$$

$$94 \qquad : \qquad 6$$

normal Markovnikov regiochemistry to a significant extent (Eq. 9) [121]. The

$$\xrightarrow[\text{2.NaBH}_4 / \text{NaOH}]{\text{1.Hg(O}_2\text{CCF}_3)_2/\text{H}_2\text{O}} \tag{9}$$

$$68 \ : \ 32$$

carboxylic acid group will also direct the incoming hydroxy group to the more remote carbon atom of the original carbon—carbon double bond (Eq. 10) [69, 87]. The trimethylsilyl group has a similar effect (Eqs. 11, 12). The

75

II. Hydroxymercuration

$$\text{(H)(CH}_3\text{)C=C(CO}_2\text{H)(CH}_3\text{)} \longrightarrow CH_3\overset{\text{OH}}{\underset{|}{C}}HCH(CH_3)CO_2H \qquad \text{(10) [87]}$$

$$H_2C=CHSi(CH_3)_3 \longrightarrow HOCH_2CH_2Si(CH_3)_3 \qquad \text{(11) [84, 89]}$$

$$\text{(12) [164]}$$

$$70 \quad : \quad 30$$

hydroxymercuration of allyltrimethylsilane, however, affords mainly silyl cleavage products [84].

A number of studies have been reported on the hydroxymercuration of cyclic olefins. The regio- and stereochemistry of the hydroxymercuration of cyclopentenes cannot be easily explained by any simple steric or conformational arguments (Eqs. 13, 14) [133]. Mixtures of regio- and stereoisomers are common.

$$\text{(13)}$$

$R = CH_3$	2%	18%	11%	69%
$R = C(CH_3)_3$	≤1%	52–71%	6–24%	11–41%

$$\text{(14)}$$

4% 6% 39% 51%

The hydroxymercuration of simple alkylcyclohexenes has been extensively studied, as seen by the many examples reported in Table 2.3. 3-Substituted alkylcyclohexenes show a high regio- and stereoselectivity for the *trans*-3-alkylcyclohexanol (Eq. 15) [133]. On the other hand, 4-substituted alkylcyclo-

$$\text{(15)}$$

$R = CH_3$	12 %	4%	6%	79 %
$R = C(CH_3)_3$	3–4 %	0–1%	2–4%	91–95%

hexenes are remarkably stereoselective, but non-regioselective (Eq. 16) [133, 157, 166–168]. The predominant products arise by trans diaxial addition to

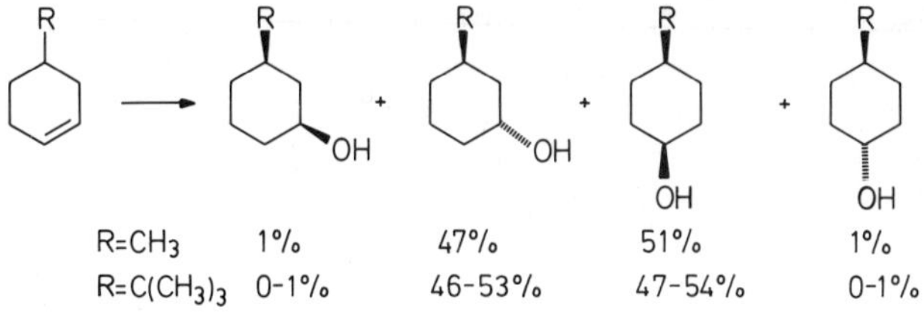

$$(16)$$

R=CH$_3$	1%	47%	51%	1%
R=C(CH$_3$)$_3$	0–1%	46–53%	47–54%	0–1%

the double bond. With 4-*tert*-butylcyclohexene, long reaction times [166] or addition of strong acids [168] lead to the more stable diequatorial products. The hydroxymercuration of 4-(hydroxymethyl)cyclohexene is reported to give totally different results from the simple 4-alkylcyclohexenes and should perhaps be reexamined (Eq. 17) [145]. Various dialkyl- and trialkylcyclohexenes have also been subjected to hydroxymercuration-demercuration (Eqs.

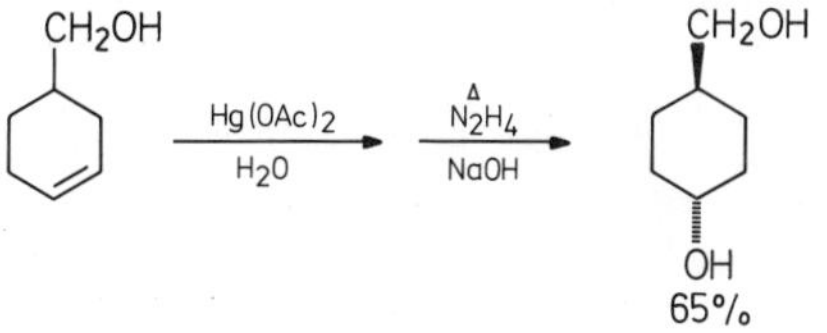

$$(17)$$

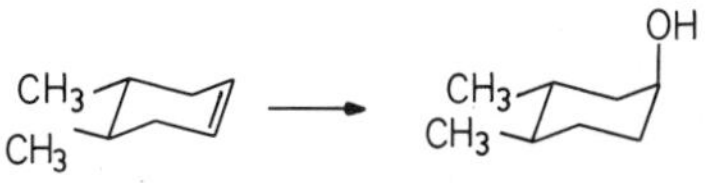

65%

$$(18)\ [155]$$

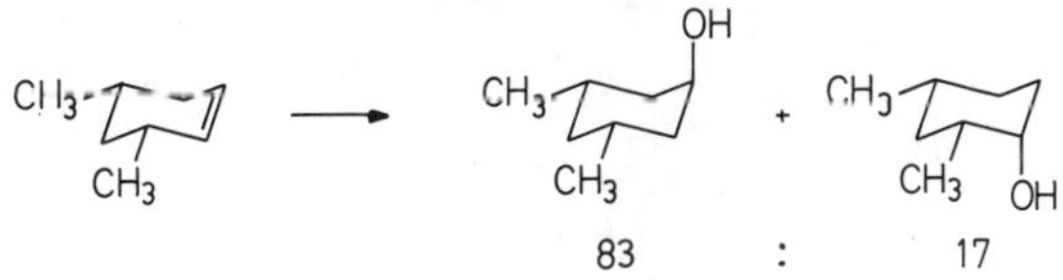

$$(19)\ [157, 158]$$

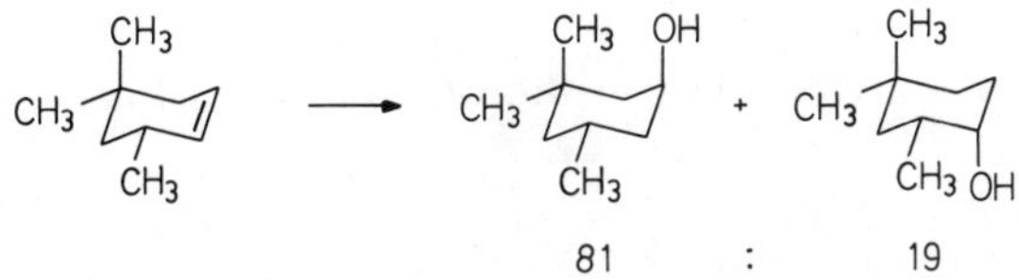

83 : 17

$$(20)\ [157, 158]$$

81 : 19

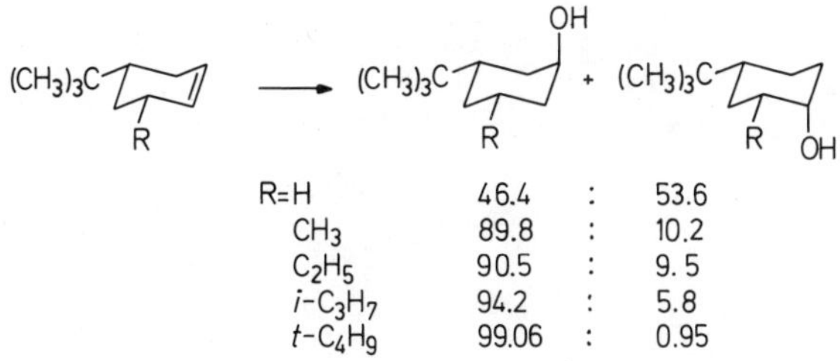

$$(21)\ [157, 158]$$

R=H	46.4	:	53.6
CH$_3$	89.8	:	10.2
C$_2$H$_5$	90.5	:	9.5
i-C$_3$H$_7$	94.2	:	5.8
t-C$_4$H$_9$	99.06	:	0.95

18–21). All products are those of trans diaxial addition. These examples clearly point out the importance of a pseudoequatorial alkyl group in the three position.

The effect of an hydroxy group on a cyclohexene ring has also been examined. Several groups have reported results for 2-cyclohexenol [39, 68, 141, 143, 144]. The product composition seems to vary with the reaction solvent, the reaction time and the procedure used to work up the diol products. The following results are representative, however (Eq. 22) [68]. Several alkyl-

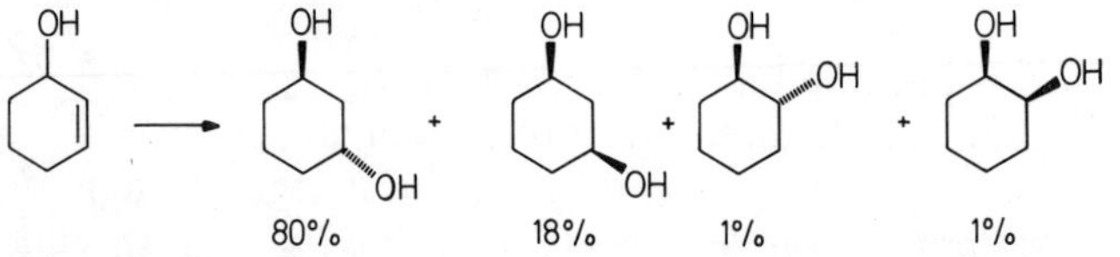

$$\text{80\%} \qquad \text{18\%} \qquad \text{1\%} \qquad \text{1\%} \tag{22}$$

substituted 2-cyclohexenols have also been hydroxymercurated and reduced (Eqs. 23, 24) [167, 170]. The results are consistent with trans diaxial addition

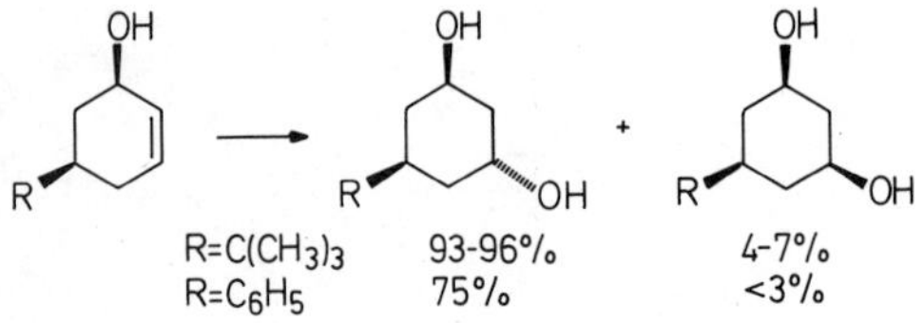

$$\begin{array}{ll} R=C(CH_3)_3 & 93\text{-}96\% \qquad 4\text{-}7\% \\ R=C_6H_5 & 75\% \qquad <3\% \end{array} \tag{23}$$

$$\begin{array}{ll} R=C(CH_3)_3 & 75\% \qquad 17\% \qquad 8\% \\ R=C_6H_5 & 60\% \qquad 5\% \qquad (?) \end{array} \tag{24}$$

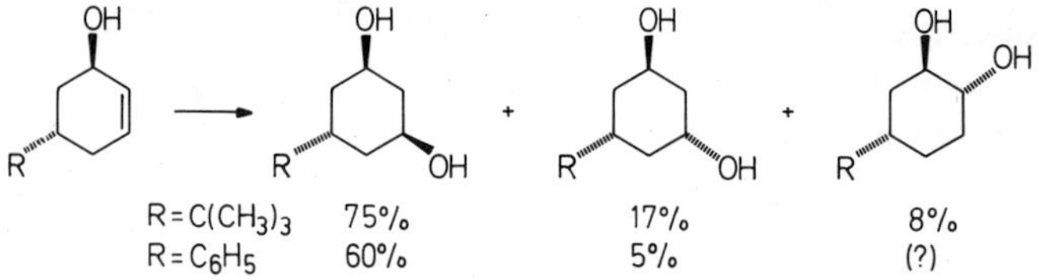

in which the new hydroxy group is introduced on the carbon furthest away from the original alcohol group.

The hydroxymercuration-demercuration of the following closely related systems is again observed to give predominantly 3-substituted products in similar stereoisomeric ratios (Eqs. 25–27).

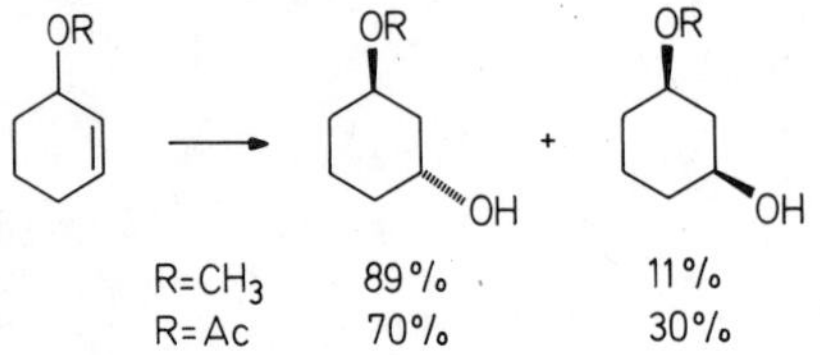

$$\begin{array}{ll} R=CH_3 & 89\% \qquad 11\% \\ R=Ac & 70\% \qquad 30\% \end{array} \tag{25} [142, 144]$$

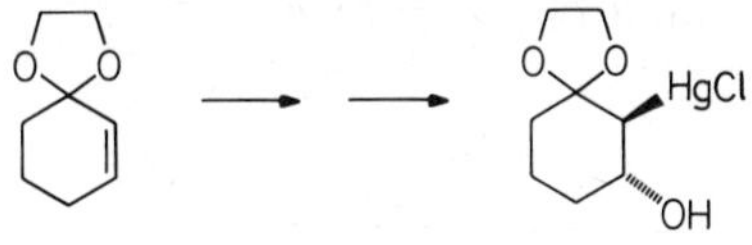

(26) [153]

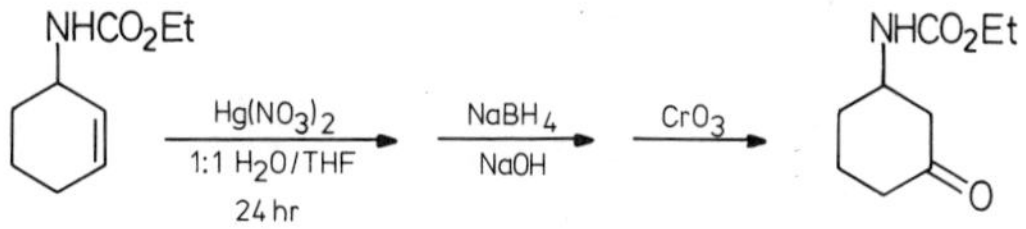

(27) [154]

3-Cyclohexenol and its methyl ether have been similarly studied (Eqs. 28, 29) [68, 145]. The results here are clearly different from the corresponding

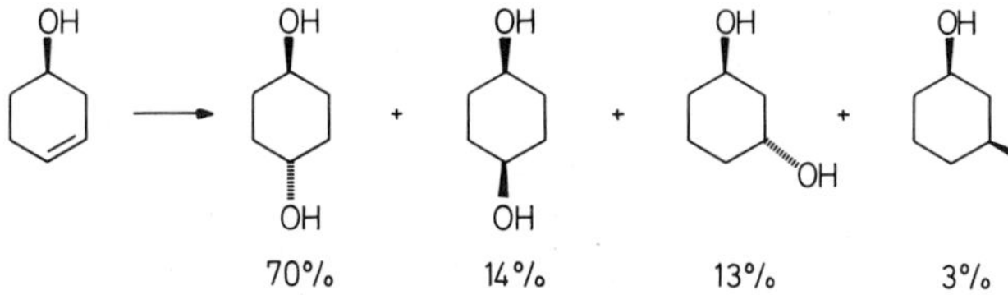

(28)

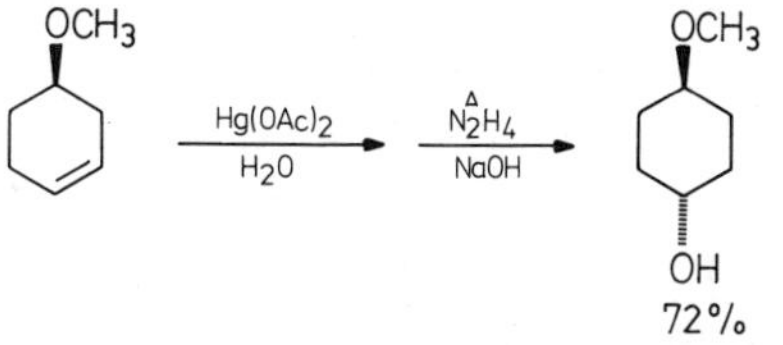

(29)

alkyl-substituted cyclohexenes and cannot be easily explained.

Larger ring allylic alcohols also tend to give the *trans*-1,3-diol as the major product (Eqs. 30, 31) [141].

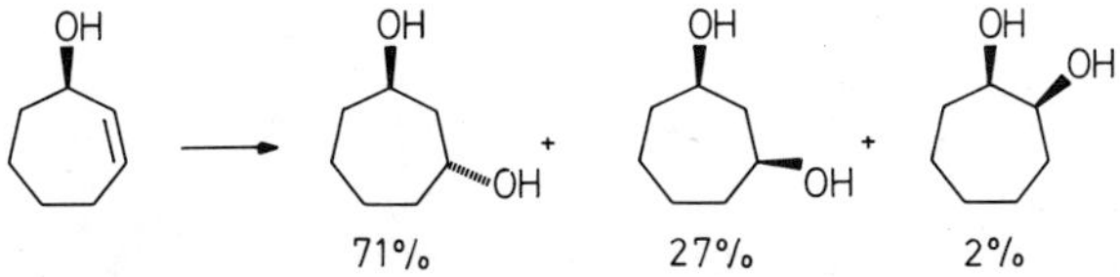

(30)

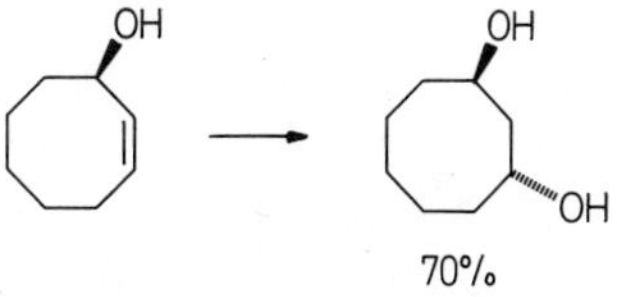

(31)

The hydroxymercuration-demercuration of several six-membered ring heterocycles has also been examined and been found to be highly regio- and stereoselective (Eqs. 32–35).

79

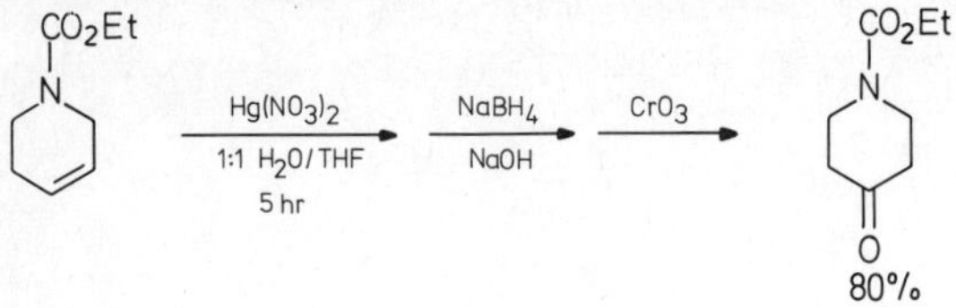

II. Hydroxymercuration

$$(32) \ [154]$$

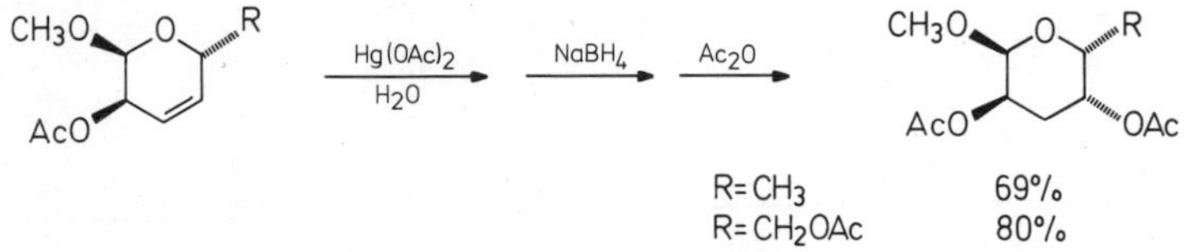

$$(33) \ [161]$$

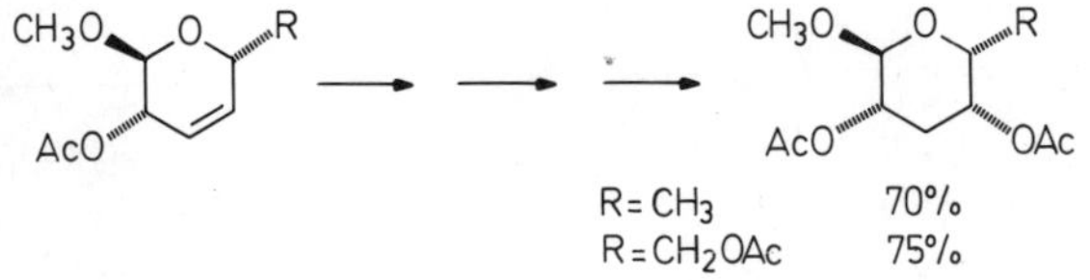

R= CH$_3$ 69%
R= CH$_2$OAc   80%

$$(34) \ [161]$$

R= CH$_3$ 70%
R = CH$_2$OAc   75%

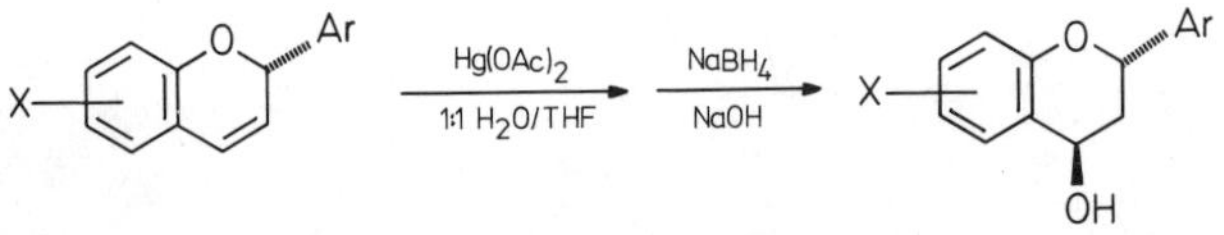

$$(35) \ [227]$$

The hydroxymercuration-demercuration of a large number of bicyclic and polycyclic alkenes has been examined. Rather than repeat the many examples already cited in Table 2.4, only a few of the more interesting cases will be discussed. Decalins and related steroid systems have been studied (Eqs. 36–38). Trans diaxial addition predominates as with simple cyclohexenes. Again

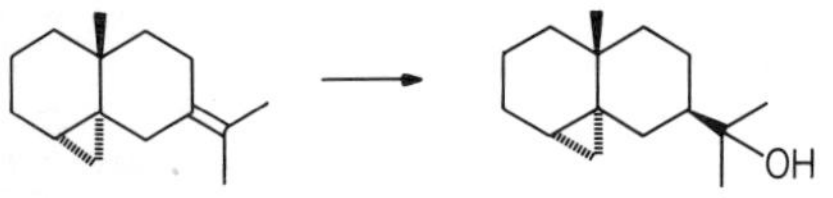

$$(36) \ [225]$$

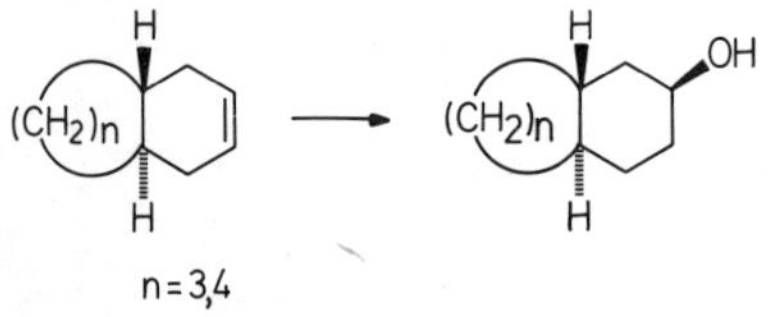

$$(37) \ [155]$$

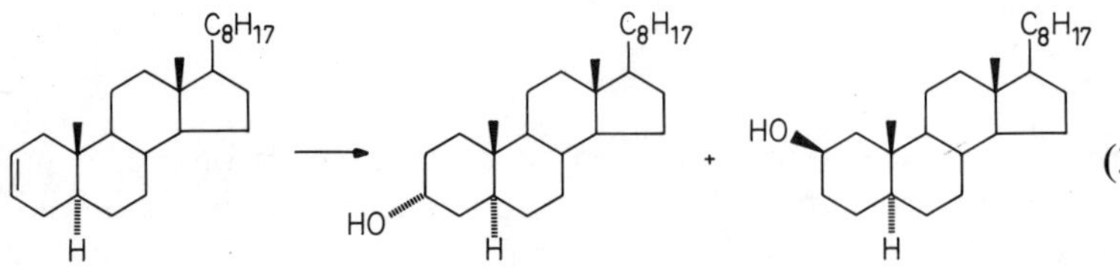

n = 3,4

$$(38) \ [230, 231]$$

80

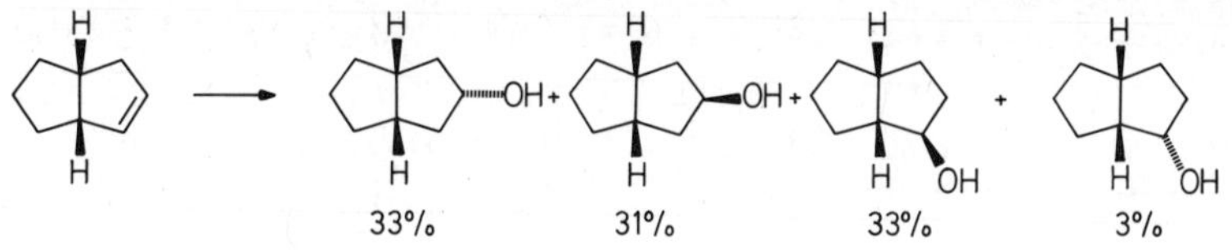

$$(39)$$

33%   31%   33%   3%

five-membered ring systems tend to give a number of products (Eq. 39)
[186, 195]. Norbornene and many substituted derivatives give cis-exo adducts
(Eq. 40) [131, 140, 182, 183, 186, 191]. Substitution about the ring fails to

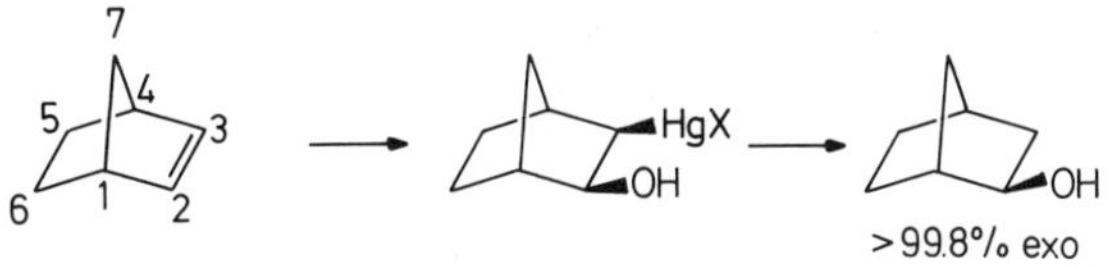

$$(40)$$

$>99.8\%$ exo

change the stereochemistry of addition [183]. Even 7,7-dimethylnorbornene
gives exclusively the cis-exo adduct [183, 185]. Unsymmetrical alkyl-substituted norbornenes usually give mixtures of regioisomers, but substitution
by certain strong electron-withdrawing groups often affords predominantly
one product as illustrated by the following examples (Eqs. 41–43). The
presence of a nitrogen moiety in a bicyclic system similarly imparts high

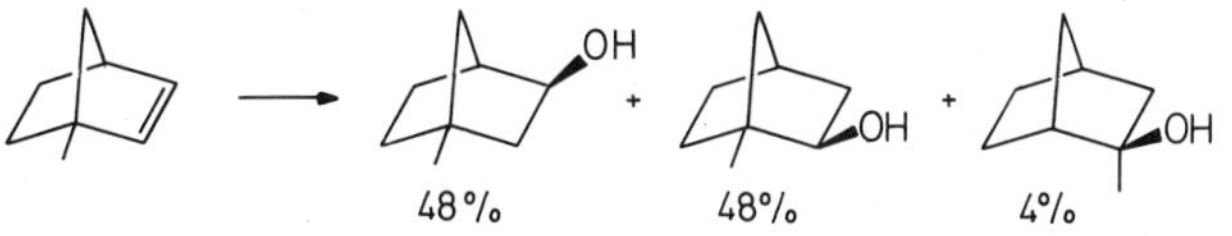

$$(41)\ [183, 185]$$

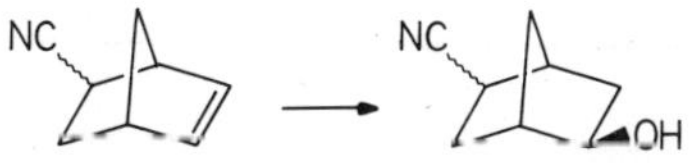

48%   48%   4%

$$(42)\ [190, 193]$$

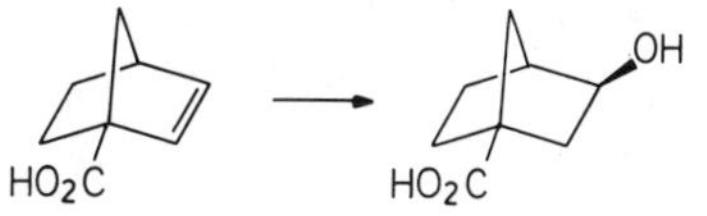

$$(43)\ [190]$$

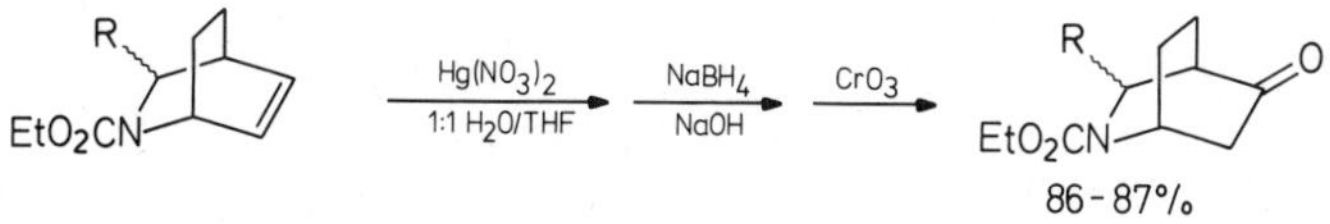

$$(44)$$

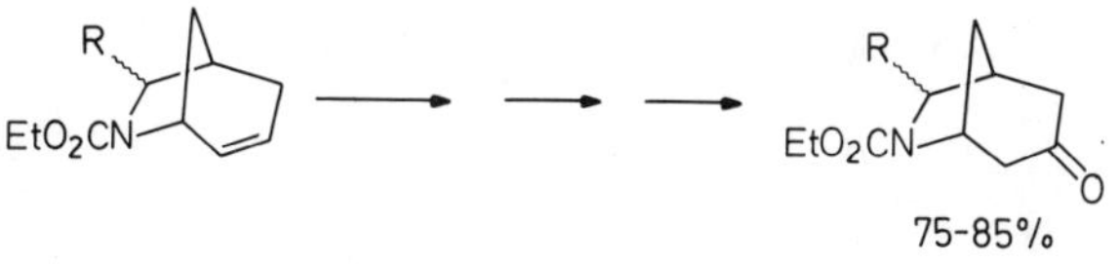

86–87%

$$(45)$$

75–85%

regioselectivity to the reaction (Eq. 44–46) [154, 207]. The regio- and stereochemistry of hydroxymercuration-demercuration of a number of other polycyclic olefins has been examined and the results are summarized in Table 2.4. Few useful generalities can be made regarding these substrates.

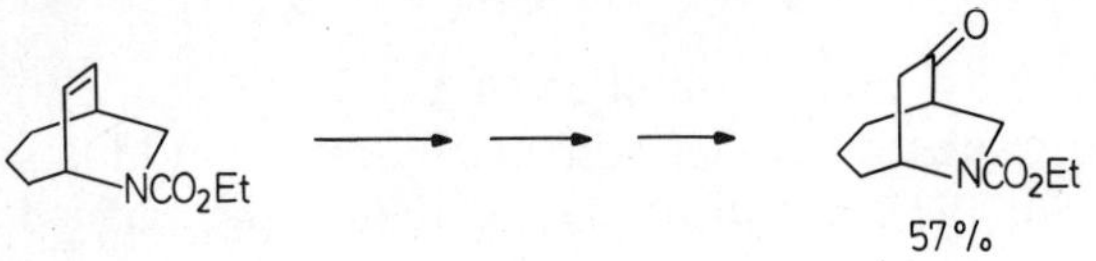

$$(46)$$

57%

Let's examine the stereochemistry of hydroxymercuration more closely [3]. Simple acyclic olefins give exclusively trans addition (Eqs. 47–49).

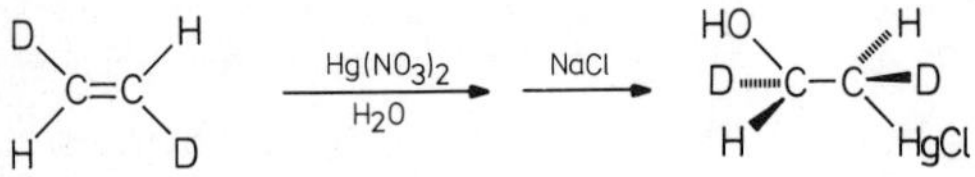

$$(47) \ [63]$$

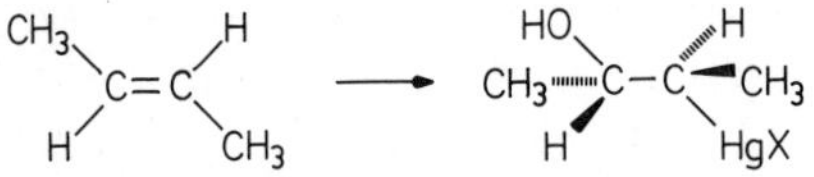

$$(48) \ [22, 28, 34, 38, 189]$$

$$(49) \ [22, 28, 34, 38, 189]$$

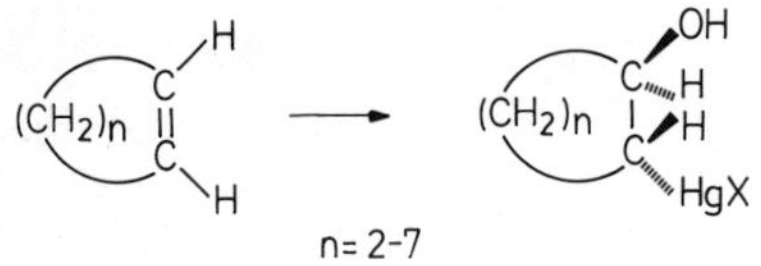

$$(50)$$

n= 2-7

Cyclobutene [127, 128], cyclopentene [34, 128, 131], cyclohexene [56, 128, 131, 189], *cis*-cycloheptene [128] and *cis*-cyclononene [128] all give trans adducts, but the more strained olefins *trans*-cyclooctene [128, 159] and *trans*-cyclononene [128] give cis addition (Eqs. 50, 51).

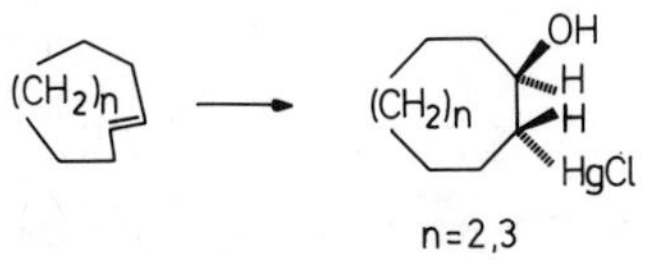

$$(51)$$

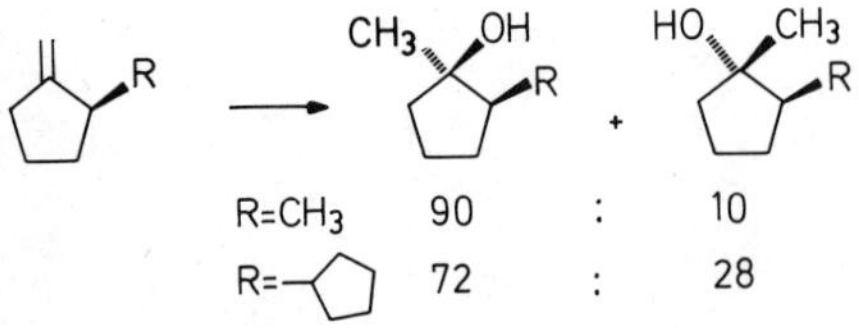

n=2,3

$$(52)$$

R=CH₃   90   :   10
R=⬠   72   :   28

A number of examples of hydroxymercuration-demercuration of methylene cycloalkanes have been reported. In 2-substituted methylene cyclopentanes, hydroxy attack occurs primarily cis to the substituent (Eq. 52, see p. 82) [147]. The opposite applies for 2-substituted methylene cyclohexanes (Eq. 53). This generalization appears to hold true for a number of 2-hydroxy [156, 160] and 2-alkoxy [156, 223] substituted methylene cyclohexanes as well, although 2-acetoxy derivatives give the opposite results [156]. 3-Substituted methylene cyclohexanes give predominantly hydroxyl attack trans to the substituent, while 4-substituents give cis attack predominantly (Eqs. 54, 55).

R	%	%	Ref.
CH_3	63–67	33–37	147,155,156
C_2H_5	74	26	155,156
$i\text{-}C_3H_7$	70	30	147
$t\text{-}C_4H_9$	97	3	147,175

(53)

R	%	%	Ref.
CH_3	54–58	42–46	147,155,156
C_2H_5	56	44	156
$t\text{-}C_4H_9$	58	42	147

(54)

R	%	%
CH_3	68	32
$t\text{-}C_4H_9$	71	29

(55) [147]

With the 2-substituted compounds, the stereochemistry is opposite that of Grignard addition to the corresponding ketones [155], while substituents in the 3 or 4 positions give results which parallel both Grignard addition to the corresponding ketone [155] and peracid epoxidation [156].

The stereochemistry of addition has been reported for a number of trisubstituted cycloalkenes. Cyclopentenes tend to add the hydroxyl group from the less hindered side of the molecule (Eqs. 56, 57) [148]. However,

73%   27%

(56)

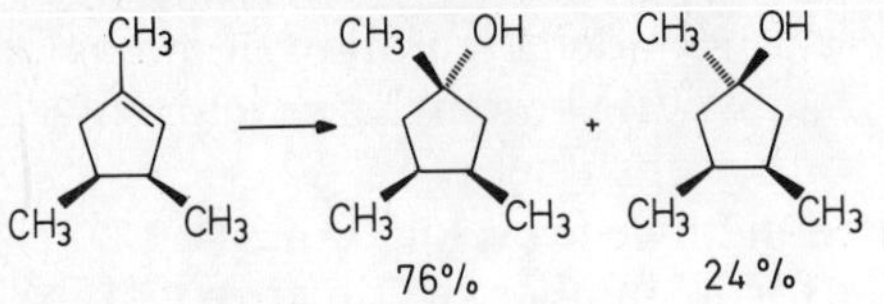

$$(57)$$

3,4-dimethylcyclopentene gives predominantly the more hindered alcohol
(Eq. 58) [133]. The stereochemistry of substituted cyclohexenes is more easily

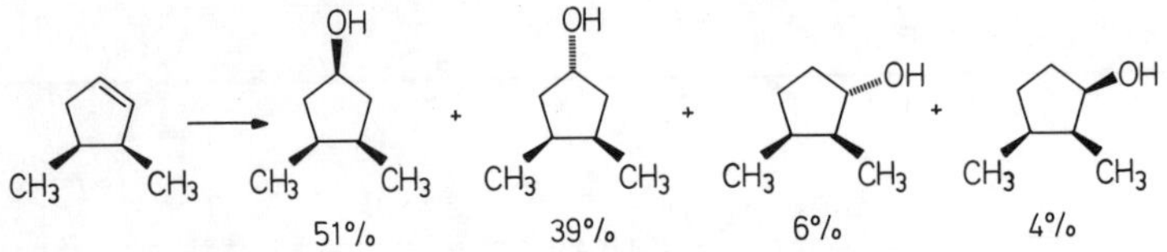

$$(58)$$

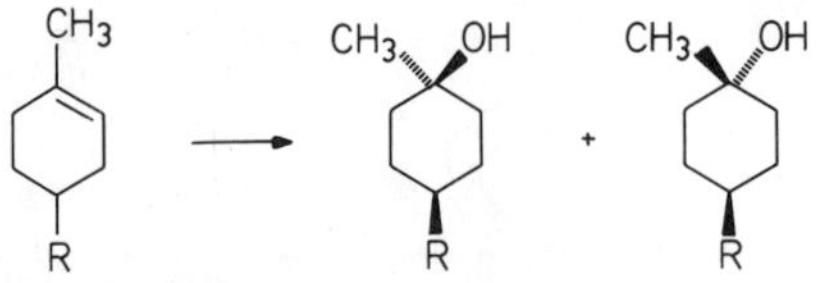

$$(59)$$

R	%	%	Ref.
t-C$_4$H$_9$	98.8-100	0-1.2	157, 166,168
i-C$_3$H$_7$	100	?	169

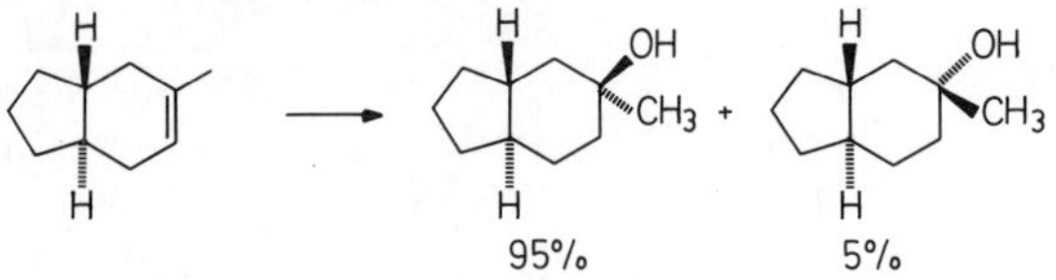

$$(60) \ [155]$$

predicted. As with the simple cyclohexenes discussed earlier, axial alcohols
are formed almost exclusively (Eqs. 59, 60).

The literature abounds with examples of the hydroxymercuration-
demercuration of bicyclic and polycyclic alkenes (see Table 2.4). A few
general comments on the stereochemistry of addition appear in order. As
noted earlier, norbornene (>99.8% exo) and all substituted norbornenes
studied to date undergo cis-exo addition no matter what the substitution
pattern [140, 183, 191]. When mercuric acetate is used as the mercurating
agent, these reactions are frequently accompanied by cis-acetoxymercuration
as well. A variety of exo-methylene norbornenes also give almost exclusive
exo hydration (Eq. 61) [148, 185, 186, 193, 195, 200]. The only exception
appears to be camphene (Eq. 62) [59]. Bicyclo[2.1.1]hex-2-ene also undergoes
cis addition (Eq. 63) [181], while both cis and trans addition are observed with
bicyclo[2.2.2]oct-2-ene (Eq. 64) [197, 226]. Less strained bicyclic and tricyclic olefins apparently give the usual trans adducts in which the hydroxyl

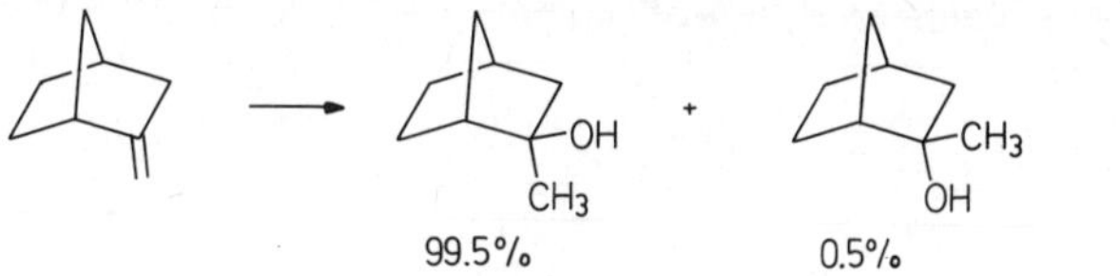

$$99.5\% \qquad 0.5\% \tag{61}$$

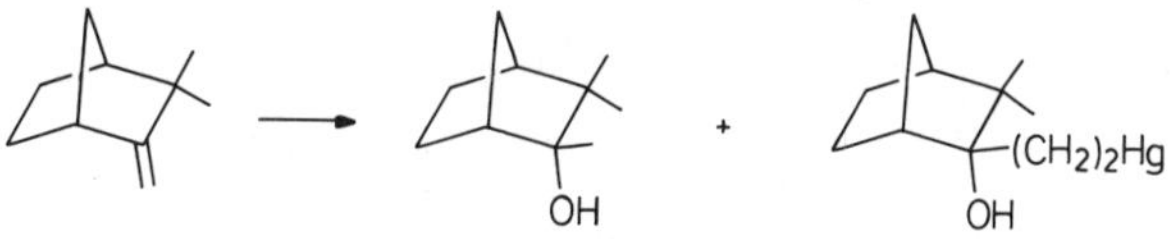

$$\tag{62}$$

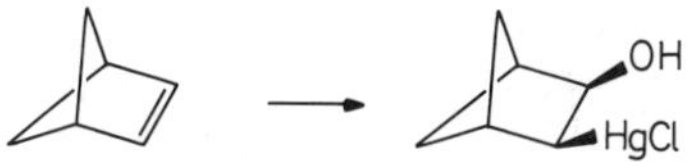

$$\tag{63}$$

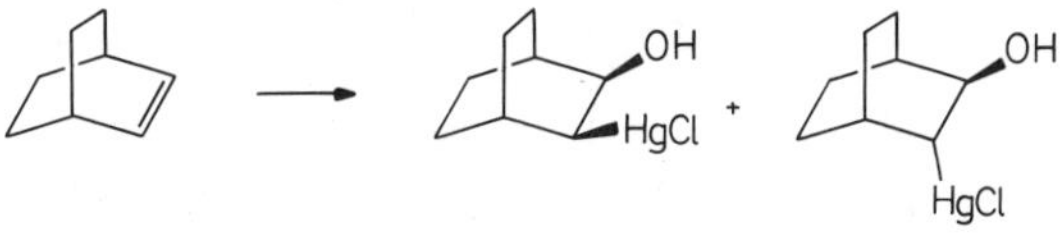

$$\tag{64}\ [197]$$

group adds primarily to the less hindered face of the double bond (Eqs. 65, 66). Table 2.4 contains a number of further examples. The carenes appear to be a notable exception to this rule (Eq. 67) [213–215].

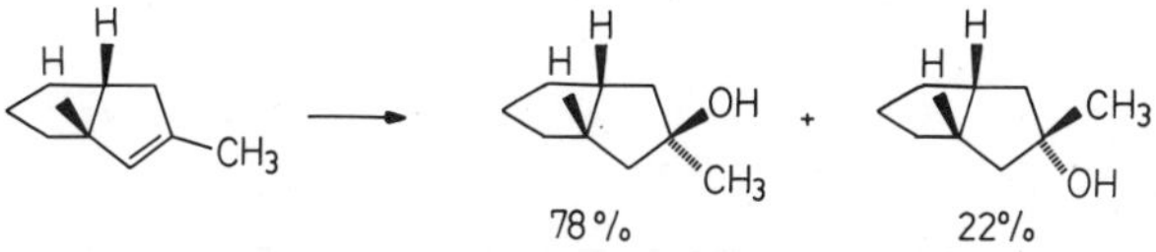

$$78\% \qquad 22\% \tag{65}\ [148,\ 195]$$

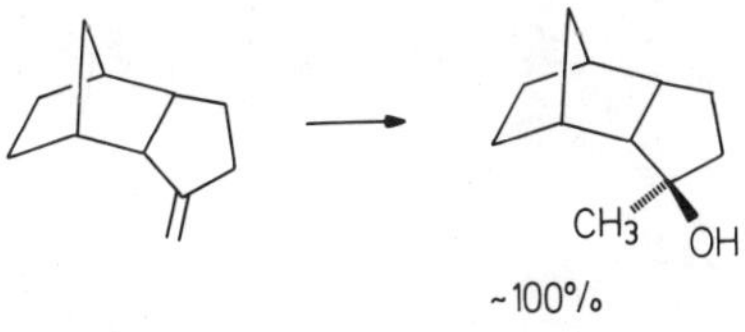

$$\sim 100\% \tag{66}\ [148,\ 186]$$

plus other products

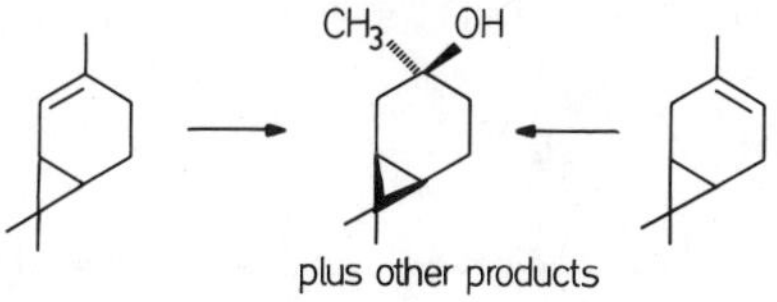

$$\tag{67}$$

Attempts to prepare optically active alcohols using optically active mercury(II) carboxylates have met with limited success [45, 51]. Best results were achieved using styrene and mercuric tartrate. Enantiomeric excesses of 25–32% were observed, but the overall yields of alcohol were less than 30% [51].

Table 2.5. Relative Rates of Hydroxymercuration of Alkenes

Alkene	Relative rate	Ref.	Alkene	Relative rate	Ref.
$H_2C=CH(CH_2)_3OH$	> 200	22	$H_2C=CHCH_2CH_2OH$	1.6	22
$H_2C=C(CH_3)_2$	> 200	22	$CH_3CH_2CH=C(CH_3)_2$	1.24	238
(methylenecyclopentane)	> 200, 59	129, 238	$H_2C=CHCH_2CHOHCH_3$	1.2	22
$H_2C=C(CH_3)(CH_2)_2CH_3$	48	54, 238	$H_2C=C(CH_3)C_6H_5$	1.18	238
$H_2C=CHCH_3$	20	22	$cis\text{-}CH_3CH=CHCH_3$	1.1	22
$H_2C=CH(CH_2)_4OH$	~20, 3.6–13	22, 92	(cyclohexene)	1.0	22, 54, 167, 238
$H_2C=CHCH_2CH_3$	16	22	$H_2C=CH_2$	1.0	22
(1-methylcyclobutene)	8.2	129	($(CH_3)_3C$-cyclohexene)	1.0	167
$H_2C=CH(CH_2)_2CH_3$	6.6	238	(cyclopentene)	0.74, 0.78	129, 238
$H_2C=C(CH_3)CH_2OH$	6.0	22	(cyclobutene)	0.69	129
$H_2C=CH(CH_2)_3CH_3$	4.8	238	($(CH_3)_3C$-norbornene, H)	0.67	220
$H_2C=CH(CH_2)_2CH\text{–}CH_2$ (epoxide)	4.4	92	$cis\text{-}CH_3CH=CHCH_2CH_3$	0.56	238
(1-methylcycloheptene)	4.1	129	(cyclohexenol, OH)	0.42, 0.08	22, 167
(cyclooctene epoxide)	4.0	92	$trans\text{-}CH_3CH=CHCH_3$	0.33	22
(norbornene, OR; R = H, Ac)	> 3.7	192	$H_2C=CHC_6H_5$	0.28	238
(norbornene)	3.7	54, 238	(CH_3, CH_3 norbornene)	0.25	183
$CH_3CH=C(CH_3)_2$	3.7	129	(AcO, H norbornene)	0.25	192
(1-methylcyclohexene)	3.7	129	$H_2C=CHCH_2OH$	0.22	22
$H_2C=CHCH(CH_3)_2$	2.5	54, 238	($(CH_3)_3C$-cyclohexenol, OH)	0.21	167
(1-methylcyclopentene)	1.86	238	(cycloheptene)	0.18	129

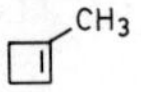

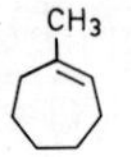

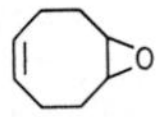

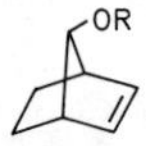

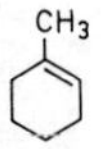

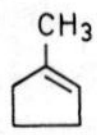

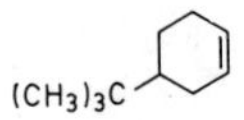

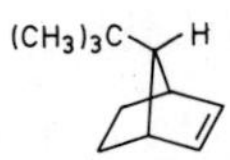

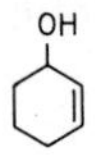

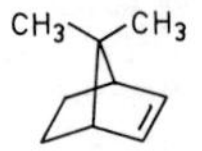

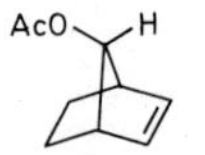

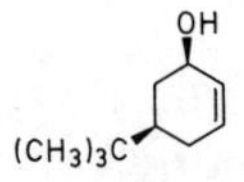

Table 2.5. (continued)

Alkene	Relative rate	Ref.
trans-$CH_3CH{=}CHCH_2CH_3$	0.17	238
$H_2C{=}CHC(CH_3)_3$	0.15	238
[structure: bicyclic with H, CH₃]	0.12	192
cis-$CH_3CH{=}CHCH(CH_3)_2$	0.09	238
[structure: bicyclooctene with cyclopropane]	0.08	92
[structure: cyclohexenol, OH]	0.08, 0.42	22, 167
cis-$CH_3CH{=}CHCH_2OH$	0.08	22
$(CH_3)_2C{=}C(CH_3)_2$	0.061	238
$(CH_3)_2CHCH{=}C(CH_3)_2$	0.056	238
$H_2C{=}CHCHOHCH_3$	0.05	22
trans-$CH_3CH{=}CHCH_2OH$	0.04	22
[structure: $(CH_3)_3C$-cyclohexenol, OH]	0.036	167
$H_2C{=}CHCHOHCH_2CH_3$	0.027	22
trans-$CH_3CH{=}CHCH(CH_3)_2$	0.026	238
$(CH_3)_3CCH{=}C(CH_3)_2$	0.02	238
cis- and trans-$CH_3CH{=}CHC_6H_5$	< 0.02	238
[structure: norbornene]	0.01	54, 238
[structure: cyclooctene]	0.002	92, 238
cis-$HOCH_2CH{=}CHCH_2OH$	0.002	22
$H_2C{=}CHCH_2Cl$	0.002	22
$H_2C{=}CHCH_2CN$	0.0008	22
trans-$CH_3CH{=}CHC(CH_3)_3$	~0.0006	39

The relative reaction rates have been determined for the hydroxymercuration of a large number of olefins. Table 2.5 summarizes these results. The relative rates reported have been taken from several different studies run under a variety of reaction conditions and have been placed on one scale. One should, therefore, not place too much weight on differences in numbers from different references, but refer to the original papers for specifics. If one dares to use the rates at which the reaction decolorizes as an indication of the rate of reaction, additional compounds could be added to Table 2.5 [92]. This has not been done, however.

A few generalities can be made regarding the relative reactivities of various olefins. The reaction rate is increased by electron donation and decreased by electron withdrawal. Disubstituted terminal olefins $R_2C=CH_2$ are more reactive than monosubstituted terminal olefins $RCH=CH_2$, which are in turn more reactive than disubstituted internal olefins $RCH=CHR$. Cis disubstituted internal olefins are approximately three times as reactive as the corresponding trans olefins. There is a relative insensitivity to the size of the ring in cyclic olefins. Increased steric hindrance at the site of hydroxyl or mercury attachment or substitution on the double bond (provided the carbonium ion stability stays the same) decreases the rate. Increased stability of the carbonium ion or decreased stability of the olefinic ground state, due either to increased cis interactions or constraint in a bicyclic ring system, also increase the reactivity of the double bond. Certain types of olefins have proven unreactive towards hydroxymercuration, particularly aryl-substituted cyclic olefins [41, 227], tetrasubstituted olefins [39, 47], and highly sterically hindered olefins, such as *syn*-7-*t*-butylnorbornene [220].

The detailed kinetics of the hydroxymercuration of alkenes have been examined by several groups. It has been observed that the reaction is first order in mercury(II) salt and alkene [22, 129, 233, 239]. A number of thermodynamic parameters have been determined for the hydroxymercuration of ethylene [22, 233]. The rate of reaction is found to be independent of the acid concentration at .001–0.1 M concentrations. The rate increases with $NaClO_4$ concentration and is unaffected by oxygen or hydrogen peroxide, but is retarded by pyridine [239] or sodium acetate [183]. The reaction apparently proceeds through a transition state with considerable carbonium ion character. The nature of this transition state has been hotly debated. Mercurinium ions have been frequently suggested as intermediates based on molecular orbital calculations [240, 241], the similarity of the stereochemical results to bromination and methoxybromination [157, 166], and the observation of such species in ion cyclotron resonance [242] and super acid NMR studies [189]. Despite initial reports [15], subsequently proved wrong [20, 236], that mercurinium ions could be observed under the usual hydroxymercuration conditions, no one has yet provided kinetic or other evidence convincingly establishing their presence [22, 233, 243]. H. C. Brown [183, 244] and others [168] have argued against mercurinium ions on the basis of kinetic and stereochemical results. They favor an unsymmetrical mercury-stabilized carbocation. The kinetics and mechanism of olefin exchange between hydroxymercurials [243], the effect of solvents on the transition state [245] and the nature of the overall equilibrium [38, 233, 246–249] have also been explored.

As noted earlier, the facile reversion of hydroxymercurial to olefin in the presence of strong acids was initially used to argue in favor of some sort of mercury-olefin complex for these compounds. This reaction has also been studied in detail to obtain information on the nature of the transition state in the hydroxymercuration reaction [14, 31, 32, 38, 189, 250]. Such studies have practical utility as well. One can separate saturated from unsaturated compounds by hydroxymercuration, separation and regeneration of the olefin [211, 251, 252]. The mercury moiety in these organomercurials also undergoes spontaneous demercuration via solvolytic processes under certain conditions, as seen in the following examples (Eqs. 68–70). This last reaction provides a useful way of distinguishing naturally occurring propenyl aroma-

$$CH_3CHCH_2HgClO_4 \longrightarrow CH_3CCH_3 + Hg \qquad (68)\ [27, 28, 36, 37, 253, 254]$$

$$\xrightarrow[H_2O]{Hg(OAc)_2/\Delta} \qquad\qquad (69)\ [211]$$

$$ArCH=CHCH_3 \xrightarrow[H_2O]{Hg(OAc)_2} ArCH-CHCH_3 \qquad (70)$$

tics from their isomeric allyl counterparts [102–105, 109, 252]. The latter compounds simply give stable hydroxymercurials from which the allyl aromatic can be regenerated.

Competing solvomercuration processes are occasionally observed during the hydroxymercuration reaction. As noted earlier, with strained bicyclic and polycyclic olefins, the hydroxymercurial is frequently accompanied by significant amounts of acetoxymercuration side product when mercuric acetate is used as the reagent [183, 184]. The hydroxymercuration of dienes and polyenes, to be discussed in the next section, frequently leads to intramolecular alkoxymercuration of the initially formed unsaturated alcohol. Numerous examples of intramolecular enol cyclizations have also been reported and will be discussed later.

Several functional groups have been observed to interfere or participate in the hydroxymercuration process. Allylic halides, unsaturated tertiary halides and secondary iodide, benzylic bromides and iodides, and unsaturated halides which can undergo neighboring group participation, all present difficulties due either to solvolysis or problems on alkaline sodium borohydride reduction [66]. However, primary alkyl chlorides, bromides and iodides; secondary alkyl chlorides; and benzylic chlorides should behave normally. The reduction problems can be circumvented by using a modified work-up procedure. Unsaturated epoxides of various sorts also encounter elimination and neighboring group participation problems [66, 92, 152, 198]. Cyclic ethers are frequent products of these reactions. Simple ethers [92], sulfides [66], nitriles [90] and ketones [152] have on occasions presented difficulties during attempted hydroxymercuration reactions. The cyclopropanes in 2- and 3-carene have also been observed to react competitively [212, 214, 215], although no difficulties with a cyclopropane were encountered in another example [225].

The hydroxymercuration of vinylic substrates provides a useful approach to a wide variety of α-mercurated carbonyl compounds as discussed in the monograph "Organomercury Compounds in Organic Synthesis" (Eq. 71). Enol acetates [71–73, 86, 106], ethers [70, 78–80, 93, 106], and silanes [255],

$$\text{C=C}\diagup^{Y} + HgX_2 + H_2O \longrightarrow -\overset{O}{\underset{HgX}{C}}-\overset{\parallel}{C}- \tag{71}$$

as well as ketene [64], have been employed in this reaction. The corresponding protodemercurated carbonyl products have also been isolated from the hydroxymercuration of enol ethers [256] and esters [85, 86], vinyl sulfides [257, 258], and vinyl halides [62, 67]. The hydroxymercuration of certain carbohydrate enol ethers provides some novel rearrangement products (Eqs. 72, 73).

$$\xrightarrow[H_2SO_4]{HgSO_4} HOCH_2CHOAcCH=CHCHO \tag{72} [146]$$

88 %

$$\xrightarrow[H_2O]{HgCl_2} \tag{73} [179, 180]$$

83%

The hydroxymercuration-demercuration process attains its great synthetic utility due to the fact that carbon skeleton rearrangements are seldom observed. Such rearrangements have been encountered, however, when highly electrophilic mercury salts are employed and the resulting hydroxymercurial is allowed to undergo solvolysis (Eqs. 74, 75). Rearrangements are also observed when the initial cationic intermediate is generated in or next to a strained ring system which can undergo facile ring opening (Eqs. 76–80). Rearrangement can also occur due to added base or during the reduction step (Eq. 81).

$$\xrightarrow[H_2O]{Hg(ClO_4)_2} \tag{74} [130]$$

(75) [229]

(76) [132]

(77) [43]

(78) [177]

(79) [102, 104, 208–210]

(80) [206]

(81) [259]

(82)

The reductive-demercuration of β-hydroxymercurials provides an exceedingly valuable method for the preparation of alcohols (Eq. 82). By far the most important reagent for this transformation is alkaline sodium borohydride. It generally gives rapid, clean demercuration. It is unnecessary to use an excess of sodium borohydride [39], but this is commonly done. The temperature of the reduction also appears to have little effect on the yield [39]. In one case where significant amounts of olefin by-product were observed during reduction, it was discovered that the use of diethyl ether as solvent instead of THF led to less olefin [151]. While there are examples of the stereospecific replacement of mercury by deuterium using alkaline NaBD$_4$ [34, 227, 230], this reaction apparently proceeds via free radicals and low stereospecificity is to be expected [34, 219]. While the presence of base apparently leads to cleaner reactions and helps to stabilize sodium borohydride, on occasions it promotes rearrangement. Thus, the demercuration of hydroxymercurials derived from allylic chlorides is best carried out in the absence of base [66]. As indicated in Tables 2.1–2.4, esters can also be saponified under the usual alkaline sodium borohydride demercuration conditions [68, 115, 192]. The reagent can also reduce ketones as well as the mercury moiety [152, 165].

A number of other reagents have been used to effect demercuration. Perhaps most important is Na(Hg)/H$_2$O. This reagent appears to be the only one which is general for demercuration with complete retention. By employing D$_2$O as the solvent, stereospecific deuteration should be possible, although this has not apparently been done yet on hydroxymercurials. While alkali metals have been utilized for demercuration [260], they appear to have no advantage over the above reagents. Zinc metal has also been used [104], but it appears to regenerate the starting olefin [103, 252]. Although hydrazine was one of the first reagents used to demercurate β-hydroxymercurials, it frequently gives low yields and olefins have been observed as side products [61, 145, 150, 169]. Finally, alkaline H$_2$S is not a general reagent for demercuration, but it is useful for the reduction of hydroxymercurials derived from α,β-unsaturated carboxylic acids (Eq. 83) [87]. Demercuration apparently proceeds with complete retention in this case.

$$\tag{83}$$

These are a number of reactions other than reductive-demercuration that the β-hydroxymercurials undergo which are of synthetic interest. Many of these, such as halogenation, oxidation, carbonylation, and free radical formation and addition to alkenes, have been discussed in detail in the monograph "Organomercury Compounds in Organic Synthesis". Many are also included in the tables in this section. A few other reactions are noteworthy. β-Hydroxymercurials can be acylated on oxygen to form the corresponding β-acyloxymercurials [16]. Certain substrates can undergo dehydration to the corresponding vinylmercurial under these conditions [60]. Treatment of the hydroxymercurials with strong bases and heat results in epoxide formation (Eq. 84) [25, 261]. Due to the low yields, however, this

$$\tag{84}$$

reaction does not appear to have any synthetic utility. The hydroxymercuration and subsequent oxidation by chromic acid [46] or palladium chloride

[44, 119] does provide a valuable method for the conversion of alkenes to ketones (Eq. 85). The palladium reaction can result in halogenation instead,

$$RCH=CH_2 \xrightarrow[\text{or } Hg(OAc)_2/Li_2PdCl_4/CuCl_2]{Hg(O_2CEt)_2/CrO_3/H_2SO_4} RCCH_3 \qquad (85)$$

$$(86)$$

if one uses a large excess of cupric chloride (Eq. 86) [63]. Note that the latter reaction proceeds with inversion of configuration. Barluenga and co-workers have recently shown that hydroxymercurials can be made to undergo transmetallation by lithium, sodium or potassium to afford the corresponding organometallics which should prove quite useful in organic synthesis (Eq. 87) [260, 262–264]. Finally, hydroxymercuration and subsequent titration of the

$$RCHCH_2HgBr \xrightarrow[\text{2. M}]{1.\,C_6H_5M} RCHCH_2M \longrightarrow RCHCH_2X \qquad (87)$$

$$M = Li, Na, K \qquad X = D, OH, CO_2R, COHR_2, SiMe_3$$

resulting acid by ammonium thiocyanate provides a useful analytical method for the determination of olefins [265].

B. Dienes and Polyenes

All dienes and polyenes which have been hydroxymercurated are summarized in Table 2.6. In acyclic systems, substrates of shorter chain length are listed first. In cyclic systems, priority is given to dienes or polyenes of smaller ring size. The general procedures for effecting hydroxymercuration are the same as those discussed in the preceding section. Much of what we have learned regarding the hydroxymercuration of simple alkenes is applicable to dienes and polyenes.

The hydroxymercuration of allenes gives several different products. The mercury-catalyzed hydration of acyclic allenes gives ketones (Eqs. 88, 89), while cyclic allenes afford allylic alcohols (Eq. 90) [294, 295]. Allenic

$$H_2C=C=CHR \xrightarrow[\text{H}_2O/H_2SO_4]{\text{cat. HgSO}_4} CH_3CCH_2R \qquad (88)\ [266, 275]$$

$$H_2C=CHCH=C=CR_2 \longrightarrow CH_3CH=CHCCHR_2 \qquad (89)\ [282]$$

$$(90)$$

Table 2.6. Hydroxymercuration of Dienes and Polyenes

Diene or polyene	Mercuric salt	Reaction conditions	Organomercurial(s) (% Yield)	Subsequent reactants	Product(s) (% Yield)	Refs.
$H_2C=C=CHCH_3$	$HgSO_4$	$H_2O/MeOH/H_2SO_4$ Δ 2 hr	—	$2,4-H_2NNHC_6H_3(NO_2)_2$	$CH_3CCH_2CH_3$ with $=NNHC_6H_3(NO_2)_2$	266
$H_2C=CHCH=CH_2$	$Hg(OAc)_2$	H_2O, NaOH/KI	—	I_2	$ICH_2CHOHCHOHCH_2I$ (2 isomers)	267
	$Hg(OAc)_2$	H_2O	$AcOHgCH_2CHOHCHOHCH_2HgOAc$ (4 – 27)	—	—	268
	$Hg(OAc)_2$	H_2O, Cl^-	$ClHgCH_2CHOHCHOHCH_2HgCl$ (60)	I_2	$ICH_2CHOHCHOHCH_2I$	53
	$Hg(OAc)_2$	—, X^-	$XHgCH_2CHOHCHOHCH_2HgX$ X = OAc, Cl, Br, I	RCOCl (or RHgOAc)	$ClHgCH_2CH(O_2CR)CH(O_2CR)CH_2HgCl$ R = CH_3, C_6H_5	269
	$Hg(O_2CR)_2$	—	$RCO_2HgCH_2CHOHCHOHCH_2HgO_2CR$ R = Et, n-Pr, n-Bu, $ClCH_2$	—	—	269
	$Hg(NO_3)_2$	H_2O	$NO_3HgCH_2CHOHCHOHCH_2HgNO_3$	—	—	270
	$Hg(NO_3)_2$	—	$NO_3HgCH_2CHOHCHOHCH_2HgNO_3$	time, KBr	(dioxane ring) $XHgCH_2$, CH_2HgX; X = NO_3, Br	269
	$Hg(NO_3)_2$	H_2O, I^-	—	I_2	$ICH_2CHOHCHOHCH_2I$ (meso ?)	267
	HgX_2 (X = ?)	H_2O, Cl^-	—	I_2	$ClCH_2CHOHCHOHCH_2I$	267
$H_2C=C=CHCH_2OH$	$Hg(OAc)_2$	1:1 H_2O/THF	—	$NaBH_4$/NaOH	$CH_3CHOHCH_2CH_2OH$ (52)	271
	$HgSO_4$	H_2O Δ 1.5 hr	—	—	$CH_3COCH=CH_2$ (15)	271
$(H_2C=CH)_2O$	$Hg(OAc)_2$	H_2O, KCl	$ClHgCH_2CHO$ (85)	$Al(O-i-C_3H_7)_3$	$ClHgCH_2CH_2OH$	78

94

Table 2.6. (continued)

Diene or polyene	Mercuric salt	Reaction conditions	Organomercurial(s) (% Yield)	Subsequent reactants	Product(s) (% Yield)	Refs.
$H_2C=C(CH_3)CH=CH_2$	$Hg(OAc)_2$	H_2O/THF 0°C 2.6 hr	—	$NaBH_4/NaOH$	$(CH_3)_2COHCH=CH_2$ (16) + $H_2C=CHCH_2CHOHCH_3$ (9) + $H_2C=C(CH_3)CHOHCH_3$ (2)	272
	$Hg(OAc)_2$	H_2O, KCl	$ClHgCH_2COH(CH_3)CHOHCH_2HgCl$ (8)	—	—	268
	$Hg(OAc)_2$	1:1 H_2O/THF 30 min	—	$NaBH_4/NaOH$	$(CH_3)_2COHCHOHCH_3$ (65)	68
trans-$H_2C=CHCH=CHCH_3$	$Hg(OAc)_2$	H_2O/THF 0°C 1hr	—	$NaBH_4/NaOH$	CH_3—CH=CH—$CHOHCH_3$ (56)	272
	$Hg(OAc)_2$	1:1 H_2O/THF 30 min	—	$NaBH_4/NaOH$	$CH_3CHOHCH_2CHOHCH_3$ (79)	68
$H_2C=CHCH_2CH=CH_2$	$Hg(OAc)_2$	1:1 H_2O/THF 45 min	—	$NaBH_4/NaOH$	$CH_3CHOHCH_2CHOHCH_3$ (83)	68
	$Hg(O_2CCF_3)_2$	1:4 H_2O/THF 25°C 8h	—	$NaBH_4/NaOH$	$CH_3CHOHCH_2CH=CH_2$ (57)	272
$H_2C=C=C(CH_3)CH_2OH$	$HgSO_4$	H_2O Δ 1.5 hr	—	—	3-methyl-2,3-dihydrofuran + $CH_3CH=C(CH_3)CHO$ (47) 85 : 15	271
$H_2C=C=CHCHOHCH_3$	$Hg(OAc)_2$	1:1 H_2O/THF	—	$NaBH_4/NaOH$	$CH_3CHOHCH_2CHOHCH_3$ (41)	271
	$HgSO_4$	H_2O Δ 1.5 hr	—	—	$CH_3COCH=CHCH_3$ (55)	271
$CH_3CH=C=CHCH_2OH$	$Hg(OAc)_2$	1:1 H_2O/THF	—	$NaBH_4/NaOH$ Δ	2-methyl-2,5-dihydrofuran (50)	271
	$HgSO_4$	H_2O Δ 1.5 hr	—	—	2-methyl-2,5-dihydrofuran (35)	271
(cyclohexadiene)	$Hg(OAc)_2$	H_2O/THF 0°C 0.5 hr	—	$NaBH_4/NaOH$	2-cyclohexen-1-ol (50) + cyclohexane-1,2-diol (16)	272

Table 2.6. (continued)

Diene or polyene	Mercuric salt	Reaction conditions	Organomercurial(s) (% Yield)	Subsequent reactants	Product(s) (% Yield)	Refs.
	$Hg(OAc)_2$	1:1 H_2O/THF	—	$NaBH_4$/NaOH	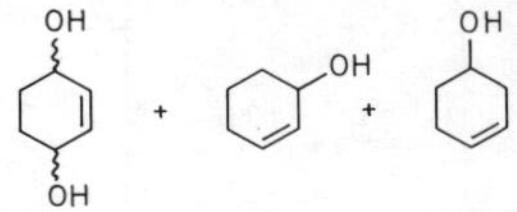 74 : 13 : 13	273
	$Hg(OAc)_2$	1:1 H_2O/THF	—	$NaBH_4$/NaOH	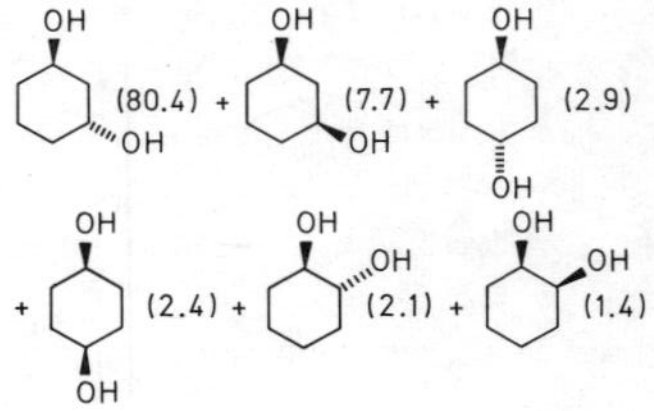 (80.4) + (7.7) + (2.9) + (2.4) + (2.1) + (1.4)	68
	$Hg(OAc)_2$	H_2O, KBr	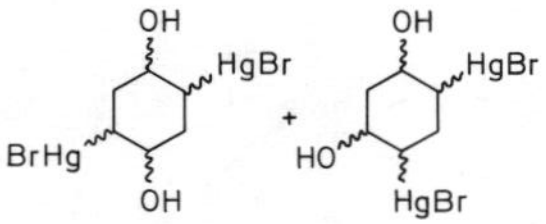 HgBr / BrHg organomercurials	$NaBH_4$	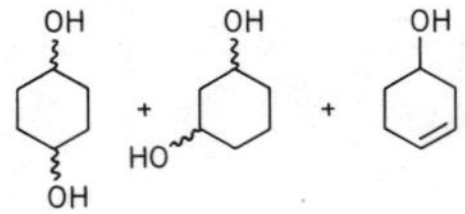	274
	$Hg(OAc)_2$	1:1 H_2O/THF	—	$NaBH_4$/NaOH	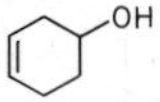 + diol, 67 : 33	273
	$Hg(OAc)_2$	1:1 H_2O/THF	—	$NaBH_4$/NaOH	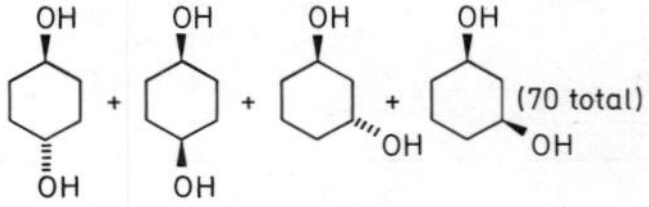 (70 total), 47 : 25 : 25 : 3	68
$H_2C{=}C(CH_3)C(CH_3){=}CH_2$	$Hg(OAc)_2$	H_2O/THF 0°C 0.5 hr	—	$NaBH_4$/NaOH	$H_2C{=}C(CH_3)COH(CH_3)_2$ (49) + $H_2C{=}C(CH_3)CH_2CHOHCH_3$ (6) + $(CH_3)_2COHCOH(CH_3)_2$ (4)	272

Table 2.6. (continued)

Diene or polyene	Mercuric salt	Reaction conditions	Organomercurial(s) (% Yield)	Subsequent reactants	Product(s) (% Yield)	Refs.
	Hg(OAc)$_2$	1:1 H$_2$O/THF 30 min	—	NaBH$_4$/NaOH	(CH$_3$)$_2$COHCOH(CH$_3$)$_2$ (88)	68
H$_2$C=C=CH(CH$_2$)$_2$CH$_3$	HgSO$_4$	H$_2$O/MeOH/H$_2$SO$_4$ 15 min	—	—	CH$_3$CO(CH$_2$)$_3$CH$_3$ (70 as semicarbazone)	275
H$_2$C=CH(CH$_2$)$_2$CH=CH$_2$	Hg(OAc)$_2$	H$_2$O, KCl	ClHgCH$_2$–O–CH$_2$HgCl (100)	Na(Hg)/H$_2$O	H$_2$C=CHCH$_2$CH$_2$CHOHCH$_3$ (90)	276
	Hg(OAc)$_2$	H$_2$O, KX	XHgCH$_2$–O–CH$_2$HgX, X = OAc, Cl, Br, I	NaBH$_4$ (on OAc)	CH$_3$–O–CH$_3$	277
	Hg(OAc)$_2$	1:1 H$_2$O/THF 30 min	—	NaBH$_4$/NaOH	CH$_3$–O–CH$_3$ (72) + CH$_3$–O–CH$_3$ (19) + H$_2$C=CH(CH$_2$)$_2$CHOHCH$_3$ (9)	68
	Hg[C(NO$_2$)$_3$]$_2$	H$_2$O RT 4 hr	(NO$_2$)$_3$HgCH$_2$–O–CH$_2$HgC(NO$_2$)$_3$ (70)	—	—	80
trans, trans–CH$_3$CH=CHCH=CHCH$_3$	Hg(OAc)$_2$	1:1 H$_2$O/THF 0°C 90 min	—	NaBH$_4$/NaOH	CH$_3$CHOHCH$_2$CHOHCH$_2$CH$_3$ (40) + CH$_3$CH=CHCHOHCH$_2$CH$_3$ (3.5)	68
H$_2$C=C=C(C$_2$H$_5$)CH$_2$OH	Hg(OAc)$_2$	1:1 H$_2$O/THF	—	NaBH$_4$/NaOH	H$_2$C=CHCOH(C$_2$H$_5$)CH$_2$OH (49)	271
	HgSO$_4$	H$_2$O Δ 1.5 hr	—	—	C$_2$H$_5$-furan + CH$_3$CH=C(C$_2$H$_5$)CHO, 90 : 9 (51 total)	271
H$_2$C=C=CHCOH(CH$_3$)$_2$	Hg(OAc)$_2$	1:1 H$_2$O/THF	—	NaBH$_4$/NaOH	CH$_3$CHOHCH$_2$COH(CH$_3$)$_2$ (39)	271
	HgSO$_4$	H$_2$O Δ 1.5 hr	—	—	CH$_3$COCH=C(CH$_3$)$_2$ (54)	271
(CH$_3$)$_2$C=C=CHCH$_2$OH	Hg(OAc)$_2$	1:1 H$_2$O/THF	—	NaBH$_4$/NaOH Δ	CH$_3$,CH$_3$-O-furan (48)	271
	HgSO$_4$	H$_2$O Δ 1.5 hr	—	—	CH$_3$,CH$_3$-O-furan (42)	271

Table 2.6. (continued)

Diene or polyene	Mercuric salt	Reaction conditions	Organomercurial(s) (% Yield)	Subsequent reactants	Product(s) (% Yield)	Refs.
$(H_2C=CHCH_2)_2O$	$Hg(OAc)_2$	$H_2O/H^+(KI)$	[1,4-dioxane ring] $XHgCH_2$—CH_2HgX cis and trans, $X = OAc, I$	I_2	[1,4-dioxane ring] ICH_2—CH_2I cis and trans	278
	$Hg(OAc)_2$	H_2O, KCl	[1,4-dioxane ring] $ClHgCH_2$—CH_2HgCl (85)	$Na(Hg)/H_2O/HOAc$	[1,4-dioxane ring] CH_3—CH_3	276
	$Hg(OAc)_2$	H_2O, Cl^-	[1,4-dioxane ring] $ClHgCH_2$—CH_2HgCl cis and trans	I_2	[1,4-dioxane ring] ICH_2—CH_2I (35) cis and trans	276, 279
	$Hg(NO_3)_2$	H_2O few min, NaX	[1,4-dioxane ring] $XHgCH_2$—CH_2HgX $X = Cl$ (31), I (23), $O_2CC_6H_5$ (70, 88)	—	—	268
	$Hg(NO_3)_2$	H_2O/HNO_3	—	I_2/KI	[1,4-dioxane ring] ICH_2—CH_2I trans > cis	278
$(H_2C=CHCH_2)_2SO_2$	$Hg(OAc)_2$	H_2O 2 hr, KI	—	I_2	[1,4-oxathiane-dioxide ring] ICH_2—CH_2I (54)	280
$(H_2C=CHCH_2)_2S$	$Hg(OAc)_2$	H_2O RT 1hr	[1,4-oxathiane ring] $AcOHgCH_2$—CH_2HgOAc (6–37)	—	—	268
	$Hg(OAc)_2$	H_2O 29 hr	[1,4-oxathiane ring] $AcOHgCH_2$—CH_2HgOAc (50)	I_2	[1,4-oxathiane ring] ICH_2—CH_2I cis >> trans	280
	$HgCl_2$	H_2O 3 hr	$(ClHgCH_2CHOHCH_2)_2S \cdot HgCl_2$ (?)	—	—	281

Table 2.6. (continued)

Diene or polyene	Mercuric salt	Reaction conditions	Organomercurial(s) (% Yield)	Subsequent reactants	Product(s) (% Yield)	Refs.
$(H_2C{=}CHCH_2)_2NH$	$Hg(OAc)_2$	H_2O, KCl	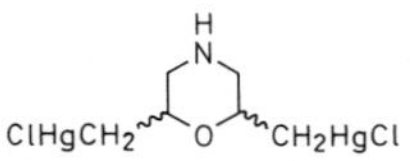 [morpholine structure] $ClHgCH_2$ … CH_2HgCl	$Na(Hg)/H_2O/HOAc$	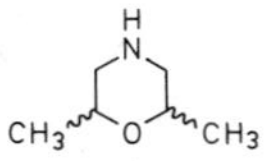 [morpholine structure] CH_3 … CH_3	276
	$Hg(OAc)_2$	H_2O 1hr, I^-	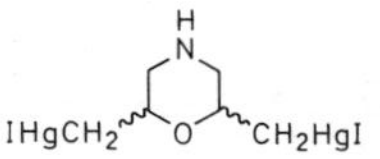 [morpholine structure] $IHgCH_2$ … CH_2HgI (27)	—	—	268
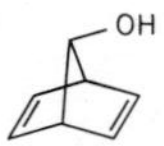 [norbornene–CH$_2$OH structure]	$Hg(OAc)_2$	1:1 H_2O/THF	—	—	C_6H_5CHO (57) + Hg	192
$H_2C{=}CHCH{=}C{=}C(CH_3)_2$	$HgSO_4$	H_2SO_4	$CH_3CH{=}CHCOCH(CH_3)_2$ (37.5–40)	—	—	282
$H_2C{=}CH(CH_2)_3CH{=}CH_2$	$Hg(OAc)_2$	1:1 H_2O/THF 30 min	—	$NaBH_4/NaOH$	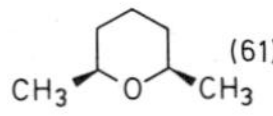 [tetrahydropyran] CH_3…CH_3 (61) 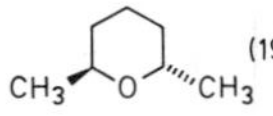 + [tetrahydropyran] CH_3…CH_3 (19) + $CH_3CHOH(CH_2)_3CHOHCH_3$ (10)	68
$H_2C{=}C{=}C(n\text{-}C_3H_7)CH_2OH$	$HgSO_4$	H_2O Δ 1.5 hr	—	—	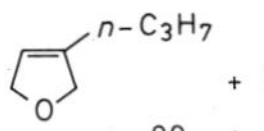 [dihydrofuran with n-C_3H_7] + $CH_3CH{=}C(n\text{-}C_3H_7)CHO$ 90 : 9 (54 total)	271
 [cyclooctatetraene structure]	$HgSO_4$	H_2O 70–80 °C	—	—	$C_6H_5CH_2CHO$ (70)	283
 [bicyclic ether structure]	$Hg(OAc)_2$	H_2O RT 216 hr	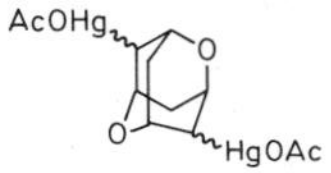 [bicyclic structure] $AcOHg$ … $HgOAc$ (43)	I_2	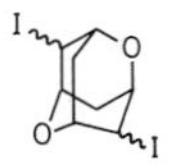 [bicyclic structure] I … I (58)	284,285
$H_2C{=}CHCH{=}C{=}C(CH_3)CH_2CH_3$	$HgSO_4$	H_2SO_4	$CH_3CH{=}CHCOCH(CH_3)CH_2CH_3$ (37.5–40)	—	—	282
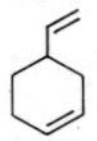 [vinylcyclohexene structure]	$Hg(O_2CCF_3)_2$	H_2O/THF 25 °C 0.5 hr	—	$NaBH_4/NaOH$	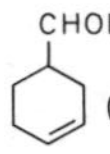 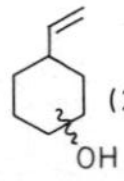 [cyclohexene–$CHOHCH_3$] (60) 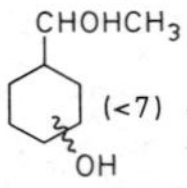 + [cyclohexane–$CHOHCH_3$] (<7) + [cyclohexane–vinyl–OH] (2) + bicyclic ethers	272

Table 2.6. (continued)

Diene or polyene	Mercuric salt	Reaction conditions	Organomercurial(s) (% Yield)	Subsequent reactants	Product(s) (% Yield)	Refs.
	$Hg(OAc)_2$	H_2O/THF 25°C 2 hr	—	$NaBH_4$/NaOH	(21) + (21) + bicyclic (10) ethers + (?) (3)	272
	$Hg(OAc)_2$	1:1 H_2O/THF	—	$NaBH_4$/NaOH	+ ether(s) + diol(s)	286
	$Hg(OAc)_2$	1:1 H_2O/THF $NaO_3S(CH_2)_{11}CH_3$	—	$NaBH_4$/NaOH	(90)	286
	$Hg(OAc)_2$	1:1 H_2O/THF	—	$NaBH_4$/NaOH	(25) + (13) + (8) + enediols (4)	68
	$Hg(OAc)_2$	1:1 H_2O/THF	—	$NaBH_4$/NaOH, Ac_2O	(49) + (41) + (3) + (3)	273
	$Hg(OAc)_2$	1:1 H_2O/THF	—	$NaBH_4$/NaOH	+ + diols 70 : 12 : 8	142
	$Hg(OAc)_2$	H_2O RT 68 hr	(68)	I_2 (or RHgI)	(64)	284, 285

Table 2.6. (continued)

Diene or polyene	Mercuric salt	Reaction conditions	Organomercurial(s) (% Yield)	Subsequent reactants	Product(s) (% Yield)	Refs.
	Hg(OAc)$_2$	1:6 H$_2$O/CH$_3$COCH$_3$, NaCl	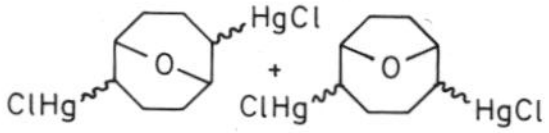 ClHg⋯(O) HgCl + ClHg⋯(O)⋯HgCl	—	—	287
	Hg(OAc)$_2$	H$_2$O/THF NaOH RT 130 min, KI	—	I$_2$/KI	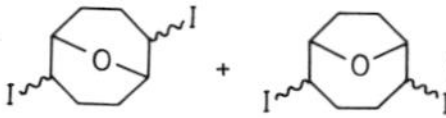 I⋯(O)⋯I (3 isomers) + I⋯(O)⋯I (3 isomers) (66 total)	288
	Hg(OAc)$_2$	1:1 H$_2$O/THF	—	NaBH$_4$/NaOH	 (O) ~55 : (O) ~45 (63 total)	273
	Hg(OAc)$_2$	1:1 H$_2$O/THF	—	NaBH$_4$/NaOH	 (O) 75 : (O) 25 (77 total)	68
	Hg(NO$_3$)$_2$·H$_2$O	H$_2$O 2 hr	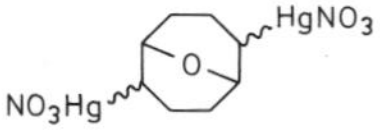 NO$_3$Hg⋯(O)⋯HgNO$_3$	NaBH$_4$/NaOH	 (O) (69 from diene)	289
	Hg(NO$_3$)$_2$·H$_2$O	80% H$_2$O$_2$/CH$_2$Cl$_2$ 10 min, KCl	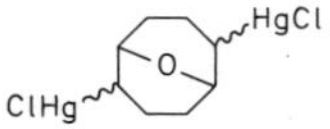 ClHg⋯(O)⋯HgCl mixture	NaBH$_4$	 (O)	287
	Hg(NO$_3$)$_2$·H$_2$O	80% H$_2$O$_2$/CH$_2$Cl$_2$ 10 min, KCl	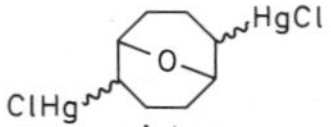 ClHg⋯(O)⋯HgCl mixture	Br$_2$	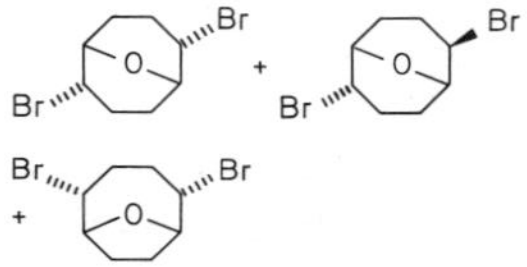 Br⋯(O)⋯Br + Br⋯(O)⋯Br + Br⋯(O)⋯Br	287
	Hg(NO$_3$)$_2$	H$_2$O	—	NaBH$_4$/NaOH	 (O)	273

Table 2.6. (continued)

Diene or polyene	Mercuric salt	Reaction conditions	Organomercurial(s) (% Yield)	Subsequent reactants	Product(s) (% Yield)	Refs.
	$Hg(NO_3)_2 \cdot H_2O$	D_2O	(bicyclic ether bearing $HgNO_3$ and NO_3Hg)	—	—	287
$H_2C{=}CH(CH_2)_4CH{=}CH_2$	$Hg(OAc)_2$	1:1 H_2O/THF 45 min	—	$NaBH_4$/NaOH	$CH_3CHOH(CH_2)_4CHOHCH_3$ (87)	68
	$Hg(OAc)_2$	1:1 H_2O/THF $NaO_3S(CH_2)_{11}CH_3$	—	$NaBH_4$/NaOH	$CH_3CHOH(CH_2)_4CH{=}CH_2$ 2 : + $CH_3CHOH(CH_2)_4CHOHCH_3$: 1	286
	$Hg(O_2CCF_3)_2$	1:4 H_2O/THF 25°C 4 h	—	$NaBH_4$/NaOH	$CH_3CHOH(CH_2)_4CH{=}CH_2$ (42) + $CH_3CHOH(CH_2)_4CHOHCH_3$ (17) + (dimethyl oxepane, CH_3 / CH_3) (1–4)	272
$H_2C{=}C{=}C(n{-}C_4H_9)CH_2OH$	$Hg(OAc)_2$	1:1 H_2O/THF	—	$NaBH_4$/NaOH	$H_2C{=}CHCOH(n{-}C_4H_9)CH_2OH$ (53)	271
	$HgSO_4$	H_2O Δ 1.5 hr	—	—	(dihydrofuran bearing $n{-}C_4H_9$) + $CH_3CH{=}C(n{-}C_4H_9)CHO$ 90 : 9 (56 total)	271
$[H_2C{=}C(CH_3)CH_2]_2SO_2$	$Hg(OAc)_2$	H_2O 45 min, KI	(dioxathiane dioxide: CH_3, CH_3, $IHgCH_2$, CH_2HgI)	I_2	(dioxathiane dioxide: CH_3, CH_3, ICH_2, CH_2I) (one isomer)	280
$[H_2C{=}C(CH_3)CH_2]_2S$	$Hg(OAc)_2$	H_2O 3 hr, KI	(dioxathiane: CH_3, CH_3, $IHgCH_2$, CH_2HgI)	I_2	(dioxathiane: CH_3, CH_3, ICH_2, CH_2I) (56) (2 isomers)	280

Table 2.6. (continued)

Diene or polyene	Mercuric salt	Reaction conditions	Organomercurial(s) (% Yield)	Subsequent reactants	Product(s) (% Yield)	Refs.
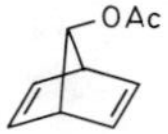 (OAc-substituted norbornadiene)	$Hg(OAc)_2$	1:1 H_2O/THF 10 min 20 °C	—	$NaBH_4$/NaOH	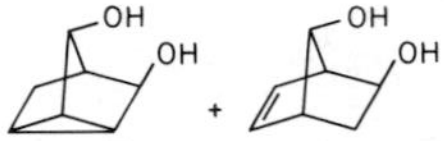 (OH, OH diol products) 58 : 42 (64 total)	192
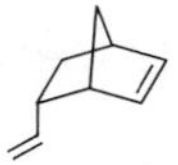 (allyl norbornene) (plus exo?)	$Hg(OAc)_2$	1:1 H_2O/THF	—	$NaBH_4$/NaOH, Ac_2O	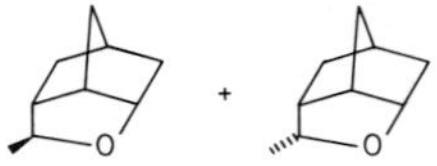 (oxa-bicyclic products) 60 : 40 (65–76); 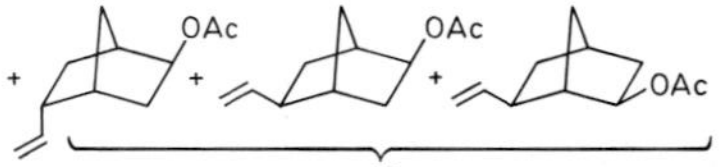 (OAc, OAc products) + diacetate, 43 : 57	196
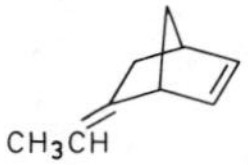 (CH_3CH ethylidene norbornene)	$Hg(OAc)_2$	1:1 H_2O/THF, NaCl	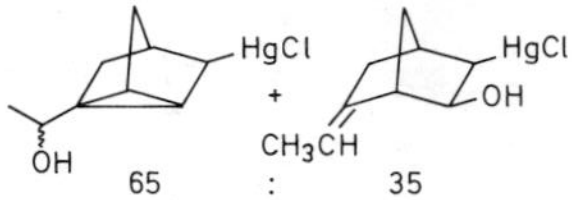 (HgCl, OH / CH_3CH, HgCl, OH organomercurials) 65 : 35	—	—	290
	$Hg(OAc)_2$	H_2O, NaCl	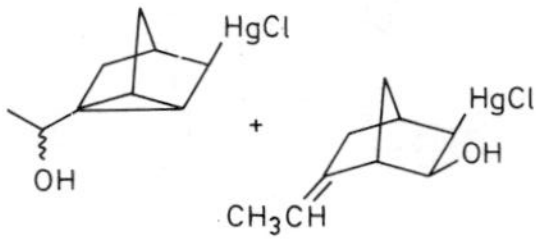 (HgCl / OH, CH_3CH, HgCl, OH organomercurials) 91 : 9	$NaBH_4$/NaOH	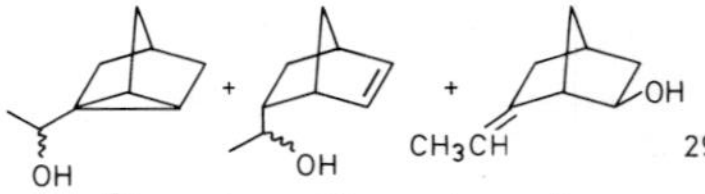 (OH / OH / CH_3CH, OH products) 53 : 38 : 9	290, 291
	$Hg(OAc)_2$	H_2O/THF	—	$NaBH_4$/NaOH	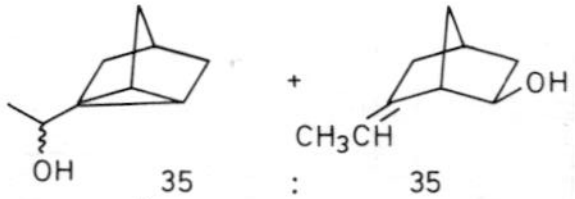 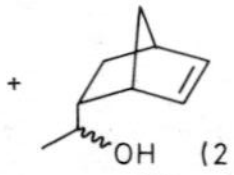 (OH / CH_3CH, OH products) 35 : 35; + (CH_3CH, OH 2 diastereomers) : 30 (67 total)	291

Table 2.6. (continued)

Diene or polyene	Mercuric salt	Reaction conditions	Organomercurial(s) (% Yield)	Subsequent reactants	Product(s) (% Yield)	Refs.
 [structure]	Hg(OAc)$_2$	H$_2$O 4 hr, NaCl	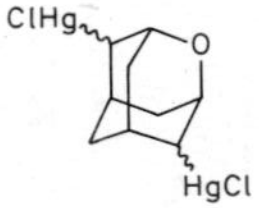 [structure] ClHg…/…HgCl	NaBH$_4$/NaOH	 [structure]	292
	—	H$_2$O 4 hr, NaCl	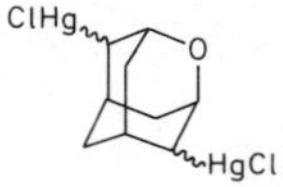 [structure] ClHg…/…HgCl	Br$_2$	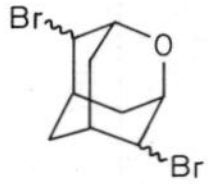 [structure] Br…/…Br	292
	Hg(OAc)$_2$	H$_2$O 10d RT	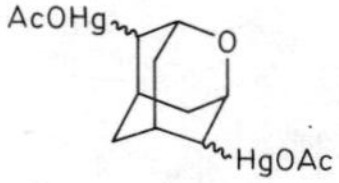 [structure] AcOHg…/…HgOAc	KI/I$_2$	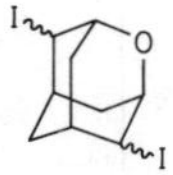 [structure] I…/…I (64)	293
H$_2$C=CHCH=C=CHCH$_2$CH(CH$_3$)$_2$	HgSO$_4$	H$_2$SO$_4$	CH$_3$CH=CHCO(CH$_2$)$_2$CH(CH$_3$)$_2$	—	—	282
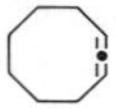 [structure]	HgSO$_4$	H$_2$O/THF/H$_2$SO$_4$ Δ 40 min – 4 hr	—	—	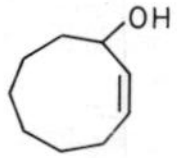 [structure] OH (74)	294, 295
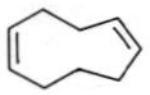 [structure]	Hg(OAc)$_2$	1:1 H$_2$O/THF	—	NaBH$_4$/NaOH	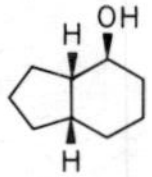 [structure] OH (68)	296
 [structure]	Hg(OAc)$_2$	1:1 H$_2$O/THF 10 min	—	NaBH$_4$/NaOH	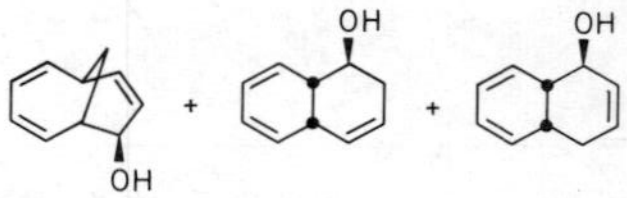 [structure] OH (major) + [structure] OH + [structure] OH	297, 298
 [structure]	Hg(OAc)$_2$	H$_2$O 0.5 hr	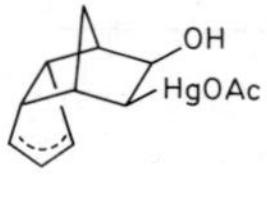 [structure] OH …HgOAc	—	—	299

Table 2.6. (continued)

Diene or polyene	Mercuric salt	Reaction conditions	Organomercurial(s) (% Yield)	Subsequent reactants	Product(s) (% Yield)	Refs.
	$Hg(OAc)_2$	H_2O 10 hr	HgOAc (42.8)	$Na(Hg)/H_2O$	OH (33.3)	300
	$Hg(OAc)_2$	1:1 H_2O/THF	—	$NaBH_4$/NaOH	OH (89)	195
	$Hg(OAc)_2$	1:1 H_2O/THF RT	—	$NaBH_4$/NaOH, $LiAlH_4$	OH (89.5) (mixture)	301
	$Hg(OAc)_2$	1:1 H_2O/THF	—	$NaBH_4$/NaOH	OH + OH (95 total) 72 : 28	302
	1:1 HgO/$HgCl_2$	1:1 H_2O/acetone overnight	OH HgCl + HgCl OH	2% $Na(Hg)/D_2O$	OH D + D OH	131
	$Hg(OAc)_2$?	H_2O, Cl^-	OH HgCl	$Na(Hg)/H_2O$	OH (12.5)	300
$H_2C=CHCH=C=$	$HgSO_4$	H_2SO_4	$CH_3CH=CHCO-$ (37.5–40)	—	—	282

Table 2.6. (continued)

Diene or polyene	Mercuric salt	Reaction conditions	Organomercurial(s) (% Yield)	Subsequent reactants	Product(s) (% Yield)	Refs.
(structure)	$Hg(OAc)_2$	H_2O/THF, NaCl	(structure) (92)	$NaBH_4$/NaOH (on RHgOAc without isolation)	(structure) (67)	291
(structure) 90:10 $E:Z$	$Hg(OAc)_2$	H_2O/THF	—	$NaBH_4$/NaOH	(structures) 68 : : 32 (72 total, based on reacted diene)	54
(structure)	$Hg(OAc)_2$	1:1 H_2O/THF 12 hr RT	—	$NaBH_4$/NaOH	(structures) (50) + (15) + (10) + (10) + (3)	165
	$Hg(OAc)_2$	H_2O, NaCl	(structure)	—	—	303
(structure)	$Hg(NO_3)_2$	1:1 H_2O/THF 168 hr	—	$NaBH_4$/NaOH, CrO_3	(structures) (29) + (23)	154

Table 2.6. (continued)

Diene or polyene	Mercuric salt	Reaction conditions	Organomercurial(s) (% Yield)	Subsequent reactants	Product(s) (% Yield)	Refs.
$(H_2C=CHCH_2)_2NCO(CH_2)_2CO_2H$	$Hg(OAc)_2$	—	$HgCH_2CHOHCH_2N(CH_2CH=CH_2)CO(CH_2)_2\overset{O}{C}{=}O$	—	—	96
$H_2C=CHCH=C=C(CH_3)C(CH_3)_3$	$HgSO_4$	H_2SO_4	$CH_3CH=CHCOCH(CH_3)C(CH_3)_3$ (37.5 – 40)	—	—	282
$H_2C=CHCH=C=C(CH_3)CH_2CH(CH_3)_2$	$HgSO_4$	H_2SO_4	$CH_3CH=CHCOCH(CH_3)CH_2CH(CH_3)_2$ (37.5 – 40)	—	—	282
$H_2C=CHCH=C=C(CH_3)(CH_2)_3CH_3$	$HgSO_4$	H_2SO_4	$CH_3CH=CHCOCH(CH_3)(CH_2)_3CH_3$ (37.5 – 40)	—	—	282
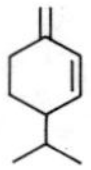	$Hg(OAc)_2$?	H_2O	—	$NaBH_4$?	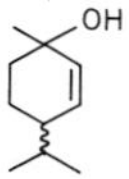 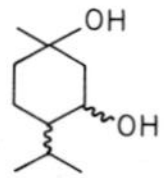 cis : trans 54.1 : 19 + 2 isomers	304
	—	—	—	—	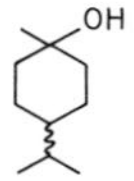 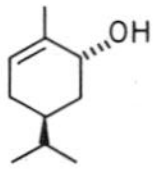 cis and trans	305
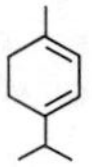	$Hg(OAc)_2$?	H_2O	—	$NaBH_4$?	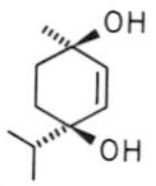 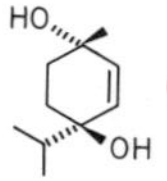 3 : 1 (63.5 total)	304
	—	—	—	—	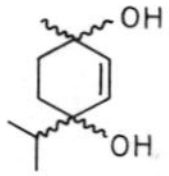	305

Table 2.6. (continued)

Diene or polyene	Mercuric salt	Reaction conditions	Organomercurial(s) (% Yield)	Subsequent reactants	Product(s) (% Yield)	Refs.
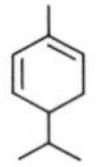	Hg(OAc)$_2$?	H$_2$O	—	NaBH$_4$?	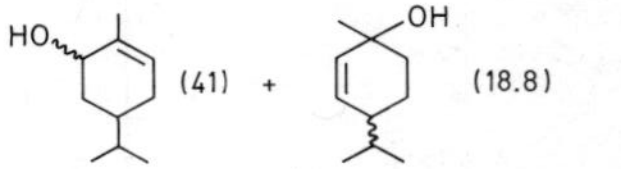 (41) + (18.8)	304
	—	—	—	—	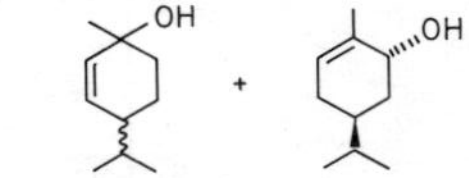 cis and trans	305
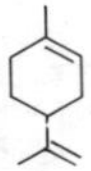	Hg(OAc)$_2$	1:1 H$_2$O/THF 2 min RT	—	NaBH$_4$/NaOH	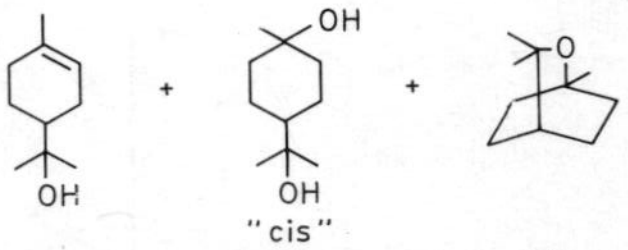 "cis" 35.1 : 26 : 8.4	172
	Hg(OAc)$_2$	1:1 H$_2$O/THF 25°C 0.5 hr	—	NaBH$_4$/NaOH	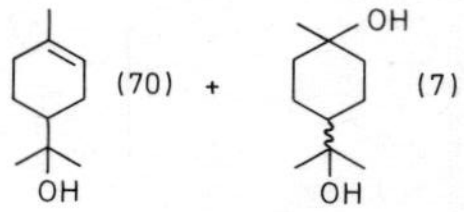 (70) + (7)	272
	Hg(OAc)$_2$	1:1 H$_2$O/THF	—	NaBH$_4$/NaOH	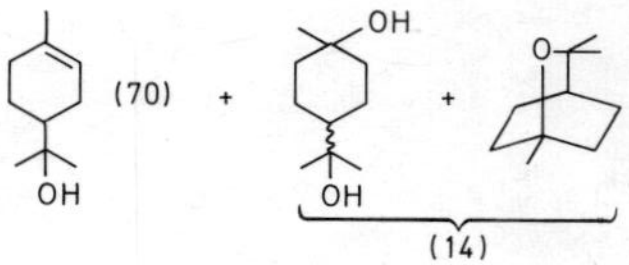 (70) + (14)	286
	Hg(OAc)$_2$	1:1 H$_2$O/THF 2 min RT	—	NaBH$_4$/NaOH	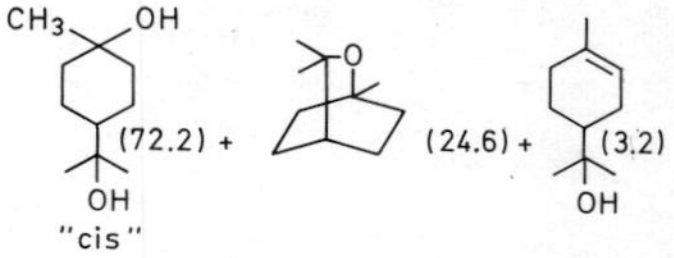 CH$_3$ (72.2) + (24.6) + (3.2) "cis"	172

Table 2.6. (continued)

Diene or polyene	Mercuric salt	Reaction conditions	Organomercurial(s) (% Yield)	Subsequent reactants	Product(s) (% Yield)	Refs.
	$Hg(OAc)_2$	1:1 H_2O/THF $NaO_3S(CH_2)_{11}CH_3$	—	$NaBH_4$/NaOH	 (97)	286
	$Hg(OAc)_2$	1:1 H_2O/THF $NaO_3S(CH_2)_{11}CH_3$	—	$NaBH_4$/NaOH	 (~95)	286
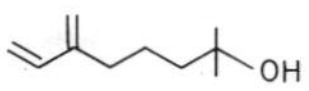	$Hg(OAc)_2$	1:1 H_2O/THF 15 hr RT	—	$NaBH_4$/NaOH	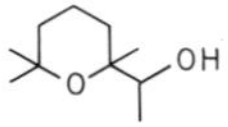 (75) (isomeric mixture)	112
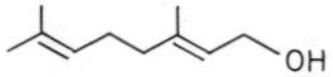	$Hg(OAc)_2$	1:1 H_2O/THF 18 hr RT	—	$NaBH_4$/NaOH	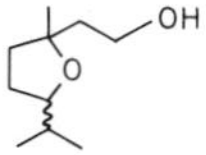 (35) + 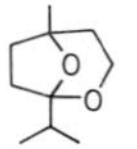 (20)	112
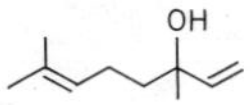	$Hg(OAc)_2$	H_2O/THF, NaCl	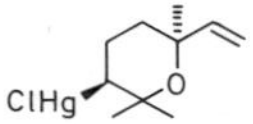 (25)	$NaBH_4$/NaOH	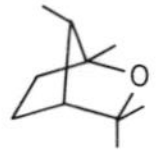 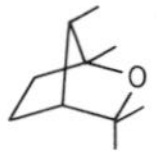 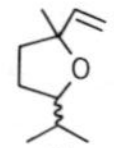 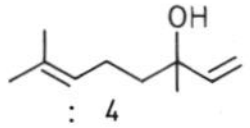 65 : 22 : 9 : 4	306

Table 2.6. (continued)

Diene or polyene	Mercuric salt	Reaction conditions	Organomercurial(s) (% Yield)	Subsequent reactants	Product(s) (% Yield)	Refs.
	Hg(OAc)₂	1:3 H₂O / THF 12 hr RT	—	NaBH₄ / NaOH	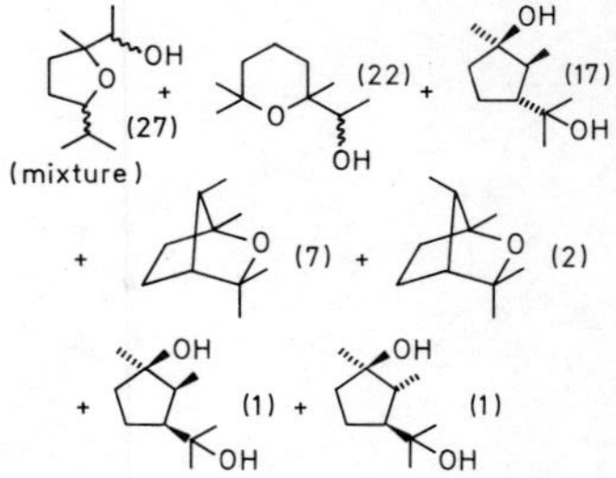 (27) (mixture) + (22) + (17) + (7) + (2) + (1) + (1)	306, 307
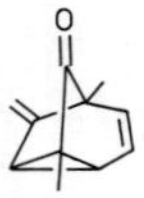	Hg(OAc)₂	1:4 H₂O / THF RT 30 min NaCl	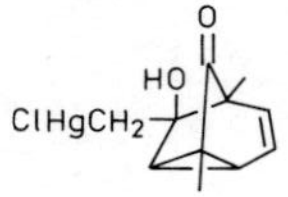 ClHgCH₂ (80)	NaBH₄ / NaOH	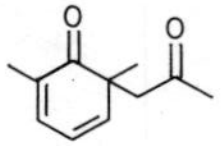 (86)	259
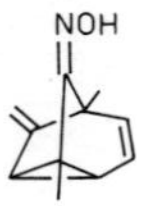	Hg(OAc)₂	1:4 H₂O / THF RT 30 min	—	NaBH₄ / NaOH	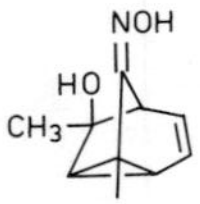 (23)	259
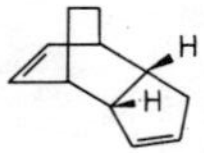	Hg(OAc)₂	1:1 H₂O / THF 80 min RT	—	NaBH₄ / NaOH	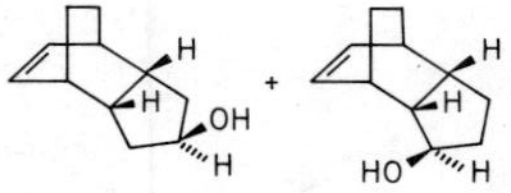 92 : 8 (96 total)	218
	Hg(OAc)₂	1:1 H₂O / THF 24 hr RT	—	NaBH₄ / NaOH	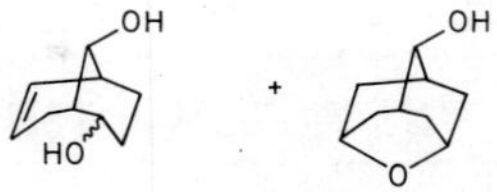 (24) 53 : 47	308

Table 2.6. (continued)

Diene or polyene	Mercuric salt	Reaction conditions	Organomercurial(s) (% Yield)	Subsequent reactants	Product(s) (% Yield)	Refs.
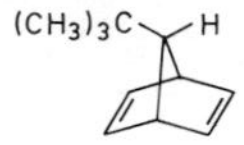 $(CH_3)_3C$—H (norbornadiene deriv.)	$Hg(OAc)_2$	1:1 H_2O / THF 15 min	—	$NaBH_4$ / NaOH, AcCl	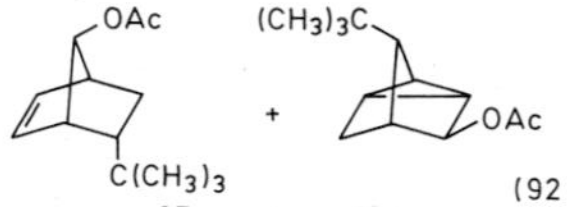 (structures) 87 : 13 (92 total)	309
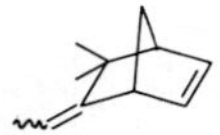 (dimethyl norbornene deriv.)	$Hg(OAc)_2$	H_2O / THF	—	$NaBH_4$ / NaOH	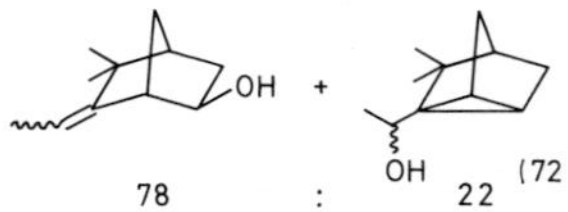 (structures) 78 : 22 (72 total)	291
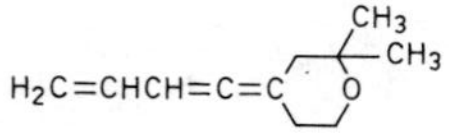 $H_2C=CHCH=C=C$ (dioxane deriv., CH_3, CH_3)	$HgSO_4$	H_2SO_4	—	—	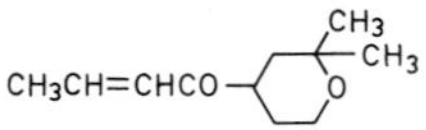 $CH_3CH=CHCO$— (tetrahydropyran deriv., CH_3, CH_3) (37.5–40)	282
$p-H_2C=CHCH_2C_6H_4CH_2CH=CH_2$	$Hg(OAc)_2$	H_2O / THF $NaO_3S(CH_2)_{11}CH_3$	—	$NaBH_4$ / NaOH	$p-CH_3CHOHCH_2C_6H_4CH_2CH=CH_2$ ~1 : + $p-CH_3CHOHCH_2C_6H_4CH_2CHOHCH_3$: 1	310
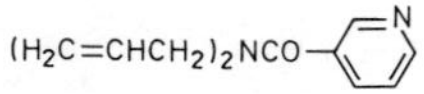 $(H_2C=CHCH_2)_2NCO$— (pyridyl)	$Hg(OAc)_2$	—	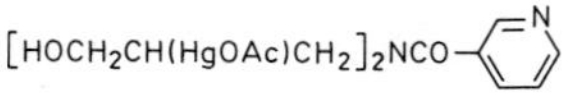 $[HOCH_2CH(HgOAc)CH_2]_2NCO$—(pyridyl) ?	—	—	98
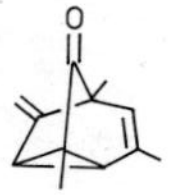 (ketone bicyclic diene)	$Hg(OAc)_2$	1:4 H_2O / THF RT 30 min	—	$NaBH_4$ / NaOH	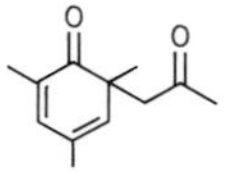 (cyclohexenedione deriv.) (90)	259
$(H_2C=CHCH_2)_2NSO_2C_6H_5$	$Hg(OAc)_2$	H_2O 29 hr RT	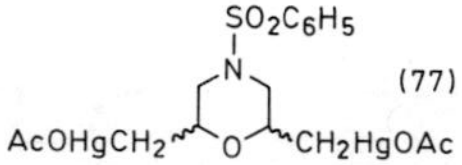 (morpholine deriv. with $SO_2C_6H_5$; $AcOHgCH_2$, CH_2HgOAc) (77) cis and trans	I_2	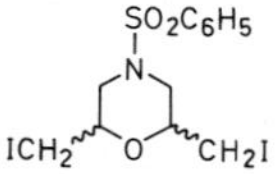 (morpholine deriv. with $SO_2C_6H_5$; ICH_2, CH_2I) (82)	311
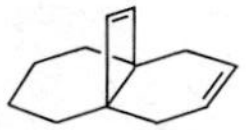 (bicyclic diene)	$Hg(OAc)_2$	1:1 H_2O / THF RT 1hr	—	$NaBH_4$ / NaOH	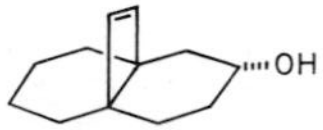 (bicyclic—OH) (70)	312

Table 2.6. (continued)

Diene or polyene	Mercuric salt	Reaction conditions	Organomercurial(s) (% Yield)	Subsequent reactants	Product(s) (% Yield)	Refs.
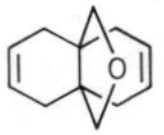 [structure]	$Hg(OAc)_2$	H_2O / THF $NaO_3S(CH_2)_{11}CH_3$	—	$NaBH_4$ / NaOH	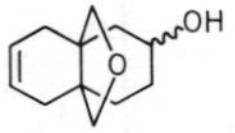 [structure] (40–83)	310
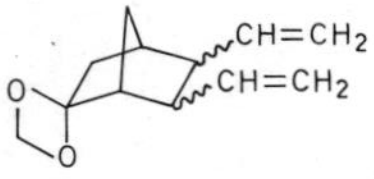 [structure] $CH=CH_2$ / $CH=CH_2$?	—	—	—	—	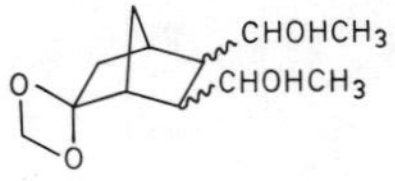 [structure] $CHOHCH_3$ / $CHOHCH_3$?	313
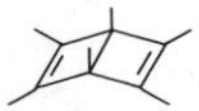 [structure]	$Hg(OAc)_2$	1:1 H_2O / THF	—	—	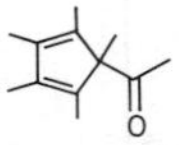 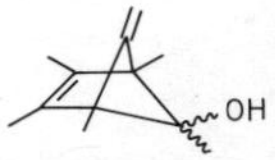 [structure] + [structure] —OH 50 : 50 overnight reaction major : minor 12 hr reaction	314, 315
	$Hg(OAc)_2$	1:1 H_2O / THF 20 °C < 1 min	—	$NaBH_4$	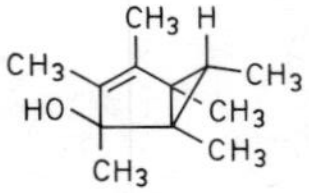 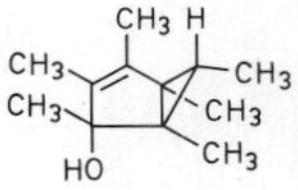 [structure] + [structure] 4 : 1	316
$(H_2C=CHCH_2)_2NCO(CH_2)_4CO_2H$	$Hg(OAc)_2$	—	$HgCH_2CHOHCH_2N(CH_2CH=CH_2)CO(CH_2)_4\overset{O}{C}=O$	—	—	96
$H_2C=CH(CH_2)_8CH=CH_2$	$Hg(OAc)_2$	1:2 H_2O / THF 60 min	—	$NaBH_4$ / NaOH	$CH_3CHOH(CH_2)_8CHOHCH_3$ (82)	68
	$Hg(O_2CCF_3)_2$	4:1 H_2O / THF 0 °C 2 hr	—	$NaBH_4$ / NaOH	$CH_3CHOH(CH_2)_8CH=CH_2$ (46) + $CH_3CHOH(CH_2)_8CHOHCH_3$ (18)	272
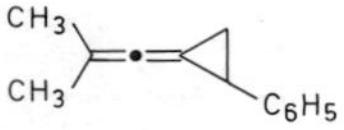 [structure] CH_3 / CH_3 / C_6H_5	$Hg(OAc)_2$	1:1 H_2O / THF 30 min 25 °C	—	$NaBH_4$ / NaOH	$(CH_3)_2C(OAc)C{\equiv}C(CH_2)_2C_6H_5$ (25) + $(CH_3)_2C=CHC(CH_2OH)=CHC_6H_5$ + $(CH_3)_2COHC{\equiv}C(CH_2)_2C_6H_5$ + $(CH_3)_2C=CHC(=CH_2)CHOHC_6H_5$	317

Table 2.6. (continued)

Diene or polyene	Mercuric salt	Reaction conditions	Organomercurial(s) (% Yield)	Subsequent reactants	Product(s) (% Yield)	Refs.
$H_2C=CH(CH_2)_7CH=C(CH_3)_2$	$Hg(OAc)_2$	H_2O / THF 25°C 1hr	—	$NaBH_4$ / NaOH	$CH_3CHOH(CH_2)_7CH=C(CH_3)_2$ (55) + $CH_3CHOH(CH_2)_8COH(CH_3)_2$ (14) + $H_2C=CH(CH_2)_8COH(CH_3)_2$ (2)	272
	$Hg(O_2CCF_3)_2$	H_2O / THF 25°C 1hr	—	$NaBH_4$ / NaOH	$CH_3CHOH(CH_2)_7CH=C(CH_3)_2$ (83) + $CH_3CHOH(CH_2)_8COH(CH_3)_2$ (2) + $H_2C=CH(CH_2)_8COH(CH_3)_2$ (1)	272
$H_2C=C(CH_3)(CH_2)_8CH=CH_2$	$Hg(OAc)_2$	1:1 H_2O / THF 25°C 1hr	—	$NaBH_4$ / NaOH	$(CH_3)_2COH(CH_2)_8CH=CH_2$ (55) + $(CH_3)_2COH(CH_2)_8CHOHCH_3$ (15) + $H_2C=C(CH_3)(CH_2)_8CHOHCH_3$ (1)	272
	$Hg(O_2CCF_3)_2$	1:5 H_2O / THF 0°C 10 min	—	$NaBH_4$ / NaOH	$(CH_3)_2COH(CH_2)_8CH=CH_2$ (43) + $(CH_3)_2COH(CH_2)_8CHOHCH_3$ (17) + $H_2C=C(CH_3)(CH_2)_8CHOHCH_3$ (4)	272
$p-CH_3C_6H_4C(CH_3)=CHCH_2CH=C(CH_3)_2$	$Hg(OAc)_2$	H_2O / dioxane 2 hr	—	$NaBH_4$ / NaOH	(35) + $p-CH_3C_6H_4COH(CH_3)(CH_2)_3COH(CH_3)_2$	318
	$Hg(OAc)_2$	1:1 H_2O / ̄HF 4 hr	—	$NaBH_4$ / NaOH	(65) + (4)	319

114

Table 2.6. (continued)

Diene or polyene	Mercuric salt	Reaction conditions	Organomercurial(s) (% Yield)	Subsequent reactants	Product(s) (% Yield)	Refs.
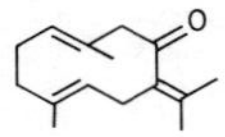 [structure]	Hg(OAc)₂	1:1 H₂O / THF 60 min	—	NaBH₄ / NaOH	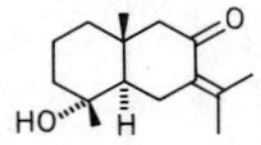 [structure] (70) 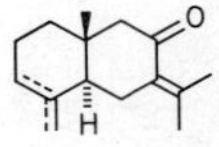 + (15)	319
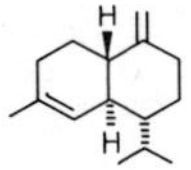 [structure]	Hg(OAc)₂	H₂O / THF 10 min RT	—	NaBH₄ / NaOH	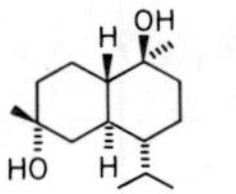 [structure] (85)	320
	Hg(OAc)₂	H₂O / THF 10 min RT	—	NaBH₄ / NaOH	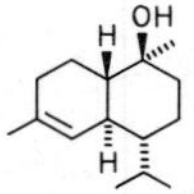 [structure] (49) 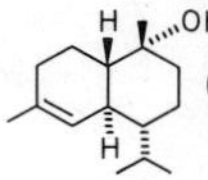 + (9) 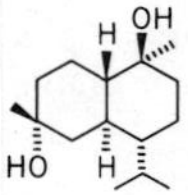 + (8)	320
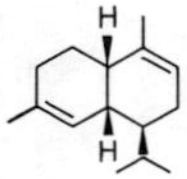 [structure]	—	—	—	—	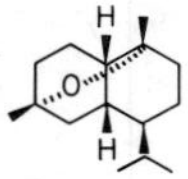 [structure]	321
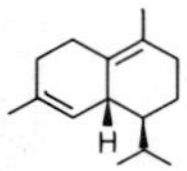 [structure]	—	—	—	NaBH₄ / NaOH	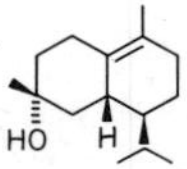 [structure]	321
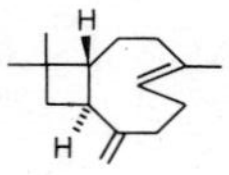 [structure]	Hg(OAc)₂	1:1 H₂O / THF 3 hr	—	NaBH₄ / NaOH	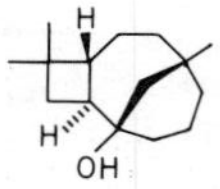 [structure] (61)	224
γ – caryophyllen (?)	Hg(OAc)₂	H₂O / NaX (X = Br, I)	C₁₅H₂₄(OH)HgX (no structure given)	—	—	202

Table 2.6. (continued)

Diene or polyene	Mercuric salt	Reaction conditions	Organomercurial(s) (% Yield)	Subsequent reactants	Product(s) (% Yield)	Refs.
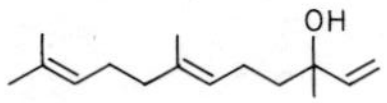	$HgCl_2$	H_2O / HOAc 30 min	—	$NaBH_4$ / NaOH	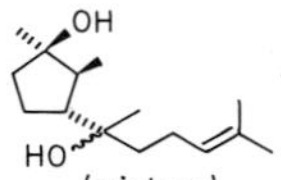 (mixture) (16)	307
	$HgCl_2$	H_2O / CH_3CN $CdCO_3$ 8 hr	—	$NaBH_4$ / NaOH	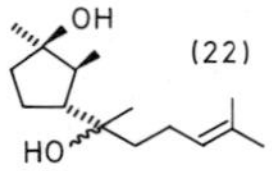 (mixture) (22) + 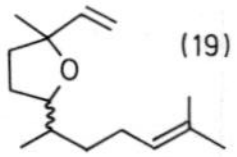 (19)	307
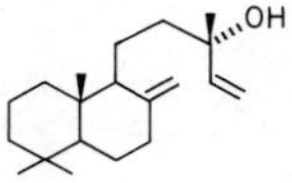	$Hg(OAc)_2$	H_2O / CH_3NO_2	—	$NaBH_4$ / NaOH	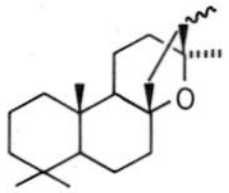 (72)	306
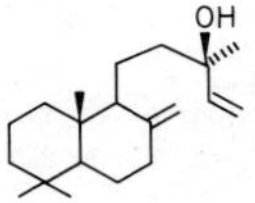	$Hg(OAc)_2$	H_2O / CH_3NO_2	—	$NaBH_4$ / NaOH	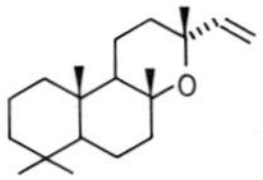	306
cis, cis-$CH_3(CH_2)_{10}CH{=}CH(CH_2)_2CH{=}CHCO_2CH_3$	$Hg(OAc)_2$	3:2 H_2O / THF 4d	—	$NaBH_4$ / NaOH	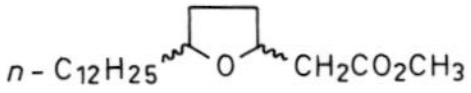 (19) + 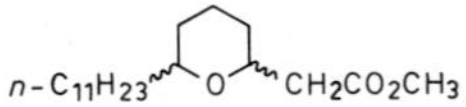 (9)	322
cis, cis-$CH_3(CH_2)_9CH{=}CH(CH_2)_2$-$CH{=}CHCH_2CO_2CH_3$	$Hg(OAc)_2$	3:2 H_2O / THF 4d	—	$NaBH_4$ / NaOH	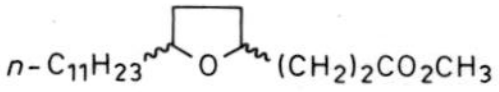 (34) + 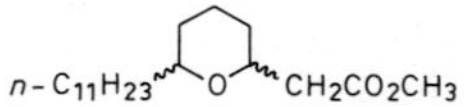 + 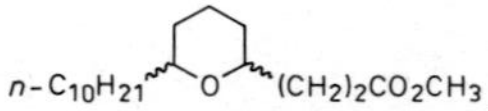 (27)	322

Table 2.6. (continued)

Diene or polyene	Mercuric salt	Reaction conditions	Organomercurial(s) (% Yield)	Subsequent reactants	Product(s) (% Yield)	Refs.
cis, cis-$CH_3(CH_2)_8CH{=}CH(CH_2)_2$-$CH{=}CH(CH_2)_2CO_2CH_3$	$Hg(OAc)_2$	3:2 H_2O/THF 4d	—	$NaBH_4$/NaOH	n-$C_{10}H_{21}$—⟨THF ring⟩—$(CH_2)_3CO_2CH_3$ (51) + n-$C_{10}H_{21}$—⟨THP ring⟩—$(CH_2)_2CO_2CH_3$ + n-C_9H_{19}—⟨THP ring⟩—$(CH_2)_3CO_2CH_3$ (16)	322
cis, cis-$CH_3(CH_2)_7CH{=}CH(CH_2)_2$-$CH{=}CH(CH_2)_3CO_2CH_3$	$Hg(OAc)_2$	3:2 H_2O/THF 4d	—	$NaBH_4$/NaOH	n-C_9H_{19}—⟨THF ring⟩—$(CH_2)_4CO_2CH_3$ (45) + n-C_9H_{19}—⟨THP ring⟩—$(CH_2)_3CO_2CH_3$ + n-C_8H_{17}—⟨THP ring⟩—$(CH_2)_4CO_2CH_3$ (24)	322
cis, cis-$CH_3(CH_2)_6CH{=}CH(CH_2)_3$-$CH{=}CH(CH_2)_3CO_2CH_3$	$Hg(OAc)_2$	3:2 H_2O/THF 4d	—	$NaBH_4$/NaOH	n-C_8H_{17}—⟨THP ring⟩—$(CH_2)_4CO_2CH_3$ (30)	322
cis, cis-$CH_3(CH_2)_6CH{=}CH(CH_2)_2$-$CH{=}CH(CH_2)_4CO_2CH_3$	$Hg(OAc)_2$	3:2 H_2O/THF 4d	—	$NaBH_4$/NaOH	n-C_8H_{17}—⟨THF ring⟩—$(CH_2)_5CO_2CH_3$ (56) + n-C_8H_{17}—⟨THP ring⟩—$(CH_2)_4CO_2CH_3$ + n-C_7H_{15}—⟨THP ring⟩—$(CH_2)_5CO_2CH_3$ (26)	322
cis, cis-$CH_3(CH_2)_5CH{=}CH(CH_2)_2$-$CH{=}CH(CH_2)_5CO_2CH_3$	$Hg(OAc)_2$	3:2 H_2O/THF 4d	—	$NaBH_4$/NaOH	n-C_7H_{15}—⟨THF ring⟩—$(CH_2)_6CO_2CH_3$ (47) + n-C_7H_{15}—⟨THP ring⟩—$(CH_2)_5CO_2CH_3$ + n-C_6H_{13}—⟨THP ring⟩—$(CH_2)_6CO_2CH_3$ (23)	322

Table 2.6. (continued)

Diene or polyene	Mercuric salt	Reaction conditions	Organomercurial(s) (% Yield)	Subsequent reactants	Product(s) (% Yield)	Refs.
trans, trans-$CH_3(CH_2)_5CH{=}CH(CH_2)_2$-$CH{=}CH(CH_2)_5CO_2CH_3$	$Hg(OAc)_2$	3:2 H_2O/THF 4d	—	$NaBH_4$/NaOH	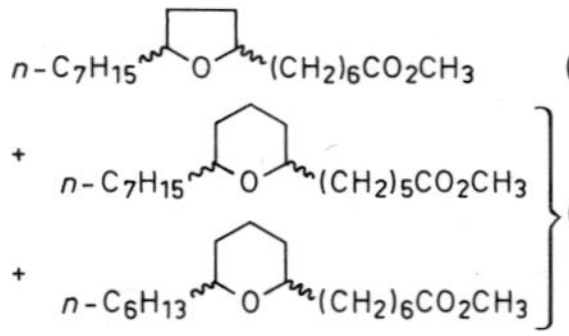 n-C_7H_{15}–[tetrahydrofuran ring]–$(CH_2)_6CO_2CH_3$ (48) + n-C_7H_{15}–[tetrahydropyran ring]–$(CH_2)_5CO_2CH_3$ + n-C_6H_{13}–[tetrahydropyran ring]–$(CH_2)_6CO_2CH_3$ (40)	322
cis, cis-$CH_3(CH_2)_4CH{=}CH(CH_2)_2$-$CH{=}CH(CH_2)_6CO_2CH_3$	$Hg(OAc)_2$	3:2 H_2O/THF 4d	—	$NaBH_4$/NaOH	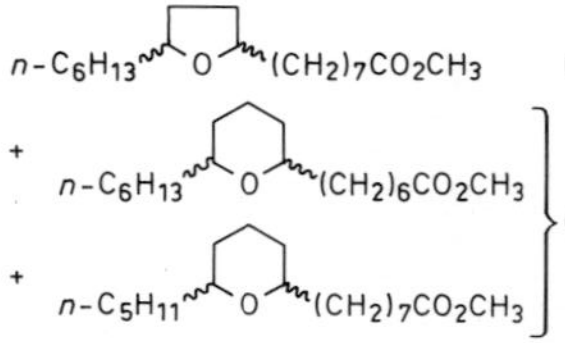 n-C_6H_{13}–[tetrahydrofuran ring]–$(CH_2)_7CO_2CH_3$ (27) + n-C_6H_{13}–[tetrahydropyran ring]–$(CH_2)_6CO_2CH_3$ + n-C_5H_{11}–[tetrahydropyran ring]–$(CH_2)_7CO_2CH_3$ (8)	322
trans, trans-$CH_3(CH_2)_4CH{=}CH(CH_2)_2$-$CH{=}CH(CH_2)_6CO_2CH_3$	$Hg(OAc)_2$	3:2 H_2O/THF 4d	—	$NaBH_4$/NaOH	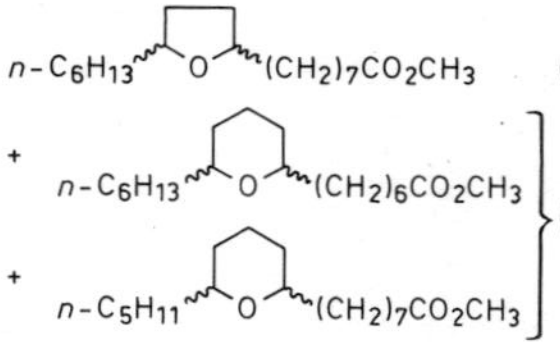 n-C_6H_{13}–[tetrahydrofuran ring]–$(CH_2)_7CO_2CH_3$ (50) + n-C_6H_{13}–[tetrahydropyran ring]–$(CH_2)_6CO_2CH_3$ + n-C_5H_{11}–[tetrahydropyran ring]–$(CH_2)_7CO_2CH_3$ (37)	322
cis, cis-$CH_3(CH_2)_3CH{=}CH(CH_2)_2$-$CH{=}CH(CH_2)_7CO_2CH_3$	$Hg(OAc)_2$	3:2 H_2O/THF 4d	—	$NaBH_4$/NaOH	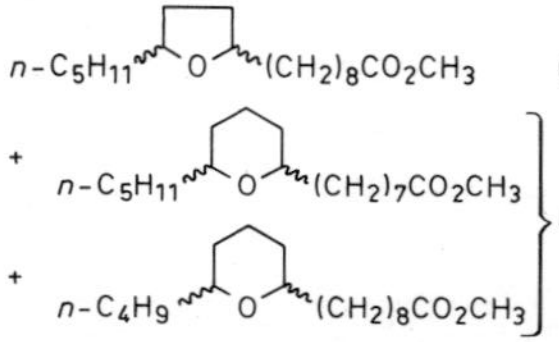 n-C_5H_{11}–[tetrahydrofuran ring]–$(CH_2)_8CO_2CH_3$ (30) + n-C_5H_{11}–[tetrahydropyran ring]–$(CH_2)_7CO_2CH_3$ + n-C_4H_9–[tetrahydropyran ring]–$(CH_2)_8CO_2CH_3$ (12)	322

Table 2.6. (continued)

Diene or polyene	Mercuric salt	Reaction conditions	Organomercurial(s) (% Yield)	Subsequent reactants	Product(s) (% Yield)	Refs.
cis, cis - $CH_3(CH_2)_2CH=CH(CH_2)_2-CH=CH(CH_2)_8CO_2CH_3$	$Hg(OAc)_2$	3:2 H_2O/THF 4d	—	$NaBH_4$/$NaOH$	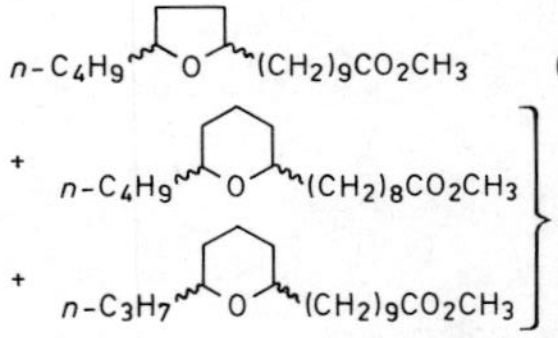 n-C_4H_9—O—$(CH_2)_9CO_2CH_3$ (73); + n-C_4H_9—O—$(CH_2)_8CO_2CH_3$, + n-C_3H_7—O—$(CH_2)_9CO_2CH_3$ (20)	322
cis, cis - $CH_3CH_2CH=CH(CH_2)_2-CH=CH(CH_2)_9CO_2CH_3$	$Hg(OAc)_2$	3:2 H_2O/THF 4d	—	$NaBH_4$/$NaOH$	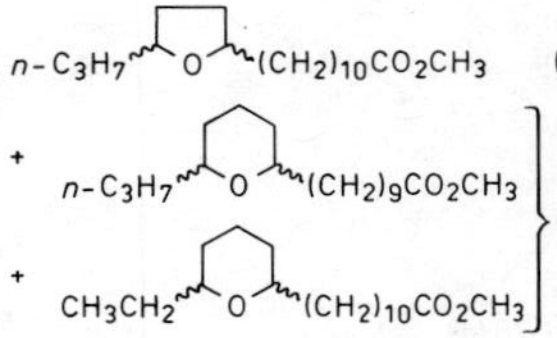 n-C_3H_7—O—$(CH_2)_{10}CO_2CH_3$ (52); + n-C_3H_7—O—$(CH_2)_9CO_2CH_3$, + CH_3CH_2—O—$(CH_2)_{10}CO_2CH_3$ (22)	322
cis, cis - $CH_3CH=CH(CH_2)_2-CH=CH(CH_2)_{10}CO_2CH_3$	$Hg(OAc)_2$	3:2 H_2O/THF 4d	—	$NaBH_4$/$NaOH$	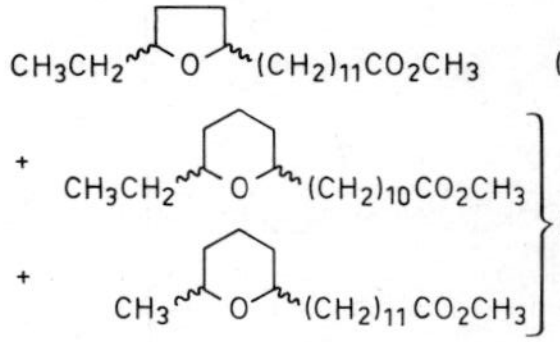 CH_3CH_2—O—$(CH_2)_{11}CO_2CH_3$ (54); + CH_3CH_2—O—$(CH_2)_{10}CO_2CH_3$, + CH_3—O—$(CH_2)_{11}CO_2CH_3$ (9)	322
cis - $H_2C=CH(CH_2)_2CH=CH(CH_2)_{11}CO_2CH_3$	$Hg(OAc)_2$	3:2 H_2O/THF 4d	—	$NaBH_4$/$NaOH$	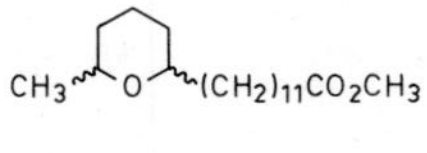 CH_3—O—$(CH_2)_{11}CO_2CH_3$ (56)	322
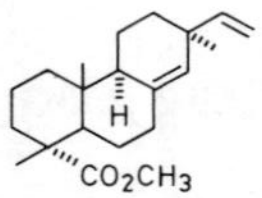 (diterpene structure, CO_2CH_3)	$Hg(OAc)_2$ or $Hg(NO_3)_2$	H_2O/t-BuOH 30 min, NaCl	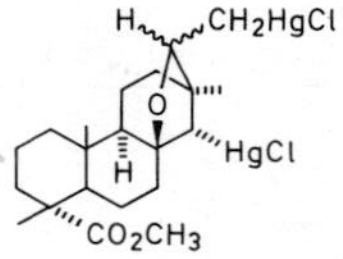 (epoxide structure, H, CH_2HgCl, $HgCl$, CO_2CH_3) (90)	$NaBH_4$/KOH	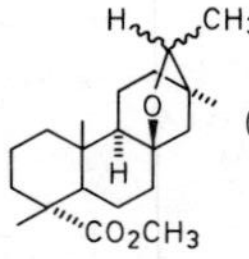 (epoxide structure, H, CH_3, CO_2CH_3) (95)	323

Table 2.6. (continued)

Diene or polyene	Mercuric salt	Reaction conditions	Organomercurial(s) (% Yield)	Subsequent reactants	Product(s) (% Yield)	Refs.
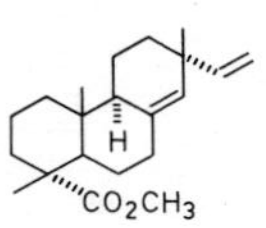 (structure)	$Hg(NO_3)_2 \cdot H_2O$	H_2O 30 min, NaCl	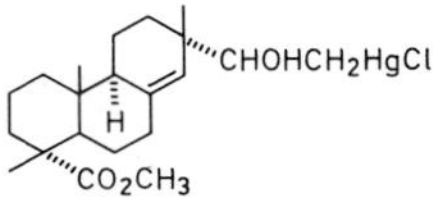 (structure) CHOHCH$_2$HgCl	$NaBH_4$ /NaOH	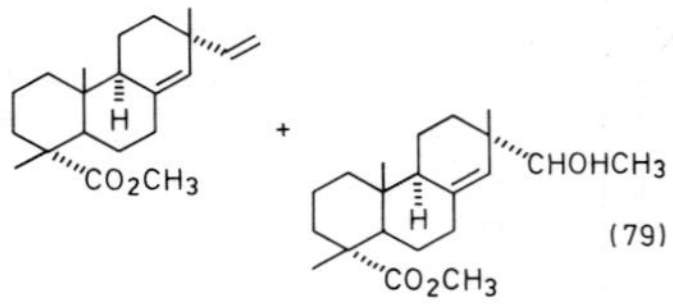 (structures) (79)	323
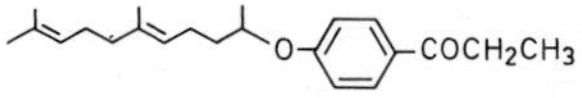 (structure)	$Hg(OAc)_2$	1:2 H_2O /dioxane RT 90 min	—	$NaBH_4$ /NaOH	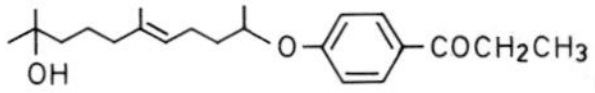 (structure) (39)	324
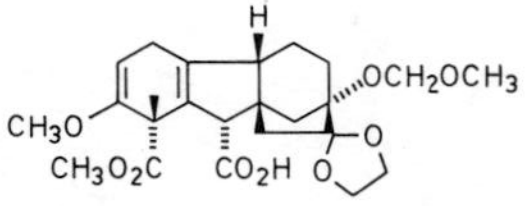 (structure)	$Hg(NO_3)_2$	1:5 H_2O /CH_3CN 24 °C 18 hr	—	—	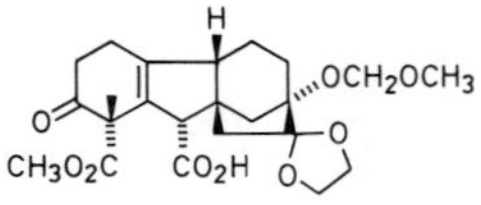 (structure)	325
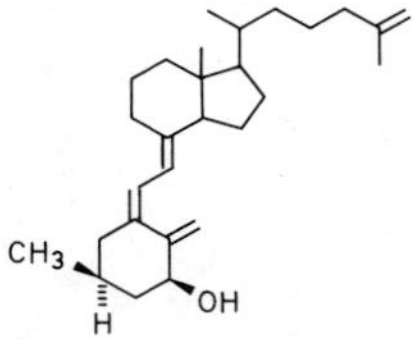 (structure)	$Hg(OAc)_2$	H_2O /THF 0 °C 2 hr	—	$NaBH_4$ /NaOH	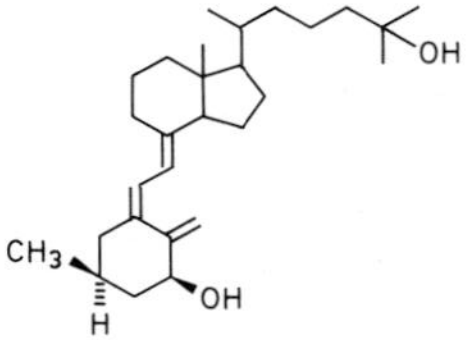 (structure) (70)	326
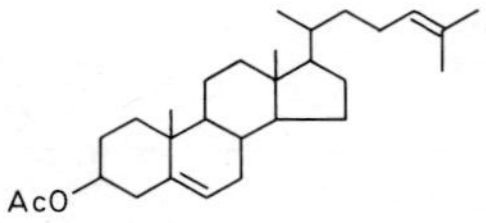 (structure)	$Hg(OAc)_2$	1:1 H_2O /THF 4 hr 0 °C 5 hr RT	—	$NaBH_4$ /NaOH	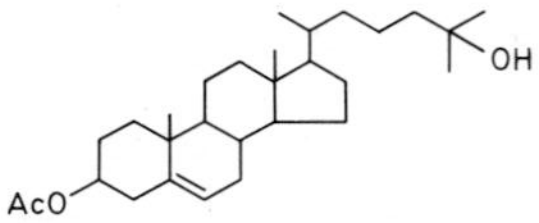 (structure) (85)	327

Table 2.6. (continued)

Diene or polyene	Mercuric salt	Reaction conditions	Organomercurial(s) (% Yield)	Subsequent reactants	Product(s) (% Yield)	Refs.
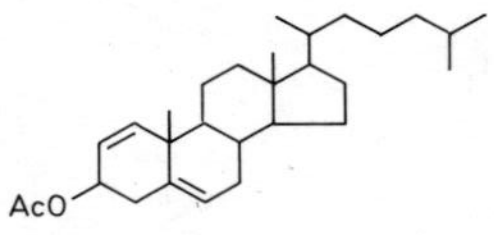	$Hg(O_2CCF_3)_2$	H_2O 3°C 8 hr 15°C 16 hr	—	$NaBH_4$ / $NaOH$	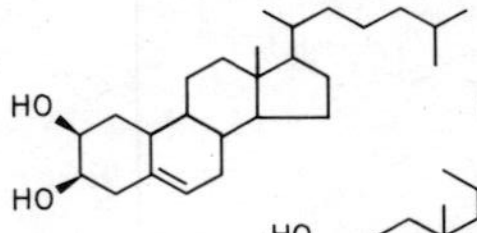 (14) 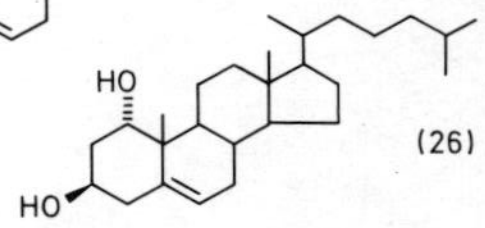 + (26)	328
	HgO	H_2O / CF_3CO_2H / THF 15 hr	—	$NaBH_4$ / $NaOH$	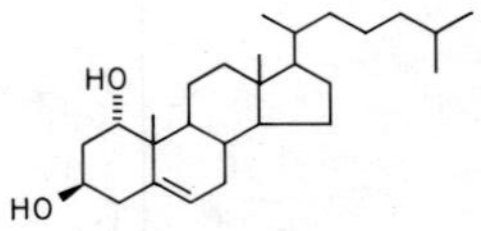 (30)	329
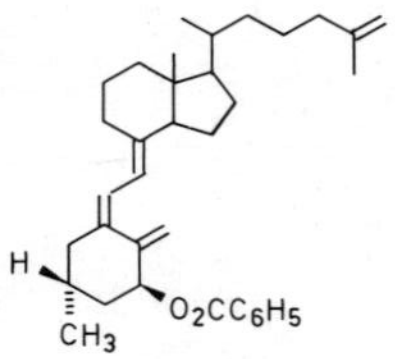	$Hg(OAc)_2$	H_2O / THF 2 hr RT	—	$NaBH_4$ / $NaOH$	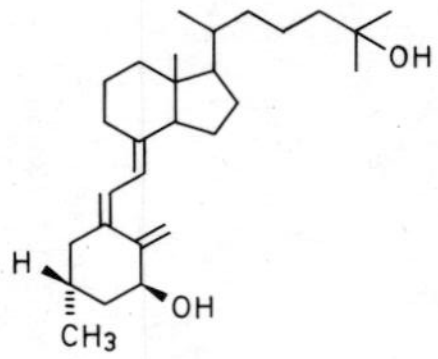 (41)	326
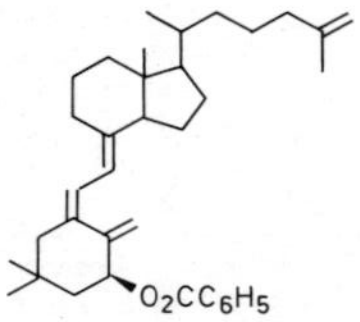	$Hg(OAc)_2$	H_2O / THF 0°C 1.5 hr	—	$NaBH_4$ / $NaOH$	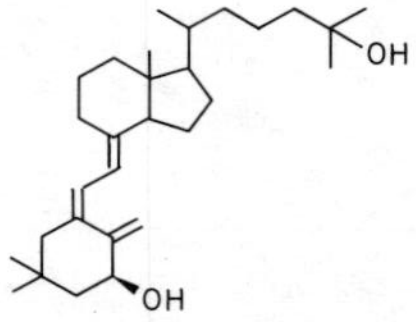 (67)	326

alcohols give a number of different types of products depending on the substitution pattern [271].

Conjugated dienes also give a variety of products depending on the substrate and reaction conditions. Simple dihydration products are frequently observed, as with 1,3-butadiene (Eq. 91) [53, 267–270]. With substituted acyclic

$$H_2C=CHCH=CH_2 \longrightarrow XHgCH_2\overset{\overset{\displaystyle HO}{|}}{CH}\overset{\overset{\displaystyle OH}{|}}{CH}CH_2HgX \longrightarrow CH_3\overset{\overset{\displaystyle HO}{|}}{CH}\overset{\overset{\displaystyle OH}{|}}{CH}CH_3 \qquad (91)$$

dienes, one can generally predict the major product by applying what one knows about the relative rates of reaction of isolated double bonds (see Table 2.5) and the directing effects of the initially introduced hydroxy group [68]. The following example illustrates the point (Eq. 92). While it is possible to

$$(92)$$

isolate the monohydration products in reasonable yield, they tend to be contaminated with starting diene and diol [272].

With cyclic conjugated dienes, the picture is more complex due to stereochemical considerations, cyclic ether formation and oxidation (Eqs. 93, 94) [273, 304, 305].

$$(93) \; [68, 273]$$

$$(94) \; [154]$$

With non-conjugated dienes and polyenes, many of the same types of products are observed. Isolated double bonds appear to be more reactive than conjugated double bonds, as indicated by the following example (Eq. 95) [326]. In non-conjugated dienes, one frequently finds that one double bond is significantly more reactive than another and one obtains only the mono-hydroxylated product [272]. The relative reactivities reported in Table 2.5 provide a good basis on which to predict such products. With double bonds of comparable reactivity, one usually obtains roughly statistical mixtures of

$$(95)$$

products [272]. However, the use of mercuric trifluoroacetate improves the yield of monohydration product [272]. The addition of sodium lauryl sulfate (SLS) to form micelles can dramatically change the situation (Eqs. 96, 97) [286, 310]. This simple solution does not work for all dienes however.

$$(96)$$

no SLS	70%	14%	19%
SLS	97%	—	—

$$(97)$$

Where five- or six-membered ring ethers can be formed by intramolecular alkoxymercuration of the initially formed enol, these products are common. Table 2.6 contains many such examples. This can be one of the best ways to make certain monocyclic and polycyclic ethers. From acyclic 1,5-dienes which can afford either five- or six-membered ring ethers, mixtures of regio- and stereoisomers are generally obtained with the tetrahydrofuran product predominating [322].

When the double bonds of a diene are appropriately situated, most commonly in a cyclic system, hydroxymercuration can also result in carbon—carbon bond formation, as the following simple example illustrates (Eq. 98) [296]. Table 2.6 contains several other examples.

$$(98)$$

Carbon skeleton rearrangements are also more prevalent among diene or polyene hydroxymercuration-demercuration products. Allenic cyclopropanes [317], cyclooctatetraene [283], bicyclo[2.2.0]hexadienes [314–316], norbornadienes [192, 309], and bullvalene [297, 298] are all observed to afford rearranged products. Rearrangement may occur during hydroxymercuration itself or upon addition of base [259], or during sodium borohydride reduction [309] which proceeds via free radicals.

The hydroxymercuration products derived from dienes and polyenes undergo many of the same reactions reported in the preceding section. For example, halogenation proceeds readily to afford polyhalogenated compounds. While alkaline sodium borohydride reduction usually goes well, rearrangements are occasionally noted. Sodium amalgam reductions of cyclic ethers can afford ring-opened products (Eq. 99) [276].

$$H_2C=CH(CH_2)_3CH=CH_2 \longrightarrow \overset{\displaystyle \bigcirc}{\underset{ClHgCH_2 \quad O \quad CH_2HgCl}{}} \xrightarrow[H_2O]{Na/Hg} H_2C=CH(CH_2)_3CHOHCH_3 \quad (99)$$

$$90\%$$

C. Alkynes

The reactions of alkynes and mercury salts which result in isolable organomercury compounds have been discussed in Chapter II, Section J of the monograph "Organomercury Compounds in Organic Synthesis". While these compounds are finding increasing utility in organic synthesis, of considerably greater preparative utility has been the mercury-promoted hydration of alkynes (Eq. 100). This reaction was last reviewed in 1963 [330].

$$RC{\equiv}CR + H_2O \xrightarrow{Hg(II)} RC\overset{\displaystyle O}{\overset{\|}{C}}CH_2R \quad (100)$$

Kucherov in 1881 first reported that the reaction of aqueous mercuric bromide and acetylene or other terminal alkynes affords acetaldehyde or the corresponding methyl ketones respectively [331, 332]. In the ensuing years, a debate raged over the nature of the organomercurials involved in these reactions [333–345]. Numerous structures were proposed, based primarily on elemental analyses and subsequent reactions, and even today the exact nature of the organomercurials formed in these reactions is unresolved.

Despite our meager understanding of the nature of the intermediates in these reactions, the mercury-promoted hydration of alkynes has proven to be a very valuable synthetic procedure. Table 2.7 summarizes the many examples of this reaction which have appeared to date. Entries are ordered according to increasing molecular formula in the usual Chemical Abstracts fashion. While every attempt has been made to include all known examples of this reaction, there no doubt remain further examples buried in the literature. The shear number of examples attests, however, to the widespread acceptance of this approach to the synthesis of carbonyl compounds.

Table 2.7. Hydroxymercuration of Alkynes

Alkyne	Mercuric salt	Reaction conditions	Product(s) (% Yield)	Ref.
$HC{\equiv}CH$	$HgBr_2$	H_2O RT	CH_3CHO	331, 332
	$Hg(OAc)_2$	H_2O/HCl	CH_3CHO	346
	$Hg(NO_3)_2$	H_2O (HCl)	CH_3CHO	334, 335
	$HgSO_4$	H_2SO_4/H_2O	CH_3CHO	347 – 349
	HgO	H_2SO_4/H_2O	CH_3CHO	345, 350
	Hg(II) – Dowex 50	H_2O RT 20 min	CH_3CHO	351
	Hg(II) – Dowex 50	$HOAc/H_2O$ Δ	CH_3CHO	352
	Hg(II) – Nafion H	$EtOH/H_2O$ 20 °C 90 min	CH_3CHO (65 as 2, 4-DNP)	353
$HC{\equiv}CCF_3$	$HgSO_4$	$H_2SO_4/MeOH$ RT 24 hr	CH_3COCF_3 + $HCOCH_2CF_3$ 84 : 16	354, 355
$HC{\equiv}CCH_3$	$HgCl_2$	H_2O	CH_3COCH_3	333
	$HgBr_2$	H_2O	CH_3COCH_3 (~100)	331
	Hg(II) – Nafion H	$EtOH/H_2O$ 20 °C 90 min	CH_3COCH_3 (67)	353
$HC{\equiv}CCH_2OH$	HgO	$Cl_3CCO_2H/HOAc/MeOH/BF_3 \cdot Et_2O$ 55 – 65 °C	CH_3COCH_2OAc (30)	356
	Hg(II) – Dowex 50	50 °C 25 min	CH_3COCH_2OH (87 as phenylhydrazone)	351
	Hg(II) – Dowex 50	$HOAc/H_2O$ Δ	CH_3COCH_2OH	352
$CH_3C{\equiv}CCF_3$	$HgSO_4$	H_2SO_4/H_2O 45 – 55 °C 20 hr	$CH_3COCH_2CF_3$	357

Table 2.7. (continued)

Alkyne	Mercuric salt	Reaction conditions	Product(s) (% Yield)	Ref.
$HC\equiv CCH=CH_2$	Hg^{2+}	H_3O^+	$CH_3COCH=CH_2$	358, 359
	$HgSO_4$ or $Hg(OAc)_2$ or $Hg_3(AsO_4)_2$ or $Hg(BF_4)_2$	H_2SO_4/H_2O	$CH_3COCH=CH_2$	360
	$HgSO_4$	H_2SO_4/H_2O	$CH_3COCH=CH_2$	361
	$HgSO_4$	$H_2SO_4/H_2O/Fe_2(SO_4)_3$	$CH_3COCH=CH_2$	362
	$HgSO_4$	$H_2SO_4/H_2O/ROH$	$CH_3COCH_2CH_2OR$ (55–81) $R=CH_3$, C_2H_5, $n\text{-}C_3H_7$, $i\text{-}C_3H_7$, $CH_2CH=CH_2$, $n\text{-}C_4H_9$, $i\text{-}C_4H_9$, $i\text{-}C_5H_{11}$, $c\text{-}C_6H_{11}$, $n\text{-}C_8H_{17}$	363
$HC\equiv CCHOHCH_3$	HgO	$Cl_3CCO_2H/HOAc/MeOH/BF_3\cdot Et_2O$ 55–65 °C	$CH_3COCHOAcCH_3$ (41)	356
$CH_3C\equiv CCH_2OH$	Hg^{2+}	$HClO_4/H_2O$	$CH_3COCH_2CH_2OH$ + $CH_3COCH=CH_2$	364
$HOCH_2C\equiv CCH_2OH$	$Hg(OAc)_2$	$HOAc/H_2O$ Δ 2 hr	$AcOCH_2COCH=CH_2$ + $AcOCH_2COCH_2CH_2OAc$	365
$CH_3C\equiv CCH=CH_2$	HgO	H_2SO_4/H_2O 6 hr	$CH_3COCH=CHCH_3$ (52)	366
$HC\equiv CCHOHCH=CH_2$	$HgSO_4$	H_2SO_4/H_2O Δ	$CH_3COCOCH_2CH_3$ (70)	367
$HC\equiv C(CH_2)_2CH_3$	$HgBr_2$	H_2O 50–60 °C	$CH_3CO(CH_2)_2CH_3$	331
	Hg–Nafion H	$EtOH/H_2O$ 20 °C 90 min	$CH_3CO(CH_2)_2CH_3$ (90)	353
$HC\equiv CCOH(CH_3)_2$	$HgCl_2$	H_2O/HCl	$CH_3COCOH(CH_3)_2$	368
	$HgSO_4$	H_2SO_4/H_2O Δ 1 hr	$CH_3COCOH(CH_3)_2$ (66, 81)	369, 370
	HgO	H_2SO_4/H_2O 60–100 °C 2 hr	$CH_3COCOH(CH_3)_2$	371
	HgO	$H_2SO_4/H_2O/MeOH$ 60–65 °C 5 hr	$CH_3COCOH(CH_3)_2$ (81)	372

Table 2.7. (continued)

Alkyne	Mercuric salt	Reaction conditions	Product(s) (% Yield)	Ref.
	HgO	HOAc/MeOH/BF$_3 \cdot$ Et$_2$O 45–55°C 3 hr	CH$_3$COC(OAc)(CH$_3$)$_2$ (49)	373
	HgO	HO$_2$CR/Ac$_2$O/MeOH/BF$_3 \cdot$ Et$_2$O	CH$_3$COC(O$_2$CR)(CH$_3$)$_2$ + CH$_3$COCOH(CH$_3$)$_2$ R = H, Me, Et, n-Pr	374
	Hg(II)–Dowex 50	H$_2$O 80°C 5 hr	CH$_3$COCOH(CH$_3$)$_2$ (68.5)	351
	Hg(II)–Dowex 50	HOAc/H$_2$O Δ	CH$_3$COCOH(CH$_3$)$_2$	352
HC≡CCH=CHCH=CH$_2$	HgO	H$_2$SO$_4$/H$_2$O	CH$_3$COCH=CHCH=CH$_2$	375
H$_2$C=CHC≡CCH=CH$_2$	HgSO$_4$	H$_2$SO$_4$/MeOH 60–65°C	H$_2$C=CHCH=CHCOCH$_3$ + RO(CH$_2$)$_2$COCH$_2$CH(OCH$_3$)CH$_3$ R = H, CH$_3$ + CH$_3$O(CH$_2$)$_2$COCH$_2$CH=CH$_2$ +	376
C$_2$H$_5$C≡CCH=CH$_2$	HgO	H$_2$SO$_4$/H$_2$O	C$_2$H$_5$COCH=CHCH$_3$ (38)	366
HC≡CCH=CHCHOHCH$_3$	HgCl$_2$	H$_2$O/EtOH Δ	(55)	367
	HgSO$_4$	H$_2$SO$_4$/H$_2$O Δ 2–3 hr	CH$_3$COCH$_2$CH$_2$COCH$_3$ (55)	367
	HgSO$_4$	H$_2$SO$_4$/H$_2$O Δ	(45)	367
	HgO	CF$_3$CO$_2$H/MeOH/BF$_3 \cdot$ Et$_2$O Δ	(25)	367
HC≡CCHOHCH=CHCH$_3$	HgSO$_4$	H$_2$SO$_4$/H$_2$O Δ	CH$_3$COCO(CH$_2$)$_2$CH$_3$ (15)	367
	HgSO$_4$	H$_2$SO$_4$/H$_2$O Δ 2–3 hr	CH$_3$COCH$_2$CH$_2$COCH$_3$	367

Table 2.7. (continued)

Alkyne	Mercuric salt	Reaction conditions	Product(s) (% Yield)	Ref.
$H_2C=CHC\equiv CCHOHCH_3$	$HgSO_4$	H_2SO_4/H_2O 80–90°C	2-methyltetrahydropyran-4-one (O-ring ketone with CH$_3$)	376, 377
	HgO	$MeOH/BF_3\cdot Et_2O$	$CH_3O(CH_2)_2COCH_2CH(OCH_3)CH_3$	378
$HC\equiv C(CH_2)_3CH_3$	$HgSO_4$	$H_2SO_4/CH_3COCH_3/H_2O$ 60°C 3 hr	$CH_3CO(CH_2)_3CH_3$ (63.6, 78.4)	379
	$HgSO_4$	$H_2SO_4/MeOH/H_2O$ 60°C 3 hr	$CH_3CO(CH_2)_3CH_3$ (79.8)	379
	$HgSO_4$	$H_2SO_4/HOAc/H_2O$ 60°C 3 hr	$CH_3CO(CH_2)_3CH_3$ (78.6)	379
	Hg(II)–Nafion H	$EtOH/H_2O$ 20°C 90 min	$CH_3CO(CH_2)_3CH_3$ (90)	353
	C_6H_5HgOH	$H_2O/HCCl_3$ 60°C 24 hr	$CH_3CO(CH_2)_3CH_3$ (67)	380
$CH_3CH_2C\equiv CCH_2CH_3$	$Hg(OAc)_2$	H_2O/THF, $NaBH_4$	$CH_3CH_2COCH_2CH_2CH_3$ (83.3)	381
$HC\equiv CCOH(CH_3)CH_2CH_3$	$HgSO_4$	H_2SO_4/H_2O Δ	$CH_3COCOH(CH_3)CH_2CH_3$ (88)	370
	HgO	H_2SO_4/H_2O 60–100°C 2–3 hr	$CH_3COCOH(CH_3)CH_2CH_3$ (40)	371, 382
	HgO	$H_2SO_4/H_2O/MeOH$ 60–65°C 3–5 hr	$CH_3COCOH(CH_3)CH_2CH_3$ (60, 80)	372, 383
$CH_3CHOHC\equiv CCHOHCH_3$	$Hg(OAc)_2$	Ac_2O 1 hr	$CH_3CHOAcCOCH=CHCH_3$ (25–30)	384
$H_2C=C(CH_3)C\equiv CCH=CH_2$	$HgSO_4$	$H_2SO_4/H_2O/MeOH$ 60°C 10d	$H_2C=C(CH_3)COCH_2CH=CH_2$ + $H_2C=C(CH_3)CO(CH_2)_3OCH_3$ + $CH_3OC(CH_3)_2COCH_2CH=CH_2$ + $CH_3OC(CH_3)_2CO(CH_2)_3OCH_3$	385
$n\text{-}C_3H_7C\equiv CCH=CH_2$	HgO	H_2SO_4/H_2O	$n\text{-}C_3H_7COCH=CHCH_3$ (43)	366

Table 2.7. (continued)

Alkyne	Mercuric salt	Reaction conditions	Product(s) (% Yield)	Ref.
$H_2C=CHC\equiv CCOH(CH_3)_2$	$HgSO_4$	H_2SO_4/H_2O Δ 3–3.5 hr	[structure] (60–65)	377, 386
	$HgSO_4$	CH_3COCH_3 RT several days	$H_2C=CHCOCH=C(CH_3)_2$ (40)	386
	$HgSO_4$	MeOH 35–40°C 12 hr	$CH_3O(CH_2)_2COCH=C(CH_3)_2$ (70–75)	387
	HgO	$ROH/BF_3\cdot Et_2O$	$RO(CH_2)_2COCH=C(CH_3)_2$ R = Me, Et, n-Pr, i-Pr, n-Bu, n-Am (40–77)	388
$H_2C=CHC\equiv CCHOHCH_2CH_3$	$HgSO_4$	H_2SO_4/H_2O 90–100°C 20 hr	[structure]	377
	HgO	$MeOH/BF_3\cdot Et_2O$	$CH_3O(CH_2)_2COCH_2CH(OCH_3)CH_2CH_3$	378
$HC\equiv C$–C(OH)(cyclopentane)	$HgSO_4/Zeo$-Karb 225	$H_2O/MeOH$ Δ 2 hr	CH_3CO, OH (cyclopentane) (80)	389
	HgO	H_2SO_4/H_2O 60°C 100 min	CH_3CO, OH (cyclopentane) (69)	390
	HgO	$MeOH/BF_3\cdot Et_2O$ 35°C 1 hr	CH_3CO, OH (cyclopentane) (40)	382
	C_6H_5HgOH	$H_2O/HCCl_3$ 50°C 3 hr	CH_3CO, OH (cyclopentane) (57)	380
$HC\equiv C(CH_2)_4CH_3$	$HgSO_4$	$H_2SO_4/HOAc/H_2O$ 70°C	$CH_3CO(CH_2)_4CH_3$ (87.1)	379

Table 2.7. (continued)

Alkyne	Mercuric salt	Reaction conditions	Product(s) (% Yield)	Ref.
	Hg(II) – Nalfion H	EtOH / H$_2$O	CH$_3$CO(CH$_2$)$_4$CH$_3$ (87–94)	353
	C$_6$H$_5$HgOH	H$_2$O / HCCl$_3$ 60 °C 24 hr	CH$_3$CO(CH$_2$)$_4$CH$_3$ (56)	380
CH$_3$C≡C(CH$_2$)$_3$CH$_3$	HgO	H$_2$SO$_4$ / MeOH 65 °C	CH$_3$CH$_2$CO(CH$_2$)$_3$CH$_3$ + CH$_3$CO(CH$_2$)$_4$CH$_3$ (68 total) 53 : 47	391
	HgO	H$_2$SO$_4$ / HOAc 105 °C	CH$_3$CH$_2$CO(CH$_2$)$_3$CH$_3$ + CH$_3$CO(CH$_2$)$_4$CH$_3$ (63 total) 53 : 47	391
HC≡CCOH(C$_2$H$_5$)$_2$	HgO	H$_2$SO$_4$ / H$_2$O	CH$_3$COCOH(C$_2$H$_5$)$_2$ (100)	392
	HgO	H$_2$SO$_4$ / H$_2$O / MeOH 60–65 °C 5 hr	CH$_3$COCOH(C$_2$H$_5$)$_2$ (88)	372
HC≡CCOH(CH$_3$)CH(CH$_3$)$_2$	HgSO$_4$	H$_2$SO$_4$ / H$_2$O	CH$_3$COCOH(CH$_3$)CH(CH$_3$)$_2$ (70)	393
	HgO	H$_2$SO$_4$ / H$_2$O Δ 3 hr	CH$_3$COCOH(CH$_3$)CH(CH$_3$)$_2$ (64)	382
HC≡CCOH(CH$_3$)CH$_2$CH$_2$CH$_3$	HgSO$_4$	H$_2$O Δ	CH$_3$COCOH(CH$_3$)CH$_2$CH$_2$CH$_3$ (75–85)	394
(CH$_3$)$_3$CC≡CCH$_2$OH	Hg(OAc)$_2$	H$_2$O	(CH$_3$)$_3$CCOCH=CH$_2$ + (CH$_3$)$_3$CCOCH$_2$CH$_2$OH	395
HOCH$_2$C≡C(CH$_2$)$_3$CH$_3$	HgO	HOAc / MeOH / BF$_3$·Et$_2$O 55–60 °C 2 hr	AcOCH$_2$CH$_2$CO(CH$_2$)$_3$CH$_3$ (45)	396
HC≡CC$_6$H$_5$	Hg(OAc)$_2$	H$_2$O / THF, NaBH$_4$	CH$_3$COC$_6$H$_5$ (55.1)	381
	Hg(ClO$_4$)$_2$	HClO$_4$ / H$_2$O / dioxane	CH$_3$COC$_6$H$_5$ (~50)	397
	Hg(II) – Nalfion H	EtOH / H$_2$O 20 °C 90 min	CH$_3$COC$_6$H$_5$ (80)	353
HC≡C–⬡	Hg(II) – Dowex 50	HOAc / H$_2$O Δ	CH$_3$CO–⬡	352
n-C$_4$H$_9$C≡CCH=CH$_2$	HgO	H$_2$SO$_4$ / H$_2$O	n-C$_4$H$_9$COCH=CHCH$_3$ (40)	366

Table 2.7. (continued)

Alkyne	Mercuric salt	Reaction conditions	Product(s) (% Yield)	Ref.
HC≡C–(cyclohexyl)	Hg(II)–Nalfion H	EtOH/H_2O 20°C 90 min	CH_3CO–(cyclohexenyl) (93)	353
	C_6H_5HgOH	H_2O/$HCCl_3$ 60°C 24 hr	CH_3CO–(cyclohexenyl) (49)	380
H_2C=$CHC≡CCOH(CH_3)CH_2CH_3$	$HgSO_4$	H_2SO_4/H_2O Δ 4–6 hr	H_2C=$CHCOCH$=$C(CH_3)CH_2CH_3$ (5–10) + H_2C=$CHC≡CC(CH_3)$=$CHCH_3$ (5–7)	386
	$HgSO_4$	MeOH 35–40°C 12 hr	$CH_3O(CH_2)_2COCH$=$C(CH_3)CH_2CH_3$ (70–75)	387
H_2C=$CHC≡CCHOHCH(CH_3)_2$	$HgSO_4$	H_2SO_4/H_2O 90–100°C 20 hr	(tetrahydropyran-4-one)–$CH(CH_3)_2$ (10)	377
	HgO	MeOH/BF_3·Et_2O	$CH_3O(CH_2)_2COCH_2CH(OCH_3)CH(CH_3)_2$	378
H_2C=$CHC≡CCHOH(CH_2)_2CH_3$	$HgSO_4$	H_2SO_4/H_2O 90–100°C 20 hr	(tetrahydropyran-4-one)–$CH_2CH_2CH_3$ (20)	377
	HgO	MeOH/BF_3·Et_2O	$CH_3O(CH_2)_2COCH_2CH(OCH_3)(CH_2)_2CH_3$	378
HC≡C–C(OH)(cyclohexyl)	$HgSO_4$	H_2SO_4/H_2O Δ	CH_3CO–C(OH)(cyclohexyl)	370
	HgO	H_2SO_4/H_2O Δ	CH_3CO–C(OH)(cyclohexyl) (65–100)	371, 390, 392, 398, 399
	HgO	H_2SO_4/H_2O/MeOH 60–65°C 5 hr	CH_3CO–C(OH)(cyclohexyl) (91)	372

Table 2.7. (continued)

Alkyne	Mercuric salt	Reaction conditions	Product(s) (% Yield)	Ref.
	HgO	Cl_3CCO_2H/HOAc/MeOH/$BF_3\cdot Et_2O$ 55–65°C	1-(CH_3CO), 1-(OAc)-cyclohexane (35)	356
	HgO	MeOH/$BF_3\cdot Et_2O$ 35°C 1hr	1-(CH_3CO), 1-(OH)-cyclohexane (73)	382
	Hg(II)–Dowex 50	H_2O/MeOH Δ 4hr	1-(CH_3CO), 1-(OH)-cyclohexane (84)	351
	Hg(II)–Dowex 50	HOAc/H_2O Δ	1-(CH_3CO), 1-(OH)-cyclohexane	352
	C_6H_5HgOH	H_2O/$HCCl_3$ 50°C 3hr	1-(CH_3CO), 1-(OH)-cyclohexane (65)	380
$HC{\equiv}C(CH_2)_5CH_3$	$Hg(OAc)_2$	H_2O/THF, $NaBH_4$	$CH_3CO(CH_2)_5CH_3$ (70)	381
	$HgSO_4$	H_2SO_4/HOAc/H_2O 70°C	$CH_3CO(CH_2)_5CH_3$ (91.2)	379
$CH_3CH_2CH_2C{\equiv}CCH_2CH_2CH_3$	Hg(II)–Dowex 50	HOAc/H_2O Δ 200min	$CH_3(CH_2)_2CO(CH_2)_3CH_3$	352
$HC{\equiv}CCOH(CH_3)C(CH_3)_3$	HgO	H_2SO_4/H_2O Δ	$CH_3COCOH(CH_3)C(CH_3)_3$ (54, 80, 100)	382, 392, 400
$HC{\equiv}CCOH(CH_3)CH_2CH(CH_3)_2$	$HgSO_4$	H_2SO_4/H_2O Δ	$CH_3COCOH(CH_3)CH_2CH(CH_3)_2$ (80, 86)	370, 393, 401
	HgO	H_2SO_4/H_2O Δ 2–3hr	$CH_3COCOH(CH_3)CH_2CH(CH_3)_2$ (58)	371, 382
	HgO	H_2SO_4/H_2O/MeOH 60–65°C 5hr	$CH_3COCOH(CH_3)CH_2CH(CH_3)_2$ (94)	372
$HC{\equiv}CCOH(CH_3)CH_2CH_2CH_2CH_3$	$HgSO_4$	H_2SO_4	$CH_3COCOH(CH_3)CH_2CH_2CH_2CH_3$ (100)	393

132

Table 2.7. (continued)

Alkyne	Mercuric salt	Reaction conditions	Product(s) (% Yield)	Ref.
$HOCH_2C{\equiv}C(CH_2)_4CH_3$	HgO	HOAc/MeOH/BF$_3\cdot$Et$_2$O 55–60°C 2 hr	$AcOCH_2CH_2CO(CH_2)_4CH_3$ (52)	396
$(CH_3)_2COHC{\equiv}CCOH(CH_3)_2$	Hg(OAc)$_2$	Ac$_2$O	$(CH_3)_2C(OAc)COCH{=}C(CH_3)_2$ (45)	384
	HgSO$_4$	H$_2$O 80°C 15 min	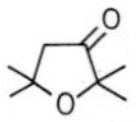 (dihydrofuranone structure) (100)	402
	HgO	Cl$_3$CCO$_2$H/MeOH or HOAc/BF$_3\cdot$Et$_2$O	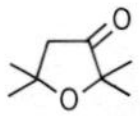 (dihydrofuranone structure) (77–78)	373
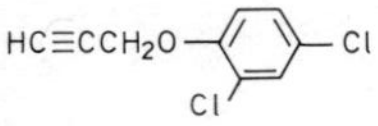 $HC{\equiv}CCH_2O$–(2,4-dichlorophenyl)	Hg(O$_2$CCF$_3$)$_2$	CH$_2$Cl$_2$ RT 2 hr	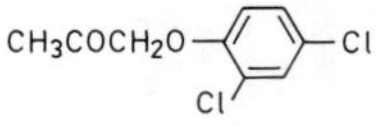 CH_3COCH_2O–(2,4-dichlorophenyl) (72)	403
p-$HC{\equiv}CCH_2OC_6H_4Cl$	Hg(O$_2$CCF$_3$)$_2$	CH$_2$Cl$_2$ RT 2 hr	p-$CH_3COCH_2OC_6H_4Cl$ (83)	403
$HC{\equiv}CCH_2OC_6H_5$	Hg(OAc)$_2$	CH$_3$OH / trace H$_2$SO$_4$	$CH_3COCH_2OC_6H_5$	403
$HC{\equiv}CCHOHC_6H_5$	HgO	Cl$_3$CCO$_2$H/HOAc/MeOH/BF$_3\cdot$Et$_2$O 55–65°C	$CH_3COCHOAcC_6H_5$ (50)	356
	HgSO$_4$	H$_2$O/H$_2$SO$_4$ 80°C	$CH_3COCHOHC_6H_5$ (80)	404
$C_6H_5C{\equiv}CCH_2OH$	Hg(OAc)$_2$	H$_2$O	$C_6H_5COCH{=}CH_2$ + $C_6H_5COCH_2CH_2OH$	395
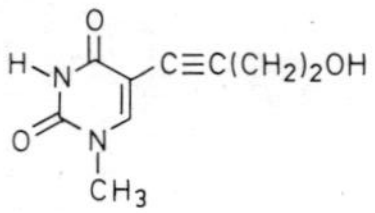 (1-methyluracil-5-yl)–$C{\equiv}C(CH_2)_2OH$	HgSO$_4$	H$_2$O 50°C 2 hr	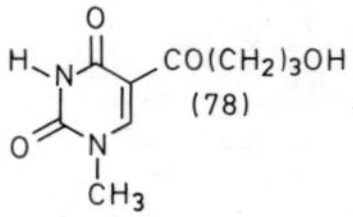 (1-methyluracil-5-yl)–$CO(CH_2)_3OH$ (78) 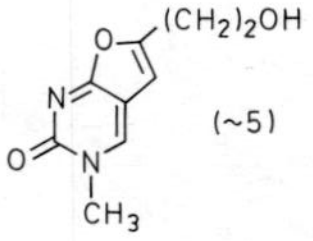 + (furo-pyrimidinone structure) (~5)	405
$H_2C{=}CHC{\equiv}CC(C_2H_5){=}CHCH_3$	HgSO$_4$	H$_2$SO$_4$/H$_2$O/MeOH Δ	$H_2C{=}CHCH_2COC(C_2H_5){=}CHCH_3$ + $CH_3O(CH_2)_3COC(C_2H_5){=}CHCH_3$	385

Table 2.7. (continued)

Alkyne	Mercuric salt	Reaction conditions	Product(s) (% Yield)	Ref.
$H_2C=CHC\equiv CCOH(C_2H_5)_2$	$HgSO_4$	CH_3COCH_3 60–80°C 3–7 hr	$H_2C=CHCOCH=C(C_2H_5)_2$	406
	$HgSO_4$	MeOH 35–40°C 12 hr	$CH_3O(CH_2)_2COCH=C(C_2H_5)_2$ (70–75)	387
	HgO	$ROH/BF_3\cdot Et_2O$	$RO(CH_2)_2COCH=C(C_2H_5)_2$ R = Me (61), n-Pr (65)	388
$H_2C=CHC\equiv CCOH(CH_3)CH_2CH_2CH_3$	$HgSO_4$	CH_3COCH_3 60–80°C 3–7 hr	$H_2C=CHCOCH=C(CH_3)CH_2CH_2CH_3$	406
	$HgSO_4$	MeOH 35–40°C 12 hr	$CH_3O(CH_2)_2COCH=C(CH_3)CH_2CH_2CH_3$ (70–75)	387
	HgO	$ROH/BF_3\cdot Et_2O$	$RO(CH_2)_2COCH=C(CH_3)CH_2CH_2CH_3$ R = Me (54), n-Pr (48)	388
1-ethynyl-2-methylcyclohexan-1-ol ($HC\equiv C$, OH, CH_3)	$HgSO_4$	$H_2SO_4/H_2O/CCl_4$ 15°C 2 hr, Δ 3 hr	CH_3CO, OH, CH_3 (cyclohexane)	407
	$HgSO_4$/Zeo-Karb 225	$H_2O/MeOH$	CH_3CO, OH, CH_3 (cyclohexane)	389
	HgO	H_2SO_4/H_2O Δ 3 hr	CH_3CO, OH, CH_3 (cyclohexane) (62)	382
1-ethynyl-3-methylcyclohexan-1-ol ($HC\equiv C$, OH, CH_3)	$HgSO_4$	H_2SO_4/H_2O Δ	CH_3CO, OH, CH_3 (cyclohexane)	389
	HgO	H_2SO_4/H_2O Δ 0.5–3 hr	CH_3CO, OH, CH_3 (cyclohexane)	382, 408

134

Table 2.7. (continued)

Alkyne	Mercuric salt	Reaction conditions	Product(s) (% Yield)	Ref.
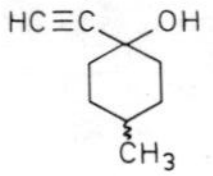 HC≡C, OH / CH$_3$ (cyclohexane ring)	HgSO$_4$/Zeo-Karb 225	H$_2$O/MeOH Δ 2 hr	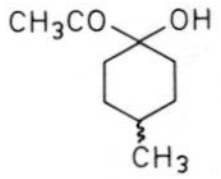 CH$_3$CO, OH / CH$_3$ (cyclohexane ring) (80)	389
	HgO	H$_2$SO$_4$/H$_2$O Δ 3 hr	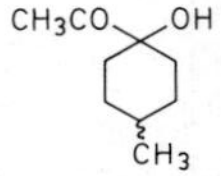 CH$_3$CO, OH / CH$_3$ (cyclohexane ring) (65)	382
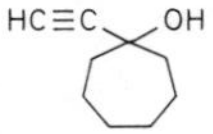 HC≡C, OH (cycloheptane ring)	HgSO$_4$	H$_2$SO$_4$/H$_2$O/CCl$_4$ <25 °C, Δ	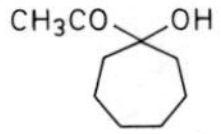 CH$_3$CO, OH (cycloheptane ring) (60)	389
	HgSO$_4$/Zeo-Karb 225	H$_2$O/MeOH Δ 2 hr	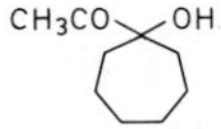 CH$_3$CO, OH (cycloheptane ring) (80)	389
(CH$_3$)$_3$CC≡CCH$_2$OAc	Hg(OAc)$_2$	H$_2$O	(CH$_3$)$_3$CCOCH$_2$CH$_2$OAc + (CH$_3$)$_3$CCOCH=CH$_2$ (?)	395
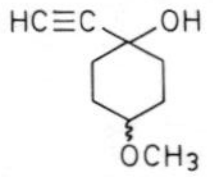 HC≡C, OH / OCH$_3$ (cyclohexane ring)	HgO	MeOH/BF$_3$·Et$_2$O 35 °C 1 hr	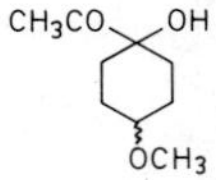 CH$_3$CO, OH / OCH$_3$ (cyclohexane ring) (34)	382
HC≡CCOH(i-C$_3$H$_7$)$_2$	HgSO$_4$	H$_2$SO$_4$/H$_2$O	CH$_3$COCOH(i-C$_3$H$_7$)$_2$ + CH$_3$COC=C(CH$_3$)$_2$ (i-C$_3$H$_7$) + HO$_2$CCH=C(i-C$_3$H$_7$)$_2$	400
	HgO	H$_2$SO$_4$/H$_2$O Δ	CH$_3$COCOH(i-C$_3$H$_7$)$_2$ (40–50)	372
HC≡CCOH(n-C$_3$H$_7$)$_2$	HgO	H$_2$SO$_4$/H$_2$O	CH$_3$COCOH(n-C$_3$H$_7$)$_2$ (100)	392
HC≡CCOH(CH$_3$)CH$_2$C(CH$_3$)$_3$	HgSO$_4$	H$_2$SO$_4$/H$_2$O	CH$_3$COCOH(CH$_3$)CH$_2$C(CH$_3$)$_3$ (90)	400

Table 2.7. (continued)

Alkyne	Mercuric salt	Reaction conditions	Product(s) (% Yield)	Ref.
$HC{\equiv}CCOH(CH_3)(CH_2)_4CH_3$	$HgSO_4$	H_2SO_4	$CH_3COCOH(CH_3)(CH_2)_4CH_3$ (80)	393
$HOCH_2C{\equiv}C(CH_2)_5CH_3$	HgO	$HOAc/MeOH/BF_3 \cdot Et_2O$ 55–60°C 2 hr	$AcOCH_2CH_2CO(CH_2)_5CH_3$ (55)	396
$HC{\equiv}CCOH(i\text{-}C_3H_7)CH(OCH_3)CH_3$	HgO	H_2SO_4 Δ 1 hr	$CH_3COCOH(i\text{-}C_3H_7)CH(OCH_3)CH_3$	409
$HC{\equiv}CCOH(CH_3)C_6H_5$	HgO	$H_2SO_4/H_2O/MeOH$ 55–65°C 1–5 hr	$CH_3COCOH(CH_3)C_6H_5$ (81, 92)	372, 410
	HgO	$HOAc/Ac_2O/BF_3 \cdot Et_2O$ RT 24 hr	$CH_3COC(OAc)(CH_3)C_6H_5$	410
$H_2C{=}CHC{\equiv}C$(1-hydroxycyclohexyl)	$HgSO_4$	$MeOH$ 35–40°C 12 hr	$CH_3O(CH_2)_2COCH{=}$(cyclohexylidene) (70–75)	387
	HgO	$ROH/BF_3 \cdot Et_2O$	$RO(CH_2)_2COCH{=}$(cyclohexylidene) R = Me (75), n-Pr (70)	388
$HC{\equiv}CCH{=}C(C_2H_5)CHOH(CH_2)_2CH_3$	$HgSO_4$	$H_2SO_4/H_2O/EtOH$ Δ 5 hr	$CH_3COCH_2CH(C_2H_5)CO(CH_2)_2CH_3$ (<25) + (furan: CH_3, C_2H_5, $(CH_2)_2CH_3$) (up to 64)	367
$HC{\equiv}CCHOHC(C_2H_5){=}CH(CH_2)_2CH_3$	$HgSO_4$	H_2SO_4/H_2O 100°C 5 hr	$CH_3COCH_2CH(C_2H_5)CO(CH_2)_2CH_3$ (40) + (furan: CH_3, C_2H_5, $(CH_2)_2CH_3$) (20–25)	367
$H_2C{=}CHC{\equiv}CCOH(CH_3)C(CH_3)_3$	HgO	$ROH/BF_3 \cdot Et_2O$	$RO(CH_2)_2COCH{=}C(CH_3)C(CH_3)_3$ R = Me (45), n-Pr (34)	388
$CH_3CHOHCH{=}CHC{\equiv}C(CH_2)_3CH_3$	HgO	$Cl_3CCO_2H/BF_3 \cdot Et_2O$ 50–55°C 3 hr, 20°C 15 hr	(furan: CH_3, $(CH_2)_4CH_3$) (30)	367
$HC{\equiv}C$(1-hydroxy-2-methylcyclohexyl), CH_3	$HgSO_4$	H_2SO_4/H_2O	CH_3CO(1-hydroxy-2-methylcyclohexyl), CH_3	389

Table 2.7. (continued)

Alkyne	Mercuric salt	Reaction conditions	Product(s) (% Yield)	Ref.
$HC{\equiv}CCOH(i\text{-}C_3H_7)C(CH_3)_3$	$HgSO_4$	H_2SO_4/H_2O 10 hr Δ	$HCOCH{=}C(i\text{-}C_3H_7)C(CH_3)_3$	400
$CH_3CH_2COH(CH_3)C{\equiv}CCOH(CH_3)CH_2CH_3$	$Hg(OAc)_2$	Ac_2O	$CH_3CH_2C(OAc)(CH_3)COCH{=}C(CH_3)CH_2CH_3$ (30)	384
$C_6H_5C{\equiv}CCH_2OAc$	$Hg(OAc)_2$	H_2O	$C_6H_5COCH{=}CH_2$ + $C_6H_5COCH_2CH_2OAc$	395
$HC{\equiv}CC(OCH_3)(CH_3)C_6H_5$	HgO	$H_2SO_4/MeOH$ 55°C > 2 hr	$CH_3COC(OCH_3)(CH_3)C_6H_5$ (81)	410
$H_2C{=}CHC{\equiv}C$—(2,2-dimethyltetrahydropyran-4-ol)	HgO	$MeOH/BF_3{\cdot}Et_2O$ 40°C 5 hr	$CH_3OCH_2CH_2COCH{=}$(2,2-dimethyltetrahydropyran)	411
$HC{\equiv}C(CH_2)_4CH{=}CH(CH_2)_2CH_3$	HgO	$Cl_3CCO_2H/EtOH/BF_3$	$CH_3CO(CH_2)_4CH{=}CH(CH_2)_2CH_3$	412
$H_2C{=}CHC{\equiv}CCOH(n\text{-}C_3H_7)_2$	$HgSO_4$	$MeOH$ 35–40°C 12 hr	$CH_3O(CH_2)_2COCH{=}C(n\text{-}C_3H_7)_2$ (70–75)	387
	HgO	$ROH/BF_3{\cdot}Et_2O$	$RO(CH_2)_2COCH{=}C(n\text{-}C_3H_7)_2$ R = Me (82), n-Pr (55)	388
$n\text{-}C_5H_{11}C{\equiv}C(CH_2)_2COCH_3$	$HgSO_4$	$H_2SO_4/H_2O/MeOH$	$n\text{-}C_6H_{13}COCH_2CH_2COCH_3$ (85)	413
	—	—	$CH_3CO(CH_2)_2CO(CH_2)_5CH_3$ (92)	414
$n\text{-}C_4H_9C{\equiv}C(CH_2)_3COCH_3$	$HgSO_4$	$H_2SO_4/H_2O/MeOH$	$n\text{-}C_5H_{11}CO(CH_2)_3COCH_3$ (>85)	413
$HC{\equiv}C$—(1-hydroxy-2,2,3-trimethylcyclohexyl)	HgO	H_2SO_4/H_2O Δ 3 hr	CH_3CO—(1-hydroxy-2,2,3-trimethylcyclohexyl) (81)	382
$CH_3C{\equiv}C(CH_2)_7CO_2H$	$Hg(OAc)_2$	$HOAc$ 90°C 8 hr HCl or H_2SO_4/H_2S	$CH_3CO(CH_2)_8CO_2H$ + $CH_3CH_2CO(CH_2)_7CO_2H$ 54 : 46	415

Table 2.7. (continued)

Alkyne	Mercuric salt	Reaction conditions	Product(s) (% Yield)	Ref.
$C_6H_5COC\equiv CC\equiv CCH_3$	$Hg(OAc)_2$	H_2SO_4/HOAc Δ 4 hr overnight RT	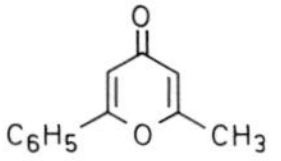 (pyranone: C_6H_5, CH_3 substituents) (70)	416
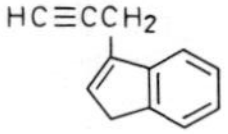 $HC\equiv CCH_2$ (indene)	$HgSO_4$	H_2SO_4/H_2O/EtOH Δ 0.5–1 hr	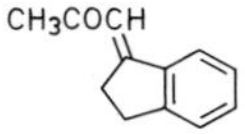 ($CH_3COCH=$ indanylidene) (25)	417
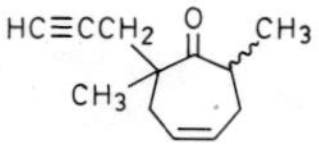 $HC\equiv CCH_2$, CH_3, CH_3 (cycloheptenone)	Hg(II)–Dowex 50	HOAc/H_2O Δ 45 min	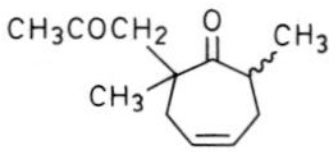 (CH_3COCH_2, CH_3, CH_3 cycloheptenone) (56)	418
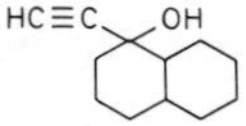 $HC\equiv C$, OH (decalin)	HgO	H_2SO_4/H_2O Δ 3 hr	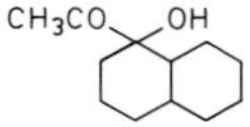 (CH_3CO, OH decalin) (68)	382
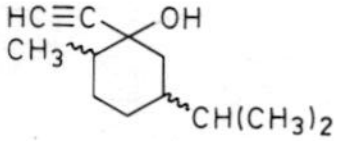 $HC\equiv C$, CH_3, OH, $CH(CH_3)_2$ (cyclohexane)	HgO	H_2SO_4/H_2O Δ 2 hr	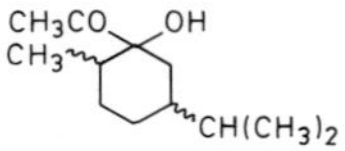 (CH_3CO, CH_3, OH, $CH(CH_3)_2$ cyclohexane)	408
$HC\equiv CCOH(CH_3)C(CH_3)_2CH_2C(CH_3)_3$	$HgSO_4$	H_2SO_4/H_2O Δ	$CH_3COCOH(CH_3)C(CH_3)_2CH_2C(CH_3)_3$	400
$C_6H_5COC\equiv CC\equiv CCH_2CH_3$	$Hg(OAc)_2$	H_2SO_4/HOAc Δ 4 hr overnight RT	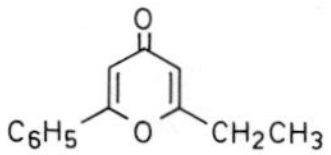 (pyranone: C_6H_5, CH_2CH_3 substituents) (50)	416
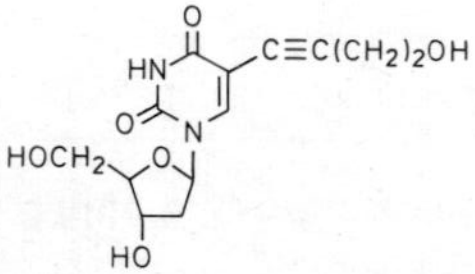 (nucleoside) $C\equiv C(CH_2)_2OH$	$HgSO_4$	H_2O	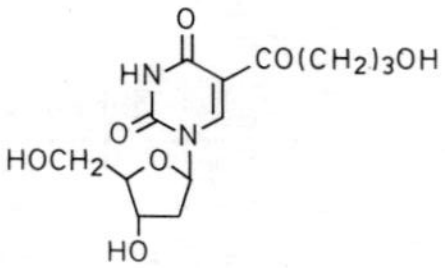 (nucleoside) $CO(CH_2)_3OH$ (64)	405

Table 2.7. (continued)

Alkyne	Mercuric salt	Reaction conditions	Product(s) (% Yield)	Ref.
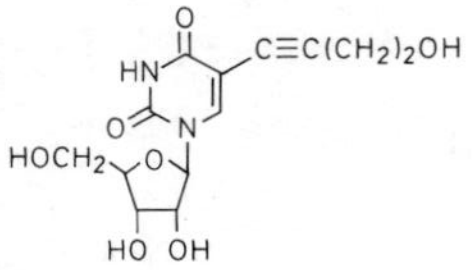 [structure: 5-(4-hydroxy-1-butynyl)uridine] $C\equiv C(CH_2)_2OH$	$HgSO_4$	H_2O	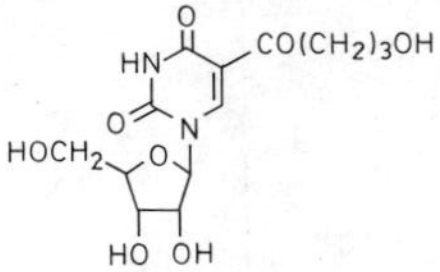 [structure] $CO(CH_2)_3OH$ (67)	405
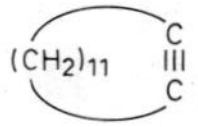 [cyclic structure] $(CH_2)_{11}$, $C\equiv C$	$Hg(OAc)_2$	H_2O/THF, $NaBH_4$	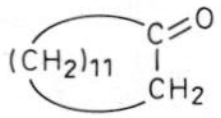 [cyclic structure] $(CH_2)_{11}$, $C=O$, CH_2 (78)	381
$HC\equiv CCOH(CH_3)(n-C_9H_{19})$	HgO	H_2SO_4/H_2O	$CH_3COCOH(CH_3)(n-C_9H_{19})$ (100)	392
$C_6H_5C\equiv CC_6H_5$	$Hg(II)-Nalfion\ H$	$EtOH/H_2O$ 40°C 90 min	$C_6H_5COCH_2C_6H_5$ (92)	353
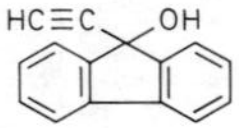 [9-ethynylfluoren-9-ol] $HC\equiv C$, OH	$HgSO_4$	$H_2SO_4/H_2O/MeOH$ 55°C >2 hr	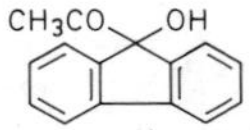 [fluorene] CH_3CO, OH (84)	410
$HC\equiv CCOH(C_6H_5)_2$	$Hg(OAc)_2$	$HOAc/Ac_2O$ RT 16 hr, H_2S	$CH_3COC(OAc)(C_6H_5)_2$ (58)	410
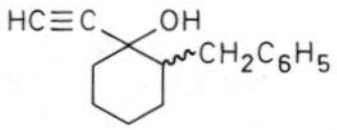 [cyclohexane] $HC\equiv C$, OH, $CH_2C_6H_5$	$HgSO_4$	H_2SO_4/H_2O 30–40°C 4–5 hr, Δ 2 hr	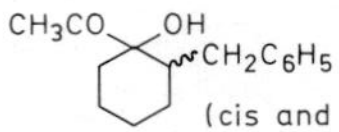 [cyclohexane] CH_3CO, OH, $CH_2C_6H_5$ (cis and trans)	419
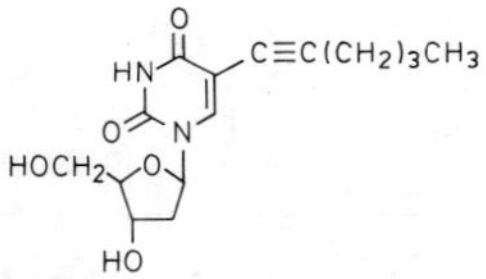 [structure: 5-(1-hexynyl)-2′-deoxyuridine] $C\equiv C(CH_2)_3CH_3$	$HgSO_4$	$H_2O/dioxane$	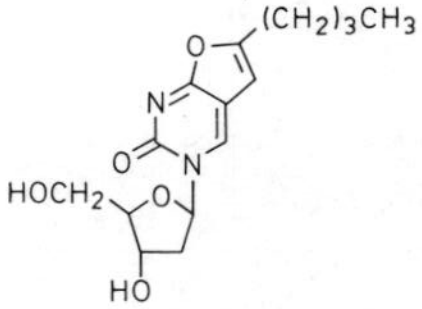 [furopyrimidine structure] $(CH_2)_3CH_3$ (36)	405
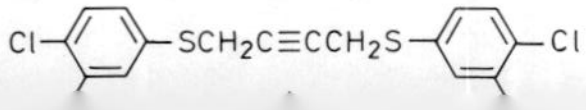 Cl—[C$_6$H$_3$Cl]—$SCH_2C\equiv CCH_2S$—[C$_6$H$_3$Cl]—Cl	HgO	$H_2SO_4/H_2O/MeOH$ Δ 12–15 hr	 Cl—[C$_6$H$_3$Cl]—$SCH_2COCH_2CH_2OCH_3$ (77.1)	420

Table 2.7. (continued)

Alkyne	Mercuric salt	Reaction conditions	Product(s) (% Yield)	Ref.
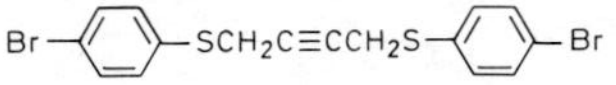 Br–C₆H₄–SCH₂C≡CCH₂S–C₆H₄–Br	HgO	H₂SO₄/H₂O/ROH Δ 12–15 hr	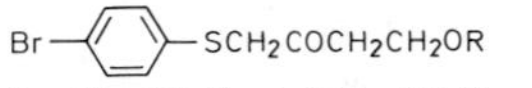 Br–C₆H₄–SCH₂COCH₂CH₂OR R = CH₃ (73.8), i–C₃H₇ (64.6)	420
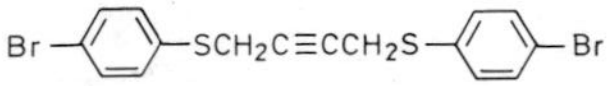 Br–C₆H₄–SCH₂C≡CCH₂S–C₆H₄–Br	HgO	H₂SO₄/HOAc 90°C 45 min, 24°C 14 hr, 100°C 1 hr	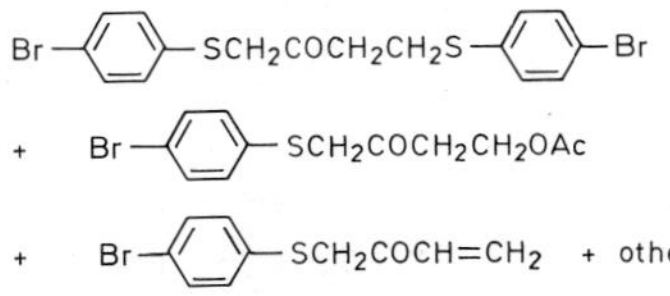 Br–C₆H₄–SCH₂COCH₂CH₂S–C₆H₄–Br + Br–C₆H₄–SCH₂COCH₂CH₂OAc + Br–C₆H₄–SCH₂COCH=CH₂ + other products	420
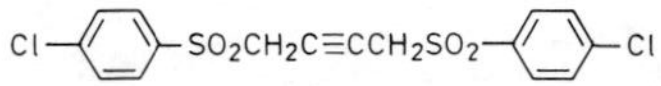 Cl–C₆H₄–SO₂CH₂C≡CCH₂SO₂–C₆H₄–Cl	HgO	H₂SO₄/HOAc 90°C 4 hr	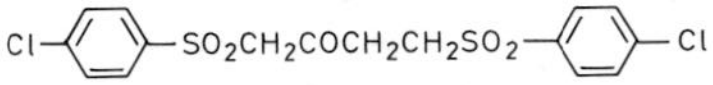 Cl–C₆H₄–SO₂CH₂COCH₂CH₂SO₂–C₆H₄–Cl (79.2)	420
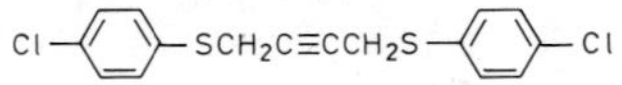 Cl–C₆H₄–SCH₂C≡CCH₂S–C₆H₄–Cl	HgO	H₂SO₄/H₂O/MeOH Δ 12–15 hr	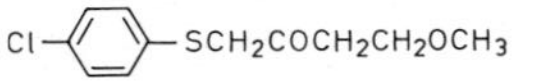 Cl–C₆H₄–SCH₂COCH₂CH₂OCH₃ (56)	420
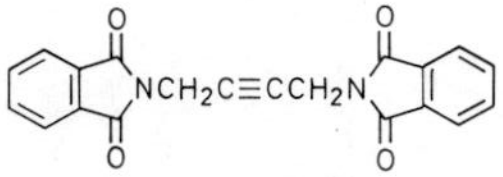 (phthalimido)–NCH₂C≡CCH₂N–(phthalimido)	Hg(OAc)₂	H₂SO₄/HOAc Δ 4 hr	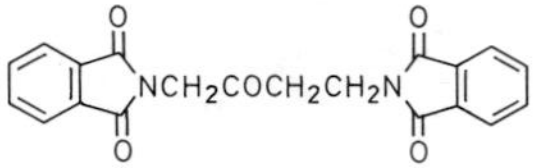 (phthalimido)–NCH₂COCH₂CH₂N–(phthalimido) (92)	421
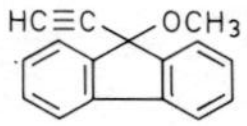 9-ethynyl-9-methoxyfluorene (HC≡C, OCH₃)	HgSO₄	MeOH RT 19 hr, steambath 10 min	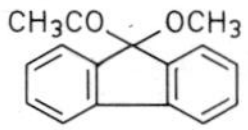 9-(acetyl/CH₃CO), 9-methoxy fluorene (CH₃CO, OCH₃)	410
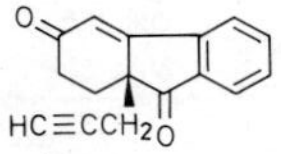 (tricyclic enone, HC≡CCH₂O substituent)	HgO	3.4 % H₂SO₄/THF 60°C 18 hr	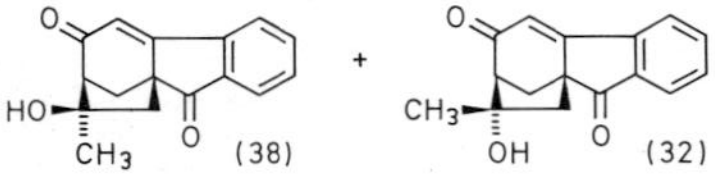 (tricyclic diketone, HO, CH₃) (38) + (tricyclic diketone, CH₃, OH) (32)	422
C₆H₅SO₂CH₂C≡CCH₂SO₂C₆H₅	HgO	H₂SO₄/HOAc 90°C 4 hr	C₆H₅SO₂CH₂COCH₂CH₂SO₂C₆H₅ (83.8)	420
C₆H₅SCH₂C≡CCH₂SC₆H₅	HgO	H₂SO₄/H₂O/MeOH Δ 12–15 hr	C₆H₅SCH₂COCH₂CH₂OCH₃ (54.6)	420

Table 2.7. (continued)

Alkyne	Mercuric salt	Reaction conditions	Product(s) (% Yield)	Ref.
$HC\equiv C$–(1-hydroxycyclohexyl)CH_2–C$_6$H$_4$–OCH_3	$HgSO_4$/Zeo-Karb 225	H_2O/MeOH 6 hr	CH_3CO–(1-hydroxycyclohexyl)CH_2–C$_6$H$_4$–OCH_3 (cis and trans)	419
$C_6H_5C\equiv CCOH(CH_3)COH(CH_3)CH_2CH(CH_3)_2$	HgO	H_2SO_4/$HClO_4$	furan: C_6H_5, CH_3, CH_3, $CH_2CH(CH_3)_2$ ClO_4^- (54.7)	423
uracil nucleoside–$C\equiv C(CH_2)_2OH$ (AcOCH$_2$, AcO)	$HgSO_4$	H_2O	uracil nucleoside–$CO(CH_2)_3OH$ (AcOCH$_2$, AcO) (59)	405
(2-CH_3C$_6$H$_4$)$SO_2CH_2C\equiv CCH_2SO_2$(2-$CH_3C_6H_4$)	HgO	H_2SO_4/HOAc 90 °C 4 hr	(2-CH_3C$_6$H$_4$)$SO_2CH_2COCH_2CH_2SO_2$(2-CH_3C$_6$H$_4$) (50)	420
(3-CH_3C$_6$H$_4$)$SO_2CH_2C\equiv CCH_2SO_2$(3-$CH_3C_6H_4$)	HgO	H_2SO_4/HOAc 90 °C 4 hr	(3-CH_3C$_6$H$_4$)$SO_2CH_2COCH_2CH_2SO_2$(3-CH_3C$_6$H$_4$) (83)	420
(4-CH_3C$_6$H$_4$)$SO_2CH_2C\equiv CCH_2SO_2$(4-$CH_3C_6H_4$)	HgO	H_2SO_4/HOAc 90 °C 4 hr	(4-CH_3C$_6$H$_4$)$SO_2CH_2COCH_2CH_2SO_2$(4-CH_3C$_6$H$_4$) (77.5)	420
(4-CH_3OC$_6$H$_4$)$SO_2CH_2C\equiv CCH_2SO_2$(4-$CH_3OC_6H_4$)	HgO	H_2SO_4/HOAc 90 °C 4 hr	(4-CH_3OC$_6$H$_4$)$SO_2CH_2COCH_2CH_2SO_2$(4-CH_3OC$_6$H$_4$) (88.5)	420
(2-CH_3C$_6$H$_4$)$SCH_2C\equiv CCH_2S$(2-CH_3C$_6$H$_4$)	HgO	H_2SO_4/H_2O/MeOH Δ 12–15 hr	(2-CH_3C$_6$H$_4$)$SCH_2COCH_2CH_2OCH_3$ (55.2)	420

Table 2.7. (continued)

Alkyne	Mercuric salt	Reaction conditions	Product(s) (% Yield)	Ref.
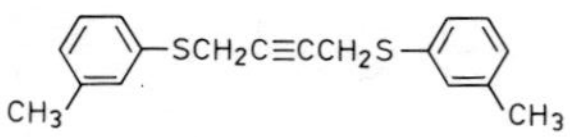 CH_3-aryl-$SCH_2C{\equiv}CCH_2S$-aryl-CH_3	HgO	$H_2SO_4/H_2O/MeOH$ Δ 12–15 hr	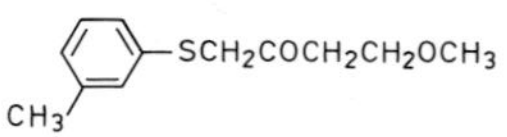 aryl-$SCH_2COCH_2CH_2OCH_3$-aryl (53.9)	420
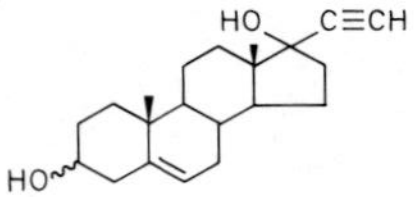 [steroid: 17-ethynyl-17-OH, Δ5-3-OH]	$HgCl_2$	H_2O/aniline/benzene 60°C 20 hr/H_2S	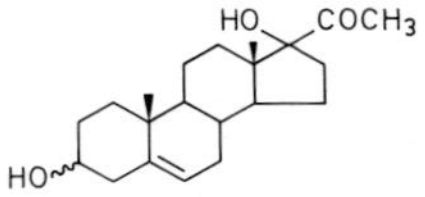 [steroid: 17-$COCH_3$, 17-OH]	424
	HgO	$HOAc/Ac_2O/BF_3{\cdot}Et_2O$ RT 1 week	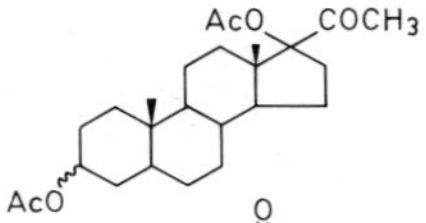 [steroid: AcO, $COCH_3$]	425
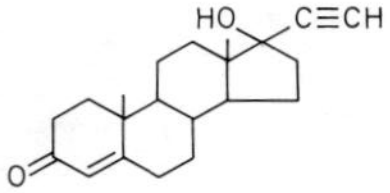	$Hg(OAc)_2$	EtOH or EtOAc 24 hr, H_2S	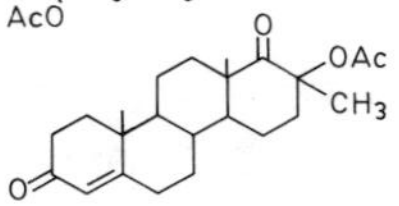 [steroid: OAc, CH_3] ?	426
[steroid: 17-ethynyl-17-OH, Δ4-3-one]	HgO	$HOAc/Ac_2O/BF_3{\cdot}Et_2O$ 20°C 15 hr	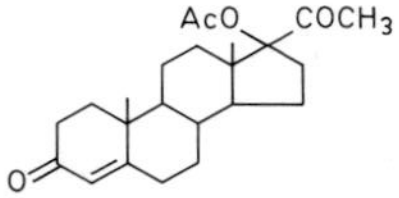 [steroid: AcO, $COCH_3$]	425, 427
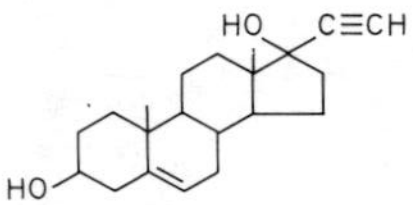	$Hg(OAc)_2$	EtOH or EtOAc RT 24 hr, H_2S	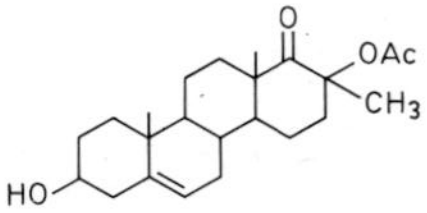 [steroid: OAc, CH_3] ?	426
[steroid: 17-ethynyl-17-OH, Δ5-3-OH]	$HgSO_4$	H_2O 110–120°C 24 hr	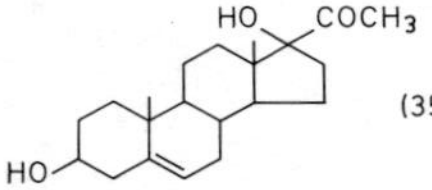 [steroid: HO, $COCH_3$] (35)	428

Table 2.7. (continued)

Alkyne	Mercuric salt	Reaction conditions	Product(s) (% Yield)	Ref.
	Hg(NHCOCH₃)₂	EtOH Δ 20 hr, H₂S	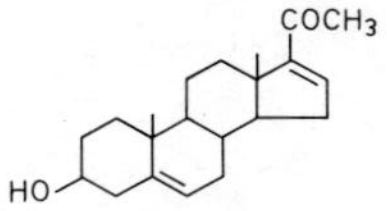 (steroid, COCH₃; HO)	429
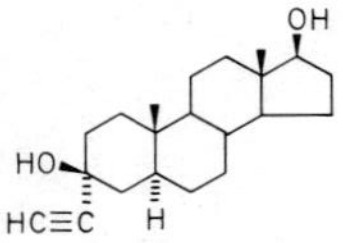 (steroid: OH; HO; HC≡C, H)	HgCl₂	piperidine 100°C 2 hr	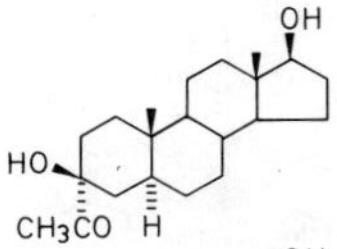 (steroid: OH; HO, CH₃CO H) (41)	430
(steroid: OH; HO; HC≡C, H)	Hg(OAc)₂	EtOAc 20°C 24 hr	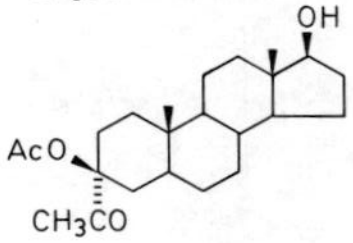 (steroid: OH; AcO, CH₃CO) (84)	430
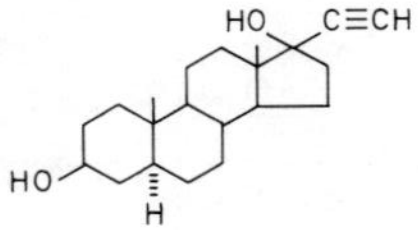 (steroid: HO, C≡CH; HO; H)	Hg(OAc)₂	EtOAc 24 hr, H₂S	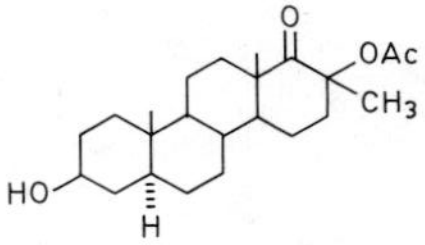 (steroid: O, OAc, CH₃; HO, H) ?	426
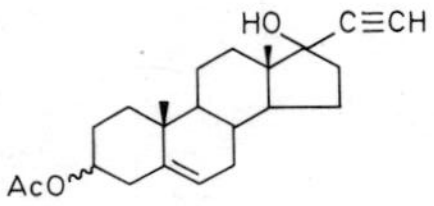 (steroid: HO, C≡CH; AcO)	HgO	HOAc / Ac₂O / BF₃·Et₂O warm, RT 1 week	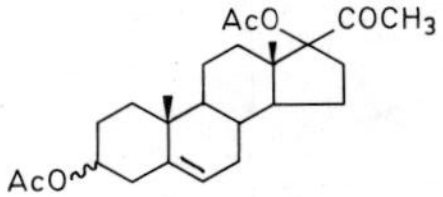 (steroid: AcO, COCH₃; AcO)	425
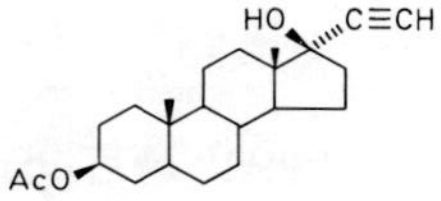 (steroid: HO, C≡CH; AcO)	HgCl₂	H₂O / aniline / benzene 60°C 20 hr	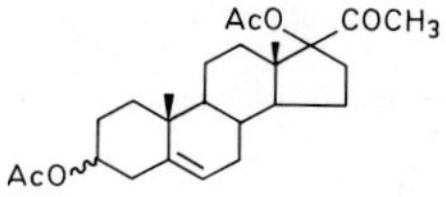 (steroid: AcO, COCH₃; AcO)	431

Table 2.7. (continued)

Alkyne	Mercuric salt	Reaction conditions	Product(s) (% Yield)	Ref.
	HgO	HOAc / Ac_2O / $BF_3 \cdot Et_2O$ RT 16 hr, NaOH	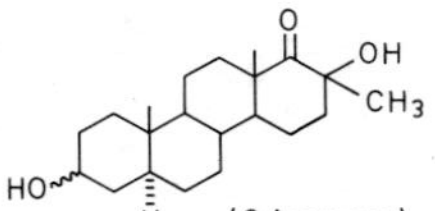 ? (2 isomers)	432
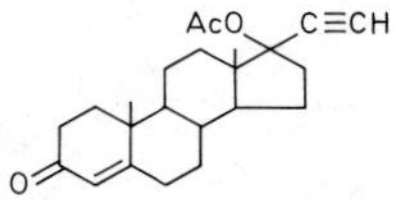	$Hg(OAc)_2$	EtOAc, H_2S	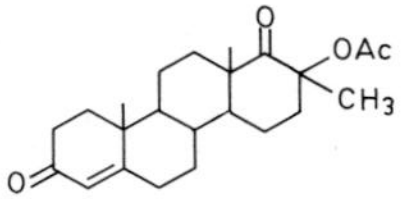 ?	426
	HgO	BF_3, H_2S	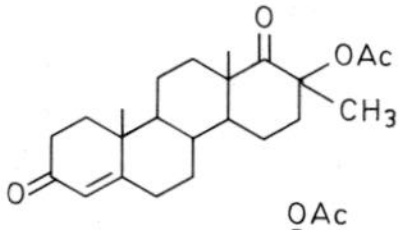 ?	426
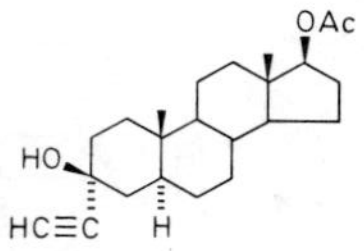	$HgCl_2$	piperidine 100 °C 2 hr, H_2S	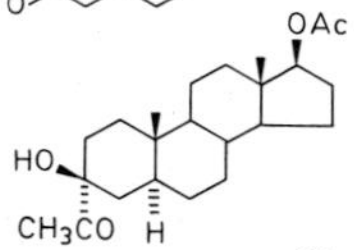 (42)	430
	$Hg(OAc)_2$	EtOAc 20 °C 24 hr	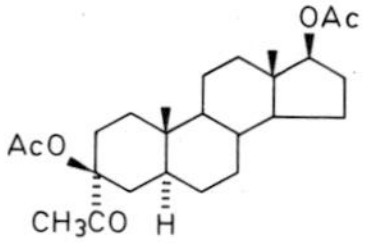 (72)	430
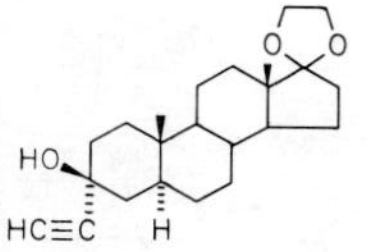	$HgCl_2$	piperidine 100 °C 2 hr	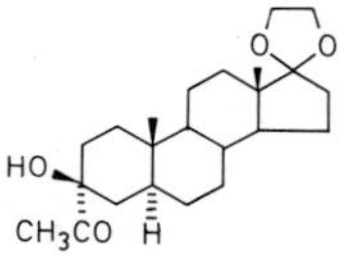 (72)	430

Table 2.7. (continued)

Alkyne	Mercuric salt	Reaction conditions	Product(s) (% Yield)	Ref.
[steroid: AcO, C≡CH; AcO, H]	Hg(OAc)$_2$	20°C 24 hr	[steroid with dioxolane; AcO, CH$_3$CO, H] (19)	430
[steroid: AcO, C≡CH; AcO, H]	HgCl$_2$	H$_2$O / aniline / benzene 60–62°C 8 hr	[steroid; AcO, COCH$_3$; AcO, H] (78)	433
[steroid: AcO, C≡CH; AcO, H]	HgO	HOAc / Ac$_2$O / BF$_3$·Et$_2$O RT overnight	[steroid; AcO, COCH$_3$; AcO]	425
(CH$_3$)$_3$C–C$_6$H$_4$–SO$_2$CH$_2$C≡CCH$_2$SO$_2$–C$_6$H$_4$	HgO	H$_2$SO$_4$ / HOAc 90°C 4 hr	(CH$_3$)$_3$C–C$_6$H$_4$–SO$_2$CH$_2$COCH$_2$CH$_2$SO$_2$–C$_6$H$_4$–C(CH$_3$)$_3$ (74.2)	420
[steroid: AcO, C≡CH; AcO]	Hg(OAc)$_2$	EtOAc 24 hr, H$_2$S	[steroid; O=, OAc, CH$_3$; AcO] ?	426
[steroid: AcO, C≡CH; AcO]	HgO	BF$_3$, H$_2$S	[steroid; O=, OAc, CH$_3$; AcO] ?	426
[steroid: C$_6$H$_5$CO$_2$, C≡CH; AcO]	HgO	HOAc / Ac$_2$O / BF$_3$·Et$_2$O RT 15 hr	[steroid; C$_6$H$_5$CO$_2$, COCH$_3$; AcO]	425

144

A wide variety of mercury reagents have proven useful in these hydration reactions. Mercuric sulfate has been the most widely used mercury salt, but each of the following mercury(II) salts have also been employed at one time or other: $Hg(OAc)_2$, $Hg(NO_3)_2$, $HgCl_2$, $HgBr_2$, $Hg_3(AsO_4)_2$, $Hg(BF_4)_2$, $Hg(ClO_4)_2$, $Hg(O_2CCF_3)_2$ and $Hg(NHCOCH_3)_2$. Mercuric oxide in the presence of a number of the corresponding inorganic acids, particularly sulfuric acid, or certain carboxylic acids has also found utility, as well as mercuric oxide plus boron trifluoride etherate in alcohol solvents. The hydration of alkynes is also promoted by organomercuric ions, RHg^+ [364], and phenylmercuric hydroxide [380]. The latter reagent is used in equivalent amounts with terminal acetylenes. It first forms the corresponding phenylmercuric acetylide, which is heated to 60 °C in aqueous chloroform for 25 hours. Yields of 55–67% of the corresponding methyl ketones are achieved. This approach is quite selective for non-conjugated terminal acetylenes, since internal acetylenes or conjugated terminal acetylenes fail to hydrolyze. The reaction conditions are also mild enough to tolerate alcohols, acetals, thioacetals, lactones, alkenes, epoxides and secondary bromides. Mercury-impregnated polymeric resins such as Dowex 50 [351, 352, 418], Zeokarb 225 [389, 419], and Nalfion H [353] in refluxing aqueous alcohol or acetic acid are also effective reagents.

The reaction as commonly run, using mercuric sulfate in aqueous sulfuric acid, is catalytic in mercury [349]. However, the catalyst can only be reused a few times before the mercury is consumed [358, 371]. The effect of various salts [349, 359], particularly iron [347, 362, 434], and manganese [434] salts, or hydrogen peroxide [435], on catalyst lifetime has been examined, but these difficulties have never been effectively overcome, thus limiting the industrial potential of this reaction.

The hydration reaction is commonly carried out in water or aqueous sulfuric acid, acetic acid or alcohol solutions, or combinations of the above. On occasion, the following reagents or solvents have also been employed either alone or in combination with the above solvents: methylene chloride, carbon tetrachloride, acetone, dioxane, ethyl acetate, benzene, aniline, piperidine and acetic anhydride. Hydrochloric acid, hydrogen sulfide or sodium borohydride have also been added to certain reactions to remove all mercury from the carbonyl products. In fact, the usual hydroxymercuration-demercuration conditions provide a convenient approach to a variety of ketones (Eq. 101) [381]. There appear to be no studies which have attempted to

$$RC\equiv CR' \quad \xrightarrow[\text{H}_2\text{O/THF}]{\text{Hg(OAc)}_2} \quad \xrightarrow{\text{NaBH}_4} \quad RC(\!\!\overset{\displaystyle O}{\overset{\|}{})\!\!CH_2R' \tag{101}$$

optimize reaction conditions for a wide variety of substrates, but a number of different procedures appear effective.

The regioselectivity of hydration has been studied for a variety of unsymmetrical acetylenes. Simple terminal alkynes always give exclusively the corresponding methyl ketones. The only exception appears to be 3,3,3-trifluoropropyne (Eq. 102) [354, 355].

$$CF_3C\equiv CH \longrightarrow CF_3\overset{\overset{\displaystyle O}{\|}}{C}CH_3 + CF_3CH_2\overset{\overset{\displaystyle O}{\|}}{C}H \qquad (102)$$
$$84 \quad : \quad 16$$

Unsymmetrical internal alkynes with simple alkyl substituents give approximately 50:50 mixtures of the two expected ketones [391, 415]. However, a number of substituents have been observed to dramatically alter the regioselectivity. Thus, neighboring fluorines [357] or specifically situated carbonyl groups [413, 414] induce high selectivity (Eqs. 103–105). The former effect must be inductive, while the carbonyl groups in the latter two examples

$$CF_3C\equiv CCH_3 \longrightarrow CF_3CH_2\overset{\overset{\displaystyle O}{\|}}{C}CH_3 \qquad (103)$$

$$n\text{-}C_5H_{11}C\equiv C(CH_2)_2\overset{\overset{\displaystyle O}{\|}}{C}CH_3 \longrightarrow n\text{-}C_6H_{13}\overset{\overset{\displaystyle O}{\|}}{C}(CH_2)_2\overset{\overset{\displaystyle O}{\|}}{C}CH_3 \qquad (104)$$

$$n\text{-}C_4H_9C\equiv C(CH_2)_3\overset{\overset{\displaystyle O}{\|}}{C}CH_3 \longrightarrow n\text{-}C_5H_{11}\overset{\overset{\displaystyle O}{\|}}{C}(CH_2)_3\overset{\overset{\displaystyle O}{\|}}{C}CH_3 \qquad (105)$$

must be involved in neighboring group participation. Vinylacetylenes specifically hydrate on the carbon further removed from the vinyl group (Eq. 106) [366]. This reaction is accompanied by olefin isomerization to the

$$H_2C=CHC\equiv CR \longrightarrow CH_3CH=CH\overset{\overset{\displaystyle O}{\|}}{C}R \qquad (106)$$

more stable conjugated ketone. The hydration of unsymmetrical dialkenylacetylenes tends to afford a variety of products and is not synthetically useful [376, 385]. Except for propargyl alcohol itself, which gives hydroxyacetone [352], substituted propargylic alcohols or acetates usually hydrate on the carbon further removed from the oxygen group and sometimes undergo elimination to the α,β-unsaturated ketone (Eq. 107) [364, 395, 396]. In enynols

$$RC\equiv CCH_2OH(OAc) \longrightarrow R\overset{\overset{\displaystyle O}{\|}}{C}CH_2CH_2OH(OAc) \longrightarrow R\overset{\overset{\displaystyle O}{\|}}{C}CH=CH_2 \qquad (107)$$

in which the directive effects of the vinyl and alcohol groups oppose each other, hydration next to the vinyl group predominates (Eq. 108) [376–378, 386–388, 406, 411]. Depending on the reaction conditions, the resulting product can undergo dehydration, cyclization or other subsequent reactions.

$$R\overset{\overset{\displaystyle OH}{|}}{C}HC\equiv CCH=CH_2 \longrightarrow R\overset{\overset{\displaystyle OH}{|}}{C}HCH_2\overset{\overset{\displaystyle O}{\|}}{C}CH=CH_2 \qquad (108)$$

Aryl groups are observed to affect hydration similarly (Eq. 109) [395, 423]. More remote hydroxy groups are observed to have an interesting directive effect as seen by comparing the following examples (Eqs. 110, 111) [405]. Diynones afford good yields of the corresponding 4-pyranones (Eq. 112) [416].

146

$$C_6H_5C\equiv CCH_2OH(OAc) \longrightarrow C_6H_5\overset{O}{\overset{\|}{C}}CH_2CH_2OH(OAc) \qquad (109)$$

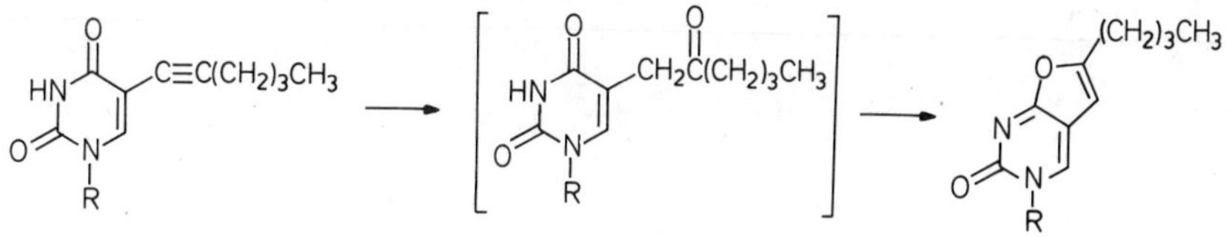

$$(110)$$

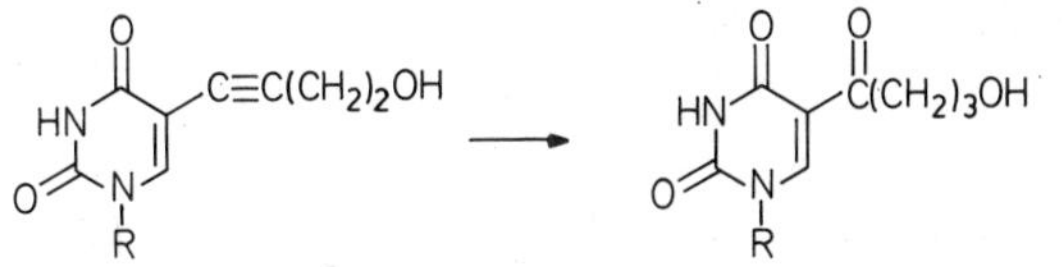

$$(111)$$

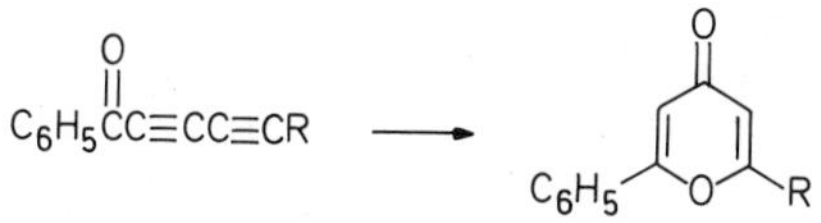

$$(112)$$

The mercury-promoted hydration of acetylenes is tolerant of a wide variety of functional groups as evident in Table 2.7. All of the following functional groups have been accomodated by this process: alkene, alcohol, conjugated diene, ester, ether, imide, α,β-unsaturated ketone, ketone, sulfide and sulfoxide. This is not to say that under the reaction conditions rearrangements and further reactions have not been observed. Remote double bonds have been observed to isomerize into conjugation with the new carbonyl group (Eqs. 113, 114) or undergo alcohol addition (Eq. 115). While numerous

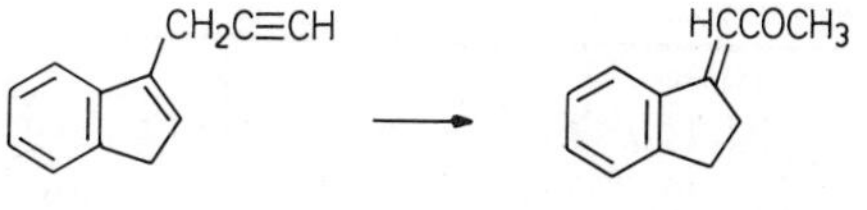

$$(113) \; [417]$$

$$RCH=CHCHC\equiv CH \longrightarrow RCH_2CH_2\overset{O}{\overset{\|}{C}}-\overset{O}{\overset{\|}{C}}CH_3 \qquad (114) \; [367]$$

$$H_2C=CHC\equiv CH \xrightarrow[ROH]{H_2SO_4\,/\,HgSO_4} ROCH_2CH_2\overset{O}{\overset{\|}{C}}CH_3 \qquad (115) \; [363]$$

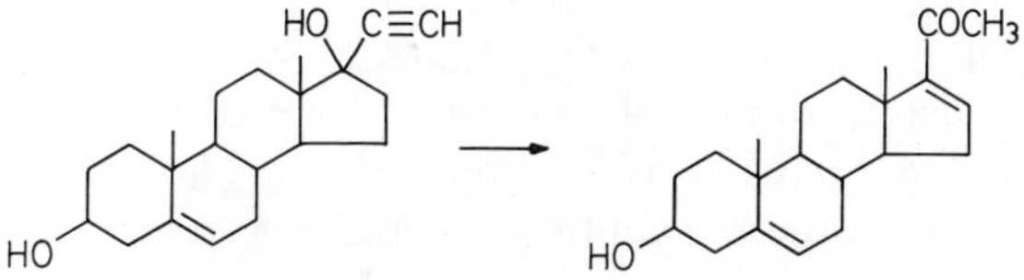

$$(116) \; [429]$$

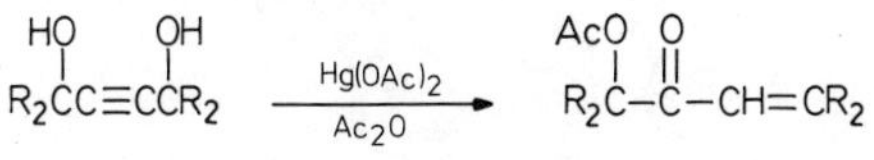

$$R_2\overset{HO}{\underset{|}{C}}C\equiv C\overset{OH}{\underset{|}{C}}R_2 \xrightarrow[Ac_2O]{Hg(OAc)_2} R_2\overset{AcO}{\underset{|}{C}}-\overset{O}{\overset{\|}{C}}-CH=CR_2 \qquad (117) \; [365,\,384]$$

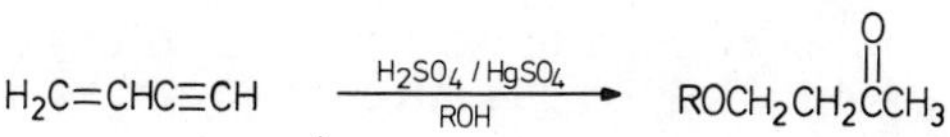

propargylic alcohols have been hydrated to hydroxyketones, dehydration to enones has also been observed (Eqs. 116, 117) [400]. As noted, esterification of such alcohols can also occur in the presence of carboxylic acid anhydrides [365, 374, 384].

Skeletal rearrangements have also been observed. Cyclization to furans and benzopyrans has been reported (Eqs. 118–120). The initial hydration

$$HC\equiv CCH=CHCHCH_3 \longrightarrow \qquad (118)\ [367]$$

$$\qquad (119)\ [403]$$

$$\qquad (120)\ [420]$$

$$\qquad (121)\ [422]$$

product can also undergo subsequent aldol condensations (Eq. 121) [413, 422] or other rearrangements [426, 432, 433, 436].

Little is known about the mechanism of these hydration reactions. The mercuric sulfate/aqueous sulfuric acid catalyzed hydration of acetylene to acetaldehyde has received the most mechanistic attention [437]. The reaction is reported to be first order in acetylene [348, 438, 439] and second order with respect to mercuric sulfate [438]. The rate-determining step is claimed to be heterogeneous hydration on the surface of a complex catalyst [348]. Mechanisms involving an initial mercury-acetylene π-complex [440], two intermediate complexes [350], and other variations [438, 441] have all been proposed, but no conclusive proof of any of these mechanisms has yet been provided. Inconclusive mechanistic studies of the hydration of phenylacetylene [397] and 2-butyn-1-ol [364] have also been reported. In spite of the paucity of mechanistic information on these reactions, the mercury-promoted hydration of acetylenes remains one of the more important applications of mercury in organic synthesis.

References

1. Wright, G. F.: Ann. N.Y. Acad. Sci., *65*, 436 (1957).
2. Chatt, J.: Chem. Rev., *48*, 7 (1951).
3. Zefirov, N. S.: Usp. Khim., *34*, 1272 (1965); Russ. Chem. Rev., *34*, 527 (1965).
4. Kitching, W.: Organometal. Chem. Rev. A, *3*, 61 (1968).
5. Fahey, R. C.: Top. Stereochem., *3*, 237 (1968).
6. Chandra, G., Muthana, M. S., Devaprabhakara, D.: J. Sci. Ind. Res. *30*, 333 (1971).
7. Kucherov, M. G.: Zh. Russ. Fiz.-Khim. Obshch., *24*, 330 (1892).
8. Hofmann, K. A., Sand, J.: Ber. Dt. Chem. Ges., *33*, 1340 (1900).
9. Sand, J., Hofmann, K. A.: Ber. Dt. Chem. Ges., *33*, 1353 (1900).
10. Sand, J., Hofmann, K. A.: Ber. Dt. Chem. Ges., *33*, 1358 (1900).
11. Aranda, V. G., Pastor, O. S.: Combustibles (Zaragoza), *10*, 183 (1950); Chem. Abstr., *46*, 423h (1952).
12. Aranda, V. G., Pastor, O. S., Cordon, J. L. M.: Combustibles (Zaragoza), *15*, 102 (1955); Chem. Zentralbl., 8102 (1959).
13. Sand, J.: J. Liebig Ann. Chem., *329*, 135 (1903).
14. Kreevoy, M. M., Stokker, G., Kretchmer, R. A., Ahmed, A. K.: J. Org. Chem., *28*, 3184 (1963).
15. Saito, Y., Matsuo, M.: J. Chem. Soc., Chem. Commun., 961 (1967).
16. Sand, J.: Ber. Dt. Chem. Ges., *34*, 1385 (1901).
17. Hofmann, K. A., Sand, J.: Ber. Dt. Chem. Ges., *33*, 2692 (1900).
18. Cotton, F. A., Leto, J. R.: J. Am. Chem. Soc., *80*, 4823 (1958).
19. Sanz Pastor, O.: Rev. Acad. Cienc. Exact., Fis.-Quim. Natur. Zaragoza, 6, 27 (1951); Chem. Abstr., *47*, 2688h (1953).
20. Kitching, W., Smith, A. J., Wells, P. R.: Austral. J. Chem., *21*, 2395 (1968).
21. Biilmann, E.: Ber. Dt. Chem. Ges., *33*, 1641 (1900).
22. Halpern, J., Tinker, H. B.: J. Am. Chem. Soc., *89*, 6427 (1967).
23. Keller, H.: Ger. Patent 967,765 (1957); Chem. Abstr., *53*, 13057c (1959).
24. Cordon, J. L. M., Aranda, V. G.: Combustibles (Zaragoza), *16*, 214 (1956); Chem. Zentralbl., 6977 (1957).
25. Arzoumanian, H., Aune, J. P., Guitard, J., Metzger, J.: J. Org. Chem., *39*, 3445 (1974).
26. Martinez Cordon, J. L.: Combustibles (Zaragoza), *14*, 229 (1954); Chem. Abstr., *49*, 8793d (1955).
27. Saito, Y., Matsuo, M.: J. Organometal. Chem., *10*, 524 (1967).
28. Shinoda, S., Saito, Y.: Bull. Chem. Soc. Japan, *47*, 1868 (1974).
29. Shinoda, S., Saito, Y.: J. Organometal. Chem., *90*, 1 (1975).
30. Sand, J., Genssler, O.: Ber. Dt. Chem. Ges., *36*, 3699 (1903).
31. Kreevoy, M. M.: J. Am. Chem. Soc., *81*, 1099 (1959).
32. Schaleger, L. L., Turner, M. A., Chamberlin, T. C., Kreevoy, M. M.: J. Org. Chem., *27*, 3421 (1962).
33. Spengler, G., Wilderotter, M., Trommer, T.: Brennstoff-Chemie, *45*, 182 (1964).
34. Pasto, D. J., Gontarz, J. A.: J. Am. Chem. Soc., *91*, 719 (1969).
35. Thomas, M. H., Wetmore, F. E. W.: J. Am. Chem. Soc., *63*, 136 (1941).
36. Matsuo, M., Saito, Y.: J. Org. Chem., *37*, 3350 (1972).
37. Shinoda, S., Saito, Y.: Bull. Chem. Soc. Japan, *47*, 2948 (1974).
38. Kreevoy, M. M., Schaleger, L. L., Ware, J. C.: Trans. Faraday Soc., *58*, 2433 (1962).
39. Brown, H. C., Geoghegan, Jr., P. J.: J. Org. Chem., *35*, 1844 (1970).
40. Brown, H. C., Geoghegan, Jr., P. J.: J. Am. Chem. Soc., *89*, 1522 (1967).

41. Brown, H. C., Geoghegan, Jr., P. J., Kurek, J. T.: J. Org. Chem., *46*, 3810 (1981).
42. Lindley, J., Mason, T. J.: Educ. Chem., *18*, 78 (1981); Chem. Abstr., *95*, 96213g (1981).
43. Nishida, S., Fujioka, T., Shimizu, N.: J. Organometal. Chem., *156*, 37 (1978).
44. Rodeheaver, G. T., Hunt, D. F.: J. Chem. Soc. D, Chem. Commun., 818 (1971).
45. Carlson, R. M., Funk, A. H.: Tetrahedron Lett., 3661 (1971).
46. Rogers, H. R., McDermott, J. X., Whitesides, G. M.: J. Org. Chem., *40*, 3577 (1975).
47. Hofman, J., Kubešová, J.: Coll. Czech. Chem. Commun., *43*, 404 (1978).
48. Levina, R. Ya., Gladshteyn, B. M.: Dokl. Akad. Nauk SSSR, *71*, 65 (1950); Chem. Abstr., *44*, 8321c (1950).
49. Lethbridge, A., Norman, R. O. C., Thomas, C. B.: J. Chem. Soc., Perkin I, 231 (1975).
50. Nesmeyanov, A. N., Freidlina, R. Kh.: Zh. Obshch. Khim., *7*, 2748 (1937); Chem. Abstr., *32*, 2912[7] (1938).
51. Sugita, T., Yamasaki, Y., Itoh, O., Ichikawa, K.: Bull. Chem. Soc. Japan, *47*, 1945 (1974).
52. Spengler, G., Weber, A.: Brennstoff-Chemie, *40*, 22 (1959).
53. Nesmeyanov, A. N., Lutsenko, I. F.: Izv. Akad. Nauk SSSR, Otdel. Khim. Nauk, 366 (1942); Chem. Abstr., *39*, 1637[1] (1945).
54. Lermontov, S. A., Belikova, N. A., Skornyakova, T. G., Pekhk, T. I., Lippmaa, É. T., Platé, A. F.: Zh. Org. Khim., *16*, 2322 (1980); J. Org. Chem. USSR, *16*, 1982 (1980).
55. Nesmeyanov, A. N., Pecherskaya, K. A.: Izv. Akad. Nauk SSSR, Otdel. Khim. Nauk, 67 (1941); Chem. Abstr., *37*, 3416[8] (1943).
56. Shearer, D. A., Wright, G. F.: Can. J. Chem., *33*, 1002 (1955).
57. Levina, R. Ya., Kostin, V. N., Gembitskii, P. A., Shostakovskii, S. M., Treshchova, E. G.: Zh. Obshch. Khim., *31*, 1185 (1961); J. Gen. Chem. USSR, *31*, 1096 (1961).
58. Olah, G. A., Westerman, P. W., Nishimura, J.: J. Am. Chem. Soc., *96*, 3548 (1974).
59. Coxon, J. M., Hartshorn, M. P., Lewis, A. J.: Tetrahedron, *26*, 3755 (1970).
60. Patai, S., Harnik, M., Hoffmann, E.: J. Am. Chem. Soc., *72*, 923 (1950).
61. Rodgman, A., Wright, G. F.: J. Am. Chem. Soc., *76*, 1382 (1954).
62. Polishchuk, V. R., German, L. S., Knunyants, I. L.: Izv. Akad. Nauk SSSR, Ser. Khim., 2024 (1971); Bull. Acad. Sci. USSR, Div. Chem. Sci., 1908 (1971).
63. Bäckvall, J. E., Åkermark, B., Ljunggren, S. O.: J. Am. Chem. Soc., *101*, 2411 (1979).
64. Tilander, B.: Arkiv för Kemi, *25*, 459 (1966).
65. Biilmann, E.: Ber. Dt. Chem. Soc., *35*, 2571 (1902).
66. Brown, H. C., Lynch, G. J.: J. Org. Chem., *46*, 930 (1981).
67. Arzoumanian, H., Metzger, J.: J. Organometal. Chem., *57*, C1 (1973).
68. Brown, H. C., Geoghegan, Jr., P. J., Kurek, J. T., Lynch, G. J.: Organometal. Chem. Syn., *1*, 7 (1970).
69. Biilmann, E.: Ber. Dt. Chem. Ges., *43*, 568 (1910).
70. Lutsenko, I. F., Badenkova, L. P., Foss, V. L.: Zh. Obshch. Khim., *27*, 3261 (1957); J. Gen. Chem. USSR, *27*, 3296 (1957).
71. Nesmeyanov, A. N., Lutsenko, I. F., Tumanova, Z. M.: Izv. Akad. Nauk SSSR, Otdel. Khim. Nauk, 601 (1949); Chem. Abstr., *44*, 7225c (1950).
72. Curtin, D. Y., Hurwitz, M. J.: J. Am. Chem. Soc., *74*, 5381 (1952).

73. Lecolier, S., Malfroot, T., Piteau, M., Senet, J. P.: Eur. Patent Appl. 2, 973 (1979); Chem. Abstr., *92*, 6698e (1980).

74. Miescher, K., Hoffmann, K.: U.S. Patent 2,156,598 (1939); Chem. Abstr., *33*, 6001^2 (1939).

75. Berkes, I., Jovanovic, V., Zivanov-Stakic, D.: Nucl.-Med., *14*, 374 (1975); Chem. Abstr., *84*, 180366q (1976).

76. Buriánek, J., Cifka, J.: J. Label. Compounds, *6*, 224 (1970).

77. Rowland, R. L.: U.S. Patent 2,635,982 (1953); Chem. Abstr., *48*, 6460e (1954).

78. Nesmeyanov, A. N., Lutsenko, I. F., Vereshchagina, N. N.: Izv. Akad. Nauk SSSR, Otdel. Khim. Nauk, 63 (1947); Chem. Abstr., *42*, 4148i (1948).

79. Lutsenko, I. F., Khomutov, R. M.: Dokl. Akad. Nauk SSSR, *102*, 97 (1955); Chem. Abstr., *50*, 4773b (1956).

80. Novikov, S. S., Tartakovskii, V. A., Godovikova, T. I., Gribov, B. G.: Izv. Akad. Nauk SSSR, Otdel. Khim. Nauk, 276 (1962); Bull. Acad. Sci. USSR, Div. Chem. Sci., 254 (1962).

81. Whitehead, C. W., Traverso, J. J.: J. Am. Chem. Soc., *80*, 2182 (1958).

82. Brown, H. C., Peters, E. N., Ravindranathan, M.: J. Am. Chem. Soc., *99*, 505 (1977).

83. Verkade, P. E.: Rec. Trav. Chim. Pays-Bas, *45*, 475 (1926).

84. Soderquist, J. A., Thompson, K. L.: J. Organometal. Chem., *159*, 237 (1978).

85. Abley, P., Byrd, J. E., Halpern, J.: J. Am. Chem. Soc., *94*, 1985 (1972).

86. Byrd, J. E., Halpern, J.: J. Chem. Soc. D, Chem. Commun., 1332 (1970)

87. Maskens, K., Polgar, N.: J. Chem. Soc., Perkin I, 109 (1973)

88. Brit. Patent 790,906 (1958); Chem. Abstr., *52*, 16213b (1958).

89. Seyferth, D., Kahlen, N.: Z. Naturforsch., *14***B**, 137 (1959).

90. Hodjat-Kachani, H., Lattes, A., Périé, J. J., Roussel, J.: J. Organometal. Chem., *96*, 175 (1975).

91. Shapiro, S. L., Parrino, V. A., Freedman, L.: J. Am. Pharm. Assoc., Sci. Ed., *46*, 689 (1957).

92. Jernow, J. L., Gray, D., Closson, W. D.: J. Org. Chem., *36*, 3511 (1971).

93. Lutsenko, I. F., Foss, V. L.: Dokl. Akad. Nauk SSSR, *98*, 407 (1954); Chem. Abstr., *49*, 12276i (1955).

94. Werner, L. H., Scholz, C. R.: J. Am. Chem. Soc., *76*, 2453 (1954).

95. Szekeres, L.: Gazz. Chim. Ital. *79*, 56 (1949); Chem. Abstr., *43*, 7921f (1949).

96. Brit. Patent 692,953 (1953); Chem. Abstr., *48*, 10056h (1954).

97. Negoro, K.: J. Chem. Soc. Japan, Pure Chem. Sec., *73*, 10 (1952); Chem. Abstr., *47*, 9929f (1953).

98. Hartmann, M., Panizzon, L.: U.S. Patent 2,136,501 (1938); Chem. Abstr., *33*, 1882^2 (1939).

99. Werner, L. H., Scholz, C. R.: U.S. Patent 2,790,816 (1957); Chem. Abstr., *51*, 16527e (1957).

100. Scholz, C. R., Werner, L. H.: Ger. Patent 946,539 (1956); Chem. Abstr., *53*, 12208c (1959).

101. Brook, A. G., Rodgman, A., Wright, G. F.: J. Org. Chem., *17*, 988 (1952).

102. Balbiano, L., Paolini, V.: Ber. Dt. Chem. Ges., *35*, 2994 (1902).

103. Balbiano, L.: Gazz. Chim. Ital., *36***I**, 237 (1906); Chem. Zentralbl. II, *77*, 119 (1906).

104. Balbiano, L., Paolini, V.: Ber. Dt. Chem. Ges., *36*, 3575 (1903).

105. Manchot, W., Bössenecker, F., Mährlein, F.: J. Liebig Ann. Chem., *421*, 316 (1920).

II. Hydroxymercuration

106. Nesmeyanov, A. N., Lutsenko, I. F., Khomutov, R. M.: Izv. Akad. Nauk SSSR, Otdel. Khim. Nauk, 217 (1960); Bull. Acad. Sci. USSR, Div. Chem. Sci., 200 (1960).

107. Ukai, T., Hayashi, M., Abe, H.: J. Pharm. Soc. Japan, *57*, 28 (1937); Chem. Abstr., *31*, 3876[4] (1937).

108. Adams, R., Roman, F. L., Sperry, W. N.: J. Am. Chem. Soc., *44*, 1781 (1922).

109. Summerbell, R. K., Kalb, G. H., Graham, E. S., Allred, A. L.: J. Org. Chem., *27*, 4461 (1962).

110. Shamshurin, A. A.: Zh. Obshch. Khim., *16*, 99 (1946); Chem. Abstr., *41*, 104d (1947).

111. Baker, R., Evans, D. A., McDowell, P. G.: J. Chem. Soc., Chem. Commun., 111 (1977).

112. Brieger, G., Burrows, E. P.: J. Agr. Food Chem., *20*, 1010 (1972).

113. Vig, O. P., Bhatia, M. S., Khurana, A. L., Matta, K. L.: J. Indian Chem. Soc., *46*, 265 (1969).

114. Iritani, N., Misumi, M., Hagiwara, Y.: Yakugaku Zasshi, *77*, 1036 (1957); Chem. Abstr., *52*, 2743e (1958).

115. Arbuzov, B. A., Shaikhutdinov, V. A., Isaeva, Z. G.: Izv. Akad. Nauk SSSR, Ser. Khim., 2124 (1972); Bull. Acad. Sci. USSR, Div. Chem. Sci., 2070 (1972).

116. Johnson, C. R., Elliott, R. C., Meanwell, N. A.: Tetrahedron Lett., 5005 (1982).

117. Ehrhardt, G., Ruschig, H., Leditschke, H.: U.S. Patent 2,695,305 (1954); Chem. Abstr., *49*, 15969g (1955).

118. Whitehead, C. W.: J. Am. Chem. Soc., *80*, 2178 (1958).

119. Yanami, T., Miyashita, M., Yoshikoshi, A.: J. Chem. Soc., Chem. Commun., 525 (1979).

120. Cann, P. F., Warren, S.: J. Chem. Soc. D, Chem. Commun., 1026 (1970).

121. Czernecki, S., Georgoulis, C., Provelenghiou, C.: Tetrahedron Lett., 2623 (1975).

122. Valcavi, U.: Swiss Patent 620,428 (1980); Chem. Abstr., *94*, 174456j (1981).

123. Arambewela, L. S. R., Khuong-Huu, F.: Phytochemistry, *20*, 349 (1981).

124. Gunstone, F. D., Inglis, R. P.: Chem. Phys. Lipids, *10*, 73 (1973).

125. Gunstone, F. D., Inglis, R. P.: Chem. Phys. Lipids, *10*, 89 (1973).

126. Ochi, K., Matsunaga, I., Shindo, M., Kaneko, C.: Jpn. Kokai Tokkyo Koho 78,144,562 (1978); Chem. Abstr., *92*, 164175m (1980).

127. Traylor, T. G.: Acct. Chem. Res., *2*, 152 (1969).

128. Waters, W. L., Traylor, T. G., Factor, A.: J. Org. Chem., *38*, 2306 (1973).

129. Abley, P., Byrd, J. E., Halpern, J.: J. Am. Chem. Soc., *95*, 2591 (1973).

130. Byrd, J. E., Cassar, L., Eaton, P. E., Halpern, J.: J. Chem. Soc. D, Chem. Commun., 40 (1971).

131. Traylor, T. G., Baker, A. W.: J. Am. Chem. Soc., *85*, 2746 (1963).

132. Babb, R. M., Gardner, P. D.: J. Chem. Soc., Chem. Commun., 1678 (1968).

133. Brown, H. C., Lynch, G. J., Hammar, W. J., Liu, L. C.: J. Org. Chem., *44*, 1910 (1979).

134. Nesmeyanov, A. N., Freidlina, R. Kh.: Ber. Dt. Chem. Ges., *69*, 1631 (1936).

135. Nesmeyanov, A. N., Freidlina, R. Kh.: Zh. Obshch. Khim., *7*, 43 (1937).

136. Rodgman, A., Shearer, D. A., Wright, G. F.: Can. J. Chem., *35*, 1377 (1957).

137. Berg, O. W., Lay, W. P., Rodgman, A., Wright, G. F.: Can. J. Chem., *36*, 358 (1958).

138. Hill, C. L., Whitesides, G. M.: J. Am. Chem. Soc., *96*, 870 (1974).

139. Georgoulis, C., Valéry, J.-M.: Bull. Soc. Chim. France, 178 (1974).
140. Anderson, M. M., Henry, P. M.: Chem. Ind., 2053 (1961).
141. Moon, S., Waxman, B. H.: J. Chem. Soc., Chem. Commun., 1283 (1967).
142. Johnson, M. R., Rickborn, B.: J. Org. Chem., *34*, 2781 (1969).
143. Moon, S., Ganz, C., Waxman, B. H.: J. Chem. Soc. D, Chem. Commun., 866 (1969).
144. Johnson, M. R., Rickborn, B.: J. Chem. Soc., Chem. Commun., 1073 (1968).
145. Henbest, H. B., Nicholls, B.: J. Chem. Soc., 227 (1959).
146. Gonzalez, F., Lesage, S., Perlin, A. S.: Carbohydr. Res., *42*, 267 (1975).
147. Senda, Y., Kamiyama, S., Imaizumi, S.: J. Chem. Soc., Perkin I, 530 (1978).
148. Brown, H. C., Hammar, W. J.: J. Am. Chem. Soc., *89*, 1524 (1967).
149. Sisti, A. J.: J. Org. Chem., *33*, 3953 (1968).
150. Park, W. R. R., Wright, G. F.: Can. J. Chem., *35*, 1088 (1957).
151. Jerkunica, J. M., Traylor, T. G.: Org. Syn., *53*, 94 (1973).
152. Barrelle, M., Apparu, M.: Bull. Soc. Chim. France, 2016 (1972).
153. Smith, R. G., Ensley, H. E., Smith, H. E.: J. Org. Chem., *37*, 4430 (1972).
154. Krow, G. R., Fan, D. M.: J. Org. Chem., *39*, 2674 (1974).
155. Jasserand, D., Granger, R., Girard, J.-P., Chapat, J.-P.: Compt. Rend. C, *272*, 1693 (1971).
156. Jasserand, D., Girard, J. P., Rossi, J. C., Granger, R.: Tetrahedron, *32*, 1535 (1976).
157. Pasto, D. J., Gontarz, J. A.: J. Am. Chem. Soc., *92*, 7480 (1970).
158. Pasto, D. J., Gontarz, J. A.: J. Am. Chem. Soc., *93*, 6909 (1971).
159. Sokolov, V. I., Troitskaya, L. L., Reutov, O. A.: Dokl. Akad. Nauk SSSR, *166*, 136 (1966); Bull. Acad. Sci. USSR, Chem. Sec., *166*, 45 (1966).
160. Carey, F. A., Frank, W. C.: J. Org. Chem., *47*, 3548 (1982).
161. Banaszek, A.: Bull. Acad. Polon. Sci., Ser. Sci. Chim., *22*, 1045 (1974).
162. Moon, S., Waxman, B. H.: J. Org. Chem., *34*, 288 (1969).
163. Fry, A. J., Simon, J. A.: J. Org. Chem., *47*, 5032 (1982).
164. Musker, W. K., Larson, G. L.: Tetrahedron Lett., 3481 (1968).
165. Misra, S. C., Chandra, G.: Indian J. Chem., *13*, 1239 (1975).
166. Pasto, D. J., Gontarz, J. A.: J. Am. Chem. Soc., *93*, 6902 (1971).
167. Chamberlain, P., Witham, G. H.: J. Chem. Soc. B, 1382 (1970).
168. Bentham, S., Chamberlain, P., Whitham, G. H.: J. Chem. Soc. D, Chem. Commun., 1528 (1970).
169. Henbest, H. B., McElhinney, R. S.: J. Chem. Soc., 1834 (1959).
170. Klein, J., Levene, R.: Tetrahedron Lett., 4833 (1969).
171. Coxon, J. M., Hartshorn, M. P., Mitchell, J. W., Richards, K. E.: Chem. Ind., 652 (1968).
172. Bambagiotti A., M., Vincieri, F. F., Coran, S. A.: J. Org. Chem., *39*, 680 (1974).
173. Brook, A. G., Wright, G. F.: J. Org. Chem., *22*, 1314 (1957).
174. Schrauth, W., Bauerschmidt, H.: Ber. Dt. Chem. Ges., *47*, 2736 (1914).
175. Suen, Y.-H., Kagan, H.: Bull. Soc. Chim. France, 2270 (1970).
176. Jarosz, S., Hicks, D. R., Fraser-Reid, B.: J. Org. Chem., *47*, 935 (1982).
177. Newman, M. S., Vander Zwan, M. C.: J. Org. Chem., *39*, 1186 (1974).
178. Misumi, S., Ohtsuka, T., Ohfune, Y., Sugita, K., Shirahama, H., Matsumoto, T.: Tetrahedron Lett., 31 (1979).
179. Ferrier, R. J.: J. Chem. Soc., Perkin I, 1455 (1979).
180. Ferrier, R. J., Prasit, P.: Carbohydr. Res., *82*, 263 (1980).
181. Bond, F. T.: J. Am. Chem. Soc., *90*, 5326 (1968).
182. Tobler, E., Foster, D. J.: Helv. Chim. Acta, *48*, 366 (1965).
183. Brown, H. C., Kawakami, J. H.: J. Am. Chem. Soc., *95*, 8665 (1973).

II. Hydroxymercuration

184. Brown, H. C., Liu, K.-T.: J. Am. Chem. Soc., *93*, 7335 (1971).
185. Brown, H. C., Kawakami, J. H., Ikegami, S.: J. Am. Chem. Soc., *89*, 1525 (1967).
186. Brown, H. C., Hammar, W. J., Kawakami, J. H., Rothberg, I., Vander Jagt, D. L.: J. Am. Chem. Soc., *89*, 6381 (1967).
187. Aberchrombie, M. J., Rodgman, A., Bharucha, K. R., Wright, G. F.: Can. J. Chem., *37*, 1328 (1959).
188. Zefirov, N. S., Prikazchikova, L. P., Yur'ev, Yu. K.: Dokl. Akad. Nauk SSSR, *152*, 869 (1963); Proc. Acad. Sci. USSR, Chem. Sec., *152*, 774 (1963).
189. Olah, G. A., Clifford, P. R.: J. Am. Chem. Soc., *95*, 6067 (1973).
190. Factor, A., Traylor, T. G.: J. Org. Chem., *33*, 2607 (1968).
191. Traylor, T. G., Baker, A. W.: Tetrahedron Lett., *19*, 14 (1959).
192. Baird, Jr., W. C., Buza, M.: J. Org. Chem., *33*, 4105 (1968).
193. Apeloig, Y., Arad, D., Lenoir, D., Schleyer, P. von R.: Tetrahedron Lett., 879 (1981).
194. Clark, R. L., Gubitz, F. W.: U.S. Patent 2,921,068 (1960); Chem. Abstr., *54*, 8867c (1960).
195. Brown, H. C., Hammar, W. J.: Tetrahedron, *34*, 3405 (1978).
196. Belikova, N. A., Lermontov, S. A., Pekhk, T. I., Lippmaa, É. T., Platé, A. F.: Zh. Org. Khim., *14*, 2273 (1978); J. Org. Chem. USSR, *14*, 2101 (1978).
197. Traylor, T. G.: J. Am. Chem. Soc., *86*, 244 (1964).
198. Ganter, C., Zwahlen, W.: Helv. Chim. Acta, *54*, 2628 (1971).
199. McNeely, K. H., Rodgman, A., Wright, G. F.: J. Org. Chem., *20*, 714 (1955).
200. Goering, H. L., Brown, C., Chang, S., Clevenger, J. V., Humski, K.: J. Org. Chem., *34*, 624 (1969).
201. Ayral-Kaloustian, S., Agosta, W. C.: J. Am. Chem. Soc., *102*, 314 (1980).
202. Deussen, E.: J. Prakt. Chem., *114*, 63 (1926).
203. Straus, F., Lemmel, L.: Ber. Dt. Chem. Ges., *54*, 25 (1921).
204. Zefirov, N. S., Kadzyaukas, P. P., Yur'ev, Yu. K., Bazanova, V. N.: Zh. Obshch. Khim., *35*, 1499 (1965); J. Gen. Chem. USSR, *35*, 1501 (1965).
205. Inman, R. C., Servé, M. P.: J. Org. Chem., *47*, 4348 (1982).
206. Corey, E. J., Glass, R. S.: J. Am. Chem. Soc., *89*, 2600 (1967).
207. Krow, G., Rodebaugh, R., Grippi, M., Carmosin, R.: Syn. Commun., *2*, 211 (1972).
208. Bluthe, N., Ecoto, J., Fetizon, M., Lazare, S.: J. Chem. Soc., Perkin I, 1747 (1980).
209. Henderson, G. G., Agnew, J. W.: J. Chem. Soc., *95*, 289 (1909).
210. Henderson, G. G., Eastburn, W. J. S.: J. Chem. Soc., *95*, 1465 (1909).
211. Balbiano, L.: Ber. Dt. Chem. Ges., *48*, 394 (1915).
212. Arbusov, B. A., Ratner, V. V., Isaeva, Z. G., Kazakova, E. Kh.: Izv. Akad. Nauk SSSR, Ser. Khim., 385 (1972); Bull. Acad. Sci. USSR, Div. Chem. Sci., 334 (1972).
213. Cocker, W., Hanna, D. P., Shannon, P. V. R.: J. Chem. Soc. C, 1302 (1969).
214. Arbusov, B. A., Ratner, V. V., Isaeva, Z. G., Belyaeva, M. G., Povodyreva, I. P.: Izv. Akad. Nauk SSSR, Ser. Khim., 1049 (1979); Bull. Acad. Sci. USSR, Div. Chem. Sci., 980 (1979).
215. Buinova, É. F., Yaremchenko, N. G., Urbanovich, T. R., Izotova, L. V.: Khim. Prir. Soedin., *5*, 646 (1979); Chem. Natural Compounds, *15*, 565 (1979).
216. Skorobogatova, E. V., Povelikina, L. N., Kartashov, V. R.: Zh. Org. Khim., *14*, 663 (1978); J. Org. Chem. USSR, *14*, 613 (1978).

217. Arbuzov, B. A., Isaeva, Z. G., Ratner, V. V.: Izv. Akad. Nauk SSSR, Ser. Khim., 1888 (1981); Chem. Abstr., *95*, 220159k (1981).
218. Takaishi, N., Fujikura, Y., Inamoto, Y.: J. Org. Chem., *40*, 3767 (1975).
219. Fujikura, Y., Inamoto, Y., Takaishi, N., Ikeda, H., Aigami, K.: J. Chem. Soc., Perkin I, 2133 (1976).
220. Baird, W. C., Jr., Surridge, J. H.: J. Org. Chem., *37*, 1182 (1972).
221. Carey, F. A., Tremper, H. S.: J. Org. Chem., *34*, 4 (1969).
222. Paquette, L. A., Bellamy, F., Wells, G. J., Böhm, M. C., Gleiter, R.: J. Am. Chem. Soc., *103*, 7122 (1981).
223. Williams, E. H., Szarek, W. A. and Jones, J. K. N.: Can. J. Chem., *47*, 4467 (1969).
224. Misra, S. C., Chandra, G.: Indian J. Chem., *11*, 613 (1973).
225. Moss, R. A., Chen, E. Y.: J. Org. Chem., *46*, 1466 (1981).
226. Sokolov, V. I.: Izv. Akad. Nauk SSSR, Ser. Khim., 1285 (1968); Bull. Acad. Sci. USSR, Div. Chem. Sci., 1213 (1968).
227. Clark-Lewis, J. W., McGarry, E. J.: Austral. J. Chem., *26*, 2447 (1973).
228. Rosenthal, A., Sprinzl, M.: Carbohydr. Res., *16*, 337 (1971).
229. Torrini, I., Romeo, A.: Tetrahedron Lett., 2605 (1975).
230. Holtmeier, W., Hobert, K., Welzel, P.: Tetrahedron Lett., 4709 (1976).
231. Herz, J. E., Gonzàlez, E.: Ciencia, *26*, 29 (1968); Chem. Abstr., *69*, 36347g (1968).
232. Holtmeier, W., Welzel, P.: Tetrahedron Lett., 3423 (1976).
233. Brandt, P., Plum, O.: Acta Chem. Scand., *7*, 97 (1953).
234. Manchot, W.: Ber. Dt. Chem. Ges., *53*, 984 (1920).
235. Manchot, W., Klüg, A.: J. Liebig Ann. Chem., *420*, 170 (1919).
236. Kitching, W., Smith, A. J., Wells, P. R.: J. Chem. Soc., Chem. Commun., 370 (1968).
237. Wells, P. R., Kitching, W.: Tetrahedron Lett., 1531 (1963).
238. Brown, H. C., Geoghegan, Jr., P. J.: J. Org. Chem., *37*, 1937 (1972).
239. Allen, E. R., Cartlidge, J., Taylor, M. M., Tipper, C. F. H.: J. Phys. Chem., *63*, 1442 (1959).
240. Sakaki, S., Kato, H., Kanai, H., Tarama, K.: Bull. Chem. Soc. Japan, *47*, 377 (1974).
241. Bach, R. D., Henneike, H. F.: J. Am. Chem. Soc., *92*, 5589 (1970).
242. Bach, R. D., Patane, J., Kevan, L.: J. Org. Chem., *40*, 257 (1975).
243. Byrd, J. E., Halpern, J.: J. Am. Chem. Soc., *92*, 6967 (1970).
244. Brown, H. C., Liu, K.-T.: J. Am. Chem. Soc., *92*, 3502 (1970).
245. Balyatinskaya, L. N.: Zh. Obshch. Khim., *47*, 198 (1977); J. Gen. Chem. USSR, *47*, 182 (1977).
246. Sand, J., Breest, F.: Z. Phys. Chem., *59*, 424 (1907).
247. Lucas, H. J., Hepner, F. R., Winstein, S.: J. Am. Chem. Soc., *61*, 3102 (1939).
248. Allen, E. R., Cartlidge, J., Taylor, M. M., Tipper, C. F. H.: J. Phys. Chem., *63*, 1437 (1959).
249. Temkin, O. N., Esikova, I. A., Mogilyanskii, A. I., Flid, R. M.: Kinet. Katal., *12*, 915 (1971); Kinetics and Catalysis (USSR), *12*, 817 (1971); Chem. Abstr., *76*, 13652h (1972).
250. Schaleger, L. L., Watamori, N.: J. Phys. Chem., *73*, 2011 (1969).
251. Thron, H., Dirscherl, W.: J. Liebig Ann. Chem., *515*, 252 (1935).
252. Balbiano, L.: Ber. Dt. Chem. Ges., *42*, 1502 (1909).
253. Kono, Y., Horioka, M.: Kyushu Yakugakkai Kaiho, *22*, 21 (1968); Chem. Abstr., *71*, 39113e (1969).
254. Kosaki, M., Shinoda, S., Saito, Y.: Bull. Chem. Soc. Japan, *48*, 3745 (1975).

255. Meyer, R., Gorrichon, L., Maroni, P.: J. Organometal. Chem., *129*, C7 (1977).
256. Mimoun, H., Charpentier, R., Mitschler, A., Fischer, J., Weiss, R.: J. Am. Chem. Soc., *102*, 1047 (1980).
257. Shostakovsky, M. F., Prilezhaeva, E. N.: Izv. Akad. Nauk SSSR, Otdel. Khim. Nauk, 517 (1954); Bull. Acad. Sci. USSR, Div. Chem. Sci., 439 (1954).
258. Shostakovsky, M. F., Prilezhaeva, E. N., Uvarova, N. I.: Izv. Akad. Nauk SSSR, Otdel. Khim. Nauk, 526 (1954); Bull. Acad. Sci. USSR, Div. Chem. Sci., 447 (1954); Chem. Abstr., *49*, 9483e (1955).
259. Nitta, M., Omata, A., Sugiyama, H.: Chem. Lett., 1615 (1980).
260. Barluenga, J., Concellon, J. M., Ara, A., Asensio, G., Yus, M.: An. Quim., *74*, 785 (1978); Chem. Abstr., *91*, 57131b (1979).
261. Kretchmer, R. A., Conrad, R. A., Mihelich, E. D.: J. Org. Chem., *38*, 1251 (1973).
262. Barluenga, J., Fañanás, F. J., Yus, M., Asensio, G.: Tetrahedron Lett., 2015 (1978).
263. Barluenga, J., Fañanás, F. J., Yus, M.: J. Org. Chem., *46*, 1281 (1981).
264. Barluenga, J., Fañanás, F. J., Yus, M.: J. Org. Chem., *44*, 4798 (1979).
265. Marquardt, R. P., Luce, E. N.: Anal. Chem., *20*, 751 (1948).
266. Jacobs, T. L., Johnson, R. N.: J. Am. Chem. Soc., *82*, 6397 (1960).
267. Summerbell, R. K., Forrester, S. R.: J. Org. Chem., *26*, 4834 (1961).
268. Goodman, L., Ross, L. O., Greene, M. O., Greenberg, J., Baker, B. R.: J. Med. Pharm. Chem., *3*, 65 (1961).
269. Aranda, V. G., Beltran, F. G., Mur, J. B.: Combustibles (Zaragoza), *25*, 3 (1967); Chem. Abstr., *68*, 114722w (1968).
270. Summerbell, R. K., Lestina, G. J.: J. Am. Chem. Soc., *79*, 3878 (1957).
271. Gelin. R., Gelin, S., Albrand, M.: Bull. Soc. Chim. France, 1946 (1972).
272. Brown, H. C., Geoghegan, P. J., Jr., Lynch, G. J., Kurek, J. T.: J. Org. Chem., *37*, 1941 (1972).
273. Moon, S., Takakis, J. M., Waxman, B. H.: J. Org. Chem., *34*, 2951 (1969).
274. Gómez-Aranda, V., Barluenga-Mur, J., Asensio-Aguilar, G.: Rev. Acad. Cienc. Exactas, Fis.-Quim. Natur. Zaragoza, *28*, 215 (1973); Chem. Abstr., *80*, 83174k (1974).
275. Hennion, G. F., Sheehan, J. J.: J. Am. Chem. Soc., *71*, 1964 (1949).
276. Nesmeyanov, A. N., Lutsenko, I. F.: Izv. Akad. Nauk SSSR, Otdel. Khim. Nauk, 296 (1943); Chem. Abstr., *38*, 5498^6 (1944).
277. Gómez-Aranda, V., Barluenga-Mur, J., Yus-Astiz, M.: Rev. Acad. Cienc. Exactas, Fis.-Quim. Natur. Zaragoza, *28*, 225 (1973); Chem. Abstr., *80*, 83164g (1974).
278. Summerbell, R. K., Lestina, G., Waite, H.: J. Am. Chem. Soc., *79*, 234 (1957).
279. Summerbell, R. K., Stephens, J. R.: J. Am. Chem. Soc., *76*, 731 (1954).
280. Summerbell, R. K., Poklacki, E. S.: J. Org. Chem., *27*, 2074 (1962).
281. Ash, A. S. F., Challenger, F., Greenwood, D.: J. Chem. Soc., 1877 (1951).
282. Badanyan, Sh. O., Pashayan, A. A., Arakelyan, S. V., Voskanyan, M. G.: Arm. Khim. Zh., *29*, 53 (1976); Chem. Abstr., *85*, 20518h (1976).
283. Reppe, W., Schlichting, O., Klager, K., Toepel, T.: J. Liebig Ann. Chem., *560*, 1 (1948).
284. Stetter, H., Meissner, H.-J., Last, W.-D.: Chem. Ber., *101*, 2889 (1968).
285. Stetter, H., Meissner, H.-J.: Tetrahedron Lett., 4599 (1966).
286. Link, C. M., Jansen, D. K., Sukenik, C. N.: J. Am. Chem. Soc., *102*, 7798 (1980).
287. Bloodworth, A. J., Khan, J. A., Loveitt, M. E.: J. Chem. Soc., Perkin I, 621 (1981).

288. Ganter, C., Wicker, K., Zwahlen, W., Schaffner-Sabba, K.: Helv. Chim. Acta, *53*, 1618 (1970).
289. Bordwell, F. G., Douglass, M. L.: J. Am. Chem. Soc., *88*, 993 (1966).
290. Belikova, N. A., Lermontov, S. A., Pekhk, T. I., Lippmaa, É. T., Platé, A. F.: Zh. Org. Khim., *14*, 884 (1978); J. Org. Chem. USSR, *14*, 823 (1978).
291. Belikova, N. A., Lermontov, S. A., Skornyakova, T. G., Pekhk, T. I., Lippmaa, É. T., Platé, A. F.: Zh. Org. Khim., *15*, 492 (1979); J. Org. Chem. USSR, *15*, 436 (1979).
292. Averina, N. V., Zefirov, N. S.: Zh. Org. Khim., *5*, 1991 (1969); J. Org. Chem. USSR, *5*, 1936 (1969).
293. Stetter, H., Schwartz, E. F.: Chem. Ber., *101*, 2464 (1968).
294. Sharma, R. K., Shoulders, B. A., Gardner, P. D.: J. Org. Chem., *32*, 241 (1967).
295. Mehta, G.: Org. Prep. Proc., *2*, 245 (1970).
296. Nagendrappa, G., Devaprabhakara, D.: Tetrahedron Lett., 4687 (1970).
297. Bergter, L., Seidl, P. R.: J. Org. Chem., *47*, 73 (1982).
298. Seidl, P., Roberts, M., Winstein, S.: J. Am. Chem. Soc., *93*, 4089 (1971).
299. Giua, M., Polla, G.: Chim. Ind. (Milan), *35*, 888 (1953).
300. Stille, J. K., Stinson, S. C.: Tetrahedron, *20*, 1387 (1964).
301. Brown, H. C., Rothberg, I., Vander Jagt, D. L.: J. Org. Chem., *37*, 4098 (1972).
302. Wilder, P., Jr., Portis, A. R., Jr., Wright, G. W., Shepherd, J. M.: J. Org. Chem., *39*, 1636 (1974).
303. Treibs, W., Lucius, G., Kögler, H., Breslauer, H.: J. Liebig Ann. Chem., *581*, 59 (1953).
304. Bambagiotti Alberti, M., Coran, S. A., Vincieri, F.: Riv. Ital. Essenze, Profumi, Piante Off., Aromat., Syndets, Saponi, Cosmet., Aerosols, *60*, 83 (1978); Chem. Abstr., *89*, 163754a (1978).
305. Bambagiotti A., M., Coran, S. A., Vincieri, F. F.: Int. Congr. Essent. Oils, [Pap.], 7th, *7*, 256 (1979); Chem. Abstr., *92*, 129091j (1980).
306. Matsuki, Y., Kodama, M., Itô, S.: Tetrahedron Lett., 4081 (1979).
307. Matsuki, Y., Kodama, M., Itô, S.: Tetrahedron Lett., 2901 (1979).
308. Moon, S., Wright, D. G., Schwartz, A. L.: J. Org. Chem., *41*, 1899 (1976).
309. Baird, W. C., Jr., Surridge, J. H.: J. Org. Chem., *37*, 304 (1972).
310. Sutter, J. K., Sukenik, C. N.: J. Org. Chem., *47*, 4174 (1982).
311. Stetter, H., Meissner, H.-J.: Chem. Ber., *96*, 2827 (1963).
312. Paquette, L. A., Thompson, G. L.: J. Am. Chem. Soc., *94*, 7118 (1972).
313. Inoue, Y., Tanimoto, F., Kitano, H.: Eur. Patent Appl., *36*, 147 (1981); Chem. Abstr., *96*, 69010c (1982).
314. Krow, G. R., Reilly, J.: Tetrahedron Lett., 3133 (1972).
315. Paquette, L. A., Lang, S. A., Jr., Short, M. R., Parkinson, B., Clardy, J.: Tetrahedron Lett., 3141 (1972).
316. Müller, E.: Chem. Ber., *106*, 3920 (1973).
317. Pasto, D. J., Miles, M. F.: J. Org. Chem., *41*, 425 (1976).
318. Mashraqui, S. H., Trivedi, G. K.: Indian J. Chem., *16B*, 849 (1978).
319. Tsankova, E., Ognyanov, I., Norin, T.: Tetrahedron, *36*, 669 (1980).
320. Agrawal, R. C., Trivedi, G. K., Bhattacharyya, S. C.: Indian J. Chem., *13*, 876 (1975).
321. Borg-Karlson, A. K., Norin, T., Wijekoon, W. M. D., Talvitie, A.: Finn. Chem. Lett., 151 (1979); Chem. Abstr., *92*, 164097n (1980).
322. Lie Ken Jie, M. S. F., Lam, C. H.: J. Chromatog., *129*, 181 (1976).
323. ApSimon, J. W., Krehm, H.: Can. J. Chem., *47*, 2865 (1969).

324. Kahovcová, J., Romanuk, M.: Coll. Czech. Chem. Commun., *46*, 1413 (1981).
325. Hook, J. M., Mander, L. N., Urech, R.: J. Am. Chem. Soc., *102*, 6629 (1980).
326. Leyes, G. A., Okamura, W. H.: J. Am. Chem. Soc., *104*, 6099 (1982).
327. Morisaki, M., Rubio-Lightbourn, J., Ikekawa, N.: Chem. Pharm. Bull., *21*, 457 (1973).
328. Morisaki, M., Bannai, K., Ikekawa, N.: Chem. Pharm. Bull., *24*, 1948 (1976).
329. Ikekawa, N., Morisaki, M., Bannai, K.: Japan. Patent 69,059 (1975); Chem. Abstr., *83*, 114757w (1975).
330. Miocque, M., Hung, N. M., Yen, V. Q.: Ann. Chim., Series 13, *8*, 157 (1963).
331. Kucherov, M.: Ber. Dt. Chem. Ges., *14*, 1540 (1881).
332. Kucherov, M.: Zh. Russ. Fiz.-Khim. Obshch., *13*, 542 (1881); Ber. Dt. Chem. Ges., *14*, 1532 (1881).
333. Kucherov, M.: Ber. Dt. Chem. Ges., *17*, 13 (1884).
334. Hofmann, K. A.: Ber. Dt. Chem. Ges., *31*, 2212 (1898).
335. Köthner, P.: Ber. Dt. Chem. Ges., *31*, 2475 (1898).
336. Hofmann, K. A.: Ber. Dt. Chem. Ges., *31*, 2783 (1898).
337. Hofmann, K. A.: Ber. Dt. Chem. Ges., *32*, 870 (1899).
338. Biltz, H., Mumm, O.: Ber. Dt. Chem. Ges., *37*, 4417 (1904).
339. Hofmann, K. A.: Ber. Dt. Chem. Ges., *37*, 4459 (1904).
340. Biltz, H.: Ber. Dt. Chem. Ges., *38*, 133 (1905).
341. Hofmann, K. A.: Ber. Dt. Chem. Ges., *38*, 1999 (1905).
342. Manchot, W., Haas, J.: J. Liebig Ann. Chem., *399*, 123 (1913).
343. Biltz, H., Reinkober, K.: J. Liebig Ann. Chem., *404*, 219 (1914).
344. Manchot, W.: J. Liebig Ann. Chem., *417*, 93 (1918).
345. Erdmann, H., Köthner, P.: Z. Anorg. Chem., *18*, 54 (1898).
346. Burkard, E., Travers, M. W.: J. Chem. Soc., *81*, 1272 (1902).
347. Johnson, G. W.: Brit. Patent 466,569 (1937); Chem. Abstr., *31*, 7892[8] (1937).
348. Schwabe, K., Berg, H.: Z. Phys. Chem., *203*, 383 (1954); Chem. Abstr., *49*, 5091f (1955).
349. Vogt, R. R., Nieuwland, J. A.: J. Am. Chem. Soc., *43*, 2071 (1921).
350. Kal'fus, M. K., Sokol'skii, D. V.: Vestnik Akad. Nauk Kazakh. SSSR, *13*, 74 (1957); Chem. Abstr., *52*, 2510c (1958).
351. Newman, M. S.: J. Am. Chem. Soc., *75*, 4740 (1953).
352. Newman, M. S.: U.S. Patent 2,853,520 (1958); Chem. Abstr., *54*, 345c (1960).
353. Olah, G. A., Meidar, D.: Synthesis, 671 (1978).
354. Henne, A. L., Nager, M.: J. Am. Chem. Soc., *74*, 650 (1952).
355. Haszeldine, R. N., Leedham, K.: J. Chem. Soc., 3483 (1952).
356. Hennion, G. F., Murray, W. S.: J. Am. Chem. Soc., *64*, 1220 (1942).
357. Haszeldine, R. N., Leedham, K.: J. Chem. Soc., 1261 (1954).
358. Arzoumanian, H., Bernard, J.-R., Coste, C., Knoche, H., Meallier, P.: Bull. Soc. Chim. France, 969 (1976).
359. Arzoumanian, H., Bernard, J.-R., Coste, C., Knoche, H., Meallier, P.: Bull. Soc. Chim. France, 974 (1976).
360. Carter, A. S.: U.S. Patent 1,896,161 (1933); Chem. Abstr., *27*, 2458 (1933).
361. Nieuwland, J. A., Calcott, W. S., Downing, F. B., Carter, A. S.: J. Am. Chem. Soc., *53*, 4197 (1931).
362. Conaway, R. F.: U.S. Patent 1,967,225 (1934); Chem. Abstr., *28*, 5834[6] (1934).
363. Nazarov, I. N., Vartanyan, S. A., Matsoyan, S. G., Zhamagortsyan, V. N.: Zh. Obshch. Khim., *23*, 1986 (1953); Chem. Abstr., *49*, 3002e (1955).
364. Szemenyei, D., Steichen, D., Byrd, J. E.: J. Mol. Catal., *2*, 105 (1977).

365. Lozac'h, N.: Bull. Soc. Chim. France, *11*, 514 (1944).
366. Kupin, B. S., Petrov, A. A.: Zh. Obshch. Khim., *29*, 2281 (1959); Chem. Abstr., *54*, 9722i (1960).
367. Heilbron, I. M., Jones, E. R. H., Smith, P., Weedon, B. C. L.: J. Chem. Soc., *54* (1946).
368. Hess, K., Munderloh, H.: Ber. Dt. Chem. Ges., *51*, 377 (1918).
369. Scheibler, H., Fischer, A.: Ber. Dt. Chem. Ges., *55*, 2903 (1922).
370. Brit. Patent 640,477 (1950); Chem. Abstr., *45*, 1622h (1951).
371. Bergmann, E. D.: U.S. Patent 2,560,921 (1951); Chem. Abstr., *46*, 3072d (1952).
372. Hennion, G. F., Watson, E. J.: J. Org. Chem., *23*, 656 (1958).
373. Froning, J. F., Hennion, G. F.: J. Am. Chem. Soc., *62*, 653 (1940).
374. Nazarov, I. N.: Izv. Akad. Nauk SSSR, Otdel. Khim. Nauk, 215 (1940); Chem. Abstr., *36*, 745[9] (1942).
375. Dolgopol'skii, I. M., Dobromil'skaya, I. M., Podkopaeva, K. N.: Zh. Obshch. Khim., *17*, 1695 (1947); Chem. Abstr., *42*, 2574c (1948).
376. Nazarov, I. N., Zaretskaya, I. I.: Izv. Akad. Nauk SSSR, Otdel. Khim. Nauk, 200 (1942); Chem. Abstr., *39*, 1619[5] (1945).
377. Nazarov, I. N., Torgov, I. B., Terekhova, L. N.: Izv. Akad. Nauk SSSR, Otdel. Khim. Nauk, 50 (1943); Chem. Abstr., *38*, 1729[7] (1944).
378. Nazarov, I. N., Okolskaya, O. V.: Izv. Akad. Nauk SSSR, Otdel. Khim. Nauk, 314 (1941); Chem. Abstr., *37*, 5369[8] (1943).
379. Thomas, R. J., Campbell, K. N., Hennion, G. F.: J. Am. Chem. Soc., *60*, 718 (1938).
380. Janout, V., Regen, S. L.: J. Org. Chem., *47*, 3331 (1982).
381. Chandra, G., Devaprabhakara, D.: Curr. Sci., *40*, 400 (1971); Chem. Abstr., *75*, 98098j (1971).
382. Papa, D., Ginsberg, H. F., Villani, F. J.: J. Am. Chem. Soc., *76*, 4441 (1954).
383. Hennion, G. F., Davis, R. B., Maloney, D. E.: J. Am. Chem. Soc., *71*, 2813 (1949).
384. Lozac'h, N.: Bull. Soc. Chim. France, Series 5, *11*, 416 (1944).
385. Nazarov, I. N., Zaretskaya, I. I.: Izv. Akad. Nauk SSSR, Otdel. Khim. Nauk, 211 (1941); Chem. Abstr., *37*, 6243[8] (1943).
386. Elizarova, A. N., Nazarov, I. N.: Izv. Akad. Nauk SSSR, Otdel. Khim. Nauk, 223 (1940); Chem. Abstr., *36*, 746[9] (1942).
387. Nazarov, I. N.: Izv. Akad. Nauk SSSR, Otdel. Khim. Nauk, 545 (1940); Chem. Abstr., *35*, 4731[9] (1941).
388. Nazarov, I. N., Romanov, V. M.: Izv. Akad. Nauk SSSR, Otdel. Khim. Nauk, 453 (1940); Chem. Abstr., *35*, 3593[9] (1941).
389. Billimoria, J. D., Maclagan, N. F.: J. Chem. Soc., 3257 (1954).
390. Stacy, G. W., Mikulec, R. A.: J. Am. Chem. Soc., *76*, 524 (1954).
391. Hennion, G. F., Pillar, C. J.: J. Am. Chem. Soc., *72*, 5317 (1950).
392. Locquin, R., Wouseng, S., Haller, M. A.: Compt. Rend., *176*, 516 (1923).
393. Leers, L.: Bull. Soc. Chim. France, Series *4*, 39, 423 (1926).
394. Locquin, R., Sung, W.: Bull. Soc. Chim. France, Series 4, *35*, 604 (1924).
395. Venus-Danilova, E. D., Danilov, S. N.: Zh. Obshch. Khim., *2*, 645 (1932); Chem. Abstr., *27*, 1624 (1933).
396. Koulkes, M.: Bull. Soc. Chim. France, 127 (1957).
397. Budde, W. L., Dessy, R. E.: J. Am. Chem. Soc., *85*, 3964 (1963).
398. Stacy, G. W., Mikulec, R. A.: Org. Syn., Coll. Vol. 4, 13 (1963).
399. Stacy, G. W., Hainley, C. A.: J. Am. Chem. Soc., *73*, 5911 (1951).
400. Hickinbottom, W. J., Hyatt, A. A., Sparke, M. B.: J. Chem. Soc., 2529 (1954).
401. Colonge, J.: Bull. Soc. Chim. France, Series 4, *49*, 441 (1931).

402. Dupont, G.: Ann. Chim., *30*, 535 (1913).
403. Bates, D. K., Jones, M. C.: J. Org. Chem., *43*, 3775 (1978).
404. Schöllkopf, U., Hänssle, P.: J. Liebig Ann. Chem., *763*, 208 (1972).
405. Robins, M. J., Barr, P. J.: J. Org. Chem., *48*, 1854 (1983).
406. Nazarov, I. N., Nagibina, T. D., Zaretskaya, I. I.: Izv. Akad. Nauk SSSR, Otdel. Khim. Nauk, 447 (1940); Chem. Abstr., *35*, 5092^6 (1941).
407. Billimoria, J. D.: J. Chem. Soc., 2626 (1953).
408. Rupe, H., Kambli, E.: J. Liebig Ann. Chem., *459*, 195 (1927).
409. Dry, L. J.: J. S. African Chem. Inst., *7*, 37 (1954); Chem. Abstr., *49*, 6825g (1955).
410. Hennion, G. F., Fleck, B. R.: J. Am. Chem. Soc., *77*, 3253 (1955).
411. Nazarov, I. N., Torgov, I. V.: Izv. Akad. Nauk SSSR, Otdel. Khim. Nauk, 129 (1943); Chem. Abstr., *38*, 4595^2 (1944).
412. Dobson, N. A., Raphael, R. A.: J. Chem. Soc., 3558 (1955).
413. Stork, G., Borch, R.: J. Am. Chem. Soc., *86*, 935 (1964).
414. Padmanabhan, S., Nicholas, K. M.: Syn. Commun., *10*, 503 (1980).
415. Sherrill, M. L., Smith, J. C.: J. Chem. Soc., 1501 (1937).
416. Kishida, Y., Hiraoka, T., Yoshimoto, M.: Chem. Pharm. Bull., *17*, 2126 (1969).
417. Hubert, A. J., Reimlinger, H.: J. Chem. Soc. C, 944 (1969).
418. Christie, J. J., Varkey, T. E., Whittle, J. A.: J. Org. Chem., *46*, 3590 (1981).
419. Billimoria, J. D.: J. Chem. Soc., 1126 (1955).
420. Thyagarajan, B. S., Majumdar, K. C., Bates, D. K.: J. Heterocyclic Chem., *12*, 59 (1975).
421. Fraser, M. M., Raphael, R. A.: J. Chem. Soc., 226 (1952).
422. Takano, S., Kasahara, C., Ogasawara, K.: J. Chem. Soc., Chem. Commun., 635 (1981).
423. Fabrycy, A., Dobrzeniecka, R.: Rocz. Chem., *41*, 1733 (1967); Chem. Abstr., *69*, 27131e (1968).
424. Stavely, H. E.: J. Am. Chem. Soc., *63*, 3127 (1941).
425. Ruzicka, L., Meldahl, H. F.: Helv. Chim. Acta, *21*, 1760 (1938).
426. Ruzicka, L., Goldberg, M. W., Hunziker, F.: Helv. Chim. Acta, *22*, 707 (1939).
427. Swiss Patent 235,916 (1945); Chem. Abstr., *43*, 7058a (1949).
428. Stavely, H. E.: J. Am. Chem. Soc., *61*, 79 (1939).
429. Goldberg, M. W., Aeschbacher, R.: Helv. Chim. Acta, *22*, 1185 (1939).
430. Kagan, H. B., Marquet, A., Jacques, J.: Bull. Soc. Chim. France, 1079 (1960).
431. Turner, R. B.: J. Am. Chem. Soc., *75*, 3484 (1953).
432. Ruzicka, L., Gätzi, K., Reichstein, T.: Helv. Chim. Acta, *22*, 626 (1939).
433. Shoppee, C. W., Prins, D. A.: Helv. Chim. Acta, *26*, 185 (1943).
434. Ger. Patent 292,818 (1914); Chem. Abstr., *11*, 1518 (1917).
435. Crosfield and Sons, J., Hilditch, T. P.: Brit. Patent 125,926 (1918); Chem. Abstr., *13*, 2218 (1919).
436. Swiss Patent 235,432 (1945); Chem. Abstr., *43*, 7054e (1949).
437. Kitamura, S., Sebe, E., Hayakawa, K., Sumino, K.: Tai-Wan K'o Hsueh, *25*, 64 (1971); Chem. Abstr., *77*, 60889w (1972).
438. Frieman, R. H., Kennedy, E. R., Lucas, H. J.: J. Am. Chem. Soc., *59*, 722 (1937).
439. Sokol'skii, D. V., Dorfman, Ya. A., Kvon, S. S.: Zh. Org. Khim., *7*, 1759 (1971); J. Org. Chem. USSR, *7*, 1825 (1971).

440. Temkin, O. N., Flid, R. M., Malakhov, A. I.: Kinet. Katal., *4*, 270 (1963); Kinetics and Catalysis (USSR), *4*, 233 (1963).
441. Hennion, G. F., Vogt, R. R., Nieuwland, J. A.: J. Org. Chem., *1*, 159 (1936).

III. Alkoxymercuration

A. Alkenes

The mercuration of alkenes in alcohol solvents, instead of water, provides a very convenient method for the addition of an alkoxy group and mercury across the double bond (Eq. 1). This type of organomercurial appears to

$$
\text{C=C} \; + \; HgX_2 \; + \; ROH \; \longrightarrow \; -\overset{RO}{\underset{HgX}{C}}-\overset{|}{C}- \; + \; HX \tag{1}
$$

have first been observed during the initial work on the hydroxymercuration of ethylene and propylene. Besides giving the anticipated hydroxymercurial, these alkenes also afford small amounts of a bis-mercurial apparently formed by alkoxymercuration of the initial hydroxymercurial (Eq. 2) [1–8]. The first direct use of an alcohol in a solvomercuration reaction appears to have been in 1910 [9].

$$
H_2C{=}CHR \; + \; H_2O \; + \; HgX_2 \; \longrightarrow \; (XHgCH_2CHR)_2O \tag{2}
$$
$$
R = H, CH_3
$$

$$
CH_3CH_2OCH_2CH_2HgCl \; + \; CH_3I \; \xrightarrow{100^\circ C} \; H_2C{=}CH_2 \tag{3}
$$

The exact nature of these organomercurials was again the center of much debate. The fact that β-ethoxyethylmercuric chloride reacts with methyl iodide to generate ethylene was used to argue for some type of olefin complex (Eq. 3) [10]. However, the preparation of separable optically active isomers from optically active alkenes was offered as proof that these compounds are addition compounds and not simple π-complexes [11, 12]. In recent years, proton [7, 13–29] and carbon-13 [30] nuclear magnetic resonance spectra, as well as x-ray crystallographic data [31–35] have firmly established the covalent nature of these compounds.

A large number of alkoxymercuration reactions have been reported as seen Tables 3.1–3.4. These tables are organized in much the same fashion as the hydroxymercuration tables. Table 3.1 covers all examples of the alkoxymercuration of simple acyclic hydrocarbons. Table 3.2 includes heteroatom-containing acyclic alkenes. Table 3.3 summarizes the work on alkenes in which the carbon—carbon double bond is part of a ring and Table

162

Table 3.1. Alkoxymercuration of Simple Acyclic Alkenes

Alkene	Alcohol	Mercuric salt	Organomercurial(s) (% Yield)	Subsequent reactants	Product(s) (% Yield)	Ref.
$H_2C=CH_2$	CH_3OH	$Hg(OAc)_2$	$AcOHgCH_2CH_2OCH_3$	X^-	$XHgCH_2CH_2OCH_3$ $X = Cl,\ Br,\ I$	14, 36
	CH_3OH	$Hg(O_2CR)_2$	$RCO_2HgCH_2CH_2OCH_3$ $R = CH_3,\ C_2H_5,\ n-C_3H_7,\ n-C_7H_{15},\ n-C_{17}H_{35},$ $ClCH_2,\ C_6H_5,\ 2-HOC_6H_4$	—	—	37
	CH_3OH	—	$XHgCH_2CH_2OCH_3$ $X = $ oxalate, tartrate, sulfate, succinate	—	—	38
	CH_3OH	$Hg(O_3SCF_3)_2$	$CF_3SO_3HgCH_2CH_2OCH_3$	—	—	39
	CH_3OH	$HgO/P_2O_5/HOAc$	$PO_3HgCH_2CH_2OCH_3$	—	—	40
	CH_3CH_2OH	$Hg(OAc)_2$	—	Cl^-	$ClHgCH_2CH_2OCH_2CH_3$	36
	$ClCH_2CH_2OH$	$Hg(OAc)_2$	—	$NaCl$	$ClHgCH_2CH_2OCH_2CH_2Cl$	41
	$n-C_3H_7OH$	$Hg(OAc)_2$	—	KCl	$ClHgCH_2CH_2OCH_2CH_2CH_3$	36, 42
	$n-C_3H_7OH$	$Hg(OAc)_2$	—	$NaBH_4$	$CH_3CH_2OCH_2CH_2CH_3$	42
	$(CH_3)_2CHOH$	—	—	Cl^-	$ClHgCH_2CH_2OCH(CH_3)_2$	43
	$n-C_6H_{13}OH$	$Hg(OAc)_2$	—	Cl^-	$ClHgCH_2CH_2O(CH_2)_5CH_3$	42
	$C_6H_5CH_2OH$	—	$IHgCH_2CH_2OCH_2C_6H_5$	$Na(Hg)/D_2O$	$DCH_2CH_2OCH_2C_6H_5$ (23)	44
	$HO(CH_2CH_2O)_mCH_2CH_2OH$ $m = 1,\ 2$	$Hg(OAc)_2$	—	$NaCl$	$ClHg(CH_2CH_2O)_nCH_2CH_2HgCl$ $n = 3,\ 4$	41
	ROH	$Hg(OAc)_2$	$AcOHgCH_2CH_2OR$ $R = Me\ (82),\ Et\ (95)$	X^-	$XHgCH_2CH_2OR$ $X = Cl,\ Br,\ I\ ;\ \ R = Me,\ Et$	45

Table 3.1. (continued)

Alkene	Alcohol	Mercuric salt	Organomercurial(s) (% Yield)	Subsequent reactants	Product(s) (% Yield)	Ref.
	ROH	$Hg(OAc)_2$	$AcOHgCH_2CH_2OR$ $R = C_6H_5(OCH_2CH_2)_n$ $n = 2, 3$	X^-	$XHgCH_2CH_2OR$ (10–30) $X = I$, $R = p\text{-}NO_2C_6H_4(OCH_2CH_2)_2$; $X = Cl$, $R = p\text{-}YC_6H_4(OCH_2CH_2)_2$, $Y = H$, CH_3, Br, OH ; $X = Cl$, $R = o\text{-}HOC_6H_4(OCH_2CH_2)_2$; $X = Cl$, $R = C_6H_5(OCH_2CH_2)_4$	46
	ROH	$Hg(OAc)_2$	—	NaCl	$ClHgCH_2CH_2OR$ (20–60) $R = Et$; $CH_3CH_2(OCH_2CH_2)_n$, $n = 1\text{-}4$; $HO(CH_2)_n$, $n = 2\text{-}4$, 6; $HO(CH_2)_2O(CH_2)_2$; $CH_3CHOHCH_2$ (?); $HOCH_2CHOHCH_2$ (?)	41
	ROH	$Hg(OAc)_2$	—	KI	$IHgCH_2CH_2OR$ $R = Et$, Ac, $HOCH_2CH_2$, $MeOCH_2CH_2$, $AcOCH_2CH_2$, F_3CCH_2, $\begin{smallmatrix}O\\O\end{smallmatrix}{>}CHCH_2$, $NCCH_2CH_2$, $i\text{-}Pr$, $n\text{-}Bu$, $i\text{-}Bu$, $sec\text{-}Bu$, $t\text{-}Bu$, $C_6H_5CH_2$	47
	—	$Hg(OAr)_2$	$ArOHgCH_2CH_2OAr$ $Ar = 2, 4, 6\text{-}C_6H_2Cl_3$; $2, 3, 4, 6\text{-}C_6HCl_4$; C_6Cl_5	—	—	48
$H_2C{=}CHCH_3$	CH_3OH	$Hg(OAc)_2$	$AcOHgCH_2CH(OCH_3)CH_3$	X^-	$XHgCH_2CH(OCH_3)CH_3$ $X = Cl$, Br	7, 14, 49
	CH_3OH	$Hg(O_2CR)_2$	$RCO_2HgCH_2CH(OCH_3)CH_3$ $R = CH_3$, C_2H_5, $n\text{-}C_3H_7$, $n\text{-}C_7H_{15}$, $n\text{-}C_{17}H_{35}$, $ClCH_2$, C_6H_5, $2\text{-}HOC_6H_4$	—	—	37
	CH_3OH	—	—	I^-	$IHgCH_2CH(OCH_3)CH_3$	50
	CH_3OH	—	$XHgCH_2CH(OCH_3)CH_3$ $X = $ oxalate	—	—	38
$H_2C{=}CHCH_2CH_3$	CH_3OH	$Hg(OAc)_2$	$AcOHgCH_2CH(OCH_3)CH_2CH_3$	—	—	14
	CH_3OH	—	—	I^-	$IHgCH_2CH(OCH_3)CH_3$	51

Table 3.1. (continued)

Alkene	Alcohol	Mercuric salt	Organomercurial(s) (% Yield)	Subsequent reactants	Product(s) (% Yield)	Ref.
	ROH	—	$XHgCH_2CH(OR)CH_2CH_3$ R = Me, X = Cl, Br, OAc, $EtCO_2$, n-$PrCO_2$, n-$C_{17}H_{35}CO_2$; R = Et, X = Cl, OAc, $EtCO_2$, n-$PrCO_2$; R = i-Pr, X = Cl, Br, OAc, $EtCO_2$, n-$PrCO_2$	—	—	52
$CH_3CH{=}CHCH_3$	CH_3OH	$Hg(OAc)_2$	—	KCl	$CH_3CH(HgCl)CH(OCH_3)CH_3$	49
cis-$CH_3CH{=}CHCH_3$	CH_3OH	$Hg(OAc)_2$	—	NaCl	$CH_3CH(HgCl)CH(OCH_3)CH_3$	53
	CH_3OH	$Hg(O_2CCF_3)_2$	$CH_3CH(HgO_2CCF_3)CH(OCH_3)CH_3$	—	—	54
	CD_3OD	$Hg(O_2CCF_3)_2$	$threo$-$CH_3CH(HgO_2CCF_3)CH(OCD_3)CH_3$ (64)	—	—	55
$trans$-$CH_3CH{=}CHCH_3$	CH_3OH	$Hg(OAc)_2$	—	NaCl	$CH_3CH(HgCl)CH(OCH_3)CH_3$	53
	CH_3OH	$Hg(O_2CCF_3)_2$	$CH_3CH(HgO_2CCF_3)CH(OCH_3)CH_3$ (80)	—	—	54
	CD_3OD	$Hg(O_2CCF_3)_2$	$erythro$-$CH_3CH(HgO_2CCF_3)CH(OCD_3)CH_3$	—	—	55
$H_2C{=}C(CH_3)_2$	CH_3OH	$Hg(OAc)_2$	$AcOHgCH_2C(OCH_3)(CH_3)_2$	X^-	$XHgCH_2C(OCH_3)(CH_3)_2$ X = Cl, Br	14, 49
	ROH	—	$XHgCH_2C(OR)(CH_3)_2$ R = Me, X = Cl, Br, OAc, $EtCO_2$, n-$PrCO_2$; R = Et, X = Cl, Br, OAc, $EtCO_2$; R = i-Pr, X = Cl, OAc, $EtCO_2$, n-$PrCO_2$	—	—	52
$H_2C{=}C(CH_3)CH_2CH_3$	CH_3OH	$Hg(OAc)_2$	—	Cl^-	$ClHgCH_2C(OCH_3)(CH_3)CH_2CH_3$	49, 56
	ROH	HgX_2	—	$NaBH_4/NaOH$	$(CH_3)_2C(OR)CH_2CH_3$ X = OAc; R = Me(100), Et(100), i-Pr(90), t-Bu(0) X = O_2CCF_3; R = Me(88), Et(86), i-Pr(58), t-Bu(0)	57

Table 3.1. (continued)

Alkene	Alcohol	Mercuric salt	Organomercurial(s) (% Yield)	Subsequent reactants	Product(s) (% Yield)	Ref.
$CH_3CH{=}C(CH_3)_2$	CH_3OH	$Hg(OAc)_2$	—	KCl	$CH_3CH(HgCl)C(OCH_3)(CH_3)_2$	49
$H_2C{=}CHCH(CH_3)_2$	CH_3OH	$Hg(OAc)_2(NaCl)$	$(Cl)AcOHgCH_2CH(OCH_3)CH(CH_3)_2$	Br_2	$BrCH_2COCH(CH_3)_2$ (72) + $BrCH_2CHBrCH(CH_3)_2$ (18)	58
	CH_3OH	—	—	I^-	$IHgCH_2CH(OCH_3)CH(CH_3)_2$	51
$CH_3CH{=}CHCH_2CH_3$	CH_3OH	$Hg(OAc)_2$	—	$NaCl$	$CH_3CH(HgCl)CH(OCH_3)CH_2CH_3$ + $CH_3CH(OCH_3)CH(HgCl)CH_2CH_3$	59
$H_2C{=}C(CH_3)CH(CH_3)_2$	CH_3OH	$Hg(OAc)_2$	—	KCl	$ClHgCH_2C(OCH_3)(CH_3)CH(CH_3)_2$	49
$H_2C{=}CHC(CH_3)_3$	CH_3OH	$Hg(OAc)_2$	$AcOHgCH_2CH(OCH_3)C(CH_3)_3$	Br_2	$BrCH_2COC(CH_3)_3$ (71) + $BrCH_2CHBrC(CH_3)_3$ (23)	58
	CH_3OH	—	—	I^-	$IHgCH_2CH(OCH_3)C(CH_3)_3$	51
	ROH	HgX_2	—	$NaBH_4/NaOH$	$CH_3CH(OR)C(CH_3)_3$ $X = OAc; R = Me(83), Et(72), i\text{-}Pr(12),$ $t\text{-}Bu(4)$ $X = O_2CCF_3; R = Me(100), Et(100),$ $i\text{-}Pr(99), t\text{-}Bu(85)$	57
$H_2C{=}CH(CH_2)_3CH_3$	CH_3OH	$Hg(O_2CR)_2$	$RCO_2HgCH_2CH(OCH_3)(CH_2)_3CH_3$ $R = Me, Et, n\text{-}Pr$	—	—	37
	ROH	HgX_2	—	$NaBH_4/NaOH$	$CH_3CH(OR)(CH_2)_3CH_3$ $X = OAc; R = Me(90), Et(98), i\text{-}Pr(91),$ $t\text{-}Bu(27)$ $X = O_2CCF_3; R = Me(100), Et(100),$ $i\text{-}Pr(100), t\text{-}Bu(100)$	57
	ROH	—	$XHgCH_2CH(OR)(CH_2)_3CH_3$ $R = Me$ or $Et; X = OAc, Cl, Br, I,$ CN, CNO, SCN	—	—	52

Table 3.1. (continued)

Alkene	Alcohol	Mercuric salt	Organomercurial(s) (% Yield)	Subsequent reactants	Product(s) (% Yield)	Ref.
$H_2C=C(CH_3)C(CH_3)_3$	CH_3OH	$Hg(OAc)_2$	$AcOHgCH_2C(OCH_3)(CH_3)C(CH_3)_3$	Br_2	$BrCH_2C(OCH_3)(CH_3)C(CH_3)_3$ (72)	58
	CH_3OH	$Hg(OAc)_2$	—	KCl	$ClHgCH_2C(OCH_3)(CH_3)C(CH_3)_3$	49, 60
$H_2C=CHCH_2C(CH_3)_3$	CH_3OH	—	—	I^-	$IHgCH_2CH(OCH_3)CH_2C(CH_3)_3$	51
$H_2C=CH(CH_2)_4CH_3$	CH_3OH	$Hg(O_2CR)_2$	$RCO_2HgCH_2CH(OCH_3)(CH_2)_4CH_3$ $R = CH_3, C_2H_5, n-C_3H_7, n-C_7H_{15},$ $n-C_{17}H_{35}, ClCH_2, C_6H_5, 2-HOC_6H_4$	—	—	37
$H_2C=CHC_6H_5$	CH_3OH	$Hg(OAc)_2$	$AcOHgCH_2CH(OCH_3)C_6H_5$ (80)	$NaBH_4$	$CH_3CH(OCH_3)C_6H_5$	61, 62
	CH_3OH	$Hg(OAc)_2$	—	$NaBH_4/NaOH$	$CH_3CH(OCH_3)C_6H_5$ (97)	57
	CH_3OH	$Hg(OAc)_2$	—	Br_2/KBr	$BrCH_2CH(OCH_3)C_6H_5$ (73)	63
	CH_3OH	$Hg(OAc)_2, Cl^-$	$ClHgCH_2CH(OCH_3)C_6H_5$ (89)	I_2/KI (on RHgOAc)	$ICH_2CH(OCH_3)C_6H_5$	29
	CH_3OH	$Hg(O_2CR)_2$	$RCO_2HgCH_2CH(OCH_3)C_6H_5$ $R = CH_3, C_2H_5, n-C_3H_7, n-C_7H_{15},$ $n-C_{17}H_{35}, ClCH_2, C_6H_5, 2-HOC_6H_4$	—	—	37
	CH_3OH	—	—	X^-	$XHgCH_2CH(OCH_3)C_6H_5$ $X = Cl$ (100), I	51, 64
	$(CH_3)_2CHOH$	—	$AcOHgCH_2CH(O-i-Pr)C_6H_5$	$NaBD_4$	$DCH_2CH(O-i-Pr)C_6H_5$	65
	ROH	HgX_2	$XHgCH_2CH(OR)C_6H_5$ $X = OAc, R = Me, Et, i-Pr, n-Bu;$ $R = Me, X = O_2CCH_2CH_3, O_2C(CH_2)_2CH_3,$ $O_2C(CH_2)_6CH_3$	Y^-	$YHgCH_2CH(OR)C_6H_5$ $R = Me$ or $Et; Y = Cl, Br, I$	66
	ROH	—	$XHgCH_2CH(OR)C_6H_5$ $R = Me, Et; X = CN, CNO, SCN,$ phosphate	—	—	52

Table 3.1. (continued)

Alkene	Alcohol	Mercuric salt	Organomercurial(s) (% Yield)	Subsequent reactants	Product(s) (% Yield)	Ref.
$H_2C=C(CH_3)CH_2C(CH_3)_3$	CH_3OH	$Hg(OAc)_2$	—	Cl^-	$ClHgCH_2C(OCH_3)(CH_3)CH_2C(CH_3)_3$	56
$H_2C=CH(CH_2)_5CH_3$	CH_3OH	$Hg(OAc)_2$	$AcOHgCH_2CH(OCH_3)(CH_2)_5CH_3$ (90)	$NaBH_4/NaOH$	$CH_3CH(OCH_3)(CH_2)_5CH_3$ (83)	67
	CH_3OH	—	$ClHgCH_2CH(OCH_3)(CH_2)_5CH_3$	—	—	23
$H_2C=C(CH_3)C_6H_5$	CH_3OH	$Hg(OAc)_2$	$AcOHgCH_2C(OCH_3)(CH_3)C_6H_5$	$NaBH_4/NaOH$	$(CH_3)_2C(OCH_3)C_6H_5$ (100)	57, 68
$H_2C=CHCH_2C_6H_5$	CH_3OH	$Hg(OAc)_2$	$AcOHgCH_2CH(OCH_3)CH_2C_6H_5$ (42)	$NaBH_4/NaOH$	$CH_3CH(OCH_3)CH_2C_6H_5$ (96)	68
	CH_3OH	$Hg(OAc)_2$	—	KI	$IHgCH_2CH(OCH_3)CH_2C_6H_5$	69
	ROH	$HgO/2\,HBF_4$	—	—	$trans$-$C_6H_5CH=CHCH_2OR$ $R = Me, Et, i\text{-}Pr, n\text{-}Bu, i\text{-}Bu, n\text{-}Am, c\text{-}Hex$	70
$CH_3CH=CHC_6H_5$	CH_3OH	$Hg(OAc)_2$	—	Br_2/KBr	$CH_3CHBrCH(OCH_3)C_6H_5$ (75)	63
cis-$CH_3CH=CHC_6H_5$	CH_3OH	$Hg(OAc)_2$ (HOAc)	$CH_3CH(HgOAc)CH(OCH_3)C_6H_5$ (69)	$NaBH_4/NaOH$	$CH_3CH_2CH(OCH_3)C_6H_5$	68
	CH_3OH	$Hg(OAc)_2, Cl^-$	$CH_3CH(OCH_3)CH(HgCl)C_6H_5$ (76) + $CH_3CH(HgCl)CH(OCH_3)C_6H_5$ (5-10)	N_2H_4	$CH_3CH(OCH_3)CH_2C_6H_5$ (7.5) + $CH_3CH_2CH(OCH_3)C_6H_5$ (11)	64
$trans$-$CH_3CH=CHC_6H_5$	CH_3OH	$Hg(OAc)_2$	$CH_3CH(HgOAc)CH(OCH_3)C_6H_5$ + $CH_3CH(OCH_3)CH(HgOAc)C_6H_5$	$NaBH_4/NaOH$	$CH_3CH_2CH(OCH_3)C_6H_5$ + $CH_3CH(OCH_3)CH_2C_6H_5$ 60 : 40	68
	CH_3OH	$Hg(OAc)_2, Cl^-$	$CH_3CH(HgCl)CH(OCH_3)C_6H_5$ (45-50) + $CH_3CH(OCH_3)CH(HgCl)C_6H_5$ (2.8)	Br_2, KOH, H^+	$CH_3CH_2COC_6H_5$ + $CH_3COCH_2C_6H_5$ 60 : 40	64
	CH_3OH	$Hg(O_2CCF_3)_2$	$CH_3CH(HgO_2CCF_3)CH(OCH_3)C_6H_5$ 65 : + $CH_3CH(OCH_3)CH(HgO_2CCF_3)C_6H_5$ (76) : 35	—	—	54

Table 3.1. (continued)

Alkene	Alcohol	Mercuric salt	Organomercurial(s) (% Yield)	Subsequent reactants	Product(s) (% Yield)	Ref.
	c-$C_6H_{11}OH$	$Hg(OAc)_2$ ($BF_3 \cdot Et_2O$) —	NaCl	$CH_3CH(HgCl)CH(O$-c-$C_6H_{11})C_6H_5$		64
	$C_6H_5CH_2OH$	$Hg(OAc)_2$ —	NaCl	$CH_3CH(HgCl)CH(OCH_2C_6H_5)C_6H_5$ (23)		64
$H_2C{=}C(CH_3)CH_2C_6H_5$	CH_3OH	$Hg(OAc)_2$	$AcOHgCH_2C(OCH_3)(CH_3)CH_2C_6H_5$	X^-	$XHgCH_2C(OCH_3)(CH_3)CH_2C_6H_5$ X = Br, I, CN	56
	CH_3OH	$Hg(OAc)_2$ —	Cl^-	$ClHgCH_2C(OCH_3)(CH_3)CH_2C_6H_5$		49, 56
$C_6H_5CH{=}C(CH_3)_2$	CH_3OH	$Hg(OAc)_2$	$C_6H_5CH(OCH_3)C(HgOAc)(CH_3)_2$ (93)	$NaBH_4/NaOH$	$C_6H_5CH(OCH_3)CH(CH_3)_2$	68
	CH_3OH	$Hg(OAc)_2$, NcCl	$C_6H_5CH(HgCl)C(OCH_3)(CH_3)_2$ (92, 100)	—	—	64, 71
cis-$(CH_3)_3CCH{=}CHC(CH_3)_3$	CH_3OH	$Hg(OAc)_2$ —	Cl^-	$threo$-$(CH_3)_3CCH(HgCl)CH(OCH_3)C(CH_3)_3$ (85)		72
	CH_3OH	$Hg(ClO_4)_2$ —	Cl^-	$threo$-$(CH_3)_3CCH(HgCl)CH(OCH_3)C(CH_3)_3$		72
$trans$-$(CH_3)_3CCH{=}CHC(CH_3)_3$	CH_3OH	$Hg(ClO_4)_2$ —	Cl^-	$threo$-$(CH_3)_3CCH(HgCl)CH(OCH_3)C(CH_3)_3$		72
$H_2C{=}C(C_6H_5)C(CH_3)_3$	CH_3OH	$Hg(OAc)_2$	$AcOHgCH_2C(OCH_3)(C_6H_5)C(CH_3)_3$	$NaBH_4/NaOH$	$CH_3C(OCH_3)(C_6H_5)C(CH_3)_3$	68
$H_2C{=}C(CH_3)CH_2$-(2,3,5-trimethylphenyl)	CH_3OH	$Hg(OAc)_2$ —	Cl^-	$ClHgCH_2C(OCH_3)(CH_3)CH_2$-(2,3,5-trimethylphenyl)		49, 56
$H_2C{=}CH(CH_2)_{10}CH_3$	CH_3OH	$Hg(O_2CR)_2$	$RCO_2HgCH_2CH(OCH_3)(CH_2)_{10}CH_3$ R = C_2H_5, n-C_3H_7, n-C_7H_{15}, n-$C_{17}H_{35}$, $ClCH_2$, C_6H_5, 2-HOC_6H_4	—	—	37
	$HOCH_2CH_2OH$	$Hg(OAc)_2$ —	$NaBH_4$	$CH_3CH(OAc)(CH_2)_{10}CH_3$ (53) + $CH_3CH(OCH_2CH_2OH)(CH_2)_{10}CH_3$ (26) + $CH_3CHOH(CH_2)_{10}CH_3$ (8)		73
$H_2C{=}C(C_6H_5)_2$	CH_3OH	$Hg(OAc)_2$	$AcOHgCH_2CH(OCH_3)(C_6H_5)_2$ (49)	—	—	74

Table 3.1. (continued)

Alkene	Alcohol	Mercuric salt	Organomercurial(s) (% Yield)	Subsequent reactants	Product(s) (% Yield)	Ref.
cis-$C_6H_5CH{=}CHC_6H_5$	CH_3OH	$Hg(OAc)_2$ (HOAc)	$C_6H_5CH(HgOAc)CH(OCH_3)C_6H_5$ (99)	$NaBH_4$ or $NaBD_4$	$C_6H_5CHXCH(OCH_3)C_6H_5$ X = H or D	68
	CH_3OH	$Hg(OAc)_2$	—	NaCl	$C_6H_5CH(HgCl)CH(OCH_3)C_6H_5$ (70)	75
	CH_3OH	$Hg(O_2CCF_3)_2$	$C_6H_5CH(HgO_2CCF_3)CH(OCH_3)C_6H_5$	KCl	$C_6H_5CH(HgCl)CH(OCH_3)C_6H_5$	54
	CH_3OH	$Hg(O_2CCF_3)_2$	—	—	$(C_6H_5)_2CHCH(OCH_3)_2$ (88)	54
trans-$C_6H_5CH{=}CHC_6H_5$	CH_3OH	$Hg(OAc)_2$ (HOAc)	$C_6H_5CH(HgOAc)CH(OCH_3)C_6H_5$ (95)	$NaBH_4$ or $NaBD_4$	$C_6H_5CHXCH(OCH_3)C_6H_5$ X = H or D	68
	CH_3OH	$Hg(OAc)_2$	—	NaCl	$C_6H_5CH(HgCl)CH(OCH_3)C_6H_5$ (6)	75
	CH_3OH	$Hg(OAc)_2$ $[(C_6H_5CO_2)_2]$	—	NaCl	$C_6H_5CH(HgCl)CH(OCH_3)C_6H_5$ (23.6)	76
	CH_3OH	$Hg(O_2CCF_3)_2$	$C_6H_5CH(HgO_2CCF_3)CH(OCH_3)C_6H_5$	—	—	54
	CH_3OH	$Hg(O_2CCF_3)_2$	—	—	$(C_6H_5)_2CHCH(OCH_3)_2$	54
$C_6H_5C(CH_3){=}CHC_6H_5$	CH_3OH	$Hg(OAc)_2$ $[(PhCO_2)_2]$, NaCl	$C_6H_5C(CH_3)(OCH_3)CH(HgCl)C_6H_5$ (74)	N_2H_4/NaOH	$C_6H_5C(CH_3)(OCH_3)CH_2C_6H_5$ (29)	77
$H_2C{=}CH(CH_2)_{12}CH_3$	CH_3OH	$Hg(O_2CR)_2$	$RCO_2HgCH_2CH(OCH_3)(CH_2)_{12}CH_3$ R = Me, Et, n-Pr	X^-	$XHgCH_2CH(OCH_3)(CH_2)_{12}CH_3$ X = Cl, Br, I	37
$H_2C{=}CH(CH_2)_{16}CH_3$	CH_3OH	$Hg(O_2CR)_2$	$RCO_2HgCH_2CH(OCH_3)(CH_2)_{16}CH_3$ R = Me, Et, n-Pr	X^-	$XHgCH_2CH(OCH_3)(CH_2)_{16}CH_3$ X = Cl, Br, I	37

Table 3.2. Alkoxymercuration of Acyclic Heteroatom-Containing Alkenes

Alkene	Alcohol	Mercuric salt	Organomercurial(s) (% Yield)	Subsequent reactants	Product(s) (% Yield)	Ref.
$D_2C{=}CD_2$	CH_3OH	$Hg(OAc)_2$	—	NaI	$IHgCD_2CD_2OCH_3$	78
$H_2C{=}CD_2$	CH_3OH	$Hg(OAc)_2$	—	Cl^-	$ClHgCH_2CD_2OCH_3$ + $ClHgCD_2CH_2OCH_3$	28
cis-$HDC{=}CHD$	CH_3OH	HgX_2	$threo$-$XHgCHDCHDOCH_3$ (X = ?)	—	—	24
$trans$-$HDC{=}CHD$	CH_3OH	HgX_2	$erythro$-$XHgCHDCHDOCH_3$ (X = ?)	—	—	24
$H_2C{=}C{=}O$	ROH	$Hg(OAc)_2/HgO$	$Hg(CH_2CO_2R)_2$ R = Me, Et, n-Pr, i-Pr, n-Bu, i-Bu, t-Bu (all ~100)	—	—	79
$F_2C{=}CHCF_3$	CH_3CH_2OH	$Hg(NO_3)_2$	—	$NaCl$	$CF_3CH(HgCl)CO_2C_2H_5$ (60)	80
$Cl_2C{=}CClOCH_3$	CH_3OH	HgX_2 (X = OAc, NO_3)	—	$NaCl$	$Hg(CCl_2CO_2CH_3)_2$ (91)	81
$H_2C{=}CHCO_2H$	CH_3OH	$Hg(OAc)_2$	(see structure) ? (96)	—	—	82
	CH_3OH	$Hg(OAc)_2$	$CH_3OCH_2CH{-}CO$ (77) (with $Hg{-}O$)	—	—	83
	CH_3OH	$Hg(OAc)_2$	$CH_3OCH_2CH(HgOAc)CO_2H$	Br_2/KBr	$CH_3OCH_2CHBrCO_2H$	84
$H_2C{=}CHOCH_3$	CH_3OH	$Hg(OAc)_2$	$AcOHgCH_2CH(OCH_3)_2$	—	—	85
	CH_3CH_2OH	$Hg(OAc)_2$	$AcOHgCH_2CH(OCH_3)OC_2H_5$ + $AcOHgCH_2CH(OC_2H_5)_2$ (1 : 1)	—	—	85
$H_2C{=}CHCH_2OH$	CH_3OH	$Hg(OAc)_2$, KBr	$BrHgCH_2CH(OCH_3)CH_2OH$ (86)	I_2	$ICH_2CH(OCH_3)CH_2OH$	86
	CH_3OH	—	—	I^-	$IHgCH_2CH(OCH_3)CH_2OH$	51

Table 3.2. (continued)

Alkene	Alcohol	Mercuric salt	Organomercurial(s) (% Yield)	Subsequent reactants	Product(s) (% Yield)	Ref.
$H_2C{=}CHSO_2CH_3$	CH_3OH	$Hg(OAc)_2\,[(PhCO_2)_2]$	$CH_3OCH_2CH(HgOAc)SO_2CH_3$	—	—	87
$H_2C{=}CHCH_2NH_2$	CH_3OH	$HgCl_2(KOAc)$	$ClHgCH_2CH(OCH_3)CH_2NH_2$	—	—	88
	CH_3OH	$Hg(OAc)_2\,(HOAc,\ H_2SO_4)$	—	NaCl, KCNO	$ClHgCH_2CH(OCH_3)CH_2NHCONH_2$	89
$BrFC{=}CFOC_2H_5$	CH_3CH_2OH	$Hg(NO_3)_2$	—	NaCl	$ClHgCBrFCO_2C_2H_5$	90
$ClFC{=}CFOC_2H_5$	CH_3CH_2OH	$Hg(NO_3)_2$	—	$HgCl_2/NaCl$ or Ph_2Hg	$XHgCFClCO_2C_2H_5$ $X = Cl\ (25.5),\ Ph$	80, 90
	CH_3CH_2OH	$Hg(NO_3)_2/HgCl_2$	$ClHgCFClCO_2C_2H_5\ +\ Hg(CFClCO_2C_2H_5)_2$	—	—	91
$F_2C{=}CFOC_2H_5$	CH_3CH_2OH	HgX_2	$XHgCF_2CO_2C_2H_5$ $X = Cl\ (83.5),\ Br$	—	—	92
	CH_3CH_2OH	$Hg(OAc)_2$	—	NaBr	$BrHgCF_2CO_2C_2H_5$ (63.5)	92
	CH_3CH_2OH	$C_6H_5HgNO_3$	$Hg(CF_2CO_2C_2H_5)_2$ (60.5)	—	—	92
	CH_3CH_2OH	$Hg(NO_3)_2$	—	NaCl	$ClHgCF_2CO_2C_2H_5$ (92.8)	92
$H_2C{=}CHCH_2CN$	CH_3OH	$HgX_2\ (X = ?),\ I^-$	$IHgCH_2CH(OCH_3)CH_2CN$	$Na(Hg)/H_2O$	$CH_3CH(OCH_3)CH_2CN$ (24)	51
$ClCH{=}CClOC_2H_5$	CH_3CH_2OH	$Hg(OAc)_2$	—	NaCl	$ClHgCHClCO_2C_2H_5$ (66)	81
$H_2C{=}C(CH_3)CHO$	CH_3OH	$Hg(OAc)_2\,(HClO_4)$	$AcOHgCH_2C(OCH_3)(CH_3)CHO$ 70 : $+\ AcOHgCH_2C(OCH_3)(CH_3)CH(OCH_3)_2$: 30 (90 total)	KBr	$BrHgCH_2C(OCH_3)(CH_3)CHO$ $+\ BrHgCH_2C(OCH_3)(CH_3)CH(OCH_3)_2$ (84 total)	93
	ROH	$Hg(OAc)_2\,(HClO_4)$	—	KBr	$BrHgCH_2C(OR)(CH_3)CHO$ $R = Me\ and/or\ t\text{-}Bu\ (?)$	94

Table 3.2. (continued)

Alkene	Alcohol	Mercuric salt	Organomercurial(s) (% Yield)	Subsequent reactants	Product(s) (% Yield)	Ref.
$H_2C=CHCOCH_3$	CH_3OH	$Hg(OAc)_2$ $(HClO_4)$	$CH_3OCH_2CH(HgOAc)COCH_3$ (~100)	KBr	$CH_3OCH_2CH(HgBr)COCH_3$ (25)	95
	CH_3OH	HgO $(BF_3 \cdot Et_2O)$	—	—	$CH_3OCH_2CH_2COCH_3$ (61)	96
$trans$-$CH_3CH=CHCHO$	CH_3OH	$Hg(OAc)_2$ $(HClO_4)$	$CH_3CH(OCH_3)CH(HgOAc)CHO$ 70 : + $CH_3CH(OCH_3)CH(HgOAc)CH(OCH_3)_2$: 30 (95 total)	KBr	$CH_3CH(OCH_3)CH(HgBr)CHO$ + $CH_3CH(OCH_3)CH(HgBr)CH(OCH_3)_2$ (75 total)	93
$H_2C=CHO_2CCH_3$	CH_3OH	$Hg(OAc)_2$	$AcOHgCH_2CH(OCH_3)O_2CCH_3$ (99)	—	—	97
	CH_3CH_2OH	$Hg(OAc)_2 / HgO$	—	KCl	$ClHgCH_2CHO$ (70)	98
$H_2C=CHCO_2CH_3$	CH_3OH	$Hg(OAc)_2$	$CH_3OCH_2CH(HgOAc)CO_2CH_3$	—	—	99
	CH_3OH	$Hg(OAc)_2$ $(HClO_4)$	$CH_3OCH_2CH(HgOAc)CO_2CH_3$ (~90)	KBr	$CH_3OCH_2CH(HgBr)CO_2CH_3$ (40)	95
	CH_3OH	$Hg(OAc)_2$, KBr	$CH_3OCH_2CH(HgBr)CO_2CH_3$	Br_2	$CH_3OCH_2CHBrCO_2CH_3$ (81-86)	100
$trans$-$CH_3CH=CHCO_2H$	CH_3OH	$Hg(OAc)_2$	—	H_2S	$CH_3CH(OCH_3)CH_2CO_2H$ (80)	101
	CH_3OH	$Hg(OAc)_2$	—	Br_2/KBr	$CH_3CH(OCH_3)CHBrCO_2H$ (88 - 93)	101
$H_2C=C(CH_3)CH_2Br$	CH_3OH	$Hg(OAc)_2$	$AcOHgCH_2C(OCH_3)(CH_3)CH_2Br$	X^-	$XHgCH_2C(OCH_3)(CH_3)CH_2Br$ X = Cl (89), I (92), CN (72)	49, 60
$H_2C=C(CH_3)CH_2Cl$	CH_3OH	$Hg(OAc)_2$	$AcOHgCH_2C(OCH_3)(CH_3)CH_2Cl$ (91)	X^-	$XHgCH_2C(OCH_3)(CH_3)CH_2Cl$ X = Cl (92), Br, I (96), CN (99)	49, 60
Z-$CH_3CH=CDCH_3$	CH_3OH	$Hg(OAc)_2$	—	Cl^-	$threo$-$CH_3CH(OCH_3)CD(HgCl)CH_3$ + $threo$-$CH_3CD(OCH_3)CH(HgCl)CH_3$	28
$H_2C=CHCH_2NHCONH_2$	CH_3OH	$HgCl_2$ (KOAc)	$ClHgCH_2CH(OCH_3)CH_2NHCONH_2$	—	—	88
	CH_3OH	$HgCl_2$ (^{197}Hg or ^{203}Hg) (NaOAc)	$ClHgCH_2CH(OCH_3)CH_2NHCONH_2$ (60-70, 75)	—	—	102, 103

Table 3.2. (continued)

Alkene	Alcohol	Mercuric salt	Organomercurial(s) (% Yield)	Subsequent reactants	Product(s) (% Yield)	Ref.
	CH_3OH	$IHgOAc$	$IHgCH_2CH(OCH_3)CH_2NHCONH_2$ (70)	—	—	13
	CH_3OH	$Hg(OAc)_2$ (^{197}Hg or ^{203}Hg)	—	$NaCl$	$ClHgCH_2CH(OCH_3)CH_2NHCONH_2$ (50–70)	104, 105
	CH_3OH	$^{203}Hg(OAc)_2$	$AcOHg^{203}CH_2CH(OCH_3)CH_2NHCONH_2$	Cl^-	$ClHg^{203}CH_2CH(OCH_3)CH_2NHCONH_2$	106
	CH_3OH	$Hg(OAc)_2 / HOAc$	$AcOHgCH_2CH(OCH_3)CH_2NHCONH_2$ (38)	NaX	$XHgCH_2CH(OCH_3)CH_2NHCONH_2$ $X = Cl$ (55), Br (41)	107
	ROH	HgX_2 $(NaOAc)$	$XHgCH_2CH(OR)CH_2NHCONH_2$ $R = Me$, $X = Cl$, Br; $R = Et$, $X = Cl$	—	—	108
	ROH	$Hg(OAc)_2$	$AcOHgCH_2CH(OR)CH_2NHCONH_2$ $R = Me$	X^-	$XHgCH_2CH(OR)CH_2NHCONH_2$ $R = Et$ (30), i-Pr, n-Bu, $X = Cl$; $R = Me$, $X = Br$, OH	109
	ROH	$Hg(OAc)_2$ $(HOAc)$	—	$NaCl$	$ClHgCH_2CH(OR)CH_2NHCONH_2$ $R = Et$ (31), i-Pr (10), n-Bu (16)	107
$H_2C=CHOC_2H_5$	CH_3OH	$Hg(OAc)_2$	$AcOHgCH_2CH(OCH_3)OC_2H_5$ (98)	—	—	97
	CH_3CH_2OH	$Hg(OAc)_2$	—	$NaCl$	$ClHgCH_2CH(OC_2H_5)_2$ (58)	15
	CH_3CH_2OH	$Hg(OAc)_2 / HgO$ (KCl)	$ClHgCH_2CH(OC_2H_5)_2$ (73)	Br_2	$BrCH_2CH(OC_2H_5)_2$ (80)	98
$H_2C=CHCH_2OCH_3$	CH_3OH	$Hg(OAc)_2$	$AcOCH_2CH(OCH_3)CH_2OCH_3$	$NaCl$ (?)	$ClHgCH_2CH(OCH_3)CH_2OCH_3$ (?)	110
	CH_3OH	—	—	I^-	$IHgCH_2CH(OCH_3)CH_2OCH_3$	51
$H_2C=C(CH_3)CH_2OH$	CH_3OH	—	—	I^-	$IHgCH_2C(OCH_3)(CH_3)CH_2OH$	60
$H_2C=CHCHOHCH_3$	CH_3OH	—	—	Cl^-	$ClHgCH_2CH(OCH_3)CHOHCH_3$	51

Table 3.2. (continued)

Alkene	Alcohol	Mercuric salt	Organomercurial(s) (% Yield)	Subsequent reactants	Product(s) (% Yield)	Ref.
$H_2C=CHCH_2SO_2CH_3$	CH_3OH	$Hg(OAc)_2$ (HOAc)	$CH_3OCH_2CH(HgOAc)CH_2SO_2CH_3$	—	—	111
$CF_3CF=CFOC_2H_5$	CH_3CH_2OH	$Hg(OAc)_2$	—	NaCl	$CF_3CF(HgCl)CO_2C_2H_5 \cdot NaCl$ (65)	80
$H_2C=C(CH_3)COCH_3$	CH_3OH	$Hg(OAc)_2$ $(HClO_4)$	$AcOHgCH_2C(OCH_3)(CH_3)COCH_3$ (~97)	KBr	$BrHgCH_2C(OCH_3)(CH_3)COCH_3$ (30)	95
	ROH	$Hg(OAc)_2$ $(HClO_4)$	—	KBr	$BrHgCH_2C(OR)(CH_3)COCH_3$ R = Me and/or t - Bu (?)	94
$CH_3CH=CHCOCH_3$	CH_3OH	$Hg(OAc)_2$ $(HClO_4)$	$CH_3CH(OCH_3)CH(HgOAc)COCH_3$ (~100)	KBr	$CH_3CH(OCH_3)CH(HgBr)COCH_3$ (73)	95
$H_2C=C(OC_2H_5)CHO$	CH_3CH_2OH	$Hg(OAc)_2$ $(HClO_4)$	$AcOHgCH_2COCH(OC_2H_5)_2$ (50)	KCl	$ClHgCH_2COCH(OC_2H_5)_2$ (62)	112
	CH_3CH_2OH	$Hg(OAc)_2$ $(HClO_4)$	—	KCl/H_2O	$ClHgCH_2COCHO$ (80)	112
$H_2C=C(CH_3)CO_2CH_3$	CH_3OH	$Hg(OAc)_2$ $(HClO_4)$	$AcOHgCH_2C(OCH_3)(CH_3)CO_2CH_3$ (60)	KBr	$BrHgCH_2C(OCH_3)(CH_3)CO_2CH_3$ (25)	94, 95
	$(CH_3)_3COH$	$Hg(OAc)_2$ $(HClO_4)$, KBr	$BrHgCH_2C(O-t-Bu)(CH_3)CO_2CH_3$	$NaBH_4$	$(CH_3)_2C(O-t-Bu)CO_2CH_3$	94
$trans$-$CH_3CH=CHCO_2CH_3$	CH_3OH	$Hg(OAc)_2$ $(HClO_4)$	$CH_3CH(OCH_3)CH(HgOAc)CO_2CH_3$ (~100)	KBr	$CH_3CH(OCH_3)CH(HgBr)CO_2CH_3$ (70)	95
	CH_3OH	$Hg(OAc)_2$	—	KBr/Br_2	$CH_3CH(OCH_3)CHBrCO_2CH_3$ (81–86)	113
H₂C=C with H and CH₃ on one carbon, CO₂H and CH₃ on other $\begin{smallmatrix}H\\CH_3\end{smallmatrix}C=C\begin{smallmatrix}CO_2H\\CH_3\end{smallmatrix}$	CH_3OH	$Hg(OAc)_2$	—	reduction	$CH_3CH(OCH_3)CH(CH_3)CO_2H$ $NaBH_4/NaOH$ ~1:1 *erythro/threo* (72) $NaBH_4$ 4:1 $H_2S/NaOH$ almost entirely *erythro* (81)	114
$\begin{smallmatrix}CH_3\\H\end{smallmatrix}C=C\begin{smallmatrix}CO_2H\\CH_3\end{smallmatrix}$	CH_3OH	$Hg(OAc)_2$	—	reduction	$CH_3CH(OCH_3)CH(CH_3)CO_2H$ $NaBH_4/NaOH$ 50:50 *erythro/threo* $H_2S/NaOH$ 100:0	114
$(CH_3)_2C=CHCO_2H$	CH_3OH	$Hg(OAc)_2$, KBr	$(CH_3)_2C(OCH_3)CH(HgBr)CO_2K$	KBr/Br_2	$(CH_3)_2C(OCH_3)CHBrCO_2H$ (53.2)	115
$H_2C=CHCH_2SCO_2CH_3$	CH_3OH	$Hg(OAc)_2$	—	KBr or NaCl	$XHgCH_2CH(OCH_3)CH_2SCO_2CH_3$ X = Cl (65), Br (60.2)	116

Table 3.2. (continued)

Alkene	Alcohol	Mercuric salt	Organomercurial(s) (% Yield)	Subsequent reactants	Product(s) (% Yield)	Ref.
$H_2C{=}C(OEt)CO_2H$	CH_3CH_2OH	$Hg(OAc)_2$	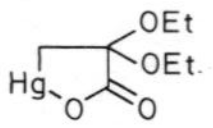 (88)	Br_2	$BrCH_2C(OEt)_2CO_2H$	82
$H_2C{=}CHCH(OCH_3)CH_2HgCl$	CH_3OH	$Hg(OAc)_2$	—	NaCl	$ClHgCH_2CH(OCH_3)CH(OCH_3)CH_2HgCl$ (59)	117
$H_2C{=}CHCH_2NHCOCH_3$	CH_3OH	$Hg(OAc)_2$ (HOAc)	—	NaCl	$ClHgCH_2CH(OCH_3)CH_2NHCOCH_3$ (65)	118
$H_2C{=}CHCH_2NHCONHCONH_2$	CH_3OH	$Hg(OAc)_2$	—	X^-	$XHgCH_2CH(OCH_3)CH_2NHCONHCONH_2$ X = OH, Cl, Br, NO_3, SO_4, O_2CPh, salicylate, glutamate	119
$H_2C{=}CHCH_2NHCONHCH_3$	CH_3OH	$Hg(OAc)_2$ (HOAc)	—	NaCl	$ClHgCH_2CH(OCH_3)CH_2NHCONHCH_3$ (25)	109, 120
$H_2C{=}C(CH_3)CH_2NHCONH_2$	CH_3OH	$Hg(OAc)_2$ (HOAc)	—	NaCl	$ClHgCH_2C(OCH_3)(CH_3)CH_2NHCONH_2$ (55)	120
$H_2C{=}CH(CH_2)_2OCH_3$	CH_3OH	$Hg(OAc)_2$	$AcOHgCH_2CH(OCH_3)(CH_2)_2OCH_3$	NaCl (?)	$ClHgCH_2CH(OCH_3)(CH_2)_2OCH_3$ (?)	110
$H_2C{=}CHCHOHCH_2OCH_3$	CH_3OH	$Hg(OAc)_2$, KI	$IHgCH_2CH(OCH_3)CHOHCH_2OCH_3$ (31)	I_2	$ICH_2CH(OCH_3)CHOHCH_2OCH_3$	86
$H_2C{=}CHSi(OCH_3)_3$	CH_3OH	$Hg(OAc)_2$	$AcOHgCH_2CH(OCH_3)Si(OCH_3)_3$?	—	—	121
	CH_3OH	$Hg(OAc)_2$	$CH_3OCH_2CH(HgOAc)Si(OCH_3)_3$? (100)	—	—	27
$H_2C{=}CHSi(CH_3)_3$	CH_3OH	$Hg(OAc)_2$	$CH_3OCH_2CH(HgOAc)Si(CH_3)_3$ (100)	—	—	27
$(CF_3)_2C{=}CFOC_2H_5$	CH_3CH_2OH	$Hg(NO_3)_2$	—	NaCl	$Hg[C(CF_3)_2CO_2C_2H_5]_2$ (21.8)	80
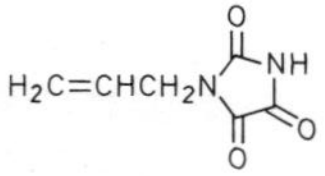	CH_3OH	$Hg(OAc)_2$	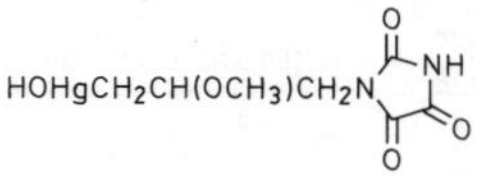	—	—	122

Table 3.2. (continued)

Alkene	Alcohol	Mercuric salt	Organomercurial(s) (% Yield)	Subsequent reactants	Product(s) (% Yield)	Ref.
$H_2C=CHCH_2N$ (hydantoin ring)	CH_3OH	$Hg(OAc)_2$	—	NaCl	$ClHgCH_2CH(OCH_3)CH_2N$ (hydantoin ring) (96)	123
$H_2C=CHCH_2N$ (hydantoin ring)	CH_3OH	$Hg(OAc)_2$	—	NaCl	$ClHgCH_2CH(OCH_3)CH_2N$ (hydantoin ring) (74)	123
$H_2C=CHCH_2$ (hydantoin ring)	CH_3OH	$Hg(OAc)_2$	—	NaCl	$ClHgCH_2CH(OCH_3)CH_2$ (hydantoin ring) (71)	123
$(CH_3)_2C=CHCOCH_3$	CH_3OH	$Hg(OAc)_2$ ($HClO_4$)	$(CH_3)_2C(OCH_3)CH(HgOAc)COCH_3$ (~100)	KBr	$(CH_3)_2C(OCH_3)CH(HgBr)COCH_3$ (83)	95
$CH_3CH=C(CH_3)COCH_3$	CH_3OH	$Hg(OAc)_2$ ($HClO_4$)	$CH_3CH(OCH_3)C(HgOAc)(CH_3)COCH_3$ (18)	—	—	95
$trans\text{-}CH_3CH=CHCO_2C_2H_5$	CH_3OH	$Hg(OAc)_2$, HBr or KBr	$CH_3CH(OCH_3)CH(HgBr)CO_2C_2H_5$	Br_2	$CH_3CH(OCH_3)CHBrCO_2C_2H_5$ (90)	124–126
$E\text{-}CH_3CH=C(CH_3)CO_2CH_3$	CH_3OH	$Hg(OAc)_2$ ($HClO_4$)	$CH_3CH(OCH_3)C(HgOAc)(CH_3)CO_2CH_3$ (~57)	KBr	$CH_3CH(OCH_3)C(HgBr)(CH_3)CO_2CH_3$ (24)	95
$(CH_3)_2C=CHCO_2CH_3$	CH_3OH	$Hg(OAc)_2$ ($HClO_4$)	$(CH_3)_2C(OCH_3)CH(HgOAc)CO_2CH_3$ (~100)	KBr	$(CH_3)_2C(OCH_3)CH(HgBr)CO_2CH_3$ (40)	95
$H_2C=CHCH_2NHCOCH_2CH_3$	CH_3OH	$Hg(OAc)_2$ (HOAc)	—	NaCl	$ClHgCH_2CH(OCH_3)CH_2NHCOCH_2CH_3$ (25)	118
$H_2C=CHCH_2NHCO_2C_2H_5$	CH_3OH	$Hg(OAc)_2$	$CH_3OCH_2CH(HgOAc)CH_2NHCO_2C_2H_5$?	—	—	127
$H_2C=CHCH_2NHCOCH_2NHCONH_2$	CH_3OH	$Hg(OAc)_2$	$AcOHgCH_2CH(OCH_3)CH_2NHCOCH_2NHCONH_2$	—	—	122
	CH_3OH	$XHgOAc$	$XHgCH_2CH(OCH_3)CH_2NHCOCH_2NHCONH_2$ $X = Cl$ (90), Br (70), SCN (62)	—	—	128
$H_2C=CHCH_2NHCONHCONHCH_3$	CH_3OH	$Hg(OAc)_2$	$AcOHgCH_2CH(OCH_3)CH_2NHCONHCONHCH_3$	—	—	119
$H_2C=CHCH_2NHCONHC_2H_5$	CH_3OH	$Hg(OAc)_2$ (HOAc)	—	NaCl	$ClHgCH_2CH(OCH_3)CH_2NHCONHC_2H_5$ (30)	109, 120

Table 3.2. (continued)

Alkene	Alcohol	Mercuric salt	Organomercurial(s) (% Yield)	Subsequent reactants	Product(s) (% Yield)	Ref.
$H_2C{=}CHO(CH_2)_3CH_3$	CH_3OH	$Hg(OAc)_2$	$AcOHgCH_2CH(OCH_3)O(CH_2)_3CH_3$ (99)	—	—	97
	$CH_3(CH_2)_3OH$	$Hg(OAc)_2/HgO,$ KBr	$BrHgCH_2CH(O{-}n{-}Bu)_2$ (70)	Br_2	$BrCH_2CH(O{-}n{-}Bu)_2$ (95)	98
$H_2C{=}CH(CH_2)_3OCH_3$	CH_3OH	$Hg(OAc)_2$	$AcOHgCH_2CH(OCH_3)(CH_2)_3OCH_3$	NaCl (?)	$ClHgCH_2CH(OCH_3)(CH_2)_3OCH_3$ (?)	110
$cis{-}CH_3CH_2CH{=}CH(CH_2)_2OH$	CH_3OH	$Hg(OAc)_2$, NaCl	$CH_3CH_2CH(OCH_3)CH(HgCl)(CH_2)_2OH$	$N_2H_4/NaOH$ Δ	$CH_3CH_2CH(OCH_3)(CH_2)_3OH$ (48) + $cis{-}CH_3CH_2CH{=}CH(CH_2)_2OH$ (40)	129
$H_2C{=}CHCH_2O(CH_2)_2SO_2CH_3$	CH_3OH	$Hg(OAc)_2$ (HOAc)	$AcOHgCH_2CH(OCH_3)CH_2O(CH_2)_2SO_2CH_3$	—	—	130
$H_2C{=}C(CH_3)CH_2N(CH_3)_2$	CH_3OH	$Hg(OAc)_2$	—	Cl^-	$ClHgCH_2C(OCH_3)(CH_3)CH_2N(CH_3)_2$	56, 60
$CH_3CH{=}CHSi(OCH_3)_3$	CH_3OH	$Hg(OAc)_2$	$CH_3CH(OCH_3)CH(HgOAc)Si(OCH_3)_3$ (100)	—	—	27
$CH_3CH{=}CHSi(CH_3)_3$	CH_3OH	$Hg(OAc)_2$	$CH_3CH(OCH_3)CH(HgOAc)Si(CH_3)_3$ 70 : + $CH_3CH(HgOAc)CH(OCH_3)Si(CH_3)_3$: 30 (85 total)	—	—	27
$H_2C{=}CHCH_2N$(barbituric acid)	CH_3OH	$Hg(OAc)_2$ (HOAc)	$AcOHgCH_2CH(OCH_3)CH_2N$(barbituric acid)	—	—	131
$H_2C{=}CHCH_2$(barbituric acid, C-5)	CH_3OH	$Hg(OAc)_2$ (HOAc)	$AcOHgCH_2CH(OCH_3)CH_2$(barbituric acid, C-5)	—	—	131
$H_2C{=}CHCH_2N$(succinimide)	CH_3OH	$Hg(OAc)_2$	—	Cl^-	$ClHgCH_2CH(OCH_3)CH_2N$(succinimide) (80)	132

Table 3.2. (continued)

Alkene	Alcohol	Mercuric salt	Organomercurial(s) (% Yield)	Subsequent reactants	Product(s) (% Yield)	Ref.
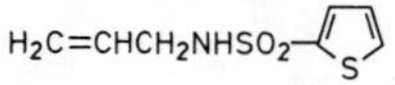 $H_2C=CHCH_2NHSO_2$—[thiophene]	CH_3OH	$Hg(OAc)_2$	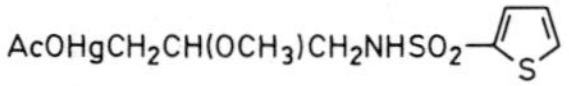 $AcOHgCH_2CH(OCH_3)CH_2NHSO_2$—[thiophene] (40)	—	—	133
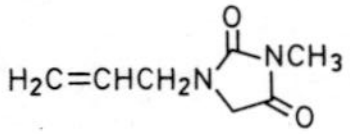 $H_2C=CHCH_2N$—[imidazolidinedione, NCH₃]	CH_3OH	$Hg(OAc)_2$	—	NaCl	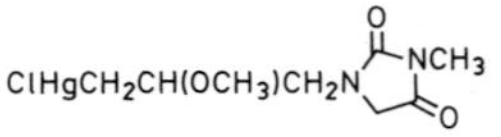 $ClHgCH_2CH(OCH_3)CH_2N$—[imidazolidinedione, NCH₃] (20)	123
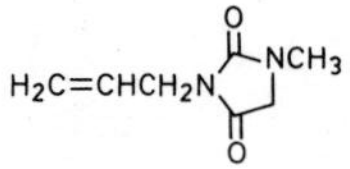 $H_2C=CHCH_2N$—[imidazolidinedione, NCH₃]	CH_3OH	$Hg(OAc)_2$	—	NaCl	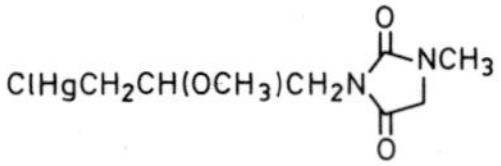 $ClHgCH_2CH(OCH_3)CH_2N$—[imidazolidinedione, NCH₃] (22)	123
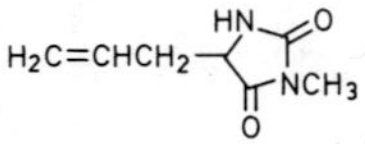 $H_2C=CHCH_2$—[hydantoin, HN / NCH₃]	CH_3OH	$Hg(OAc)_2$	—	NaCl	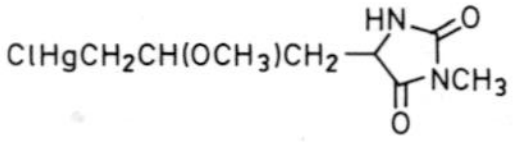 $ClHgCH_2CH(OCH_3)CH_2$—[hydantoin, HN / NCH₃] (55)	123
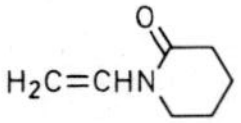 $H_2C=CHN$—[2-piperidinone]	CH_3OH	$Hg(OAc)_2$	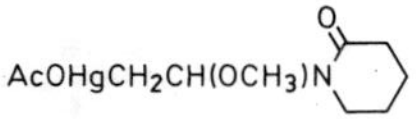 $AcOHgCH_2CH(OCH_3)N$—[2-piperidinone]	—	—	134
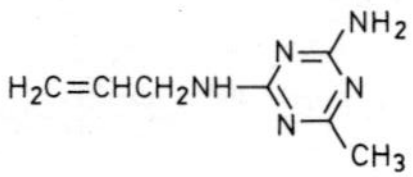 $H_2C=CHCH_2NH$—[triazine, NH₂ / CH₃]	CH_3OH	$Hg(OAc)_2$	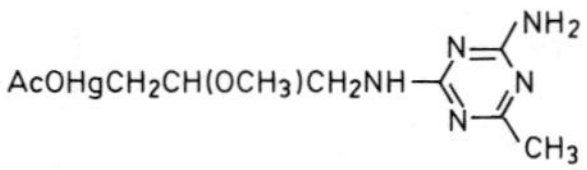 $AcOHgCH_2CH(OCH_3)CH_2NH$—[triazine, NH₂ / CH₃]	Cl^-	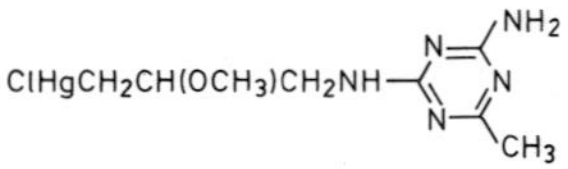 $ClHgCH_2CH(OCH_3)CH_2NH$—[triazine, NH₂ / CH₃]	135
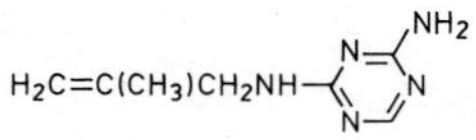 $H_2C=C(CH_3)CH_2NH$—[triazine, NH₂]	CH_3OH	$Hg(OAc)_2$	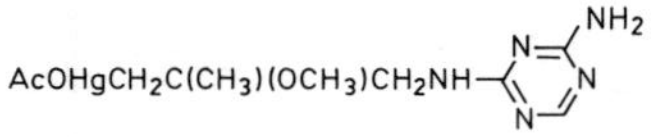 $AcOHgCH_2C(CH_3)(OCH_3)CH_2NH$—[triazine, NH₂]	—	—	135
$(CH_3)_2C=CHCO_2C_2H_5$	CH_3OH	$Hg(OAc)_2$, KBr	$(CH_3)_2C(OCH_3)CH(HgBr)CO_2C_2H_5$ (70)	Br_2	$(CH_3)_2C(OCH_3)CHBrCO_2C_2H_5$ (70)	136
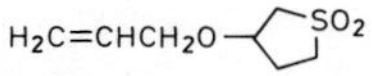 $H_2C=CHCH_2O$—[tetrahydrothiophene-SO₂]	CH_3OH	$Hg(OAc)_2$ (HOAc)	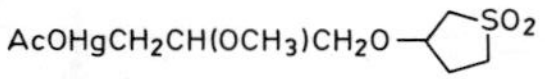 $AcOHgCH_2CH(OCH_3)CH_2O$—[tetrahydrothiophene-SO₂] (91)	—	—	137
$H_2C=CHCH_2NHCO(CH_2)_2CH_3$	CH_3OH	$Hg(OAc)_2$ (HOAc)	—	NaCl	$ClHgCH_2CH(OCH_3)CH_2NHCO(CH_2)_2CH_3$ (20)	118
$H_2C=CHCH_2NHCONHCONHC_2H_5$	CH_3OH	$Hg(OAc)_2$	$AcOHgCH_2CH(OCH_3)CH_2NHCONHCONHC_2H_5$	—	—	119

Table 3.2. (continued)

Alkene	Alcohol	Mercuric salt	Organomercurial(s) (% Yield)	Subsequent reactants	Product(s) (% Yield)	Ref.
$H_2C{=}CHCH_2NHSO_2N{-}$(morpholino)	CH_3OH	$ClHgOAc$	$ClHgCH_2CH(OCH_3)CH_2NHSO_2N{-}$(morpholino) (30)	—	—	128
$H_2C{=}CHCH_2NHCONH(CH_2)_2CH_3$	CH_3OH	$Hg(OAc)_2$ (HOAc)	—	NaCl	$ClHgCH_2CH(OCH_3)CH_2NHCONH(CH_2)_2CH_3$ (50)	109, 120
$H_2C{=}CH(CH_2)_4OCH_3$	CH_3OH	$Hg(OAc)_2$	$AcOHgCH_2CH(OCH_3)(CH_2)_4OCH_3$	NaCl (?)	$ClHgCH_2CH(OCH_3)(CH_2)_4OCH_3$ (?)	110
$D_2C{=}CDC_6H_5$	CH_3OH	—	—	I^-	$IHgCD_2CD(OCH_3)C_6H_5$	51, 138
$D_2C{=}CHC_6H_5$	CH_3OH	—	—	I^-	$IHgCD_2CH(OCH_3)C_6H_5$	51, 138
$H_2C{=}CDC_6H_5$	CH_3OH	—	—	I^-	$IHgCH_2CD(OCH_3)C_6H_5$	51, 138
$o{-}H_2C{=}CHC_6H_4NO_2$	CH_3OH	$Hg(OAc)_2$	—	NaCl	$o{-}ClHgCH_2CH(OCH_3)C_6H_4NO_2$ (92)	139
$p{-}H_2C{=}CHC_6H_4NO_2$	CH_3OH	$Hg(OAc)_2$, Cl^-	$p{-}ClHgCH_2CH(OCH_3)C_6H_4NO_2$ (90, 95)	I_2/KI (on RHgOAc)	$p{-}ICH_2CH(OCH_3)C_6H_4NO_2$	29, 139
$p{-}H_2C{=}CHOC_6H_4NO_2$	CH_3OH	$Hg(OAc)_2$	$p{-}AcOHgCH_2CH(OCH_3)OC_6H_4NO_2$ (98.3)	—	—	97
$H_2C{=}CHOC_6H_5$	CH_3OH	$Hg(OAc)_2$	$AcOHgCH_2CH(OCH_3)OC_6H_5$ (98.3)	—	—	97
	C_6H_5OH	$Hg(NHCOCH_3)_2$	$CH_3CONHHgCH_2CH(OC_6H_5)_2$ (87)	—	—	98
$H_2C{=}CHCH_2N{-}$(5,5-dimethylhydantoin-3-yl)	CH_3OH	$Hg(OAc)_2$	—	NaCl	$ClHgCH_2CH(OCH_3)CH_2N{-}$(5,5-dimethylhydantoin-3-yl) (17)	123
$H_2C{=}CHCH_2NHCONHCO{-}$ $(CH_2)_2CO_2H$	CH_3OH	$Hg(OAc)_2$	—	NaOH ?	$HOHgCH_2CH(OCH_3)CH_2NHCONHCOCH_2{-}$ CH_2CO_2Na ?	122

Table 3.2. (continued)

Alkene	Alcohol	Mercuric salt	Organomercurial(s) (% Yield)	Subsequent reactants	Product(s) (% Yield)	Ref.
	CH_3OH	$Hg(OAc)_2$ $(NaHCO_3-HOAc)$	$HOHgCH_2CH(OCH_3)CH_2NHCONHCO-(CH_2)_2CO_2H$ (90)	NaX or Br$_2$/KBr	$RHgX$ (X = Cl, Br, I) or RBr	140
	$HOCH_2CH_2OH$	$Hg(OAc)_2$ (OH$^-$)	$HOHgCH_2CH(OCH_2CH_2OH)CH_2-NHCONHCO(CH_2)_2CO_2Na$ (60)	—	—	141
(cyclopentyl)$-CH=CHCO_2H$	CH_3OH	$Hg(OAc)_2$	(cyclopentyl)$-CH(OCH_3)CH(HgOAc)CO_2H$	Br$_2$/KBr hν	(cyclopentyl)$-CH(OCH_3)CHBrCO_2H$	142
$H_2C=CHN$ (azocane-2-one)	CH_3OH	$Hg(OAc)_2$	$AcOHgCH_2CH(OCH_3)N$ (azocane-2-one)	—	—	134
$H_2C=CHCH_2NHCON$ (morpholino)	CH_3OH	$Hg(OAc)_2$ (HOAc)	—	NaCl	$ClHgCH_2CH(OCH_3)CH_2NHCON$ (morpholino) (40)	109, 120
$H_2C=CHO-$(cyclohexyl)	CH_3OH	$Hg(OAc)_2$	$AcOHgCH_2CH(OCH_3)O-$(cyclohexyl) (99)	—	—	97
$H_2C=CHCH_2NHCOC(CH_3)_3$	CH_3OH	$Hg(OAc)_2$ (HOAc)	—	NaCl	$ClHgCH_2CH(OCH_3)CH_2NHCOC(CH_3)_3$ (55)	118
$H_2C=CHCH_2NHCO(CH_2)_3CH_3$	CH_3OH	$Hg(OAc)_2$ (HOAc)	—	NaCl	$ClHgCH_2CH(OCH_3)CH_2NHCO(CH_2)_3CH_3$ (40)	118
$H_2C=CHCH_2NHCON(C_2H_5)_2$	CH_3OH	$Hg(OAc)_2$	—	NaCl	$ClHgCH_2CH(OCH_3)CH_2NHCON(C_2H_5)_2$	109
$H_2C=CHCH_2NHCONH(CH_2)_3CH_3$	CH_3OH	$Hg(OAc)_2$ (HOAc)	—	NaCl	$ClHgCH_2CH(OCH_3)CH_2NHCONH(CH_2)_3CH_3$ (30)	109, 120
$H_2C=CH(CH_2)_5OCH_3$	CH_3OH	$Hg(OAc)_2$	$AcOHgCH_2CH(OCH_3)(CH_2)_5OCH_3$	NaCl (?)	$ClHgCH_2CH(OCH_3)(CH_2)_5OCH_3$ (?)	110
cis-$C_6H_5CH=CHCN$	CH_3OH	$Hg(OAc)_2$ (HNO$_3$)	$C_6H_5CH(OCH_3)CH(HgOAc)CN$ (26)	Br$_2$	$C_6H_5CH(OCH_3)CHBrCN$	143
	CH_3OH	$Hg(OAc)_2$ (BF$_3$·Et$_2$O)	—	NaCl	$C_6H_5CH(OCH_3)CH(HgCl)CN$ (78)	76
trans-$C_6H_5CH=CHCN$	CH_3OH	$Hg(OAc)_2$ (HNO$_3$)	$C_6H_5CH(OCH_3)CH(HgOAc)CN$ (16)	Br$_2$	$C_6H_5CH(OCH_3)CHBrCN$	143
	CH_3OH	$Hg(OAc)_2$ (BF$_3$·Et$_2$O)	—	NaCl	$C_6H_5CH(OCH_3)CH(HgCl)CN$ (47)	76

Table 3.2. (continued)

Alkene	Alcohol	Mercuric salt	Organomercurial(s) (% Yield)	Subsequent reactants	Product(s) (% Yield)	Ref.
p-$H_2C{=}CHCOC_6H_4NO_2$	CH_3OH	$Hg(OAc)_2$	p-$CH_3OCH_2CH(HgOAc)COC_6H_4NO_2$ (63)	KBr/Br_2	$CH_3OCH_2CHBrCOC_6H_4NO_2$ (74, 78)	144, 145
$trans$-o-$HO_2CCH{=}CHC_6H_4NO_2$	CH_3OH	$Hg(OAc)_2$	o-$OC{-}CHCH(OCH_3)C_6H_4NO_2$ $\;\;\mid\;\;\mid$ $O{-}Hg$	$H_2S/NaOH$	o-$HO_2CCH_2CH(OCH_3)C_6H_4NO_2$	146
$trans$-m-$HO_2CCH{=}CHC_6H_4NO_2$	CH_3OH	$Hg(OAc)_2$	m-$OC{-}CHCH(OCH_3)C_6H_4NO_2$ $\;\;\mid\;\;\mid$ $O{-}Hg$	$H_2S/NaOH$	m-$HO_2CCH_2CH(OCH_3)C_6H_4NO_2$	146
$H_2C{=}CH{-}\!\!\bigwedge\!\!_{Fe(CO)_3}$	CH_3OH	$Hg(OAc)_2$	$AcOHgCH_2\overset{OCH_3}{\underset{}{CH}}\!\!\bigwedge\!\!_{Fe(CO)_3}$ (4.5)	—	—	147
$trans$-$C_6H_5CH{=}CHCHO$	CH_3OH	$Hg(OAc)_2$ $(HClO_4)$	$C_6H_5CH(OCH_3)CH(HgOAc)CHO$ (70)	—	—	93
$H_2C{=}CHCOC_6H_5$	CH_3OH	$Hg(OAc)_2$ (KBr)	$CH_3OCH_2CH(HgBr)COC_6H_5$ (77)	Br_2	$CH_3OCH_2CHBrCOC_6H_5$ (78)	144, 145
$C_6H_5CH{=}CHCO_2H$	CH_3OH	$Hg(OAc)_2$	$C_6H_5CH(OCH_3)CH(HgOAc)CO_2H$ (85)	Br_2/KBr $h\nu$	$C_6H_5CH(OCH_3)CHBrCO_2H$ (80)	63
cis-$C_6H_5CH{=}CHCO_2H$	CH_3OH	$Hg(OAc)_2$	$C_6H_5CH(OCH_3)CH{-}CO$ $\;\;\mid\;\;\;\;\mid$ $Hg{-}O$ (100)	—	—	148
$trans$-$C_6H_5CH{=}CHCO_2H$	CH_3OH	$Hg(OAc)_2$ $(HOAc)$	$C_6H_5CH(OCH_3)CH{-}CO$ $\;\;\mid\;\;\;\;\mid$ $Hg{-}O$ (79)	$Br_2(KBr)$	$C_6H_5CH(OCH_3)CHBrCO_2H$ 1 or 2 diastereomers	82, 146, 149
	CH_3OH	$Hg(OAc)_2$	$C_6H_5CH(OCH_3)CH{-}CO$ $\;\;\mid\;\;\;\;\mid$ $Hg{-}O$ (85)	$NaCl/HCCl_3$	$C_6H_5CH(OCH_3)CH(HgCl)CO_2H \cdot HCCl_3$ (27)	148
	CH_3OH	$Hg(OAc)_2$	$C_6H_5CH(OCH_3)CH{-}CO$ $\;\;\mid\;\;\;\;\mid$ $Hg{-}O$	$HOAc/KBr$	$C_6H_5CH(OCH_3)CH(HgBr)CO_2H$	149
	CH_3OH	$Hg(OAc)_2$, KBr	$C_6H_5CH(OCH_3)CH(HgBr)CO_2H$ (92)	Br_2	$C_6H_5CH(OCH_3)CHBrCO_2H$ (63.6)	136
	ROH	$Hg(OAc)_2$	$C_6H_5CH(OR)CH{-}CO$ $\;\;\mid\;\;\;\;\mid$ $Hg{-}O$ $R = Me$ (95), Et (95), n-Bu(75), n-Am(80), n-$C_{16}H_{33}$	—	—	150

Table 3.2. (continued)

Alkene	Alcohol	Mercuric salt	Organomercurial(s) (% Yield)	Subsequent reactants	Product(s) (% Yield)	Ref.
$trans$-o-$HO_2CCH{=}CHC_6H_4OH$	CH_3OH	$Hg(OAc)_2$	o-$OC{-}CHCH(OCH_3)C_6H_4OH$ (82), $O{-}Hg$	$NaOH/H_2S$	o-$HO_2CCH_2CH(OCH_3)C_6H_4OH$ (25)	151
$H_2C{=}CHCH_2N$ (pyridinone ring, CO_2H)	CH_3OH	—	$CH_3OCH_2CH(HgOH)CH_2N$ (pyridinone ring, CO_2Na) ?	—	—	152
$H_2C{=}CHCH_2OC_6H_5$	CH_3OH	—	—	I^-	$IHgCH_2CH(OCH_3)CH_2OC_6H_5$	51
p-$H_2C{=}CHC_6H_4OCH_3$	CH_3OH	$Hg(OAc)_2$, Cl^-	p-$ClHgCH_2CH(OCH_3)C_6H_4OCH_3$ (87)	I_2/KI (on RHgOAc)	p-$ICH_2CH(OCH_3)C_6H_4OCH_3$	29
o-$H_2C{=}CHCH_2C_6H_4OH$	CH_3OH	$Hg(OAc)_2$	o-$AcOHgCH_2CH(OCH_3)CH_2C_6H_4OH$ 12 : + (benzofuran-CH_2HgOAc) : 7	$NaBH_4$	o-$CH_3CH(OCH_3)CH_2C_6H_4OH$ (62) + o-$H_2C{=}CHCH_2C_6H_4OH$ (23) + (benzofuran-CH_3) (13)	153
$trans$-$HOCH_2CH{=}CHC_6H_5$	CH_3OH	$Hg(OAc)_2$	—	Cl^-	$HOCH_2CH(HgCl)CH(OCH_3)C_6H_5$ (66) mainly erythro	22, 151
$HO_2CCH{=}CH{-}$ (aryl, NO_2, CH_3O)	CH_3OH	$Hg(OAc)_2$	$HO_2CCH(HgOAc)CH(OCH_3){-}$ (aryl, NO_2, CH_3O)	$H_2S/NaOH$	$HO_2CCH_2CH(OCH_3){-}$ (aryl, NO_2, CH_3O)	160
$H_2C{=}CHCH_2NHCO{-}$ (pyridine, HO_2C)	CH_3OH	$Hg(OAc)_2$	—	$NaOH$	$HOHgCH_2CH(OCH_3)CH_2NHCO{-}$ (pyridine, NaO_2C)	122
cis-$C_6H_5CH{=}CHCOCH_3$	CH_3OH	$Hg(OAc)_2$	$C_6H_5CH(OCH_3)CH(HgOAc)COCH_3$	—	—	159
$trans$-$C_6H_5CH{=}CHCOCH_3$	CH_3OH	$Hg(OAc)_2$	$C_6H_5CH(OCH_3)CH(HgOAc)COCH_3$ (66) erythro / threo = 85 : 15	X_2	$C_6H_5CH(OCH_3)CHXCOCH_3$ (erythro and threo, only threo isolated) X = Br (88), I	22, 159
$H_2C{=}CHCH_2{-}$ (methylenedioxyphenyl)	CH_3OH	$Hg(OAc)_2$	$AcOHgCH_2CH(OCH_3)CH_2{-}$ (methylenedioxyphenyl) (94)	$NaCl$	$ClHgCH_2CH(OCH_3)CH_2{-}$ (methylenedioxyphenyl) (90)	150

Table 3.2. (continued)

Alkene	Alcohol	Mercuric salt	Organomercurial(s) (% Yield)	Subsequent reactants	Product(s) (% Yield)	Ref.
$H_2C=C(C_6H_5)CO_2CH_3$	ROH	$Hg(OAc)_2$ $(HClO_4)$	$AcOHgCH_2C(OR)(C_6H_5)CO_2CH_3$ $R = Me$ (~50)	KBr	$BrHgCH_2C(OR)(C_6H_5)CO_2CH_3$ $R = Me$ and/or $t-Bu$ (?)	94, 95
$cis-C_6H_5CH=CHCO_2CH_3$	CH_3OH	$Hg(OAc)_2$	$C_6H_5CH(OCH_3)CH(HgOAc)CO_2CH_3$	—	—	159
$trans-C_6H_5CH=CHCO_2CH_3$	CH_3OH	$Hg(OAc)_2$	$C_6H_5CH(OCH_3)CH(HgOAc)CO_2CH_3$ (64–65)	$NaOH,\ H_2SO_4$	$C_6H_5CH(OCH_3)CH-CO$ $\quad\quad\quad\quad\quad\ \ |\quad\ \ |$ $\quad\quad\quad\quad\quad\ Hg-O$	9, 159
	CH_3OH	$Hg(OAc)_2$	$C_6H_5CH(OCH_3)CH(HgOAc)CO_2CH_3$	NH_3/H_2S	$C_6H_5CH(OCH_3)CH_2CO_2CH_3$	161
	CH_3OH	$Hg(OAc)_2,\ KBr$	$C_6H_5CH(OCH_3)CH(HgBr)CO_2CH_3$	Br_2	$C_6H_5CH(OCH_3)CHBrCO_2CH_3$ (91)	136
	CH_3OH	$Hg(OAc)_2$ $(HClO_4)$	$C_6H_5CH(OCH_3)CH(HgOAc)CO_2CH_3$ (90)	KBr	$C_6H_5CH(OCH_3)CH(HgBr)CO_2CH_3$ (74)	95
	CH_3CH_2OH	$Hg(OAc)_2$	$C_6H_5CH(OC_2H_5)CH(HgOAc)CO_2CH_3$ (31)	—	—	9
	ROH	$Hg(OAc)_2$	$C_6H_5CH(OR)CH(HgOAc)CO_2CH_3$ $R = Et,\ n-Pr$ (71), $i-Pr$ (55), $i-Bu$ (35)	—	—	162
$H_2C=CH(CH_2)_2CH(OCH_3)CH_2-$ $HgOAc$	CH_3CH_2OH	$Hg(OAc)_2$	$AcOHgCH_2CH(OC_2H_5)(CH_2)_2CH(OCH_3)-$ CH_2HgOAc	$NaBH_4$	$CH_3CH(OC_2H_5)(CH_2)_2CH(OCH_3)CH_3$	154
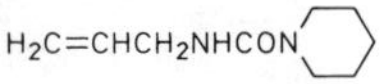 $H_2C=CHCH_2NHCON\bigcirc$	CH_3OH	$Hg(OAc)_2$ $(HOAc)$	—	NaCl	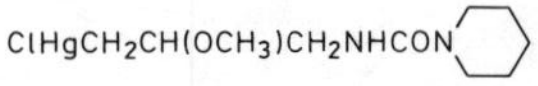 $ClHgCH_2CH(OCH_3)CH_2NHCON\bigcirc$ (55)	109, 120
$n-C_6H_{13}CH=CHCO_2H$	$CH_3(CH_2)_3OH$	$Hg(OAc)_2$	$n-C_6H_{13}CH(O-n-Bu)CH(Hg^+)CO_2^-$ (72)	—	—	141
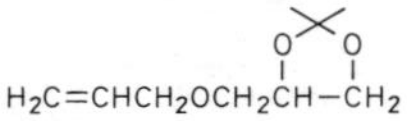 $H_2C=CHCH_2OCH_2CH-CH_2$ (dioxolane)	CH_3OH	$Hg(OAc)_2$	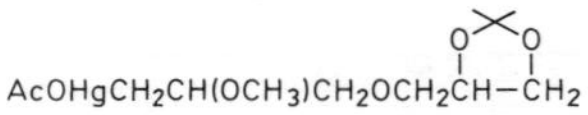 $AcOHgCH_2CH(OCH_3)CH_2OCH_2CH-CH_2$ (dioxolane)	—	—	155
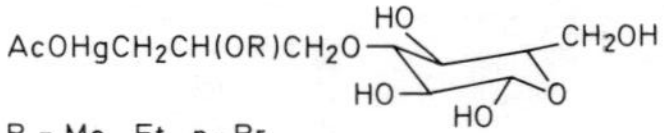 $H_2C=CHCH_2O$ (sugar) $R = Me,\ Et,\ n-Pr$	ROH	$Hg(OAc)_2$	$AcOHgCH_2CH(OR)CH_2O$ (sugar)	—	—	155

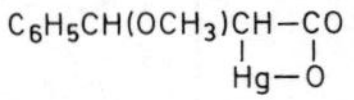

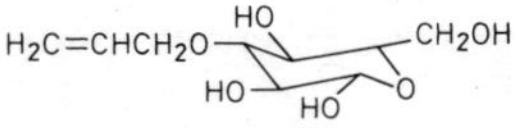

Table 3.2. (continued)

Alkene	Alcohol	Mercuric salt	Organomercurial(s) (% Yield)	Subsequent reactants	Product(s) (% Yield)	Ref.
$H_2C=CHCH_2OCH(CHOHCH_2OH)$- $CHOHCHOHCH_2OH$	ROH	$Hg(O_2CR')_2$	$R'CO_2HgCH_2CH(OR)CH_2OCH(CHOHCH_2OH)CHOHCHOHCH_2OH$ $R' = CH_3, \ R = CH_3, \ C_2H_5, \ HOCH_2CH_2;$ $R' = C_6H_5, \ R = CH_3;$ $R' = HO_2CCHOHCHOH, \ R = CH_3$	—	—	155, 156
$H_2C=CHCH_2OCH_2(CHOH)_4CH_2OH$	CH_3OH	$Hg(OAc)_2$	$AcOHgCH_2CH(OCH_3)CH_2OCH_2(CHOH)_4CH_2OH$	—	—	155
$H_2C=CHCH_2Si(OC_2H_5)_3$	CH_3CH_2OH	$Hg(OAc)_2$	$AcOHgCH_2CH(OC_2H_5)CH_2Si(OC_2H_5)_3$	—	—	27
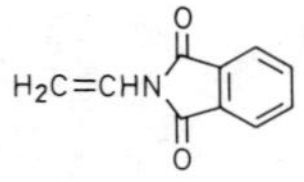 $H_2C=CHN$(phthalimide)	CH_3OH	$Hg(OAc)_2$	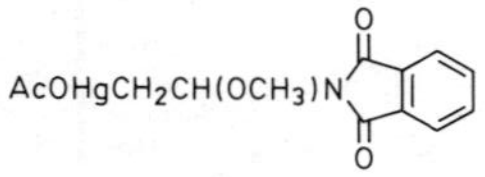 $AcOHgCH_2CH(OCH_3)N$(phthalimide)	KBr/Br_2	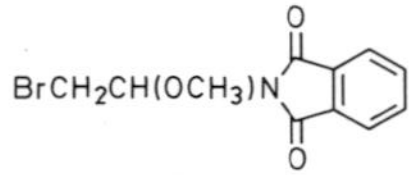 $BrCH_2CH(OCH_3)N$(phthalimide)	157
$H_2C=C(CO_2H)NHCOC_6H_5$	CH_3OH	$Hg(OAc)_2$	$AcOHgCH_2C(OCH_3)(CO_2H)NHCOC_6H_5$ (~100)	$NaBH_4/NaOH,$ CH_2N_2	$CH_3C(OCH_3)(CO_2CH_3)NHCOC_6H_5$ (20)	158
$trans-p-$ $NO_2C_6H_4CH=CHCOCH_3$	CH_3OH	$Hg(OAc)_2$	$p-NO_2C_6H_4CH(OCH_3)CH(HgOAc)COCH_3$	—	—	159
$trans-p-$ $NO_2C_6H_4CH=CHCO_2CH_3$	CH_3OH	$Hg(OAc)_2$	$p-NO_2C_6H_4CH(OCH_3)CH(HgOAc)CO_2CH_3$	—	—	159
$p-H_2C=C(CH_3)C_6H_4OCH_3$	CH_3OH	$Hg(OAc)_2$	—	KCl	$p-ClHgCH_2C(OCH_3)(CH_3)C_6H_4OCH_3$	49
$p-CH_3CH=CHC_6H_4OCH_3$	CH_3OH	HgX_2 (X = ?)	$p-CH_3CH(HgX)CH(OCH_3)C_6H_4OCH_3$	—	—	166
$m-H_2C=CHCH_2OC_6H_4OCH_3$	CH_3OH	$Hg(OAc)_2$	$m-AcOHgCH_2CH(OCH_3)CH_2OC_6H_4OCH_3$ $12 : AcOHg$ $+ \ H_2C=CHCH_2O$—(aryl, HgOAc, OCH_3) $2 :$ $+ \ m-AcOHgCH_2CHOAcCH_2OC_6H_4OCH_3$ $: 1$ 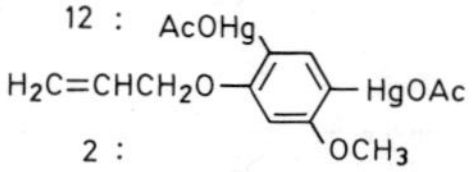	$NaBH_4$	$m-CH_3CH(OCH_3)CH_2OC_6H_4OCH_3$ (57) $+ \ m-CH_3CHOAcCH_2OC_6H_4OCH_3$? (~5)	153
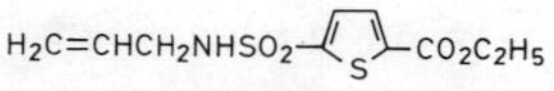 $H_2C=CHCH_2NHSO_2$-(thiophene)-$CO_2C_2H_5$	CH_3OH	$Hg(OAc)_2$	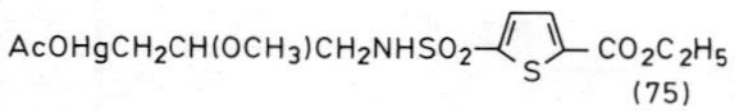 $AcOHgCH_2CH(OCH_3)CH_2NHSO_2$-(thiophene)-$CO_2C_2H_5$ (75)	—	—	133

Table 3.2. (continued)

Alkene	Alcohol	Mercuric salt	Organomercurial(s) (% Yield)	Subsequent reactants	Product(s) (% Yield)	Ref.
$H_2C=CHCH_2NHCO$ 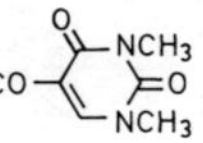 (1,3-dimethyluracil)	ROH	XHgOAc	$XHgCH_2CH(OR)CH_2NHCO$ 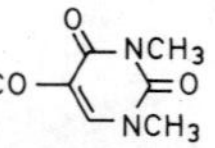 (1,3-dimethyluracil); R = Me, X = Cl (62), Br (90), I (95), SCN (75); R = Et, X = Cl (83); R = $MeOCH_2CH_2$, X = Cl (90)	—	—	128
$H_2C=C(CO_2H)NHCO$–(cyclohexyl) 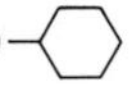	CH_3OH	$Hg(OAc)_2$	$AcOHgCH_2C(OCH_3)(CO_2H)NHCO$–(cyclohexyl) (~100) 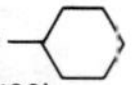	$NaBH_4/NaOH$, CH_2N_2	$CH_3C(OCH_3)(CO_2CH_3)NHCO$–(cyclohexyl) (12) 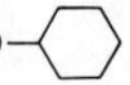	158
$H_2C=CHCH_2N$ (5,5-diethylbarbituric) 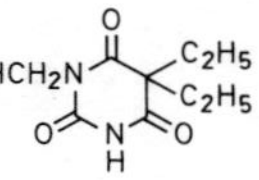	ROH	$Hg(OAc)_2$ (HOAc)	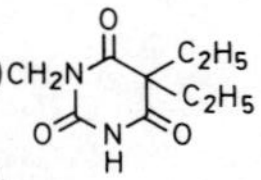 $AcOHgCH_2CH(OR)CH_2N$ (5,5-diethylbarbituric); R = Me (97), Et (98), n-Pr (94), i-Pr (83), n-Bu (8), i-Bu (40)	—	—	167
$HO_2CCH=CH(CH_2)_4CO_2C_2H_5$	CH_3OH	$Hg(OAc)_2$	$HO_2CCH(HgOAc)CH(OCH_3)(CH_2)_4CO_2C_2H_5$	KBr/Br_2 $h\nu$	$HO_2CCHBrCH(OCH_3)(CH_2)_4CO_2C_2H_5$	168
$H_2C=CHCH_2NHCONH$–(cyclohexyl) 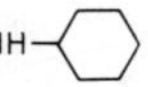	CH_3OH	$Hg(OAc)_2$ (HOAc)	—	NaCl	$ClHgCH_2CH(OCH_3)CH_2NHCONH$–(cyclohexyl) (75) 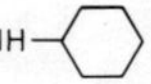	120
$C_6H_5CH=CHCO_2CH_3$	ROH	$Hg(OAc)_2$	$C_6H_5CH(OR)CH(HgX)CO_2CH_3$; X = OH, R = CH_2CH_2OH (78); X = OAc, R = CH_2CH_2OMe (80) and R = CH_2CH_2Cl (75)	—	—	141
p-$H_2C=CHCH_2OC_6H_4CO_2H$	ROH	$Hg(OAc)_2$	p-$HOHgCH_2CH(OR)CH_2OC_6H_4CO_2H$; R = Me (68), CH_2CH_2OH (73)	—	—	141
cis-o-$HO_2CCH=CHC_6H_4OCH_3$	CH_3OH	$Hg(OAc)_2$	o-OC—$CHCH(OCH_3)C_6H_4OCH_3$ (83.7), O—Hg	$H_2S/NaOH$	o-$HO_2CCH_2CH(OCH_3)C_6H_4OCH_3$	151
$trans$-o-$HO_2CCH=CHC_6H_4OCH_3$	CH_3OH	$Hg(OAc)_2$	o-OC—$CHCH(OCH_3)C_6H_4OCH_3$ (78.6), O—Hg	$H_2S/NaOH$	o-$HO_2CCH_2CH(OCH_3)C_6H_4OCH_3$	151

Table 3.2. (continued)

Alkene	Alcohol	Mercuric salt	Organomercurial(s) (% Yield)	Subsequent reactants	Product(s) (% Yield)	Ref.
o-HO$_2$CCH=CHC$_6$H$_4$OCH$_3$	CH$_3$OH	Hg(OAc)$_2$	o-HO$_2$CCH(HgOAc)CH(OCH$_3$)C$_6$H$_4$OCH$_3$	H$_2$S/NaOH	o-HO$_2$CCH$_2$CH(OCH$_3$)C$_6$H$_4$OCH$_3$	160
$trans$-p- HO$_2$CCH=CHC$_6$H$_4$OCH$_3$	CH$_3$OH	Hg(OAc)$_2$	HO$_2$CCH(HgOAc)CH(OCH$_3$)—(ring: HgOAc, OCH$_3$, HgOAc) 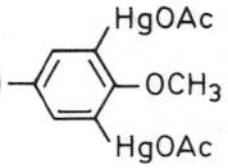	H$_2$S/NaOH	p-HO$_2$CCH$_2$CH(OCH$_3$)C$_6$H$_4$OCH$_3$	146
p-H$_2$C=C(CH$_3$)CH$_2$C$_6$H$_4$Cl	CH$_3$OH	Hg(OAc)$_2$	—	Cl$^-$	p-ClHgCH$_2$C(OCH$_3$)(CH$_3$)CH$_2$C$_6$H$_4$Cl	56
H$_2$C=CHCH$_2$NHCOC$_6$H$_5$	CH$_3$OH	Hg(OAc)$_2$ (HOAc)	AcOHgCH$_2$CH(OCH$_3$)CH$_2$NHCOC$_6$H$_5$ (80)	NaCl	ClHgCH$_2$CH(OCH$_3$)CH$_2$NHCOC$_6$H$_5$ (10)	118
o- or p- H$_2$C=CHCH$_2$NHC$_6$H$_4$CO$_2$H	CH$_3$OH	Hg(OAc)$_2$	o- or p-AcOHgCH$_2$CH(OCH$_3$)CH$_2$C$_6$H$_4$CO$_2$H	—	—	163
m-H$_2$C=CHCH$_2$NHSO$_2$C$_6$H$_4$CO$_2$H	HOCHRCH$_2$OH	Hg(OAc)$_2$	m-HOHgCH$_2$CH(OCH$_2$CHOHR) – CH$_2$NHSO$_2$C$_6$H$_4$CO$_2$H anhydride ? R = H, Me (88), CH$_2$OH (52)	—	—	164
H$_2$C=CHCH$_2$NHCONHC$_6$H$_5$	CH$_3$OH	Hg(OAc)$_2$ (HOAc)	AcOHgCH$_2$CH(OCH$_3$)CH$_2$NHCONHC$_6$H$_5$ (40)	NaCl	ClHgCH$_2$CH(OCH$_3$)CH$_2$NHCONHC$_6$H$_5$ (30)	120
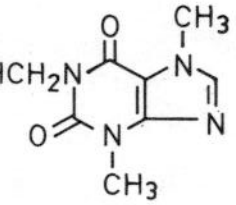 H$_2$C=CHCH$_2$– (theophylline ring) ?	CH$_3$CH$_2$OH	Hg(OAc)$_2$	AcOHgCH$_2$CH(OC$_2$H$_5$)CH$_2$– (theophylline ring) ?	X$^-$	XHgCH$_2$CH(OC$_2$H$_5$)CH$_2$– (theophylline ring) ? X = Cl, Br	165
(CH$_3$)$_2$C=CH–(cyclobutane: OH, CH$_3$, CH$_3$) 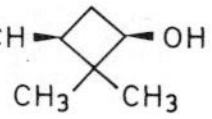	CH$_3$OH	Hg(OAc)$_2$	—	NaBH$_4$	(CH$_3$)$_2$C(OCH$_3$)CH$_2$–(cyclobutane: OH, CH$_3$, CH$_3$) 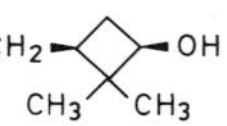 + E-(CH$_3$)$_2$C(OCH$_3$)CH=CHCH$_2$CHOHCH(CH$_3$)$_2$	169
H$_2$C=CHCH$_2$SCO$_2$(CH$_2$)$_5$CH$_3$	CH$_3$CH$_2$OH	Hg(OAc)$_2$	—	NaCl	ClHgCH$_2$CH(OC$_2$H$_5$)CH$_2$SCO$_2$(CH$_2$)$_5$CH$_3$ (70.3)	116
H$_2$C=CHCH$_2$NHCONH(CH$_2$)$_5$CH$_3$	CH$_3$OH	Hg(OAc)$_2$ (HOAc)	—	NaCl	ClHgCH$_2$CH(OCH$_3$)CH$_2$NHCONH(CH$_2$)$_5$CH$_3$ (60, 65)	109, 120
H$_2$C=CHCH$_2$N(phthalimide) 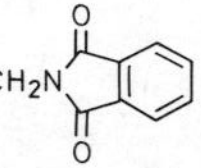	CH$_3$OH	Hg(OAc)$_2$	AcOHgCH$_2$CH(OCH$_3$)CH$_2$N(phthalimide) 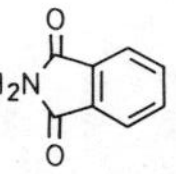	—	—	170, 171

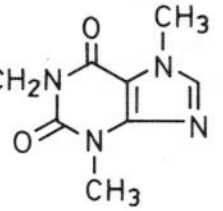

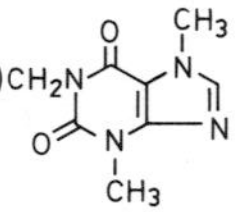

Table 3.2. (continued)

Alkene	Alcohol	Mercuric salt	Organomercurial(s) (% Yield)	Subsequent reactants	Product(s) (% Yield)	Ref.
	CH$_3$OH	Hg(OAc)$_2$	CH$_3$OCH$_2$CH(HgOAc)CH$_2$N(phthalimide) ? (60)	—	—	172, 173
	CH$_3$OH	Hg(OAc)$_2$	o-AcOHgCH$_2$CH(OCH$_3$)CH$_2$NHCOC$_6$H$_4$CO$_2$H ?	—	—	173, 174
	CH$_3$OH	Hg(OAc)$_2$ (KBr)	XHgCH$_2$CH(OCH$_3$)CH$_2$N(phthalimide); X = OAc (78), Br (12) + CH$_3$OCH$_2$CH(HgX)CH$_2$N(phthalimide); X = Br (3)	Y$_2$	YCH$_2$CH(OCH$_3$)CH$_2$N(phthalimide); Y = Br (55), I (88) + CH$_3$OCH$_2$CHYCH$_2$N(phthalimide); Y = I (86)	175
	CH$_3$OH	Hg(OAc)$_2$ /HgCl$_2$	ClHgCH$_2$CH(OCH$_3$)CH$_2$N(phthalimide) (87) + CH$_3$OCH$_2$CH(HgCl)CH$_2$N(phthalimide)	I$_2$ (on separated RHgCl's)	ICH$_2$CH(OCH$_3$)CH$_2$N(phthalimide) (81) or CH$_3$OCH$_2$CHICH$_2$N(phthalimide) (100)	175
	ROH	Hg(OAc)$_2$	AcOHgCH$_2$CH(OR)CH$_2$N(phthalimide); R = CH$_2$CH$_2$OH (95), CH$_2$CH$_2$OCH$_3$ (88)	—	—	141
p-H$_2$C=C(CH$_3$)CH$_2$C$_6$H$_4$CF$_3$	CH$_3$OH	Hg(OAc)$_2$	—	Cl$^-$	p-ClHgCH$_2$C(OCH$_3$)(CH$_3$)CH$_2$C$_6$H$_4$CF$_3$	56
H$_2$C=C(CO$_2$H)NHCOCH$_2$C$_6$H$_5$	ROH	Hg(OAc)$_2$	AcOHgCH$_2$C(OR)(CO$_2$H)NHCOCH$_2$C$_6$H$_5$ (~100)	NaBH$_4$/NaOH, CH$_2$N$_2$	CH$_3$C(OR)(CO$_2$CH$_3$)NHCOCH$_2$C$_6$H$_5$; R = Me (82), Et (16)	158
H$_2$C=C(CO$_2$H)NHCOCH$_2$OC$_6$H$_5$	CH$_3$OH	Hg(OAc)$_2$	AcOHgCH$_2$C(OCH$_3$)(CO$_2$H)NHCOCH$_2$OC$_6$H$_5$ (~100)	NaBH$_4$/NaOH, CH$_2$N$_2$	CH$_3$C(OCH$_3$)(CO$_2$CH$_3$)NHCOCH$_2$OC$_6$H$_5$ (29)	158

Table 3.2. (continued)

Alkene	Alcohol	Mercuric salt	Organomercurial(s) (% Yield)	Subsequent reactants	Product(s) (% Yield)	Ref.
$H_2C=CHCH_2NHCONHCOC_6H_5$	CH_3OH	$Hg(OAc)_2$ (HOAc)	$AcOHgCH_2CH(OCH_3)CH_2NHCONHCOC_6H_5$ (80)	NaCl	$ClHgCH_2CH(OCH_3)CH_2NHCONHCOC_6H_5$ (70)	120
trans-$C_6H_5CH=CHCO_2C_2H_5$	CH_3OH	$Hg(OAc)_2$	$C_6H_5CH(OCH_3)CH(HgOAc)CO_2C_2H_5$ (44.5)	—	—	162
$C_6H_5CH=CHCO_2C_2H_5$	CH_3OH	$Hg(OAc)_2$, X^-	$C_6H_5CH(OCH_3)CH(HgX)CO_2C_2H_5$ X = OAc, Cl, Br, I	—	—	66
	CH_3OH	$Hg(OAc)_2$	—	Br_2/KBr	$C_6H_5CH(OCH_3)CHBrCO_2C_2H_5$ (96)	63
trans-p- $CH_3OC_6H_4CH=CHCOCH_3$	CH_3OH	$Hg(OAc)_2$	$p-CH_3OC_6H_4CH(OCH_3)CH(HgOAc)COCH_3$	—	—	159
$C_6H_5(CH_2)_2CH=CHCO_2H$	CH_3OH	$Hg(OAc)_2$	$C_6H_5(CH_2)_2CH(OCH_3)CH(Hg^+)CO_2^-$ (75)	—	—	141
o-$H_2C=CHCH_2C_6H_4OCH_2CO_2H$	CH_3OH	$Hg(OAc)_2$	$o-HOHgCH_2CH(OCH_3)CH_2C_6H_4OCH_2CO_2H$	—	—	176
trans-p- $CH_3OC_6H_4CH=CHCO_2CH_3$	CH_3OH	$Hg(OAc)_2$	$p-CH_3OC_6H_4CH(OCH_3)CH(HgOAc)CO_2CH_3$	—	—	159
$H_2C=CHCH_2NHCONHCH_2C_6H_5$	CH_3OH	$Hg(OAc)_2$ (HOAc)	—	NaCl	$ClHgCH_2CH(OCH_3)CH_2NHCONHCH_2C_6H_5$ (65)	120
(xanthine derivative) $H_2C=CHCH_2O-$ 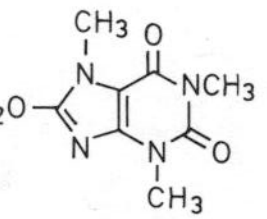	CH_3OH	$Hg(OAc)_2$ (HOAc)	$AcOHgCH_2CH(OCH_3)CH_2O-$ 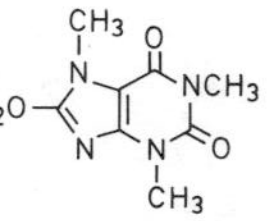 (xanthine)	H_2O Δ	$HOHgCH_2CH(OCH_3)CH_2O-$ 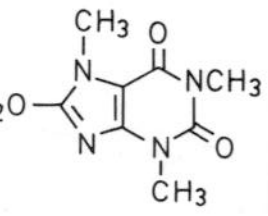 (xanthine)	177
(xanthine derivative) $H_2C=CHCH_2N-$ 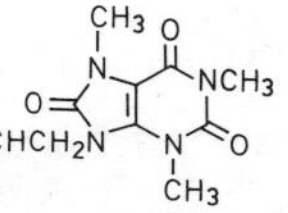	CH_3OH	$Hg(OAc)_2$	$AcOHgCH_2CH(OCH_3)CH_2N-$ 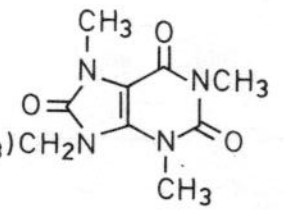 (xanthine) (83)	H_2O Δ	$HOHgCH_2CH(OCH_3)CH_2N-$ 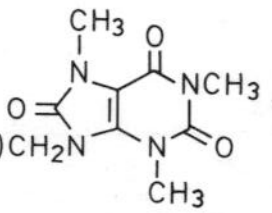 (xanthine)	177
$p-H_2C=C(CH_3)CH_2C_6H_4OCH_3$	CH_3OH	$Hg(OAc)_2$	—	Cl^-	$p-ClHgCH_2C(OCH_3)(CH_3)CH_2C_6H_4OCH_3$	49, 56
$C_6H_5CH=CHSi(OCH_3)_3$	CH_3OH	$Hg(OAc)_2$	$C_6H_5CH(OCH_3)CH(HgOAc)Si(OCH_3)_3$ (100)	—	—	27

Table 3.2. (continued)

Alkene	Alcohol	Mercuric salt	Organomercurial(s) (% Yield)	Subsequent reactants	Product(s) (% Yield)	Ref.
$C_6H_5CH{=}CHSi(CH_3)_3$	CH_3OH	$Hg(OAc)_2$	$C_6H_5CH(HgOAc)CH(OCH_3)Si(CH_3)_3$ 50 : + $C_6H_5CH(OCH_3)CH(HgOAc)Si(CH_3)_3$: 50 (72 total)	—	—	27
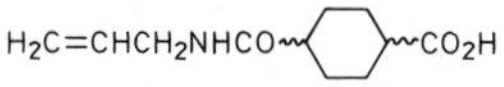 $H_2C{=}CHCH_2NHCO$—⬡—CO_2H	CH_3OH	$Hg(OAc)_2$	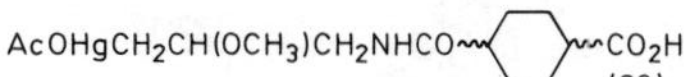 $AcOHgCH_2CH(OCH_3)CH_2NHCO$—⬡—$CO_2H$ (63)	—	—	178, 179
$H_2C{=}CH(CH_2)_8CN$	CH_3OH	$Hg(OAc)_2$ (?)	$AcOHgCH_2CH(OCH_3)(CH_2)_8CN$	—	—	180
$H_2C{=}CH(CH_2)_9OH$	CH_3OH	$Hg(OAc)_2$ (?)	$AcOHgCH_2CH(OCH_3)(CH_2)_9OH$	—	—	181
$H_2C{=}CH(CH_2)_9NH_2$	CH_3OH	$Hg(OAc)_2$ (?)	$AcOHgCH_2CH(OCH_3)(CH_2)_9NH_2$	—	—	182
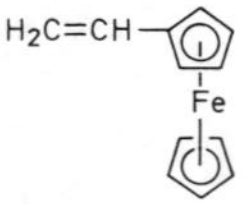 $H_2C{=}CH$-ferrocenyl	CH_3OH	$Hg(OAc)_2$	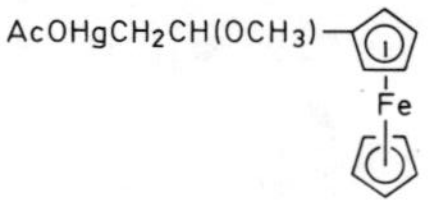 $AcOHgCH_2CH(OCH_3)$-ferrocenyl	$NaBH_4/NaOH$	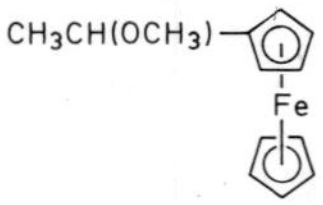 $CH_3CH(OCH_3)$-ferrocenyl	61, 183
$o\text{-}H_2C{=}CHCH_2NHCOC_6H_4$ - OCH_2CO_2Na	CH_3OH	—	$o\text{-}HOHgCH_2CH(OCH_3)CH_2NHCOC_6H_4OCH_2CO_2Na$	—	—	184
$p\text{-}CH_3CH{=}CHCONHC_6H_4$ - CH_2CO_2H	$HOCH_2CH_2OH$	$Hg(OAc)_2$	$p\text{-}CH_3CH(HgOH)CH(OCH_2CH_2OH)CONHC_6H_4CH_2CO_2H$? (50)	—	—	141
$o\text{-}H_2C{=}CHCH_2NHCOC_6H_4$ - OCH_2CO_2H	CH_3OH	$Hg(OAc)_2$	$o\text{-}AcOHgCH_2CH(OCH_3)CH_2NHCOC_6H_4OCH_2CO_2H$ (50)	—	—	133
	CH_3OH	$Hg(OAc)_2$	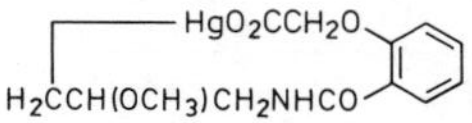 $H_2CCH(OCH_3)CH_2NHCO$—⬡—OCH_2CCO_2gH (?)	—	—	122
	$ROCH_2CH_2OH$	$Hg(OAc)_2$	—	$NaOH$ (H_2SO_4)	$o\text{-}HOHgCH_2CH(OCH_2CH_2OR)CH_2NHCOC_6H_4$ - $OCH_2CO_2H(Na)$ $R = H,\ HOCH_2CH_2$	185, 186

Table 3.2. (continued)

Alkene	Alcohol	Mercuric salt	Organomercurial(s) (% Yield)	Subsequent reactants	Product(s) (% Yield)	Ref.
	$ROCH_2CH_2OH$	HgO	(structure) HgO_2CCH_2O—; $CH_2CH(OCH_2CH_2OR)CH_2NHCO$—; $R = H, Me, Et, n\text{-}Bu, Et_2CHCH_2$	—	—	186
	ROH	HgO	$o\text{-}{}^+HgCH_2CH(OR)CH_2NHCOC_6H_4OCH_2CO_2^-$; $R = CH_2CH_2OH\ (80),\ CH_2CH_2OCH_3\ (52),\ n\text{-}Bu\ (90),\ CH_2CH_2OEt\ (54),\ CH_2CH_2OCH_2CH_2OH\ (55),\ CH_2CH_2O\text{-}n\text{-}Bu\ (70),\ CH_2CH_2OC_6H_5\ (88),\ CH_2CH_2OCH_2CH(CH_2CH_3)_2\ (70),\ n\text{-}C_{10}H_{21}\ (39)$	—	—	141
$m\text{-}H_2C{=}CHCH_2NHCOC_6H_4\text{-}OCH_2CO_2H$	$ROCH_2CH_2OH$	$Hg(OAc)_2$	$m\text{-}{}^+HgCH_2CH(OCH_2CH_2OR)CH_2NHCOC_6H_4OCH_2CO_2^-$; $R = CH_3\ (71),\ H\ (73)$ (71)	—	—	141
$p\text{-}H_2C{=}CHCH_2NHCOC_6H_4\text{-}OCH_2CO_2H$	$HOCH_2CH_2OH$	$Hg(OAc)_2$	$p\text{-}AcOHgCH_2CH(OCH_2CH_2OH)CH_2NHCOC_6H_4OCH_2CO_2H$ (63)	—	—	141
$H_2C{=}CHCH_2NHCOCH_2NHCOC_6H_5$	CH_3OH	$Hg(OAc)_2$	$AcOHgCH_2CH(OCH_3)CH_2NHCOCH_2NHCOC_6H_5$ (30)	—	—	187
	CH_3OH	$ClHgOAc$	$ClHgCH_2CH(OCH_3)CH_2NHCOCH_2NHCOC_6H_5$ (98)	—	—	128
$p\text{-}H_2C{=}CHCH_2OC_6H_4CO_2C_2H_5$	ROH	$Hg(OAc)_2$	$p\text{-}AcOHgCH_2CH(OR)CH_2OC_6H_4CO_2C_2H_5$; $R = Me\ (70),\ Et\ (75),\ CH_2CH_2OCH_3\ (78)$	—	—	141
$H_2C{=}CHCH_2\text{-}$ (3-HO-4-$CO_2C_2H_5$ phenyl)	ROH	$Hg(OAc)_2$	$AcOHgCH_2CH(OR)CH_2\text{-}$ (HO, $CO_2C_2H_5$ phenyl); $R = Me\ (50),\ CH_2CH_2OH\ (77)$	—	—	141
$trans\text{-}p\text{-}CH_3OC_6H_4CH{=}CHCO_2C_2H_5$	CH_3OH	$Hg(OAc)_2,\ KX$	$p\text{-}CH_3OC_6H_4CH(OCH_3)CH(HgX)CO_2C_2H_5$; $X = Br\ (90),\ I$	Y_2	CH_3O—(phenyl, Z)—$CH(OCH_3)CHYCO_2C_2H_5$; $Z = H,\ Y = I\ (71);\ Z = Y = Br\ (93)$	136
$H_2C{=}CHCH_2\text{-}$ (4-OCH_2CO_2H-3-OCH_3 phenyl)	CH_3OH	$Hg(OAc)_2$	$AcOHgCH_2CH(OCH_3)CH_2\text{-}$ (phenyl)—OCH_2CO_2H, OCH_3 (87)	—	—	188

Table 3.2. (continued)

Alkene	Alcohol	Mercuric salt	Organomercurial(s) (% Yield)	Subsequent reactants	Product(s) (% Yield)	Ref.
p-$H_2C=C(CH_3)COH(CH_3)C_6H_4OCH_3$	CH_3OH	HgX_2	—	KI	p-$IHgCH_2C(CH_3)(COCH_3)C_6H_4OCH_3$ $X = OAc\,(35,94);\ NO_3\,(35,78);\ ClO_4\,(40,80)$	189
$H_2C=CHCH_2N\!\!<^{CO}_{CO}\!\!>(CH_2)_7$	CH_3OH	$Hg(OAc)_2$	$AcOHgCH_2CH(OCH_3)CH_2N\!\!<^{CO}_{CO}\!\!>(CH_2)_7$	—	—	174
cis-$CH_3O_2CCH=CH(CH_2)_7CH_3$	CH_3OH	$Hg(OAc)_2$	—	$NaBH_4/NaOH$	$CH_3O_2CCH_2CH(OCH_3)(CH_2)_7CH_3$ (95)	190
cis-$CH_3O_2CCH_2CH=CH(CH_2)_6CH_3$	CH_3OH	$Hg(OAc)_2$	—	$NaBH_4/NaOH$	$CH_3O_2CCH_2CH(OCH_3)(CH_2)_7CH_3$ 5 : + $CH_3O_2C(CH_2)_2CH(OCH_3)(CH_2)_6CH_3$: 95	190
cis-$CH_3O_2C(CH_2)_2CH=CH(CH_2)_5CH_3$	CH_3OH	$Hg(OAc)_2$	—	$NaBH_4/NaOH$	$CH_3O_2C(CH_2)_3CH(OCH_3)(CH_2)_5CH_3$ 74 : + $CH_3O_2C(CH_2)_2CH(OCH_3)(CH_2)_6CH_3$: 26	190
cis-$CH_3O_2C(CH_2)_3CH=CH(CH_2)_4CH_3$	CH_3OH	$Hg(OAc)_2$	—	$NaBH_4/NaOH$	$CH_3O_2C(CH_2)_3CH(OCH_3)(CH_2)_5CH_3$ 50 : + $CH_3O_2C(CH_2)_4CH(OCH_3)(CH_2)_4CH_3$: 50	190
cis-$CH_3O_2C(CH_2)_4CH=CH(CH_2)_3CH_3$	CH_3OH	$Hg(OAc)_2$	—	$NaBH_4/NaOH$	$CH_3O_2C(CH_2)_4CH(OCH_3)(CH_2)_4CH_3$ 50 : + $CH_3O_2C(CH_2)_5CH(OCH_3)(CH_2)_3CH_3$: 50	190
cis-$CH_3O_2C(CH_2)_5CH=CH(CH_2)_2CH_3$	CH_3OH	$Hg(OAc)_2$	—	$NaBH_4/NaOH$	$CH_3O_2C(CH_2)_5CH(OCH_3)(CH_2)_3CH_3$ 50 : + $CH_3O_2C(CH_2)_6CH(OCH_3)(CH_2)_2CH_3$: 50	190
cis-$CH_3O_2C(CH_2)_6CH=CHCH_2CH_3$	CH_3OH	$Hg(OAc)_2$	—	$NaBH_4/NaOH$	$CH_3O_2C(CH_2)_7CH(OCH_3)CH_2CH_3$ 64 : + $CH_3O_2C(CH_2)_6CH(OCH_3)(CH_2)_2CH_3$: 36	190

Table 3.2. (continued)

Alkene	Alcohol	Mercuric salt	Organomercurial(s) (% Yield)	Subsequent reactants	Product(s) (% Yield)	Ref.
cis- $CH_3O_2C(CH_2)_7CH=CHCH_3$	CH_3OH	$Hg(OAc)_2$	—	$NaBH_4/NaOH$	$CH_3O_2C(CH_2)_8CH(OCH_3)CH_3$ 97 : + $CH_3O_2C(CH_2)_7CH(OCH_3)CH_2CH_3$: 3	190
$CH_3O_2C(CH_2)_8CH=CH_2$	CH_3OH	$Hg(OAc)_2$	—	$NaBH_4/NaOH$	$CH_3O_2C(CH_2)_8CH(OCH_3)CH_3$ (91)	73, 190
$H_2C=CHCH_2NHCON(n-C_4H_9)_2$	CH_3OH	$Hg(OAc)_2$ (HOAc)	—	$NaCl$	$ClHgCH_2CH(OCH_3)CH_2NHCON(n-C_4H_9)_2$ (35)	109, 120
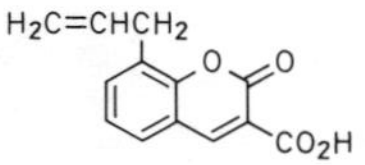 $H_2C=CHCH_2$ (8-substituted coumarin-3-CO_2H)	CH_3OH	$Hg(OAc)_2$	—	$NaOH$?	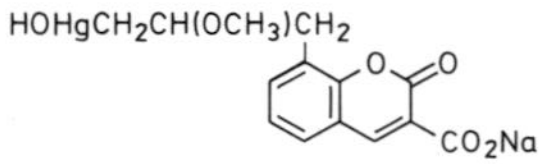 $HOHgCH_2CH(OCH_3)CH_2$ (8-substituted coumarin-3-CO_2Na)	122
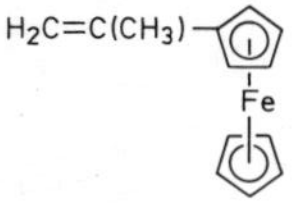 $H_2C=C(CH_3)$-ferrocenyl	CH_3OH	$Hg(OAc)_2$	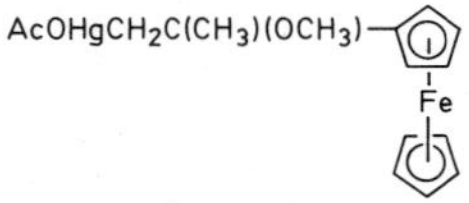 $AcOHgCH_2C(CH_3)(OCH_3)$-ferrocenyl	$NaBH_4/NaOH$	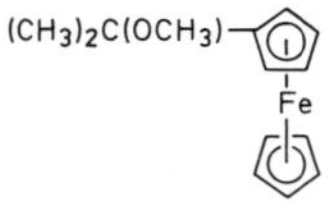 $(CH_3)_2C(OCH_3)$-ferrocenyl	61, 183
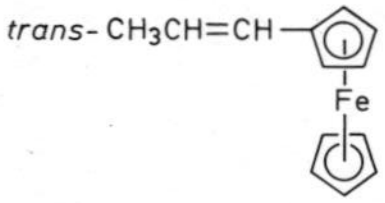 $trans$-$CH_3CH=CH$-ferrocenyl	CH_3OH	$Hg(OAc)_2$	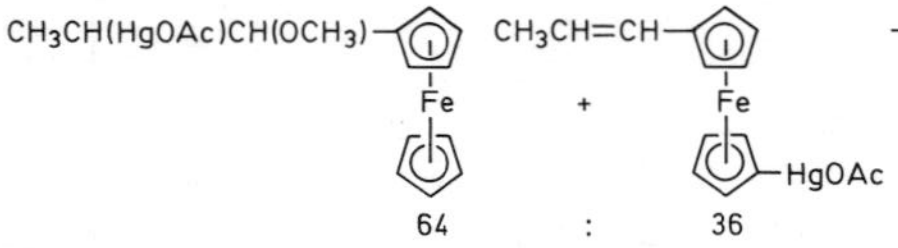 $CH_3CH(HgOAc)CH(OCH_3)$-ferrocenyl + $CH_3CH=CH$-(HgOAc-ferrocenyl) 64 : 36	—	—	61
$H_2C=CHCH_2NHCOCH(CH_3)$-$NHCOC_6H_5$	CH_3OH	$Hg(OAc)_2$	$AcOHgCH_2CH(OCH_3)CH_2NHCOCH(CH_3)$-$NHCOC_6H_5$ (48)	—	—	187
p-$H_2C=CHCH_2NHSO_2C_6H_4$-$NHCOCH_2CH_2CO_2H$	ROH	$Hg(OAc)_2$	p-$HOHgCH_2CH(OR)CH_2NHSO_2C_6H_4$-$NHCOCH_2CH_2CO_2H$ anhydride? R = $HOCH_2CH_2$ (96), $CH_3CHOHCH_2$ (74), $HOCH_2CHOHCH_2$ (77), $CH_3CHOHCH_2CH_2$ (93)	—	—	164
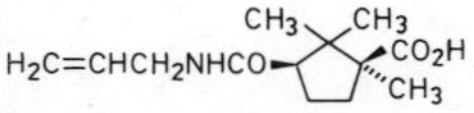 $H_2C=CHCH_2NHCO$-(cyclopentane-CH_3, CH_3, CO_2H, CH_3)	CH_3OH	$Hg(OAc)_2$	—	NaX	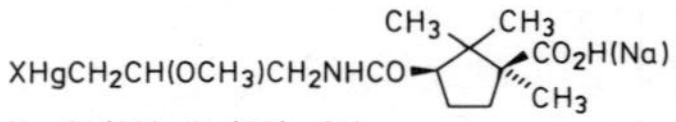 $XHgCH_2CH(OCH_3)CH_2NHCO$-(cyclopentane-CH_3, CH_3, $CO_2H(Na)$, CH_3) X = Cl (89), Br (72), OH	122, 191, 192

Table 3.2. (continued)

Alkene	Alcohol	Mercuric salt	Organomercurial(s) (% Yield)	Subsequent reactants	Product(s) (% Yield)	Ref.
$H_2C=CHCH_2NHCO$-[2,2,3-trimethyl-cyclopentane-CO_2H]	ROH	HgX_2 (X = Cl, OAc)	$HOHgCH_2CH(OR)CH_2NHCO$-[...CO_2H] $R = Me, Et, n\text{-}Pr, i\text{-}Pr$	—	—	193
$H_2C=CH$-N(carbazole)	ROH	HgX_2	$XHgCH_2CHOR$-N(carbazole) $R = Me,\ X = OAc,\ n\text{-}C_6H_{13}CO_2$; $R = n\text{-}Bu,\ X = OAc$	—	—	134
$H_2C=CHCH_2NHCO$-(1-naphthyl)	CH_3OH	$Hg(OAc)_2$	$AcOHgCH_2CH(OCH_3)CH_2NHCO$-(1-naphthyl) (35)	NaCl	$ClHgCH_2CH(OCH_3)CH_2NHCO$-(1-naphthyl) (40)	118
$H_2C=CHCH_2NHCONH$-(1-naphthyl)	CH_3OH	$Hg(OAc)_2$ (HOAc)	—	NaCl	$ClHgCH_2CH(OCH_3)CH_2NHCONH$-(1-naphthyl) (40)	120
$H_2C=CHCH_2NHCO$-[benzene, HO_2CCH_2O-, -$NHCOCH_3$]	ROH	$Hg(OAc)_2$	$AcOHgCH_2CH(OR)CH_2NHCO$-[...$NHCOCH_3$] $R = Me, Et, i\text{-}Pr$	—	—	194
$m\text{-}H_2C=CHCH_2NHCOC_6H_4\text{-}OCH_2CO_2C_2H_5$	ROH	$Hg(OAc)_2$	$m\text{-}AcOHgCH_2CH(OR)CH_2NHCOC_6H_4OCH_2CO_2C_2H_5$ $R = CH_2CH_2OMe$ (67), CH_2CH_2OEt (75), $CH_2CH_2O\text{-}n\text{-}Bu$ (70)	—	—	141
	ROH	$Hg(OAc)_2$	$m\text{-}HOHgCH_2CH(OR)CH_2NHCOC_6H_4OCH_2CO_2C_2H_5$ $R = CH_2CH_2Cl$ (98), $CH(CH_3)CO_2Et$ (55)	—	—	141
$trans\text{-}$ $CH_3CH=CHCO_2$-(menthyl)	CH_3OH	$Hg(OAc)_2$ (HOAc or $BF_3 \cdot Et_2O$)	$CH_3CH(OCH_3)CH(HgOAc)CO_2R$	$NaBH_4/NaOH$ or H_2S/C_5H_5N	$CH_3CH(OCH_3)CH_2CO_2R$ % R : HOAc-12, $BF_3 \cdot Et_2O$-15.5	195
$p\text{-}C_6H_5CH=CHCOC_6H_4Br$	CH_3OH	$Hg(OAc)_2$	$p\text{-}C_6H_5CH(OCH_3)CH(HgOAc)COC_6H_4Br$	Br_2	$p\text{-}C_6H_5CH(OCH_3)CHBrCOC_6H_4Br$	196

Table 3.2. (continued)

Alkene	Alcohol	Mercuric salt	Organomercurial(s) (% Yield)	Subsequent reactants	Product(s) (% Yield)	Ref.
p-C_6H_5CH=$CHCOC_6H_4Cl$	CH_3OH	$Hg(OAc)_2$	p-$C_6H_5CH(OCH_3)CH(HgOAc)COC_6H_4Cl$	—	—	197
m-$NO_2C_6H_4CH$=$CHCOC_6H_5$	CH_3OH	$Hg(OAc)_2$	m-$NO_2C_6H_4CH(OCH_3)CH(HgOAc)COC_6H_5$	Br_2	m-$NO_2C_6H_4CH(OCH_3)CHBrCOC_6H_5$	196
$trans$-p-$NO_2C_6H_4CH$=$CHCOC_6H_5$	CH_3OH	$Hg(OAc)_2$	p-$NO_2C_6H_4CH(OCH_3)CH(HgOAc)COC_6H_5$	Br_2	p-$NO_2C_6H_4CH(OCH_3)CHBrCOC_6H_5$	159, 196
cis-C_6H_5CH=$CHCOC_6H_5$	CH_3OH	$Hg(OAc)_2$	$C_6H_5CH(OCH_3)CH(HgOAc)COC_6H_5$ $S, R / R, R = {\sim}7:3$	—	—	159, 198
$trans$-C_6H_5CH=$CHCOC_6H_5$	CH_3OH	$Hg(OAc)_2$	$C_6H_5CH(OCH_3)CH(HgOAc)COC_6H_5$ erythro / threo = 81 : 19 $S, R / R, R = {\sim}70 : 30$	—	—	22, 159, 198
C_6H_5CH=$CHCOC_6H_5$	CH_3OH	$Hg(OAc)_2$	$C_6H_5CH(OCH_3)CH(HgOAc)COC_6H_5$	X_2	$C_6H_5CH(OCH_3)CHXCOC_6H_5$ X = Br, I	197
	CH_3OH	$Hg(OAc)_2$ $(HClO_4)$	$C_6H_5CH(OCH_3)CH(HgOAc)COC_6H_5$ (77)	KBr	$C_6H_5CH(OCH_3)CH(HgBr)COC_6H_5$ (74)	95
	CH_3CH_2OH	$Hg(OAc)_2$	$C_6H_5CH(OEt)CH(HgOAc)COC_6H_5$	X_2	$C_6H_5CH(OEt)CHXCOC_6H_5$ X = Br, I	197
H_2C=$CHCH_2NHCOCH$-$(CH_2CH_2SCH_3)NHCOC_6H_5$	CH_3OH	$Hg(OAc)_2$	$AcOHgCH_2CH(OCH_3)CH_2NHCOCH$-$(CH_2CH_2SCH_3)NHCOC_6H_5$ (58)	—	—	187
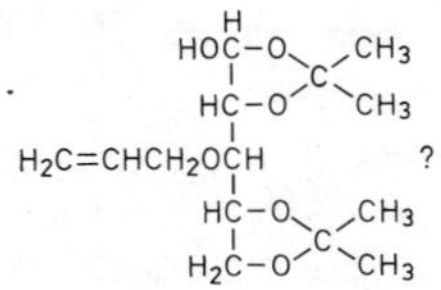 H_2C=$CHCH_2OCH[HC(O-C(CH_3)_2-O-CH)]$?	CH_3OH	$Hg(OAc)_2$	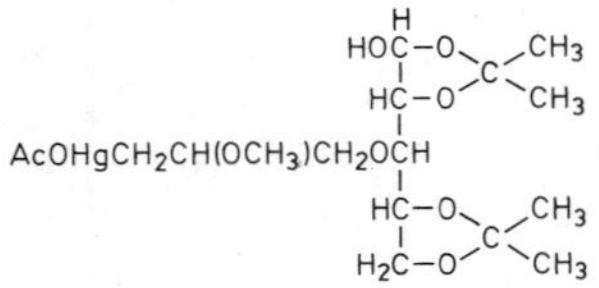 $AcOHgCH_2CH(OCH_3)CH_2OCH[HC(O-C(CH_3)_2-O-CH)]$?	—	—	155
$C_6H_5C(OCH_3)$=$CHCOC_6H_5$	CH_3OH	$Hg(OAc)_2$	$C_6H_5COC(HgOAc)_2COC_6H_5$ (80)	—	—	197
$trans$-p-$CH_3OC_6H_4CH$=$CHCOC_6H_5$	CH_3OH	$Hg(OAc)_2$	p-$CH_3OC_6H_4CH(OCH_3)CH(HgOAc)COC_6H_5$	—	—	159

Table 3.2. (continued)

Alkene	Alcohol	Mercuric salt	Organomercurial(s) (% Yield)	Subsequent reactants	Product(s) (% Yield)	Ref.
trans- $C_6H_5CH=CHCH_2O_2CC_6H_5$	CH_3OH	$Hg(OAc)_2$ (HNO_3), KCl	$C_6H_5CH(OCH_3)CH(HgX)CH_2O_2CC_6H_5$ $X = OAc,\ Cl\ (>95)$	KSCN	$C_6H_5CH(OCH_3)CH=CH_2\ +\ HO_2CC_6H_5$ (~90)	199
trans- $C_6H_5CH=CHCO_2CH_2C_6H_5$	CH_3OH	$Hg(OAc)_2$	$C_6H_5CH(OCH_3)CH(HgOAc)CO_2CH_2C_6H_5$ (52.7)	—	—	162
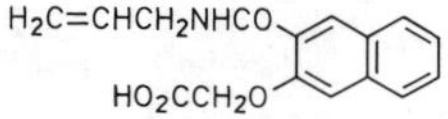 $H_2C=CHCH_2NHCO$ …naphthalene… HO_2CCH_2O	CH_3CH_2OH	$Hg(OAc)_2$	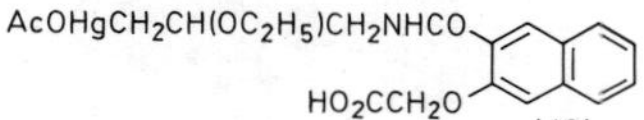 $AcOHgCH_2CH(OC_2H_5)CH_2NHCO$ …naphthalene… HO_2CCH_2O (40)	—	—	133
$H_2C=CHCH_2COH(C_6H_5)_2$	ROH	$Hg(OAc)_2$	$AcOHgCH_2CH(OR)CH_2COH(C_6H_5)_2$ $R = CH_3$ (70), CH_2CH_2OH (41)	—	—	141
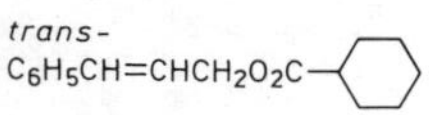 trans- $C_6H_5CH=CHCH_2O_2C$—cyclohexyl	CH_3OH	$Hg(OAc)_2$ (HNO_3), KCl	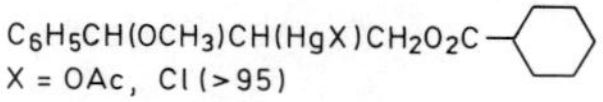 $C_6H_5CH(OCH_3)CH(HgX)CH_2O_2C$—cyclohexyl $X = OAc,\ Cl\ (>95)$	KSCN (on RHgOAc)	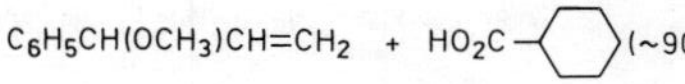 $C_6H_5CH(OCH_3)CH=CH_2\ +\ HO_2C$—cyclohexyl (~90)	199
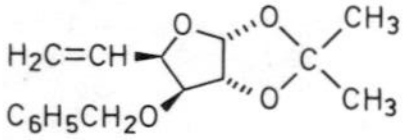 $H_2C=CH$ …(furanose acetonide) $C_6H_5CH_2O$	ROH	$Hg(O_2CCF_3)_2$	—	$NaBH_4/NaOH$	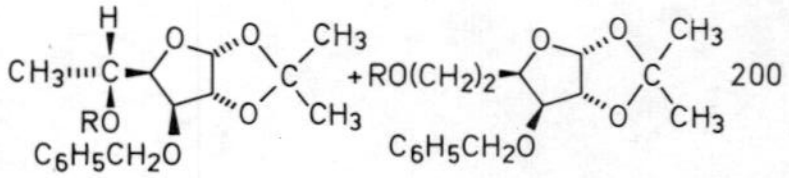 (furanose products) $R = i\text{-}Pr$ (55) $R = t\text{-}Bu$ (29); $+RO(CH_2)_2$ … $R = i\text{-}Pr$ (30) (R=H, 15) $R = t\text{-}Bu$ (59) (R=H, 12)	200
$H_2C=CHCH_2NHCOCH-$ $(i\text{-}C_4H_9)NHCOC_6H_5$	CH_3OH	$Hg(OAc)_2$	$AcOHgCH_2CH(OCH_3)CH_2NHCOCH-$ $(i\text{-}C_4H_9)NHCOC_6H_5$ (88)	—	—	187
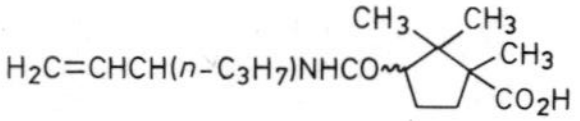 $H_2C=CHCH(n\text{-}C_3H_7)NHCO$ …(cyclopentane, CH_3, CH_3, CO_2H)	ROH	HgX_2 $(X=Cl, OAc)$	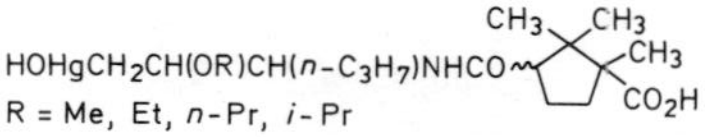 $HOHgCH_2CH(OR)CH(n\text{-}C_3H_7)NHCO$ …(cyclopentane, CH_3, CH_3, CO_2H) $R = Me,\ Et,\ n\text{-}Pr,\ i\text{-}Pr$	—	—	193
$C_6H_5C(OEt)=CHCOC_6H_5$	CH_3CH_2OH	$Hg(OAc)_2$	$C_6H_5COC(HgOAc)_2COC_6H_5$	—	—	197
trans- $C_6H_5CH=CHCH_2O_2CCH_2C_6H_5$	CH_3OH	$Hg(OAc)_2$ (HNO_3), KCl	$C_6H_5CH(OCH_3)CH(HgX)CH_2O_2CCH_2C_6H_5$ $X = OAc,\ Cl\ (>95)$	KSCN	$C_6H_5CH(OCH_3)CH=CH_2\ +\ HO_2CCH_2C_6H_5$ (~90)	199

Table 3.2. (continued)

Alkene	Alcohol	Mercuric salt	Organomercurial(s) (% Yield)	Subsequent reactants	Product(s) (% Yield)	Ref.
$trans-p-$ $C_6H_5CH{=}CHCH_2O_2CC_6H_4OCH_3$	CH_3OH	$Hg(OAc)_2$ (HNO_3), KCl	$p-C_6H_5CH(OCH_3)CH(HgX)CH_2O_2CC_6H_4OCH_3$ X = OAc, Cl (>95)	KSCN	$C_6H_5CH(OCH_3)CH{=}CH_2$ + $p-HO_2CC_6H_4OCH_3$ (~90)	199
$trans-C_6H_5CH{=}CHCO_2$⟨ribofuranose diacetonide⟩ 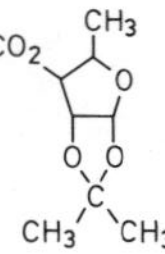	CH_3OH	$Hg(OAc)_2$ (HNO_3)	—	$NaBH_4$	$C_6H_5CH(OCH_3)CH_2CO_2H$ (89) 22% S	201
$H_2C{=}CHCH(n-C_4H_9)NHCO$⟨cyclopentane⟩ 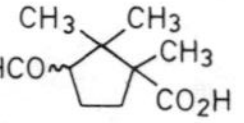	ROH	HgX_2 (X=Cl, OAc)	$HOHgCH_2CH(OR)CH(n-C_4H_9)NHCO$⟨cyclopentane⟩ R = Me, Et, n-Pr, i-Pr 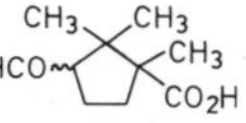	—	—	193
$H_2C{=}CHCH(n-C_5H_{11})NHCO$⟨cyclopentane⟩ 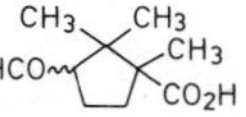	ROH	HgX_2 (X=Cl, OAc)	$HOHgCH_2CH(OR)CH(n-C_5H_{11})NHCO$⟨cyclopentane⟩ R = Me, Et, n-Pr, i-Pr 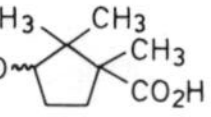	—	—	193
$trans-$ $CH_3(CH_2)_{11}CH{=}CH(CH_2)_3CO_2H$	CH_3OH	$Hg(OAc)_2$	—	$NaBH_4$, methylation	$CH_3(CH_2)_{11}CHCH(CH_2)_3CO_2CH_3$ (64) $H{\frown}OCH_3$ + $CH_3(CH_2)_{12}CHOH(CH_2)_3CO_2CH_3$ (23)	202
$cis-$ $CH_3(CH_2)_7CH{=}CH(CH_2)_7CO_2H$	CH_3OH	$Hg(OAc)_2$	—	$NaBH_4$	$CH_3(CH_2)_7CHCH(CH_2)_7CO_2H$ $H{\frown}OCH_3$	202
$H_2C{=}CHCH_2NHCOCH-$ $(CH_2C_6H_5)NHCOC_6H_5$	CH_3OH	$Hg(OAc)_2$	$AcOHgCH_2CH(OCH_3)CH_2NHCOCH-$ $(CH_2C_6H_5)NHCOC_6H_5$ (77)	—	—	187
$C_6H_5CH{=}CHCO_2$⟨bornyl⟩ 	CH_3OH	$Hg(OAc)_2$	$C_6H_5CH(OCH_3)CH(HgOAc)CO_2R$	NaBr	$C_6H_5CH(OCH_3)CH(HgBr)CO_2R$ 2 isomers	12
$trans-C_6H_5CH{=}CHCO_2$⟨bornyl⟩ 	CH_3OH	$Hg(OAc)_2$ (HNO_3)	—	$NaBH_4$	$C_6H_5CH(OCH_3)CH_2CO_2H$ (67) 4% R	201

Table 3.2. (continued)

Alkene	Alcohol	Mercuric salt	Organomercurial(s) (% Yield)	Subsequent reactants	Product(s) (% Yield)	Ref.
$trans$-$C_6H_5CH=CHCO_2$ (cyclohexyl with CH_3 and $(CH_3)_2CH$) (l)	CH_3OH	$Hg(OAc)_2$	$C_6H_5CH(OCH_3)CH(HgOAc)CO_2R$ (94)	NaX	$C_6H_5CH(OCH_3)CH(HgX)CO_2R$ X = Cl, Br, I	11
	CH_3OH	$Hg(OAc)_2$ $(BF_3 \cdot Et_2O)$	$C_6H_5CH(OCH_3)CH(HgOAc)CO_2R$	H_2S / C_5H_5N	$C_6H_5CH(OCH_3)CH_2CO_2R$ 2.3% S	195
	CH_3OH	$Hg(OAc)_2$ (HNO_3)	—	$NaBH_4$	$C_6H_5CH(OCH_3)CH_2CO_2H$ (36) 19% S	201
p-$(CH_3)_2C=CH(CH_2)_2CH(CH_3)$-$(CH_2)_2OC_6H_4COC_2H_5$	ROH	$Hg(OAc)_2$	—	$NaBH_4 / NaOH$	p-$(CH_3)_2C(OR)(CH_2)_3CH(CH_3)(CH_2)_2$-$OC_6H_4COC_2H_5$ R = Et (22), n-Pr (18), $\triangleright$—CH_2 (15)	203
cis-n-$C_7H_{15}CH=CH(CH_2)_3$-$C\equiv C(CH_2)_3CO_2CH_3$	CH_3OH	$Hg(OAc)_2$	—	$NaBH_4$	methyl 5 or 6-hydroxy, 10 or 11-methoxy-octadecanoate (mixture) (48)	190
cis-$CH_3(CH_2)_4CH$—$CHCH_2$- (epoxide) $CH=CH(CH_2)_7CO_2CH_3$	CH_3OH	$Hg(OAc)_2$	—	$NaBH_4$	$CH_3(CH_2)_4CH$—$CHCH_2CHCH(CH_2)_7CO_2CH_3$ (78) (epoxide, $H\cdots OCH_3$) + $CH_3(CH_2)_4$ (tetrahydrofuran, CH_3O) $(CH_2)_8CO_2CH_3$ (22)	202
cis-$CH_3(CH_2)_{14}CH=CHCO_2CH_3$	CH_3OH	$Hg(OAc)_2$	—	$NaBH_4$	$CH_3(CH_2)_{14}CH(OCH_3)CH_2CO_2CH_3$ (78)	73
$trans$-$CH_3(CH_2)_{14}CH=CHCO_2CH_3$	CH_3OH	$Hg(OAc)_2$	—	$NaBH_4$	$CH_3(CH_2)_{14}CH(OCH_3)CH_2CO_2CH_3$ (94)	73
cis-$CH_3(CH_2)_{13}CH=CHCH_2CO_2CH_3$	CH_3OH	$Hg(OAc)_2$	—	$NaBH_4$	$CH_3(CH_2)_{13}CH(OCH_3)(CH_2)_2CO_2CH_3$ (96) + $CH_3(CH_2)_{14}CH(OCH_3)CH_2CO_2CH_3$ (4)	73
$trans$-$CH_3(CH_2)_{13}CH=CHCH_2CO_2CH_3$	CH_3OH	$Hg(OAc)_2$	—	$NaBH_4$	$CH_3(CH_2)_{13}CH(OCH_3)(CH_2)_2CO_2CH_3$ (90) + $CH_3(CH_2)_{14}CH(OCH_3)CH_2CO_2CH_3$ (7)	73

Table 3.2. (continued)

Alkene	Alcohol.	Mercuric salt	Organomercurial(s) (% Yield)	Subsequent reactants	Product(s) (% Yield)	Ref.
cis-$CH_3(CH_2)_{12}CH=CH(CH_2)_2CO_2CH_3$	CH_3OH	$Hg(OAc)_2$	—	$NaBH_4$	$CH_3(CH_2)_{12}CH(OCH_3)(CH_2)_3CO_2CH_3$ (80) + $CH_3(CH_2)_{13}CH(OCH_3)(CH_2)_2CO_2CH_3$ (20)	73
trans-$CH_3(CH_2)_{12}CH=CH(CH_2)_2CO_2CH_3$	CH_3OH	$Hg(OAc)_2$	—	$NaBH_4$	$CH_3(CH_2)_{12}CH(OCH_3)(CH_2)_3CO_2CH_3$ (79) + $CH_3(CH_2)_{13}CH(OCH_3)(CH_2)_2CO_2CH_3$ (20)	73
cis-$CH_3(CH_2)_{11}CH=CH(CH_2)_3CO_2CH_3$	CH_3OH	$Hg(OAc)_2$	—	$NaBH_4$	$CH_3(CH_2)_{11}CH(OCH_3)(CH_2)_4CO_2CH_3$ (50) + $CH_3(CH_2)_{12}CH(OCH_3)(CH_2)_3CO_2CH_3$ (50)	73
cis-$CH_3(CH_2)_{10}CH=CH(CH_2)_4CO_2CH_3$	CH_3OH	$Hg(OAc)_2$	—	$NaBH_4$	$CH_3(CH_2)_{10}CH(OCH_3)(CH_2)_5CO_2CH_3$ (50) + $CH_3(CH_2)_{11}CH(OCH_3)(CH_2)_4CO_2CH_3$ (50)	73
cis-$CH_3(CH_2)_7CH=CH(CH_2)_7CO_2CH_3$	CH_3OH	$Hg(OAc)_2$	$CH_3(CH_2)_7CH(OCH_3)CH(HgOAc)(CH_2)_7CO_2CH_3$ Cl^- + isomer ?		$CH_3(CH_2)_7CH(OCH_3)CH(HgCl)(CH_2)_7CO_2CH_3$ + isomer ?	204
	CH_3OH	$Hg(OAc)_2$	—	$NaBH_4$	$CH_3(CH_2)_7CH(OCH_3)(CH_2)_8CO_2CH_3$ (50) + $CH_3(CH_2)_8CH(OCH_3)(CH_2)_7CO_2CH_3$ (50)	73
	ROH	$Hg(OAc)_2$	—	$NaBH_4$	$CH_3(CH_2)_7CHCH(CH_2)_7CO_2CH_3$ (H OR) R = Me (98), Et (95), $HOCH_2CH_2$ (53), n-Bu (0), t-Bu (0), C_6H_5 (0), n-Hex (0)	73
cis-$CH_3(CH_2)_5CHOHCH_2-CH=CH(CH_2)_7CO_2CH_3$	CH_3OH	$Hg(OAc)_2$	—	$NaBH_4$	$CH_3(CH_2)_5CHOH(CH_2)_2CH(OCH_3)(CH_2)_7CO_2CH_3$ (66) + $CH_3(CH_2)_5CHOHCH_2CH(OCH_3)(CH_2)_8CO_2CH_3$ (18) + $CH_3(CH_2)_5$—(O)—$(CH_2)_7CO_2CH_3$ 2 isomers : 8% and 2%	202
methyl cis- 5 and 6-hydroxy-10-octadecenoate	CH_3OH	$Hg(OAc)_2$	—	$NaBH_4$	$n-C_8H_{17}$—(O)—$(CH_2)_4CO_2CH_3$ + methyl 5-hydroxy-10 and/or 11-methoxyoctadecanoate	190
(I)	CH_3OH	$Hg(OAc)_2$ ($BF_3 \cdot Et_2O$)	$C_6H_5C(CH_3)(OCH_3)CH(HgOAc)CO_2R$	H_2S/C_5H_5N	$C_6H_5C(CH_3)(OCH_3)CH_2CO_2R$ 4.1% S	195

Table 3.2. (continued)

Alkene	Alcohol	Mercuric salt	Organomercurial(s) (% Yield)	Subsequent reactants	Product(s) (% Yield)	Ref.
cis - CH$_3$(CH$_2$)$_7$CH=CH - (CH$_2$)$_7$CO$_2$C$_2$H$_5$	CH$_3$CH$_2$OH	Hg(OAc)$_2$	CH$_3$(CH$_2$)$_7$CH(OEt)CH(HgOAc)(CH$_2$)$_7$CO$_2$C$_2$H$_5$ + isomer ?	Cl$^-$	CH$_3$(CH$_2$)$_7$CH(OEt)CH(HgCl)(CH$_2$)$_7$CO$_2$C$_2$H$_5$ + isomer ?	204
cis - CH$_3$(CH$_2$)$_5$CH(OCH$_3$)CH$_2$ - CH=CH(CH$_2$)$_7$CO$_2$CH$_3$	CH$_3$OH	Hg(OAc)$_2$	—	NaBH$_4$	CH$_3$(CH$_2$)$_5$CH(OCH$_3$)(CH$_2$)$_2$CH(OCH$_3$)(CH$_2$)$_7$CO$_2$CH$_3$ (79) + CH$_3$(CH$_2$)$_5$CH(OCH$_3$)CH$_2$CH(OCH$_3$)(CH$_2$)$_8$CO$_2$CH$_3$ (18)	73
trans - CH$_3$(CH$_2$)$_5$CH(OCH$_3$)CH$_2$ - CH=CH(CH$_2$)$_7$CO$_2$CH$_3$	CH$_3$OH	Hg(OAc)$_2$	—	NaBH$_4$	CH$_3$(CH$_2$)$_5$CH(OCH$_3$)(CH$_2$)$_2$CH(OCH$_3$)(CH$_2$)$_7$CO$_2$CH$_3$ (70) + CH$_3$(CH$_2$)$_5$CH(OCH$_3$)CH$_2$CH(OCH$_3$)(CH$_2$)$_8$CO$_2$CH$_3$ (18)	73
cis - CH$_3$(CH$_2$)$_4$CH=CH(CH$_2$)$_2$ - CH(OCH$_3$)(CH$_2$)$_7$CO$_2$CH$_3$	CH$_3$OH	Hg(OAc)$_2$	—	NaBH$_4$	CH$_3$(CH$_2$)$_4$CH(OCH$_3$)(CH$_2$)$_3$CH(OCH$_3$)(CH$_2$)$_7$CO$_2$CH$_3$ (56) + CH$_3$(CH$_2$)$_5$CH(OCH$_3$)(CH$_2$)$_2$CH(OCH$_3$)(CH$_2$)$_7$CO$_2$CH$_3$ (41)	73
trans - CH$_3$(CH$_2$)$_4$CH=CH(CH$_2$)$_2$ - CH(OCH$_3$)(CH$_2$)$_7$CO$_2$CH$_3$	CH$_3$OH	Hg(OAc)$_2$	—	NaBH$_4$	CH$_3$(CH$_2$)$_4$CH(OCH$_3$)(CH$_2$)$_3$CH(OCH$_3$)(CH$_2$)$_7$CO$_2$CH$_3$ (58) + CH$_3$(CH$_2$)$_5$CH(OCH$_3$)(CH$_2$)$_2$CH(OCH$_3$)(CH$_2$)$_7$CO$_2$CH$_3$ (38)	73
trans-C$_6$H$_5$CH=CHCO$_2$ [structure]	CH$_3$OH	Hg(OAc)$_2$ (HNO$_3$)	—	NaBH$_4$	C$_6$H$_5$CH(OCH$_3$)CH$_2$CO$_2$H (77) 22% S	201
cis - CH$_3$(CH$_2$)$_5$CH(OAc)CH$_2$ - CH=CH(CH$_2$)$_7$CO$_2$CH$_3$	CH$_3$OH	Hg(OAc)$_2$	—	NaBH$_4$	CH$_3$(CH$_2$)$_5$CH(OAc)(CH$_2$)$_2$CH(OCH$_3$)(CH$_2$)$_7$CO$_2$CH$_3$ (78) + CH$_3$(CH$_2$)$_5$CH(OAc)CH$_2$CH(OCH$_3$)(CH$_2$)$_8$CO$_2$CH$_3$ (20)	73
trans - CH$_3$(CH$_2$)$_5$CH(OAc)CH$_2$ - CH=CH(CH$_2$)$_7$CO$_2$CH$_3$	CH$_3$OH	Hg(OAc)$_2$	—	NaBH$_4$	CH$_3$(CH$_2$)$_5$CH(OAc)(CH$_2$)$_2$CH(OCH$_3$)(CH$_2$)$_7$CO$_2$CH$_3$ (60) + CH$_3$(CH$_2$)$_5$CH(OAc)CH$_2$CH(OCH$_3$)(CH$_2$)$_8$CO$_2$CH$_3$ (10)	73

Table 3.2. (continued)

Alkene	Alcohol	Mercuric salt	Organomercurial(s) (% Yield)	Subsequent reactants	Product(s) (% Yield)	Ref.
cis-$CH_3(CH_2)_4CH{=}CH(CH_2)_2$-$CH(OAc)(CH_2)_7CO_2CH_3$	CH_3OH	$Hg(OAc)_2$	—	$NaBH_4$	$CH_3(CH_2)_4CH(OCH_3)(CH_2)_3CH(OAc)(CH_2)_7CO_2CH_3$ (77) + $CH_3(CH_2)_5CH(OCH_3)(CH_2)_2CH(OAc)(CH_2)_7CO_2CH_3$ (20)	73
$trans$-$CH_3(CH_2)_4CH{=}CH(CH_2)_2$-$CH(OAc)(CH_2)_7CO_2CH_3$	CH_3OH	$Hg(OAc)_2$	—	$NaBH_4$	$CH_3(CH_2)_4CH(OCH_3)(CH_2)_3CH(OAc)(CH_2)_7CO_2CH_3$ (64) + $CH_3(CH_2)_5CH(OCH_3)(CH_2)_2CH(OAc)(CH_2)_7CO_2CH_3$ (32)	73
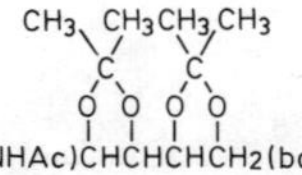 $EtO_2CC(OEt){=}CHCH(OCH_3)CH(NHAc)CHCHCHCH_2$ (both C-4 epimers)	CH_3CH_2OH	$Hg(OAc)_2$ $(HClO_4)$	—	$NaBH_4/NaOH$	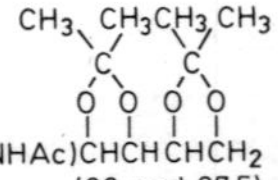 $EtO_2C(OEt)_2CH_2CH(OCH_3)CH(NHAc)CHCHCHCH_2$ (92 and 97.5)	205
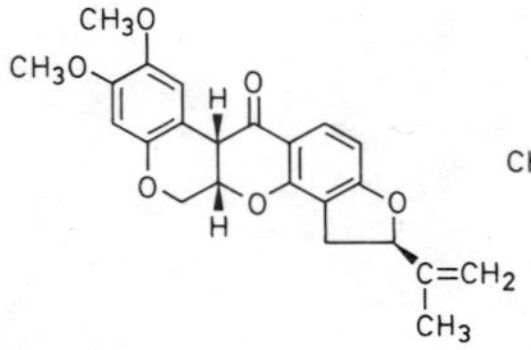	CH_3OH	$Hg(OAc)_2$	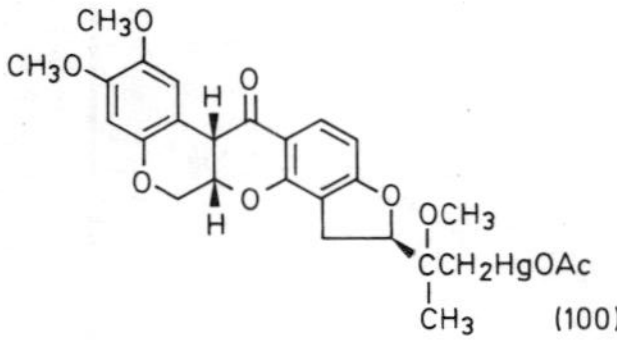 CH_2HgOAc … OCH_3, CH_3 (100)	—	—	206
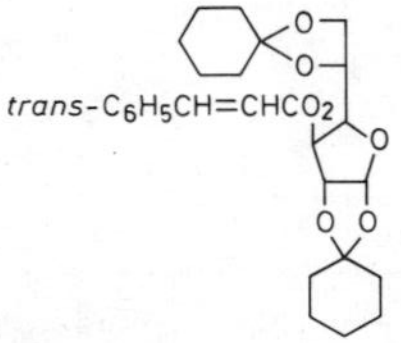 $trans$-$C_6H_5CH{=}CHCO_2$	CH_3OH	$Hg(OAc)_2$ (HNO_3)	—	$NaBH_4$	$C_6H_5CH(OCH_3)CH_2CO_2H$ (81) 27% S	201
$H_2C{=}C(OCH_3)CH(OBz)CH(OBz)$-$CH(OTs)CH(OCH_3)(OAc)$	CH_3OH	$Hg(OAc)_2$	$AcOHgCH_2C(OCH_3)_2CH(OBz)CH(OBz)CH(OTs)CH(OCH_3)(OAc)$ (75)	—	—	207

Table 3.3. Alkoxymercuration of Cyclic Alkenes

Alkene	Alcohol	Mercuric salt	Organomercurial(s) (% Yield)	Subsequent reactants	Product(s) (% Yield)	Ref.
cyclobutene	CH_3OH	$Hg(OAc)_2$, NaCl	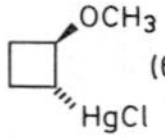 cyclobutane with OCH$_3$ and HgCl (65)	$NaBH_4$	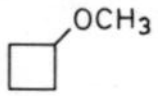 cyclobutane with OCH$_3$	23, 208
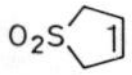 O$_2$S-cyclopentene	CH_3OH	$Hg(OAc)_2$ [HOAc, $(PhCO_2)_2$]	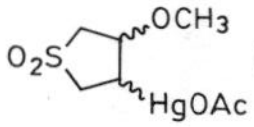 O$_2$S-ring with OCH$_3$ and HgOAc (63)	—	—	137
cyclopentene	CH_3OH	$Hg(OAc)_2$	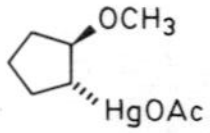 cyclopentane with OCH$_3$ and HgOAc (92)	—	—	30, 209
	CH_3OH	$Hg(OAc)_2$	—	Na X	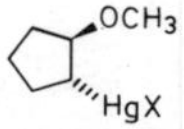 cyclopentane with OCH$_3$ and HgX; X = Cl (88), Br (90), I (65)	32
	CH_3OH	$Hg(OAc)_2$, NaCl	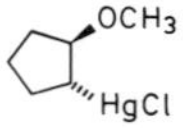 cyclopentane with OCH$_3$ and HgCl	$NaBH_4$	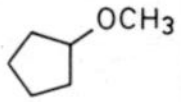 cyclopentane with OCH$_3$	208
 3,4-dihydro-2H-pyran	CH_3OH	$Hg(OAc)_2$	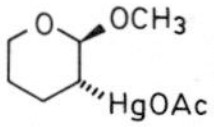 pyran ring with OCH$_3$ and HgOAc (74)	NaCl	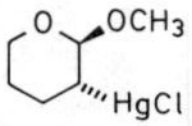 pyran ring with OCH$_3$ and HgCl	16, 210
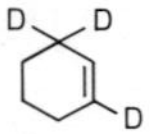 D,D,D-labeled cyclohexene	CH_3OH	$Hg(OAc)_2$	—	NaCl	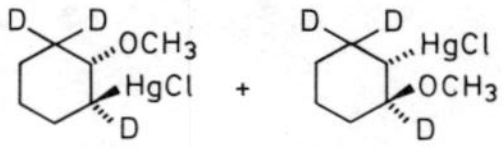 D,D-cyclohexane with OCH$_3$ and HgCl, D + D,D-cyclohexane with HgCl and OCH$_3$, D	15
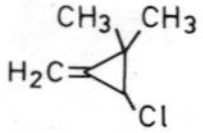 H$_2$C=cyclopropane(CH$_3$,CH$_3$)Cl	CH_3CH_2OH	HgX_2, NaCl	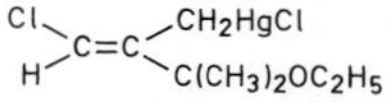 Cl-CH=C=C(CH$_2$HgCl)(C(CH$_3$)$_2$OC$_2$H$_5$), H; X = Cl (29), OAc (93)	$NaBH_4$	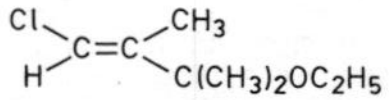 Cl-CH=C=C(CH$_3$)(C(CH$_3$)$_2$OC$_2$H$_5$), H	211
cyclohexene	CH_3OH	$Hg(OAc)_2$	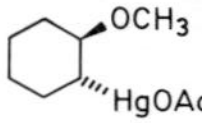 cyclohexane with OCH$_3$ and HgOAc (90)	—	—	30, 209, 212

Table 3.3. (continued)

Alkene	Alcohol	Mercuric salt	Organomercurial(s) (% Yield)	Subsequent reactants	Product(s) (% Yield)	Ref.
	CH_3OH	$Hg(OAc)_2$, NaCl	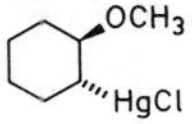 cyclohexane with OCH₃ and HgCl (81, 89)	$NaBH_4$	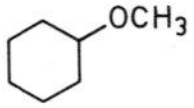 cyclohexane with OCH₃	208, 210, 213
	CH_3OH	$Hg(OAc)_2$	—	KBr	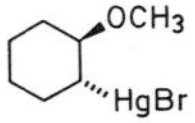 cyclohexane with OCH₃ and HgBr (76)	214
	CH_3OH	$Hg(O_2CR)_2$	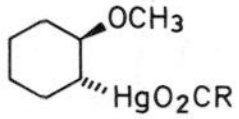 cyclohexane with OCH₃ and HgO₂CR; R = CH_3, C_2H_5, n-C_3H_7, n-C_7H_{15}, n-$C_{17}H_{35}$, $ClCH_2$, C_6H_5, 2-HOC_6H_4	—	—	37
	CH_3OH	$Hg(O_2CCF_3)_2$	—	NaCl	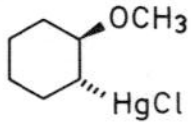 cyclohexane with OCH₃ and HgCl (58)	215
	CH_3OH	$Hg(II)$-d, l-mandelate	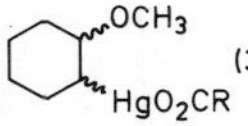 cyclohexane with OCH₃ and HgO₂CR (39)	Br^-	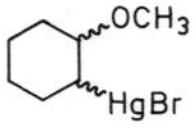 cyclohexane with OCH₃ and HgBr (83)	216
	CH_3OH	$Hg(II)$ (+)-mandelate	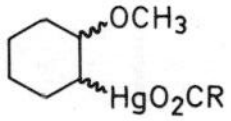 cyclohexane with OCH₃ and HgO₂CR (54)	Br^-	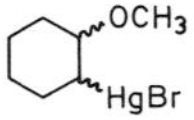 cyclohexane with OCH₃ and HgBr, optically inactive	216
	CH_3OH	$Hg(NO_3)_2$ (HNO_3)	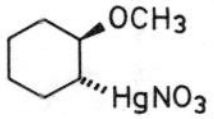 cyclohexane with OCH₃ and HgNO₃	NaCl	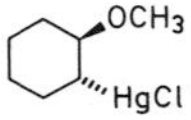 cyclohexane with OCH₃ and HgCl (80)	217
	ROH	$Hg(OAc)_2$	—	NaX	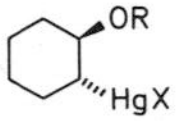 cyclohexane with OR and HgX; X = Cl R = Me (80), Et (79), n-Pr, i-Pr (80), t-Bu, $CH_3OCH_2CH_2$ (60); X = Br, I R = Me	215, 218, 219
	ROH	HgX_2.	—	$NaBH_4$/NaOH	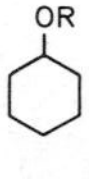 cyclohexane with OR; X = OAc, R = Me (100), Et (100), i-Pr (73), t-Bu (18); X = O_2CCF_3, R = Me (100), Et (100), i-Pr (98), t-Bu (90)	57

204 **Table 3.3.** (continued)

Alkene	Alcohol	Mercuric salt	Organomercurial(s) (% Yield)	Subsequent reactants	Product(s) (% Yield)	Ref.
	CH_3CH_2OH	$Hg(OAc)_2$	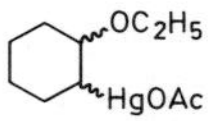 cyclohexane, 1-OC_2H_5, 2-HgOAc	—	—	220, 221
	$CH_3(CH_2)_2OH$	$Hg(OAc)_2$	—	NaCl	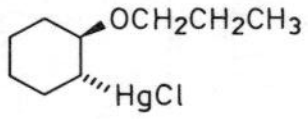 cyclohexane, 1-$OCH_2CH_2CH_3$, 2-HgCl (71)	219
	$(CH_3)_3COH$	$Hg(OAc)_2$ ($BF_3 \cdot Et_2O$)	—	Cl^-	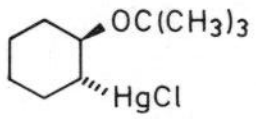 cyclohexane, 1-$OC(CH_3)_3$, 2-HgCl	222
	$C_6H_5CH_2OH$	$Hg(OAc)_2$	—	KBr	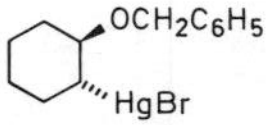 cyclohexane, 1-$OCH_2C_6H_5$, 2-HgBr (79)	214
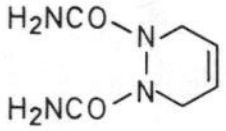 dihydropyridazine, 1,2-bis(H_2NCO)	CH_3OH	$Hg(OAc)_2$ (HNO_3)	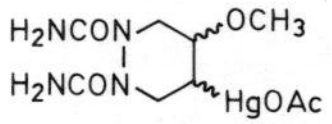 H_2NCON, H_2NCON ring, OCH_3, HgOAc (78)	NaCl	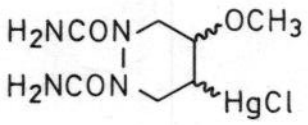 H_2NCON, H_2NCON ring, OCH_3, HgCl (40)	223
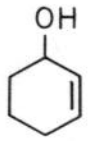 cyclohex-2-en-1-ol	CH_3OH	$Hg(OAc)_2$	—	$NaBH_4$	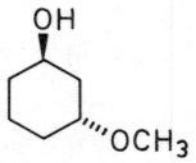 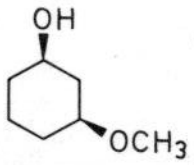 OH, OCH_3 cyclohexanes + (92 total) 84.4 : 15.6	224,225
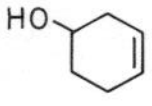 4-hydroxycyclohexene (HO-)	CH_3OH	$Hg(OAc)_2$, NaCl	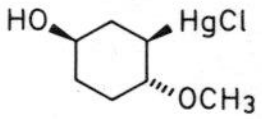 HO-cyclohexane, HgCl, OCH_3 (95)	$N_2H_4/NaOH$ Δ	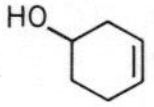 HO-cyclohexene (52) 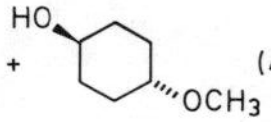 + HO-cyclohexane OCH_3 (40)	129, 226
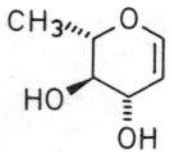 glycal (CH_3, O, HO, OH)	CH_3OH	$Hg(OAc)_2$	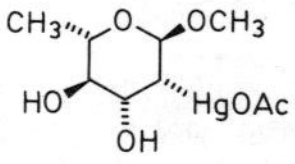 pyranose, CH_3, OCH_3, HO, OH, HgOAc (81)	$NaBH_4$ or KBH_4	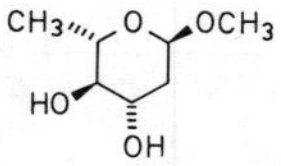 pyranose, CH_3, OCH_3, HO, OH (92)	227, 228

Table 3.3. (continued)

Alkene	Alcohol	Mercuric salt	Organomercurial(s) (% Yield)	Subsequent reactants	Product(s) (% Yield)	Ref.
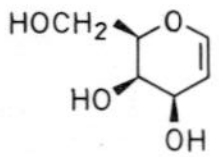 (structure)	CH_3OH	$Hg(OAc)_2$, NaCl	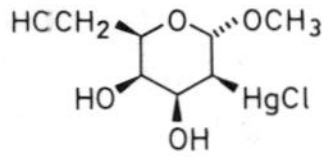 (structure)	$NaBH_4$	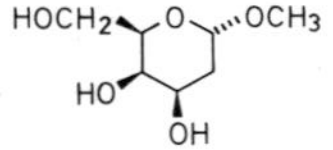 (structure)	35
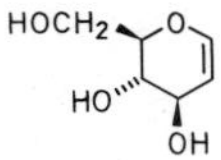 (structure)	CH_3OH	$Hg(OAc)_2$	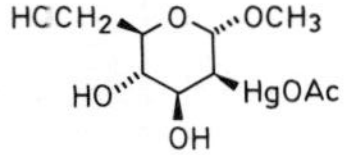 (structure) (60–70)	KBH_4/OH^-	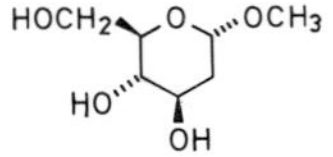 (structure) (73)	229
(structure)	CH_3OH	$Hg(OAc)_2$, Cl^-	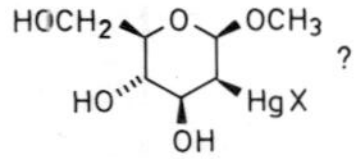 (structure) ? X = OAc (≧72), Cl, Br	Br_2	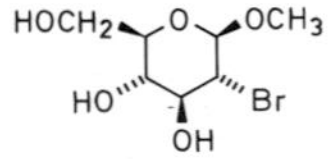 (structure) (60 from RHgOAc)	210
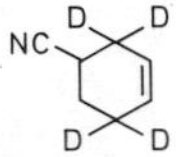 (structure)	ROH	$Hg(OAc)_2$	—	NaCl	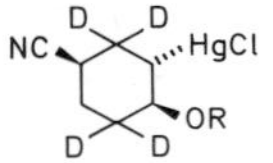 (structure) R = CH_3, CD_3	25, 26
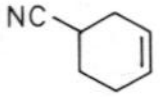 (structure)	CH_3OH	$Hg(OAc)_2$, NaCl	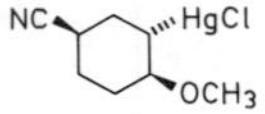 (structure)	$NaBH_4$	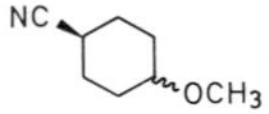 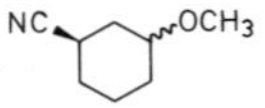 (structure) + (structure)	25
	CH_3OH	$Hg(OAc)_2$, NaCl	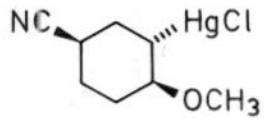 (structure)	Br_2	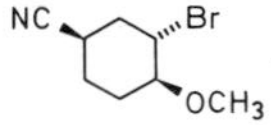 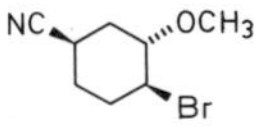 (structure) + (structure) 69 : 31	25
	CH_3OH	$Hg(OAc)_2$, NaCl	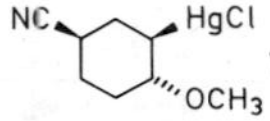 (structure) ?	$N_2H_4/NaOH$ Δ	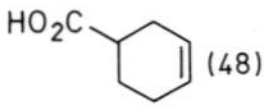 (structure) (48) 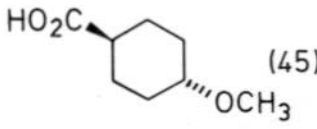 + (structure) (45)	129
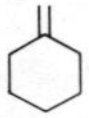 (structure)	CH_3OH	$Hg(OAc)_2$, Cl^-	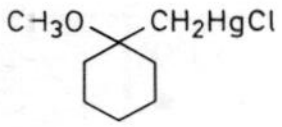 (structure) (91)	$NaHB(OCH_3)_3$	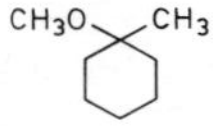 (structure) (71)	59

Table 3.3. (continued)

Alkene	Alcohol	Mercuric salt	Organomercurial(s) (% Yield)	Subsequent reactants	Product(s) (% Yield)	Ref.
(1-methylcyclohexene)	CH_3OH	$Hg(OAc)_2$	—	NaCl	(cyclohexane, CH_3, OCH_3, $HgCl$) (75)	230
(3-methylcyclohexene)	CH_3OH	$Hg(OAc)_2$, NaCl	(CH_3, OCH_3, $HgCl$) (3.2) + (CH_3, OCH_3, $HgCl$) (2.9) + (CH_3, $HgCl$, OCH_3) (0.3) (wrong structures)	—	—	230
(cycloheptene)	CH_3OH	$Hg(OAc)_2$	(OCH_3, $HgOAc$) (86)	—	—	209
(cycloheptene)	CH_3OH	$Hg(OAc)_2$, NaCl	(OCH_3, $HgCl$) (70, 80)	$NaBH_4$	(cycloheptane, OCH_3)	208, 231
(1-methoxycyclohexene)	CH_3OH	$Hg(OAc)_2$	—	NaCl	(cyclohexanone, $HgCl$)	232
(CH_3O-cyclohexene)	CH_3OH	$Hg(OAc)_2$, NaCl	(CH_3O, $HgCl$, OCH_3)	N_2H_4/NaOH Δ	(CH_3O, OCH_3) (38) + (CH_3O-cyclohexene) (35)	129, 226
($HOCH_2$-cyclohexene)	CH_3OH	$Hg(OAc)_2$, NaCl	($HOCH_2$, $HgCl$, OCH_3)	N_2H_4/NaOH Δ	($HOCH_2$-cyclohexene) (59) + ($HOCH_2$, OCH_3) (32)	129
(NC, NC, D, D dicyanocyclohexene)	CD_3OH	$Hg(OAc)_2$, NaCl	(NC, NC, D's, $HgCl$, OCD_3)	Br_2	(NC, NC, D's, Br, OCD_3) 86 : (NC, NC, D's, Br, Br) 14	25, 26
(NC, NC dicyanocyclohexene)	CH_3OH	$Hg(OAc)_2$, NaCl	(NC, NC, $HgCl$, OCH_3)	$NaBH_4$ or Br_2	(NC, NC, X, OCH_3) X = H, Br (56)	25

Table 3.3. (continued)

Alkene	Alcohol	Mercuric salt	Organomercurial(s) (% Yield)	Subsequent reactants	Product(s) (% Yield)	Ref.
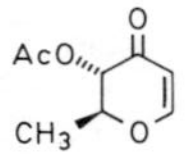 [structure: AcO-/CH₃ pyranone]	CH_3OH	HgX_2 (?)	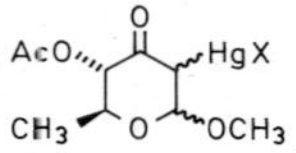 [structure: AcO-/CH₃ HgX, OCH₃]	—	—	233
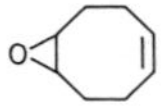 [structure: epoxycyclooctene]	CH_3OH	$Hg(OAc)_2$	—	$NaBH_4/NaOH$	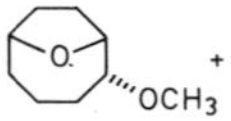 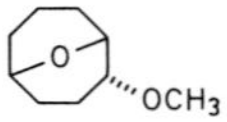 [structures: O, OCH₃ + O, OCH₃]	234
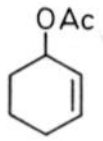 [structure: cyclohexenyl OAc]	CH_3OH	$Hg(OAc)_2$	—	$NaBH_4$	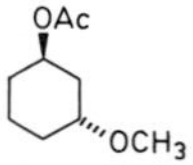 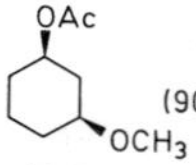 [structures: OAc, OCH₃ + OAc, OCH₃ (90 total); 74.5 : 25.5]	224
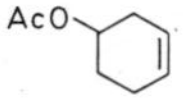 [structure: AcO cyclohexene]	CH_3OH	$Hg(OAc)_2$	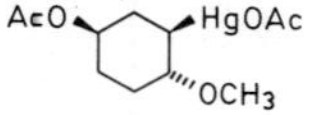 [structure: AcO, HgOAc, OCH₃ (90)]	$N_2H_4/NaOH$	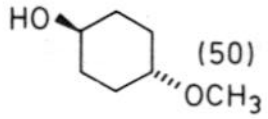 [structures: HO, OCH₃ (50) + HO (40)] 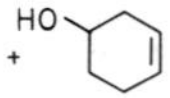	129
	CH_3OH	$Hg(OAc)_2$	—	NaCl	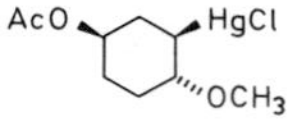 [structure: AcO, HgCl, OCH₃]	226
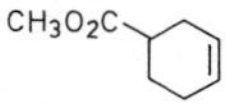 [structure: CH_3O_2C cyclohexene]	CH_3OH	$Hg(OAc)_2$, NaCl	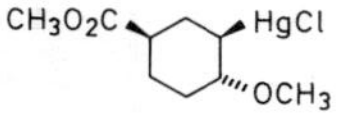 [structure: CH_3O_2C, HgCl, OCH₃]	$N_2H_4/NaOH$ Δ	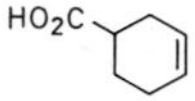 [structures: HO_2C (58) + HO_2C, OCH₃ (30)] 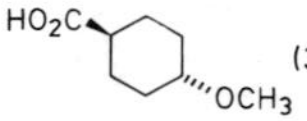	129
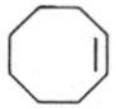 [structure: cyclooctene]	CH_3OH	$Hg(OAc)_2$	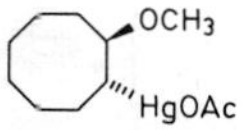 [structure: OCH₃, HgOAc (93)]	—	—	209, 235
	CH_3OH	$Hg(OAc)_2$, NaCl	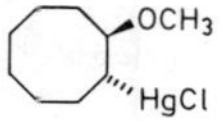 [structure: OCH₃, HgCl (65, 70)]	$NaBH_4$	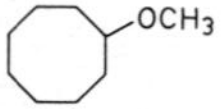 [structure: OCH₃ cyclooctane]	208, 231

Table 3.3. (continued)

Alkene	Alcohol	Mercuric salt	Organomercurial(s) (% Yield)	Subsequent reactants	Product(s) (% Yield)	Ref.
 (cyclooctene, bridged)	CH₃OH	Hg(OAc)₂	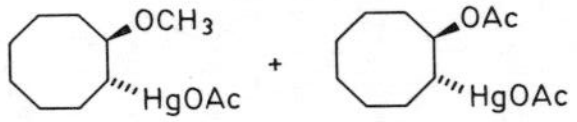 OCH₃ / HgOAc + OAc / HgOAc, 54 : 46	—	—	235
	CH₃OH	Hg(OAc)₂	—	NaCl	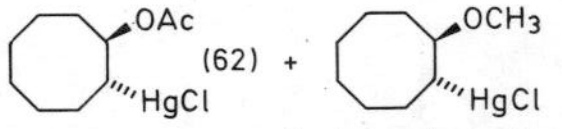 OAc / HgCl (62) + OCH₃ / HgCl (only product after 6 days)	236
	CH₃OH	Hg(OAc)₂ , NaCl	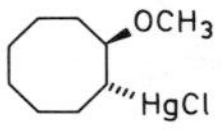 OCH₃ / HgCl (70)	NaBH₄	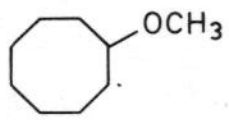 OCH₃	208
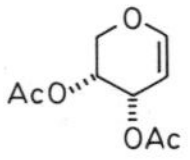 (AcO-pyran, OAc)	CH₃OH	Hg(OAc)₂	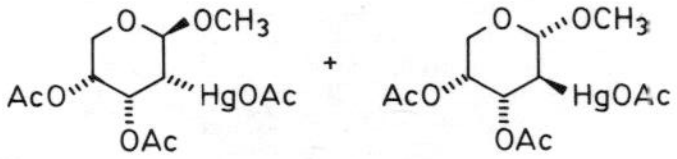 OCH₃ / HgOAc / OAc (AcO) + OCH₃ / HgOAc / OAc (AcO), 65 : 35	NaBH₄ / NaOCH₃	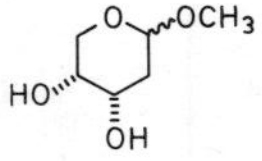 OCH₃ / HO / OH (> 90 overall)	237
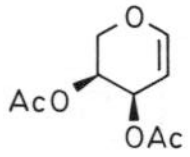 (AcO-pyran, OAc)	CH₃OH	Hg(OAc)₂	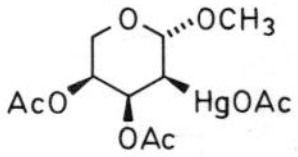 OCH₃ / HgOAc / OAc (AcO) (58)	NaBH₄	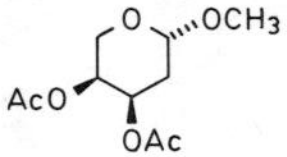 OCH₃ / OAc (AcO) (73)	238
	CH₃OH	Hg(OAc)₂	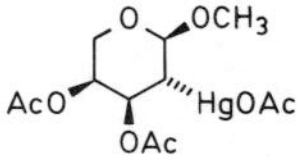 OCH₃ / HgOAc / OAc (AcO) (35)	H₂NCSNH₂ [NaBH₄]	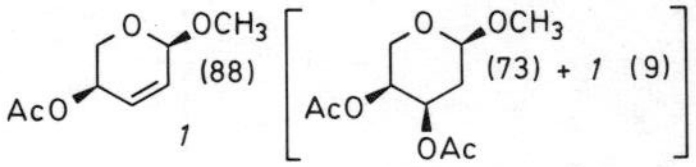 OCH₃ (AcO) (88) 1 [OCH₃ / OAc (AcO) (73) + 1 (9)]	238
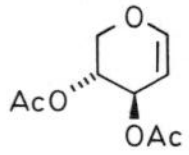 (AcO-pyran, OAc)	CH₃OH	Hg(OAc)₂	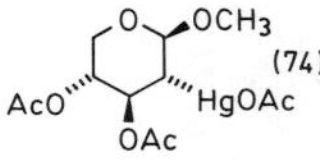 OCH₃ / HgOAc / OAc (AcO) (74) plus diastereomer (see below)	H₂NCSNH₂ [NaBH₄]	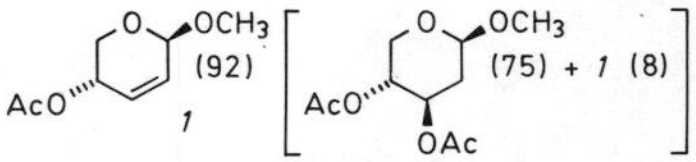 OCH₃ (AcO) (92) 1 [OCH₃ / OAc (AcO) (75) + 1 (8)]	238
	CH₃OH	Hg(OAc)₂	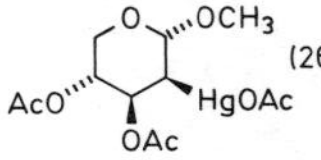 OCH₃ / HgOAc / OAc (AcO) (26) plus diastereomer (see above)	H₂NCSNH₂ [NaBH₄]	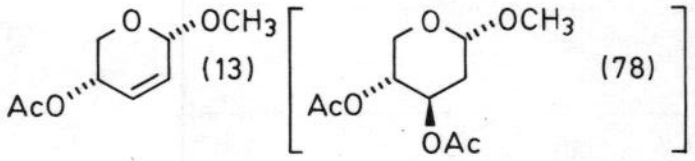 OCH₃ (AcO) (13) [OCH₃ / OAc (AcO) (78)]	238
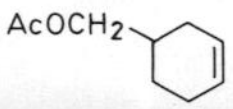 AcOCH₂ (cyclohexene)	CH₃OH	Hg(OAc)₂ , NaCl	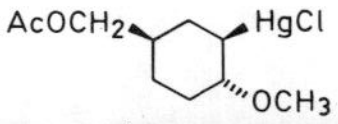 AcOCH₂ / HgCl / OCH₃	N₂H₄ / NaOH Δ	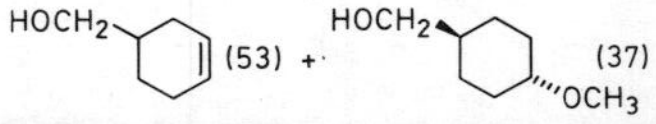 HOCH₂ (53) + HOCH₂ / OCH₃ (37)	129

Table 3.3. (continued)

Alkene	Alcohol	Mercuric salt	Organomercurial(s) (% Yield)	Subsequent reactants	Product(s) (% Yield)	Ref.
cis- and *trans*-$(CH_2)_7$ with $CH=CH$	CH_3OH	$Hg(OAc)_2$	—	NaCl	$(CH_2)_7$—$CHOCH_3$ / $CHHgCl$ (40)	231
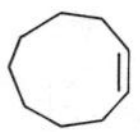 [cyclooctene]	CH_3OH	$Hg(OAc)_2$, NaCl	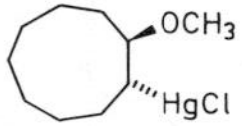 OCH_3 / HgCl (54)	$NaBH_4$	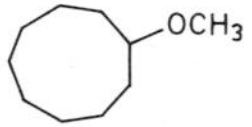 OCH_3	208
 [bicyclic alkene]	CH_3OH	$Hg(OAc)_2$, NaCl	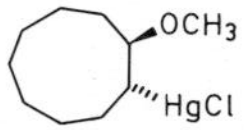 OCH_3 / HgCl (65)	$NaBH_4$	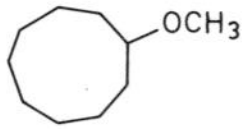 OCH_3	208
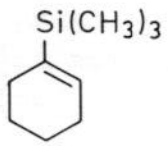 $Si(CH_3)_3$ [cyclohexene]	CH_3OH	$Hg(OAc)_2$?	—	$NaBH_4$ / NaOH	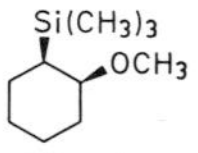 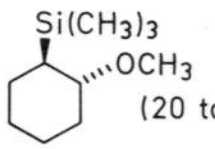 $Si(CH_3)_3$ OCH_3 + $Si(CH_3)_3$ OCH_3 (20 total) 65 : 35	239
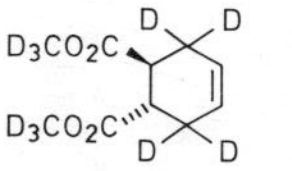 D_3CO_2C ... D_3CO_2C [tetradeutero cyclohexene]	D_3COH	$Hg(OAc)_2$	—	NaCl	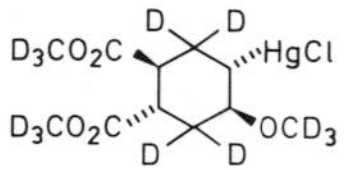 D_3CO_2C / HgCl, D_3CO_2C / OCD_3	20
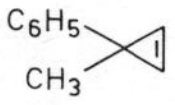 C_6H_5 / CH_3 [cyclopropene]	CH_3OH	$Hg(OAc)_2$	—	Cl^-	*cis*-$ClHgCH=CHC(OCH_3)(CH_3)C_6H_5$ (~35) 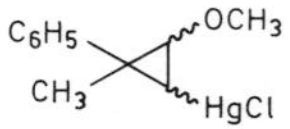 + C_6H_5, CH_3 / OCH_3, HgCl (15–20) + *cis*-$[C_6H_5C(OCH_3)CH_3CH=CH]_2Hg$ (~5)	240
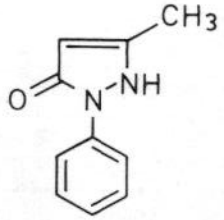 [pyrazolone, CH_3, NH]	CH_3OH	$Hg(OAc)_2$	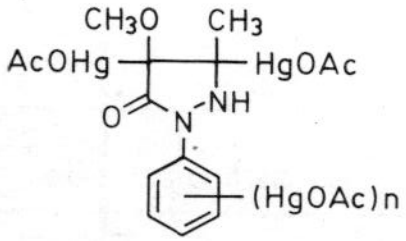 CH_3O CH_3 / AcOHg, HgOAc, NH / $(HgOAc)n$; n = 1 (~90) or 2 (~90)	—	—	241

Table 3.3. (continued)

Alkene	Alcohol	Mercuric salt	Organomercurial(s) (% Yield)	Subsequent reactants	Product(s) (% Yield)	Ref.
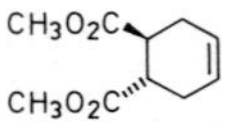 [dimethyl cyclohexene-dicarboxylate: CH_3O_2C, CH_3O_2C]	CH_3OH	$Hg(OAc)_2$	—	NaCl	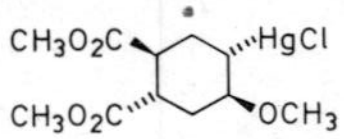 [CH_3O_2C, CH_3O_2C, HgCl, OCH_3]	20
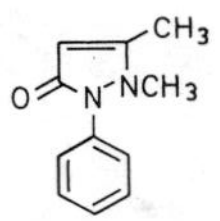 [pyrazolone ring with CH_3, N-NCH_3, phenyl, O]	CH_3OH	$Hg(OAc)_2$	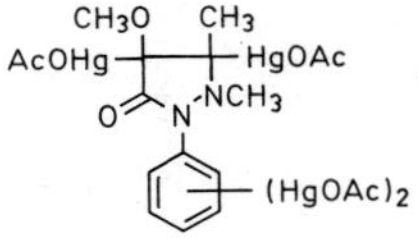 [CH_3O, CH_3, AcOHg, HgOAc, N-NCH_3, O, phenyl-$(HgOAc)_2$] (97.4)	—	—	241
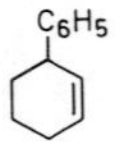 [phenylcyclohexene: C_6H_5]	CH_3OH	$Hg(OAc)_2$	—	NaCl	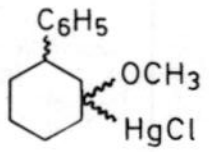 [C_6H_5, OCH_3, HgCl] (93)	242
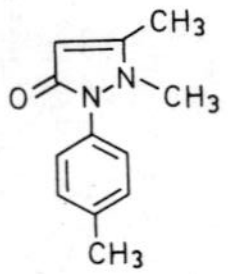 [pyrazolone ring with CH_3, N-NCH_3, O, tolyl-CH_3]	CH_3OH	$Hg(OAc)_2$	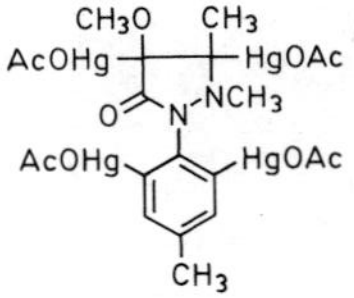 [CH_3O, CH_3, AcOHg, HgOAc, N-NCH_3, O, AcOHg, HgOAc, CH_3] (~90)	—	—	241
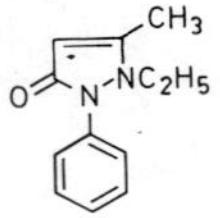 [pyrazolone ring with CH_3, N-NC_2H_5, O, phenyl]	ROH	$Hg(OAc)_2$	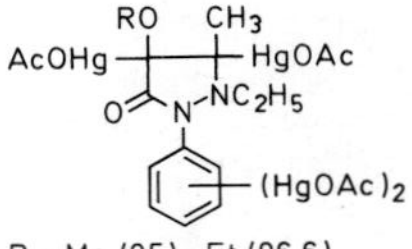 [RO, CH_3, AcOHg, HgOAc, N-NC_2H_5, O, phenyl-$(HgOAc)_2$] R = Me (95), Et (96.6)	—	—	241
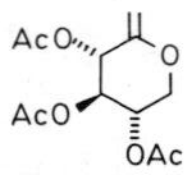 [glycal: AcO, AcO, OAc, O]	CH_3OH	$Hg(OAc)_2$	—	NaCl	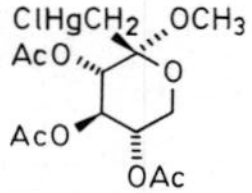 [$ClHgCH_2$, OCH_3, AcO, AcO, OAc, O]	243
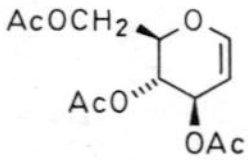 [glycal: $AcOCH_2$, AcO, OAc, O]	CH_3OH	$Hg(OAc)_2$	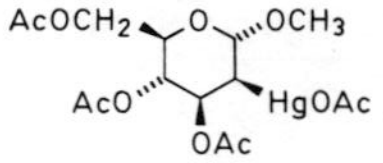 [$AcOCH_2$, O, OCH_3, AcO, HgOAc, OAc] (93)	$NaBH_4$	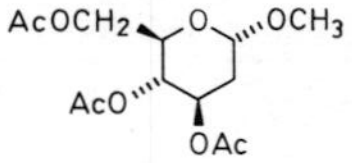 [$AcOCH_2$, O, OCH_3, AcO, OAc] (90)	238

210

Table 3.3. (continued)

Alkene	Alcohol	Mercuric salt	Organomercurial(s) (% Yield)	Subsequent reactants	Product(s) (% Yield)	Ref.
	CH_3OH	$Hg(OAc)_2$	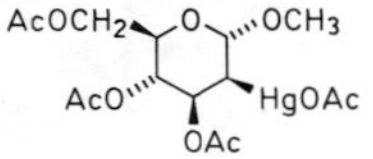 (53)	H_2NCSNH_2 [$NaBH_4$]	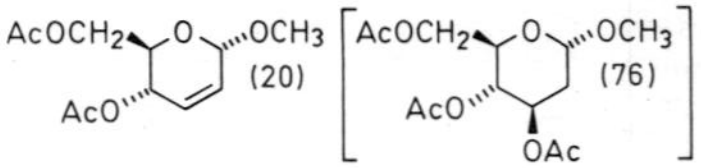 (20) (76)	238
	CH_3OH	$Hg(OAc)_2$	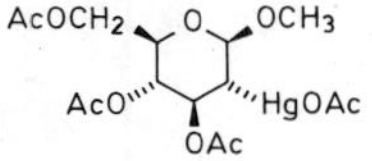 (47)	H_2NCSNH_2 [$NaBH_4$]	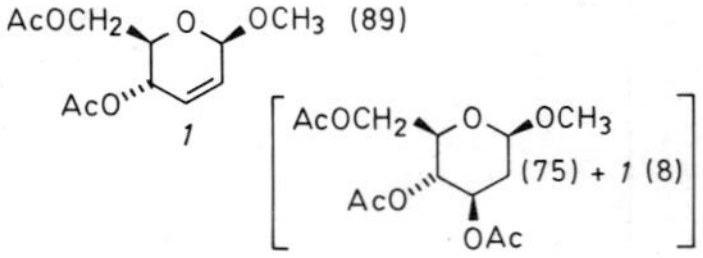 (89); (75) + 1 (8)	238
	CH_3OH	$Hg(OAc)_2$ (Cl^-)	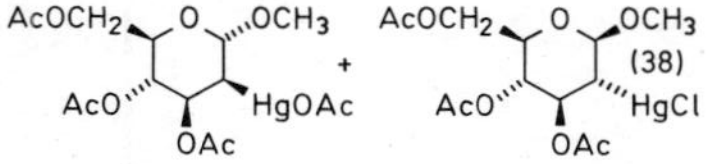 (38) (major) + (minor)	KBH_4/OH^-	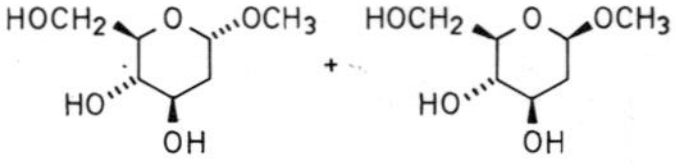 +	229
	CH_3OH	$Hg(ClO_4)_2$	—	$NaCl$	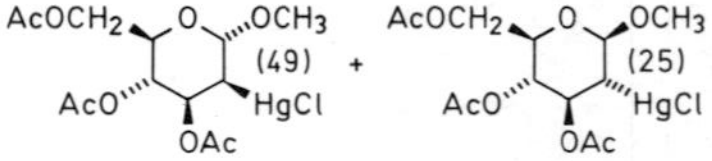 (49) + (25)	244
	ROH	$Hg(OAc)_2$	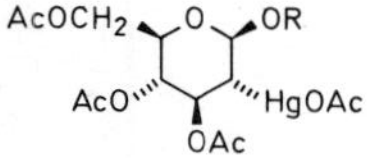 $R = Me$, Et (33)	NaI	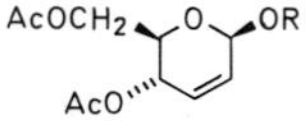 $R = Me$, Et (85)	245
	ROH	$Hg(OAc)_2$, X^-	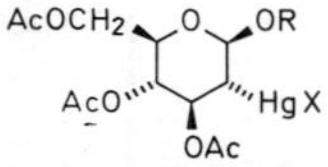 $X = OAc$, $R = Me$ (45), Et (39), $i\text{-}Pr$ (40); $X = Br$, $R = Me$	Br_2 (or $RHgCl$ or $RHgOAc$)	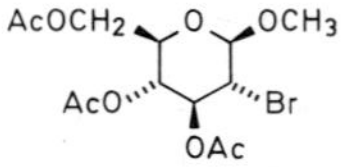 (~100)	210
	ROH	$Hg(OAc)_2$	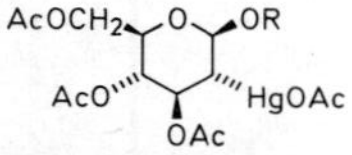 $R = Me$, $n\text{-}Pr$ (25), $i\text{-}Pr$, $c\text{-}C_5H_9$ (15), $n\text{-}Hex$ (7.5)	—	—	246

Table 3.3. (continued)

Alkene	Alcohol	Mercuric salt	Organomercurial(s) (% Yield)	Subsequent reactants	Product(s) (% Yield)	Ref.
	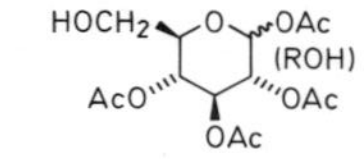 (structure)	$Hg(ClO_4)_2$	—	$NaBH_4$	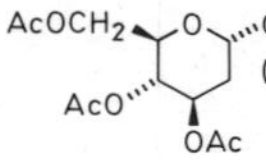 (23) 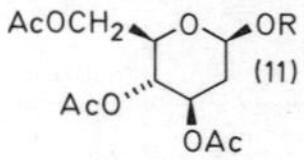 + (11)	244
	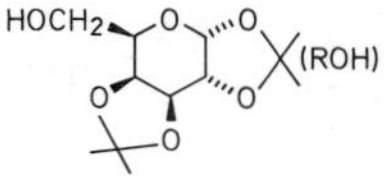 (structure)	$Hg(ClO_4)_2$	—	$NaBH_4$	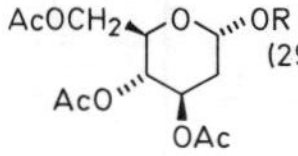 (29) 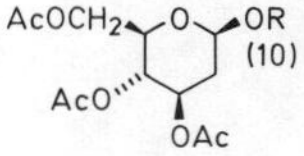 + (10)	244
 (cis or trans ?)	CH_3OH	$Hg(OAc)_2$	—	NaCl	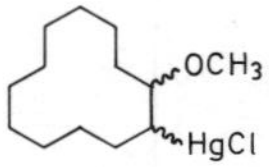 (structure)	247
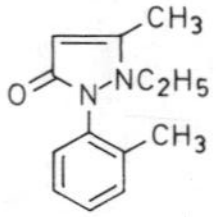 (structure)	CH_3OH	$Hg(OAc)_2$	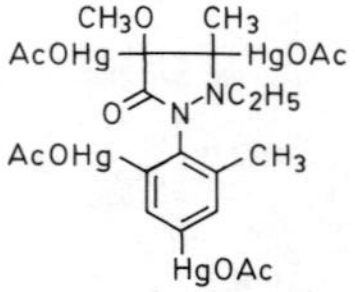 (~90)	—	—	241
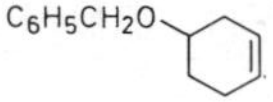 $C_6H_5CH_2O$— (structure)	CH_3OH	$Hg(OAc)_2$, NaCl	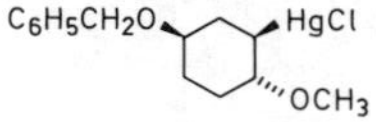 (structure)	N_2H_4/NaOH Δ	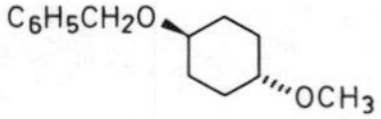 (>32)	129
(structure)	CH_3OH	$Hg(OAc)_2$	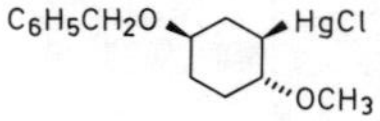 (structure)	N_2H_4, Li/EtNH$_2$	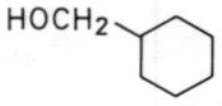 (60) 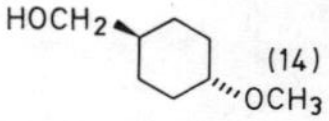 + (14)	129
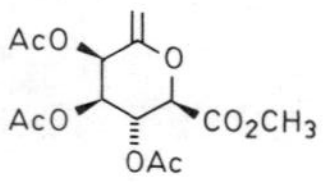 (structure)	CH_3OH	$Hg(OAc)_2$	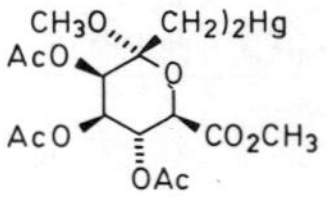 (80)	—	—	248

Table 3.3. (continued)

Alkene	Alcohol	Mercuric salt	Organomercurial(s) (% Yield)	Subsequent reactants	Product(s) (% Yield)	Ref.
(pyranose: AcO, AcO, OAc, CO$_2$CH$_3$)	CH$_3$OH	Hg(OAc)$_2$	(CH$_3$O, (CH$_2$)$_2$Hg, AcO, AcO, CO$_2$CH$_3$, OAc) (80)	—	—	248
(pyranose: AcOCH$_2$, AcO, OAc, OAc)	CH$_3$OH	Hg(OAc)$_2$	(AcOCH$_2$, OCH$_3$, OAc ?, AcO, HgOAc, OAc) (18)	—	—	246
(pyranose: AcO, AcO, OAc, CH$_2$OAc)	CH$_3$OH	Hg(OAc)$_2$	(CH$_3$O, (CH$_2$)$_2$Hg, AcO, AcO, CH$_2$OAc, OAc) (74)	—	—	248
(pyranose: AcO, AcO, OAc, CH$_2$OAc)	CH$_3$OH	Hg(OAc)$_2$	(CH$_3$O, (CH$_2$)$_2$Hg, AcO, AcO, CH$_2$OAc, OAc) (80)	—	—	248
(cyclopentene)–(CH$_2$)$_{12}$CO$_2$H	CH$_3$CH$_2$OH	Hg(OAc)$_2$ (HOAc)	(cyclopentane: (CH$_2$)$_{12}$CO–O–Hg, OC$_2$H$_5$) and/or regioisomer	—	—	249
(cyclopentene)–(CH$_2$)$_{12}$CO$_2$C$_2$H$_5$	CH$_3$CH$_2$OH	Hg(OAc)$_2$	(cyclopentane: CH$_2$)$_{12}$CO$_2$C$_2$H$_5$, HgOAc, OC$_2$H$_5$) and/or regioisomer	—	—	249
(dihydropyridazine: C$_2$H$_5$O$_2$CN, C$_2$H$_5$O$_2$CN, C$_6$H$_5$, C$_6$H$_5$)	CH$_3$OH	Hg(OAc)$_2$ (HClO$_4$)	(C$_2$H$_5$O$_2$CN, C$_2$H$_5$O$_2$CN, C$_6$H$_5$, OCH$_3$, HgX, C$_6$H$_5$) X = OAc (27), ClO$_4$ (92)	KBr / Br$_2$ (on RHgOAc)	(C$_2$H$_5$O$_2$CN, C$_2$H$_5$O$_2$CN, C$_6$H$_5$, OCH$_3$, Br, C$_6$H$_5$) (77)	250

Table 3.3. (continued)

Alkene	Alcohol	Mercuric salt	Organomercurial(s) (% Yield)	Subsequent reactants	Product(s) (% Yield)	Ref.
(BzO, BzO, OTs, OCH₃ pyranose)	CH₃OH	Hg(OAc)₂	(CH₃O, BzO, BzO, OTs, OCH₃, CH₂HgOAc pyranose) (59)	NaBH₄	(CH₃O, BzO, BzO, OTs, OCH₃, CH₃ pyranose) (97)	207
	CH₃OH	C₆H₅HgOAc	(CH₃O, BzO, BzO, OTs, OCH₃, (CH₂)₂Hg pyranose) (73)	CH₃COCl	(BzO, BzO, OTs, OCH₃ pyranose) (35) + (BzO, BzO, OTs, OCH₃, CH(OCH₃)(OAc) pyranose) (47)	207
(BzO, BzO, OTs, OCH₃ pyranose)	CH₃OH	Hg(OAc)₂	AcOHgCH₂C(OCH₃)₂CH(OBz)CH(OBz)CH(OTs)CH(OCH₃)(OAc)	—	—	207

Table 3.4. Alkoxymercuration of Polycyclic Alkenes

Alkene	Alcohol	Mercuric salt	Organomercurial(s) (% Yield)	Subsequent reactants	Product(s) (% Yield)	Ref.
	CH_3OH	$Hg(OAc)_2$ (HNO_3)	(74)	NaCl	$\cdot\,H_2O$	223
	$CH_3(CH_2)_2OH$	$Hg(OAc)_2$ (HNO_3)	—	NaCl	(47)	223
	CH_3OH	$Hg(OAc)_2$ (HNO_3)	—	NaCl	(60)	223
	CH_3OH	$Hg(OAc)_2$	60–80 : 20–40	—	—	251
	CH_3OH	$Hg(OAc)_2$	94 : 6	—	—	235
	CH_3OH	$Hg(OAc)_2$	—	NaCl	(50)	252, 253
	CH_3OH	$Hg(OAc)_2$, NaCl	? (95)	Na(Hg)/H_2O or N_2H_4	? (50)	254

Table 3.4. (continued)

Alkene	Alcohol	Mercuric salt	Organomercurial(s) (% Yield)	Subsequent reactants	Product(s) (% Yield)	Ref.
	ROH	HgX$_2$	—	NaBH$_4$/NaOH	X = OAc, R = Me (89), Et (60), i-Pr (15), t-Bu (1); X = O$_2$CCF$_3$, R = Me (100), Et (100), i-Pr (95), t-Bu (90)	57
	(CH$_3$)$_3$COH	Hg(OAc)$_2$ (HNO$_3$ or HClO$_4$), Cl$^-$		Cl$_2$/C$_5$H$_5$N	71% exo, 13% endo	255
	CH$_3$OH	HgO(HOAc)		—	—	256
	CH$_3$OH	Hg(OAc)$_2$, NaCl	2.5 : 1 (85 total)	Na/Hg/NaOH	68 : 32	257
	ROH	Hg(OAc)$_2$ (HNO$_3$)	R = Me (61), Et, i-Pr, n-Bu, HO(CH$_2$)$_2$, EtO(CH$_2$)$_2$, HO(CH$_2$)$_2$O(CH$_2$)$_2$ (all ?)	NaCl	(80)	223, 258
	CH$_3$OH	Hg(OAc)$_2$	77 : 13 : 7 : 3	—	—	259

Table 3.4. (continued)

Alkene	Alcohol	Mercuric salt	Organomercurial(s) (% Yield)	Subsequent reactants	Product(s) (% Yield)	Ref.
	CH₃OH (plus solvent)	Hg(NO₃)₂	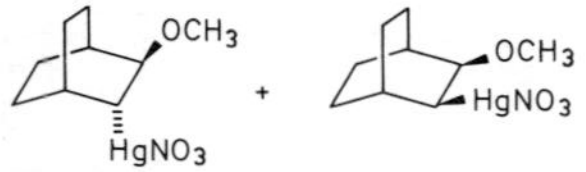 (bicyclo, OCH₃/HgNO₃) + (OCH₃/HgNO₃) solvent: CH₃OH 81 : 19; CH₂Cl₂ <1 : >99; dioxane >99 : <1	—	—	259
	CH₃OH	—	(OCH₃/HgCl)	—	—	23
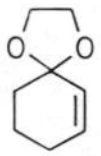	CH₃OH	Hg(OAc)₂	—	NaCl	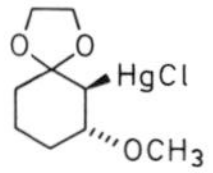 (HgCl / OCH₃ spiro structure) (78)	22
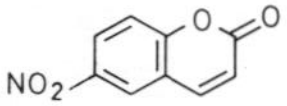	CH₃OH	Hg(OAc)₂	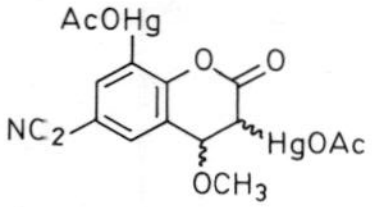 (AcOHg, HgOAc, OCH₃, NO₂ coumarin)	Br₂	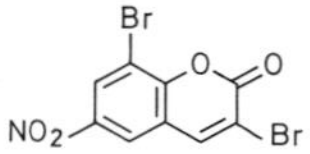 (Br / NO₂ coumarin)	260
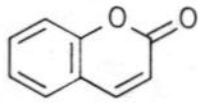	CH₃OH	Hg(OAc)₂	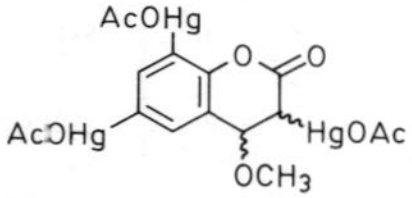 (AcOHg, HgOAc, OCH₃, AcOHg coumarin)	Br₂ [or H₂S]	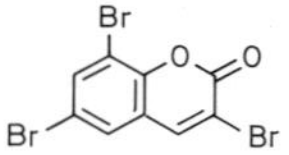 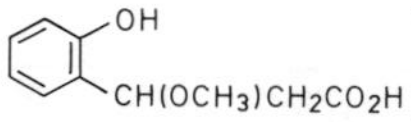 (Br / Br coumarin) or [OH / CH(OCH₃)CH₂CO₂H]	260
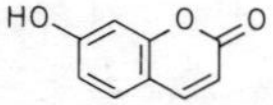	CH₃OH	Hg(OAc)₂	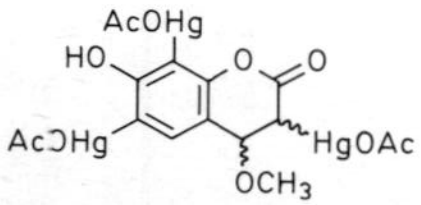 (AcOHg, HgOAc, OCH₃, HO, AcOHg coumarin)	HNO₃	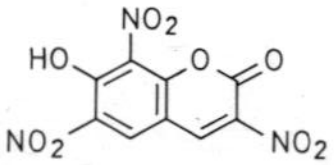 (NO₂ / NO₂ / HO coumarin)	261

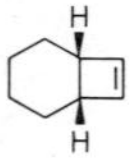

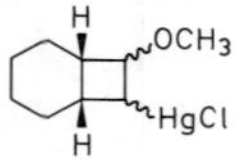

217

Table 3.4. (continued)

Alkene	Alcohol	Mercuric salt	Organomercurial(s) (% Yield)	Subsequent reactants	Product(s) (% Yield)	Ref.
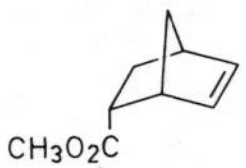	CH_3OH	$Hg(OAc)_2$	—	NaCl	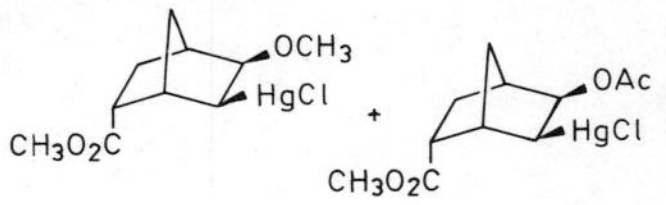 5 : 1	17
	CH_3OH	$Hg(OAc)_2$	—	NaCl	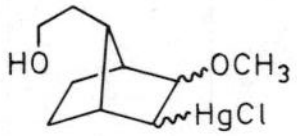	262
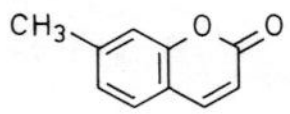	CH_3OH	$Hg(OAc)_2$	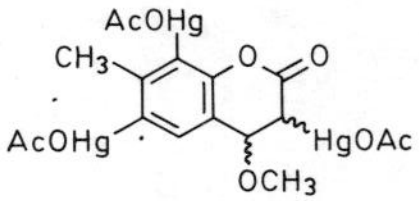	Br_2	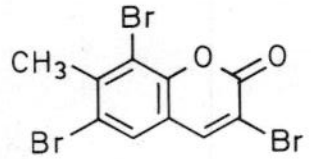	260
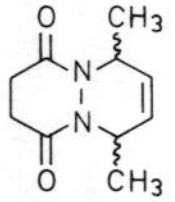	ROH	$Hg(OAc)_2$	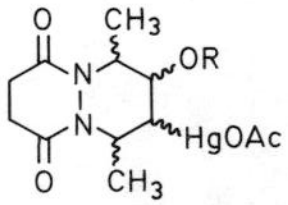 $R = Me,\ Et,\ i\text{-}Pr,\ n\text{-}Bu,\ HO(CH_2)_2,$ $EtO(CH_2)_2,\ HO(CH_2)_2O(CH_2)_2$ (all ?)	—	—	258
	CH_3OH	$HgO/Hg(NO_3)_2$	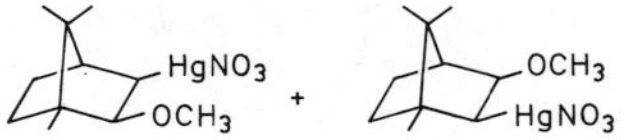 2 : 1	—	—	263
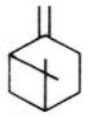	CH_3CH_2OH	$Hg(OAc)_2$ (HOAc)	—	NaX	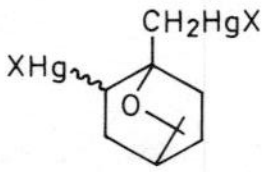 X = Cl, Br, I	264
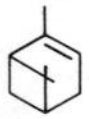	ROH (R = Me, Et)	$HgCl_2$	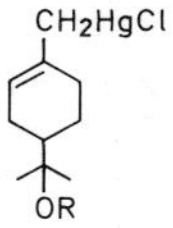	on standing (HCl) or $NaBH_4/NaOH$	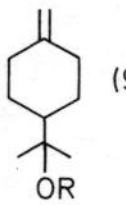 (95) or 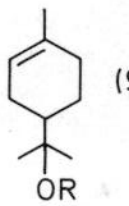 (95)	265

Table 3.4. (continued)

Alkene	Alcohol	Mercuric salt	Crganomercurial(s) (% Yield)	Subsequent reactants	Product(s) (% Yield)	Ref.
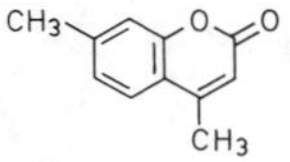 (7-methyl-4-methylcoumarin structure)	CH₃OH	Hg(OAc)₂	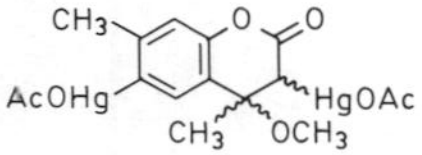 (coumarin–AcOHg/HgOAc/OCH₃ structure)	Br₂	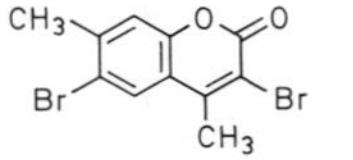 (6,8-dibromo-7-methyl-4-methylcoumarin structure)	261
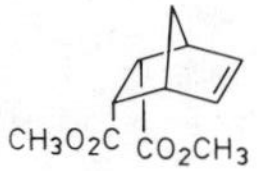 (norbornene dicarboxylate structure, CH₃O₂C CO₂CH₃)	CH₃OH	Hg(OAc)₂	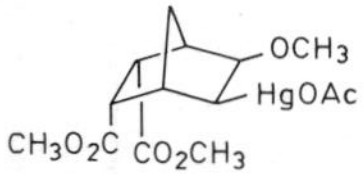 (structure, OCH₃ HgOAc; CH₃O₂C CO₂CH₃)	—	—	266
	CH₃OH	Hg(OAc)₂ (BF₃·Et₂O)	—	NaCl	(two structures: OCH₃ HgCl, CH₃O₂C CO₂CH₃) ? (10) + (OAc HgCl, CH₃O₂C CO₂CH₃)	254
 (pinene-type bicyclic alkene structure)	CH₃OH	Hg(OAc)₂	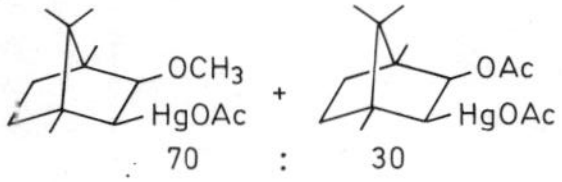 (OCH₃ HgOAc) + (OAc HgOAc) 70 : 30	—	—	251
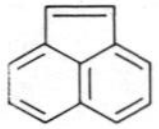 (acenaphthylene structure)	CH₃OH	Hg(OAc)₂	—	NaCl	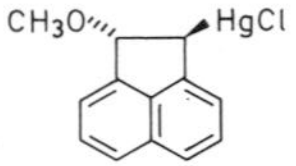 (CH₃O HgCl acenaphthene structure) (100)	19
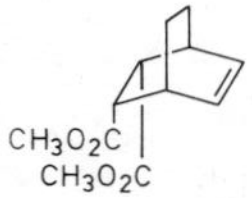 (bicyclic diester alkene structure, CH₃O₂C CH₃O₂C)	CH₃OH	Hg(OAc)₂	—	NaCl	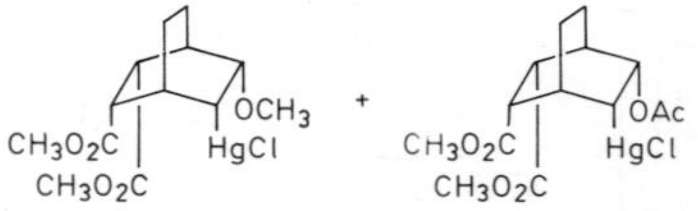 (OCH₃ HgCl, CH₃O₂C CH₃O₂C) + (OAc HgCl, CH₃O₂C CH₃O₂C) (62 total)	267
(decalin-type bicyclic alkene structure)	CH₃OH	—	(CH₃O HgCl bicyclic structure)	—	—	23

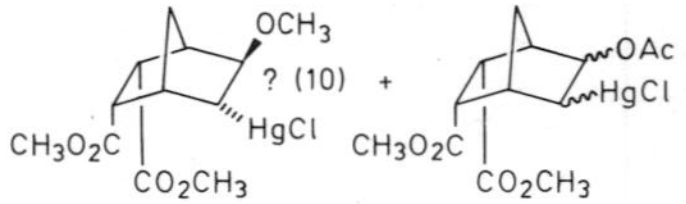

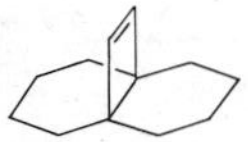

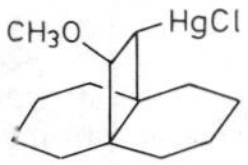

Table 3.4. (continued)

Alkene	Alcohol	Mercuric salt	Organomercurial(s) (% Yield)	Subsequent reactants	Product(s) (% Yield)	Ref.
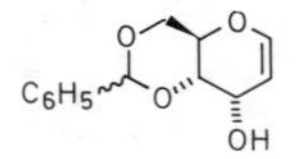	CH_3OH	$Hg(OAc)_2$	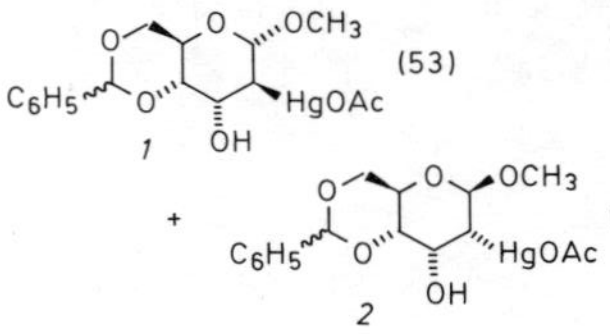 (53) + (40)	$NaBH_4/NaOH$	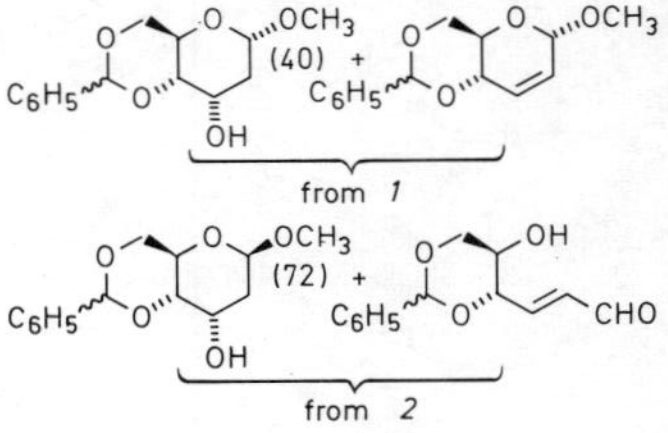 (40) + from 1 (72) + from 2	268
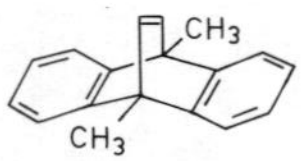	CH_3OH	$Hg(OAc)_2$	—	$NaBH_4$	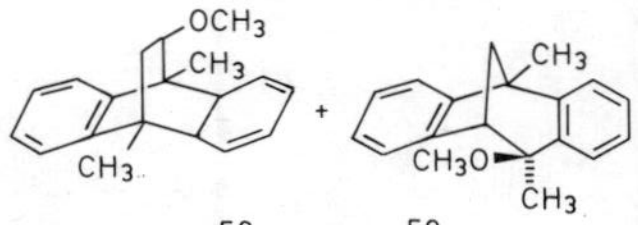 ~50 : ~50	269
	CH_3OH	$Hg(OAc)_2$ (HOAc) [Cl^-]	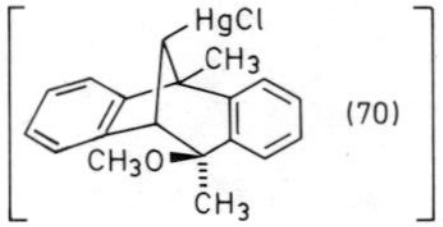 (70)	$NaBH_4$	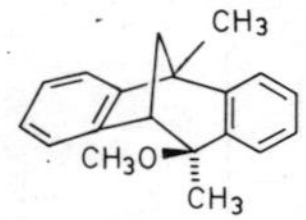	269
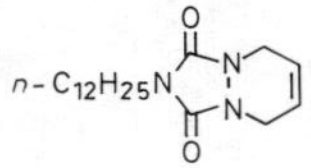 $n-C_{12}H_{25}N$	CH_3OH	$Hg(OAc)_2$ (HNO_3), Cl^-	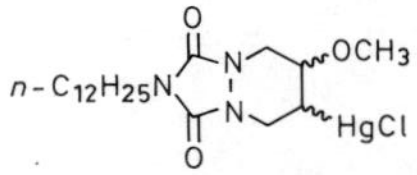 $n-C_{12}H_{25}N$ (25)	—	—	223
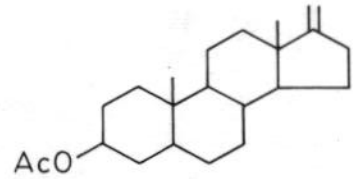 AcO	CH_3OH	$Hg(OAc)_2$	—	$NaBH_4[NaBD_4]/$ NaOH, Ac_2O	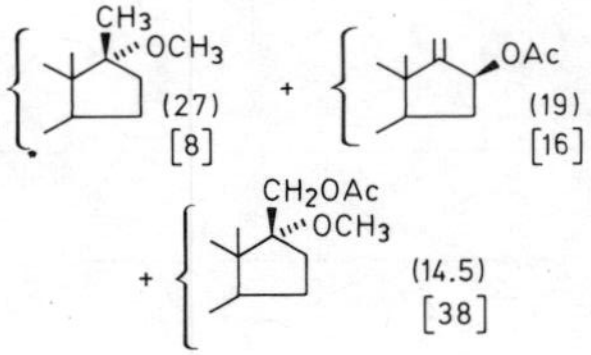 (27) [8] + (19) [16] + (14.5) [38]	270

3.4 contains examples of bicyclic or polycyclic alkenes which have been alkoxy-mercurated.

Although the vast majority of alkoxymercuration reactions have utilized mercuric acetate, a wide variety of other mercury salts can be employed, including mercuric chloride, mercuric bromide, mercuric nitrate, mercuric perchlorate, mercuric triflate and a wide variety of mercuric carboxylates. Combinations of these salts have also been used, presumably generating salts of the type XHgOAc (X=Cl, Br, I, SCN) [13, 128]. While mercuric oxide has been employed in the alkoxymercuration of unsaturated carboxylic acids, it is most commonly utilized in combination with other mercuric salts or inorganic or organic acids. Optically active mercuric carboxylates have been used in an attempt to induce optical activity into the resulting products [212, 216]. Mercuric acetate and mercuric chloride containing ^{197}Hg and ^{203}Hg afford radiolabelled organomercurials of interest for their diuretic properties [102–104, 106, 108]. Phenylmercuric acetate and nitrate also participate in alkoxymercuration reactions [92, 207]. Mercuric methoxide adds to ketene to generate the corresponding dialkylmercurial ester or one can more conveniently use mercuric acetate in methanol [79].

There are very few studies on the relative merits of different mercury salts for alkoxymercuration. H. C. Brown and M.-H. Rei have reported the most comprehensive study, in which they compare the reaction of mercuric acetate or mercuric trifluoroacetate with a variety of alkenes and alcohols [57]. They conclude that for most alkenes and primary or secondary alcohols, mercuric acetate works well, but mercuric trifluoroacetate generally gives faster reactions and higher yields. With tertiary alcohols, like *tert*-butanol, mercuric acetate gives poor results and one should use mercuric trifluoroacetate. However, with certain alkenes, such as 2-methyl-1-butene, neither salt works with *tert*-butanol. The major problem in many solvomercuration reactions appears to be competing addition of the mercury salt across the olefin double bond. A number of examples of this can be found in the tables, particularly with strained, relatively hindered alkenes. Mercuric salts containing weakly nucleophilic anions, such as mercuric trifluoroacetate or mercuric nitrate, often prove to be a convenient solution to this problem. In some cases, it has been observed that if one simply allows the reaction to proceed for a longer time, the initial mercury salt adduct is converted to the desired alkoxymercurial [57, 236]. Alternatively, hydroxy- or acetoxymercurials can react with strong acids in alcohol to afford the corresponding alkoxymercurials (Eqs. 4, 5).

(4) [271]

(5) [255]

On occasions, certain additional reagents have been added to the alkoxymercuration reaction to promote reaction. Inorganic acids, such as nitric and perchloric acids, are commonly added, as is acetic acid or boron trifluoride etherate. It has also been claimed that benzoyl peroxide facilitates alkoxymercuration [76, 212]. Sodium and potassium acetate have been added to mercuric halides to promote reaction. The effects of pH or reactant concentrations have received little attention [52]. While a wide range of reaction temperatures have been utilized for alkoxymercurations, it appears that most reactions can be effected at room temperature.

The alkoxymercuration reaction is most commonly carried out in an excess of the desired alcohol as the solvent. Only isolated examples of benzene [46], DMF [73], acetonitrile [244], methylene chloride [259], and dioxane [259] as solvents have been reported. The solvent appears to be rather important. It can in fact play a vital role in determining the stereochemistry of the product as shown in the following interesting example (Eq. 6) [259]. There appear to be

$$\tag{6}$$

very few examples of alkoxymercuration employing only stoichiometric amounts of the alcohol. To accomplish the following stoichiometric alkoxy-mercuration-demercuration reactions, equivalent amounts of alkene, alcohol, mercuric perchlorate and collidine in acetonitrile solvent were utilized (Eq. 7) [244].

$$\tag{7}$$

(34 % total)

While methanol has been the most widely used alcohol for alkoxy-mercuration, a careful examination of the tables reveals that a considerable variety of alcohols can be accommodated by this reaction. Thus, alcohols containing ester [47, 141, 244], ether [47], cyano [47], phenol [46], halogen [47, 141], and acetal [47, 150, 244] groups have all been utilized in alkoxymercuration reactions. As shown above (Eq. 7) carbohydrate alcohols have also been employed. Deuterated methanol has proven useful for certain

NMR studies [20, 25, 26, 55]. Diols and triols have been mono- [41, 47, 73, 141, 155, 164, 185, 186, 223, 258] or dimercurated [41] (Eq. 8). While

$$XHg-C-C-O-C_n-OH \quad \longleftarrow \quad HO-C_n-OH \quad \longrightarrow \quad XHg-C-C-O-C_n-O-C-C-HgX \qquad (8)$$

$$(9)$$

there are relatively few reports of alcohols which do not react, H. C. Brown has indicated that with mercuric acetate yields drop off substantially as the alcohol becomes more hindered [57]. Mercuric trifluoroacetate circumvents this problem. Similar problems are observed in the above example involving competitive intramolecular phenoxymercuration (Eq. 9) [141]. This type of intramolecular cyclization will be discussed in greater detail in the next section of this chapter.

Relatively few examples of intermolecular phenoxymercuration have been reported (Eqs. 10, 11).

$$H_2C=CHOC_6H_5 \quad \xrightarrow[C_6H_5OH]{Hg(NHCOCH_3)_2} \quad CH_3CONHHgCH_2CH(OC_6H_5)_2 \qquad (10)\ [98]$$

$$H_2C=CH_2 \quad \xrightarrow{Hg(OAr)_2} \quad ArOCH_2CH_2HgOAr \qquad (11)\ [48]$$

As noted earlier, the overall nature of the alkoxymercurials remained a puzzle for some time. Even today certain structural questions remain. For example, the alkoxymercurials derived from α,β-unsaturated acids have been written as four-membered ring carboxylate salts [9, 82, 115, 146, 148–151, 267], ionic salts [141], or polymers [149] (Eq. 12), but no clear-cut evidence for any of these structures has ever been provided. A variety of other unsaturated acids have also been reported to undergo alkoxymercuration to generate either carboxylate salts [141], cyclic mercury carboxylates [186] or hydroxy-mercuric compounds [122, 142, 176, 193] (Eq. 13, see p. 224). Their exact nature has never been established.

$$(12)$$

While the vast majority of alkoxymercuration reactions lead directly to alkylmercuric salts, there are several examples of reactions which afford dialkylmercurials directly. Phenylmercuric acetate [207] and nitrate [92] have been observed to react with olefins to afford the corresponding dialkylmercurials directly (Eqs. 14, 15). These reactions presumably proceed by dis-

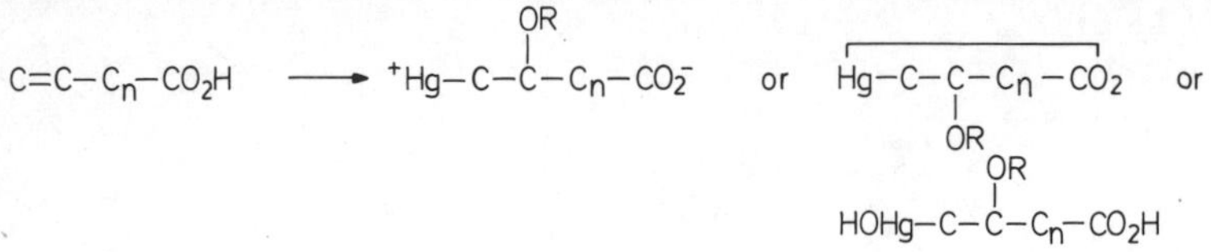

$$C=C-C_n-CO_2H \longrightarrow {}^+Hg-C-\overset{\overset{\displaystyle OR}{|}}{C}-C_n-CO_2^- \quad or \quad Hg-C-\overset{}{C}-C_n-CO_2 \quad or$$

$$HOHg-C-\overset{\overset{\displaystyle OR}{|}}{C}-C_n-CO_2H \tag{13}$$

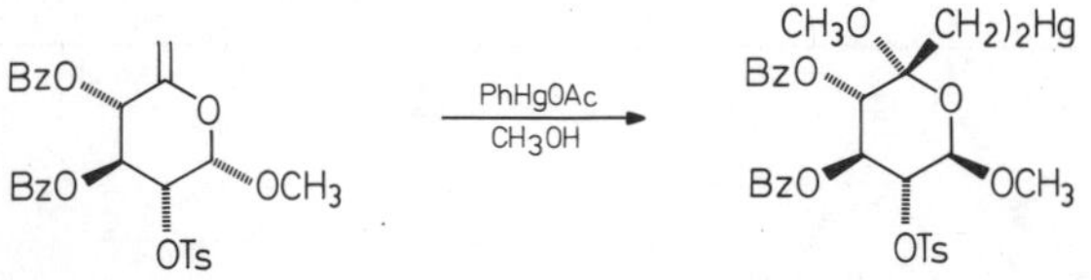

$$\tag{14}$$

$$F_2C=CFOC_2H_5 \xrightarrow[C_2H_5OH]{PhHgNO_3} Hg(CF_2CO_2C_2H_5)_2 \tag{15}$$

proportionation of the initial phenylmercurial. The product is apparently also a function of the olefin involved, as seen by the following examples (Eqs. 16–18). These few examples appear to be the only instances in which dialkylmercurials form directly.

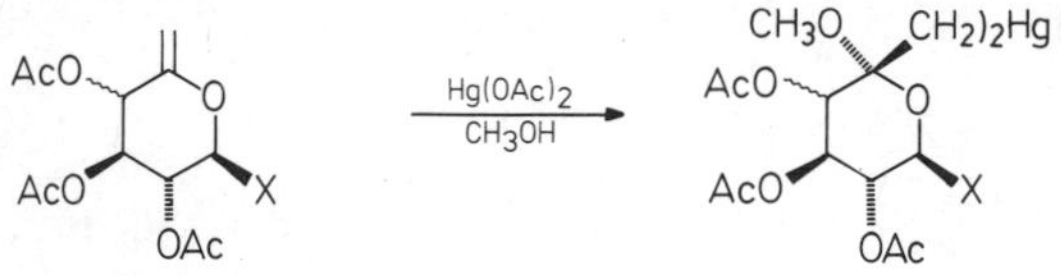

$$X=CO_2CH_3,CH_2OAc \tag{16}\ [248]$$

$$X_2C=CXOR \xrightarrow[ROH]{Hg(NO_3)_2} Hg(CX_2CO_2R)_2 \tag{17}\ [81,\ 91]$$
$$X=halogen$$

$$H_2C=C=O \xrightarrow[ROH]{Hg(OAc)_2} Hg(CH_2CO_2R)_2 \tag{18}\ [79]$$

The regioselectivity of alkoxymercuration is essentially identical to that of hydroxymercuration. Exclusive Markovnikov addition is reported for all simple alkenes and alcohols except the following (Eqs. 19, 20). This appears to be a steric phenomena.

$$RCH=CH_2 \xrightarrow[Hg(O_2CCX_3)_2]{R'OH} \xrightarrow[NaOH]{NaBH_4} \overset{\overset{\displaystyle OR'}{|}}{RCHCH_3} + RCH_2CH_2OR'$$

R	R'	X			
$(CH_3)_3C$	CH_3	H	98	:	2
$CH_3(CH_2)_3$	$(CH_3)_3C$	H	88	:	12
$CH_3(CH_2)_3$	$(CH_3)_3C$	F	~88	:	~12

$$\tag{19}\ [57]$$

$$ (20) \ [200] $$

$$
\begin{array}{ccc}
R=i\text{-}Pr & 55 & : & 30 \\
R=t\text{-}Bu & 29 & : & 59
\end{array}
$$

Little work on the regioselectivity of alkoxymercuration of simple unsymmetrical internal alkenes has appeared. 2-Pentene [59], 1,1-dideutero-ethylene [28], z-2-deutero-2-butene [28], 1-phenylnorbornene [263] and bornylene [263] all give regio mixtures of products. Arene substitution on the olefin double bond has only a weak directing effect (Eqs. 21, 22).

$$ (21) $$

$$
\begin{array}{llccc}
\textit{trans} & X=OAc & 60 & : & 40 & [64,68] \\
 & X=O_2CCF_3 & 65 & : & 35 & [54] \\
\textit{cis} & X=OAc & 75\text{-}80 & : & 5\text{-}10 & [64]
\end{array}
$$

$$ (22) \ [64, 68, 71] $$

However, certain functional groups have a pronounced effect on the regiochemistry of alkoxymercuration. α,β-Unsaturated carboxylic acids [63, 82–84, 101, 114, 115, 136, 141, 142, 146, 148, 150, 160, 168, 272], esters [9, 11, 12, 63, 100, 113, 124, 136, 141, 159, 161, 162, 190, 195, 201, 267], and amides [12] of almost all substitution patterns give the α-mercurated carbonyl product (Eq. 23), with the following exceptions (Eqs. 24, 25). The structure

$$ (23) $$

$$ (24) \ [94] $$

$$ R=CH_3, C_6H_5 \quad (R'=Me, t\text{-}Bu) $$

$$p\text{-}CH_3CH{=}CHCONHC_6H_4CH_2CO_2H \xrightarrow[HOCH_2CH_2OH]{Hg(OAc)_2} p\text{-}CH_3\overset{\underset{|}{HOHg}}{C}H\overset{\underset{|}{OCH_2CH_2OH}}{C}HCONHC_6H_4CH_2CO_2H \quad (25)\ [141]$$

of the product from the latter reaction (Eq. 25) seems highly questionable, however. Coumarins undergo Markovnikov addition as well, but the aromatic ring is also substituted under the conditions required to effect reaction (Eq. 26, see below) [260, 261]. As the distance between the functional group

$$\text{(26)}$$

and the double bond increases, its directional effect diminishes, A systematic examination of the methoxymercuration-demercuration of all possible methyl *cis*-undecenoates indicates considerable selectivity for the more remote methoxy derivate for the Δ^2-through Δ^4-isomers. A 50:50 mixture of regio-isomers is not observed until one reaches the Δ^5-isomer [190]. Interestingly cnough, the Δ^8- and Δ^9-isomers also show a strong preference (74% and 97% respectively) for the more remote methyl ethers. A through-space effect seems likely. Similar directive effects are observed for the Δ^2-through Δ^6-isomers of methyl octadecenoate [73]. The Δ^9-isomer shows no selectivity however. As anticipated, oleic acid and its esters, elaidic esters and other long-chain, unsaturated esters in which the double bond is remote from the ester functionality exhibit no regioselectivity [73, 202, 273]. The carbomethoxy group has also shown a substantial directing effect in the norbornene system (Eq. 27) [17].

$$\text{(27)}$$

$$5 \quad : \quad 1$$

$$\text{(28)}$$

Other carbonyl compounds exert directive effects similar to carboxylic acids and their derivatives. Thus, α,β-unsaturated aldehydes and ketones generally undergo mercury addition alpha to the carbonyl group (Eq. 28) [22, 93, 95, 144, 145, 159, 196, 197, 233], except where the double bond is terminal and disubstituted (Eq. 29) [93, 94]. Since such olefins are generally

$$H_2C=\overset{\overset{\textstyle CH_3}{|}}{C}COR \xrightarrow[CH_3OH]{Hg(OAc)_2} AcOHgCH_2\underset{\underset{\textstyle OCH_3}{|}}{\overset{\overset{\textstyle CH_3}{|}}{C}}COR \qquad (29)$$

$$R=H, CH_3$$

rather unreactive, alkoxymercuration is often effected by adding catalytic amounts of strong acids to the reaction mixture. With α,β-unsaturated aldehydes, acetal formation can become a major side reaction under these conditions [93].

As noted in the monograph "Organomercury Compounds in Organic Synthesis", numerous vinyl ethers and related derivatives undergo hydroxymercuration with formation of the corresponding α-mercurated carbonyl compounds. The alkoxymercuration of such compounds affords mercurated acetals directly, as shown by the following two simple examples (Eqs. 30, 31). This reaction has been widely employed for the functionalization of glycals (Eq. 32) [34, 35, 210, 227–229, 237, 238, 244–246, 268, 274, 275]. As anticipated, β-alkoxy enones afford the expected regioisomers (Eq. 33) [233]. With α-alkoxy carbonyl systems, the product is not so easily predicted. It is reported, however, that the alkoxy group exhibits the dominant directing effect (Eq. 34) [82, 112, 205]. The amido group behaves similarly (Eq. 35) [158]. The alkoxymercuration of halo ethers leads to α-mercurated esters (Eq. 36) [81, 90–92].

$$H_2C=CHOC_2H_5 \longrightarrow ClHgCH_2CH(OC_2H_5)_2 \qquad (30)\ [15]$$

$$(31)\ [16, 210]$$

$$(32)\ [35]$$

$$(33)$$

$$\underset{R=H, OH, OC_2H_5}{}\overset{OC_2H_5}{\underset{COR}{C=C}} \xrightarrow[C_2H_5OH]{Hg(OAc)_2} XHg-\underset{\underset{\textstyle COR}{|}}{C}-\underset{\underset{\textstyle OC_2H_5}{|}}{\overset{\overset{\textstyle OC_2H_5}{|}}{C}}-OC_2H_5 \qquad (34)$$

$$H_2C=\overset{\overset{\textstyle HNCOR}{|}}{C}CO_2H \xrightarrow[R'OH]{Hg(OAc)_2} AcOHgCH_2\underset{\underset{\textstyle OR'}{|}}{\overset{\overset{\textstyle HNCOR}{|}}{C}}CO_2H \qquad (35)$$

$$X_2C=CXOR \longrightarrow XHgCX_2CO_2R \quad or \quad Hg(CX_2CO_2R)_2 \qquad (36)$$

III. Alkoxymercuration

The directive effects of simple acetoxy, hydroxy and alkoxy groups in alkoxymercuration have been only briefly examined, and have been found to be essentially identical to those observed in hydroxymercuration. While the

$$CH_3(CH_2)_4CH=CHCH_2CH_2\overset{\overset{\displaystyle X}{|}}{C}H(CH_2)_7CO_2CH_3 \longrightarrow CH_3(CH_2)_4\overset{\overset{\displaystyle CH_3O}{|}}{C}H(CH_2)_3\overset{\overset{\displaystyle X}{|}}{C}H(CH_2)_7CO_2CH_3 \qquad (37)$$

X = Ac, OH, OCH_3

$$CH_3(CH_2)_5\overset{\overset{\displaystyle X}{|}}{C}HCH_2CH=CH(CH_2)_7CO_2CH_3 \longrightarrow CH_3(CH_2)_5\overset{\overset{\displaystyle X}{|}}{C}H(CH_2)_2\overset{\overset{\displaystyle OCH_3}{|}}{C}H(CH_2)_7CO_2CH_3 \qquad (38)$$

X = Ac, OH, OCH_3

(39) [22, 151, 199]

X = OH, O_2CR

(40) [22, 224, 225]

X = OH, OAc, OCH_2CH_2O   74–85 : 15–26

substitution pattern of the alkene is the most important factor in determining the regiochemistry of substitution, these functional groups do exert a powerful directive effect. For example, the methoxymercuration-demercuration of 9-substituted methyl 12-octadecenoates and 12-substituted methyl 9-octadecenoates affords predominantly, but not exclusively, the products of substitution on the carbon more remote from the substituent (Eqs. 37, 38) [73, 202]. Unsaturated alcohols capable of undergoing intramolecular alkoxymercuration to form five- or six-membered ring ethers do so readily. This reaction will be discussed in more detail later. Allylic alcohols, esters or ketals all undergo substitution to place the new alkoxy group on the more remote carbon (Eqs. 39, 40). Similar effects are observed for a wide variety of

(41)

X = HO, OAc, OCH_3, OCH_2Ph, CH_2OH, CH_2OAc, CO_2CH_3, CN

$$H_2C=C=O \longrightarrow Hg(CH_2CO_2R)_2 \qquad (42)$$

$$C_6H_5CH=CHCN \longrightarrow C_6H_5\overset{\overset{\displaystyle OCH_3}{|}}{\underset{\underset{\displaystyle HgOAc}{|}}{C}}HCHCN \qquad (43)$$

$$\text{H}_2\text{C=CHCH}_2\text{N} \underset{\text{O}}{\overset{\text{O}}{\diamond}} \text{R} \longrightarrow \text{AcOHgCH}_2\overset{\text{OCH}_3}{\text{CHCH}_2}\text{N}\underset{\text{O}}{\overset{\text{O}}{\diamond}}\text{R} \qquad (44)$$

$$R = C_6H_4, (CH_2)_7$$

4-substituted cyclohexenes (Eq. 41, see p. 228) [25, 26, 129, 226]. These reactions are highly regio- and stereoselective.

The directive effects of a number of other functional groups have also been reported. The alkoxymercuration of ketene [79], *cis-* and *trans-*styryl cyanide [76, 143] and N-allyl imides [170, 171, 173–175] proceeds in the anticipated fashion (Eqs. 42–44). The regiochemistry of this last reaction has been disputed, but it appears that the major product is that shown and only small amounts of the other regioisomer are formed. Similar confusion exists with allyl amides (Eqs. 45, 46). The latter product (Eq. 46) seems unlikely in view of the many related alkenes which alkoxymercurate in a Markovnikov fashion. The sulfone and trimethylsilyl groups appear to afford primarily *anti-*Markovnikov products (Eqs. 47–49). Conflicting results are reported for alkenyl trimethoxysilanes [27, 121]. The regiochemistry of the alkoxymercuration of alkyl-substituted cyclohexenes has not been examined with the care afforded the analogous hydroxymercuration reactions [230], but one can fairly safely assume that the regioselectivity of the two processes is likely to be very similar. The tables also contain a number of examples of the alkoxymercuration of internal alkenes with remote functional groups which are reported as if the reaction affords only one regioisomer. These results seem highly questionable and are included in the tables followed by a question mark. The interest in many of these compounds stems from their potential utility as germicides or weed killers.

$$\text{H}_2\text{C=CHCH}_2\text{NHCOC}_6\text{H}_4\text{OCH}_2\text{CO}_2\text{H} \longrightarrow \text{AcOHgCH}_2\overset{\text{OR}}{\text{CHCH}_2}\text{NHCOC}_6\text{H}_4\text{OCH}_2\text{CO}_2\text{H} \qquad (45)\ [141]$$

$$\text{H}_2\text{C=CHCH}_2\text{NHCO}_2\text{C}_2\text{H}_5 \longrightarrow \text{CH}_3\text{OCH}_2\overset{\text{HgOAc}}{\text{CHCH}_2}\text{NHCO}_2\text{C}_2\text{H}_5 \qquad (46)\ [127]$$

$$\text{H}_2\text{C=CH(CH}_2)_n\text{SO}_2\text{CH}_3 \longrightarrow \text{CH}_3\text{OCH}_2\overset{\text{HgOAc}}{\text{CH(CH}_2)_n}\text{SO}_2\text{CH}_3 \qquad (47)\ [87, 111]$$

$$n = 0,1$$

$$(48)\ [239]$$

$$65 \quad : \quad 35$$

229

III. Alkoxymercuration

$$RCH{=}CHSi(CH_3)_3 \longrightarrow \underset{\underset{HgOAc}{|}}{RCHCHSi(CH_3)_3}\overset{\overset{OCH_3}{|}}{} + \underset{\underset{OCH_3}{|}}{RCHCHSi(CH_3)_3}\overset{\overset{HgOAc}{|}}{}$$

$$
\begin{array}{ccc}
R = H & 100 & : & 0 \\
CH_3 & 70 & : & 30 \\
C_6H_5 & 50 & : & 50
\end{array}
$$

(49) [27]

$$(CH_3)_3C\text{—}CH{=}CH\text{—}C(CH_3)_3 \longrightarrow \longleftarrow (CH_3)_3C\text{—}CH{=}CH\text{—}C(CH_3)_3$$

(50)

Let's take a closer look at the stereochemistry of the alkoxymercuration process. Valuable NMR spectroscopic techniques have been worked out to help in the assignment of stereochemistry in methoxymercurials [18, 21, 23]. Thus, by NMR spectroscopy the methoxymercuration of both *cis*- and *trans*-1,2-dideuteroethylene has been shown to involve exclusively anti addition [24]. Similar results are reported for *cis*- and *trans*-2-butene [54, 55]. On the other hand, *cis*-1,2-di-*t*-butylethylene gives an anti adduct, but the *trans* isomer gives a syn adduct (Eq. 50) [72]. This appears to be the only example of syn addition to an unstrained alkene. The low reactivity of the *trans*-isomer and the syn addition are attributed to steric effects. The methoxymercuration of *cis*- and *trans*-stilbenes [54, 68] and styryl cyanides [76, 143] have been reported to yield two different diastereomers, but no assignment of stereochemistry has been made. *trans*-Cinnamyl alcohol [22] and *cis*- and *trans*-cinnamic acids and derivatives [159] reportedly afford predominantly the anti diastereomers, but *cis*- and *trans*-styryl and substituted styryl ketones give mixtures of diastereomers [22, 159, 198]. The methoxymercuration of tiglic and angelic acids

(51)

(52)

$$R = i\text{-}Pr,\ t\text{-}Bu$$

(53)

and subsequent demercuration with hydrogen sulfide in alkali produces mainly the erythro product, suggesting that the former demercuration proceeds with almost complete retention and the latter with inversion (Eq. 51, see p. 230) [114].

Several studies have examined the possibility of asymmetric induction by neighboring chiral centers in the alkoxymercuration process. The above carbohydrate is reported to afford the indicated stereoisomer (Eq. 52, see p. 230) [200]. No mention is made of the epimer. The stereochemistry of methoxymercuration of a variety of chiral cinnamic esters [11, 12, 195, 201] and amides [12] has been examined, but enantiomeric excesses of only 27% or less have been obtained (Eq. 53, see 230) [201]. Methoxymercuration-demercuration of several similar menthyl esters also gave low ($<16\%$) asymmetric induction [195].

$$\text{(bicyclic alkene)} \xrightarrow[\text{CH}_3\text{OH}]{1.1\,\text{Hg(NO}_3)_2/\text{HgO}} \xrightarrow{n\text{-C}_6\text{H}_{13}\text{CH}=\text{CH}_2} \xrightarrow{\text{NaBH}_4} n\text{-C}_6\text{H}_{13}\overset{*}{\underset{}{\text{C}}}\text{H}(\text{OCH}_3)\text{CH}_3 \qquad (54)$$

An alternative, quite novel approach to the preparation of optically active ethers involves the alkoxymercuration of an optically active alkene, followed by transfer alkoxymercuration to the desired alkenes and subsequent demercuration (Eq. 54) [263]. Transfer of the methoxymercury moiety must be proceeding directly from the chiral olefin adduct to the octene or else all optical activity would be lost. While optical yields as high as 36% are reported, this approach to chiral ethers is not really practical since the exchange reaction cannot be allowed to proceed to completion and the actual yields are therefore quite low.

The stereochemistry of alkoxymercuration of simple cyclic alkenes is identical to that of hydroxymercuration. While the alkoxymercuration of a cyclopropene has been observed, no stereochemistry was reported [240]. Alkoxymercuration of cyclobutene[208], cyclopentene [32, 208], cyclohexene [31, 33, 208, 212, 215, 216], *cis*-cycloheptene [208, 231], *cis*-cyclooctene [208, 231, 235], *cis*-cyclononene [208] and acenaphthylene [19] affords anti adducts, while *trans*-cyclooctene [208, 235, 236] and *trans*-cyclononene [208] form syn addition compounds. The methoxymercuration of cyclohexene using mercury(II) salts derived from (+)-lactic [212] and (+)-mandelic [216] acids affords equal mixtures of two diastereomers. The addition of sodium chloride to one of the diastereomers of the former reaction is said to afford an optically active organomercurial of unknown optical purity.

Relatively little work has been reported on the alkoxymercuration of substituted cyclohexenes, but it appears safe to assume that the results would

$$\text{(substituted cyclohexene)} \longrightarrow \longrightarrow \text{(trans adduct, OCH}_3\text{)} + \text{(cis adduct, OCH}_3\text{)} \qquad (55)$$

| X = OH | 85 | : | 15 | [224, 225] |
| X = OAc | 74.5 | : | 25.5 | [224] |

III. Alkoxymercuration

$$(56) \ [22]$$

be nearly identical to those found for hydroxymercuration. Both 1- and 4-methylcyclohexene have been methoxymercurated, but the stereochemistry of the adducts was incorrectly assigned [230]. A wide variety of 3-oxygenated cyclohexenes undergo alkoxymercuration to form predominantly the *trans*-1,3-products (Eqs. 55, see p. 231; 56). Homoallylic derivatives of all types have been reported to give the following adducts (Eq. 57) [129, 226]. However, conflicting results have been reported for cyano- and carbomethoxy-substituted cyclohexenes (Eqs. 58, 59) [20, 25, 26]. These last results are con-

$$(57)$$

$$X = OH, OAc, OCH_3, OCH_2C_6H_5, CH_2OH, CH_2OAc, CH_2OCH_2C_6H_5$$

$$(58)$$

$$(59)$$

$$X = CN, CO_2CH_3$$

$$(60)$$

$$65 \quad : \quad 35$$

$$(61)$$

$$(62)$$

major   minor

sistent with trans diaxial addition to the more stable conformers. 1-Trimethyl-silylcyclohexene gives a mixture of β-methoxysilanes (Eq. 60, see above) [239].

Conflicting work has been reported on the alkoxymercuration of D-glucals. D-Glucal itself is reported to yield both *syn*- and *anti*-methoxymercurials (Eq. 61, see p. 232) [210, 229]. The syn adduct is most likely incorrect, since the anti adduct (whose stereochemistry was only assumed) corresponds to the major isomer formed in the alkoxymercuration of D-glucal triacetate (Eq. 62, see p. 232) [210, 229, 238, 244–246]. Confusion apparently arises here, because the minor isomer is more easily isolated. Its structure has been confirmed by x-ray crystallographic analysis [34]. 6-Deoxy-L-glucal affords an analogous product in 81% yield (Eq. 63) [227, 228].

(63)

(64)

(65) [237]

65 : 35

(66) [238]

58 : 35

A number of other glycals have been subjected to methoxymercuration. D-Galactal and its triacetate give only one stereoisomer confirmed by x-ray crystallography (Eq. 64) [35, 238]. Other glycals are reported to give mixtures of products (Eqs. 65–68). All carbohydrates bearing an exo-methylene group

(67) [238]

74 : 26

(68) [268]

53 : 40

233

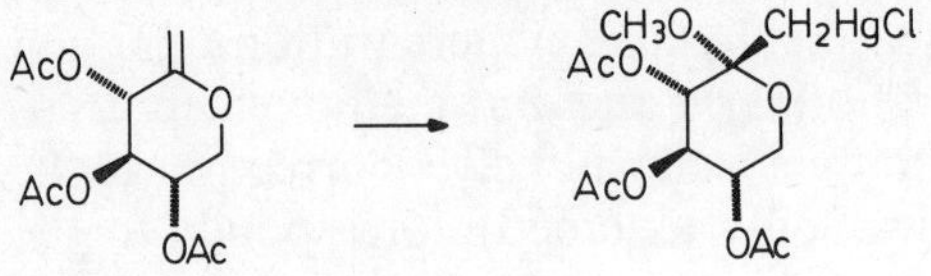

III. Alkoxymercuration

(69) [243]

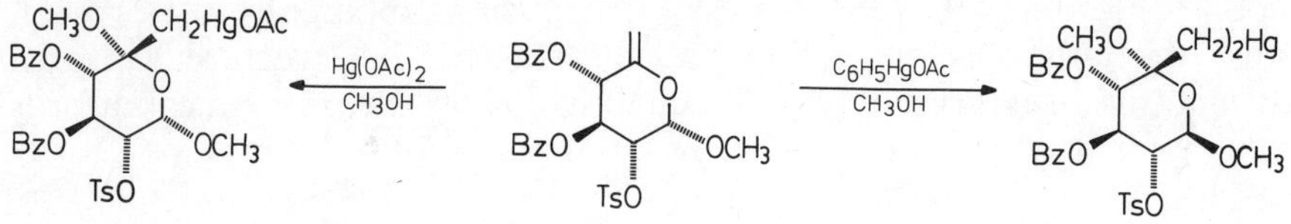

(70) [207]

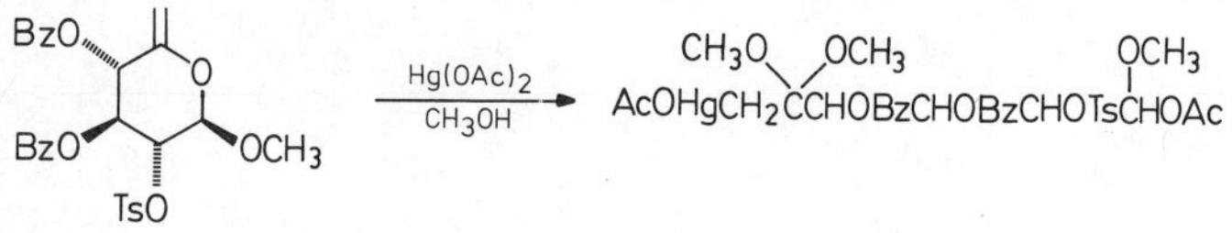

(71) [207]

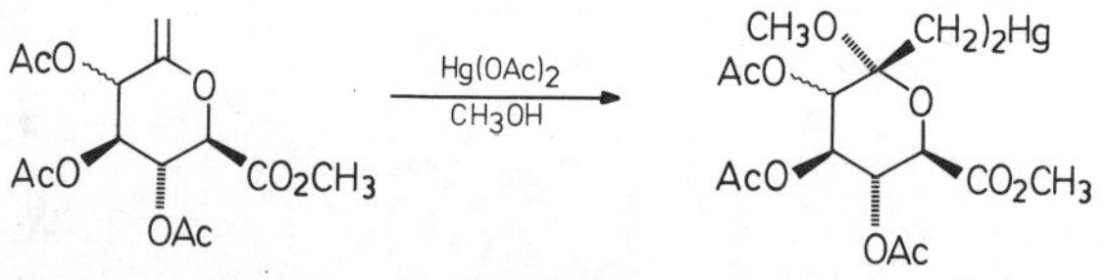

$$AcOHgCH_2CCHOBzCHOBzCHOTsCHOAc$$

(72) [248]

seem to lead to a single stereoisomer, but both alkylmercuric salts and dialkylmercurials are reportedly formed in these reactions (Eqs. 69–73). Note the acetal inversion in equation 70 and ring opening in equation 71. Axial attack of the methoxyl group is preferred in these reactions.

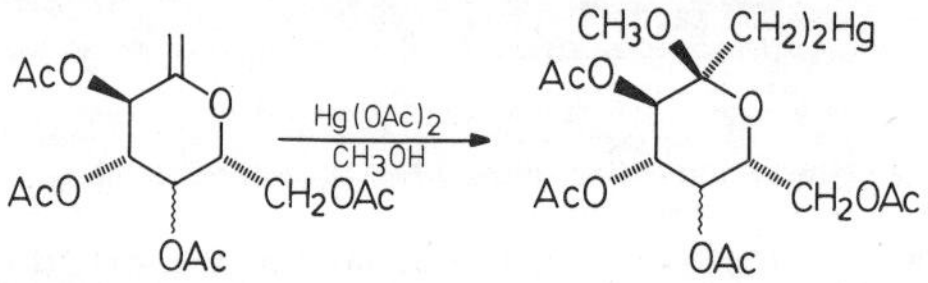

(73) [248]

(74)

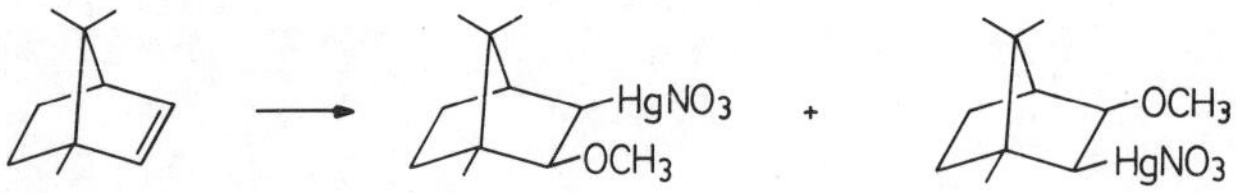

(75) [263]

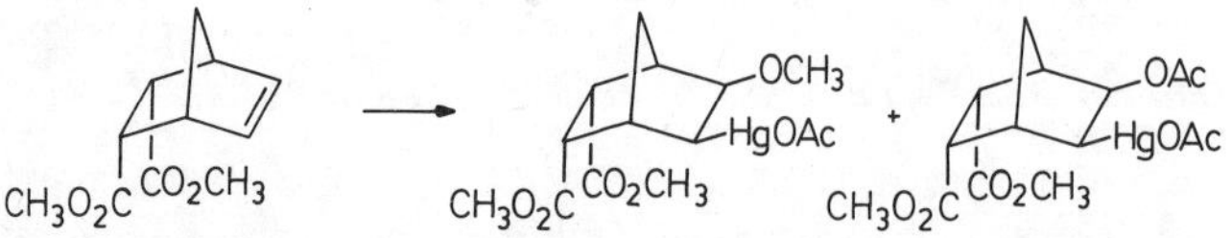

(76) [254, 266]

234

The stereochemistry of the alkoxymercuration of a limited number of bicyclic alkenes has been reported. The methoxymercuration of norbornene was incorrectly first reported to proceed by anti addition [254]. It is now known to proceed syn with accompanying syn acetoxymercuration (Eq. 74, see p. 234) [235, 253]. Similar observations have been reported for other norbornene derivatives (Eqs. 75, 76, see p. 234). While the corresponding [2.2.2] bicyclic diester has been reported to afford a syn endo adduct, this result needs confirmation (Eq. 77) [267].

$$\text{(77)}$$

$$HO-C_n-C{=}C \longrightarrow C_n{-}\overset{O}{\overbrace{\quad}}{-}C{-}C{-}HgX \tag{78}$$

$$HOCH_2CH{=}CH_2 \longrightarrow HOCH_2\overset{OCH_3}{\underset{|}{C}}HCH_2HgX \tag{79}$$

$$HO_2C-C_n-CH{=}CH_2 \longrightarrow HO_2C-C_n-\overset{OR}{\underset{|}{C}}H-CH_2HgOH \tag{80}$$

As noted in Tables 3.1–3.4, virtually every important organic functional group can be accommodated by the alkoxymercuration reaction. An unusually wide range of functional groups have been examined, because of the interest in the resulting alkoxymercurials as diuretics. Olefins containing alcohol groups can of course present difficulties due to intramolecular alkoxymercuration (Eq. 78). This reaction will be discussed in detail later. Nevertheless, even simple allylic alcohols which have been observed to form dioxane derivatives can, under the right conditions, give the desired alkoxymercurial (Eq. 79) [86].

Certain functional groups can present difficulties however. For example, unsaturated carboxylic acids are observed in some cases to afford β-alkoxy alkylmercuric hydroxides (eq. 80) [164, 194]. Epoxide groups can undergo

$$\text{(81) [234]}$$

$$\underset{ROCSCH_2CH{=}CH_2}{\overset{O}{\overset{\|}{}}} \longrightarrow ROCSCH_2\overset{OR'}{\underset{|}{C}}HCH_2HgX + XHgSCSCH_2\overset{OR'}{\underset{|}{C}}HCH_2HgX \tag{82}$$

intramolecular participation (Eq. 81) [202]. Thiocarbonates can be cleaved (Eq. 82) [116] and vinyl halo ethers afford mercurated esters (Eq. 83, see p. 236)

[80]. Activated aromatic compounds can undergo competitive aromatic substitution (Eqs. 84–86). With certain alkenes, allylic mercurials apparently form

$$X_2C=CXOR \longrightarrow XHgCX_2CO_2R \tag{83}$$

$$X=OCH_3,\ OAc \tag{84) [153]}$$

$$R^1=t\text{-Bu},\ R^2=R^3=H;\quad R^1=H,\ R^2=R^3=Me \tag{85) [61]}$$

competitively and undergo solvolysis (Eqs. 87, 88). While a similar product has been formed from allylbenzene, a different mechanism was suggested (Eq. 89) [70]. With certain strained bicyclic compounds like norbornene, acyloxymercuration products are formed alongside the usual alkoxymercurials (Eq. 90) [251].

$$\tag{86) [241]}$$

$$\xrightarrow[\text{CH}_3\text{OH}]{\text{Hg(OAc)}_2\ /\Delta} \tag{87) [276]}$$

$$\xrightarrow[\text{CH}_3\text{OH}]{\text{Hg(OAc)}_2/\Delta}\ \xrightarrow{\text{NaBH}_4}\ \xrightarrow{\text{Ac}_2\text{O}} \tag{88) [270]}$$

236

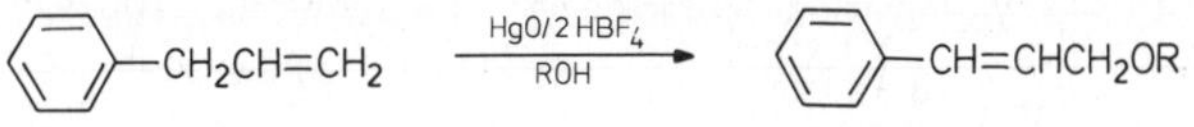

$$C_6H_5-CH_2CH=CH_2 \xrightarrow[ROH]{HgO/2\ HBF_4} C_6H_5-CH=CHCH_2OR \tag{89}$$

$$\text{(norbornene)} \longrightarrow \text{(2-OCH}_3\text{, 3-HgOAc)} + \text{(2-OAc, 3-HgOAc)} \tag{90}$$

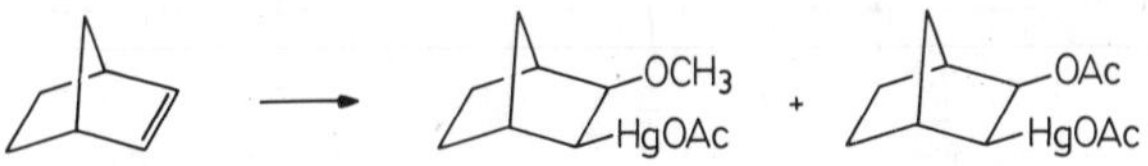

There are relatively few cases of carbon skeleton rearrangements during alkoxymercuration. However, aryl groups have been noted to undergo migration in the following reactions (Eqs. 91, 92). These rearrangements are not observed in the presence of less ionic mercury salts. Unsaturated cyclopropanes and cyclobutanes are observed to undergo ring-opening reactions (Eqs. 93–95; Eqs. 96, 97, see p. 238). One other bicyclic alkene has also been reported to rearrange (Eq. 98, see p. 238) [269].

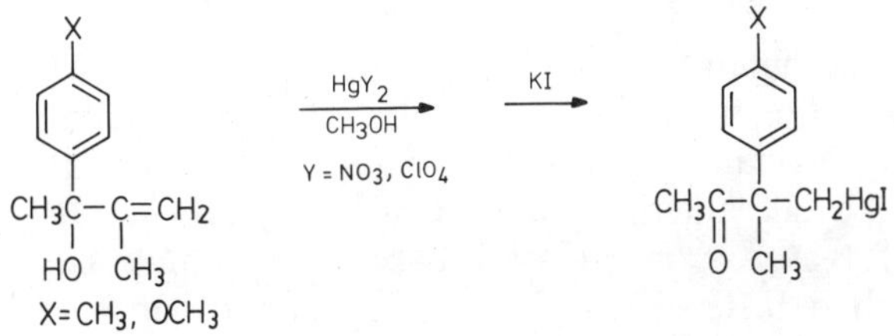

$$\begin{array}{c} X-C_6H_4-\underset{\underset{CH_3}{|}}{\underset{\underset{HO}{|}}{C}}-C=CH_2 \end{array} \xrightarrow[CH_3OH]{HgY_2} \xrightarrow{KI} X-C_6H_4-\underset{\underset{O}{||}}{C}-\underset{\underset{CH_3}{|}}{C}-CH_2HgI \tag{91}\ [189]$$

$$Y = NO_3,\ ClO_4 \qquad X = CH_3,\ OCH_3$$

$$C_6H_5CH=CHC_6H_5 \xrightarrow[CH_3OH]{2\ Hg(O_2CCF_3)_2} (C_6H_5)_2CHCH(OCH_3)_2 \tag{92}\ [54]$$

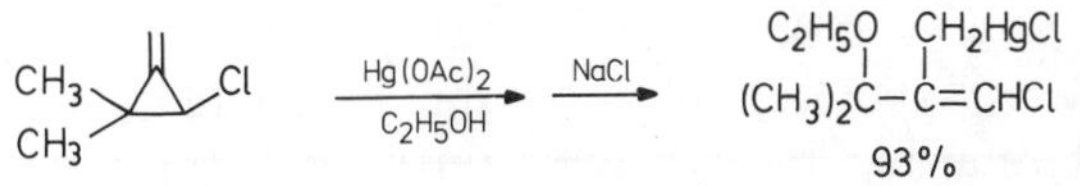

cis and *trans*

$$\underset{CH_3}{\overset{CH_3}{>}}\!\!\!\triangle\!\!-Cl \xrightarrow[C_2H_5OH]{Hg(OAc)_2} \xrightarrow{NaCl} \underset{C_2H_5O\ \ CH_2HgCl}{(CH_3)_2C-C=CHCl} \tag{93}\ [211]$$

93%

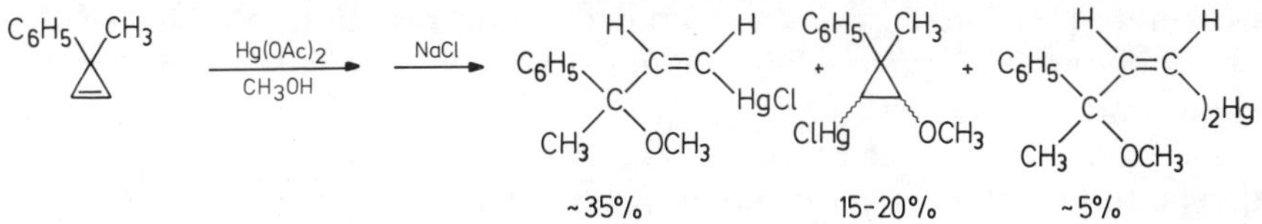

$$\tag{94}\ [240]$$

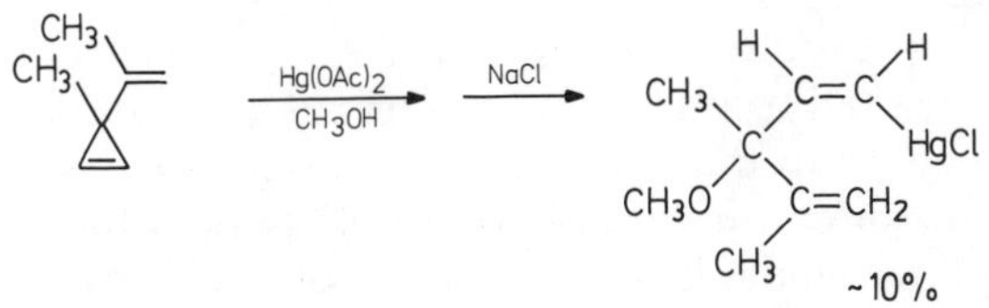

~35%   15–20%   ~5%

$$\tag{95}\ [240]$$

~10%

The kinetics of alkoxymercuration have been studied for both mechanistic and practical reasons. The alkoxymercuration of α,β-unsaturated acids [277], esters [99, 227–279], aldehydes [277, 279], ketones [280] and nitriles [76, 143], as well as 5-vinyl-2-picoline [281] are reportedly accelerated by the

III. Alkoxymercuration

(96) [264]

(97) [265]

(98)

addition of acids, generally in the order $HClO_4 > HNO_3 > HOAc$ [277, 279, 281]. On the other hand, the methoxymercuration of acrylonitrile is accelerated by $HClO_4$, but appreciably retarded by HOAc [282], and the reaction of simple acyclic and cyclic alkenes is also retarded by HOAc [219, 283]. While the addition of NaOAc or nitrogen bases to the methoxymercuration of acrylonitrile had little effect [282], NaOAc or other bases have been reported to retard the alkoxymercuration of simple alkenes [219, 283] and all the other substrates mentioned above [99, 278–281, 284]. In the methoxymercuration of 5-vinyl-2-picoline, the effect of other added salts was slight [281]. The methoxymercuration of cyclohexene and *trans*-stilbene is said to be promoted by the addition of peroxides [76, 212], but this is not true for styryl cyanide [76]. Methoxymercuration of the latter alkene is accelerated by $BF_3 \cdot Et_2O$, but retarded by pyridine and acetonitrile [76].

In general, the more electrophilic mercury salts such as mercuric trifluoroacetate [57], mercuric nitrate and mercuric perchlorate [72] give the fastest reactions, often succeeding where mercuric acetate fails. Second order rate constants for the methoxymercuration of cyclooctene and 3,3-dimethyl-1-butene have recently been reported for a variety of mercuric carboxylates [285]. The following relative reactivity is observed for HgX_2: $X = CF_3CO_2 > F_2CHCO_2 > ClCH_2CO_2 > Cl_2CHCO_2 > CH_3CO_2 > Me_2CHCO_2 > Me_3CCO_2 > n\text{-}BuCO_2 > Cl$.

Surprisingly, little work has appeared on the effect of varying the nature of the alcohol. In a study of the alkoxymercuration of α,β-unsaturated carboxylic acids, it was observed that the relative rate of reaction was methanol > glycols > ethanol through pentanol [284]. In a similar study, it was reported that shortening the chain length of the alcohol *decreases* the rate of reaction, but the figures presented in that paper seem to say the opposite [277]. In alkoxymercurations using mercuric trifluoroacetate, it is found that with more hindered alcohols and certain olefins, trifluoroacetoxymercuration preceeds alkoxymercuration and only after long reaction times is the desired alkoxymercurial formed [57].

238

Actual kinetic data have been determined for a number of alkoxymercuration reactions [52]. Competition studies of the relative rates of methoxymercuration of 28 simple olefins have been correlated with the Taft equation to give $\varrho^* = -1.00$ and $\delta = 0.91$ [286]. It has been argued that the low carbonium ion character of this reaction suggests a transition state similar to an unsymmetrically bridged mercurinium ion. Second order kinetics are reported for the alkoxymercuration of a number of simple alkenes [212, 219, 285, 287], and the rates of formation of electron donor-acceptor complexes between 10 such olefins and mercuric chloride or mercuric acetate have been reported [285]. While the rates of bromination and alkoxymercuration correlate with the acceptor ability of the electrophile, they show divergent trends with the donor ability of the olefin. However, when steric considerations are taken into account, the relative reactivities of various olefins towards electrophilic bromination and solvomercuration are essentially identical. Simply stated, steric effects in the mercuration reaction are significantly larger in the transition state than in bromination. This is borne out in the methoxymercuration of ring-substituted styrenes [29, 288, 289]. Second order rate constants and a number of thermodynamic parameters have been determined for this reaction. Rho values ranging from -1.59 to -3.16 have been offered as evidence for an unsymmetrically bridged or mercurinium-like intermediate. The following relative reactivities for alkyl-substituted styrenes have been reported: $ArCR=CH_2 > ArCH=CH_2 > ArCH=CHR$ [68].

The kinetics of methoxymercuration of a number of α,β-unsaturated carbonyl systems have been examined. The following relative reactivities are observed: $C_6H_5CH=CH_2 > H_2C=CHCO_2H > H_2C=CHCO_2R > XC_6H_4CH=CHCO_2H > C_6H_5CH=CHCO_2R > C_6H_5CH=CHCHO > H_2C=C(CH_3)CO_2R$ [277, 279, 284]. The methoxymercuration of methyl acrylate and methyl methacrylate exhibits second order kinetics with activation energies of 9 and 15 kcal/mole respectively [99, 279]. *trans*-Cinnamyl alcohol is more reactive than cinnamic acid [277]. Detailed kinetics for cinnamic acid [290], substituted cinnamic acids [291] and cinnamate esters [159, 278], have been published. A rho value of -1.57 and other thermodynamic data have been used to support a mercurinium ion-like intermediate in the methoxymercuration of various cinnamic acids [290]. With the corresponding esters, rho is approximately -0.9 (using σ^+ constants), leading the authors to conclude that most of the positive charge is located on mercury not carbon [159]. The corresponding substituted styryl ketones are approximately ten times more reactive than the cinnamic esters and have a rho value of -0.96, suggesting a similar mechanism holds for these substrates [159]. Thermodynamic data for the second order methoxymercuration of various substituted chalcones have also been reported [280].

The second order methoxymercuration of acrylonitrile is approximately 100 times slower than that of the acrylate esters [282]. This observation and the fact that the reaction is retarded by acetic acid, but relatively unaffected by sodium acetate or nitrogen bases, have been suggested to support a non-ionic mechanism presumably proceeding via $CH_3OHgOAc$.

Unlike all the other alkoxymercuration reactions which have been reported to exhibit second order kinetics, the methoxymercuration of 5-vinyl-2-picoline is not first order in mercuric acetate [281]. This is presumably due to mercury coordination to the basic nitrogen.

Besides the kinetic data reported above, a great deal of information is now available on the relative reactivity of various alkenes towards alkoxymercuration. In general, cis disubstituted alkenes are more reactive than the corresponding trans isomers. This has been reported for unsaturated fatty acid methyl esters [292], *cis*- and *trans*-stilbene [75, 293], and *cis*- and *trans*-cinnamonitriles [143], and a number of other simple alkenes to be discussed shortly. The opposite is true for the *cis*- and *trans*-di-*t*-butylethylenes, however [72]. The cis isomer is approximately 1000 times less reactive than cyclohexene due to steric hindrance and the trans isomer is still less reactive. Remember that this latter compound undergoes syn addition. Steric hindrance to methoxymercuration is also reported to change the nature of the products from alkenylferrocenes (Eq. 99) [183]. In the methoxymercuration of nine pure *n*-undecenes, the rates are reported to fall as the double bond is displaced toward the center of the carbon chain [294].

$$\text{(structure: } R\text{-}CCH_2HgOAc,\ OCH_3,\ Fe) \xleftarrow{R=H,Me} \text{(structure: } R\text{-}C{=}CH_2,\ Fe) \xrightarrow{R=t\text{-}Bu} \text{(structure: } R\text{-}C{=}CH_2,\ Fe\text{-}HgOAc) \tag{99}$$

As reported earlier for the hydroxymercuration of alkenes (Table 2.5), the relative rates of methoxymercuration of a wide variety of alkenes (excluding most of the functionally substituted alkenes discussed immediately above) have been summarized in Table 3.5 with all data adjusted relative to cyclohexene. Unlike Table 2.5, the present data is broken down according to reference. Where conflicts in data are apparent, the reader is advised to check the original references for details. One significant complication in determining the relative rates of methoxymercuration by competition studies is the observation that reactions allowed to proceed for some time afford equilibrium distributions of products and not the desired kinetic results [209]. This appears to be the source of some of the discrepancies in reference [295].

Most of the generalizations made earlier in discussing the relative rates of hydroxymercuration of alkenes apply to the methoxymercuration reaction as well. For example, the rate generally decreases as the carbon—carbon double bond is moved further towards the center of the alkene. This is nicely illustrated by the following relative rates for a series of *n*-undecenes [297].

1 >	cis-2 >	cis-3 >	cis-4 >	cis-5 >	trans-2 >	trans-3 >	trans-4 >	trans-5
1000	86.3	56.2	39.1	35.8	21.5	10.0	6.17	5.37

However, there are exceptions to this trend. For example, the 3-hexenes are more reactive than the 2-hexenes [37], and it is reported that *cis-* and *trans-*2-pentadecene are more reactive towards mercuric propionate in ethanol/dioxane than 1-pentadecene [37]. Under these conditions one can apparently separate the internal olefins from the terminal isomer leaving almost pure 1-pentadecene. The reason for the unusual selectivity in this case is not clear. Note also that *cis-*alkenes are generally more reactive than their *trans-*isomers. In the above system, *cis-* and *trans-*2-pentadecene can be separated sufficiently to afford 95% pure *trans-*2-pentadecene [37]. The *cis/trans* relative rate ratios for 3-hexene, 4-octene and 2-undecene are 5.6, 6.7 and 4.0, respectively [37, 272]. In one further study of the relative rates of reaction of a variety of alkenes, the following results have been reported with no actual rate constants presented [37]. From this data and that of Table 3.5, it

```
      C                              C                        C
      |                              |                        |
C-C-C=C  ≃  C-C-C-C-C=C  >  C-C-C-C=C  >  C-C-C=C-C  >  C-C-C=C-C-C >

                   C
                   |
C-C-C-C=C-C  >  C-C-C=C-C

                                  C                              C
                                  |                              |
C-C-C-C-C-C-C=C  ≃  C-C-C-C-C-C-C=C  ≃  C-C-C-C-C-C=C  >

              C                 C       C
              |                 |       |
C-C-C-C-C-C=C  ≃  C-C-C-C-C-C=C  >  C-C-C-C-C-C-C=C-C  ≃
                                C
                                |
C-C-C-C-C-C=C-C-C  ≃  C-C-C-C-C-C=C-C  >  C-C-C-C-C=C-C-C
```

is quite obvious that the alkoxymercuration reaction is quite sensitive to steric factors. It is also clear that a general increase in rate with the strain energy of the double bond is not observed [296].

As noted earlier, the mechanism of the solvomercuration reaction has been the center of considerable controversy. The slow rate at which mercuric acetate adds to alkenes, and alcohols replace the acetate moiety in the resulting β-acetoxymercuricals rules out such a mechanism for alkoxymercuration [212, 219, 298]. The fact that acetic acid retards the rate of alkoxymercuration of simple alkenes was used by Wright to rule out an ionic mechanism in favor of the direct addition of an alkoxymercuric acetate across the carbon—carbon double bond [219, 298]. However, this mechanism has almost [282] been completely discarded in favor of an ionic process involving mercury-stabilized carbocations. The nature of this intermediate remains the center of controversy. The considerable difference noted in the characteristics of alkoxymercuration versus other electrophilic addition reactions [272, 286], has most recently been explained by arguing that steric effects in solvomercuration are significantly larger in the transition state for alkoxymercuration than for these other reactions [283, 285]. Bach has argued that the preferred mechanism is one involving fast reversible formation of a

Table 3.5. Relative Rates of Methoxymercuration of Alkenes

Alkene	Relative rates						
	Ref.: 283[a]	295	272	296	285	286	72
(cyclopropyl)$_2$C=CH$_2$	700 (11,000)						
cyclopropyl–CH=CH$_2$	700 (1,700)						
CH$_3$(cyclopropyl)C=CH$_2$	350 (8,700)						
methylenecyclopentane		170	17				
norbornadiene				27			
(CH$_3$CH$_2$)$_2$C=CH$_2$		26					
CH$_3$CH$_2$CH$_2$C(CH$_3$)=CH$_2$		21	5.9				
C$_6$H$_5$(cyclopropyl)C=CH$_2$		15					
CH$_3$(CH$_2$)$_3$C(CH$_3$)=CH$_2$		14					
methylenecyclohexane			13				
CH$_3$(CH$_2$)$_3$CH=CH$_2$	(4.17)	3.2	4.1		12		
CH$_3$(CH$_2$)$_5$CH=CH$_2$	7.4		3.9		11		9.5

Alkene	Relative rates						
	Ref.: 283[a]	295	272	296	285	286	72
CH$_3$(CH$_2$)$_2$CH=CH$_2$					10		
cyclopentene							10.1
(CH$_3$)$_2$C=CH$_2$						8.7	
(CH$_3$)$_2$CHCH$_2$CH=CH$_2$		8.2					
CH$_3$CH$_2$C(CH$_3$)=CH$_2$			7.3				
CH$_3$CH$_2$CH(CH$_3$)CH=CH$_2$		6.8			1.3		
CH$_3$CH$_2$CH=CH$_2$						6.2	
(CH$_3$)$_2$CHC(CH$_3$)=CH$_2$		5.9	5.2				
norbornene	4.6 (10.8)		1.0				4.5
CH$_3$(CH$_2$)$_4$CH=CH$_2$						3.95	
(CH$_3$)$_2$C=CHCH$_3$			1.6		3.71		
cis-cyclopropyl–CH=CHCH$_3$	(2.79)						
1-methylcyclohexene			1.5	2.2			
H$_2$C=CH$_2$	(2.08)						

Table 3.5. (continued)

Alkene	Relative rates						
	Ref.: 283[a]	295	272	296	285	286	72
$CH_3OCH_2C(CH_3)=CH_2$						1.68	
$C_6H_5CH=CH_2$					1.48		
trans-(cyclopropyl)$CH=CHCH_3$	(1.29)						
$CH_3CH_2CH=C(CH_3)_2$		1.1				1.05	
[benzobicyclic alkene structure]				1.1			
[1-methylcyclopentene structure]		1.0	0.45				
[cyclohexene structure]	1.0	1.0	1.0	1.0	1.0	1.0	1.0
[4-tert-butylcyclohexene structure]				0.96			
[cyclooctadiene structure]			0.92				
[cyclopentene structure]			0.77, 0.73				
[cycloheptene structure]			0.25		0.70		
cis-$CH_3CH_2C(CH_3)=CHCH_3$		0.59					

Alkene	Relative rates						
	Ref.: 283[a]	295	272	296	285	286	72
$(CH_3)_3CC(CH_3)=CH_2$	0.50 (1.0)						
[1-methylcyclopentene structure]		0.50					
$ClCH_2CH_2CH=CH_2$						0.47	
$CH_3OCH_2CH=CH_2$						0.46	
trans-$CH_3CH_2C(CH_3)=CHCH_3$		0.45					
$C_6H_5CH_2CH=CH_2$						0.41	
[cyclobutene structure]				0.40			0.36
$(CH_3)_3CCH_2C(CH_3)=CH_2$						0.24	
cis-$CH_3(CH_2)_2CH=CHCH_3$		0.23					
cis-$CH_3(CH_2)_2CH=CH(CH_2)_2CH_3$	0.22		0.10				
trans-$CH_3CH_2CH=CHCH_2CH_3$		0.18	0.044				
[bicyclo[4.2.0] structure]						0.17	
$CH_3(CH_2)_2CH=C(CH_3)_2$		0.14					
$ClCH_2C(CH_3)=CH_2$						0.12	
trans-$CH_3OCH_2CH=CHCH_3$						0.11	

Table 3.5. (continued)

244

Alkene	Relative rates						
	Ref.: 283[a]	295	272	296	285	286	72
cis-$CH_3CH_2CH{=}CHCH_2CH_3$	0.58	0.27	0.25, 0.065				
cis-$(CH_3)_2CHCH{=}CHCH_3$		0.09					
$(CH_3)_3CCH{=}CH_2$	0.067 (1.17)	0.91	0.07		0.09		
$(CH_3CH_2)_2C{=}CHCH_3$			0.08				
 [bicyclo[2.2.1] alkene structure]				0.065			
$ClCH_2CH{=}CH_2$						0.06	
cis- and $trans$-$(CH_3)_2CHCH{=}CHCH_2CH_3$		0.05					
$trans$-$(CH_3)_2CHCH{=}CHCH_3$		0.05					
 [bicyclo[2.2.2] alkene structure]			0.03	0.03			0.03
$trans$-$ClCH_2CH{=}CHCH_3$						0.027	

[a] Numbers in parentheses are obtained by direct UV measurement.

Alkene	Relative rates						
	Ref.: 283[a]	295	272	296	285	286	72
$trans$-$CH_3(CH_2)_2CH{=}CHCH_3$		0.09					
$(CH_3)_3CCH{=}C(CH_3)_2$						0.019	
$trans$-$CH_3(CH_2)_2CH{=}CH(CH_2)_2CH_3$		0.015					
$(CH_3)_2C{=}C(CH_3)_2$		0.009			0.006	0.007	
$[(CH_3)_3C]_2C{=}CH_2$	0.005						
cis-$C_6H_5CH{=}CHCH_3$					0.004		
 [cyclooctene structure]			0.004	0.004	0.003		
cis-$(CH_3)_3CCH{=}CHC(CH_3)_3$							0.001
cis- and $trans$-$(CH_3)_3CCH{=}CHCH_3$					0.0006		
cis-$C_6H_5CH{=}CHBr$					0.00003		

[a] Numbers in parentheses are obtained by direct UV measurement.

bridged π-complex (k_1) with rate-limiting attack by solvent (k_2) on this mercurinium ion (Eq. 100) [72, 235, 259, 296]. Others have argued from

$$\ce{>C=C<} \quad \underset{k_{-1}}{\overset{\overset{\text{+HgX}}{k_1}}{\rightleftharpoons}} \quad \overset{\text{HgX}}{\ce{>C\cdots C<}} \quad \underset{k_{-2}}{\overset{\overset{\text{ROH}}{k_2}}{\rightleftharpoons}} \quad \underset{\text{RO}}{\overset{\text{HgX}}{-\ce{C}-\ce{C}-}} \tag{100}$$

correlations with the Taft equation [286] and deuterium isotope effects [28] that the intermediate is unsymmetrical and is accompanied by considerable carbon—oxygen bond formation [271]. Perhaps Tidwell is correct when he argues for a continuum of ions ranging from bridged to open ions depending on the substrate involved [283].

The acid-promoted dealkoxymercuration of a wide variety of alkoxymercurials has also been studied in detail in the hopes of shedding some light on the mechanism of the alkoxymercuration reaction. Besides several overall studies of the dealkoxymercuration reaction [43, 218, 299], detailed kinetic studies of solvent effects [300], solvent isotope effects [301], variation in the olefin structure [51, 302, 303], including deuterium isotope effects [138], variation in the alcohol structure [47, 303] and other thermodynamic parameters [304] have been reported. All results are consistent with a reversal of the alkoxymercuration mechanism discussed earlier (Eq. 100).

Relatively little work has been done on the stereochemistry of dealkoxymercuration. The methoxymercuration of unsaturated fatty acids and esters provides a convenient method of separating these compounds from their saturated analogs. By treating the methoxymercurials with dilute acid, commonly hydrochloric acid, one can regenerate the olefins in high purity [305–312]. It is claimed that this overall process proceeds with no isomerization of the carbon—carbon double bond [292, 313]. There is one notable exception to this generalization. The *threo*-methoxymercurial derived from either *cis*- or *trans*-di-*t*-butylethylene affords the *trans*-olefin upon treatment with dilute hydrochloric acid [72]. This appears to be the only known example of a *syn*-dealkoxymercuration.

Several other dealkoxymercuration studies have also appeared. For example, the addition of super acids to alkoxymercurials reportedly generates mercurinium ions observable by NMR techniques [314]. Transalkoxymercuration between various olefins has also been studied [209, 263, 315, 316]. The mechanism of this reaction appears different from the alkoxy- or dealkoxymercuration reactions so far discussed, since alkoxymercurials derived from optically active olefins are capable of chirality transfer [263].

By far the most important reaction of the β-alkoxymercurials is replacement of the mercury moiety by hydrogen, since the two step sequence of alkoxymercuration-demercuration provides a valuable method for the Markovnikov addition of alcohols to alkenes. The most important reagent for demercuration is alkaline sodium borohydride. It reacts rapidly and cleanly with most alkoxymercurials to afford high yields of the corresponding ethers. Sodium borodeuteride behaves similarly, providing a valuable approach to deuterated ethers [65, 169, 270]. Numerous functional groups are accommodated by this reaction and relatively few difficulties have been encountered with these reagents. However, when the mercury is next to a *trans*-hydroxy group or a four-membered ring, side products are observed (Eqs. 101, 102). In the latter reaction (Eq. 102), the rearrangement has been shown to occur during reduction, not during the alkoxymercuration process. In the presence of oxygen, sodium borohydride affords oxidation products instead of the usual reduction products (Eq. 103) [270].

$$(101) \ [268]$$

$$(102) \ [169]$$

$$(103) \ [214]$$

$$(104)$$

$$(105) \ [114]$$

Alkaline hydrogen sulfide has proven useful for the demercuration of a variety of α-mercurated esters and acids (Eq. 104) [101, 146, 151, 160, 161, 317]. However, methoxymercurated coumarins apparently undergo elimination to the unsaturated, ring-opened acid under these conditions [260, 261]. This approach reportedly gives higher yields with mercurated menthyl esters [195] and greater stereochemical control with methoxymercurials derived from E- and z-2-methyl-2-butenoic acids than sodium borohydride (Eq. 105). In this last example, the E-acid is reduced with almost complete retention while the z-acid affords only inverted product. Reduction with alkaline sodium borohydride gives equal amounts of the two diastereomers. With α-amidoacrylic acids, hydrogen sulfide reductions gave lower yields than sodium borohydride [158].

Several other reducing agents appear useful for demercuration. These include potassium borohydride [228, 229]; sodium trimethoxyborohydride

[59, 232]; calcium metal in methanol/THF [318]; Li, Na, K or Mg/THF/ anisidine [319]; N-benzyl-1,4-dihydronicotinamide [320] and sodium-mercury amalgam [44, 254]. The last reagent is useful for deuterium labeling when employed in D_2O. Alkaline hydrazine has also been employed for demercuration, but the yields of products are usually quite low and side products abound [64, 71, 77, 321]. Finally, one novel photochemical demercuration has been reported (Eq. 106) [322]. Photolysis of the corresponding organo-mercury chloride gave mercury and a different unidentified product.

$$\text{(106)}$$

Numerous β-alkoxymercurials have been halogenated as seen by the many examples in Tables 3.1–3.4. Only the more interesting examples will be discussed here. The methoxymercuration-bromination of acrylic [100] and crotonic [100, 124, 125] acids and esters affords bromo derivatives used in the synthesis of the amino acids serine and threonine respectively (Eq. 107). The stereochemistry of this sequence has been examined on *trans*-cinnamic

$$RCH=CHCO_2R' \longrightarrow \overset{CH_3O\ \ Br}{\underset{|\ \ \ \ \ |}{RCHCHCO_2R'}} \longrightarrow \longrightarrow \overset{HO\ \ NH_2}{\underset{|\ \ \ \ \ |}{RCHCHCO_2H}} \qquad \text{(107)}$$

acid and its methyl ester. By appropriate choice of the bromination conditions, the acid can be converted predominantly into one diastereomer [149]. Surprisingly, different isomers are reported for the acid and the ester [136]. As noted in Chapter III of the monograph "Organomercury Compounds in Organic Synthesis", the reaction pathway for these halogenations is highly dependent on the reaction conditions employed.

A number of other interesting alkoxymercurials have been halogenated. α,β-Unsaturated ketones undergo methoxymercuration-bromination. This sequence has been employed in the synthesis of a key intermediate for chloramphenicol (Eq. 108, see p. 248) [144]. As noted earlier, the mercuration of styryl ketones affords predominantly ($\sim 80\%$) the erythro isomer, but only the threo bromides could be isolated in low yield from subsequent bromination and iodination reactions [22]. Methoxymercuration-bromination of *cis*- and *trans*-styryl nitriles gives mixtures of the corresponding bromides [143]. The iodination of methoxymercurials derived from glycals gives epimeric iodides [323], while the products of bromination depend on the reaction conditions [210].

While intermolecular phenoxymercuration is difficult, phenoxyiodination can be accomplished simply by treating the olefin with iodine, phenol and mercuric sulfate, followed by alkaline potassium borohydride (Eq. 109) [324].

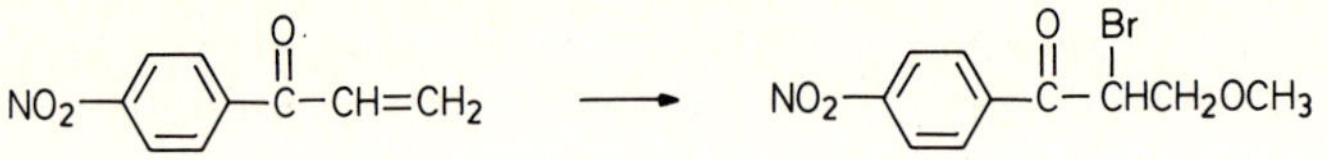

III. Alkoxymercuration

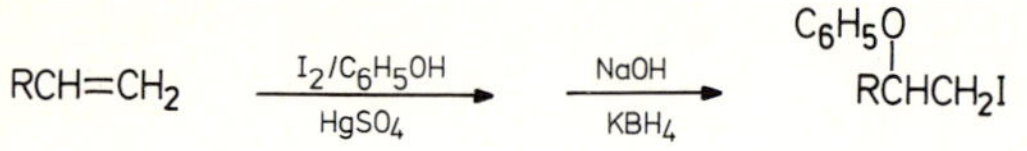

$$NO_2\text{-}C_6H_4\text{-}\overset{O}{\overset{\|}{C}}\text{-}CH{=}CH_2 \longrightarrow NO_2\text{-}C_6H_4\text{-}\overset{O}{\overset{\|}{C}}\text{-}\overset{Br}{\overset{|}{C}}HCH_2OCH_3 \qquad (108)$$

$$RCH{=}CH_2 \xrightarrow[HgSO_4]{I_2/C_6H_5OH} \xrightarrow[KBH_4]{NaOH} \overset{C_6H_5O}{\overset{|}{R}}CHCH_2I \qquad (109)$$

$$CH_3OC_6H_4CH{=}CHCO_2C_2H_5 \longrightarrow CH_3O\text{-}C_6H_3\text{-}\overset{Br}{\overset{|}{C}}H\overset{Br}{\overset{|}{C}}HCO_2C_2H_5 \text{ (}OCH_3\text{)} \qquad (110)$$

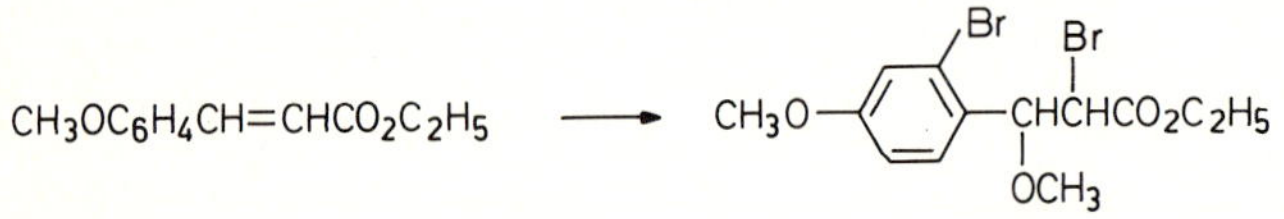

In a few cases, halogenation has not lead to the expected organic halide products. Attempted bromination of a β-methoxy organomercuric perchlorate failed where the corresponding organomercuric acetate or bromide afforded the anticipated bromide [250]. Ethyl p-methoxycinnamate yields a dibromide after methoxymercuration-bromination (Eq. 110) [136]. Iodination proceeds as expected to give the mono iodide. Dibromides, α-bromoketones and re-arranged carbonyl compounds are observed in other halogenation reactions (Eqs. 111–113). Nevertheless, by an appropriate choice of reaction conditions, it appears that most β-alkoxymercurials can be halogenated to afford the desired halo ethers.

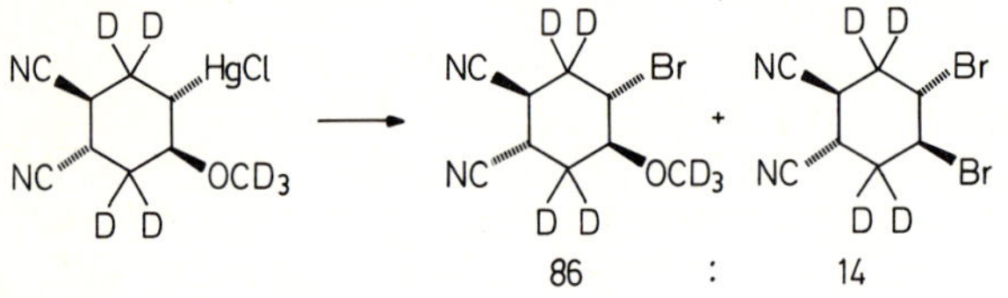

$$(111)\ [25]$$

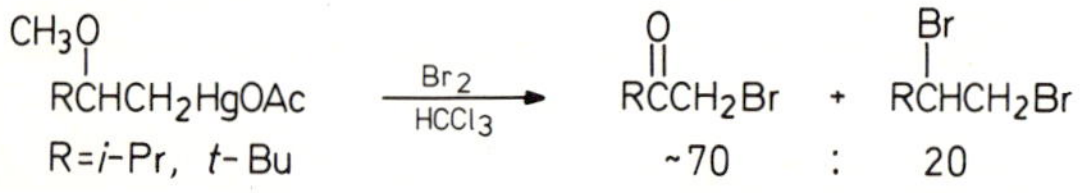

$$86 \quad : \quad 14$$

$$\overset{CH_3O}{\overset{|}{R}}CHCH_2HgOAc \xrightarrow[HCCl_3]{Br_2} \overset{O}{\overset{\|}{R}}CCH_2Br + \overset{Br}{\overset{|}{R}}CHCH_2Br \qquad (112)\ [58]$$
$$R={i}\text{-}Pr,\ {t}\text{-}Bu \qquad \sim 70 \quad : \quad 20$$

$$\overset{ClHg}{\overset{|}{C_6H_5}}\overset{OCH_3}{\overset{|}{C}}HC(CH_3)_2$$
$$\xrightarrow[CCl_4]{Br_2} C_6H_5CH(CH_3)COCH_3$$
$$\xrightarrow[CH_3OH]{Br_2} C_6H_5C(CH_3)_2CHO \qquad (113)\ [71]$$
$$\xrightarrow[CH_3OH]{I_2} C_6H_5C(CH_3)_2CHO + C_6H_5CH(OCH_3)C(OCH_3)(CH_3)_2$$

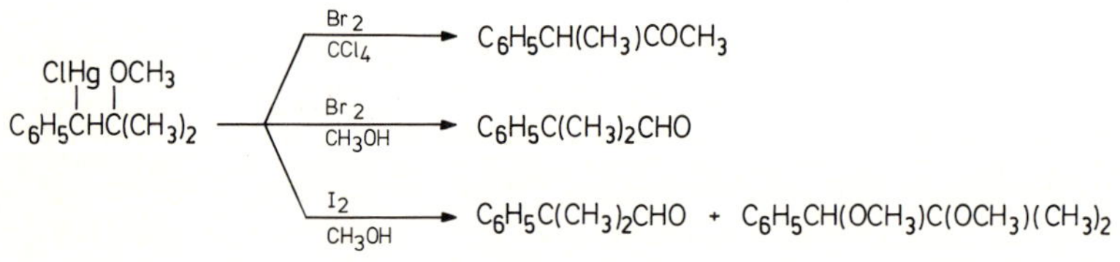

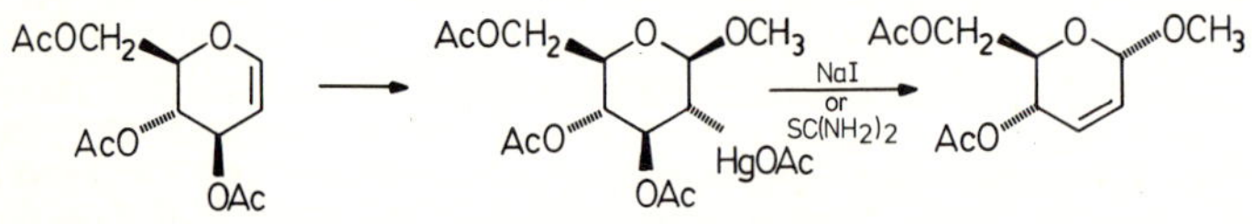

$$(114)$$

$$C_6H_5CH=CHCH_2O_2CR \longrightarrow C_6H_5\overset{CH_3O}{\underset{}{C}}H\overset{HgOAc}{\underset{}{C}}HCH_2O_2CR \xrightarrow{KSCN} C_6H_5CH(OCH_3)CH=CH_2 \; + \; RCO_2H$$

(115)

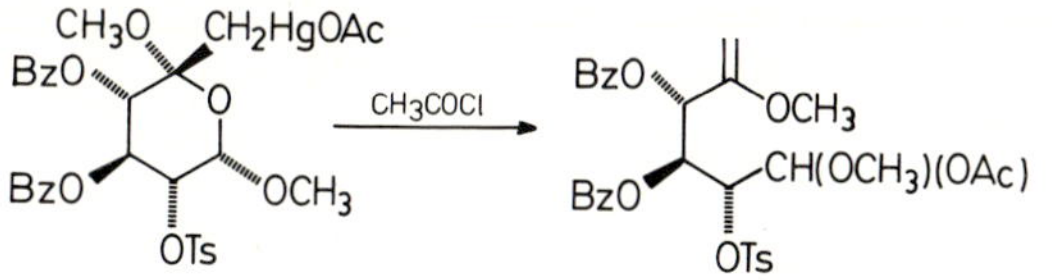

(116)

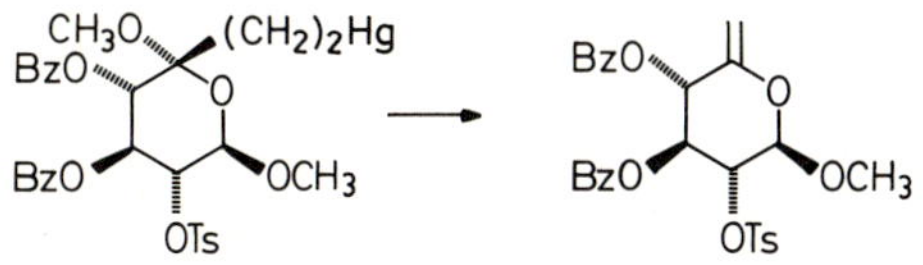

(117)

Several other potentially useful reactions of β-alkoxymercurials have been reported. As noted in Chapters VII and VIII of the monograph "Organomercury Compounds in Organic Synthesis", these mercurials undergo borohydride-induced free radical additions to alkenes [325–327] and carbonylation [36, 42, 328] reactions which can be quite useful. Alkoxymercuration of olefins in the presence of ethylene glycol and subsequent palladium-promoted rearrangement to ethylene ketals has been discussed earlier in Chapter IV of that monograph [329]. Certain elimination reactions of these mercurials also appear synthetically useful. For example, the methoxymercuration of glycals [238, 245, 330] and cinnamyl esters [199] leads to β-acyloxymercurials which undergo facile mercuric carboxylate eliminations (Eqs. 114, see p. 248; 115). In the reaction of glycals and thiourea, only the *anti*-acetate undergoes elimination in this manner. In the latter reaction (Eq. 115), the cinnamyl group has been suggested as a useful protecting group for carboxylic acids. Two other interesting glycal eliminations have been reported using acetyl chloride (Eqs. 116, 117) [207]. It seems likely that the conformation of the mercurial is the determining factor in the direction of elimination in these reactions. Other reactions of potential interest are the thermal isomerization of cyclic *anti*-alkoxymercurials to their *syn* isomers [32] and the stannous chloride [331] and sodium hydrosulfite [332] disproportionation to dialkylmercurials.

Mercury salts also participate in several other very important synthetic transformations in which intermediate β-alkoxymercurials are no doubt intermediates although they are not commonly isolated. For example, vinyl ethers readily undergo alkoxymercuration to mercurated acetals (Eq. 118) [85, 98, 333]. Halogenation affords the corresponding halo acetals [98]. Using only catalytic amounts of mercury salts (mercuric acetate [333–337], mercuric acetate/sodium acetate [338], mercuric acetate/mercuric oxide [334], mercuric acetate/sulfuric acid [339] or mercuric sulfate [340]), the predominant reaction becomes vinyl ether exchange (Eq. 119). This reaction is particularly

$$H_2C=CHOR \xrightarrow[ROH]{Hg(OAc)_2} AcOHgCH_2CH(OR)_2$$

(118)

III. Alkoxymercuration

$$H_2C\!=\!CHOR \quad \xrightarrow[R'OH]{Hg(II)} \quad H_2C\!=\!CHOR' \tag{119}$$

important for the preparation of allyl vinyl ethers which are thermally rearranged to γ,δ-unsaturated carbonyl compounds by the Claisen rearrangement. Mechanistic studies suggest that this reaction proceeds by alkoxymercuration followed by acid-promoted dealkoxymercuration [85, 333]. Equilibrium constants for the reaction with mercuric acetate have been determined [341]. In the presence of amino alcohols or diamines, this reaction generates heterocycles (Eq. 120) [342].

$$\tag{120}$$

Enol esters have also been employed in these reactions to provide vinyl ethers [343, 344], alkoxy esters [343, 345], and acetals or ketals [343, 346, 347] (Eqs. 121–124). Sulfur and nitrogen variations on these last two reactions have also been reported [346, 347].

$$H_2C\!=\!CHOAc + ROH \xrightarrow{Hg(II)} H_2C\!=\!CHOR \tag{121}$$

$$H_2C\!=\!CHOAc + ROH \longrightarrow \underset{\underset{OAc}{|}}{CH_3CHOR} \tag{122}$$

$$\tag{123}$$

$$H_2C\!=\!CHOAc + ROH \longrightarrow CH_3CH(OR)_2 \tag{124}$$

Allyl alcohol, allyl ethyl ether and diallyl ether also undergo exchange with alcohols in the presence of mercuric acetate and boron trifluoride etherate to generate the corresponding allyl ethers (Eq. 125) [348]. An alkoxymercuration-dealkoxymercuration sequence has been suggested for this process.

$$H_2C\!=\!CHCH_2OR \quad \xrightarrow[R'OH]{Hg(II)} \quad H_2C\!=\!CHCH_2OR' \tag{125}$$

Besides the many very important synthetic applications of the alkoxymercuration reaction, this reaction has found considerable analytical utility. For example, the percent of mercuric acetate in a mixture of metal acetates can be ascertained by first titrating with perchloric acid to determine the total acetate concentration and then adding a methanol solution of styrene and titrating again [62]. Alkoxymercuration followed by various titrimetric, spectrophotometric, complexometric or other analytical techniques has provided a valuable method for determining the degree of unsaturation in a hydrocarbon mixture or for checking the purity of olefins [349–360]. These methods have their limitations however.

The alkoxymercuration of alkenes also provides adducts of sufficiently different physical properties that they can usually be readily separated from closely related hydrocarbons [309, 311, 361]. This approach has found greatest utility in the separation of unsaturated fatty acids and esters from their saturated counterparts [287, 292, 305, 308, 310, 362]. As discussed earlier, dilute hydrochloric acid regenerates the original alkenes. Due to the difference in the rates of reaction, it is also possible to separate *cis* and *trans*-isomers in this fashion [273, 287, 292, 313]. Regioisomers and olefins of differing degrees of unsaturation have also been separated by complete alkoxymercuration and subsequent chromatography [292, 306, 307, 312, 313, 363, 364]. The position of unsaturation in naturally-occurring olefins and polyenes can also be determined by mass spectral analysis of their methoxymercuration-demercuration products [365–368].

B. Alkenols

The mercuration of unsaturated alcohols (Table 3.6) and phenols (Table 3.7) provides a very useful method for the synthesis of cyclic and polycyclic five- and six-membered ring ethers. The first reports in 1900 of the solvomercuration of an unsaturated alcohol, namely allyl alcohol, claimed a variety of products whose structures were based solely on elemental analyses [449–451]. Subsequent work has shown that, depending on the reaction conditions, one can obtain either the simple hydroxymercuration product [373, 374] or the *cis-* [370] or *trans-* [371, 372]-dimercurated dioxane (Eq. 126). The unimolecular cyclization of terpineol [374, 408] (Eq. 127) and two acyclic enols [400, 408] was reported shortly thereafter.

$$H_2C=CHCH_2OH \xrightarrow{HgX_2} XHgCH_2\overset{OH}{\underset{|}{C}}HCH_2OH \quad \text{or} \quad \tag{126}$$

$$\tag{127}$$

As noted above, the reaction conditions can be critical to the formation of the desired mercurial. A wide variety of mercury salts have been employed

Table 3.6. Intramolecular Alkoxymercuration of Alkenols

Alkenol	Mercuric salt	Reaction conditions	Organomercurial(s) (% Yield)	Subsequent reactants	Product(s) (% Yield)	Ref.
$H_2C=CHCH_2OH$	HgO/HNO_3	H_2O 0 °C few min	(dioxane) NO_3HgCH_2–/–CH_2HgNO_3 (64–72)	—	—	83
	$Hg(NO_3)_2$	HNO_3/H_2O 50 °C	(dioxane) NO_3HgCH_2–/–CH_2HgNO_3	—	—	369
	$Hg(NO_3)_2$	HNO_3/H_2O 6 hr 0 °C, KI	(dioxane) $IHgCH_2$–/–CH_2HgI	I_2/KI	(dioxane) ICH_2–/–CH_2I (trans 64, 90 ?) (cis > trans ?)	370–374
$H_2C=CH(CH_2)_3OH$	$Hg(OAc)_2$	H_2O (KCl)	(tetrahydrofuran) –CH_2HgX X = OAc (70) X = Cl (82, 90)	$NaBH_4/O_2$ (on RHgCl)	(tetrahydrofuran) –CH_2OH (10)	375, 376
	$Hg(OAc)_2$	CH_3OH (NaCl)	(tetrahydrofuran) –$CH_2HgOAc(Cl)$	$N_2H_4/NaOH$ Δ	(tetrahydrofuran) –CH_3	110
	$Hg(OAc)_2$	1:1 H_2O/THF 30 min	—	$NaBH_4/NaOH$	(tetrahydrofuran) –CH_3 (94)	377
$H_2C=CHCH_2OCH_2CH_2OH$	$Hg(OAc)_2$	Δ H_2O 15 min NaOH, KI	(dioxane) –CH_2HgI	I_2	(dioxane) –CH_2I	378–380
$H_2C=CH(CH_2)_2CHOHCH_3$	HgX_2 (X = Cl, OAc)	H_2O	(tetrahydrofuran) CH_3–/–CH_2HgX X = Cl (30) 58:42 cis/trans X = OAc (65) 55:45	$NaBH_4/NaOH$ (on RHgOAc)	(tetrahydrofuran) CH_3–/–CH_3 67:33 cis/trans	381
	$Hg(OAc)_2$	1:1 H_2O/THF	(tetrahydrofuran) CH_3–/–CH_2HgOAc	$NaBH_4/NaOH$ (phase transfer)	(tetrahydrofuran) CH_3–/–CH_3 (98)	382

Table 3.6. (continued)

Alkenol	Mercuric salt	Reaction conditions	Organomercurial(s) (% Yield)	Subsequent reactants	Product(s) (% Yield)	Ref.
	Hg(OAc)$_2$	H$_2$O, KCl	CH$_3$~⟨O⟩~CH$_2$HgCl (80)	Na(Hg)/H$_2$O/HOAc	CH$_3$~⟨O⟩~CH$_3$ + H$_2$C=CH(CH$_2$)$_2$CHOHCH$_3$	376
	Hg(OAc)$_2$	H$_2$O (KCl)	CH$_3$~⟨O⟩~CH$_2$HgX X = OAc (65) 55 : 45 cis/trans X = Cl (90) 58 : 42	NaBH$_4$/O$_2$ (on RHgCl)	CH$_3$~⟨O⟩~CH$_2$OH (20) 57 : 43 cis/trans	375
H$_2$C=CH(CH$_2$)$_4$OH	Hg(OAc)$_2$	CH$_3$OH (NaCl)	⟨pyran⟩–CH$_2$HgOAc(Cl)	N$_2$H$_4$/NaOH Δ	⟨pyran⟩–CH$_3$	110
	Hg(OAc)$_2$	1:1 H$_2$O/THF 30 min	—	NaBH$_4$/NaOH	⟨pyran⟩–CH$_3$ + CH$_3$CHOH(CH$_2$)$_4$OH + H$_2$C=CH(CH$_2$)$_4$OH 93 : 5 : 2 (98 total)	377, 383
	Hg(ClO$_4$)$_2$	1 NaOH/ 0.01 N HClO$_4$, NaCl	⟨pyran⟩–CH$_2$HgCl (90)	—	—	384
trans-CH$_3$CH$_2$CH=CH(CH$_2$)$_2$OH	CH$_3$OH	Hg(OAc)$_2$, NaCl	ClHg–⟨tetrahydrofuran⟩–CH$_2$CH$_3$	N$_2$H$_4$/NaOH Δ	CH$_3$CH$_2$–⟨tetrahydrofuran⟩ (42) + trans-CH$_3$CH$_2$CH=CH(CH$_2$)$_2$OH (42)	129
H$_2$C=CHCH$_2$OCH$_2$CHOHCH$_2$OH	Hg(OAc)$_2$	H$_2$O overnight	HOCH$_2$~⟨dioxane⟩~CH$_2$HgOAc	—	—	385
	Hg(OAc)$_2$ (NaI)	—	HOCH$_2$~⟨dioxane⟩~CH$_2$HgX X = OAc, I	—	—	156, 380
	Hg(OAc)$_2$	Δ H$_2$O 1 hr, Ac$_2$O, KI	AcOCH$_2$~⟨dioxane⟩~CH$_2$HgI (42)	I$_2$	AcOCH$_2$~⟨dioxane⟩~CH$_2$I	378

Table 3.6. (continued)

Alkenol	Mercuric salt	Reaction conditions	Organomercurial(s) (% Yield)	Subsequent reactants	Product(s) (% Yield)	Ref.
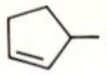 (CH₂)₂OH (cyclopentene)	Hg(OAc)₂	H₂O/THF 0.5 hr	—	NaBH₄/NaOH	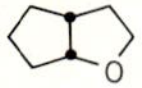 (75)	386
HO 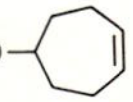 (cycloheptenol)	Hg(OAc)₂	NaOAc/H₂O 30 min, KI	IHg (>100 crude)	I₂	(94)	387
OCHOHCCl₃	Hg(O₂CCF₃)₂	THF (Cl⁻)	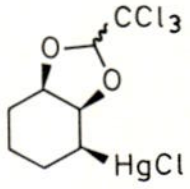 HgCl (46)	NaBH₄/NaOH (on RHgO₂CCF₃)	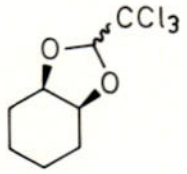 (80)	388, 389
HOCH₂	HgCl₂	H₂O/CH₃COCH₃	HgCl	Na(Hg)/D₂O	(>80) + (10)	17
	Hg(OAc)₂	CH₃OH 2–7d, NaCl	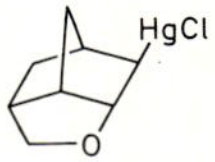 HgCl	N₂H₄/NaOH		129
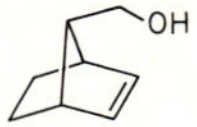 OH	Hg(OAc)₂	CH₃OH 90 min, NaCl	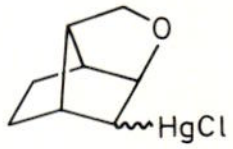 HgCl (64)	—	—	17, 390
HO,  S	Hg(OAc)₂	—	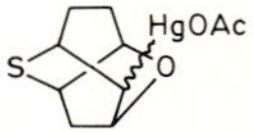 HgOAc	—	—	391
 HO, O	Hg(OAc)₂	H₂O (KI)	 HgX (X = OAc, I)	NaBH₄ [KI/I₂]	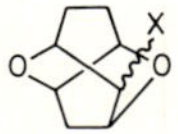 X = H, I (85) (4:1 mixture of isomers)	391–393

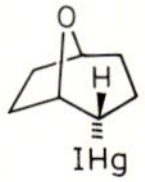

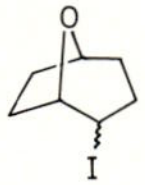

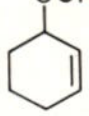

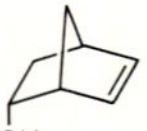

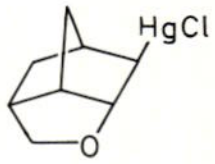

Table 3.6. (continued)

Alkenol	Mercuric salt	Reaction conditions	Organomercurial(s) (% Yield)	Subsequent reactants	Product(s) (% Yield)	Ref.
cis-$CH_3CH_2CH=CH(CH_2)_2$-OCHOHCCl$_3$	Hg(O$_2$CCF$_3$)$_2$	THF	—	NaBH$_4$/NaOH acetal removal	$CH_3CH_2CHOH(CH_2)_3OH$ + $CH_3(CH_2)_2CHOH(CH_2)_2OH$ 1 : 0.82	388
trans-$CH_3(CH_2)_2CH=CHCH_2$-OCHOHCCl$_3$	Hg(O$_2$CCF$_3$)$_2$	THF	—	NaBH$_4$/NaOH	[1,3-dioxane, $CH_3(CH_2)_2$, CCl$_3$] 56 + [1,3-dioxolane, $CH_3(CH_2)_3$, CCl$_3$] 44 (79)	388, 389
trans-$CH_3(CH_2)_2CH=CHCH_2$-OCHOHCF$_3$	Hg(O$_2$CCF$_3$)$_2$	—	—	NaBH$_4$/NaOH	[1,3-dioxane, $CH_3(CH_2)_2$, CF$_3$] (52)	389
[H_2N, HO-cyclooctene]	Hg(OAc)$_2$	H$_2$O/THF/H$_2$SO$_4$	—	NaBH$_4$/NaOH	[H_2N, O-bicyclic] (30) + [H_2N, O-bicyclic] (10)	394
[4-methylenecyclohexyl]-CH$_2$OH	Hg(OAc)$_2$	t-BuOH 48 hr NaCl	[O-bicyclic, ClHgCH$_2$] (67)	NaBH$_4$/NaOH	[O-bicyclic, CH$_3$] (80)	395
[cyclohexenyl]-(CH$_2$)$_2$OH	Hg(OAc)$_2$	H$_2$O/THF 45 min	—	NaBH$_4$/NaOH	[bicyclic O] (67)	386
HO-[cyclooctene]	Hg(OAc)$_2$	H$_2$O/NaOAc 8 min	[O-bicyclic, HgOAc] (78)	NaBH$_4$/NaOH	[O-bicyclic] (77)	396
	Hg(NO$_3$)$_2$·H$_2$O	H$_2$O/KNO$_3$ 14 hr	[O-bicyclic, HgNO$_3$] (78)	NaBH$_4$	HO-[cyclooctene]	396
	Hg(OAc)$_2$	H$_2$O/NaOAc	[O-bicyclic, HgOAc]	NaBH$_4$ [I$^-$/I$_2$]	[O-bicyclic, X] + [O-bicyclic, X] + [O-bicyclic, X] X = H — : — : none X = I 49.5–64.5 : 30.5–33 : 5–17.5	397, 398

Table 3.6. (continued)

Alkenol	Mercuric salt	Reaction conditions	Organomercurial(s) (% Yield)	Subsequent reactants	Product(s) (% Yield)	Ref.
	$Hg(OAc)_2$	1:1 H_2O/THF	—	$NaBH_4$/NaOH	 (63)  + 15 ~85 : 15	399
	$Hg(NO_3)_2$	H_2O, KNO_3	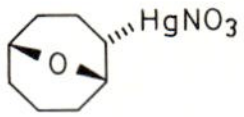 HgNO$_3$	$NaBH_4$ [I_2 on RHgI]	   X = H — : — : none X = I 56–79 : 17–21 : 4–24	397, 398
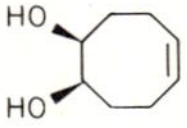 (cyclooctenediol)	$Hg(OAc)_2$	1:1 H_2O/THF 25 °C	—	$NaBH_4$/NaOH	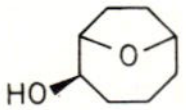 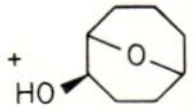 88 : 12 (assumed) (60 total)	234
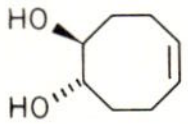 (cyclooctenediol)	$Hg(OAc)_2$	1:1 H_2O/THF 25 °C	—	$NaBH_4$/NaOH	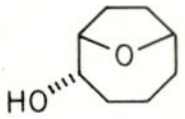 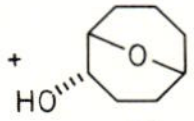 77 : 23 (55 total)	234
	$Hg(NO_3)_2$	H_2O 25 °C 2 hr	—	$NaBH_4$/NaOH	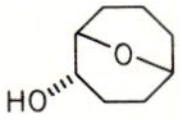 (82)	234
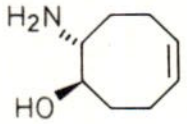 (amino-cyclooctenol)	$Hg(OAc)_2$	H_2O/THF/H_2SO_4	—	$NaBH_4$/NaOH	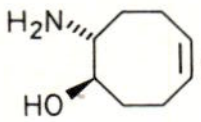 (60) 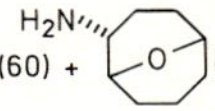 + (30) 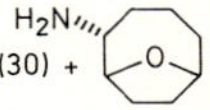 + (10)	394
$H_2C{=}CH(CH_2)_2COH(CH_3)(C_2H_5)$	$Hg(OAc)_2$	KOH, KI	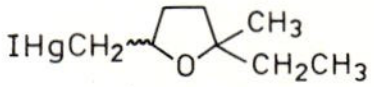 $IHgCH_2$–(furan)–CH_3, CH_2CH_3 (~100) cis and trans	I_2	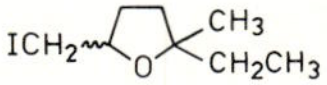 ICH_2–(furan)–CH_3, CH_2CH_3	374, 400
$(CH_3)_2C{=}CH(CH_2)_2CHOHCH_3$	HgX_2 (X = Cl, OAc)	1:1 H_2O/THF	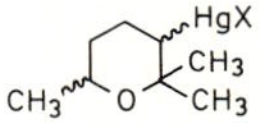 HgX, CH_3, CH_3, CH_3 (pyran)	$NaBH_4$/NaOH (phase transfer)	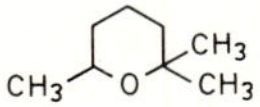 CH_3–(pyran)–CH_3, CH_3 (98)	382

Table 3.6. (continued)

Alkenol	Mercuric salt	Reaction conditions	Organomercurial(s) (% Yield)	Subsequent reactants	Product(s) (% Yield)	Ref.
$H_2C{=}CHCH_2OCH(CHOHCH_2OH)_2$	$Hg(OAc)_2$	—	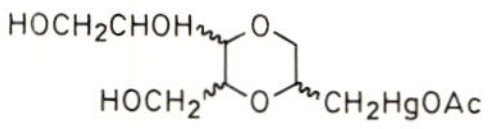 [structure: $HOCH_2CHOH$ / $HOCH_2$ dioxane ring with CH_2HgOAc]	—	—	156, 380, 385
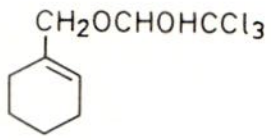 [cyclohexene with $CH_2OCHOHCCl_3$]	$Hg(O_2CCF_3)_2$	THF	—	$NaBH_4/NaOH$	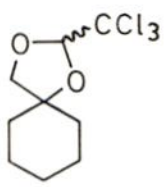 [spiro dioxolane with CCl_3] (90)	388, 389
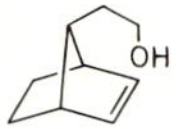 [bicyclic alkenol with OH]	$Hg(OAc)_2$	CH_3OH 25 °C 15 min, NaCl	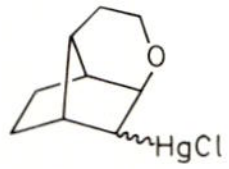 [bicyclic ether with $HgCl$] (87) (2 isomers)	—	—	262
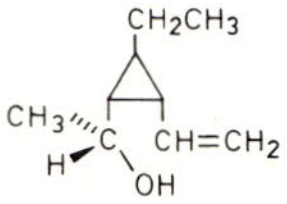 [cyclopropyl carbinol, CH_2CH_3, CH_3, $CH{=}CH_2$, OH]	$Hg(OAc)_2$	THF 20 min	—	$NaBH_4/NaOH$	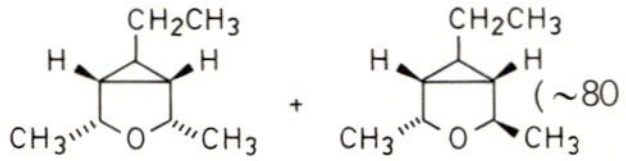 [tetrahydropyran CH_2CH_3, CH_3 75] + [isomer 25] (~80 total)	401
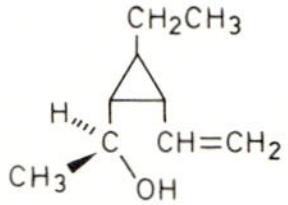 [cyclopropyl carbinol, CH_2CH_3, CH_3, $CH{=}CH_2$, OH]	$Hg(OAc)_2$	THF 20 min	—	$NaBH_4/NaOH$	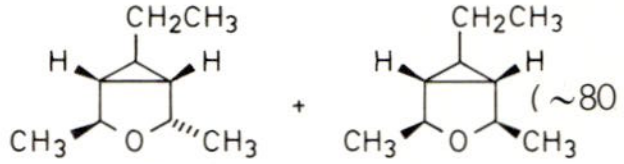 [tetrahydropyran 85] + [isomer 15] (~80 total)	401
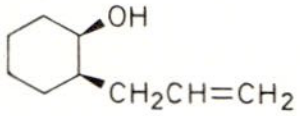 [cyclohexane, OH, $CH_2CH{=}CH_2$]	HgX_2 (X = Cl, OAc)	H_2O (THF)	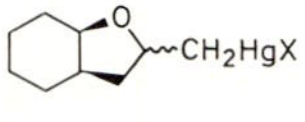 [bicyclic furan CH_2HgX] (20–60)	$NaBH_4/NaOH$	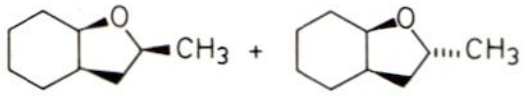 [CH_3 50–75] + [CH_3 25–50]	402
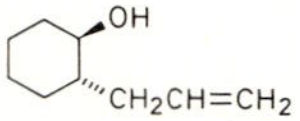 [cyclohexane, OH, $CH_2CH{=}CH_2$]	HgX_2 (X = Cl, OAc)	H_2O (THF)	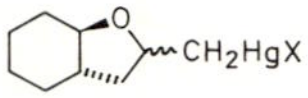 [bicyclic furan CH_2HgX] (10–75)	$NaBH_4/NaOH$	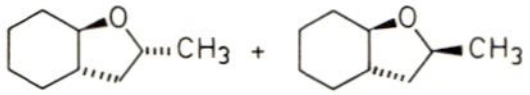 [CH_3 25–89] + [CH_3 11–75]	402
[cyclohexane, OH, $CH_2CH{=}CH_2$]	$Hg(OAc)_2$	H_2O (KCl)	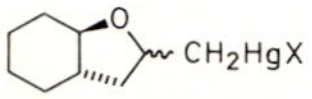 [bicyclic furan CH_2HgX] X = OAc (75) 26 : 74 "cis/trans" X = Cl (95) 26 : 74	$NaBH_4/O_2$ (on $RHgCl$)	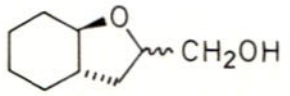 [bicyclic furan CH_2OH] (60) 36 : 64 (2 diastereomers)	375

Table 3.6. (continued)

Alkenol	Mercuric salt	Reaction conditions	Organomercurial(s) (% Yield)	Subsequent reactants	Product(s) (% Yield)	Ref.
	Hg(OAc)$_2$	1:1 H$_2$O/THF	 [structure] CH$_2$HgOAc	NaBH$_4$/NaOH (phase transfer)	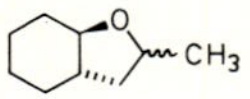 [structure] CH$_3$ (83) 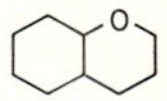 + [structure] (10)	382
H$_2$C=CHCH$_2$OCH(CHOHCHO) – CHOHCHOHCH$_2$OH	Hg(OAc)$_2$	—	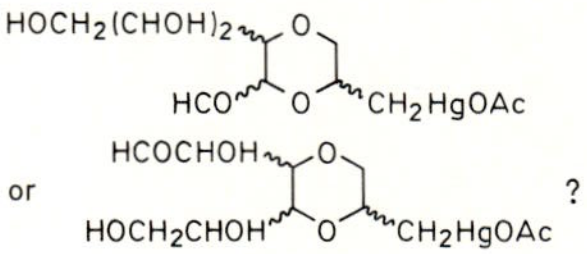 HOCH$_2$(CHOH)$_2$ [structure] CH$_2$HgOAc or HCOCHOH [structure] HOCH$_2$CHOH CH$_2$HgOAc ?	—	—	380
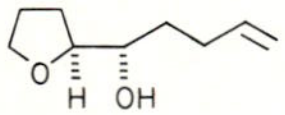 [structure] H OH	HgX$_2$ (X = OAc, Cl)	H$_2$O/THF	—	NaBH$_4$/NaOH	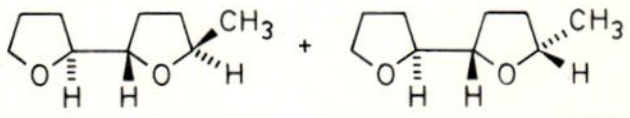 [structures] X = OAc 84 : 16 (85 total); X = Cl 50 : 50 (88 total)	403, 404
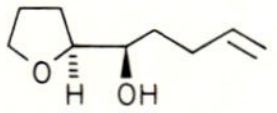 [structure] H OH	HgX$_2$ (X = OAc, Cl)	H$_2$O/THF	—	NaBH$_4$/NaOH	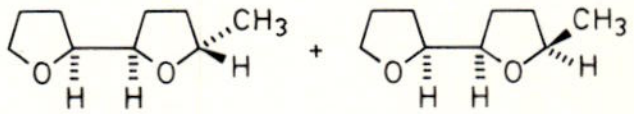 [structures] X = OAc 81 : 19 (85 total); X = Cl 56 : 44 (88 total)	403, 404
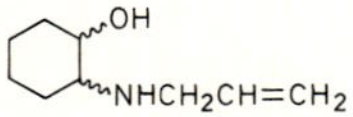 [structure] OH NHCH$_2$CH=CH$_2$	HgCl$_2$	1:1 H$_2$O/THF	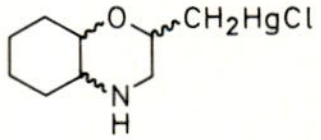 [structure] CH$_2$HgCl	NaBH$_4$/NaOH (phase transfer)	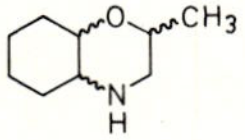 [structure] CH$_3$ (58)	405
	Hg(OAc)$_2$	1:1 H$_2$O/THF	—	NaBH$_4$/NaOH	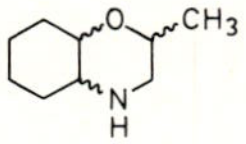 [structure] CH$_3$ (50)	406
(H$_2$C=CHCH$_2$)$_2$NCH$_2$CHOHCH$_3$	Hg(OAc)$_2$	1:1 H$_2$O/THF	—	NaBH$_4$/NaOH	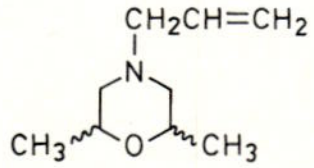 [structure] CH$_2$CH=CH$_2$ ... CH$_3$ ... CH$_3$ (45)	406

Table 3.6. (continued)

Alkenol	Mercuric salt	Reaction conditions	Organomercurial(s) (% Yield)	Subsequent reactants	Product(s) (% Yield)	Ref.
$H_2C=CH(CH_2)_2CHOHC(CH_3)_3$	$HgCl_2$	H_2O	$(CH_3)_3C$–[tetrahydrofuran ring]–CH_2HgCl (75) 68 : 32 cis/trans	$NaBH_4/O_2$	$(CH_3)_3C$–[tetrahydrofuran ring]–CH_2OH (40) 60 : 40 cis/trans	375
	HgX_2 (X = Cl, OAc)	H_2O	$(CH_3)_3C$–[tetrahydrofuran ring]–CH_2HgX X = Cl (75) 64 : 36 cis/trans X = OAc (45) 24 : 76 cis/trans	$NaBH_4/NaOH$ (on $RHgOAc$)	$(CH_3)_3C$–[tetrahydrofuran ring]–CH_3 22 : 78 cis/trans	381
	$Hg(OAc)_2$	1:1 H_2O/THF	$(CH_3)_3C$–[tetrahydrofuran ring]–CH_2HgOAc	$NaBH_4/NaOH$ (phase transfer)	$(CH_3)_3C$–[tetrahydrofuran ring]–CH_3 (100)	382
$(CH_3)_2C=CH(CH_2)_2COH(CH_3)_2$	$Hg(OAc)_2$	DME 6.5 hr RT, NaCl	[tetrahydropyran ring, CH_3, CH_3, CH_3, CH_3, $HgCl$] (60)	$N_2H_4/NaOH$	[tetrahydropyran ring, CH_3, CH_3, CH_3, CH_3]	407
	$Hg(OAc)_2$	no H_2O 3 hr, X^-	[tetrahydropyran ring, CH_3, CH_3, CH_3, CH_3, HgX] X = Cl (91) I (63)	—	—	407
	$Hg(OAc)_2$	H_2O, X^-	[tetrahydropyran ring, CH_3, CH_3, CH_3, CH_3, HgX] X = Cl (25) I (21) + $(CH_3)_2COHCH(HgX)(CH_2)_2COH(CH_3)_2$ X = Cl (17) I (25)	—	—	400, 407, 408
	—	–, I^-	[tetrahydropyran ring, CH_3, CH_3, CH_3, CH_3, HgI]	—	—	374
$H_2C=CHCH_2OCH_2(CHOH)_4CH_2OH$	$Hg(OAc)_2$	—	$HOCH_2(CHOH)_3$–[dioxane ring]–CH_2HgOAc	—	—	156, 385

Table 3.6. (continued)

Alkenol	Mercuric salt	Reaction conditions	Organomercurial(s) (% Yield)	Subsequent reactants	Product(s) (% Yield)	Ref.
$H_2C=CHCH_2OCH(CH_2OH)$ – $(CHOH)_3CH_2OH$	$Hg(OAc)_2$	—	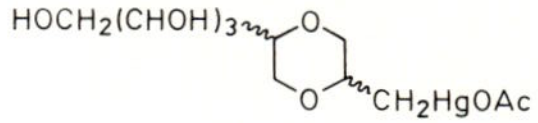 $HOCH_2(CHOH)_3$ [ring] CH_2HgOAc	—	—	380
$H_2C=CHCH_2OCH(CHOHCH_2OH)$ – $CHOHCHOHCH_2OH$	$Hg(OAc)_2$	H_2O	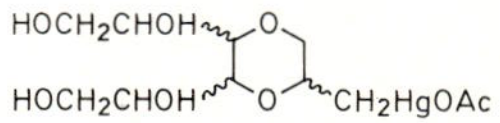 $HOCH_2CHOH$ / $HOCH_2CHOH$ [ring] CH_2HgOAc	Ac_2O, KI	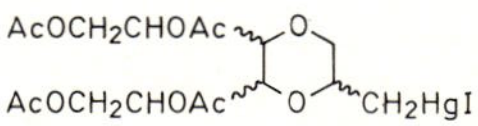 $AcOCH_2CHOAc$ / $AcOCH_2CHOAc$ [ring] CH_2HgI	378, 385
	$Hg(OAc)_2$	$H_2O/HOAc$ 16 hr	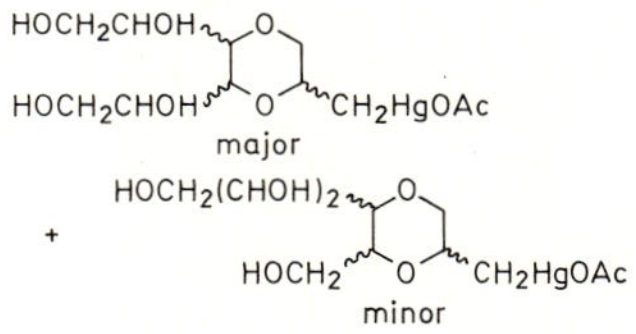 $HOCH_2CHOH$ / $HOCH_2CHOH$ [ring] CH_2HgOAc (major) + $HOCH_2(CHOH)_2$ / $HOCH_2$ [ring] CH_2HgOAc (minor)	—	—	156, 380
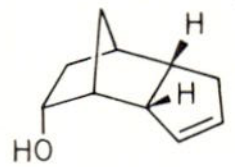 [bicyclic alkenol, HO]	$Hg(OAc)_2$	—	—	$NaBH_4/NaOH$	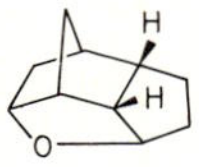 [oxa-tricyclic product, H]	409
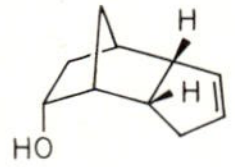 [bicyclic alkenol, HO]	$Hg(OAc)_2$	—	—	$NaBH_4/NaOH$	 [oxa-tricyclic product]	409
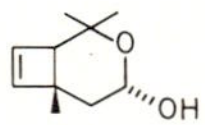 [pyran alkenol, OH]	$Hg[O_2CC(CH_3)_3]_2$	THF 20 hr	—	$NaBH_4/NaOH$	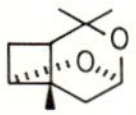 [oxa-bicyclic product] (42)	410
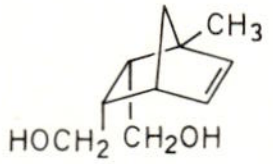 [terpenoid alkenol, CH_3, $HOCH_2$ CH_2OH]	$Hg(OAc)_2$	CH_3OH	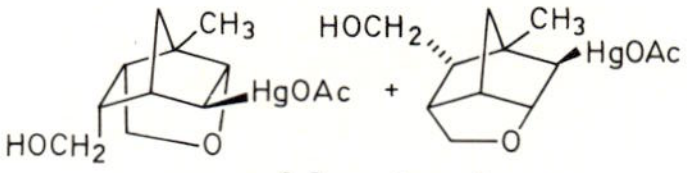 [CH_3, $HOCH_2$, $HgOAc$ isomer] + [$HOCH_2$, CH_3, $HgOAc$ isomer], 2.3 : 1	$NaBH_4$	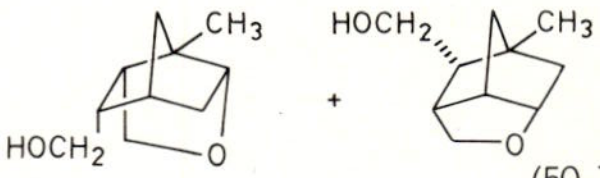 [CH_3, $HOCH_2$ product] + [$HOCH_2$, CH_3 product] (50–70 total)	411

Table 3.6. (continued)

Alkenol	Mercuric salt	Reaction conditions	Organomercurial(s) (% Yield)	Subsequent reactants	Product(s) (% Yield)	Ref.
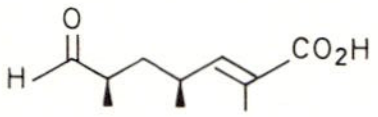 [structure: aldehyde–CO₂H alkenol]	Hg(OAc)₂	CH₃OH	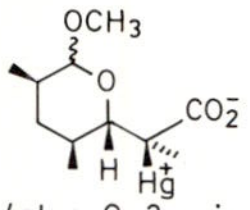 [structure, OCH₃ lactone] (70) (plus C-2 epimer)	Na₂CS₃ / NaOH −60 °C	[two lactone structures] 78 : 22 (with NaBH₄ 95 %)	412
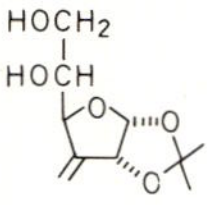 [structure: HOCH₂–HOCH dioxolane]	Hg(OAc)₂ / NaN₃	1:1 H₂O/THF 85 °C 90 hr	—	NaBH₄ / NaOH	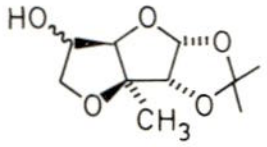 [bicyclic structure] (37.5)	413
 [structure: methylcyclohexene with OH]	Hg(OAc)₂	40–45 °C 36 hr, NaCl	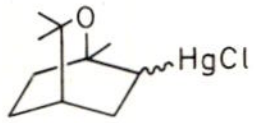 [structure, –HgCl] (40)	N₂H₄ / NaOH	 [bicyclic structure] (45)	414
	Hg(OAc)₂	THF 24 hr 55 °C	—	NaBH₄ / NaOH	 [bicyclic structure]	415
	Hg(OAc)₂	1:1 H₂O/THF 2 min RT	—	NaBH₄ / NaOH	 [cyclohexane diol structure] (68.6) "cis"  + [bicyclic structure] (29.2)	416
	Hg(NO₃)₂	H₂O / Et₂O, KOH / KI	 [bicyclic –HgI structure] (8) 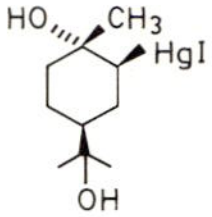 + [HO, CH₃ –HgI structure]	—	—	417
	Hg(NO₃)₂, KI	—	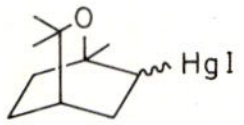 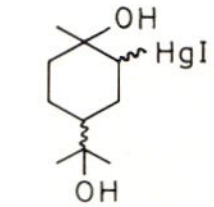 [bicyclic –HgI structure] + [OH –HgI structure]	Na / Hg	 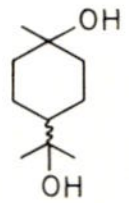 [bicyclic structure] + [cyclohexane diol structure]	374, 408

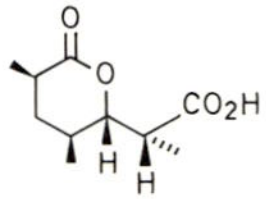

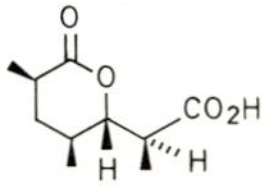

Table 3.6. (continued)

Alkenol	Mercuric salt	Reaction conditions	Organomercurial(s) (% Yield)	Subsequent reactants	Product(s) (% Yield)	Ref.
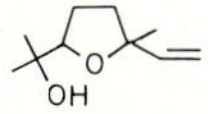 (1:1 cis /trans)	Hg(OAc)₂	1:1 H₂O/THF RT	—	NaBH₄/NaOH	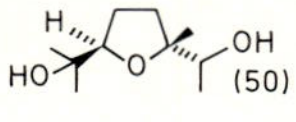 (50) 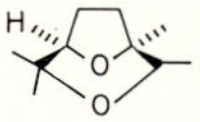 + (40)	418
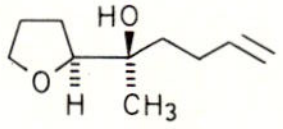	HgX₂ (X = OAc, Cl)	H₂O/THF	—	NaBH₄/NaOH	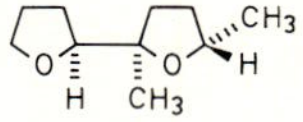 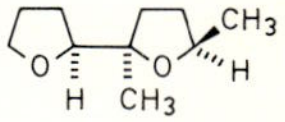 X = OAc 74 : 26 (80–90); X = Cl 35 : 65	404
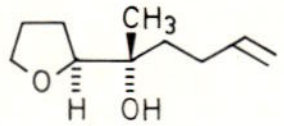	HgX₂ (X = OAc, Cl)	H₂O/THF	—	NaBH₄/NaOH	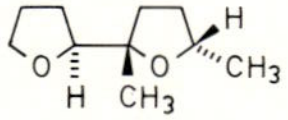 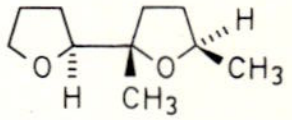 X = OAc 64 : 36 (80–90); X = Cl 33 : 67	404
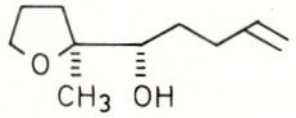	HgX₂ (X = OAc, Cl)	H₂O/THF	—	NaBH₄/NaOH	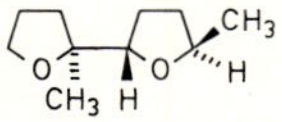 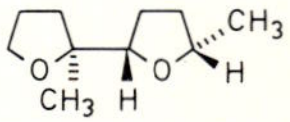 X = OAc 73 : 27 (80–90); X = Cl 43 : 57	404
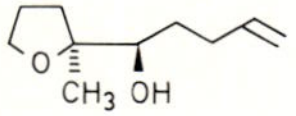	HgX₂ (X = OAc, Cl)	H₂O/THF	—	NaBH₄/NaOH	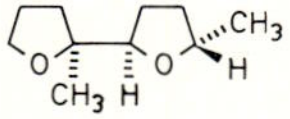 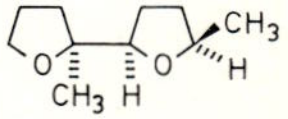 X = OAc 87 : 13 (80–90); X = Cl 52 : 48	404
	Hg(OAc)₂	CH₃OH	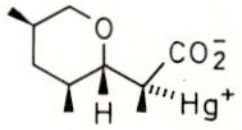 (86)	NaBH₄/NaOH	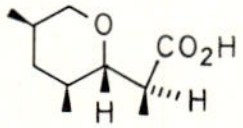	412
	Hg(OAc)₂	CH₃OH	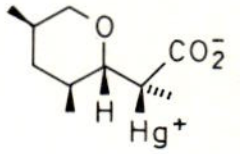 (92)	NaBH₄/NaOH or H₂S/C₅H₅N	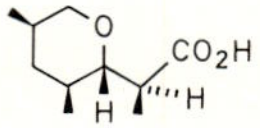	412

262

Table 3.6. (continued)

Alkenol	Mercuric salt	Reaction conditions	Organomercurial(s) (% Yield)	Subsequent reactants	Product(s) (% Yield)	Ref.
$H_2C=CH(CH_2)_2CHOHCH_2NHC(CH_3)_3$	$Hg(OAc)_2$	H_2O / THF	—	$NaBH_4$ / NaOH	CH_3 ⟶ $CH_2NHC(CH_3)_3$ (tetrahydrofuran ring) cis and trans	394
$H_2C=CH(CH_2)_2CHOHC_6H_5$	$HgCl_2$	H_2O	C_6H_5 ⟶ CH_2HgCl (60) (tetrahydrofuran ring) 92 : 8 cis/trans	$NaBH_4$ / O_2	C_6H_5 ⟶ CH_2OH (25) (tetrahydrofuran ring) 90 : 10 cis/trans	375
	HgX_2 (X = Cl, OAc)	H_2O	C_6H_5 ⟶ CH_2HgX (tetrahydrofuran ring) X = Cl (60) 93 : 7 cis/trans; X = OAc (75) 48 : 52 cis/trans	$NaBH_4$ / NaOH	C_6H_5 ⟶ CH_3 (tetrahydrofuran ring) X = Cl 90 : 10 cis/trans; X = OAc 47 : 53 cis/trans	381
	$Hg(OAc)_2$	1:1 H_2O/THF	C_6H_5 ⟶ CH_2HgOAc (tetrahydrofuran ring)	$NaBH_4$ / NaOH (phase transfer)	C_6H_5 ⟶ CH_3 (100) (tetrahydrofuran ring)	382
(bicyclic structure with CH_2CH_3, $HOCH_2$ CH_2OH)	$Hg(OAc)_2$	CH_3OH	(two bicyclic organomercurial structures with CH_2CH_3, $HOCH_2$, $HgOAc$) 5.0 : 1	$NaBH_4$	(two bicyclic product structures with CH_2CH_3, $HOCH_2$) (50–70 total)	411
$H_2C=CHCH$– $(CH_2CO_2C_2H_5)$ (cyclobutanol) –OH	$Hg(O_2CCF_3)_2$	C_6H_6 1.5 hr	—	I_2	(bicyclic structure with $CH_2CO_2C_2H_5$, CH_2I) (40)	419
(tetrahydrofuran structure with OH, CH_3 CH_3, allyl)	HgX_2 (X = OAc, Cl)	H_2O / THF	—	$NaBH_4$ / NaOH	(two bis-tetrahydrofuran structures with CH_3, CH_3 CH_3, H) X = OAc 82 : 18; X = Cl 54 : 46 (80–90)	404
(tetrahydrofuran structure with CH_3, CH_3 OH, allyl)	HgX_2 (X = OAc, Cl)	H_2O / THF	—	$NaBH_4$ / NaOH	(two bis-tetrahydrofuran structures with CH_3, CH_3 CH_3, H) X = OAc major : minor; X = Cl minor : major (80–90)	404

Table 3.6. (continued)

Alkenol	Mercuric salt	Reaction conditions	Organomercurial(s) (% Yield)	Subsequent reactants	Product(s) (% Yield)	Ref.
$H_2C=C(CH_3)CH_2CCH_2C(CH_3)=CH_2$ (with CH_2OH and CH_2NH_2 substituents)	$Hg(OAc)_2$	1:1 H_2O/THF 25°C 3 hr	—	$NaBH_4$/NaOH	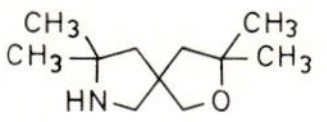 (pyrrolidine–tetrahydrofuran spiro structure) (80)	420
trans-$CH_3CH=CH(CH_2)_2CHOHC_6H_5$	$Hg(OAc)_2$	—	—	$NaBH_4$	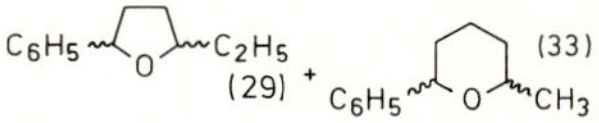 C_6H_5–(tetrahydrofuran)–C_2H_5 (29) + C_6H_5–(tetrahydropyran)–CH_3 (33) 53 : 47 cis/trans	421
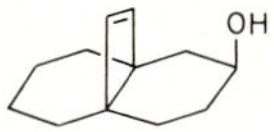 (bicyclic alkenol with OH)	$Hg(OAc)_2$	1:1 H_2O/THF	—	$NaBH_4$/NaOH	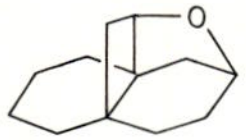 (tricyclic ether) (100)	422
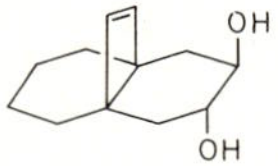 (bicyclic alkenol with two OH)	$Hg(OAc)_2$	H_2O/THF 25°C 30 min, NaCl	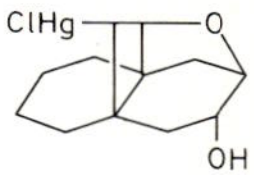 (ClHg tricyclic ether–OH) (92)	—	—	422
	$Hg(OAc)_2$	1:1 H_2O/THF 20 min	—	$NaBH_4$/NaOH	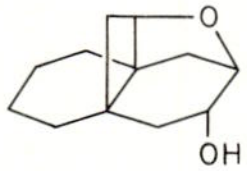 (tricyclic ether–OH) (100 crude)	422
	$Hg(OAc)_2$	1:1 H_2O/THF 30 min RT	—	$NaBH_4$/NaOH	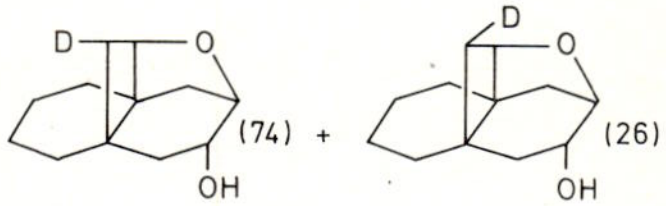 (D-labeled tricyclic ether–OH) (74) + (D-labeled tricyclic ether–OH) (26)	422
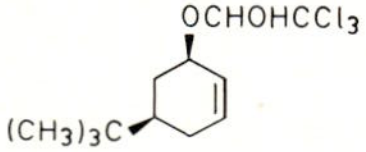 (cyclohexenyl $OCHOHCCl_3$, $(CH_3)_3C$)	$Hg(O_2CCF_3)_2$	THF	—	$NaBH_4$/NaOH	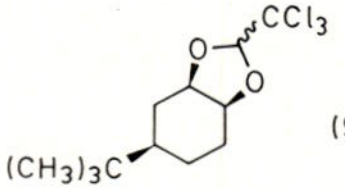 (dioxole–CCl_3, $(CH_3)_3C$ cyclohexane) (92, 94)	388, 389

264

Table 3.6. (continued)

Alkenol	Mercuric salt	Reaction conditions	Organomercurial(s) (% Yield)	Subsequent reactants	Product(s) (% Yield)	Ref.
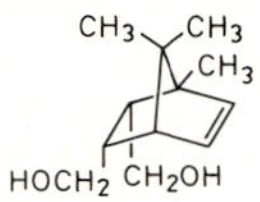 $HOCH_2$ CH_2OH (bicyclic, gem-dimethyl)	$Hg(OAc)_2$	CH_3OH	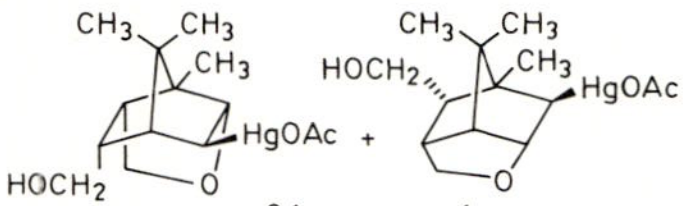 (two organomercurials, $HgOAc$) 3.1 : 1	$NaBH_4$	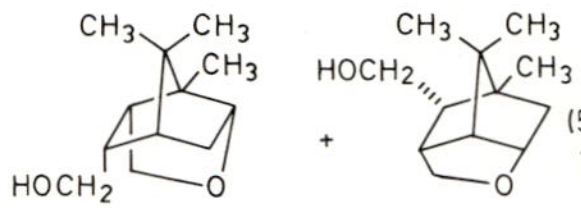 (two products) (50–70 total)	411
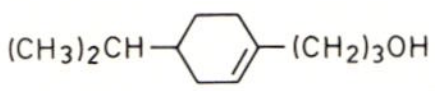 $(CH_3)_2CH$—$(CH_2)_3OH$	$Hg(OAc)_2$	1:1 H_2O/THF (KCl)	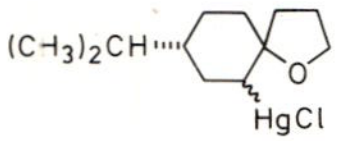 $(CH_3)_2CH$··· (HgCl) (80)	$NaBH_4$/NaOH (on RHgOAc) [O_2 on RHgCl]	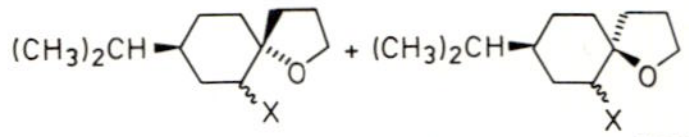 $(CH_3)_2CH$— + $(CH_3)_2CH$— (X) X = H 68 : 32 (65) X = OH — : — (45)	423
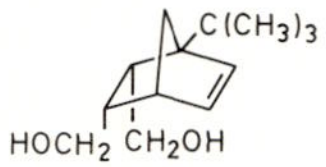 $HOCH_2$ CH_2OH (bicyclic, $C(CH_3)_3$)	$Hg(OAc)_2$	CH_3OH	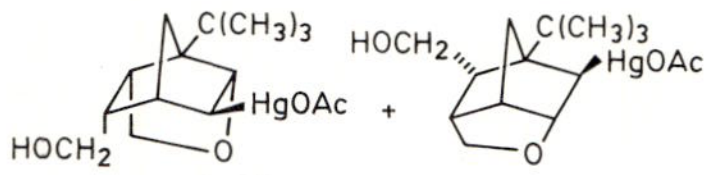 (two organomercurials, $HgOAc$) ≧ 50 : 1	$NaBH_4$	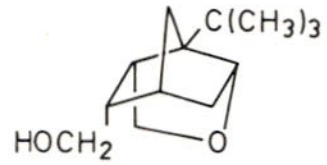 (70)	411
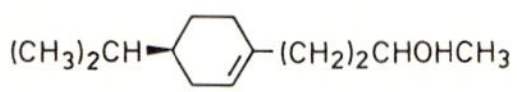 $(CH_3)_2CH$—$(CH_2)_2CHOHCH_3$	$Hg(OAc)_2$	1:1 H_2O/THF (KCl)	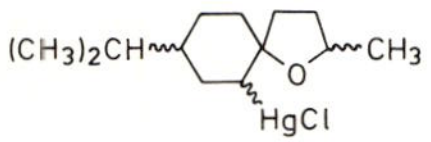 $(CH_3)_2CH$ ~~~CH_3 (HgCl) (90)	$NaBH_4$/NaOH (on RHgOAc) [O_2 on RHgCl]	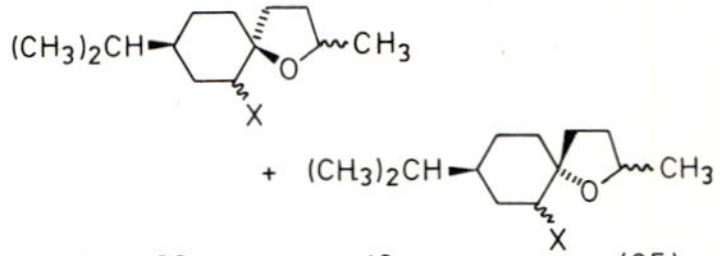 $(CH_3)_2CH$— CH_3 (X) + $(CH_3)_2CH$— CH_3 (X) X = H 90 : 10 (85) X = OH — : — (65)	423
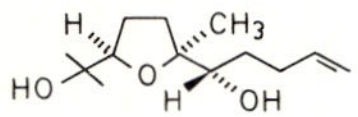 (tetrahydrofuran diol, CH_3, alkenyl)	$Hg(OAc)_2$	H_2O/THF 1 hr	—	$NaBH_4$/NaOH	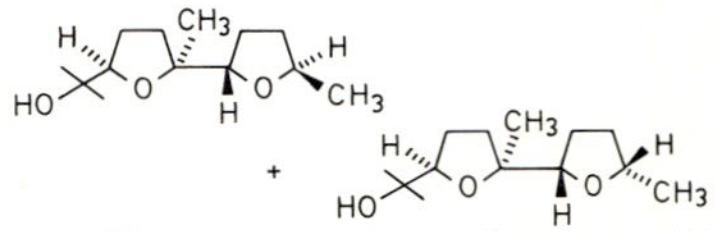 (bis-THF products) 90 : 10 (74)	424
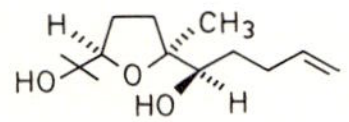 (tetrahydrofuran diol, CH_3, alkenyl)	$Hg(OAc)_2$	H_2O/THF 1 hr	—	$NaBH_4$/NaOH	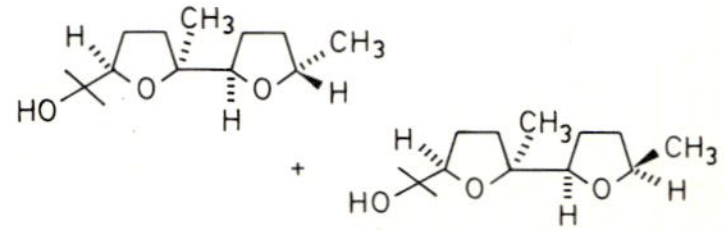 (bis-THF products) 90 : 10 (72)	424

Table 3.6. (continued)

Alkenol	Mercuric salt	Reaction conditions	Organomercurial(s) (% Yield)	Subsequent reactants	Product(s) (% Yield)	Ref.
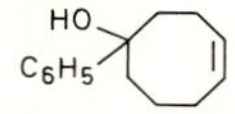 HO / C_6H_5 (cyclooctenol)	$Hg(OAc)_2$	$H_2O/NaOAc$ 1hr	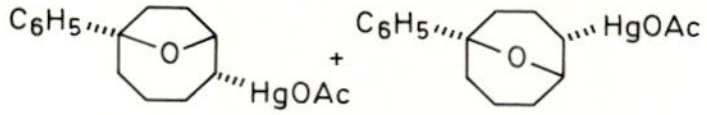 C_6H_5···O···$HgOAc$ + C_6H_5···O···$HgOAc$	$NaBH_4/NaOH$	 C_6H_5—O (70) : C_6H_5—O (30)	425
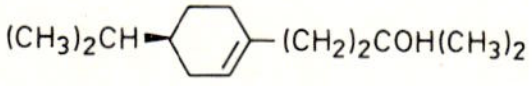 $(CH_3)_2CH$—(cyclohexene)—$(CH_2)_2COH(CH_3)_2$	$Hg(OAc)_2$	1:1 H_2O/THF (KCl)	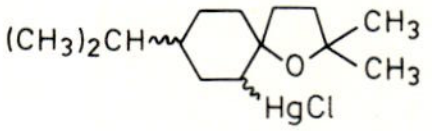 $(CH_3)_2CH$···(bicyclic) CH_3 CH_3 O, HgCl (75)	$NaBH_4/NaOH$ (on RHgOAc) [O_2 on RHgCl]	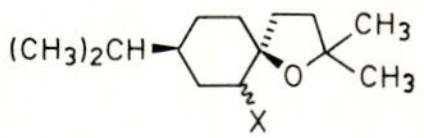 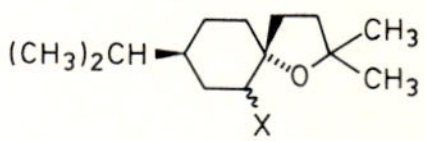 $(CH_3)_2CH$···(bicyclic) CH_3 CH_3 O, X + $(CH_3)_2CH$···(bicyclic) CH_3 CH_3 O, X; X = H 95 : 5 (80); X = OH — : — (60)	423
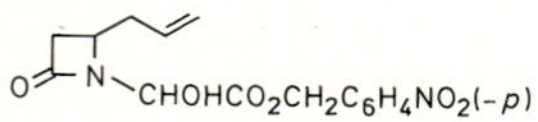 O=(lactam)N—CHOHCO$_2$CH$_2$C$_6$H$_4$NO$_2$(-p) (allyl)	$Hg(OAc)_2$	1:4 H_2O/THF 0°C 3hr, NaCl	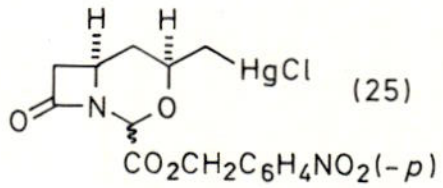 (bicyclic) H H, CH_2HgCl, O, $CO_2CH_2C_6H_4NO_2$(-p) (25)	Br_2	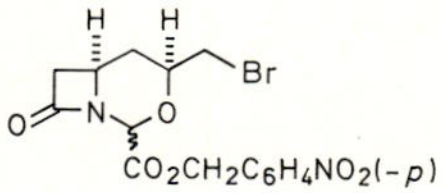 (bicyclic) H H, CH_2Br, O, $CO_2CH_2C_6H_4NO_2$(-p)	426
p-$(CH_3)_2C$=$CH(CH_2)_2$-$COH(CH_3)C_6H_4CH_3$	$Hg(OAc)_2$	$H_2O/$dioxane	—	$NaBH_4/NaOH$	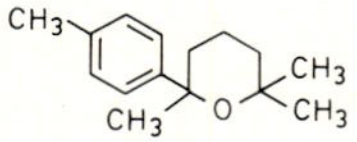 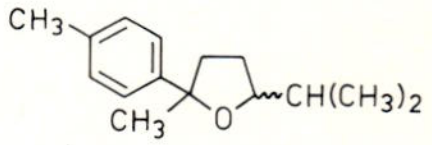 CH_3—C$_6$H$_4$—(pyran) CH_3 CH_3 CH_3 + CH_3—C$_6$H$_4$—(furan) CH_3 O $CH(CH_3)_2$; 1 : 1 (90)	427
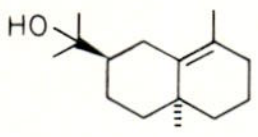 HO—(decalin, bicyclic with CH_3 groups)	$Hg(OAc)_2$	1:1 H_2O/THF	—	$NaBH_4/NaOH$	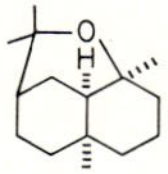 (tricyclic ether) (60) 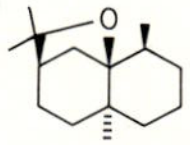 + (tricyclic ether) (20)	428
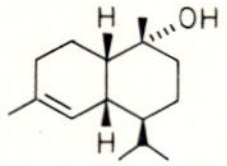 H, OH (decalin with isopropyl)	—	—	—	—	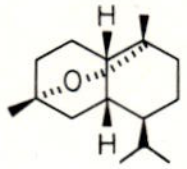 (tricyclic ether)	429

Table 3.6. (continued)

Alkenol	Mercuric salt	Reaction conditions	Organomercurial(s) (% Yield)	Subsequent reactants	Product(s) (% Yield)	Ref.
$H_2C=CHCH_2OCH$ [cyclic acetonide/orthoester structure]	$Hg(OAc)_2$ (I^-)	—	[cyclic structure with CH_2HgX, $X = OAc, I$]	I_2	[cyclic structure with CH_2I]	378, 380, 430
[aryl structure, CH_3O...CH_3] $H_2C=CH(CH_2)_2COH[CH(CH_3)_2]$	$Hg(OAc)_2$	1:1 H_2O/THF 2 hr	—	$NaBH_4$/NaOH	[tetrahydrofuran structure $(CH_3)_2CH$, CH_3, OCH_3] (66.6)	431
$C_2H_5O_2CC(OC_2H_5)=CHCH(OCH_3)$-$CH(NHAc)(CHOH)_3CH_2OH$	$Hg(O_2CCF_3)_2$	THF 0°C 6 hr	—	$NaBH_4$/NaOH	[pyran structure, OC_2H_5, CO_2H, OCH_3, $AcNH$, $HOCH_2$, HO] (88)	432
$H_2C=CHCH_2C(C_6H_5)_2CH_2OH$	$Hg(OAc)_2$	CH_3OH 25°C 20d, NaCl	$ClHgCH_2$ [tetrahydrofuran, C_6H_5, C_6H_5] (50)	—	—	252
[naphthalene structure, CH_3O, OCH_3, $CHOHCH_3$, $CH_2CH=CH_2$, OCH_3]	$Hg(OAc)_2$	—	—	$NaBH_4$	[pyran/naphthalene structure, CH_3O, OCH_3, CH_3, OCH_3, CH_3] (93) 1:1 cis/trans	433
trans-$CH_3(CH_2)_{13}CH=CH(CH_2)_2OH$	$Hg(OAc)_2$	DMF	—	$NaBH_4$	$CH_3(CH_2)_{13}$ [tetrahydrofuran] (100)	434
cis-$CH_3(CH_2)_{12}CH=CH(CH_2)_3OH$	$Hg(OAc)_2$	DMF	—	$NaBH_4$	$CH_3(CH_2)_{13}$ [tetrahydrofuran] (92)	434

Table 3.6. (continued)

Alkenol	Mercuric salt	Reaction conditions	Organomercurial(s) (% Yield)	Subsequent reactants	Product(s) (% Yield)	Ref.
trans- $CH_3(CH_2)_{12}CH=CH(CH_2)_3OH$	$Hg(OAc)_2$	DMF	—	$NaBH_4$	$CH_3(CH_2)_{13}$ (furan ring) $+ CH_3(CH_2)_{12}$ (pyran ring) 88 : 12	434
cis- $CH_3(CH_2)_{11}CH=CH(CH_2)_4OH$	$Hg(OAc)_2$	DMF	—	$NaBH_4$	$CH_3(CH_2)_{12}$ (pyran ring) (100)	434
trans- $CH_3(CH_2)_{11}CH=CH(CH_2)_4OH$	$Hg(OAc)_2$	DMF	—	$NaBH_4$	$CH_3(CH_2)_{12}$ (pyran ring) (100)	434
erythro- (bicyclic structure with Br, OH, $CO_2C(CH_3)_3$)	$Hg(O_2CCF_3)_2$	CH_3NO_2	—	$C_5H_5HBr_3$	(bicyclic structure with Br, OH, $CO_2C(CH_3)_3$)	435
cis- $CH_3(CH_2)_5CHOHCH_2-CH=CH(CH_2)_7CO_2CH_3$	$Hg(OAc)_2$	H_2O	—	$NaBH_4$	$CH_3(CH_2)_5CHOHCH_2CHOH(CH_2)_8CO_2CH_3$ (9) $+ CH_3(CH_2)_5CHOH(CH_2)_2CHOH(CH_2)_7CO_2CH_3$ (70) $+ CH_3(CH_2)_5$ (furan) $(CH_2)_7CO_2CH_3$ (17)	202
	$Hg(OAc)_2$	CH_3OH	—	$NaBH_4$	$CH_3(CH_2)_5CHOH(CH_2)_2CH(OCH_3)(CH_2)_7CO_2CH_3$ (66) $+ CH_3(CH_2)_5CHOHCH_2CH(OCH_3)(CH_2)_8CO_2CH_3$ (18) $+ CH_3(CH_2)_5$ (furan) $(CH_2)_7CO_2CH_3$ 2 isomers : 8% and 2%	202
trans- $CH_3(CH_2)_5CHOHCH_2-CH=CH(CH_2)_7CO_2CH_3$	$Hg(OAc)_2$	CH_3OH	—	$NaBH_4$	$CH_3(CH_2)_5$ (furan) $(CH_2)_7CO_2CH_3$ 2 isomers : 20% and 79%	202
cis- $CH_3(CH_2)_6CH=CH(CH_2)_3-CHOH(CH_2)_4CO_2CH_3$	$Hg(OAc)_2$	CH_3OH 4 d	—	$NaBH_4$	$CH_3(CH_2)_7$ (pyran) $(CH_2)_4CO_2CH_3$	83

Table 3.6. (continued)

Alkenol	Mercuric salt	Reaction conditions	Organomercurial(s) (% Yield)	Subsequent reactants	Product(s) (% Yield)	Ref.
cis-$CH_3(CH_2)_4CH=CH(CH_2)_2$-$CHOH(CH_2)_7CO_2CH_3$	$Hg(OAc)_2$	CH_3OH	—	$NaBH_4$	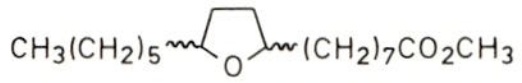 $CH_3(CH_2)_5$⋯O⋯$(CH_2)_7CO_2CH_3$ 2 isomers : 67% and 33%	202
$trans$-$CH_3(CH_2)_4CH=CH(CH_2)_2$-$CHOH(CH_2)_7CO_2CH_3$	$Hg(OAc)_2$	CH_3OH	—	$NaBH_4$	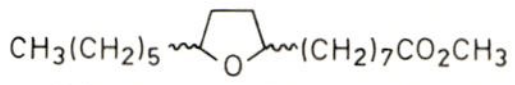 $CH_3(CH_2)_5$⋯O⋯$(CH_2)_7CO_2CH_3$ 2 isomers : 57% and 35% 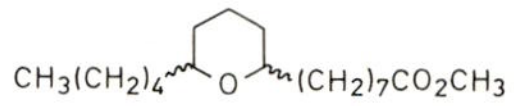 + $CH_3(CH_2)_4$⋯O⋯$(CH_2)_7CO_2CH_3$ (8)	202
cis-$CH_3(CH_2)_4CHOHCHOHCH_2$-$CH=CH(CH_2)_7CO_2CH_3$	$Hg(OAc)_2$	CH_3OH	—	$NaBH_4$	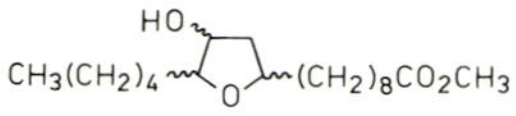 HO⋯ $CH_3(CH_2)_4$⋯O⋯$(CH_2)_8CO_2CH_3$ (92)	202
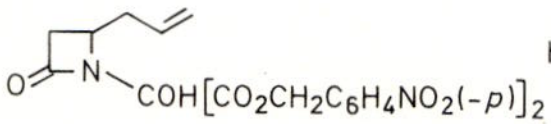 (structure) ...$COH[CO_2CH_2C_6H_4NO_2(-p)]_2$	$Hg(OAc)_2$	1:4 H_2O/THF 0°C 3hr, NaCl	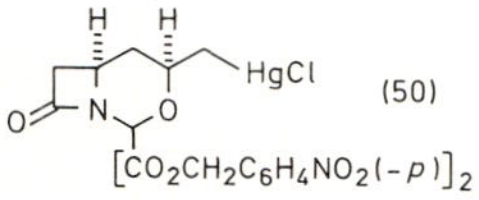 (structure) HgCl (50) $[CO_2CH_2C_6H_4NO_2(-p)]_2$	Br_2	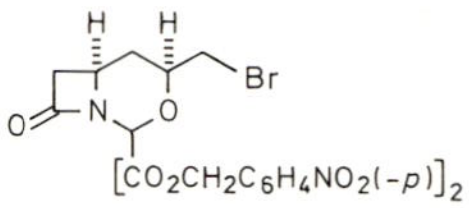 (structure) Br $[CO_2CH_2C_6H_4NO_2(-p)]_2$	426
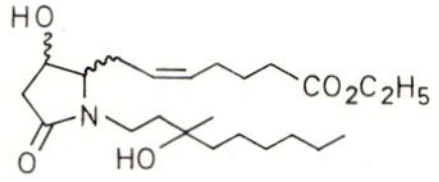 (structure) ...$CO_2C_2H_5$	$Hg(O_2CCF_3)_2$	$CaCO_3$/THF 1hr	—	$NaBH_4$	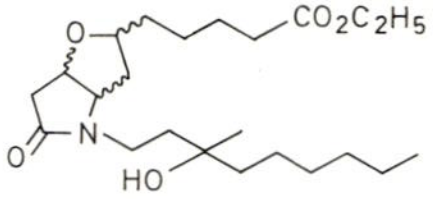 (structure) $CO_2C_2H_5$ (47)	436
$erythro$- 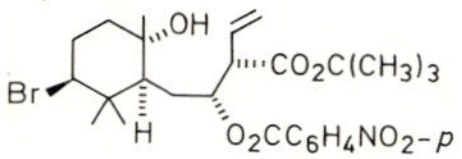 (structure) $CO_2C(CH_3)_3$ $O_2CC_6H_4NO_2$-p	$Hg(O_2CCF_3)_2$	CH_3NO_2, KBr	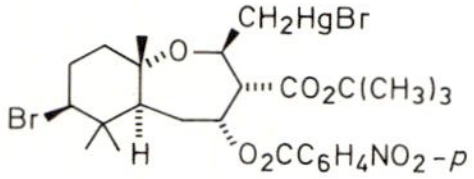 (structure) CH_2HgBr $CO_2C(CH_3)_3$ $O_2CC_6H_4NO_2$-p	$C_5H_5NHBr_3$	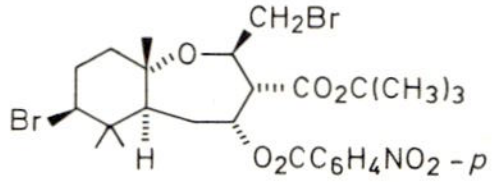 (structure) CH_2Br $CO_2C(CH_3)_3$ (74) $O_2CC_6H_4NO_2$-p	435
$threo$- (structure) $CO_2C(CH_3)_3$ $O_2CC_6H_4NO_2$-p 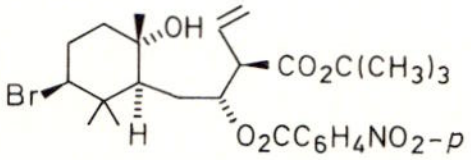	$Hg(O_2CCF_3)_2$	CH_3NO_2, KBr	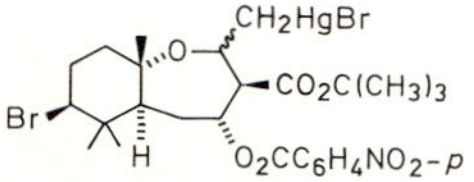 (structure) CH_2HgBr $CO_2C(CH_3)_3$ $O_2CC_6H_4NO_2$-p (5:1 mixture)	$C_5H_5NHBr_3$	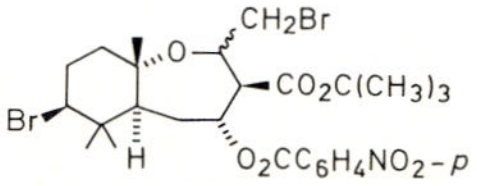 (structure) CH_2Br $CO_2C(CH_3)_3$ $O_2CC_6H_4NO_2$-p (mixture 69% and 15%)	435

Table 3.6. (continued)

Alkenol	Mercuric salt	Reaction conditions	Organomercurial(s) (% Yield)	Subsequent reactants	Product(s) (% Yield)	Ref.
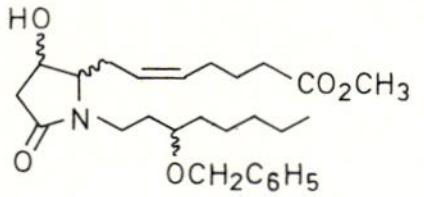 HO… CO$_2$CH$_3$ / OCH$_2$C$_6$H$_5$ (structure)	—	—	—	—	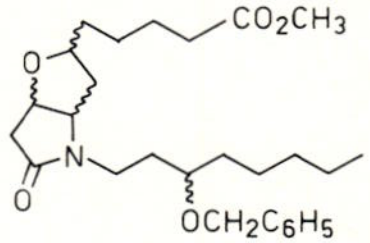 CO$_2$CH$_3$ / OCH$_2$C$_6$H$_5$ (structure) (≥ 50)	436
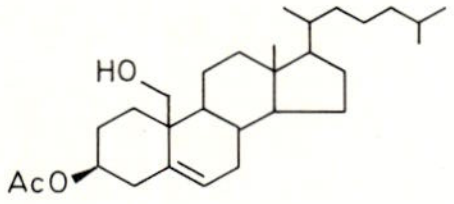 HO… AcO… (steroid structure)	Hg(O$_2$CCF$_3$)$_2$	—	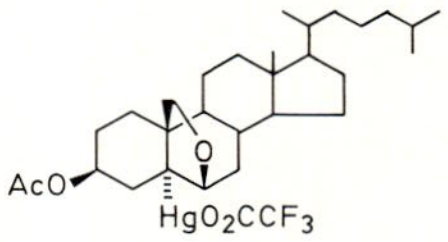 AcO… HgO$_2$CCF$_3$ (structure)	NaBH$_4$/NaOH	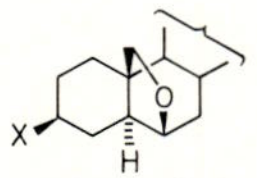 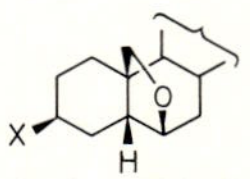 X = OAc (27), OH + X = OAc (18), OH	437
H$_2$C=CH(CHO$_2$CC$_6$H$_5$)$_3$ – CHOHCH$_2$O$_2$CC$_6$H$_5$	Hg(OAc)$_2$	THF, KCl	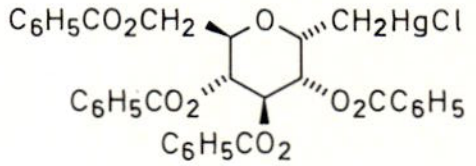 C$_6$H$_5$CO$_2$CH$_2$—O—CH$_2$HgCl / C$_6$H$_5$CO$_2$ / O$_2$CC$_6$H$_5$ / C$_6$H$_5$CO$_2$ (structure) (98)	NaBH$_4$ [O$_2$]	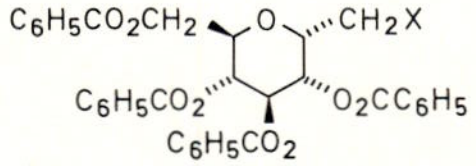 C$_6$H$_5$CO$_2$CH$_2$—O—CH$_2$X / C$_6$H$_5$CO$_2$ / O$_2$CC$_6$H$_5$ / C$_6$H$_5$CO$_2$ (structure) X = H (73) [X = OH (81)]	438

Table 3.7. Intramolecular Phenoxymercuration of Alkenylphenols

Phenol	Mercuric salt	Organomercurial(s) (% Yield)	Subsequent reactants	Product(s) (% Yield)	Ref.
$H_2C=CHCH_2$-phenol (HO, Br)	HgX_2 (X = Cl, OAc)	5-Br-2-(CH_2HgX)-2,3-dihydrobenzofuran (~100)	—	—	439
o-$H_2C=CHCH_2C_6H_4OH$	$HgCl_2$	2-(CH_2HgCl)-2,3-dihydrobenzofuran (81)	—	—	375
	$Hg(OAc)_2$	2-(CH_2HgOAc)-2,3-dihydrobenzofuran (74)	$NaBH_4$/NaOH	2-CH_3-2,3-dihydrobenzofuran	440
	$Hg(OAc)_2$	2-(CH_2HgOAc)-2,3-dihydrobenzofuran (98)	$NaBH_4$	2-CH_3-2,3-dihydrobenzofuran (36) + $H_2C=CHCH_2$-(2-HO-phenyl) (62)	153
	HgX_2 (X = Cl, OAc)	2-(CH_2HgX)-2,3-dihydrobenzofuran; X = Cl, OAc (100)	I_2 (on RHgI)	2-CH_2I-2,3-dihydrobenzofuran	441
	$Hg[(S)$-valinate$]_2$	—	$NaBH_4$/NaOH	2-CH_3-2,3-dihydrobenzofuran (< 5% enantiomeric excess)	442
$H_2C=CHCH_2$-benzene (HO, HO)	$HgCl_2$	4-OH-2-(CH_2HgCl)-2,3-dihydrobenzofuran	—	—	443
$H_2C=CHCH_2$-benzene (HO, OH)	$HgCl_2$	6-HO-2-(CH_2HgCl)-2,3-dihydrobenzofuran (100)	—	—	443
$H_2C=CHCH_2$-benzene (HO, CO_2H)	HgX_2 (X = Cl, OAc)	7-CO_2H-2-(CH_2HgX)-2,3-dihydrobenzofuran (~100)	—	—	439

Table 3.7. (continued)

Phenol	Mercuric salt	Organomercurial(s) (% Yield)	Subsequent reactants	Product(s) (% Yield)	Ref.
	Hg(OAc)$_2$	—	NaOH ?	[2,3-dihydrobenzofuran with CO$_2$Na; 2-CH$_2$HgOH]	122
H$_2$C=CHCH$_2$–C$_6$H$_3$(HO)(CO$_2$H)	HgX$_2$ (X = Cl, OAc)	[HO$_2$C-2,3-dihydrobenzofuran-2-CH$_2$HgX] (~100)	—	—	439
o-H$_2$C=C(CH$_3$)CH$_2$C$_6$H$_4$OH	HgCl$_2$	[2-CH$_3$-2,3-dihydrobenzofuran-2-CH$_2$HgCl]	—	—	444
	Hg(OAc)$_2$	[2-CH$_3$-2,3-dihydrobenzofuran-2-CH$_2$HgOAc] (75)	NaBH$_4$/NaOH	[2,2-(CH$_3$)$_2$-2,3-dihydrobenzofuran]	440
o-H$_2$C=CHCH(CH$_3$)C$_6$H$_4$OH	Hg(OAc)$_2$	[3-CH$_3$-2,3-dihydrobenzofuran-2-CH$_2$HgOAc] (91)	NaBH$_4$/NaOH	[2-CH$_3$-3-CH$_3$-2,3-dihydrobenzofuran]	440
trans-o-CH$_3$CH=CHCH$_2$C$_6$H$_4$OH	Hg(OAc)$_2$	[2,3-dihydrobenzofuran-2-CH(HgOAc)CH$_3$] + [2-CH$_3$-chroman-3-HgOAc]; 52 : 48 (78 total)	NaBH$_4$/NaOH	[2,3-dihydrobenzofuran-2-CH$_2$CH$_3$] + [2-CH$_3$-chroman]; 52 : 48	440
H$_2$C=CHCH$_2$–C$_6$H$_3$(HO)(CH$_3$)	HgX$_2$ (X = Cl, OAc)	[CH$_3$-2,3-dihydrobenzofuran-2-CH$_2$HgX] (~100)	—	—	439
	Hg(OAc)$_2$	[CH$_3$-2,3-dihydrobenzofuran-2-CH$_2$HgOAc] (70)	NaBH$_4$/NaOH	[CH$_3$-2-CH$_3$-2,3-dihydrobenzofuran]	440

Table 3.7. (continued)

Phenol	Mercuric salt	Organomercurial(s) (% Yield)	Subsequent reactants	Product(s) (% Yield)	Ref.
$H_2C=CHCH_2$— (HO, CH_3 cresol)	HgX_2 ($X = Cl, OAc$)	CH_3—[benzofuran]—CH_2HgX (~100)	—	—	439
$H_2C=CHCH_2$— (HO, CH_3 cresol)	HgX_2 ($X = Cl, OAc$)	CH_3—[benzofuran]—CH_2HgX (~100)	—	—	439
(cyclopentenyl phenol, OH)	HgX_2 ($X = Cl, OAc$)	[tricyclic]—HgX; $X = Cl$ (65), $X = OAc$ (60)	—	—	445
$H_2C=CHCH_2$— (HO, CO_2CH_3)	HgX_2 ($X = Cl, OAc$)	CO_2CH_3—[benzofuran]—CH_2HgX (~100)	—	—	439
$H_2C=C(CH_3)CH_2$— (HO, CH_3)	$HgCl_2$	CH_3—[benzofuran](CH_3)—CH_2HgCl	—	—	444
$H_2C=C(CH_3)CH_2$— (HO, CH_3)	$HgCl_2$	CH_3—[benzofuran](CH_3)—CH_2HgCl	—	—	444
$H_2C=CHCH(CH_3)$— (HO, CH_3)	$Hg(OAc)_2$	CH_3—[benzofuran](CH_3)—CH_2HgOAc (67)	$NaBH_4/NaOH$	CH_3—[benzofuran](CH_3)—CH_3	440
trans- $CH_3CH=CHCH_2$— (HO, CH_3)	$Hg(OAc)_2$	CH_3—[benzofuran]—$CH(HgOAc)CH_3$ + CH_3—[chromane](CH_3)—$HgOAc$ (85 total) 51 : 49	$NaBH_4/NaOH$	CH_3—[benzofuran]—CH_2CH_3 + CH_3—[chromane]—CH_3 48 : 52	440

Table 3.7. (continued)

Phenol	Mercuric salt	Organomercurial(s) (% Yield)	Subsequent reactants	Product(s) (% Yield)	Ref.
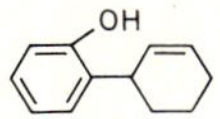 [2-(2-cyclohexenyl)phenol structure]	Hg(OAc)$_2$	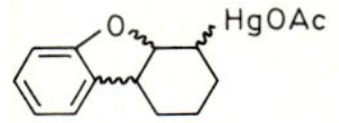 [benzofuran-fused HgOAc structure]	NaBH$_4$/NaOH (phase transfer)	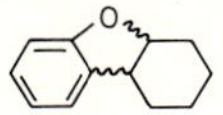 [benzofuran-fused structure] (90)	382
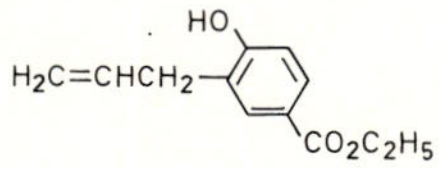 [H$_2$C=CHCH$_2$, HO, CO$_2$C$_2$H$_5$ phenol structure]	Hg(OAc)$_2$	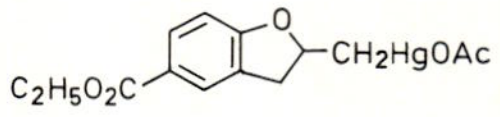 [C$_2$H$_5$O$_2$C dihydrobenzofuran CH$_2$HgOAc structure] (14–50)	—	—	141
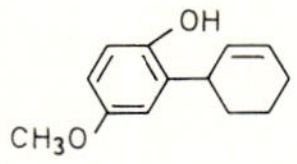 [CH$_3$O, OH, cyclohexenyl phenol structure]	Hg(OAc)$_2$	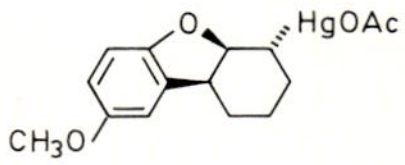 [CH$_3$O benzofuran-fused HgOAc structure] (59)	NaBH$_4$	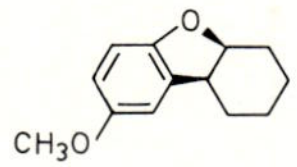 [CH$_3$O benzofuran-fused structure]	445
	HgX$_2$ (X = Cl, NO$_3$, ClO$_4$)	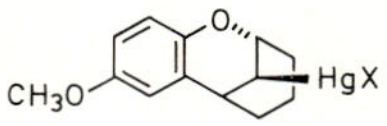 [CH$_3$O bridged HgX structure] [X = Cl (50)]	NaBH$_4$	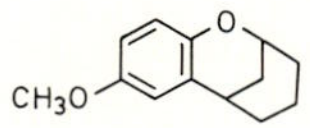 [CH$_3$O bridged structure]	445
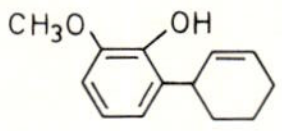 [CH$_3$O, OH, cyclohexenyl phenol structure]	Hg(OAc)$_2$ (KCl)	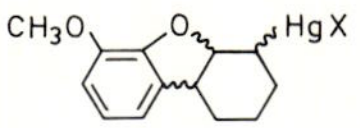 [CH$_3$O benzofuran-fused HgX structure] X = OAc (76), X = Cl (75)	NaBH$_4$ or NaBD$_4$ (on RHgOAc)	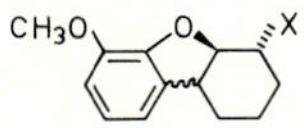 [CH$_3$O benzofuran-fused X structure] X = H (45), D (40)	375
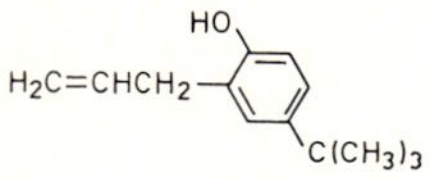 [H$_2$C=CHCH$_2$, HO, C(CH$_3$)$_3$ phenol structure]	Hg(OAc)$_2$, KI	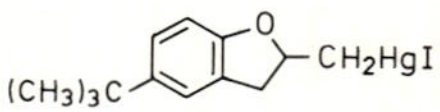 [(CH$_3$)$_3$C dihydrobenzofuran CH$_2$HgI structure]	I$_2$/KI	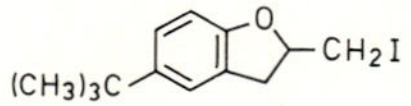 [(CH$_3$)$_3$C dihydrobenzofuran CH$_2$I structure]	446
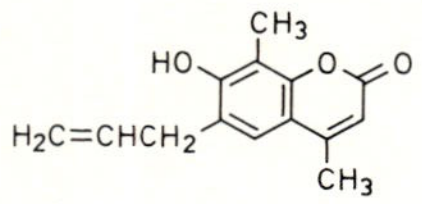 [HO, CH$_3$, H$_2$C=CHCH$_2$, CH$_3$ coumarin structure]	HgCl$_2$	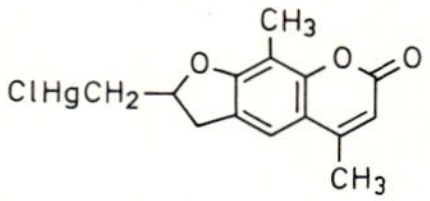 [ClHgCH$_2$ furo-coumarin structure]	I$_2$/KI	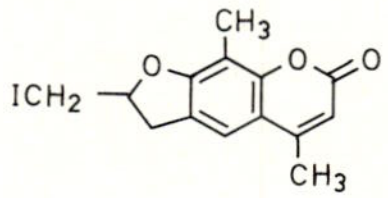 [ICH$_2$ furo-coumarin structure]	447
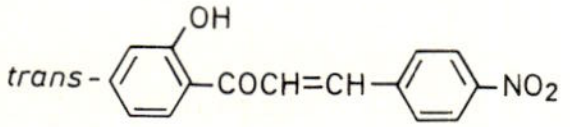 _trans_-[HO, COCH=CH, NO$_2$ chalcone structure]	Hg(OAc)$_2$	—	—	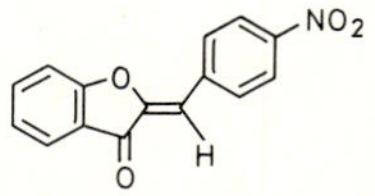 [aurone (4-nitrobenzylidene) structure] (~100)	448

Table 3.7. (continued)

Phenol	Mercuric salt	Organomercurial(s) (% Yield)	Subsequent reactants	Product(s) (% Yield)	Ref.
o-$C_6H_5CH=CHCOC_6H_4OH$	$Hg(OAc)_2$	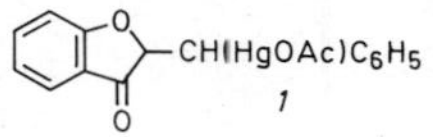 [structure] CHIHgOAc)C_6H_5 + [structure] *1* 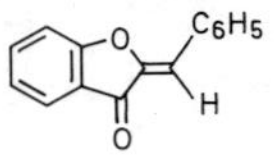	$NaBH_4$ (on *1*)	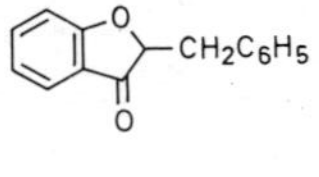 [structure] CH$_2C_6H_5$	448
	$Hg(OAc)_2$	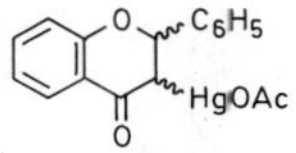 [structure] C_6H_5, HgOAc — cis and trans	$NaBH_4$/NaOH	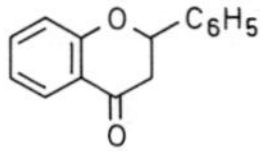 [structure] C_6H_5	448
o-$H_2C=CHCH(C_6H_5)C_6H_4OH$	$Hg(OAc)_2$	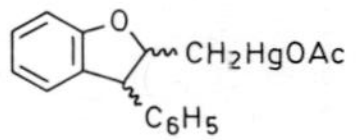 [structure] CH_2HgOAc (85) C_6H_5	$NaBH_4$/NaOH	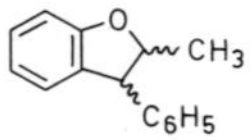 [structure] CH_3 C_6H_5	440
trans-CH_3O—[structure, OH]—$COCH=CHC_6H_5$ 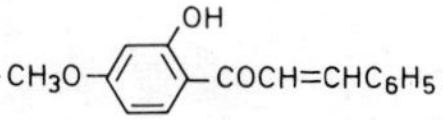	$Hg(OAc)_2$	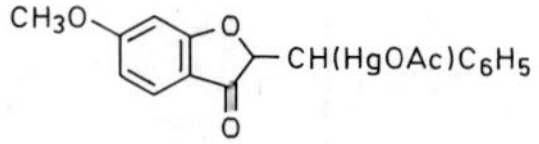 CH_3O—[structure]—CH(HgOAc)C_6H_5	CaO/DMSO	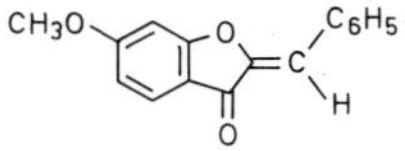 CH_3O—[structure]—C_6H_5, H (54 overall)	448

in these cyclization reactions, including mercuric chloride, mercuric acetate, mercuric nitrate, mercuric sulfate, mercuric perchlorate and several different mercuric carboxylates. Both the electrophilicity of the mercury salt and the acidity of the resulting acid appear to be important in determining the regio- [396, 445] and stereochemistry [370–372] of the resulting organomercurial. In a number of reactions, excess acid or buffers, such as sodium acetate, calcium carbonate or cadmium carbonate, have been added to the reaction. Where competing hydroxymercuration may be a problem, the pH has been shown to affect the ratio of products [417]. While numerous solvents have proven effective in these reactions, including water, H_2O/THF, methanol, *tert*-butanol, DMF, benzene, THF, 1,2-dichloroethane and nitromethane, the protic solvents can of course participate in these reactions and should be avoided in those cases where intramolecular cyclization is more difficult.

The intramolecular alkoxymercuration reaction has proven to be a very valuable method for the synthesis of five- and six-membered ring ethers. There appears to be only one example of any other ring size ever having been generated by this reaction [435]. Olefinic alcohols which can form five- and six-membered ring ethers preferentially cyclize even under the usual hydroxy- mercuration conditions, while other unsaturated alcohols simply afford diols [377]. However, with *o*-allylphenol, mercuric acetate and methanol, a slight preference for methoxymercuration over intramolecular phenoxymercu- ration has been observed [153].

Let's examine more closely the regiochemistry of these cyclization reactions, keeping in mind that only five- and six-membered ring ethers can generally be formed. We should also realize that two studies have noted a regiochemical dependence on the mercury salt employed in the reaction (Eqs. 128, 129). Our discussion will be divided into acyclic, monocyclic and poly- cyclic enols.

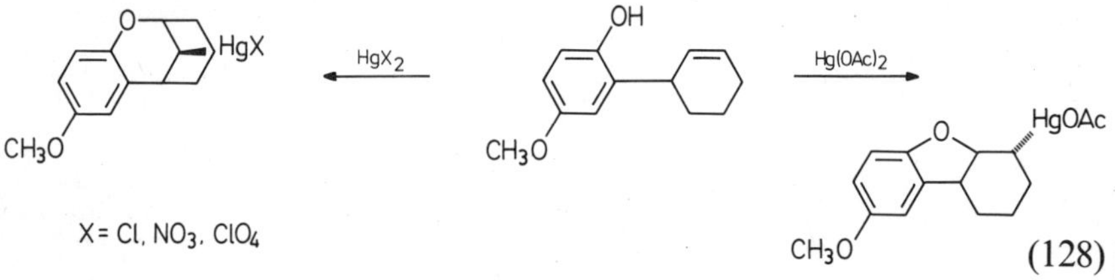

$$(128) \ [445]$$

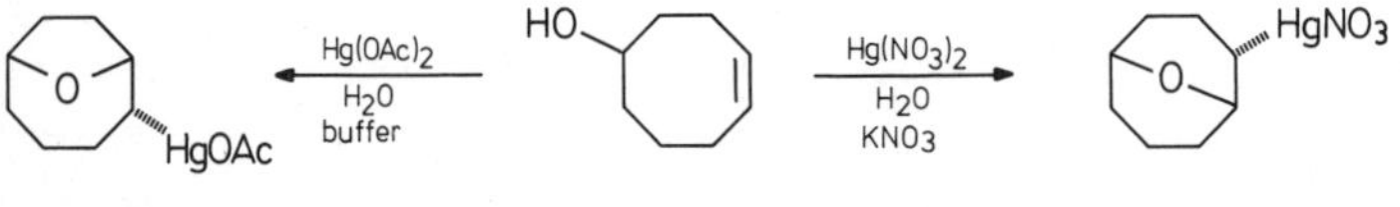

$$(129) \ [396]$$

$$R-C=C-C-C-OH \longrightarrow \longrightarrow R-\overset{\diagup\diagdown}{\underset{O}{\diagdown\diagup}} \qquad (130)$$

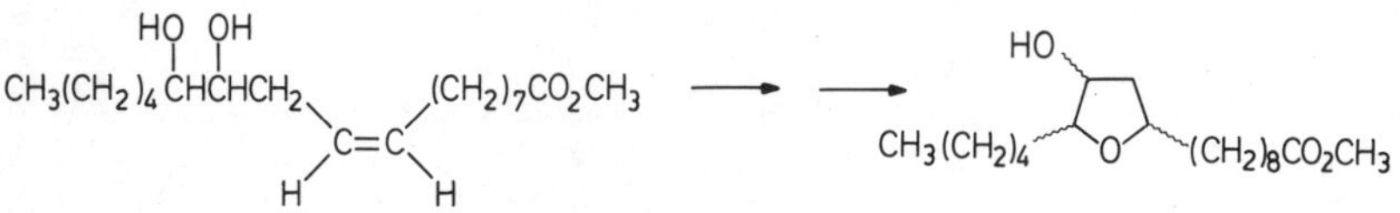

$$(131)$$

$$H_2C\!=\!\overset{\underset{\displaystyle |}{R}}{C}\!-\!C\!-\!C\!-\!C\!-\!OH \longrightarrow \longrightarrow \quad\text{[cyclic ether]} \tag{132}$$

R = H, alkyl

$$\text{[}o\text{-allylic phenol]} \longrightarrow \longrightarrow \text{[2,2-disubstituted dihydrobenzofuran]} \tag{133}$$

A number of publications have appeared in which the preference for five-versus six-membered ring formation has been studied as a function of the position, substitution pattern and stereochemistry of the carbon—carbon double bond [383]. Homoallylic alcohols have been cyclized to tetrahydrofurans using mercuric acetate in methanol or DMF (Eq. 130) [202, 434]. Given a choice between a homoallylic alcohol or a Δ^4-alkenol, cyclization to the more remote alcohol is observed (Eq. 131) [202]. Δ^4-Alkenols and o-allylic phenols with a terminal double bond also afford exclusively tetrahydrofurans (Eqs. 132, 133) [383, 400, 440]. cis-Disubstituted Δ^4-alkenols afford exclusively the five-membered ring ethers (Eq. 134) [202, 383, 434, 436], while the

$$\text{[}cis\text{-alkenol]} \longrightarrow R\!-\!CH_2\!-\!\text{[tetrahydrofuran]} \tag{134}$$

$$\text{[}trans\text{-alkenol]} \longrightarrow R\!-\!CH_2\!-\!\text{[tetrahydrofuran]} + R\!-\!\text{[tetrahydropyran]} \tag{135}$$

$$\text{[}o\text{-allylic phenol]} \longrightarrow \text{[dihydrobenzofuran-CH}_2\text{CH}_3\text{]} + \text{[chroman-CH}_3\text{]} \tag{136}$$

$$\text{[chromanone]} \underset{\substack{HClO_4 \\ CH_2Cl_2}}{\overset{Hg(OAc)_2}{\longleftarrow}} \text{[}o\text{-hydroxyphenyl enone]} \overset{Hg(OAc)_2}{\underset{DMSO}{\longrightarrow}} \text{[benzofuranone]} \tag{137}$$

cis and trans

corresponding $trans$-isomers, be they alcohols [202, 383, 421, 434] or phenols [440, 448], give mixtures of five- and six-membered ring ethers (Eqs. 135, 136). The six-membered ring ether predominates (58:42) in this last reaction (Eq. 136) when either mercuric acetate or mercuric chloride are used, but addition of pyridine favors formation of the five-membered ring (67:33) [440]. While the presence of a carbonyl group under one set of reaction conditions will direct the cyclization to the six-membered ring product, under other conditions this reaction gives only the five-membered ring product (Eq. 137) [448]. The latter product reacts with excess mercuric acetate or calcium oxide in DMSO to give aurones. Δ^4-Alkenols in which the double bond is trisubstituted are reported to give exclusively the tetrahydropyran [383, 400, 407, 408], but one exception has been noted [427] (Eqs. 138, 139). Δ^5-Alkenols containing either a terminal double bond or a *cis*- or *trans*-disubstituted double bond cyclize cleanly to the tetrahydropyran (Eq. 140) [110, 202, 377, 383, 384, 434]. No seven-membered ring ether is observed here. Where more than one alcohol group is in a position to cyclize to a tetrahydropyran ring system as in certain carbohydrates, it appears that little selectivity exists (Eq. 141) [156, 378, 380, 385]. The intramolecular alkoxymercuration of

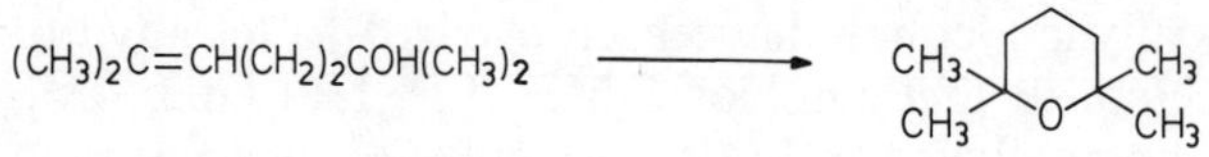
(CH3)2C=CH(CH2)2COH(CH3)2
CH3
CH3
O
CH3
CH3

(138)

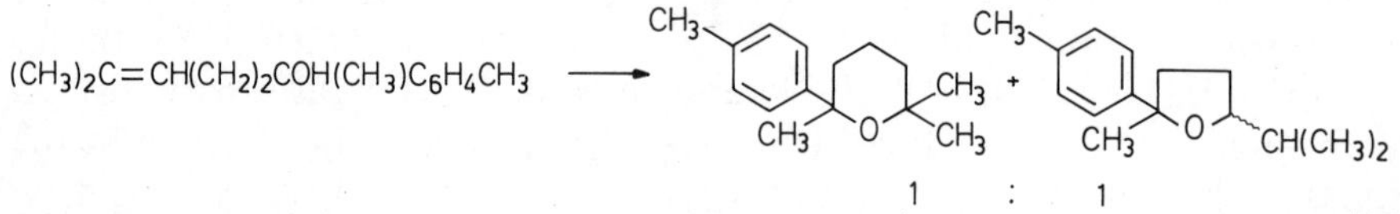
(CH3)2C=CH(CH2)2COH(CH3)C6H4CH3
CH3
CH3
O
CH3
CH3
+
CH3
CH3
O
CH(CH3)2
1 : 1

(139)

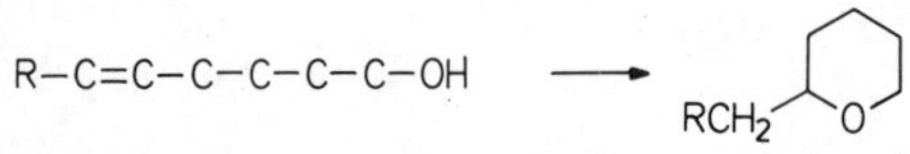
R–C=C–C–C–C–C–OH
RCH2
O

(140)

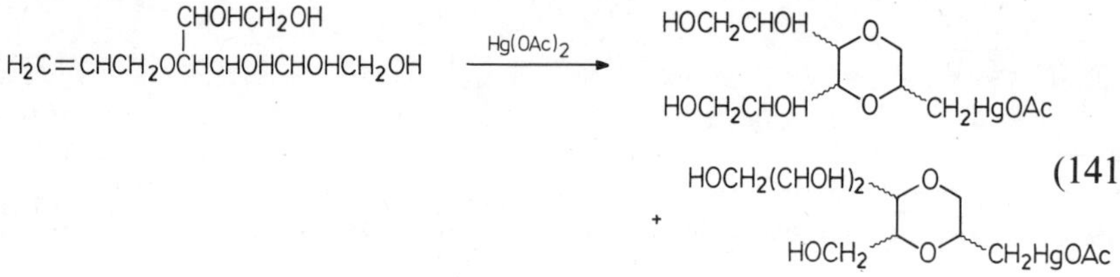
CHOHCH2OH
H2C=CHCH2OCHCHOHCHOHCH2OH
Hg(OAc)2
HOCH2CHOH
O
HOCH2CHOH
O
CH2HgOAc
HOCH2(CHOH)2
O
+
HOCH2
O
CH2HgOAc

(141) [380]

acyclic Δ^3-, Δ^4-, and Δ^5-alkenols, and acids, esters, aldehydes and ketones reduceable to such alcohols, has been suggested as a useful analytical technique for the identification, isolation and estimation of these compounds [434, 452]. This approach has been used on selected seed and fish oils, as well as rat liver lipids.

The rules of the game become somewhat more complex when cyclic systems are subjected to intramolecular alkoxymercuration. When the carbon—carbon double bond is not part of a ring, the preceeding generalizations apply.

The only reported cyclization to a seven-membered ring ether involves the following unsaturated cyclic alcohol (Eq. 142) [435]. The erythro isomer here affords a single diastereomer, while the threo isomer gives a 5:1 mixture of diastereomers. Removal of the *p*-nitrobenzoyl group affords a diol which preferentially cyclizes to the tetrahydrofuran.

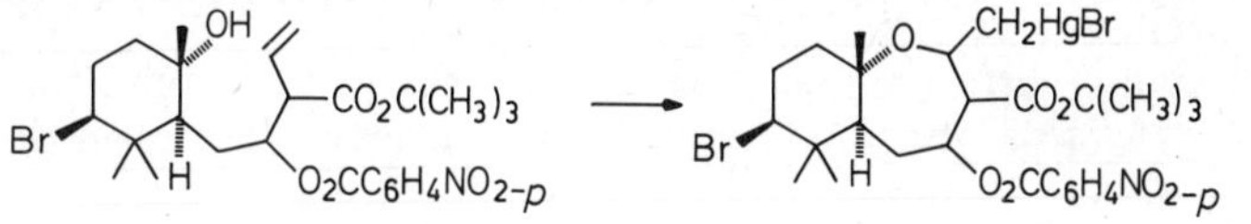

$$(142)$$

When the double bond is within a ring, steric factors and ring strain become important considerations. All of the following cyclizations proceed cleanly in the direction indicated (Eqs. 143–146). The last two examples,

$$(143)\ [386]$$

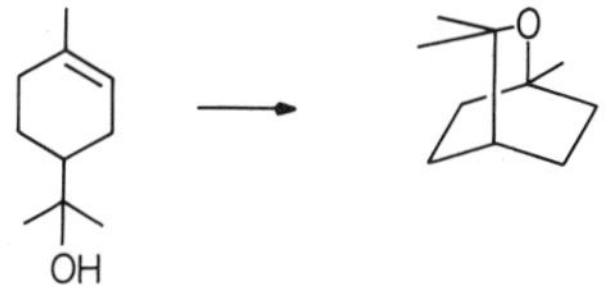

$$(144)\ [408,\ 414\text{--}416]$$

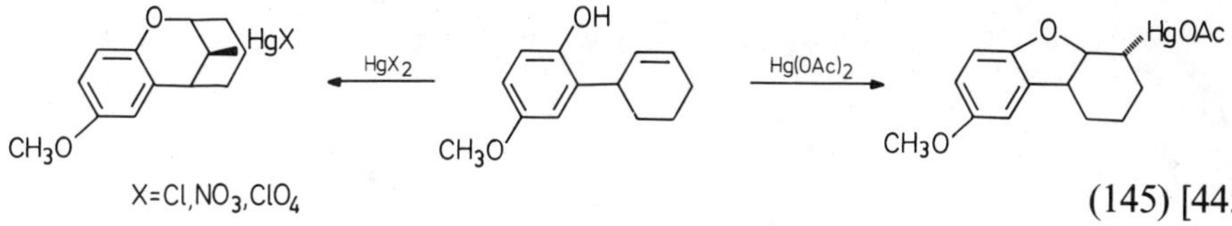

$$(145)\ [445]$$

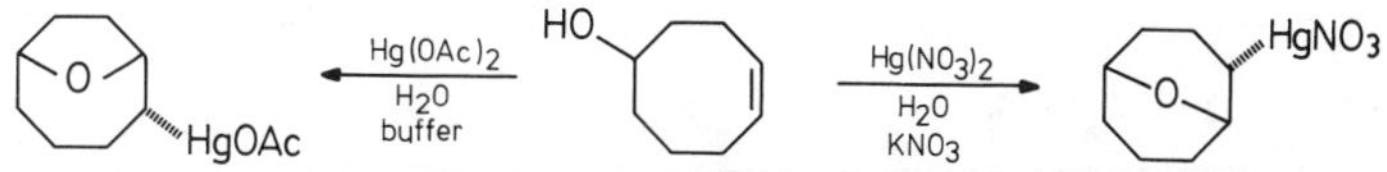

$$(146)\ [396]$$

discussed briefly earlier, indicate how easily the direction of cyclization can be altered. On the other hand, 5-phenylcycloocten-5-ol gives a mixture of regioisomers with mercuric acetate (Eq. 147) [425].

A unique application of this type of cyclization is reported by Overman [388, 389]. By first reacting allylic alcohols with chloral and then effecting intramolecular alkoxymercuration, cyclic acetals are formed (Eq. 148). Acetal removal with sodium affords a stereospecific 1,2-diol synthesis. While

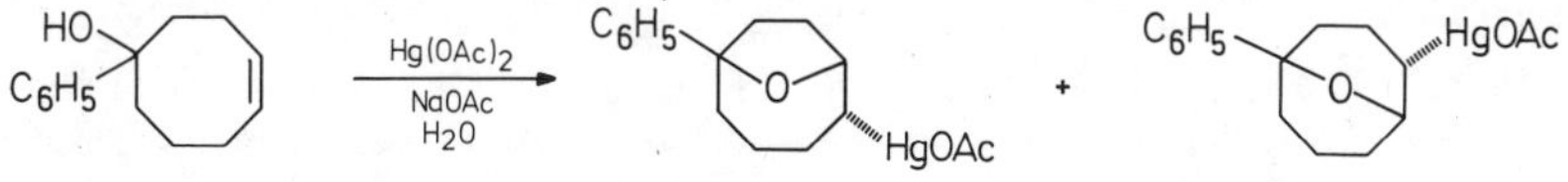

$$(147)$$

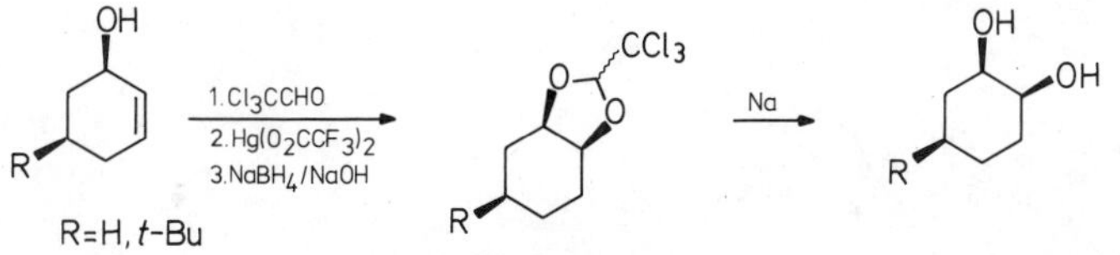

$$(148)$$

this approach works well with certain cyclic alcohols, it fails on a number of others. In acyclic systems, little regioselectivity is observed, eliminating any advantage this approach might have over direct hydroxymercuration of the allylic alcohol.

In polycyclic systems, intramolecular alkoxymercuration frequently affords only one major product in high yield, although mixtures have been reported as seen in the last two examples below (Eqs. 149–158). Note that in

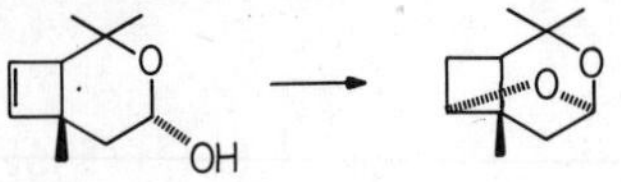

(149) [410]

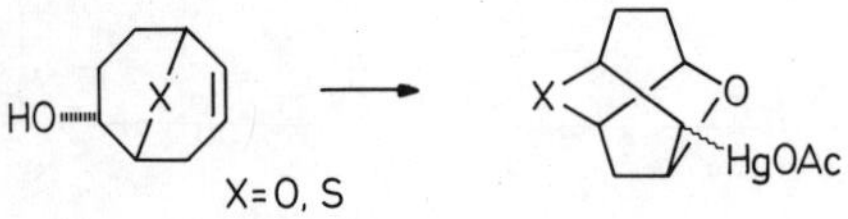

(150) [391, 393, 397]

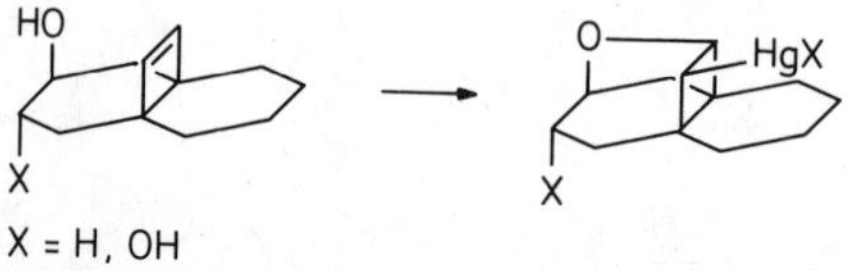

(151) [422]

(152) [409]

(153) [409]

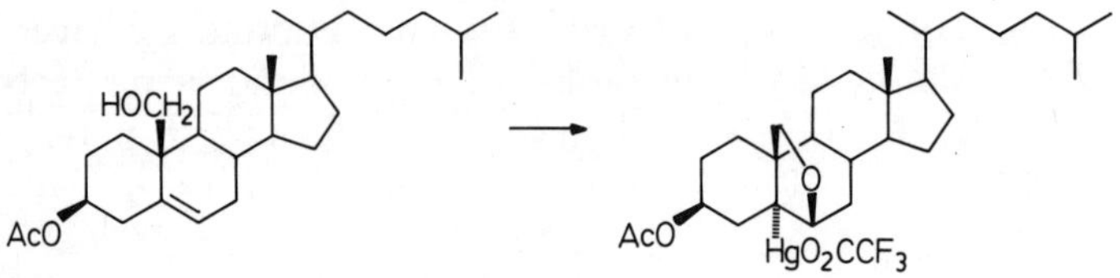

(154) [437]

(155) [17, 390]

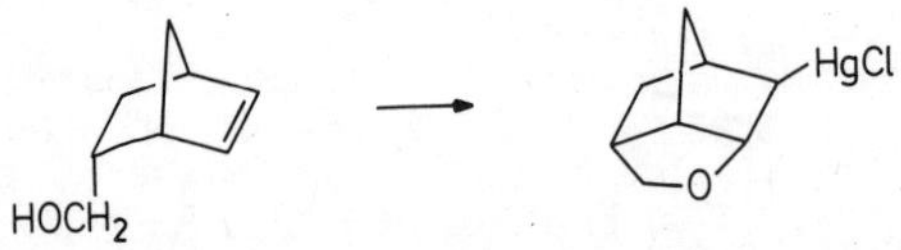

(156) [17, 129]

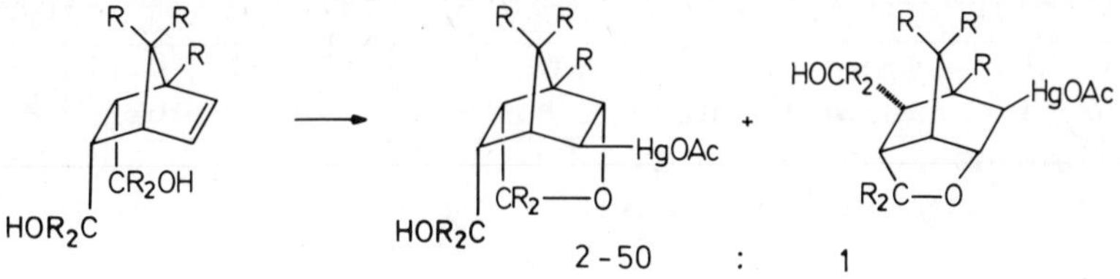

(157) [411]

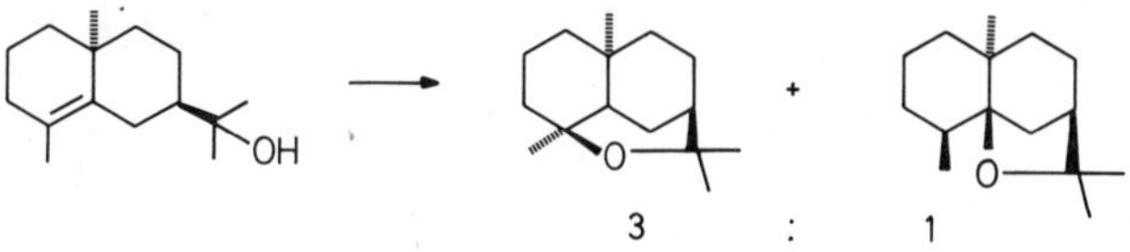

(158) [428]

the bicyclo[2.2.1]heptyl system, all additions apparently proceed in an anti fashion. The steroid example (Eq. 154) is interesting in that the product arises by an anti-Markovnikov addition rather than form a four-membered ring ether.

Let's take a closer look at the stereochemistry of these cyclization reactions. The cyclodimerization of allyl alcohol apparently can be controlled so as to give predominantly either the cis or the trans disubstituted dioxane after iodination (Eq. 159) [370, 377]. The cyclization of *cis*- and *trans*-disubstituted Δ^3-alkenols [202] and terminal [375, 381] or *cis*- and *trans*-disubstituted [202, 421] Δ^4-alkenols gives mixtures of *cis*- and *trans*-tetrahydrofurans (Eqs. 160, 161). The ratio varies with the substituents present, the

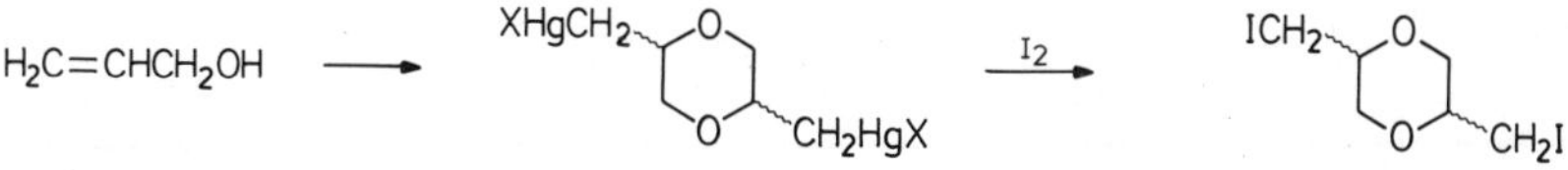

(159)

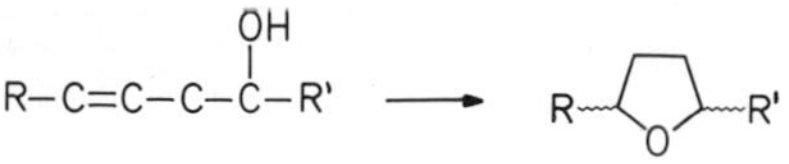

(160)

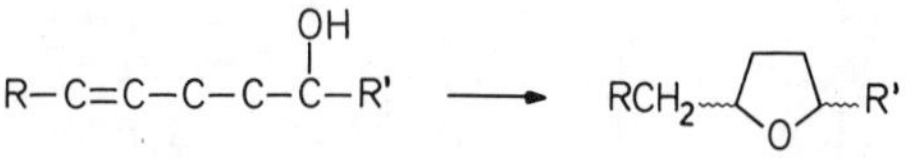

(161)

stereochemistry of the double bond and the mercury salt employed. Cyclization to a tetrahydropyran has also been reported to afford a mixture of stereoisomers (Eq. 162) [433].

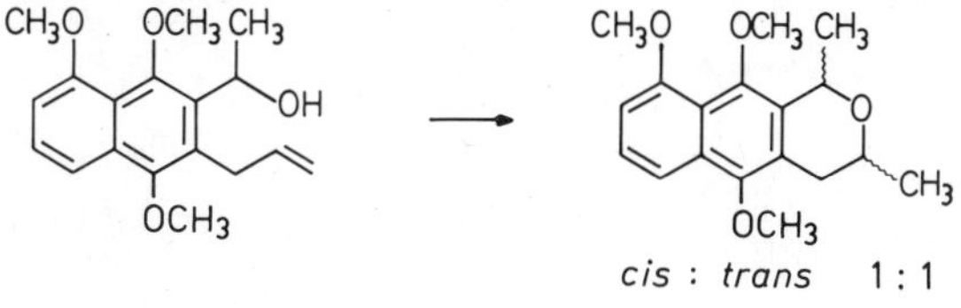

(162)

cis : trans   1 : 1

The cyclization of acyclic alkenols containing chiral centers often proceeds with high stereospecificity as shown by the following examples (Eqs. 163–167). Is is noteworthy that all these cyclizations involve tetrahydropyran formation. The cyclization of chiral alkenols to tetrahydrofurans tends to proceed with lower stereospecificity [401, 403, 404].

(163) [412]

> 95% stereoselective

(164) [412]

plus C-2 epimer

(165) [412]

> 95% stereoselective

(166) [438]

(167) [432]

Some work has been reported on the stereoselectivity of cyclization of cyclic alcohols. The following stereoselective cyclization was a key step in the synthesis of a thromboxane A$_2$ analog (Eq. 168) [419]. *Cis-* and *trans-*2-allylcyclohexanols afford mixtures of bicyclic tetrahydrofurans, the ratio of which varies with the substrate, the mercury salt and the reaction conditions (Eq. 169) [375, 402]. While the intramolecular alkoxymercuration

(168)

(169)

of 3-(2-hydroxyethyl)cyclopentene and -cyclohexene (Eq. 143) [386] and
o-(3-cyclopentenyl)phenol and 2-(3-cyclohexenyl)-4-methoxyphenol (Eq. 145)
[445] apparently proceed stereospecifically, the following cyclization yields
mixtures in which the ratio is dependent on the degree of substitution about
the alcohol (Eq. 170) [423].

$$
\begin{array}{lccc}
& R^1=R^2=H & 68 & : & 32 \\
& R^1=H, \; R^2=CH_3 & 10 & : & 90 \\
& R^1=R^2=CH_3 & 5 & : & 95 \\
\end{array}
\tag{170}
$$

In the cyclization of polycyclic alkenols, we noted earlier a high degree
of regioselectivity (Eqs. 149–156). It appears that in all cases where the
stereochemistry of the mercury moiety has been established, cyclization
proceeds by anti addition of mercury and the alcohol. This appears true in
bicyclo[2.2.1]heptyl systems also, although the intermolecular solvomercuration of these compounds affords syn exo adducts.

There has been only one attempt to induce chirality in an intramolecular
alkoxymercuration reaction (Eq. 171) [442]. Unfortunately, enantiomeric
excesses of less than 5% were achieved.

$$\tag{171}$$

$$\tag{172}$$

Relatively little work has been reported on the kinetics of intramolecular alkoxymercuration reactions. As noted earlier, under typical
hydroxymercuration conditions, Δ^4- and Δ^5-alkenols cyclize preferentially to
tetrahydrofurans and tetrahydropyrans [377]. While 1-penten-5-ol cyclizes
too rapidly to determine accurate kinetics [384], it is estimated that endo-5-
hydroxymethylnorbornene reacts at least 10^4 times faster than norbornene
itself [453]. The reaction of cis-3-hexen-1-ol and mercuric acetate in methanol
gives only methoxymercuration, while the trans isomer undergoes cyclization
(Eq. 172) [129]. Under similar conditions, 2-allylphenol gives a 12:7 ratio of
methoxymercurial and cyclized mercurial after 5 minutes [153]. For the
cyclization of o-allylphenol by mercuric chloride, the equilibrium constants
at three different temperatures, the heat of reaction and the overall kinetics
have been determined [454]. The reaction appears quite complex. The effect of
substituents on the relative rates of cyclization have also been established
for the following reaction (Eq. 173) [448]. The data is inconsistent with
electrophilic attack by $HgOAc^+$ and $ArOHgOAc$ addition to the double bond
has been proposed.

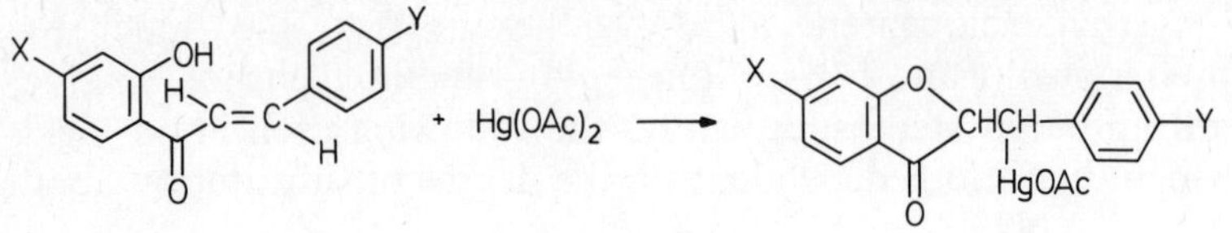

III. Alkoxymercuration

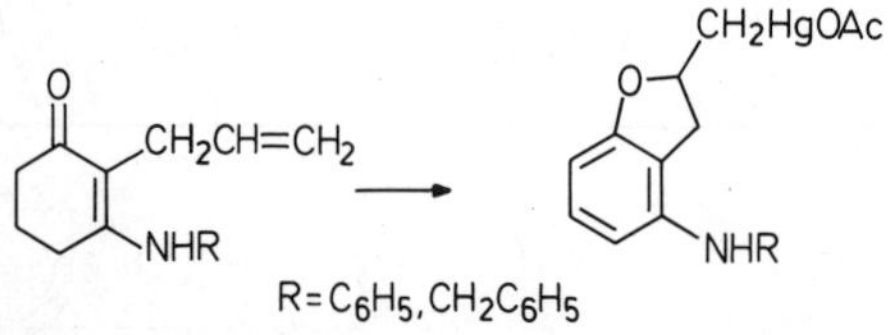

$$(173)$$

$$H_2C{=}CH(CH_2)_2\overset{\overset{\displaystyle OH}{|}}{C}HCH_2NHC(CH_3)_3 \longrightarrow CH_3{-}\!\!\!\!\!\overset{O}{\frown}\!\!\!\!\!{-}CH_2NHC(CH_3)_3 \qquad (174)$$

$$(175)$$

$$R = C_6H_5, CH_2C_6H_5$$

A close examination of Tables 3.6 and 3.7 confirms the versatility of the intramolecular alkoxy- and phenoxymercuration reaction. Virtually every important organic functional group is accommodated. Note that even allylresorcinols cyclize rather than undergo aromatic substitution [443]. Even aminoalkenols, which tend to undergo preferential intramolecular aminomercuration, can be induced to cyclize via oxygen if the reaction is run in the presence of acid (Eq. 174) [394]. Even some allyl enaminoketones undergo preferential intramolecular phenoxymercuration in the presence of an amine when heated with mercuric acetate (Eq. 175) [455].

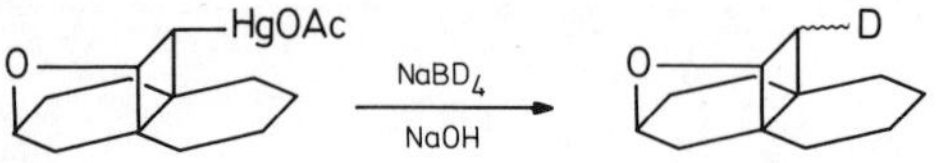

$$\xrightarrow[\text{NaOH}]{\text{NaBD}_4} \qquad (176) \ [422]$$

$$26:74 \ \ cis/trans$$

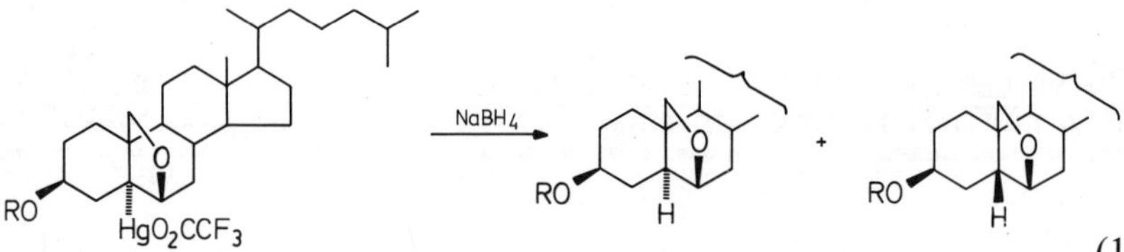

$$\xrightarrow{\text{NaBH}_4} \qquad (177) \ [437]$$

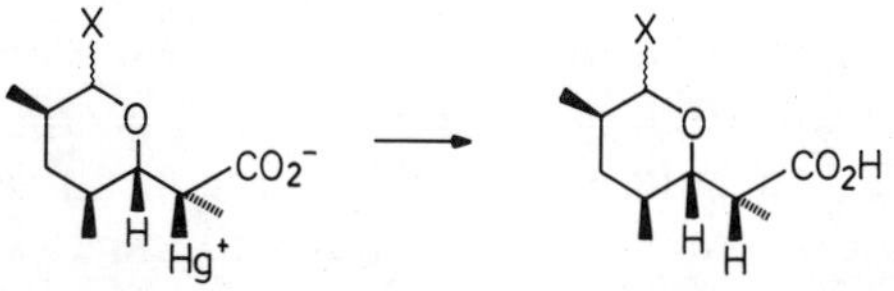

$$(178)$$

The demercuration of the intramolecular alkoxymercuration products presents certain difficulties. While reduction with sodium borohydride is generally successful, it is not without its problems. As expected, loss of stereochemistry has been observed (Eqs. 176, 177). In the synthesis of

Prelog-Djerassi lactone via intramolecular solvomercuration, it was important to demercurate the above intermediate with retention (Eq. 178) [412]. A number of reducing reagents were examined in this reaction, including sodium borohydride, hydrogen sulfide, sodium sulfide, sodium-mercury amalgam, sodium trithiocarbonate and hydrogen over Wilkinson's catalyst. Sodium borohydride gave almost complete inversion, while the others gave epimeric mixtures. Sodium trithiocarbonate gave the best results, affording the desired isomer is a 3.5:1 ratio. In one unusual case, the stereochemistry at oxygen is reported to change during sodium borohydride demercuration (Eq. 179) [381]. No satisfactory explanation for this phenomenon is available.

$$CH_3 - [\text{ring}] - CH_2HgCl \xrightarrow[\text{NaOH}]{\text{NaBH}_4} CH_3 - [\text{ring}] - CH_3 \qquad (179)$$

cis : *trans*   55 : 45   *cis* : *trans*   67 : 33

$$HO - [\text{ring}] \longleftarrow [\text{ring}] - HgX \longrightarrow [\text{ring}] \qquad (180)$$

One other problem common to these demercuration reactions is elimination to the starting enol (Eq. 180) [396]. The success of such reductions is dependent on the mercury ligand X, the pH of the reaction and the solvent employed, best results being obtained in aqueous alkali. To overcome this problem a phase-transfer approach employing sodium borohydride, aqueous sodium hydroxide, methylene chloride and tetraalkylammonium chlorides has recently been introduced [382, 405, 438]. Faster reductions, fewer rearrangements or by-products, and higher yields are evident using this approach.

Several other reagents have been used in these demercuration reactions. Reduction of intramolecular phenoxymercuration products with N-benzyl-1,4-dihydronicotinamide also gives good results, better than simple sodium borohydride reductions [320]. Sodium-mercury amalgam reductions do not appear to offer any advantages over sodium borohydride. While successes have been reported with this reagent [374, 408], elimination to the starting enol appears to be more common [17, 376, 407, 422]. Furthermore, mercurials derived from intramolecular phenoxymercuration of *o*-allylic phenols afford predominantly dialkylmercurials [439, 441]. Successful alkaline hydrazine reductions have been reported [110, 407, 417], but the yields appear to be low and this reaction is also accompanied by ring opening to the enol [129].

Numerous examples of the halogenation of these mercurated cyclic ethers have been reported as seen in Tables 3.6 and 3.7. Apparently both alcohol- and phenol-derived mercurials undergo this reaction smoothly. Relatively little work on the stereochemistry of these reactions has been reported. However, it has been shown that the iodination of bicyclic and tricyclic mercurials gives mixtures of epimers, as one would expect [391–393, 397, 398]. It has also been observed that the 9-oxabicyclo[4.2.1]octylmercurial derived from 5-hydroxycyclooctene affords rearranged bicyclo[3.3.1] products upon iodination and vice versa [397, 398].

One other reaction of some importance is the oxidative demercuration of these mercurials using sodium borohydride and oxygen in DMF (Eq. 181) [375, 423]. Yields of 10–65% have been reported for this reaction and it has found use in the synthesis of a carbohydrate (Eq. 182) [438].

$$C_6H_5 \underset{O}{\diagdown}\!-\!CH_2HgCl \xrightarrow[O_2]{NaBH_4} C_6H_5 \underset{O}{\diagdown}\!-\!CH_2OH \tag{181}$$

$$\tag{182}$$

C. Dienes and Polyenes

A number of dienes and polyenes have been subjected to alkoxymercuration (Table 3.8). Much of what we have learned about the alkoxymercuration of simple alkenes proves applicable to understanding the alkoxymercuration of dienes and polyenes. Our discussion will be divided into the alkoxymercuration of allenes; conjugated dienes and polyenes; non-conjugated acyclic, monocyclic and polycyclic dienes and polyenes; and polyunsaturated alcohols.

Allenes undergo mono- or dialkoxymercuration depending on the substitution pattern of the allene. The first report of this type of reaction appeared in 1949 with the observation that 1,2-hexadiene reacts with mercuric oxide and boron trifluoride etherate in methanol to afford a ketal (Eq. 183) [464]. In the 1950's the first monomercuration products from allenes were isolated, but their structures were not established [500, 501]. More recent work has indicated that allene itself affords a dimercurated ketal most easily isolated as the corresponding ketone (Eq. 184) [456]. 1,2-Butadiene yields a mixture of mercurials (Eq. 185) [456]. More highly substituted allenes give exclusively monomercuration products in which the mercury adds to the central carbon and the ether group predominantly to the more highly sub-

$$H_2C{=}C{=}CH(CH_2)_2CH_3 \longrightarrow \overset{CH_3O}{\underset{}{\diagup}}\!\!\!\overset{}{\underset{}{\diagdown}}\!\!\!\overset{OCH_3}{\underset{}{}}\atop CH_3\overset{}{C}(CH_2)_3CH_3 \tag{183}$$

$$H_2C{=}C{=}CH_2 \xrightarrow[CH_3OH]{Hg(OAc)_2} \overset{CH_3O}{\underset{}{\diagup}}\!\!\!\overset{}{\underset{}{\diagdown}}\!\!\!\overset{OCH_3}{\underset{}{}}\atop AcOHgCH_2\overset{}{C}CH_2HgOAc \xrightarrow{H_2O} AcOHgCH_2\overset{\overset{O}{\|}}{C}CH_2HgOAc \tag{184}$$

$$H_2C{=}C{=}CHCH_3 \longrightarrow AcOHgCH_2\overset{\overset{O}{\|}}{C}CH(HgOAc)CH_3 \; + \; H_2C{=}\overset{\overset{HgOAc}{|}}{C}CH(OCH_3)CH_3 \tag{185}$$
60%   35%

Table 3.8. Alkoxymercuration of Dienes and Polyenes

Diene or polyene	Alcohol	Mercuric salt	Organomercurial(s) (% Yield)	Subsequent reactants	Product(s) (% Yield)	Ref.
$H_2C=C=CH_2$	CH_3OH	$Hg(OAc)_2$	$AcOHgCH_2C(OCH_3)_2CH_2HgOAc$	H_2O	$AcOHgCH_2COCH_2HgOAc$ (95)	456
$H_2C=C=CHCH_3$	CH_3OH	$Hg(OAc)_2$, H_2O (KCl)	$H_2C=C(HgX)CH(OCH_3)CH_3$ X = OAc (60), Cl + $XHgCH_2COCH(HgX)CH_3$ X = OAc (35), Cl	I_2 on dimercurial	$ICH_2COCHICH_3$	456
$H_2C=CHCH=CH_2$	CH_3OH	$Hg(OAc)_2$, NaCl	$ClHgCH_2CH(OCH_3)CH=CH_2$ (91)	—	—	117
	CH_3OH	$Hg(OAc)_2$	—	$H_2C=CXY$ / $NaHB(OCH_3)_3$	$H_2C=CHCH(OCH_3)CH_2CH_2CHXY$ X Y H CN (47) H CO_2CH_3 (34) H $COCH_3$ (22) CH_3 CN (24) Cl CN (60)	326
	CH_3OH	$Hg(OAc)_2$	$AcOHgCH_2CH(OCH_3)CH(OCH_3)CH_2HgOAc$ 4:1 meso (~70) / racemic (~10)	Br_2 or I^-/I_2	$XCH_2CH(OCH_3)CH(OCH_3)CH_2X$ X = Br, I	457
	CH_3CH_2OH	$Hg(OAc)_2$	meso-$AcOHgCH_2CH(OC_2H_5)CH(OC_2H_5)$-$CH_2HgOAc$	KI/I_2	meso-$ICH_2CH(OC_2H_5)CH(OC_2H_5)CH_2I$	458
	$HOCH_2CH_2OH$	HgO/HNO_3	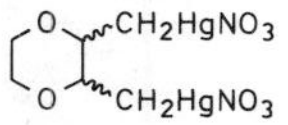 (dioxane ring with CH_2HgNO_3, CH_2HgNO_3) (61)	X^-	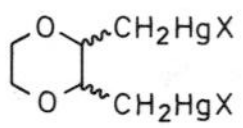 (dioxane ring with CH_2HgX, CH_2HgX) X = I, $PhCO_2$	83
	$HOCH_2CH_2OH$	$Hg(NO_3)_2$ (KI)	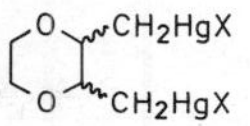 (dioxane ring with CH_2HgX, CH_2HgX) X = NO_3, I	I_2 (on RHgI)	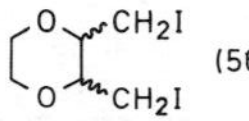 (dioxane ring with CH_2I, CH_2I) (56) cis and trans	459
$NO_3HgCH_2CHOHCHOHCH_2HgNO_3$		$Hg(NO_3)_2$	—	I^-/I_2	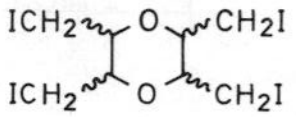 (ICH_2, ICH_2 / CH_2I, CH_2I dioxane) (?)	459

Table 3.8. (continued)

Diene or polyene	Alcohol	Mercuric salt	Organomercurial(s) (% Yield)	Subsequent reactants	Product(s) (% Yield)	Ref.
	ROH	$Hg(OAc)_2$	$AcOHgCH_2CH(OR)CH(OR)CH_2HgOAc$ R = Me (61), Et (54), *n*-Pr, *n*-Bu	X^-	$XHgCH_2CH(OR)CH(OR)CH_2HgX$ R = Me, X = Cl, Br, I, $PhCO_2$, n-$C_7H_{15}CO_2$, n-$C_{17}H_{35}CO_2$, SCN; R = *n*-Pr, X = Cl, $PhCO_2$; R = *i*-Pr, X = Cl; R = *n*-Bu, X = Cl, $PhCO_2$	83
$H_2C{=}C{=}C(CH_3)_2$	CH_3OH	$Hg(OAc)_2$	$H_2C{=}C(HgOAc)C(OCH_3)(CH_3)_2$ (84)	KI/I_2	$H_2C{=}C(I)C(OCH_3)(CH_3)_2$	456
$H_2C{=}C(CH_3)CH{=}CH_2$	CH_3CH_2OH	—	$XHgCH_2C(CH_3)(OC_2H_5)CH(OC_2H_5)CH_2HgX$ X = OAc, Cl, Br, I, CN, CNO, SCN	—	—	52
	ROH	$Hg(OAc)_2$	$AcOHgCH_2C(CH_3)(OR)CH(OR)CH_2HgOAc$ R = Et	X^-	$XHgCH_2C(CH_3)(OR)CH(OR)CH_2HgX$ R = Me, X = Cl, $PhCO_2$; R = Et, X = Cl	83
trans-$H_2C{=}CHCH{=}CHCH_3$	CH_3OH	$Hg(OAc)_2$	—	KCl	$ClHgCH_2CH(OCH_3)CH{=}CHCH_3$	460
	CH_3OH	$Hg(OAc)_2$	—	$H_2C{=}C(Cl)CN$ / $NaHB(OCH_3)_3$	$CH_3CH{=}CHCH(OCH_3)CH_2CH_2CH(Cl)CN$ (59)	326
	CH_3OH	$Hg(NO_3)_2 \cdot H_2O$	—	KCl	$ClHgCH_2CH{=}CHCH(OCH_3)CH_3$ + $ClHgCH_2CH(OCH_3)CH{=}CHCH_3$ 2 : 1 (89 total)	460
$CH_3CH{=}C{=}CHCH_3$	CH_3OH	$Hg(OAc)_2$	$CH_3CH{=}C(HgOAc)CH(OCH_3)CH_3$ *E* (18) and *Z* (75)	—	—	456
	CH_3CH_2OH	$HgCl_2$	$CH_3CH{=}C(HgCl)CH(OC_2H_5)CH_3$	—	—	461
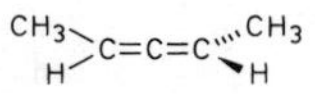	CH_3OH	$Hg(OAc)_2$	—	NaCl	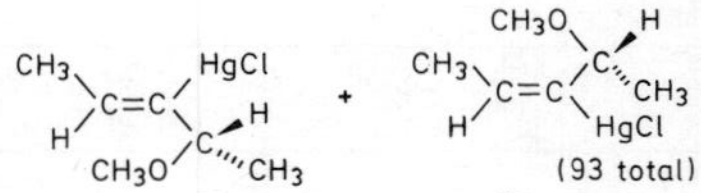 83 : 17 (93 total)	462

Table 3.8. (continued)

Diene or polyene	Alcohol	Mercuric salt	Organomercurial(s) (% Yield)	Subsequent reactants	Product(s) (% Yield)	Ref.
	CH_3OH	$Hg(OAc)_2$	—	$NaCl$	[two isomeric allylic mercurials] (19 % ee) 95 : 5	463
[cyclohexadiene]	CH_3OH	$Hg(OAc)_2$	—	$H_2C=C(Cl)CN/$ $NaHB(OCH_3)_3$	[cyclohexene with $CH_2CH(Cl)CN$ and OCH_3] (27)	326
[4-methylene-1,3-dioxane]	$HOCH_2CH_2OH$	$Hg(OAc)_2/HOAc$	—	KI/I_2	[bis-dioxane with CH_2I groups]	459
$H_2C=C(CH_3)C(CH_3)=CH_2$	CH_3OH	$Hg(OAc)_2$	—	$H_2C=C(Cl)CN/$ $NaHB(OCH_3)_3$	$H_2C=C(CH_3)C(CH_3)(OCH_3)CH_2CH_2CH(Cl)CN$ (24)	326
$CH_3CH=C=C(CH_3)_2$	CH_3OH	$Hg(OAc)_2$	$CH_3(H)C=C(HgOAc)C(CH_3)_2OCH_3$ (72) $+ (CH_3)_2C=C(HgOAc)CH(OCH_3)CH_3$ (18)	—	—	456
$H_2C=C=CH(CH_2)_2CH_3$	CH_3OH	$HgO/BF_3 \cdot Et_2O$	—	—	$CH_3C(OCH_3)_2(CH_2)_3CH_3$	464
$H_2C=CH(CH_2)_2CH=CH_2$	CH_3OH	HgX_2 $(X = Cl, OAc)$	$H_2C=CH(CH_2)_2CH(OCH_3)CH_2HgX$	—	—	154
	CH_3OH	$Hg(OAc)_2$	$AcOHgCH_2CH(OCH_3)(CH_2)_2CH(OCH_3) -$ CH_2HgOAc (52:8)	I^-	$IHgCH_2CH(OCH_3)(CH_2)_2CH(OCH_3)CH_2HgI$	83
	CH_3OH	$Hg(OAc)_2$ (KX)	$XHgCH_2CH(OCH_3)(CH_2)_2CH(OCH_3)CH_2HgX$ $X = OAc, Cl, Br, I$	$NaBH_4$	$CH_3CH(OCH_3)(CH_2)_2CH(OCH_3)CH_3$	154

Table 3.8. (continued)

Diene or polyene	Alcohol	Mercuric salt	Organomercurial(s) (% Yield)	Subsequent reactants	Product(s) (% Yield)	Ref.
$(H_2C=CHCH_2)_2O$	CH_3OH	$Hg(OAc)_2$	—	KI	$IHgCH_2CH(OCH_3)CH_2OCH_2CH(OCH_3)CH_2HgI$ (42)	83
	$(HOCH_2CH_2OCH_2CH_2)_2O$	$Hg(OAc)_2/HClO_4/$ $KClO_4,\ KCl$	—	$NaBH_4/NaOH$	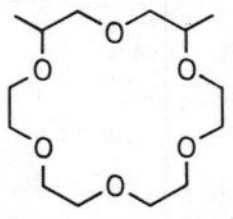 [crown ether structure] (11) 2:1 meso/(±)	465
 [norbornadiene]	CH_3OH	$HgCl_2$	 [CH_3O…HgCl bicyclic structure]	—	—	466
	CH_3OH	$Hg(OAc)_2$	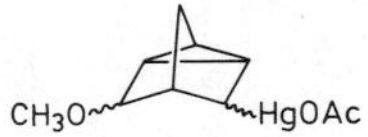 [CH_3O…HgOAc bicyclic structure]	—	—	467
	CH_3CH_2OH	$HgCl_2$	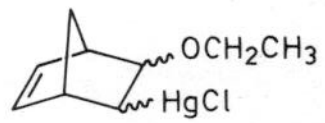 [OCH_2CH_3 / HgCl bicyclic structure] ? (83)	—	—	468
	ROH	$HgCl_2$	 [RO…HgCl bicyclic structure] R = Me, Et	—	—	469
	ROH	$HgCl_2$	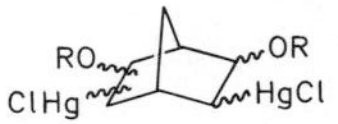 [RO…OR / ClHg…HgCl bicyclic structure] R = Me, Et	—	—	468
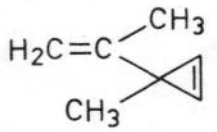 [2-methyl-1-methylenecyclopropene, $H_2C=C(CH_3)$/CH_3]	CH_3OH	$Hg(OAc)_2$	—	Cl^-	cis-$ClHgCH=CHC(OCH_3)(CH_3)C(CH_3)=CH_2$ (~10)	240
$(CH_3)_2C=C=C(CH_3)_2$	CH_3OH	$Hg(OAc)_2$	$(CH_3)_2C=C(HgOAc)C(OCH_3)(CH_3)_2$ (90)	—	—	456
(S)-$(+)$-$CH_3CH_2CH=C=CHCH_2CH_3$	CH_3OH	$Hg(OAc)_2,\ NaCl$	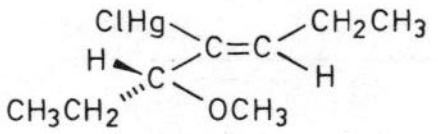 [allene addition structure] (94) (33.1% ee)	—	—	463

Table 3.8. (continued)

Diene or polyene	Alcohol	Mercuric salt	Organomercurial(s) (% Yield)	Subsequent reactants	Product(s) (% Yield)	Ref.
$(R)-(-)-CH_3CH_2CH=C=CHCH_2CH_3$	CH_3OH	$Hg(OAc)_2$, NaCl	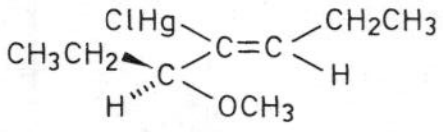 (96) (21.6% ee)	—	—	463
	CH_3OH	$EtHgOAc / BF_3 \cdot Et_2O$	—	—	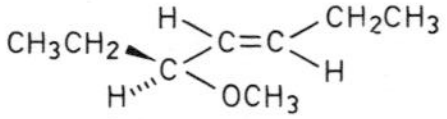 (>90) (25% ee)	463
$(H_2C=CHCH_2)_2NCONH_2$	CH_3OH	$Hg(OAc)_2/HOAc$	—	NaCl	$[ClHgCH_2CH(OCH_3)CH_2]_2NCONH_2$ (60)	178
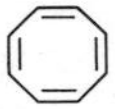	CH_3OH	$Hg(OAc)_2$	—	—	$C_6H_5CH_2CH(OCH_3)_2$ (78)	470
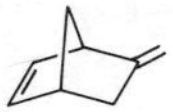	ROH	$Hg(OAc)_2$	—	$NaBH_4$	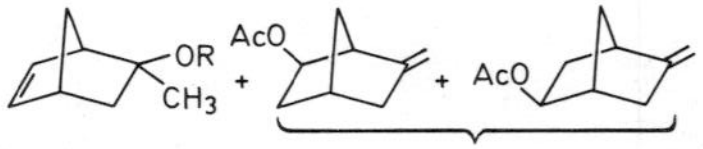 R = Me (59) 100 : 0; Et (83) 95 : 5; n-Pr 65 : 35; i-Pr 36 : 64	471
	CH_3OH	$Hg(OAc)_2$, NaCl	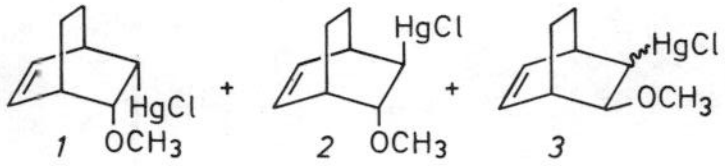 1 : 2 : 3 minor; 1.6 : 1 : minor 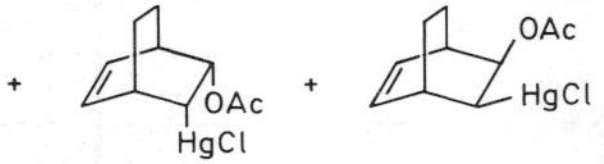 (+ OAc/HgCl minor, minor)	Na/Hg/NaOH	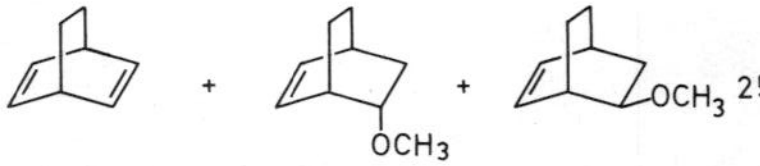 from 1+3 40 : 56.7 : 3.3; from 2+3 22 : 72.3 : 5.7	257
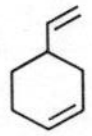	CH_3OH	$Hg(OAc)_2$	—	KCl	$ClHgCH_2CH(OCH_3)-$⟨OCH3/HgCl⟩ 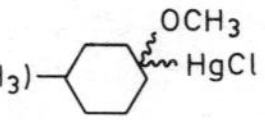	83

Table 3.8. (continued)

Diene or polyene	Alcohol	Mercuric salt	Organomercurial(s) (% Yield)	Subsequent reactants	Product(s) (% Yield)	Ref.
	CH_3OH	—	HgCl, OCH_3	$NaBH_4/NaOH$ (phase transfer)	OCH_3 (98) + (2)	405
	CH_3OH	$Hg(OAc)_2$	OCH_3, HgOAc (100)	$NaBH_4/NaOH$	OCH_3 (70)	472, 473
$CH=CH_2$, $Fe(CO)_3$	CH_3OH	$Hg(OAc)_2$	OCH_3 $-CHCH_2HgOAc$ (4.5), $Fe(CO)_3$	—	—	147
	CH_3OH	$Hg(OAc)_2$, NaCl	**1** OCH_3, HgCl + OCH_3, HgCl + **2** HgCl, OCH_3 ~1.55 : ~1 : + HgCl, OAc + HgCl, OAc : ~0.33	$Na/Hg/NaOH$ (on **1** + **2**)	OCH_3 + OCH_3 17 – 25 : 1	257
$CHCH_3$	CH_3OH	$Hg(OAc)_2$, NaCl	CH_3O, $CHCH_3$, ClHg + OCH_3, $CHCH_3$, ClHg	$NaBH_4$ or $Na(Hg)$	CH_3O, $CHCH_3$ 66 : OCH_3, $CHCH_3$ 34	474
	CH_3OH	$Hg(OAc)_2$	—	$NaBH_4/NaOH$	CH_3O, $CHCH_3$ 66 : OCH_3, $CHCH_3$ 25 : + OCH_3, $CHCH_3$ 9 (49 total)	475

Table 3.8. (continued)

Diene or polyene	Alcohol	Mercuric salt	Organomercurial(s) (% Yield)	Subsequent reactants	Product(s) (% Yield)	Ref.
 (structure)	CH_3OH	$Hg(OAc)_2$	—	$NaBH_4/NaOH$	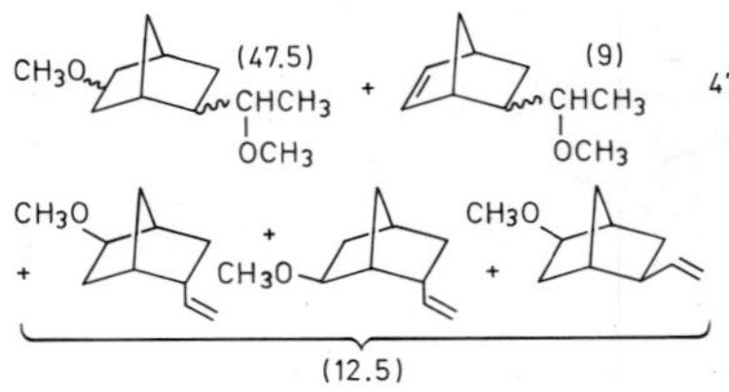 (structures: (47.5) + (9); + (12.5))	476
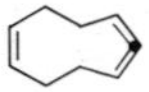 (structure)	CH_3CH_2OH	$HgCl_2$	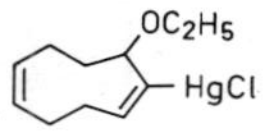 (structure, OC_2H_5 / $HgCl$)	—	—	461
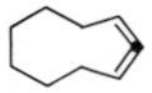 (structure)	CH_3OH	$HgCl_2$	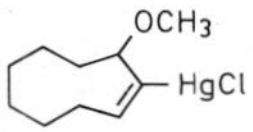 (structure, OCH_3 / $HgCl$)	$NaBH_4$	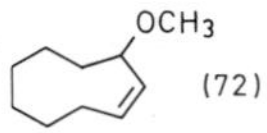 (structure, OCH_3) (72)	477
	CH_3OH	—	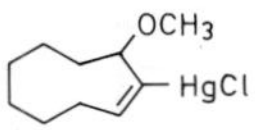 (structure, OCH_3 / $HgCl$)	Na/NH_3	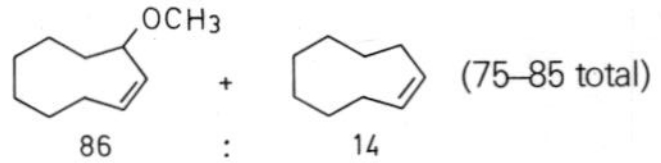 (structures, OCH_3) 86 : 14 (75–85 total)	478
	CH_3OH	$HgSO_4/H_2SO_4$	—	—	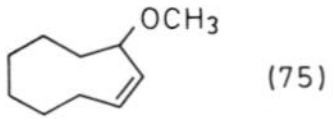 (structure, OCH_3) (75)	479
	CH_3CH_2OH	$Hg(OAc)_2/NaCl$ or $HgCl_2$	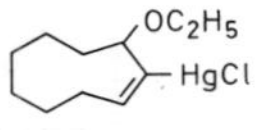 (structure, OC_2H_5 / $HgCl$)	—	—	480
	CH_3CH_2OH	$HgCl_2$	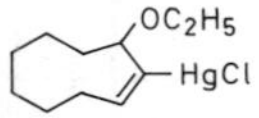 (structure, OC_2H_5 / $HgCl$)	$Na/NH_3/EtOH$	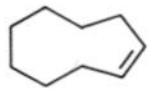 (structure)	481, 482
	CH_3CH_2OH	$HgCl_2$	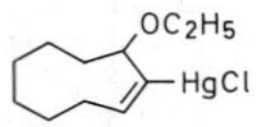 (structure, OC_2H_5 / $HgCl$)	Na/NH_3	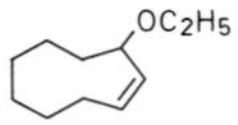 (structure, OC_2H_5)	461

Table 3.8. (continued)

Diene or polyene	Alcohol	Mercuric salt	Organomercurial(s) (% Yield)	Subsequent reactants	Product(s) (% Yield)	Ref.
	CH_3CH_2OH	$HgSO_4/BF_3$	—	—	(structure, OC_2H_5) (74)	479
	CH_3CH_2OH	HgX_2 (X = NO_3, ClO_4, SO_4 or $OAc/BF_3\cdot Et_2O$) or $RHgOAc/BF_3\cdot Et_2O$ (R = Et, Ph)	—	—	(structure, OC_2H_5)	480
	ROH	$HgSO_4/H_2SO_4$	—	—	(structure, OR) R = Me, Et	461
	ROH	HgO/BF_3	—	—	(structure, OR) R = Me(69), Et(80), i-Pr(81), t-Bu(63), $PhCH_2$(85)	479
(structure, $(CH_2)_5$ diene)	CH_3OH	$Hg(OAc)_2$, NaCl	(organomercurial structure) (100) (8.4% ee)	—	—	463
(structure, $(CH_2)_5$ diene)	ROH	$HgO/BF_3\cdot Et_2O$	—	—	(structure, OR) R = Me, Et (optically active)	479
(structure)	CH_3OH	$Hg(OAc)_2$, NaCl or KI	(structure, OCH_3, HgX) X = Cl(82), I(80)	$NaBH_4$ [or Na/NH_3] on RHgI	(structure, OCH_3) [or (structure) (65–85)]	483
(structure)	CH_3OH	$Hg(OAc)_2$	(structure, HgOAc, OCH_3) (75)	—	—	484

Table 3.8. (continued)

Diene or polyene	Alcohol	Mercuric salt	Organomercurial(s) (% Yield)	Subsequent reactants	Product(s) (% Yield)	Ref.
	CH_3OH	$Hg(OAc)_2$, NaCl or KI	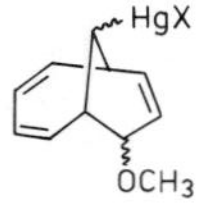 X = Cl (77), I (75)	$NaBH_4$ [or Na/NH_3] on RHgI	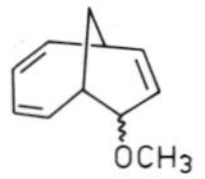 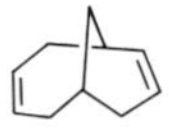 (65–85)	483
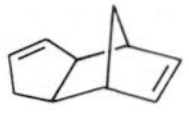	CH_3OH	HgX_2	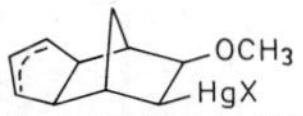 X = Cl (50), OAc (50)	$Na/Hg/H_2O$ (on RHgOAc)	 (50.2)	485
	CH_3OH	$HgCl_2$	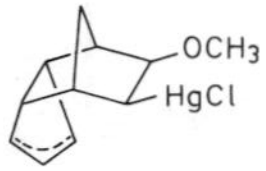 (79)	$Na/Hg/H_2O$ (D_2O)	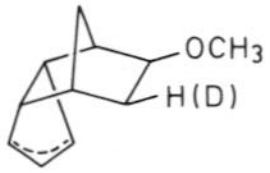 (D, 36.4)	485
$(H_2C=CHCH_2)_2$ 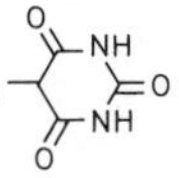	CH_3OH	$Hg(OAc)_2/HOAc$	$[AcOHgCH_2CH(OCH_3)CH_2]_2$ 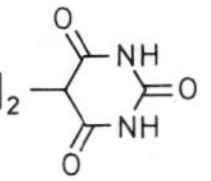	—	—	131
$H_2C=CHCH_2$, 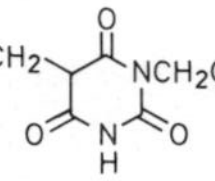 $NCH_2CH=CH_2$	CH_3OH	$Hg(OAc)_2/HOAc$	$AcOHgCH_2CH(OCH_3)CH_2$, 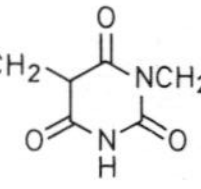 $NCH_2CH(OCH_3)CH_2HgOAc$	—	—	131
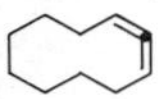	CH_3OH	$HgCl_2$	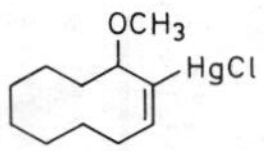	$NaBH_4$	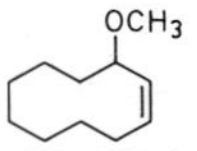 (54) 78 : 22 cis/trans	477
	CH_3OH	—	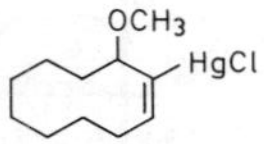	Na/NH_3	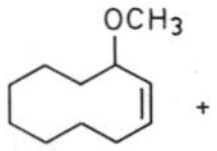 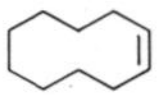 (75–85 total) 85 : 15	478
	CH_3CH_2OH	$HgCl_2$	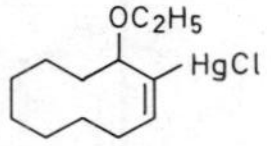 (72)	$Na/NH_3/EtOH$	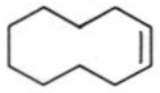	481, 482

Table 3.8. (continued)

Diene or polyene	Alcohol	Mercuric salt	Organomercurial(s) (% Yield)	Subsequent reactants	Product(s) (% Yield)	Ref.
$(CH_3)_2C=C=$ cyclopropane $(CH_3)_2$ with CH_3 groups	CH_3OH	$Hg(OAc)_2$, NaCl	ClHg–C= with $(CH_3)_2C$, OCH_3 (cyclopropyl CH_3 groups)	Br_2	Br–C= with $(CH_3)_2C$, OCH_3 (cyclopropyl CH_3 groups)	486
cyclic $(CH_2)_8$ with $CH=C=CH$ (cyclodeca)	CH_3OH	$HgCl_2$	$trans$-$(CH_2)_8$ ring, $CHOCH_3$, $C–HgCl$, CH	$NaBH_4$	$(CH_2)_8$ ring, $CHOCH_3$, CH, CH (80); 15 : 85 cis/trans	477, 487
	CH_3OH	—	$trans$-$(CH_2)_8$ ring, $CHOCH_3$, $C–HgCl$, CH	Na/NH_3	$trans$-$(CH_2)_8$ ring $CHOCH_3$ + $(CH_2)_8$ ring CH_2, CH (75–85 total); cis : trans, 74 : 5 : 21	478
chlorinated bicyclic (aldrin-type, Cl...)	ROH	$HgCl_2$	chlorinated bicyclic with OCH_3, $HgCl$	—	—	488
	ROH	$Hg(OAc)_2$	chlorinated bicyclic with OR, $HgOAc$; R = Me, Et	—	—	489
$trans$-$C_6H_5CH=CHCO_2CH_2CH=CH_2$	CH_3OH	$Hg(OAc)_2$	—	NaCl	$C_6H_5CH(OCH_3)CH(HgCl)CO_2CH_2CH(OCH_3)CH_2HgCl$	162
$H_2C=CHCH_2O$–triazine–$OCH_2CH=CH_2$ with $OCH_2CH=CH_2$	ROH	$Hg(OAc)_2$	—	NaOH	$HOHgCH_2CH(OR)CH_2O$–triazine–$OCH_2CH(OR)CH_2HgOH$ with $OCH_2CH(OR)CH_2HgOH$; R = Me (97.5), Et (51)	135

Table 3.8. (continued)

Diene or polyene	Alcohol	Mercuric salt	Organomercurial(s) (% Yield)	Subsequent reactants	Product(s) (% Yield)	Ref.
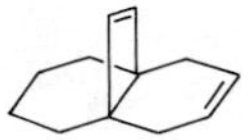	CH_3OH	$Hg(OAc)_2$	—	$NaBH_4/NaOH$	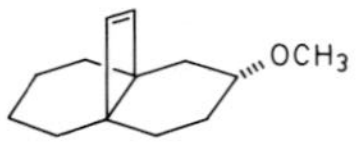 (64)	422
	CH_3OH	$Hg(OAc)_2$	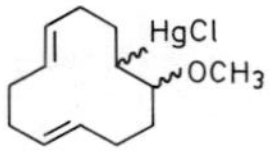 (83.5)	—	—	490
	CH_3OH	$Hg(OAc)_2$	—	NaX	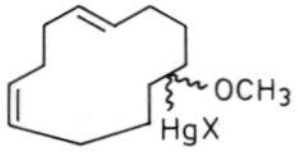 X = Cl(56), Br	247, 490
	ROH	$Hg(OAc)_2$	—	NaCl	$CH_2CH=CHCH_2CH_2CHOR$ $CH_2CH=CHCH_2CH_2CHHgCl$ R = Me, Et	247
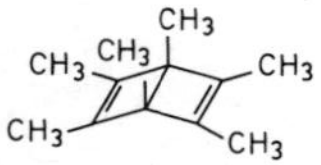	CH_3OH	$Hg(OAc)_2$	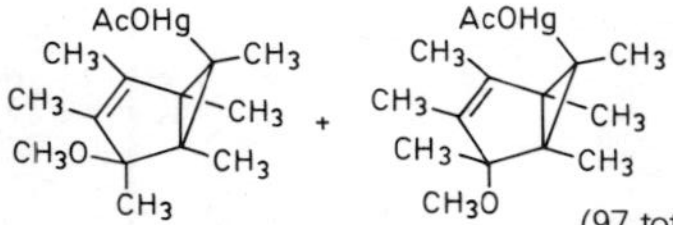 80 : 20 (97 total)	$NaBH_4/NaOH$	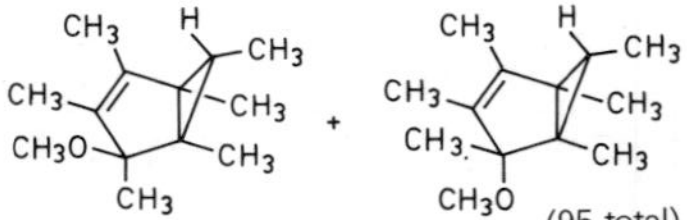 4 : 1 (95 total)	491
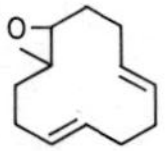 (stereochemistry ?)	CH_3OH	$Hg(OAc)_2$	—	NaCl	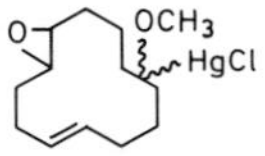 (stereochemistry ?)	247
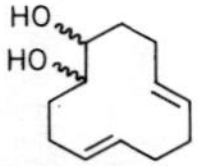 (stereochemistry ?)	CH_3OH	$Hg(OAc)_2$	—	NaCl	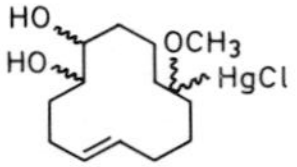 (stereochemistry ?)	247

Table 3.8. (continued)

Diene or polyene	Alcohol	Mercuric salt	Organomercurial(s) (% Yield)	Subsequent reactants	Product(s) (% Yield)	Ref.
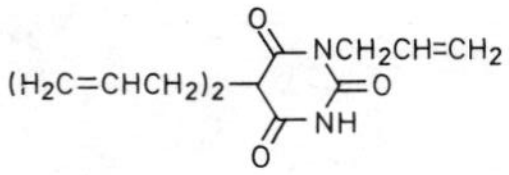 $(H_2C=CHCH_2)_2$ barbituric acid with $NCH_2CH=CH_2$	CH_3OH	$Hg(OAc)_2/HOAc$	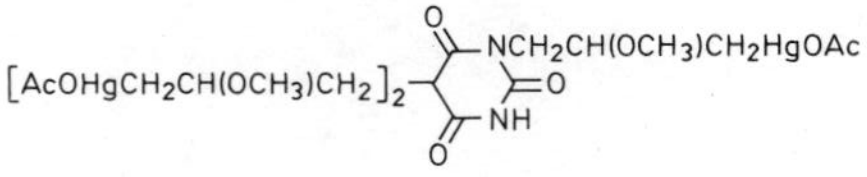 $[AcOHgCH_2CH(OCH_3)CH_2]_2$ barbituric acid with $NCH_2CH(OCH_3)CH_2HgOAc$	—	—	131
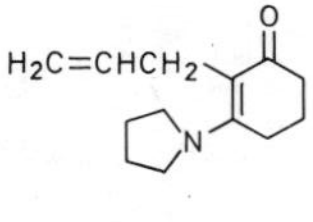 $H_2C=CHCH_2$-cyclohexenone-pyrrolidine	—	$Hg(OAc)_2$	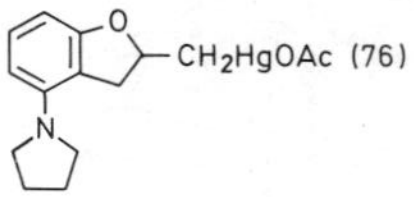 dihydrobenzofuran-pyrrolidine with CH_2HgOAc (76)	$NaBH_4/NaOH$	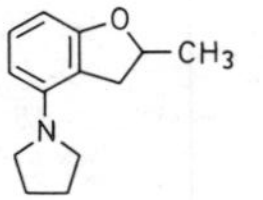 dihydrobenzofuran-pyrrolidine with CH_3 (42)	455
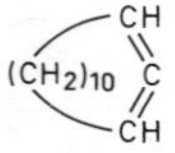 $(CH_2)_{10}$ cyclic diene	CH_3OH	$HgCl_2$	$trans$-$(CH_2)_{10}$ ring, $CHOCH_3$ / C—$HgCl$ 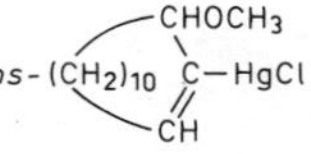	$NaBH_4$	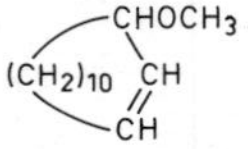 $(CH_2)_{10}$ ring, $CHOCH_3$ / CH (76) 15 : 85 cis/trans	477, 487
	CH_3OH	—	$trans$-$(CH_2)_{10}$ ring, $CHOCH_3$ / C—$HgCl$ 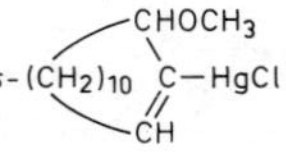	Na/NH_3	$trans$-$(CH_2)_{10}$ ring, $CHOCH_3$ / CH + $(CH_2)_{10}$ ring, CH_2 / CH cis trans (75–85 total) 40 : 16 : 44 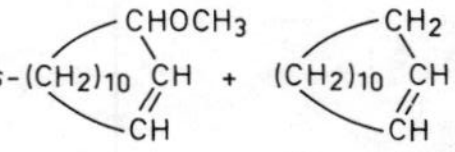	478
	CH_3OH	$HgSO_4/H_2SO_4$	—	—	$trans$-$(CH_2)_{10}$ ring, $CHOCH_3$ / CH (65) 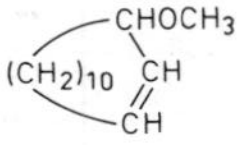	479
	CH_3CH_2OH	$HgCl_2$	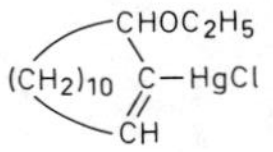 $(CH_2)_{10}$ ring, $CHOC_2H_5$ / C—$HgCl$ (68)	$Na/NH_3/EtOH$	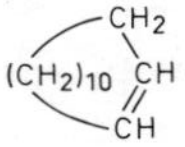 $(CH_2)_{10}$ ring, CH_2 / CH	482
	CH_3CH_2OH	$HgO/BF_3 \cdot Et_2O$	—	—	$trans$-$(CH_2)_{10}$ ring, $CHOC_2H_5$ / CH (78) 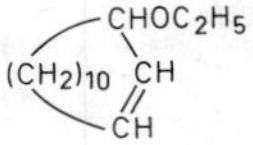	479

Table 3.8. (continued)

Diene or polyene	Alcohol	Mercuric salt	Organomercurial(s) (% Yield)	Subsequent reactants	Product(s) (% Yield)	Ref.
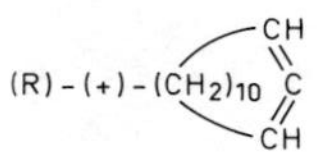 $(R)-(+)-(CH_2)_{10}$ [structure]	CH_3OH	$Hg(OAc)_2$, $NaCl$	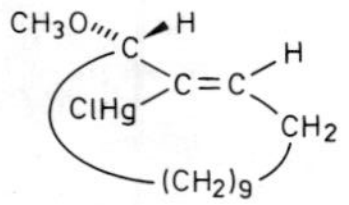 [structure]	—	—	463
o- or p-$(H_2C=CHCH_2)_2NCOC_6H_4CO_2H$	CH_3OH	$Hg(OAc)_2$	o- or p-$[AcOHgCH_2CH(OCH_3)CH_2]_2NCOC_6H_4CO_2H$	—	—	163
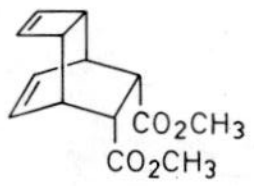 [structure] CO_2CH_3 CO_2CH_3	CH_3OH	$Hg(OAc)_2$, $NaCl$	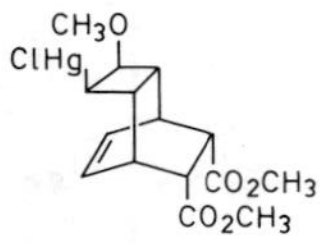 [structure] CH_3O $ClHg$ CO_2CH_3 CO_2CH_3 (95)	—	—	492
	CH_3CH_2OH	$Hg(OAc)_2$, $NaCl$	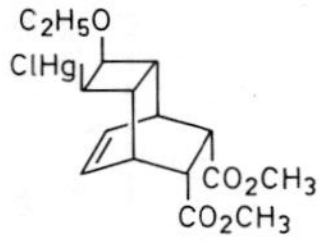 [structure] C_2H_5O $ClHg$ CO_2CH_3 CO_2CH_3	$NaBH_4$	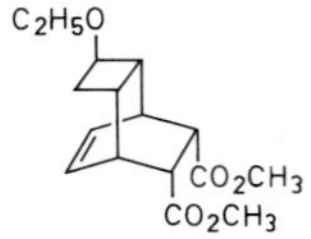 [structure] C_2H_5O CO_2CH_3 CO_2CH_3	493
$trans$- [structure] $-CH=CH-$ [structure] 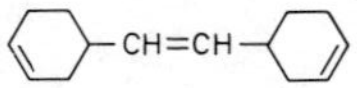	CH_3OH	$Hg(OAc)_2$, $NaCl$	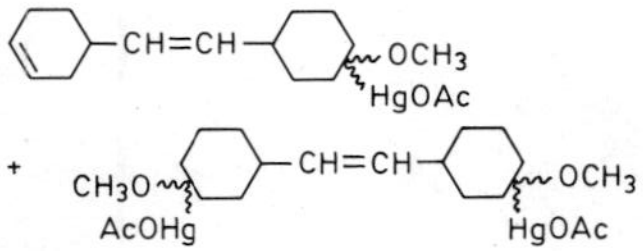 [structure] $-CH=CH-$ OCH_3 $HgOAc$ + CH_3O $AcOHg$ $-CH=CH-$ OCH_3 $HgOAc$	—	—	494
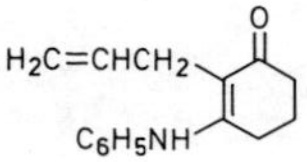 $H_2C=CHCH_2$ [structure] C_6H_5NH	—	$Hg(OAc)_2$	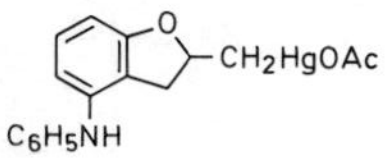 [structure] CH_2HgOAc (71) C_6H_5NH	$NaBH_4/NaOH$	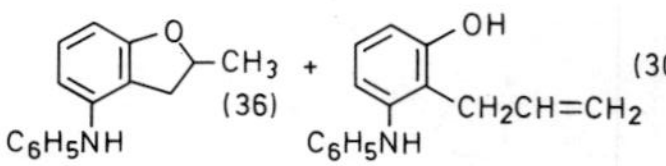 [structure] CH_3 C_6H_5NH (36) + [structure] OH $CH_2CH=CH_2$ C_6H_5NH (30)	455
o-$(H_2C=CHCH_2)_2-$ $NCOC_6H_4OCH_2CO_2H$	CH_3OH	$Hg(OAc)_2$, $NaCl$	o-$[ClHgCH_2CH(OCH_3)CH_2]_2-$ $NCOC_6H_4OCH_2CO_2H$ (86)	—	—	178, 495

Table 3.8. (continued)

Diene or polyene	Alcohol	Mercuric salt	Organomercurial(s) (% Yield)	Subsequent reactants	Product(s) (% Yield)	Ref.
p-$(H_2C{=}CHCH_2)_2$-$NCOC_6H_4OCH_2CO_2H$	CH_3OH	$Hg(OAc)_2/HOAc$, NaCl	p-$[ClHgCH_2CH(OCH_3)CH_2]$-$N(CH_2CH{=}CH_2)COC_6H_4OCH_2CO_2H$ (39)	—	—	178
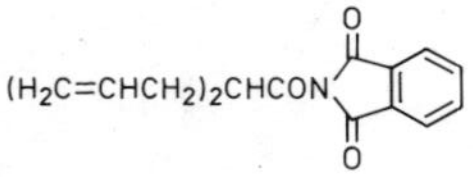 $(H_2C{=}CHCH_2)_2CHCON$⟨phthalimido⟩	CH_3OH	$Hg(OAc)_2$	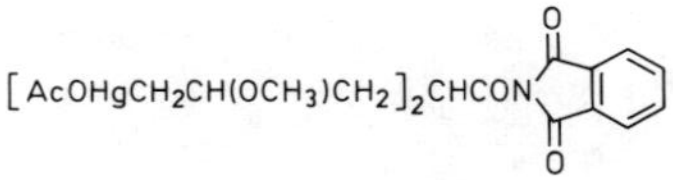 $[AcOHgCH_2CH(OCH_3)CH_2]_2CHCON$⟨phthalimido⟩	—	—	163
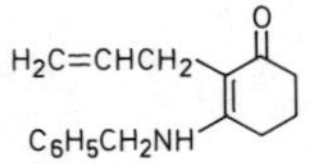 $H_2C{=}CHCH_2$-⟨3-(benzylamino)cyclohex-2-enone⟩, $C_6H_5CH_2NH$	—	$Hg(OAc)_2$	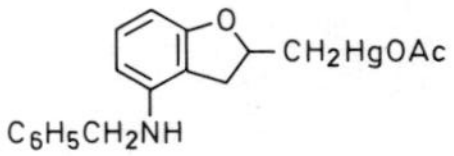 ⟨2-(AcOHgmethyl)-4-(benzylamino)-2,3-dihydrobenzofuran⟩–CH_2HgOAc (75), $C_6H_5CH_2NH$	$NaBH_4/NaOH$	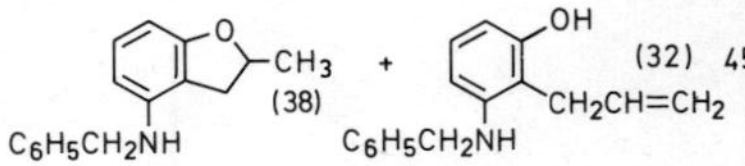 ⟨2-methyl-4-(benzylamino)-2,3-dihydrobenzofuran⟩–CH_3 (38), $C_6H_5CH_2NH$ + ⟨2-allyl-3-(benzylamino)phenol⟩ OH, $CH_2CH{=}CH_2$, $C_6H_5CH_2NH$ (32)	455
$(C_6H_5CH{=}CH)_2CO$	ROH	$Hg(OAc)_2$	$[C_6H_5CH(OR)CH(HgOAc)]_2CO$ R = Me, Et	—	—	197
p-$(CH_3)_2C{=}CH(CH_2)_2$-$C(CH_3){=}CHCH_2OC_6H_4COC_2H_5$	ROH	$Hg(OAc)_2$	—	$NaBH_4/NaOH$	p-$(CH_3)_2C(OR)(CH_2)_3C(CH_3){=}CHCH_2$-$OC_6H_4COC_2H_5$ R = Me (63), Et (42), n-Pr (37), i-Pr (25), ⟨cyclopropyl⟩–CH_2 (10), $C_6H_5CH_2$ (23), F_3CCH_2 (17), H_2C⟨epoxide⟩$CHCH_2$ (7), Cl_3CCH_2 (15), $BrCH_2CH_2$ (26), $ClCH_2CH_2$ (30), $NCCH_2CH_2$ (11), $HOCH_2CH_2$ (29), $CH_3OCH_2CH_2$ (26), $C_2H_5OCH_2CH_2$ (28) + p-$(CH_3)_2C(OR)(CH_2)_3C(CH_3)(OR)(CH_2)_2$-$OC_6H_4COC_2H_5$ R = Me, Et	203
$CH_3(CH_2CH{=}CH)_3(CH_2)_7CO_2CH_3$	CH_3OH	$Hg(OAc)_2$	$CH_3(CH_2CH$⟨AcOHg, OCH₃⟩$CH)_3(CH_2)_7CO_2CH_3$	—	—	496
cis,cis-$CH_3(CH_2)_4CH{=}CH(CH_2)_2$-$CH{=}CH(CH_2)_6CO_2CH_3$	CH_3OH	$Hg(OAc)_2$	—	$NaBH_4$	various dimethoxy esters (96)	73

Table 3.8. (continued)

Diene or polyene	Alcohol	Mercuric salt	Organomercurial(s) (% Yield)	Subsequent reactants	Product(s) (% Yield)	Ref.
9-*cis*, 11-*trans*-$CH_3(CH_2)_5$-$CH=CHCH=CH(CH_2)_7CO_2CH_3$	CH_3OH	$Hg(OAc)_2$	—	$NaBH_4$	monomethoxy (30) + $CH_3(CH_2)_5$-$CH(OCH_3)(CH_2)_2CH(OCH_3)(CH_2)_7CO_2CH_3$ (8) + 10, 12- and 9, 11- dimethoxy (62)	73
cis, *cis*-$CH_3(CH_2)_4CH=CHCH_2$-$CH=CH(CH_2)_7CO_2CH_3$	CH_3OH	$Hg(OAc)_2$	$CH_3(CH_2)_4CH(CH_3O)(HgOAc)CHCH_2CH(CH_3O)(HgOAc)CH(CH_2)_7CO_2CH_3$	$NaBH_4$	$CH_3(CH_2)_5CH(OCH_3)CH_2CH(OCH_3)(CH_2)_8$-$CO_2CH_3$ (11) + other dimethoxy (79) + monomethoxy (7)	73, 496
$H_2C=CH(CH_2)_8CO(CH_2)_8CH=CH_2$	ROH	$Hg(OAc)_2$	$AcOHgCH_2CH(OR)(CH_2)_8CO(CH_2)_8CH=CH_2$ R = Me, *n*-Bu (?)	Cl^-	$ClHgCH_2CH(O$-*n*-$Bu)(CH_2)_8CO(CH_2)_8CH=CH_2$	497
(iron porphyrin structure)	ROH	$Hg(O_2CCF_3)_2$	—	$NaBH_4$/NaOH	(iron porphyrin product structure) R = Me, *n*-Bu	498
(tetraphenyl di-tert-butyl bicyclic structure)	ROH	$Hg(OAc)_2$ [or Hg_2Cl_2]	—	—	(product structure) R = Me (68) [63], Et (31), Ph (46)	499

stituted terminus of the allene (Eq. 186). Note that alkoxymercuration of the cyclopropyl allene leaves the cyclopropane ring untouched, unlike many other electrophilic additions to this compound. 2-Methyl-2,3-pentadiene gives a mixture of regioisomers (Eq. 187) [456], while 2,3-pentadiene [456, 462, 463] and 3,4-heptadiene [463] (Eq. 188) afford stereoisomeric mixtures. The

$$\text{(186)}$$

R^1	R^2	R^3	R^4	X	%Yield	Ref.
H	H	CH_3	CH_3	OAc	84	456
C_2H_5	H	H	C_2H_5	OAc	>90	463
CH_3	CH_3	CH_3	CH_3	OAc	90	456
$(CH_3)_2C - C(CH_3)_2$		CH_3	CH_3	Cl	—	486

$$\text{(187)}$$

72 %   18 %

$$\text{(188)}$$

$R = CH_3$	81–86	:	14–19
$R = C_2H_5$	19	:	1

methoxymercuration of optically active 2,3-pentadiene and 3,4-heptadiene at $-78\ ^\circ C$ proceeds by high, perhaps complete, trans addition at least for the major isomer [462, 463]. The reaction of (R)-(−)-3,4-heptadiene, ethylmercuric acetate and boron trifluoride etherate in methanol affords the corresponding optically active allylic ether directly (Eq. 189) [463]. Finally, it has been noted that tetramethylallene is approximately 10^4 times more reactive than tetramethylethylene.

$$\text{(189)}$$

Numerous cyclic allenes form monomercurials using either mercuric chloride or mercuric acetate in an alcohol solvent [461, 477, 480–482, 487]. The nine- and ten-membered ring allenes give vinylmercurials in which the mercury and vinyl hydrogen are cis, while the eleven- and thirteen-membered

302

$$(190)$$

ring allenes give trans products (Eq. 190) [463, 477]. Attack of the mercury electrophile on the hydrogen-containing side of the allene affords the cis product, while attack on the side of the ring yields the trans mercurial. Methoxymercuration of the chiral nine- and thirteen-membered ring allenes again proceeds with high, if not complete, trans stereospecificity [463]. If the acidity of the reaction media is increased by using mercuric sulfate/sulfuric acid or a variety of organomercuric or mercuric salts plus boron trifluoride etherate, the cyclic allyl ethers minus mercury are obtained directly (Eq. 191) [461, 463, 479, 480]. These ethers retain the stereochemistry of the correspond-

$$(191)$$

$$(192)$$

ing mercurials and can be obtained optically active if chiral allenes are employed [463, 479, 480]. The isolation of these optically active products rules out an allylic cation as the sole intermediate and suggests the presence of a mercurinium ion-like intermediate.

Both sodium borohydride and sodium in liquid ammonia have been used to demercurate these allene-derived mercurials. Using sodium borohydride the cyclic mercurials are reportedly reduced to the corresponding allylic ethers with some loss of stereospecificity [477]. A phase transfer approach gives higher yields of ether, but the stereochemistry of demercuration has not been established [405]. The earlier mentioned cyclopropyl-containing vinylmercurial (Eq. 186) affords only the corresponding divinylmercurial upon treatment with sodium borohydride [486]. With this compound, lithium aluminum hydride produces only the starting allene. Bromination, however, proceeds as expected. Sodium in liquid ammonia reacts with the mercurials derived from cyclic allenes to give either the allylic ethers [461] with some loss of double bond stereochemistry [478], or the corresponding simple cyclic olefins in which the ether group is entirely removed [482]. Finally, iodination of one acyclic allene-derived product has been reported (Eq. 192) [456].

Conjugated dienes can be mono- or dimercurated. 1,3-Butadiene and 1,3-pentadiene give exclusively the 1,2-adduct upon treatment with mercuric acetate in methanol (Eq. 193) [117, 460]. These mercurials, as well as the 1,2-adducts from 2,3-dimethyl-1,3-butadiene and 1,3-cyclohexadiene, react with electron-deficient olefins and sodium trimethoxyborohydride to give the expected alkene addition products (Eq. 194) [326]. If the methoxymercuration of 1,3-pentadiene is carried out using mercuric nitrate, a 2:1 mixture of 1,4- and 1,2-adducts is obtained instead (Eq. 195) [460]. A unique approach to the monomethoxymercuration of 1,3,5-hexatriene involves the protection of two of the double bonds as an iron tricarbonyl complex and subsequent mercuration (Eq. 196) [147]. Unfortunately, the yield of mercurial is very low and not synthetically useful at this juncture. The methoxymercuration of cyclooctatetraene has also been examined, but only rearranged product is observed (Eq. 197) [470].

$$H_2C=CHCH=CHR \longrightarrow AcOHgCH_2\overset{\displaystyle OCH_3}{\overset{|}{C}}HCH=CHR \tag{193}$$

$$H_2C=CHCH=CH_2 \xrightarrow[CH_3OH]{Hg(OAc)_2} \xrightarrow[NaHB(OCH_3)_3]{H_2C=CHCN} H_2C=CH\overset{\displaystyle OCH_3}{\overset{|}{C}}HCH_2CH_2CH_2CN \tag{194}$$

$$H_2C=CHCH=CHCH_3 \longrightarrow ClHgCH_2CH=\overset{\displaystyle OCH_3}{\overset{|}{C}}HCHCH_3 \; + \; ClHgCH_2\overset{\displaystyle OCH_3}{\overset{|}{C}}HCH=CHCH_3 \tag{195}$$

$$2 \qquad : \qquad 1$$

$$H_2C=CHCH=CHCH=CH_2 \longrightarrow \underset{Fe(CO)_3}{\diagup\!\!\!\!\diagdown}\!\!-CH=CH_2 \longrightarrow \underset{Fe(CO)_3}{\diagup\!\!\!\!\diagdown}\!\!-\overset{\displaystyle OCH_3}{\overset{|}{C}}HCH_2HgOAc \tag{196}$$

$$4.5\%$$

$$\bigodot \xrightarrow[CH_3OH]{Hg(OAc)_2} C_6H_5CH_2CH(OCH_3)_2 \tag{197}$$

The dialkoxymercuration of conjugated dienes has also been accomplished. 1,3-Butadiene reportedly gives primarily the meso adduct from two 1,2-additions (Eq. 198) [457, 458]. However, significant quantities of the

$$H_2C=CHCH=CH_2 \xrightarrow[ROH]{Hg(OAc)_2} AcOHgCH_2\overset{\displaystyle RO\;\;OR}{\overset{|\;\;\;|}{C}}HCHCH_2HgOAc \tag{198}$$

$$R=CH_3, C_2H_5$$

$$H_2C=CHCH=CH_2 \xrightarrow[HOCH_2CH_2OH]{Hg(NO_3)_2} \overset{O \diagup CH_2HgNO_3}{\underset{O \diagdown CH_2HgNO_3}{\bigcirc}} \tag{199}$$

racemate are also obtained. Halogenation affords the two halides as expected. Employing diols in this reaction provides a unique approach to dioxanes (Eqs. 199–201) [459, 502].

$$H_2C=CHCH=CH_2 \xrightarrow[H_2O]{Hg(NO_3)_2} NO_3HgCH_2\overset{\underset{|}{OH}}{CH}\overset{\underset{|}{OH}}{CH}CH_2HgNO_3 \xrightarrow[C_4H_6]{Hg(NO_3)_2}$$

$$\text{(200)}$$

$$\text{(201)}$$

Acyclic non-conjugated dienes can be mono- or dimercurated. By employing modest amounts of the mercuric salt, mono adducts have been isolated from the following compounds (Eqs. 202, 203).

$$H_2C=CH(CH_2)_2CH=CH_2 \longrightarrow XHgCH_2\overset{\underset{|}{OCH_3}}{CH}(CH_2)_2CH=CH_2 \qquad \text{(202) [154]}$$

$$(H_2C=CHCH_2)_2N\!\!-\!\!\langle\ \rangle\!\!-\!\!OCH_2CO_2H \longrightarrow \qquad \text{(203) [178]}$$

$$(H_2C=CHCH_2)_2O \ + \ HOCH_2(CH_2OCH_2)_3CH_2OH \longrightarrow \longrightarrow \qquad \text{(204)}$$

A number of acyclic dienes and polyenes, including simple and functionally substituted dienes, fatty acid esters [73, 496], protohemin IX [498] and divinylbenzene polymers [503], have been successfully di- or polymercurated. One interesting example is the above crown ether synthesis (Eq. 204) [465]. The rate constant for the first methoxymercuration of methyl linoleate has been determined, as well as that for the second methoxymercuration [496]. Unfortunately, it has proven difficult to separate fatty acid ester dienes from polyenes by methoxymercuration and subsequent column chromatography [496].

The monomercuration of 1,5-cyclooctadiene [473], and *trans, trans, trans*- and *cis, trans, trans*-1,5,9-cyclododecatriene and certain derivatives [49, 247] have been accomplished using mercuric acetate in methanol. Neither the stereo- nor regiochemistry of addition to these compounds has been established, although it appears that the addition to *cis, trans, trans*-1,5,9-cyclododecatriene occurs at one of the trans double bonds [490]. The methoxymercuration of *trans*-bis(4-cyclohexenyl)ethylene using 2:1 mercuric acetate to triene gives a 10% yield of the six-membered ring monoadducts and major amounts of six-membered ring diadducts [494]. Finally, in a rather remarkable reaction, 2-allyl-3-aminocyclohexenones appear to undergo oxidation to the corresponding phenol and cyclization to the tetrahydrofuran even in the presence of an aniline moiety which might be expected to undergo aromatic substitution or at least intramolecular aminomercuration (Eq. 205) [455]. However, competitive aminomercuration is observed in some cases to be discussed later in Chapter VI.

$$\tag{205}$$

$$\tag{206}$$

$$\tag{207}$$

The mono- and polymercuration of bi- and polycyclic dienes has been examined by a number of different groups. Norbornadiene has been reported to give a simple norbornenylmercurial after three days at 60 °C (Eq. 206) [468]. Bis adducts are also reported. However, other workers report that nortricyclyl products are formed using either mercuric acetate or mercuric chloride in methanol or ethanol at room temperature (Eq. 207) [466, 467, 469]. The stereochemistry of all these mercurials has yet to be established. The methoxymercuration-demercuration of *endo*-5-vinylnorbornene gives a mixture of products containing 47.5% dimethyl ether, 12.5% of methanol addition only to the bicyclic double bond, and 9% addition only to the vinyl group [476]. Alkoxymercuration-demercuration of 5-methylenenorbornene gives exclusive alcohol addition to the exocyclic double bond, alongside some endocyclic acetic acid adducts (Eq. 208) [471]. The ratio varies with the alcohol. On the other hand, 5-ethylidenenorbornene undergoes methoxymercuration exclusively on the endocyclic double bond to yield a mixture of bi- and tricyclic products (Eq. 209) [475]. The alkoxymercuration of *exo*- and *endo*-dicyclopentadiene affords norbornyl mono adducts of unknown regio- or stereochemistry (Eq. 210) [485, 504]. The exo compound reacts faster than the endo diene. Sodium amalgam reduction in water or D_2O affords cis exo ethers plus a substantial amount of the starting dienes. Cis exo adducts analogous to these have also been prepared from aldrin [488] and related polychlorinated dienes [489] (Eq. 211).

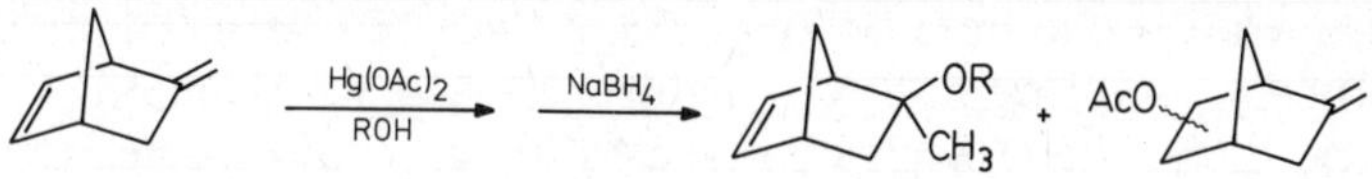

$$R = Me \quad\quad 100 \quad : \quad 0$$
$$Et \quad\quad 95 \quad : \quad 5$$
$$n\text{-Pr} \quad\quad 65 \quad : \quad 35$$
$$i\text{-Pr} \quad\quad 36 \quad : \quad 64$$

(208)

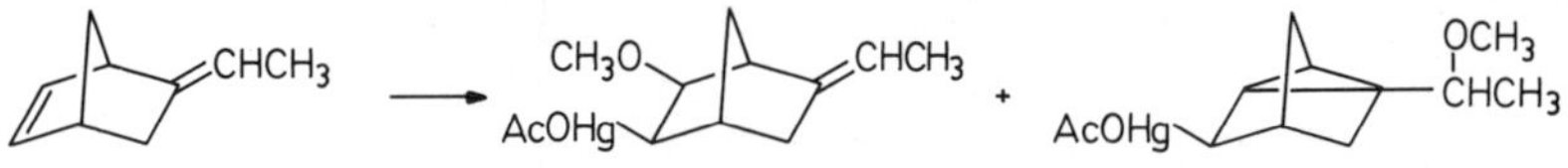

$$66 \quad : \quad 34$$

(209)

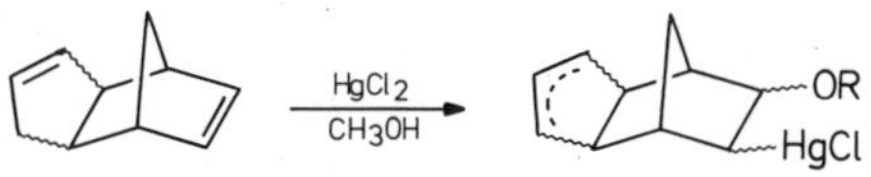

(210)

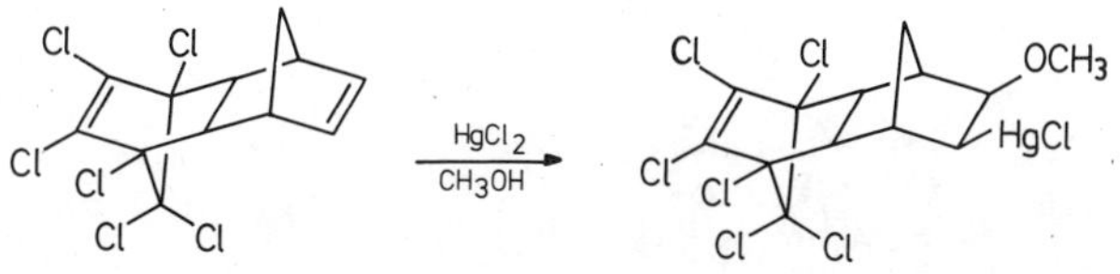

(211)

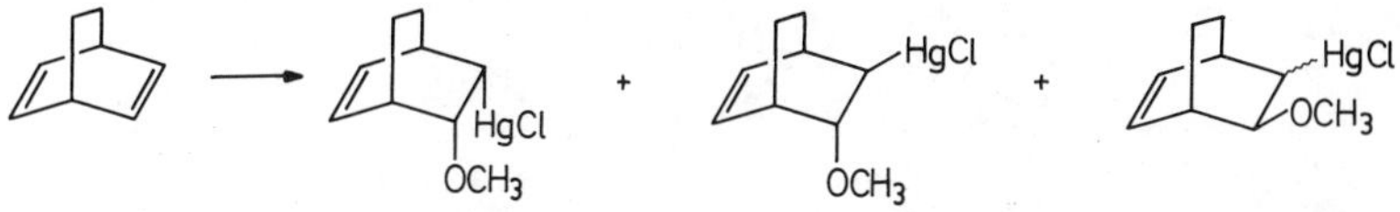

(212)

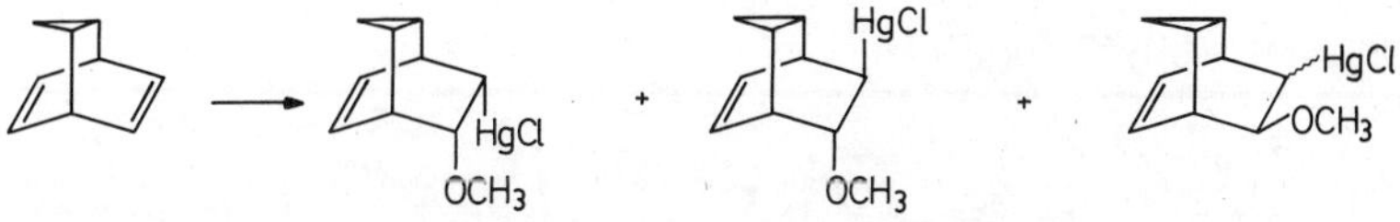

(213)

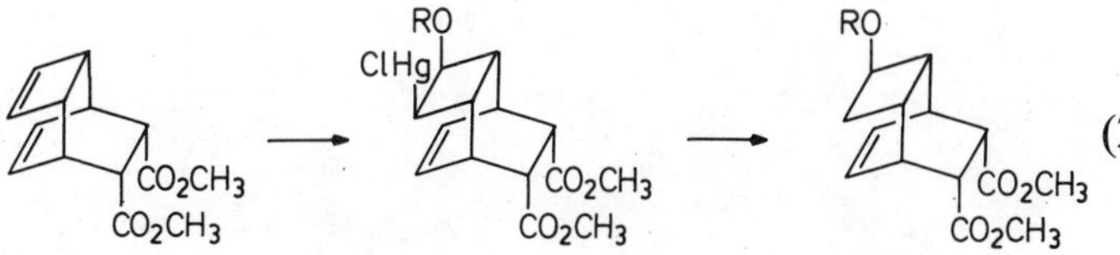

(214) [492, 493, 506]

(215) [422]

Alkoxymercuration results for a number of other polycyclic olefins have been reported. The bicyclo[2.2.2]octadiene system affords stereoisomeric mixtures of alkoxymercurials, plus minor amounts of acetoxymercurials (Eqs. 212, 213) [257, 505]. While sodium amalgam reduction of these mercurials proceeds as expected and sodium borohydride reduction of the bicyclic mercurials proceeds normally, sodium borohydride reduction of the tricyclic mercurials in methylene chloride affords mainly bicyclo[3.2.1]octadienyl ethers. Other tricyclic dienes have been reported to undergo more selective alkoxymercuration and reduction reactions (Eqs. 214, 215). In some systems, rearrangements and addition reactions have been observed (Eqs. 216–219). In the latter reaction (Eq. 219), two one-electron oxidations to a cation are suggested.

$$(216) \quad [491]$$

$$(217) \quad [483]$$

$$(218) \quad [483, 484]$$

$$(219) \quad [499]$$

$$\text{NuH} = \text{ROH, ArOH, RNH}_2, \text{RCO}_2\text{H}$$

$$(220) \quad [507]$$

$$(221) \quad [506]$$

$$(222) \quad [511]$$

Finally, let us examine the alkoxymercuration of dienes and polyenes which contain an alcohol group (Table 3.9). Allenic alcohols have been reported to undergo a variety of reactions with mercury salts, including intramolecular alkoxymercuration (Eqs. 220–223). The reaction products from this last reaction (p. 315) are highly dependent on the reaction conditions and solvent employed.

Table 3.9. Alkoxymercuration of Dienols

Dienol	Alcohol	Mercuric salt	Organomercurial(s) (% Yield)	Subsequent reactants	Product(s) (% Yield)	Ref.
$CH_3CH=C=CHCH_2OH$	CH_3OH	$Hg(OAc)_2$	—	$NaBH_4$	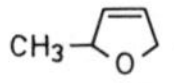 CH_3–[furan] + $CH_3CH(OCH_3)CH=CHCH_2OH$ 91 : 9 (45 total)	507
$(CH_3)_2C=C=CHCH_2OH$	CH_3OH	$Hg(OAc)_2$	—	$NaBH_4$	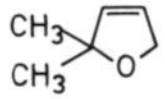 [dimethyl furan structure] (49)	507
$H_2C=C=C\!\!\begin{smallmatrix}(CH_2)_3CH_3\\ CH_2OH\end{smallmatrix}$	CH_3OH	$Hg(OAc)_2$	—	$NaBH_4$	$H_2C=CHC(OCH_3)(CH_2OH)(CH_2)_3CH_3$ (47)	507
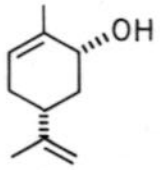 [cyclohexene diol structure]	—	$Hg(OAc)_2$	—	$NaBH_4/NaOH$	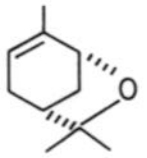 [bicyclic ether structure] (> 85)	508
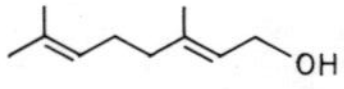 [dienol structure with OH]	—	$Hg(OAc)_2$	—	$NaBH_4/NaOH$	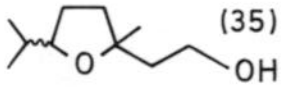 [tetrahydrofuran structure] (35) 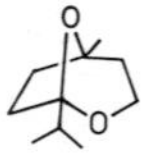 + [bicyclic ether structure] (20)	418
	CH_3CH_2OH	$Hg(OAc)_2$	—	$NaBH_4/NaOH$	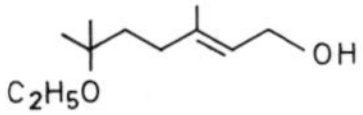 [C_2H_5O substituted alkenol] OH (50) 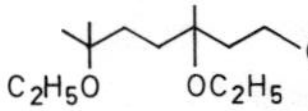 + [C_2H_5O, OC_2H_5 substituted alcohol] OH (20)	418
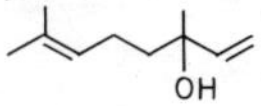 [dienol structure with OH]	—	$Hg(OAc)_2$	—	$NaBH_4/NaOH$	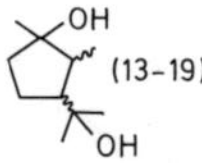 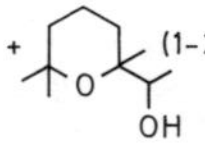 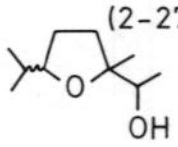 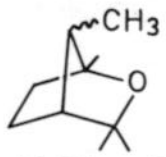 [cyclopentane diol] (13–19) + [tetrahydropyran] (1–22) + [tetrahydrofuran] (2–27) + [bicyclic ether] cis (2–16) and trans (7–52) 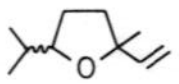 + [furan] (0–10) + [cyclopentanol structure] 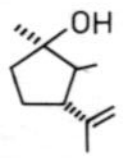	509, 510

Table 3.9. (continued)

Dienol	Alcohol	Mercuric salt	Organomercurial(s) (% Yield)	Subsequent reactants	Product(s) (% Yield)	Ref.
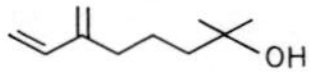 [structure]	—	Hg(OAc)$_2$	—	NaBH$_4$/NaOH	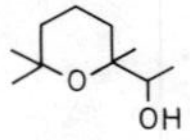 [structure] (75) or [structure] 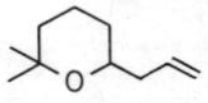	418
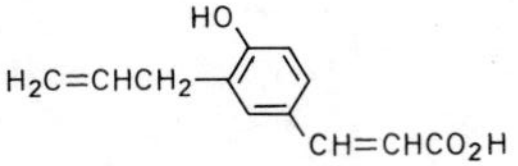 $H_2C=CHCH_2$–[phenol, CH=CHCO$_2$H]	—	HgX$_2$ (X = Cl, OAc)	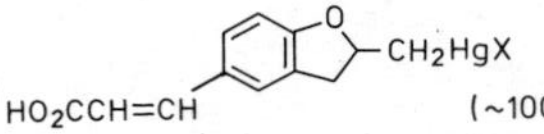 HO$_2$CCH=CH–[benzofuran]–CH$_2$HgX (~100)	—	—	439
CH$_3$(CH$_2$)$_3$, CH$_3$CH$_2$ C=C=C (CH$_2$)$_4$OH, H	—	Hg(O$_2$CCF$_3$)$_2$	—	NaBH$_4$/NaOH	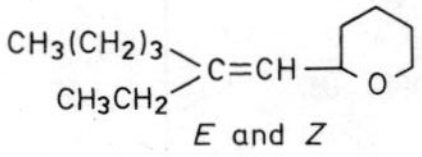 CH$_3$(CH$_2$)$_3$, CH$_3$CH$_2$ C=CH–[pyran] (80) E and Z	511
CH$_3$(CH$_2$)$_3$, CH$_3$ C=C=C (CH$_2$)$_4$OH, CH$_3$	—	Hg(O$_2$CCF$_3$)$_2$	—	NaBH$_4$/NaOH	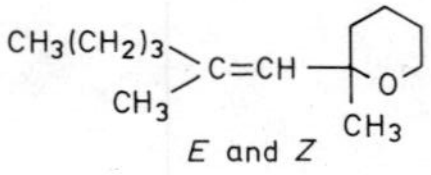 CH$_3$(CH$_2$)$_3$, CH$_3$ C=CH–[pyran, CH$_3$] (70) E and Z	511
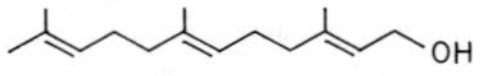 [structure] OH	CH$_3$CH$_2$OH	Hg(OAc)$_2$	—	NaBH$_4$/NaOH	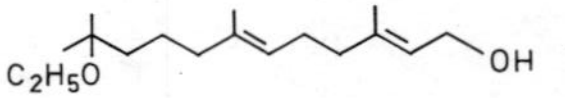 C$_2$H$_5$O–[structure]–OH (30)	418
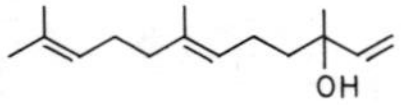 [structure] OH	—	HgCl$_2$	—	NaBH$_4$/NaOH	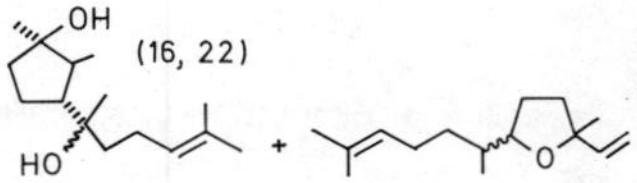 [structure] (16, 22) + [structure] (19)	509
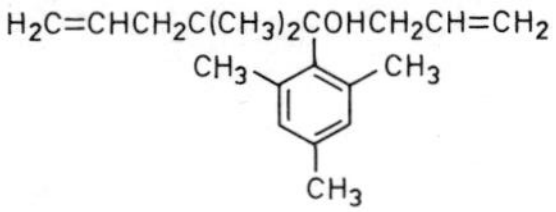 $H_2C=CHCH_2C(CH_3)_2COHCH_2CH=CH_2$ [mesityl]	CH$_3$OH	Hg(OAc)$_2$	—	KI	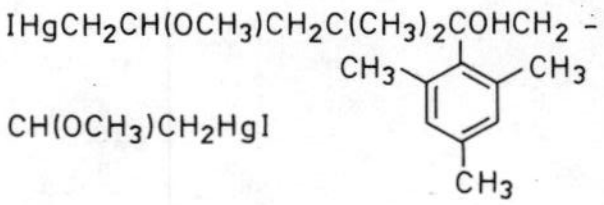 IHgCH$_2$CH(OCH$_3$)CH$_2$C(CH$_3$)$_2$COHCH$_2$– [mesityl] CH(OCH$_3$)CH$_2$HgI	69
cis, cis -CH$_3$CH$_2$CH=CH(CH$_2$)$_2$– CHOHCH$_2$CH=CH(CH$_2$)$_7$CO$_2$CH$_3$	—	Hg(OAc)$_2$ DMF	—	NaBH$_4$	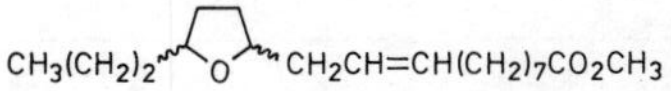 CH$_3$(CH$_2$)$_2$–[tetrahydrofuran]–CH$_2$CH=CH(CH$_2$)$_7$CO$_2$CH$_3$ (86)	202
	CH$_3$OH	Hg(OAc)$_2$	—	NaBH$_4$	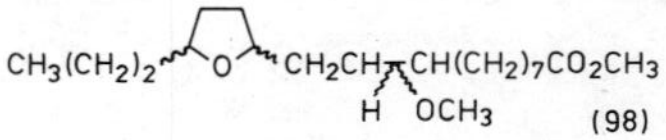 CH$_3$(CH$_2$)$_2$–[tetrahydrofuran]–CH$_2$CH–CH(CH$_2$)$_7$CO$_2$CH$_3$, H OCH$_3$ (98)	202

Table 3.9. (continued)

Dienol	Alcohol	Mercuric salt	Organomercurial(s) (% Yield)	Subsequent reactants	Product(s) (% Yield)	Ref.
all cis-$CH_3(CH_2)_4(CH{=}CHCH_2)_4(CH_2)_3OH$	—	$Hg(OAc)_2$	—	$NaBH_4$	all cis-$CH_3(CH_2)_4(CH{=}CHCH_2)_3CH_2$— 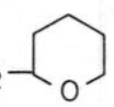 (structure)	202, 434
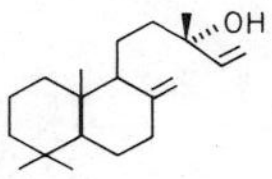 (structure)	—	$Hg(OAc)_2$	—	$NaBH_4/NaOH$	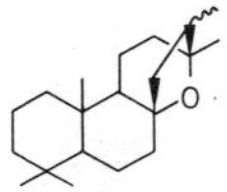 (structure) (72)	510
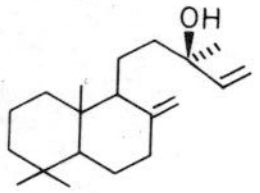 (structure)	—	$Hg(OAc)_2$	—	$NaBH_4/NaOH$	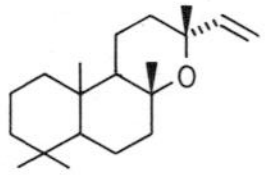 (structure)	510
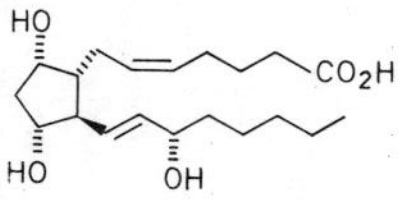 (structure)	—	$Hg(OAc)_2$	—	$NaBH_4/NaOH$	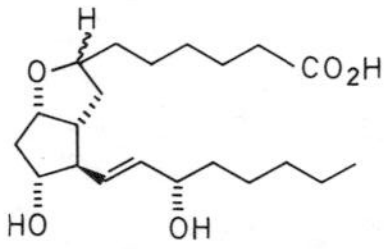 (structure) (both C-6 epimers)	512
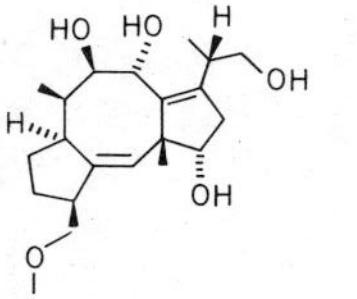 (structure)	—	—	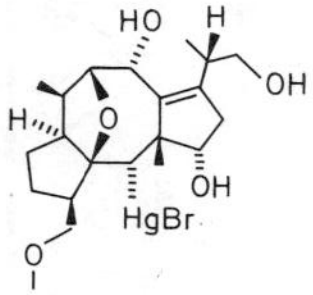 (structure, HgBr)	$NaBH_4$	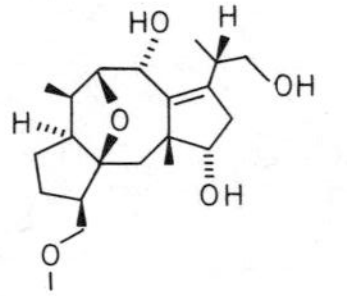 (structure)	513, 514
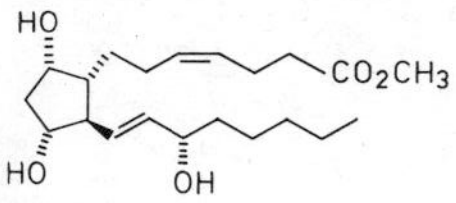 (structure)	—	$Hg(OAc)_2$	—	$NaBH_4$	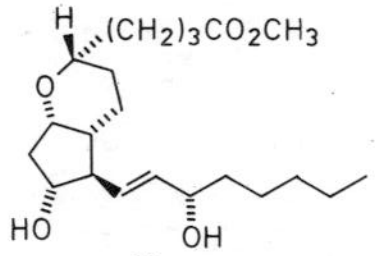 (structure) (41) 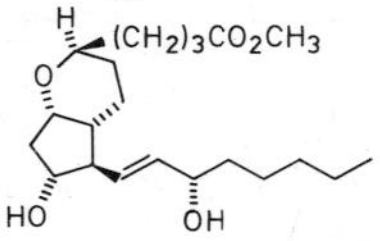 + (structure) (22)	515

Table 3.9. (continued)

Dienol	Alcohol	Mercuric salt	Organomercurial(s) (% Yield)	Subsequent reactants	Product(s) (% Yield)	Ref.
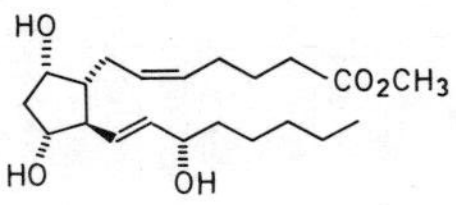	—	Hg(OAc)$_2$, NaCl	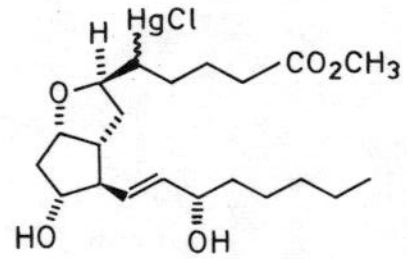 (51, 58) 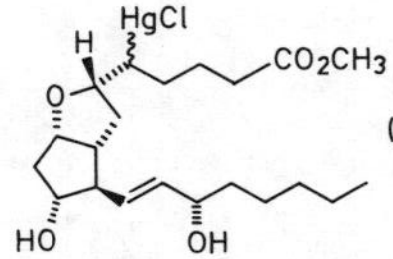 + (31, 24)	NaBH$_4$	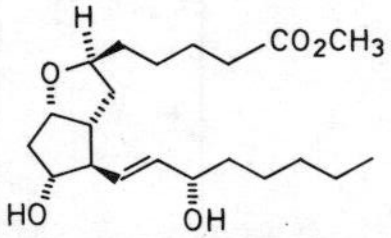 (only product direct from PGF$_{2\alpha}$ methyl ester) 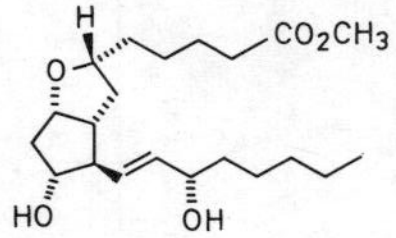 + (48)	516, 517
	—	Hg(OAc)$_2$	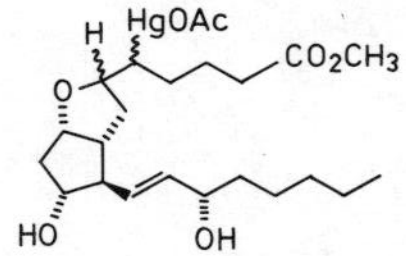	NaBH$_4$ / KOH	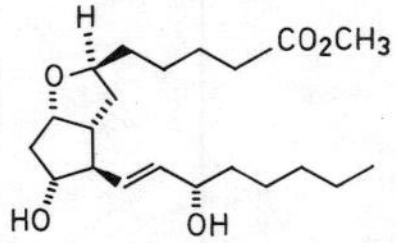 (24) 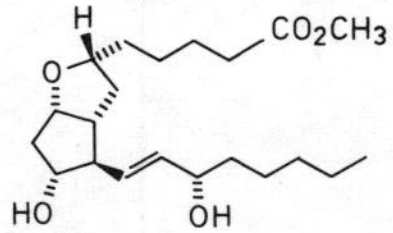 + (13)	512

Table 3.9. (continued)

Dienol	Alcohol	Mercuric salt	Organomercurial(s) (% Yield)	Subsequent reactants	Product(s) (% Yield)	Ref.
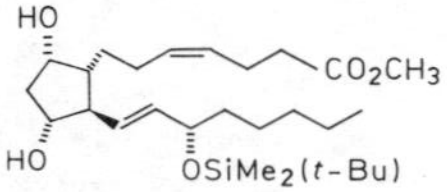 [structure: cyclopentane dienol with CO₂CH₃, OSiMe₂(t-Bu), two HO]	—	Hg(OAc)₂	—	NaBH₄	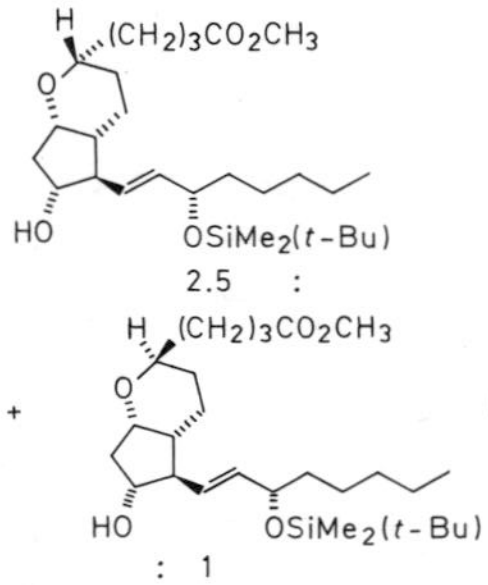 [structures: bicyclic products, ratio 2.5 : 1, (65 total)]	515
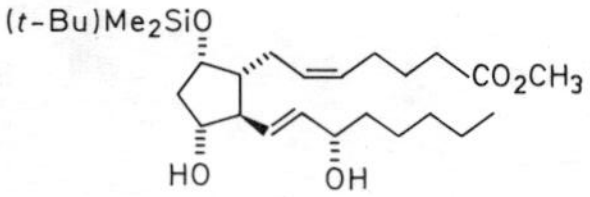 [structure: (t-Bu)Me₂SiO cyclopentane dienol with CO₂CH₃, HO, OH]	—	Hg(OAc)₂, NaCl	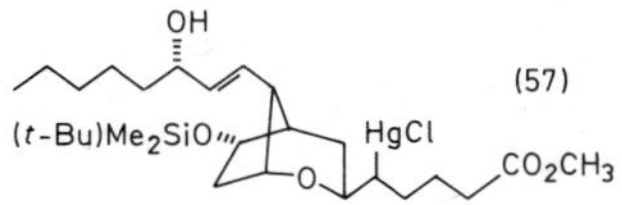 [structure with OH, (t-Bu)Me₂SiO, HgCl, CO₂CH₃] (57)	Ac₂O, NaBH₄/O₂	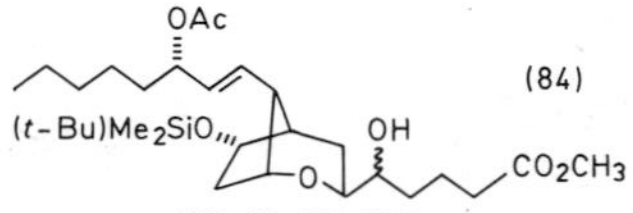 [structure with OAc, (t-Bu)Me₂SiO, OH, CO₂CH₃] (84) 52:48 5R/5S	518
(C₆H₅)₂C=C=C(Br)C(OH)(C₆H₅)₂	—	Hg(OAc)₂	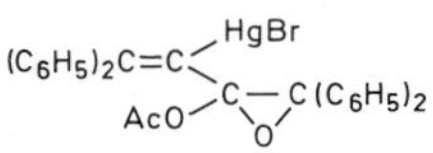 [structure: (C₆H₅)₂C=C, HgBr, AcO–C–C(C₆H₅)₂ epoxide] (92)	Δ	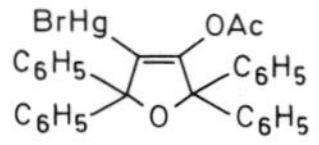 [furan structure: BrHg, OAc, C₆H₅ groups] (100)	519
	—	Hg(OAc)₂	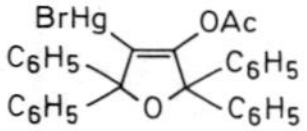 [furan structure: BrHg, OAc, C₆H₅ groups] (67)	NaBH₄	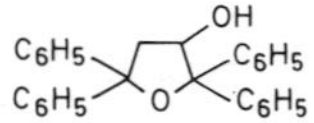 [furan structure: OH, C₆H₅ groups] (91)	519
	—	Hg(OAc)₂	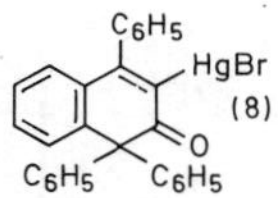 [naphthalenone structure: C₆H₅, HgBr] (8) 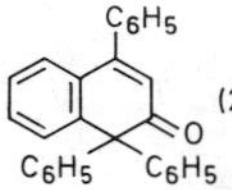 + [naphthalenone: C₆H₅] (27)	NaBH₄	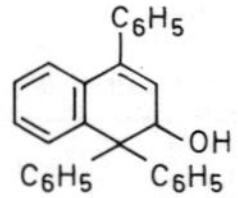 [naphthalenol structure: C₆H₅, OH] (90)	519

Table 3.9. (continued)

Dienol	Alcohol	Mercuric salt	Organomercurial(s) (% Yield)	Subsequent reactants	Product(s) (% Yield)	Ref.
[structure]	ROH	$Hg(OAc)_2$	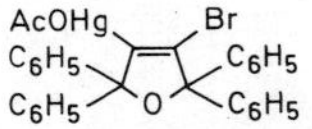 [structure] $R = Me(14)$, $Et(18)$ + $(C_6H_5)_2C(OR)C\equiv CC(OH)(C_6H_5)_2$ $R = Me(54)$, $Et(41)$	—	—	519
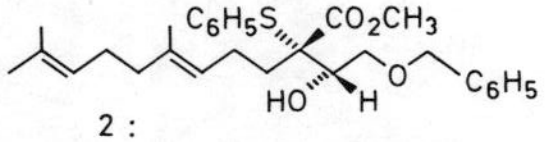 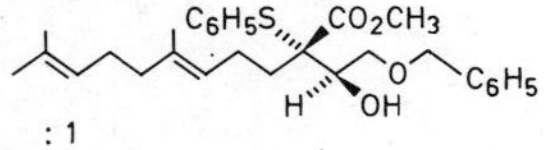 [structure] 2 :	—	$Hg(O_2CCF_3)_2$	—	$Br_2/LiBr/O_2/$ C_5H_5N	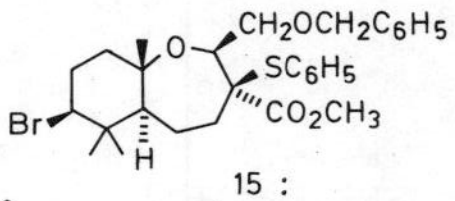 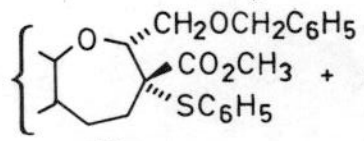 [structures] 15 : 14 : 11 : 15	435
[structure] : 1						
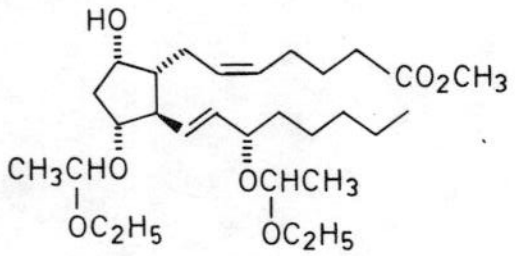 [structure] CH_3CHO, OC_2H_5	—	$Hg(OAc)_2$	—	$NaBH_4/NaOH$, H_3O^+, OH^-	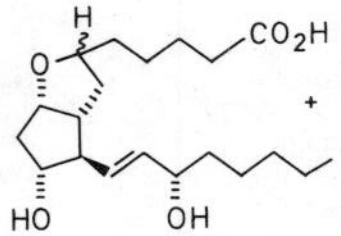 [structures] (31) + (50) 1:1 mixture	517
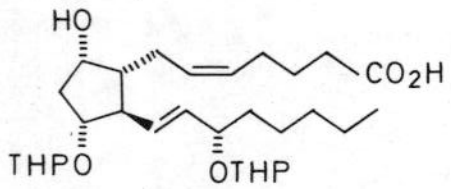 [structure] THPO, OTHP	—	$Hg(O_2CCF_3)_2$	—	$NaBH_4$, HOAc	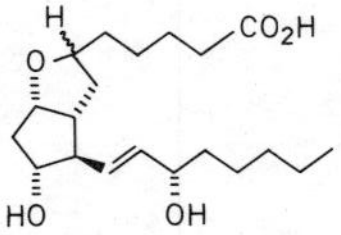 [structures] (90) (2 C-6 epimers)	520

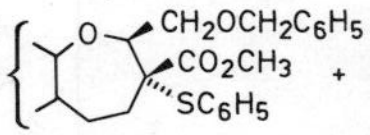

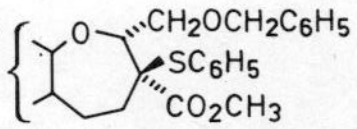

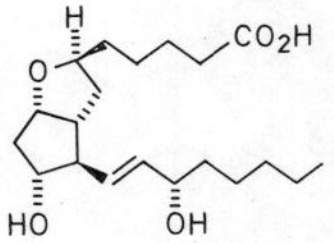

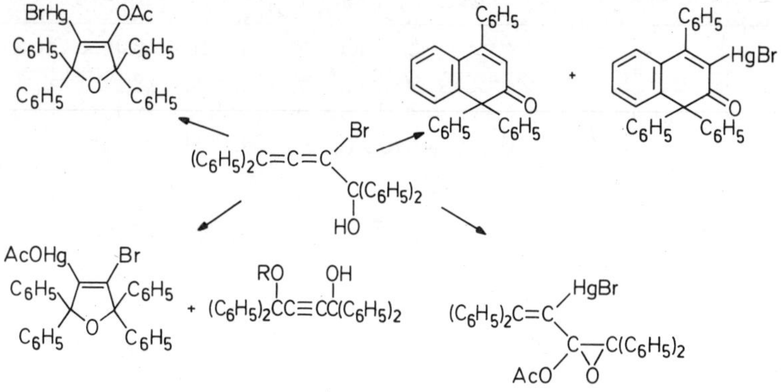

$$(223) \ [519]$$

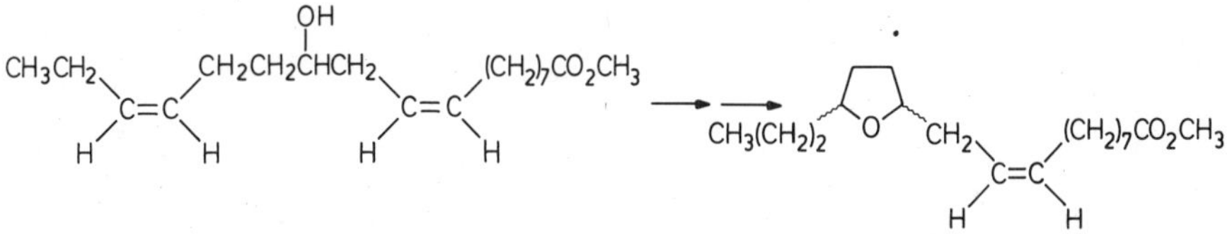

$$(224) \ [202]$$

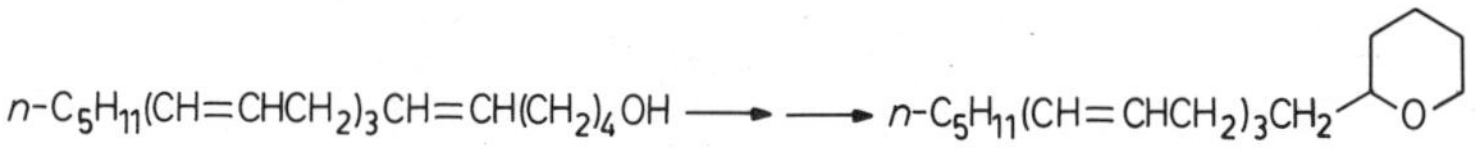

$$n\text{-}C_5H_{11}(CH{=}CHCH_2)_3CH{=}CH(CH_2)_4OH \longrightarrow \longrightarrow n\text{-}C_5H_{11}(CH{=}CHCH_2)_3CH_2 \qquad (225) \ [202, 434]$$

A number of other examples of intramolecular alkoxymercuration of polyunsaturated alcohols are also known. Acyclic alcohols cyclize readily (Eqs. 224, 225). This can be a useful method of locating double bonds in polyenoic fatty esters [521].

Several naturally-occurring cyclic dienols have been successfully cyclized via intramolecular alkoxymercuration (Eqs. 226–228). Most important

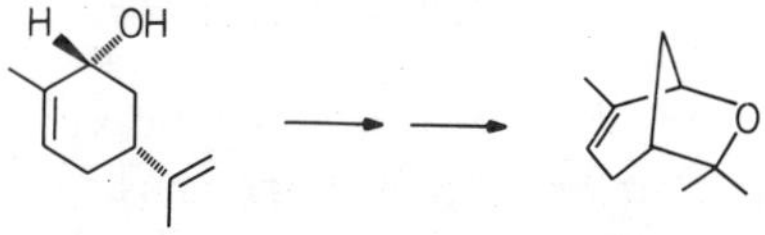

$$(226) \ [508]$$

$$(227) \ [510]$$

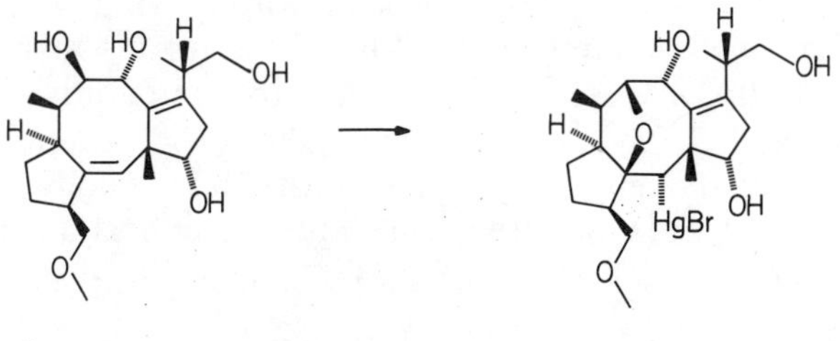

$$(228) \ [513, 514]$$

315

of these are the prostaglandins (Eqs. 229–231). Note that all of these compounds cyclize with the regiochemistry expected from our earlier discussion of acyclic alkenols. Besides being successfully demercurated by alkaline sodium borohydride, some of these mercurials have also been oxidized to alcohols using oxygen and sodium borohydride (Eqs. 229, 231).

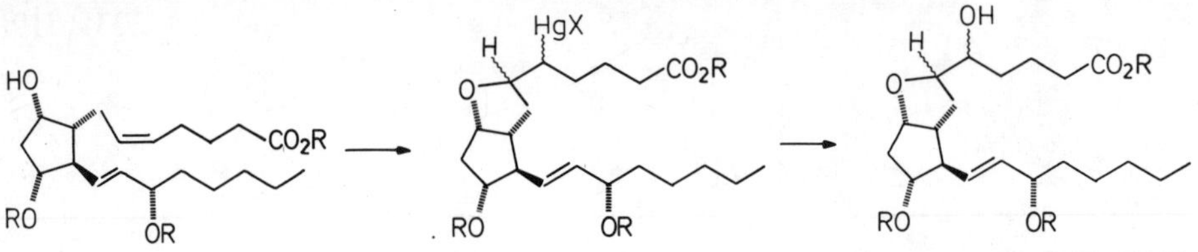

(229) [512, 516, 517, 520]

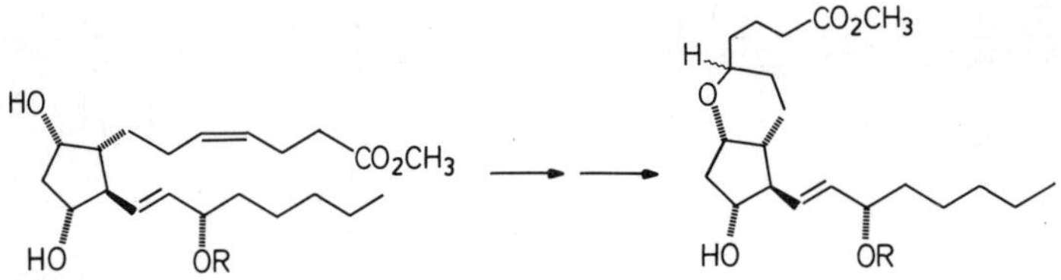

(230) [515]

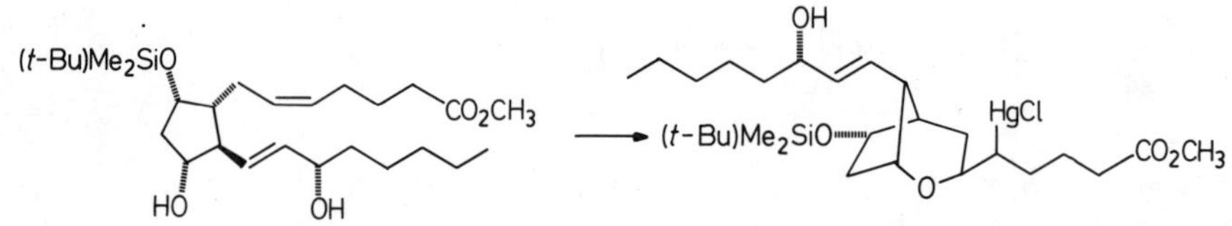

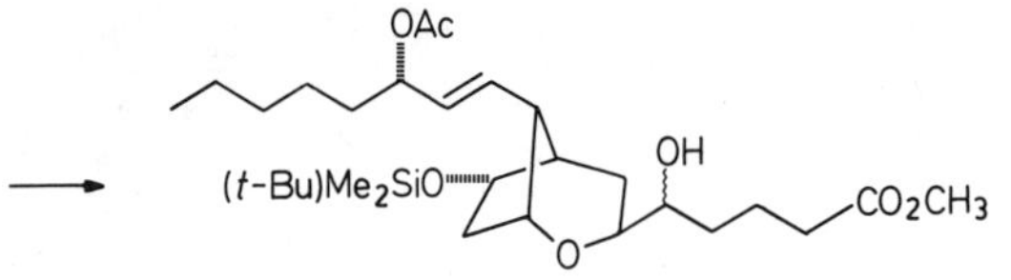

(231) [518]

With certain terpenoid alcohols, cyclization has proven difficult and rearrangement and carbomercuration products have been observed. Thus, linalool affords only a low yield of the anticipated mercurial (Eq. 232),

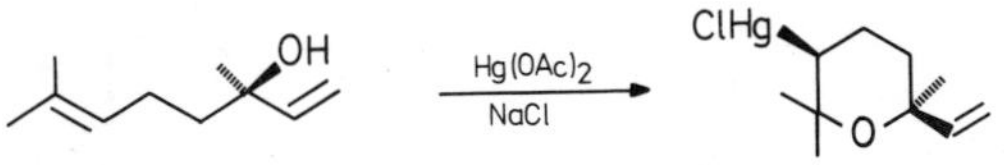

(232)

which upon reaction with alkaline sodium borohydride affords a variety of rearranged products and none of the expected ether [510]. Geraniol and myrcenol also give surprising products under the usual hydroxymercuration-demercuration conditions (Eqs. 233, 234) [418]. In ethanol, however, geraniol and farnesol have been regioselectively converted into the corresponding mono ethers in low yields (Eq. 235) [418]. Other unsaturated alcohols have undergone preferential carbocyclization (Eqs. 236, 237).

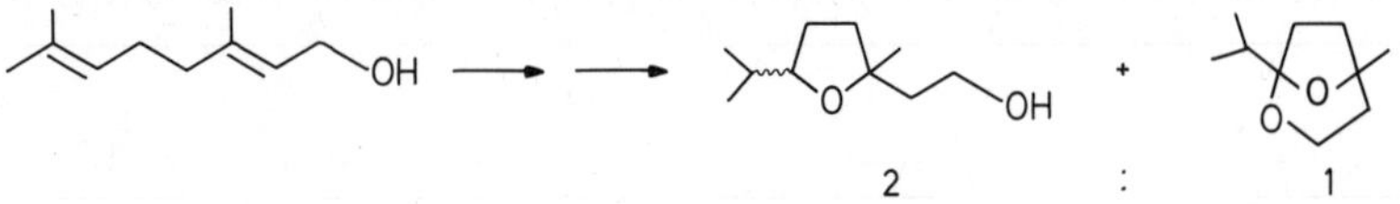

(233)

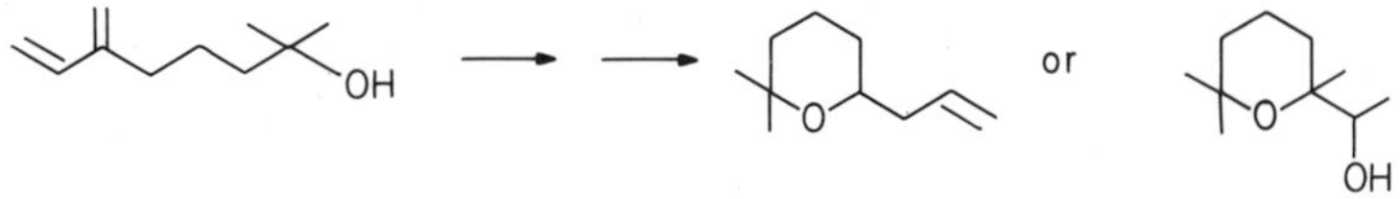

(234)

(235)

(236) [509]

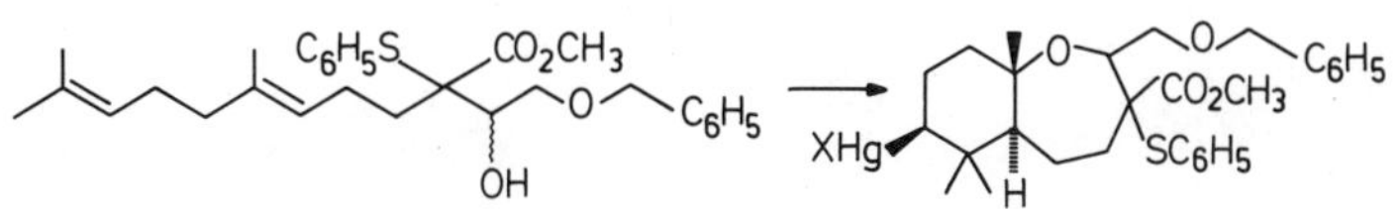

(237) [435]

D. Alkynes

The alkoxymercuration of alkynes provides a convenient approach to vinyl ethers, acetals or ketals, dioxanes or furans depending on the nature of the acetylene and the reaction conditions employed. Table 3.10 (see p. 319) summarizes all results reported to date.

Mercury salts promote the addition of alcohols to the carbon—carbon triple bond of acetylenes. However, it is almost impossible to stop this reaction at the vinylmercurial stage. The following reaction appears to be the only case in which a vinylmercurial has actually been isolated from such a reaction (Eq. 238) [544]. More commonly the initial vinylmercurial is rapidly protodemercurated by the acid generated in the initial alkoxymercuration step (Eqs. 239–242). Note the interesting regioselectivity in the last two examples. By a careful choice of reaction conditions, one can intercept the desired vinylmercurials (Eqs. 243, 244). It is not clear at present how general these reactions really are.

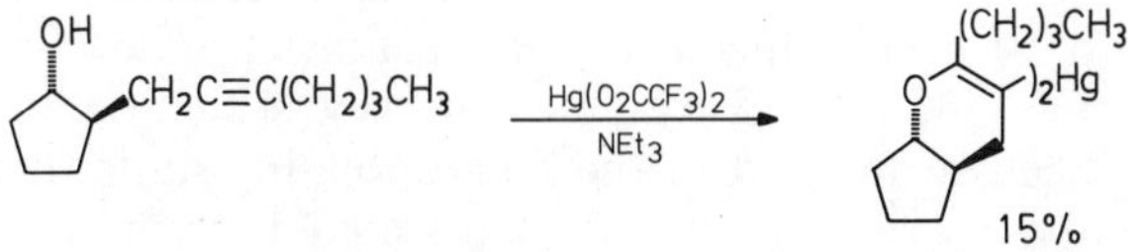

(238)

III. Alkoxymercuration

$$HC\equiv CH \quad + \quad C_2H_5OH \quad \xrightarrow{Hg_3(PO_4)_2} \quad H_2C=CHOC_2H_5 \qquad \text{(239) [526]}$$

$$HC\equiv CCH_2CH_2OH \quad + \quad CH_3OH \quad \xrightarrow[\substack{BF_3\cdot Et_2O \\ Cl_3CCO_2H}]{HgO} \quad H_2C=\overset{\overset{\displaystyle OCH_3}{|}}{C}CH_2CH_2OH \qquad \text{(240) [531]}$$

(241) [544]

(242) [544]

$$n\text{-}C_6H_{13}C\equiv CH \quad + \quad ROH \quad \xrightarrow{Hg(OAc)_2} \quad \xrightarrow[NaOH]{NaBH_4} \quad n\text{-}C_6H_{13}\overset{\overset{\displaystyle OR}{|}}{C}=CH_2 \qquad \text{(243) [545]}$$

(244) [544]

X = Cl, Br, I

In the majority of cases, no attempt has been made to isolate the vinyl ethers. Instead, acidic mercury catalysts such as $HgO/BF_3\cdot Et_2O\,(Cl_3CCO_2H)$, $HgO/BF_3/CH_3OH$, HgO/H_2SO_4, $HgSO_4/H_2SO_4$, $HgSO_4$, $Hg_3(PO_4)_2$ or $Hg(OAc)_2$ have been employed in an excess of the alcohol and the corresponding acetal or ketal has been isolated. The HgO/BF_3 catalysts appear to be more effective than HgO/H_2SO_4 and accommodate certain acid sensitive alcohols not tolerated by the latter catalyst [525, 549]. When using mercuric acetate in methanol, the corresponding ketones or alcohols can be obtained upon aqueous hydrochloric acid or alkaline sodium borohydride work-up respectively [190, 306].

A wide variety of alcohols have been utilized in these reactions, ranging from simple primary, secondary and tertiary alcohols to diols, triols and polyols. With $HgO/BF_3\cdot Et_2O/Cl_3CCO_2H$, it is reported that branched alcohols are much less reactive than their straight chain counterparts [541]. The reaction conditions are generally mild enough that even hydroxy esters can be employed without hydrolysis of the esters [525, 528]. A number of 1,2-, 1,3- and even 1,4-diols have been utilized in the preparation of cyclic acetals and ketals [525, 527, 533, 550, 551]. Five-, six- and seven-membered ring acetals and ketals have been formed from the corresponding diols. In competition studies, six-membered ring dioxanes are formed more easily

Table 3.10. Alkoxymercuration of Alkynes

Alkyne	Alcohol	Reagent(s)	Product(s) (% Yield)	Ref.
HC≡CH	CH_3OH	$HgSO_4$	$CH_3CH(OCH_3)_2$	522
	CH_3OH	$HgSO_4 / H_2SO_4$	$CH_3CH(OCH_3)_2$ (25 – 30)	523
	CH_3OH	$HgO / BF_3 \cdot Et_2O$	$CH_3CH(OCH_3)_2$	524
	$HOCH_2CH_2Cl$	$HgO / BF_3 / CH_3OH$	$CH_3CH(OCH_2CH_2Cl)_2$ (71)	525
	CH_3CH_2OH	$HgSO_4 / H_2SO_4$	$CH_3CH(OC_2H_5)_2$ (30 – 35)	523
	CH_3CH_2OH	$Hg_3(PO_4)_2$	$H_2C{=}CHOC_2H_5$	526
	$HOCH_2CH_2OH$	$HgSO_4 / H_2SO_4$	[1,3-dioxolane, 2-CH_3] (75)	527
	$HOCH_2CH_2OH$	$HgO / BF_3 / CH_3OH$	[1,3-dioxolane, 2-CH_3] (62)	525
	$HOCH_2CO_2CH_3$	$Hg_3(PO_4)_2$	$CH_3CH(OCH_2CO_2CH_3)_2$ (56)	528
	$HOCH_2CO_2CH_3$	$HgO / BF_3 / CH_3OH$ $[Hg_3(PO_4)_2]$	$CH_3CH(OCH_2CO_2CH_3)_2$ (81) [56]	528, 529
	$BrCH_2CHOHCH_2OH$	$HgSO_4 / H_2SO_4$	[1,3-dioxolane, 2-CH_3, 4-CH_2Br] (40)	527
	$HO(CH_2)_3Cl$	$HgO / BF_3 / CH_3OH$	$CH_3CH(OCH_2CH_2CH_2Cl)_2$ (38)	525
	$CH_3CHOHCH_2OH$	$HgSO_4 / H_2SO_4$	[1,3-dioxolane, 2-CH_3, 4-CH_3] (70)	527
	$HO(CH_2)_3OH$	$HgSO_4 / H_2SO_4$	[1,3-dioxane, 2-CH_3] (75)	527

Table 3.10. (continued)

Alkyne	Alcohol	Reagent(s)	Product(s) (% Yield)	Ref.
	$HO(CH_2)_3OH$	HgO / BF_3 / CH_3OH	(45)	525
	$HOCH_2CHOHCH_2OH$	HgO / BF_3 / CH_3OH	[78 : 22] (65–75 total) (41.7)	525
	$HO(CH_2)_4OH$	$HgSO_4$ / H_2SO_4	(20)	527
	$C(CH_2OH)_4$	HgO / BF_3 / CH_3OH	(90)	525
	$(CH_3)_2COHCHOHCH_3$	$HgSO_4$ / H_2SO_4	(71)	527
	$RO_2CCHOHCHOHCO_2R$	HgO / BF_3 / CH_3OH	$R = Me$ (81), Et (74)	525
	$HOCH_2CO_2CH_2CH(CH_3)_2$	HgO / BF_3 /CH_3OH [$Hg_3(PO_4)_2$]	$CH_3CH[OCH_2CO_2CH_2CH(CH_3)_2]_2$ (70) [56]	528, 529
	$(CH_3)_2COHCOH(CH_3)_2$	$HgSO_4$ / H_2SO_4	(61)	527
	$(CH_3)_2COHCOH(CH_3)_2$	HgO / BF_3 / CH_3OH		525

Table 3.10. (continued)

Alkyne	Alcohol	Reagent(s)	Product(s) (% Yield)	Ref.
	$[CH_3CH_2C(CH_3)OH]_2$	HgO / BF$_3$ / CH$_3$OH	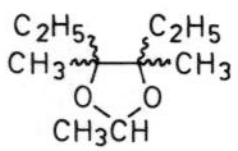 (dioxolane structure)	525
	HOCH$_2$CH$_2$OR ?	HgO / BF$_3$ / CH$_3$OH	CH$_3$CH(OCH$_2$CH$_2$OR)$_2$? R = Me (44), Et (74), n-Bu (52), CH$_2$CH$_2$OEt	525
	ROH	HgO / BF$_3$ / CH$_3$OH	CH$_3$CH(OR)$_2$ R = Me (37), Et (40), n-Pr (37), i-Pr (40), n-Bu (33.6), sec-Bu (35), i-Bu (43.4), n-Am (51.5), CH$_3$CH$_2$CH(CH$_3$)CH$_2$ (43.6), CH$_3$(CH$_2$)$_2$CHCH$_3$ (30), (CH$_3$)$_2$CHCHCH$_3$ (30), (CH$_3$CH$_2$)$_2$CH (40), CH$_3$CH$_2$C(CH$_3$)$_2$ (53.7), (CH$_3$CH$_2$)$_3$C, CH$_3$(CH$_2$)$_3$C(CH$_3$)(C$_2$H$_5$), H$_2$C=CHCH$_2$, (tetrahydrofuranyl-CH$_2$) 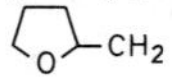 (31.6), c-C$_6$H$_{11}$, C$_6$H$_5$CH$_2$ (45.5), (C$_6$H$_5$)$_2$CH , C$_6$H$_5$(CH$_2$)$_2$ (50.4), C$_6$H$_5$CHCH$_3$, CH$_3$CH$_2$CHC$_6$H$_5$ (40), CH$_3$CH$_2$C(CH$_3$)C$_6$H$_5$, (C$_6$H$_5$)$_3$C (43)	549
	HOCH$_2$CO$_2$H	HgO / HgSO$_4$ / H$_2$SO$_4$ / Na$_2$SO$_4$	(dioxolanone structure)	530
	CH$_3$CHOHCO$_2$H	HgO / BF$_3$ / CH$_3$OH	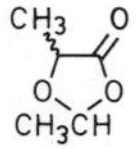 (dioxolanone structure) (61)	525
	HO$_2$CCHOHCHOHCO$_2$H	HgO / BF$_3$ / CH$_3$OH	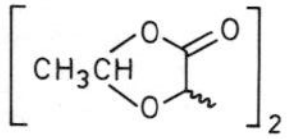 (bis-dioxolanone structure)	525
	(CH$_3$)$_2$COHCO$_2$H	HgO / BF$_3$ / CH$_3$OH	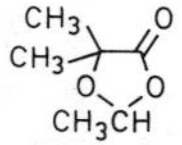 (dioxolanone structure) (71)	525
	C$_6$H$_5$CHOHCO$_2$H	HgO / BF$_3$ / CH$_3$OH	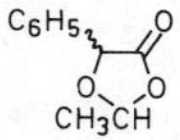 (dioxolanone structure) (64)	525

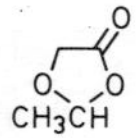

Table 3.10. (continued)

Alkyne	Alcohol	Reagent(s)	Product(s) (% Yield)	Ref.
	$(C_6H_5)_2COHCO_2H$	$HgO / BF_3 / CH_3OH$	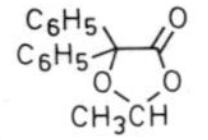 (66)	525
$HC{\equiv}CCH_3$	CH_3OH	$HgO / BF_3 \cdot Et_2O$	$CH_3C(OCH_3)_2CH_3$	524
$HC{\equiv}CCH_2OH$	CH_3OH	$HgO / BF_3 \cdot Et_2O / Cl_3CCO_2H$	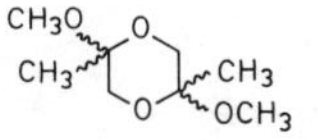 (5)	531
$HC{\equiv}CCH{=}CH_2$	CH_3OH	$HgO/BF_3 \cdot Et_2O / Cl_3CCO_2H$	$CH_3C(OCH_3)_2CH_2CH_2OCH_3$ (65)	532
	$HOCH_2CH_2OH$	$HgO / BF_3 \cdot Et_2O / Cl_3CCO_2H$	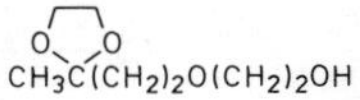 + 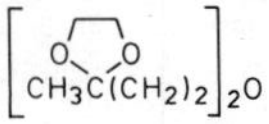	533
$HC{\equiv}CCH_2CH_3$	CH_3OH	$HgO / BF_3 \cdot Et_2O$	$CH_3C(OCH_3)_2CH_2CH_3$	524
$HC{\equiv}CCHOHCH_3$	CH_3OH	$HgO / BF_3 \cdot Et_2O / Cl_3CCO_2H$	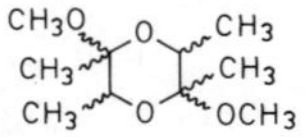 (41)	531
	$HOCH_2CH_2OH$	$HgO / BF_3 \cdot Et_2O / Cl_3CCO_2H$	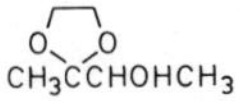 (67)	531
$HC{\equiv}CCH_2CH_2OH$	CH_3OH	$HgO / BF_3 \cdot Et_2O / Cl_3CCO_2H$	$H_2C{=}C(OCH_3)CH_2CH_2OH$ (47) + $CH_3C(OCH_3)_2CH_2CH_2OH$ (10)	531
$HC{\equiv}CC(CH_3){=}CH_2$	ROH	$Hg(OAc)_2$, $NaBH_4 / NaOH$	$CH_3C(OR)_2C(CH_3){=}CH_2$ R = Me (81.5), Et (58.2)	534
$HC{\equiv}C(CH_2)_2CH_3$	CH_3OH	$HgO / BF_3 \cdot Et_2O$	$CH_3C(OCH_3)_2(CH_2)_2CH_3$	524
$HC{\equiv}CCOH(CH_3)_2$	CH_3OH	$HgO / BF_3 \cdot Et_2O / Cl_3CCO_2H$	$CH_3C(OCH_3)_2COH(CH_3)_2$ (80) + 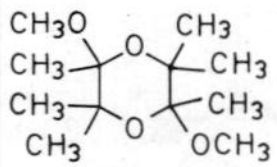 (4.4)	535

Table 3.10. (continued)

Alkyne	Alcohol	Reagent(s)	Product(s) (% Yield)	Ref.
	CH_3CH_2OH	$HgO/BF_3 \cdot Et_2O/Cl_3CCO_2H$	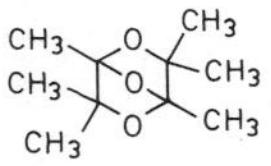 major + 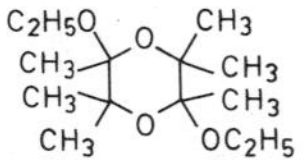 minor	536
	$HOCH_2CH_2OH$	$HgC/BF_3 \cdot Et_2O/Cl_3CCO_2H$	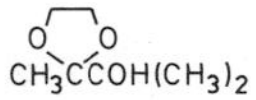 $CH_3CCOH(CH_3)_2$ (57)	531, 537
	ROH	$HgO/BF_3 \cdot Et_2O/Cl_3CCO_2H$	$CH_3C(OR)_2COH(CH_3)_2$ + major, 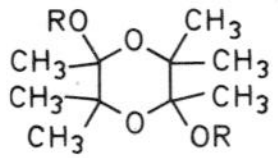 minor; $R = Me, Et, n-Pr, n-Bu, n-Am$	537
$CH_3C{\equiv}CCH_2OCH_3$	CH_3OH	$HgO/BF_3 \cdot Et_2O/Cl_3CCO_2H$	$CH_3C(OCH_3)_2CH_2CH_2OCH_3$ (57)	538
$H_2C{=}CHC{\equiv}CCH_2CH_3$	CH_3OH	$HgO/BF_3 \cdot Et_2O/Cl_3CCO_2H$	$CH_3O(CH_2)_2C(OCH_3)_2(CH_2)_2CH_3$	96
$HC{\equiv}CCHOHCH{=}CHCH_3$	CH_3OH	$HgO/BF_3 \cdot Et_2O/Cl_3CCO_2H$	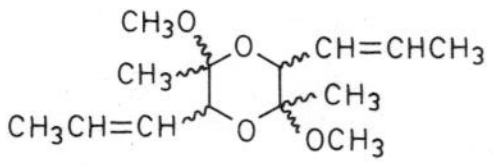 (22)	539
$HC{\equiv}C(CH_2)_3CH_3$	CH_3OH	$HgO/BF_3 \cdot Et_2O$	$CH_3C(OCH_3)_2(CH_2)_3CH_3$ (70)	524, 540
	$HOCH_2CH_2OH$	$HgO/BF_3 \cdot Et_2O/Cl_3CCO_2H$	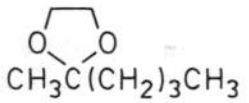 $CH_3C(CH_2)_3CH_3$	533
	$ClCH_2CHOHCH_2OH$	$HgO/BF_3 \cdot Et_2O/Cl_3CCO_2H$	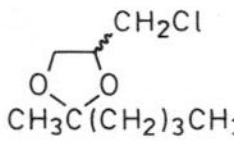 $CH_3C(CH_2)_3CH_3$	533

Table 3.10. (continued)

Alkyne	Alcohol	Reagent(s)	Product(s) (% Yield)	Ref.
	$CH_3O_2CCHOHCHOHCO_2CH_3$	$HgO/BF_3 \cdot Et_2O/Cl_3CCO_2H$	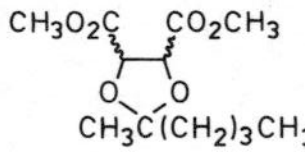	533
	$HOCH_2(CHOH)_4CH_2OH$	$HgO/BF_3 \cdot Et_2O/Cl_3CCO_2H$	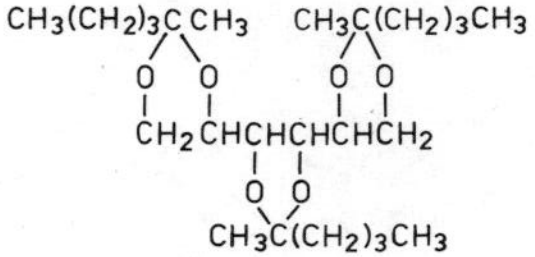	533
	ROH	$HgO/BF_3 \cdot Et_2O/Cl_3CCO_2H$	$CH_3C(OR)_2(CH_2)_3CH_3$ R = Et, *n*-Pr, *n*-Bu, *n*-Am, *n*-Hex (62)	541
	$(CH_3)_2COHCO_2H$	$HgO/BF_3 \cdot Et_2O/Cl_3CCO_2H$	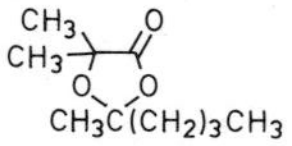	533
	$C_6H_5CHOHCO_2H$	$HgO/BF_3 \cdot Et_2O/Cl_3CCO_2H$	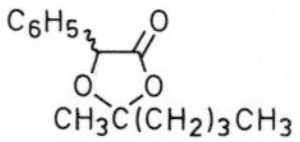	533
$HC{\equiv}CCOH(CH_3)CH_2CH_3$	CH_3OH	$HgO/BF_3 \cdot Et_2O$	$CH_3C(OCH_3)_2COH(CH_3)CH_2CH_3$	542
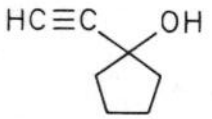	$HOCH_2CH_2OH$	$HgO/BF_3 \cdot Et_2O$	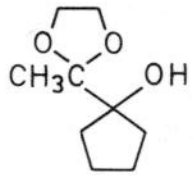 (60)	543
$HC{\equiv}CC(CH_3)_2O_2CCH_3$	CH_3OH	$HgO/BF_3 \cdot Et_2O/Cl_3CCO_2H$	$CH_3C(OCH_3)_2COH(CH_3)_2$	537
$HC{\equiv}C(CH_2)_4CH_3$	CH_3OH	$HgO/BF_3 \cdot Et_2O$	$CH_3C(OCH_3)_2(CH_2)_4CH_3$	524

Table 3.10. (continued)

Alkyne	Alcohol	Reagent(s)	Product(s) (% Yield)	Ref.
	$HOCH_2CH_2OH$	$HgO / BF_3 \cdot Et_2O$	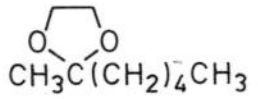 $CH_3C(CH_2)_4CH_3$ (cyclic acetal) (75)	540
	ROH	$HgO / BF_3 \cdot Et_2O / Cl_3CCO_2H$	$CH_3C(OR)_2(CH_2)_4CH_3$ $R = Et, n-Pr, n-Bu, n-Am, n-Hex$	541
$HC\equiv CCOH(C_2H_5)_2$	CH_3OH	$HgO / BF_3 \cdot Et_2O$	$CH_3C(OCH_3)_2COH(C_2H_5)_2$	542
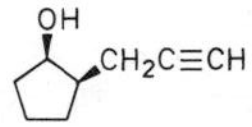 (2-(prop-2-ynyl)cyclopentanol) $CH_2C\equiv CH$, OH	—	$HgCl_2 / NEt_3$	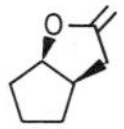 (bicyclic enol ether) (89)	544
	—	$HgCl_2 / NBS / DMAP$	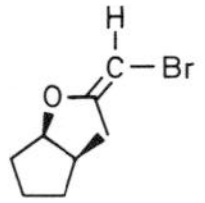 (bicyclic bromomethylene ether) (93)	544
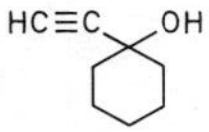 $HC\equiv C$, OH (1-ethynylcyclohexanol)	$HOCH_2CH_2OH$	$HgO / BF_3 \cdot Et_2O / Cl_3CCO_2H$	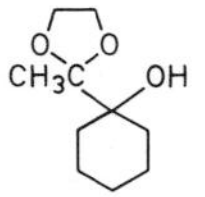 CH_3C (dioxolane) OH (63)	531
$HC\equiv C(CH_2)_5CH_3$	CH_3OH	$Hg(OAc)_2, NaBH_4 / NaOH$	$H_2C=C(OCH_3)(CH_2)_5CH_3$ (65)	545
	CH_3CH_2OH	$Hg(OAc)_2, NaBH_4 / NaOH$	$H_2C=C(OC_2H_5)(CH_2)_5CH_3$ (36)	545
$CH_3C\equiv C(CH_2)_4CH_3$	CH_3OH	$HgO / BF_3 \cdot Et_2O / Cl_3CCO_2H$	$CH_3CH_2C(OCH_3)_2(CH_2)_4CH_3$ (55)	538
$(CH_3)_2COHC\equiv CCOH(CH_3)_2$	CH_3OH	$HgO / BF_3 \cdot Et_2O / Cl_3CCO_2H$	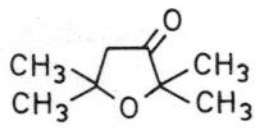 (tetrahydrofuranone with 4 CH_3) (77)	535

Table 3.10. (continued)

Alkyne	Alcohol	Reagent(s)	Product(s) (% Yield)	Ref.
$HC{\equiv}CCHOHC_6H_5$	CH_3OH	$HgO/BF_3{\cdot}Et_2O/Cl_3CCO_2H$	CH_3O, CH_3, C_6H_5, C_6H_5, CH_3, OCH_3, CH_3O (36)	531
$H_2C{=}CHCH_2C{\equiv}C(CH_2)_3CH_3$	CH_3OH	$HgO/BF_3{\cdot}Et_2O/Cl_3CCO_2H$	$H_2C{=}CH(CH_2)_2C(OCH_3)_2(CH_2)_3CH_3$	96
$H_2C{=}CHCH_2C{\equiv}C(CH_2)_4CH_3$	CH_3OH	$HgO/BF_3{\cdot}Et_2O/Cl_3CCO_2H$	$H_2C{=}CH(CH_2)_2C(OCH_3)_2(CH_2)_4CH_3$	96
$H_2C{=}CHCH_2C{\equiv}CC_6H_5$	CH_3OH	$HgO/BF_3{\cdot}Et_2O/Cl_3CCO_2H$	$H_2C{=}CH(CH_2)_2C(OCH_3)_2C_6H_5$ (80)	96
$C_6H_5C{\equiv}CCOH(CH_3)CHOHCH_3$	—	$HgCl_2$	CH_3, $HgCl$, CH_3, C_6H_5 (furan)	546
(cyclopentane-OH)$CH_2C{\equiv}C(CH_2)_3CH_3$	—	$Hg(O_2CCF_3)_2/NEt_3$ (NX succinimide)	$(CH_2)_3CH_3$, $C{-}X$ X = H, Cl (32), Br (88), I	544
(cyclopentane-OH)$CH_2C{\equiv}C(CH_2)_3CH_3$	—	$Hg(O_2CCF_3)_2/NEt_3$ (NX succinimide)	$(CH_2)_3CH_3$, X X = H (86), Cl (82), Br (76), I (88)	544
$HC{\equiv}C(CH_2)_8CO_2CH_3$	CH_3OH	$Hg(OAc)_2$, HCl	$CH_3CO(CH_2)_8CO_2CH_3$ (94)	190
	CH_3OH	$Hg(OAc)_2$, $NaBH_4$	$CH_3CHOH(CH_2)_8CO_2CH_3$ (58)	190
$CH_3C{\equiv}C(CH_2)_7CO_2CH_3$	CH_3OH	$Hg(OAc)_2$, HCl	$CH_3CO(CH_2)_8CO_2CH_3$ + $CH_3CH_2CO(CH_2)_7CO_2CH_3$ (96 total) 53 : 47	190
	CH_3OH	$Hg(OAc)_2$, $NaBH_4$	$CH_3CHOH(CH_2)_8CO_2CH_3$ + $CH_3CH_2CHOH(CH_2)_7CO_2CH_3$ (96 total) 54 : 46	190

Table 3.10. (continued)

Alkyne	Alcohol	Recgent(s)	Product(s) (% Yield)	Ref.
$C_6H_5C{\equiv}CCOH(i\text{-}C_3H_7)CHOHCH(CH_3)_2$	—	$HgCl_2$	(furan product) (38.8)	547
$C_6H_5C{\equiv}CCOH(n\text{-}C_3H_7)CHOH(CH_2)_2CH_3$	—	$HgCl_2$	(furan product)	547
$cis\text{-}CH_3(CH_2)_6CH{=}CH(CH_2)_3\text{-}C{\equiv}C(CH_2)_3CO_2CH_3$	CH_3OH	$Hg(OAc)_2$, HCl	$cis\text{-}CH_3(CH_2)_6CH{=}CH(CH_2)_3CO(CH_2)_4CO_2CH_3$ + $cis\text{-}CH_3(CH_2)_6CH{=}CH(CH_2)_4CO(CH_2)_3CO_2CH_3$ (44.8 total)	190
	CH_3OH	$Hg(OAc)_2$, $NaBH_4$	$CH_3(CH_2)_6CH(OCH_3)CH(CH_2)_3CH(OH)CH(CH_2)_3CO_2CH_3$ (48)	190
$C_6H_5C{\equiv}CCOH(C_6H_5)CHOHC_6H_5$	—	HgC_2	(furan product)	546
$[CH_3O_2C(CH_2)_8C{\equiv}C]_2$	CH_3OH	$Hg(OAc)_2$, HCl	$CH_3O_2C(CH_2)_8COCH_2CO(CH_2)_9CO_2CH_3$ (70)	548

than the five-membered ring dioxolanes, which in turn are preferred over the seven-membered ring diethers [550, 551]. While mixtures of five- and six-membered ring products are observed with glycerol, the major product appears to be the dioxolane [525, 527, 550]. When α-hydroxy acids are employed, the corresponding lactones are formed, generally in good yield (Eq. 245) [525, 533].

$$RC \equiv CH \; + \; RCHOHCO_2H \; \longrightarrow \tag{245}$$

$$RC \equiv CH \; \longrightarrow \; R\overset{RO \quad OR}{\underset{}{C}}CH_3 \tag{246}$$

$$CH_3OCH_2C \equiv CCH_3 \; \longrightarrow \; CH_3OCH_2CH_2\overset{CH_3O \quad OCH_3}{\underset{}{C}}CH_3 \tag{247}$$

Relatively little work has been reported on the regiochemistry of these alkoxymercuration reactions. Terminal acetylenes always afford products of alcohol addition to the more highly substituted carbon (Eq. 246). Very few unsymmetrical internal acetylenes have been subjected to this reaction, but it is reported that methyl 9-undecynoate shows a slight preference (53:47) for alcohol addition to carbon number ten [190]. Only one ketal product is reported from both 1-methoxy-2-butyne and 2-octyne (Eqs. 247, 248) [538], but a mixture of regioisomers seems likely in this last reaction. Alkoxymercuration-hydrolysis of a symmetrical conjugated diyne has been reported to yield the corresponding beta diketone (Eq. 249) [548].

$$CH_3(CH_2)_4C \equiv CCH_3 \; \longrightarrow \; CH_3(CH_2)_4\overset{CH_3O \quad OCH_3}{\underset{}{C}}CH_2CH_3 \tag{248}$$

$$[CH_3O_2C(CH_2)_8C \equiv C]_2 \; \longrightarrow \; CH_3O_2C(CH_2)_8\overset{O}{\overset{\|}{C}}CH_2\overset{O}{\overset{\|}{C}}(CH_2)_9CO_2CH_3 \tag{249}$$

$$H_2C = CHCH_2C \equiv CR \; \longrightarrow \; H_2C = CHCH_2CH_2\overset{CH_3O \quad OCH_3}{\underset{}{C}}R \tag{250}$$

$$\underset{CH_3}{\underset{|}{H_2C = CC \equiv CH}} \; \longrightarrow \; H_2C = \underset{CH_3}{\underset{|}{C}} - \overset{RO \quad OR}{\underset{}{C}} - CH_3 \tag{251}$$

$$H_2C = CHC \equiv CR \; \longrightarrow \; CH_3OCH_2CH_2\overset{CH_3O \quad OCH_3}{\underset{}{C}}CH_2R \tag{252}$$

Several enynes have been subjected to alkoxymercuration. Enynes in which the double and triple bonds are distant apparently give products of both alkene and alkyne addition, with little chemoselectivity [190]. Treatment with methanolic hydrochloric acid regenerates the carbon—carbon double bond, affording a mixture of ketones from alkyne attack. On the other hand, 1,4-enynes reportedly undergo clean chemo- and regioselective alcohol addition to the more remote carbon of the triple bond (Eq. 250) [96]. The results with conjugated enynes are mixed. Both exclusive addition to the triple bond [534] and complete saturation [96, 532, 533] of the enyne have been reported on different acetylenes (Eqs. 251, 252).

$$(253)\ [543]$$

$$(254)$$

$$(255)$$

$$(256)$$

$$(257)$$

$$(258)$$

The alkoxymercuration of propargylic alcohols gives a variety of products. In many cases, particularly with ethylene glycol as the alcohol, the anticipated ketal can be isolated in good yield (Eq. 253) [531, 535, 537, 542]. In many examples, 1,4-dioxanes have been reported either as minor products or sometimes the major product (Eq. 254) [531, 535, 537, 539]. In one instance a bicyclic ether appears to be the major product (Eq. 255) [536]. Acetylenic diols behave differently and afford furanones [535] and furans [546, 547] (Eqs. 256, 257). In acetone as the solvent, this latter reaction affords the protodemercurated furans directly. Finally, propargylic acetates appear to behave normally except that the acetate group is removed during reaction (Eq. 258) [537].

References

1. Hofmann, K. A., Sand, J.: Ber. Dt. Chem. Ges., *33*, 1340 (1900).
2. Sand, J., Genssler, O.: Ber. Dt. Chem. Ges., *36*, 3699 (1903).
3. Aranda, V. G., Pastor, O. S.: Combustibles (Zaragoza), *10*, 183 (1950); Chem. Abstr., *46*, 423h (1952).
4. Aranda, V. G., Pastor, O. S., Cordon, J. L. M.: Combustibles (Zaragoza), *15*, 102 (1955); Chem. Zentralbl., 8102 (1959).
5. Pastor, O. Sanz: Rev. Acad. Cienc. Exact., Fis.-Quim. Natur. Zaragoza, *6*, 27 (1951); Chem. Abstr., *47*, 2688h (1953).
6. Kitching, W., Smith, A. J., Wells, P. R.: J. Chem. Soc., Chem. Commun., 370 (1968).
7. Kitching, W., Smith, A. J., Wells, P. R.: Austral. J. Chem., *21*, 2395 (1968).
8. Sand, J.: Ber. Dt. Chem. Ges., *34*, 2906 (1901).
9. Schrauth, W., Schoeller, W., Struensee, R.: Ber. Dt. Chem. Ges., *43*, 695 (1910).
10. Manchot, W.: Ber. Dt. Chem. Ges., *53*, 984 (1920).
11. Sandborn, L. T., Marvel, C. S.: J. Am. Chem. Soc., *48*, 1409 (1926).
12. Griffith, E., Marvel, C. S.: J. Am. Chem. Soc., *53*, 789 (1931).
13. Cotton, F. A., Leto, J. R.: J. Am. Chem. Soc., *80*, 4823 (1958).
14. Brownstein, S.: Diss. Faraday Soc., *34*, 25 (1962).
15. Kreevoy, M. M., Schaeffer, J. F.: J. Organometal. Chem., *6*, 589 (1966).
16. Zefirov, N. S., Shekhtman, N. M.: Dokl. Akad. Nauk SSSR, *177*, 842 (1967); Proc. Acad. Sci. USSR, Chem. Sec., *177*, 1110 (1967).
17. Factor, A., Traylor, T. G.: J. Org. Chem. *33*, 2607 (1968).
18. Waters, W. L.: Tetrahedron Lett., 3769 (1969).
19. Sokolov, V. I., Troitskaya, L. L., Reutov, O. A.: Zh. Org. Khim., *5*, 174 (1969); J. Org. Chem. USSR, *5*, 169 (1969).
20. Zefirov, N. S., Chekulaeva, V. N., Antonova, N. D., Gurvich, L. G.: Dokl. Akad. Nauk SSSR, *192*, 567 (1970); Proc. Acad. Sci. USSR, Chem. Sec., *192*, 363 (1970).
21. Schaumburg, K.: Lipids, *5*, 505 (1970).
22. Smith, R. G., Ensley, H. E., Smith, H. E.: J. Org. Chem., *37*, 4430 (1972).
23. Richter, R. F., Philips, C., Bach, R. D.: Tetrahedron Lett., 4327 (1972).
24. Ibusuki, T., Saito, Y.: J. Organometal. Chem., *56*, 103 (1973).
25. Zefirov, N. S., Gurvich, L. G.: J. Organometal. Chem., *81*, 309 (1974).
26. Zefirov, N. S., Gurvich, L. G.: Zh. Org. Khim., *10*, 127 (1974); J. Org. Chem. USSR, *10*, 126 (1974).
27. Voronkov, M. G., Chernov, N. F., Kalikhman, I. D.: Dokl. Akad. Nauk SSSR, *233*, 361 (1977); Proc. Acad. Sci. USSR, Chem. Sec., *233*, 138 (1977).
28. Shinoda, S., Isemura, M., Saito, Y.: Bull. Chem. Soc. Japan, *52*, 1855 (1979).
29. Lewis, A., Azoro, J.: J. Org. Chem., *46*, 1764 (1981).
30. Parker, R. G., Roberts, J. D.: J. Am. Chem. Soc., *92*, 743 (1970).
31. Brook, A. G., Wright, G. F.: Acta Cryst., *4*, 50 (1951).
32. Brook, A. G., Donovan, R., Wright, G. F.: Can. J. Chem., *31*, 536 (1953).
33. Wright, G. F.: Ann. N.Y. Acad. Sci., *65*, 436 (1957).
34. Ehrlich, H. W. W.: J. Chem. Soc., 509 (1962).
35. Bain, J., Harding, M. M.: J. Chem. Soc., 4025 (1965).
36. Nefedov, B. K., Sergeeva, N. S., Éidus, Ya. T.: Izv. Akad. Nauk SSSR, Ser. Khim., 1753 (1972); Bull. Acad. Sci. USSR, Div. Chem. Sci., *1697* (1972).
37. Spengler, G., Frömmel, H., Schäff, R., Faul, Ph., Lonsky, P.: Brennstoff-Chemie, *37*, 47 (1953).
38. Brit. Patent 411,507 (1934); Chem. Abstr., *28*, 6947³ (1934).

39. Peringer, P.: J. Inorg. Nucl. Chem., *42*, 1501 (1980).
40. Keller, H.: Ger. Patent 967,765 (1957); Chem. Abstr., *53*, 13057c (1959).
41. Shukis, A. J., Tallman, R. C.: J. Am. Chem. Soc., *65*, 2365 (1943).
42. Nefedov, B. K., Sergeeva, N. S., Éidus, Ya. T.: Izv. Akad. Nauk SSSR, Ser. Khim., 2494 (1972); Bull. Acad. Sci. USSR, Div. Chem. Sci., 2426 (1972).
43. Kreevoy, M. M., Stokker, G., Kretchmer, R. A., Ahmed, A. K.: J. Org. Chem., *28*, 3184 (1963).
44. Slama, J. T., Smith, H. W., Willson, C. G., Rapoport, H.: J. Am. Chem. Soc., *97*, 6556 (1975).
45. Schoeller, W., Schrauth, W., Essers, W.: Ber. Dt. Chem. Ges., *46*, 2864 (1913).
46. Shukis, A. J., Tallman, R. C.: J. Am. Chem. Soc., *66*, 1462 (1944).
47. Kreevoy, M. M., Turner, M. A.: J. Org. Chem., *30*, 373 (1965).
48. Urbschat, E.: Ger. Patent 912,452 (1952); Chem. Zentralbl., 444 (1955).
49. Kiefer, E. F., Waters, W. L.: J. Am. Chem. Soc., *87*, 4401 (1965).
50. Kreevoy, M. M.: J. Am. Chem. Soc., *81*, 1099 (1959).
51. Schaleger, L. L., Turner, M. A., Chamberlin, T. C., Kreevoy, M. M.: J. Org. Chem., *27*, 3421 (1962).
52. Spengler, G., Wilderotter, M., Trommer, T.: Brennstoff-Chemie, *45*, 182 (1964).
53. Thomas, M. H., Wetmore, F. E. W.: J. Am. Chem. Soc., *63*, 136 (1941).
54. Bloodworth, A. J., Griffin, I. M.: J. Chem. Soc., Perkin I, 195 (1975).
55. Bloodworth, A. J., Griffin, I. M.: J. Chem. Soc., Perkin II, 531 (1975).
56. Kiefer, E. F., Waters, W. L., Carlson, D. A.: J. Am. Chem. Soc., *90*, 5127 (1968).
57. Brown, H. C., Rei, M.-H.: J. Am. Chem. Soc., *91*, 5646 (1969).
58. Bilke, H., Collin, G., Duschek, Ch., Höbold, W., Höhn, R., Pritzkow, W., Schmidt, H., Schnurpfeil, D.: J. Prakt. Chem., *311*, 1037 (1969).
59. Robson, J. H., Wright, G. F.: Can. J. Chem., *38*, 21 (1960).
60. Kiefer, E. F., Gericke, W.: J. Am. Chem. Soc., *90*, 5131 (1968).
61. Floris, B., Illuminati, G.: J. Organometal. Chem., *168*, 203 (1979).
62. Kundu, K. K., Das, M. N.: Anal. Chem., *31*, 1358 (1959).
63. Levi, M.: Farmatsiya (Bulgaria), *8*, 31 (1958); Chem. Abstr., *54*, 10934f (1960).
64. Park, W. R. R., Wright, G. F.: J. Org. Chem., *19*, 1435 (1954).
65. Uemura, S., Tabata, A., Okano, M.: J. Chem. Soc., Chem. Commun., 517 (1972).
66. Spengler, G., Weber, A.: Brennstoff-Chemie, *40*, 22 (1959).
67. Lethbridge, A., Norman, R. O. C., Thomas, C. B.: J. Chem. Soc., Perkin I, 1929 (1974).
68. Bassetti, M., Floris, B., Illuminati, G.: J. Organometal. Chem., *202*, 351 (1980).
69. Geissman, T. A., Horowitz, R. M.: J. Am. Chem. Soc., *73*, 5759 (1951).
70. Barluenga, J., Alonso-Cires, L., Asensio, G.: Tetrahedron Lett., 2239 (1981).
71. Berman, L., Hall, R. H., Pyke, R. G., Wright, G. F.: Can. J. Chem., *30*, 541 (1952).
72. Bach, R. D., Richter, R. F.: J. Org. Chem., *38*, 3442 (1973).
73. Gunstone, F. D., Inglis, R. P.: Chem. Phys. Lipids, *10*, 73 (1973).
74. Patai, S., Harnik, M., Hoffmann, E.: J. Am. Chem. Soc., *72*, 923 (1950).
75. Downing, D. C., Wright, G. F.: J. Am. Chem. Soc., *68*, 141 (1946).
76. Birks, A. M., Wright, G. F.: J. Am. Chem. Soc., *62*, 2412 (1940).
77. Rodgman, A., Wright, G. F.: J. Am. Chem. Soc., *76*, 1382 (1954).
78. Kreevoy, M. M., Ditsch, L. T.: J. Am. Chem. Soc., *82*, 6124 (1960).

79. Lutsenko, I. F., Foss, V. L., Ivanova, I. L.: Dokl. Akad. Nauk SSSR, *141*, 1107 (1961); Proc. Acad. Sci. USSR, Chem. Sec., 1270 (1961).

80. Polishchuk, V. R., German, L. S., Knunyants, I. L.: Izv. Akad. Nauk SSSR, Ser. Khim., 2024 (1971); Bull. Acad. Sci. USSR, Div. Chem. Sci., 1908 (1971).

81. Seyferth, D., Woodruff, R. A.: J. Organometal. Chem., *71*, 335 (1974).

82. Keiko, N. A., Musorina, T. P., Tatarinova, A. A., Voronkov, M. G.: Izv. Akad. Nauk SSSR, Ser. Khim., 1617 (1975); Bull. Acad. Sci. USSR, Div. Chem. Sci., 1503 (1975).

83. Goodman, L., Ross, L. O., Greene, M. O., Greenberg, J., Baker, B. R.: J. Med. Pharm. Chem., *3*, 65 (1961).

84. Izumiya, N.: J. Chem. Soc. Japan, Pure Chem. Sec., *71*, 214 (1950); Chem. Abstr., *45*, 7011h (1951).

85. Yuki, H., Hatada, K., Nagata, K.: Bull. Chem. Soc. Japan, *43*, 1817 (1970).

86. Blicke, F. F., Biel, J. H.: J. Am. Chem. Soc., *76*, 3163 (1954).

87. Bertin, D., Girault, P.: Fr. Patent 2,063,601 (1971); Chem. Abstr., *76*, 153936d (1972).

88. Delmar, G. S., Macallum, E. N.: Ger. Patent 959,907 (1957); Chem. Abstr., *53*, 18865d (1959).

89. Danish Patent 84,183 (1958); Chem. Abstr., *53*, 11238g (1959).

90. Seyferth, D., Woodruff, R. A.: J. Org. Chem., *38*, 4031 (1973).

91. Seyferth, D., Woodruff, R. A.: J. Fluorine Chem., *2*, 214 (1972/73).

92. Polishchuk, V. R., German, L. S., Knunyants, I. L.: Izv. Akad. Nauk SSSR, Ser. Khim., 795 (1971); Bull. Acad. Sci. USSR, Div. Chem. Sci., 711 (1971).

93. Bloodworth, A. J., Bunce, R. J.: J. Organometal. Chem., *60*, 11 (1973).

94. Bloodworth, A. J., Bunce, R. J.: J. Chem. Soc. D, Chem. Commun., 753 (1970).

95. Bloodworth, A. J., Bunce, R. J.: J. Chem. Soc. C, 1453 (1971).

96. Killian, D. B., Hennion, G. F., Nieuwland, J. A.: J. Am. Chem. Soc., *58*, 892 (1936).

97. Popova, M. N., Stepanov, D. E., Yakovenko, V. S., Stepuk, E. S., Komarovskaya, I. A., Donskikh, V.: Dokl. Vses. Konf. Khim. Atsetilena, 4th, *2*, 360 (1972); Chem. Abstr., *79*, 78910x (1973).

98. Nesmeyanov, A. N., Lutsenko, I. F., Khomutov, R. M.: Izv. Akad. Nauk. SSSR, Otdel. Khim. Nauk, 942 (1957); Bull. Acad. Sci. USSR, Div. Chem. Sci., 971 (1957).

99. Mallik, K. L., Das, M. N.: J. Am. Chem. Soc., *82*, 4269 (1960).

100. Schiltz, L. R., Carter, H. E.: J. Biol. Chem., *116*, 793 (1940).

101. West, H. D., Carter, H. E.: J. Biol. Chem., *119*, 103 (1937).

102. Lisewska, I., Potocki, J., Pasternak, A.: J. Label. Compounds, *8*, 561 (1972); Chem. Abstr., *78*, 159776q (1973).

103. Hallaba, E., El-Asrag, H., Abou Zeid, Y.: Int. J. Appl. Radiat. Isotop., *20*, 195 (1969); Chem. Abstr., *71*, 13188h (1969).

104. Mani, R. S., Desai, C. N., Raghavan, S. V.: Indian J. Chem., *3*, 415 (1965).

105. Anghileri, L. J.: J. Nucl. Med., *4*, 193 (1964); Chem. Abstr., *62*, 11005e (1965).

106. Caro, R. A., De Paoli, T., Hager, A. A., Chueco, C. A., Marquez, R., Nicolini, J. O., Radicella, R.: Int. J. Appl. Radiat. Isotop., *23*, 384 (1972); Chem. Abstr., *77*, 152302c (1972).

107. Rowland, R. L., Perry, W. L., Foreman, E. L., Friedman, H. L.: J. Am. Chem. Soc., *72*, 3595 (1950).

108. Barza, S. A.: Brit. Patent 792,992 (1958); Chem. Abstr., *52*, 18222e (1958).

109. Rowland, R. L.: U.S. Patent 2,635,982 (1953); Chem. Abstr., *48*, 6460e (1954).
110. Beyer, D., Duschek, Ch., Franz, H. J., Höhn, R., Höbold, W., Kluge, P., Pritzkow, W., Schmidt, H.: J. Prakt. Chem., *313*, 956 (1971).
111. Bertin, D., Girault, P.: Fr. Demande 2,106,745 (1972); Chem. Abstr., *78*, 30002e (1973).
112. Keiko, N. A., Musorina, T. N., Kalikhman, I. D., Brodskaya, É. I., Tatarinova, A. A., Kopylovskaya, B. Kh., Voronkov, M. G.: Zh. Obshch. Khim., *47*, 421 (1977); J. Gen. Chem. USSR, *47*, 389 (1977).
113. Carter, H. E., West, H. D.: Org. Syn., *20*, 81 (1940).
114. Maskens, K., Polgar, N.: J. Chem. Soc., Perkin I, 109 (1973).
115. Tatsuoka, S., Murakami, M., Ueno, T., Morimoto, A.: Ann. Repts. Takeda Res. Lab., *9*, 7 (1950); Chem. Abstr., *46*, 2498b (1952).
116. Freidlina, R. Kh., Chukovskaya, E. Ts.: Izv. Akad. Nauk SSSR, Otdel. Khim. Nauk, 187 (1957); Bull. Acad. Sci. USSR, Div. Chem. Sci., 197 (1957).
117. McNeely, K. H., Wright, G. F.: J. Am. Chem. Soc., *77*, 2553 (1955).
118. Rowland, R. L., Perry, W. L., Gerstein, S.: J. Am. Chem. Soc., *73*, 91 (1951).
119. Brit. Patent 790,906 (1958); Chem. Abstr., *52*, 16213b (1958).
120. Rowland, R. L., Perry, W. L., Gerstein, S.: J. Am. Chem. Soc., *73*, 3691 (1951).
121. Viventi, R. V.: U.S. Patent 3,532,726 (1970); Chem. Abstr., *74*, 3728g (1971).
122. Werner, L. H., Scholz, C. R.: J. Am. Chem. Soc., *76*, 2453 (1954).
123. Gubitz, F. W., McKeon, Jr., W. B.: J. Med. Pharm. Chem., *5*, 168 (1962).
124. Mayeda, S.: Proc. Imp. Acad. Tokyo, *11*, 258 (1935).
125. Abderhalden, E., Heyns, K.: Ber. Dt. Chem. Ges., *67*, 530 (1934).
126. Wood, M. L., Madden, R. J., Carter, H. E.: J. Biol. Chem., *117*, 1 (1937).
127. Miescher, K., Hoffmann, K.: U.S. Patent 2,156,598 (1939); Chem. Abstr., *33*, 6001^2 (1939).
128. Whitehead, C. W., Traverso, J. J.: J. Am. Chem. Soc., *80*, 2182 (1958).
129. Henbest, H. B., Nicholls, B.: J. Chem. Soc., 227 (1959).
130. Fr. Patent 2,063,602 (1971); Chem. Abstr., *76*, 113376q (1972).
131. Nadkarni, M. V., Jones, J. W.: J. Am. Pharm. Assoc., Sci. Ed., *39*, 297 (1950).
132. Brit. Patent 782,050 (1957); Chem. Abstr., *52*, 2100d (1958).
133. Foye, W. O., Kotak, H. M., Hefferren, J. J.: J. Am. Pharm. Assoc., Sci. Ed., *41*, 273 (1952).
134. Friederich, H. H., Lausberg, D.: Ger. Patent 1,005,964 (1957); Chem Abstr., *53*, 18058g (1959).
135. Shapiro, S. L., Parrino, V. A., Freedman, L.: J. Am. Pharm. Assoc., Sci. Ed., *46*, 689 (1957).
136. Schrauth, W., Geller, H.: Ber. Dt. Chem. Ges., *55*, 2783 (1922).
137. Loev, B.: J. Org. Chem., *26*, 4394 (1961).
138. Kreevoy, M. M., Eisen, B. M.: J. Org. Chem., *28*, 2104 (1963).
139. Shaborov, Yu. S., Mochalov, S. S., Oretskaya, T. S., Karpova, V. V.: J. Organometal. Chem., *150*, 7 (1978).
140. Pearson, D. E., Sigal, Jr., M. V.: J. Org. Chem., *15*, 1055 (1950).
141. Whitehead, C. W.: J. Am. Chem. Soc., *80*, 2178 (1958).
142. Roth, R., Erlenmeyer, H.: Helv. Chim. Acta, *38*, 1276 (1955).
143. Brown, W. H., Wright, G. F.: J. Am. Chem. Soc., *62*, 1991 (1940).
144. Kao, Y.-S., Pan, P.-C., Loh, S.-H., Chen, C.-H., Hsü, H. Y.: Sci. Sinica, *7*, 738 (1958); Chem. Abstr., *53*, 13097c (1959).

III. Alkoxymercuration

145. Kao, Y.-S., Pan, P.-C., Loh, S.-H., Chen, C.-H., Hsu, H.-Y.: Hua Hsüeh Hsüeh Pao, *24*, 162 (1958); Chem. Abstr., *53*, 7088e (1959).
146. Rangaswami, S., Rao, V. S., Seshadri, T. R.: Proc. Ind. Acad. Sci. A, *7*, 296 (1938); Chem. Zentralbl. II, 3078 (1938).
147. Johnson, B. F. G., Lewis, J., Parker, D. G., Postle, S. R.: J. Chem. Soc., Dalton, 794 (1977).
148. Park, W. R. R., Wright, G. F.: J. Org. Chem., *19*, 1325 (1954).
149. Van Loon, E. J., Carter, H. E.: J. Am. Chem. Soc., *59*, 2555 (1937).
150. Matejka, K.: Ber. Dt. Chem. Ges., *69*, 274 (1936).
151. Biilmann, E.: J. Liebig Ann. Chem., *388*, 259 (1912).
152. Cahen, R. L.: J. Am. Pharm. Assoc., Sci. Ed., *36*, 139 (1947).
153. Lethbridge, A., Norman, R. O. C., Thomas, C. B.: J. Chem. Soc., Perkin I, 2465 (1975).
154. Gómez-Aranda, V., Barluenga-Mur, J., Yus-Astiz, M.: Rev. Acad. Cienc. Exactas, Fis.-Quim. Natur. Zaragoza, *28*, 225 (1973); Chem. Abstr., *80*, 83164g (1974).
155. Werner, L. H., Scholz, C. R.: U.S. Patent 2,790,816 (1957); Chem. Abstr., *51*, 16527e (1957).
156. Brit. Patent 776,776 (1957); Chem. Abstr., *52*, 423f (1958).
157. Kato, K.: Nippon Kagaku Zasshi, *81*, 829 (1960); Chem. Abstr., *56*, 493a (1962).
158. Gallina, C., Maneschi, M., Romeo, A.: J. Chem. Soc., Perkin I, 1134 (1973).
159. Cabaleiro, M. C., Ayala, A. D., Johnson, M. D.: J. Chem. Soc., Perkin II, 1207 (1973).
160. Rangaswami, S., Rao, V. S., Seshadri, T. R.: Proc. Ind. Acad. Sci. A, *7*, 312 (1938); Chem. Zentralbl. I, 104 (1939).
161. Sah, P. P. T., Tseu, C.-Z.: J. Chin. Chem. Soc., *5*, 134 (1937).
162. Schrauth, W., Schoeller, W., Struensee, R.: Ber. Dt. Chem. Ges., *44*, 1048 (1911).
163. Dominikiewicz, M., Kijewska, M.: Arch. Chem. Farm., *4*, 8 (1939); Chem. Abstr., *34*, 740^2 (1940).
164. Ehrhart, G., Ruschig, H., Leditschke, H.: U.S. Patent 2,704,767 (1955); Chem. Abstr., *50*, 15586b (1956).
165. Ukai, T., Hayashi, M., Abe, H.: J. Pharm. Soc. Japan, *57*, 28 (1937); Chem. Abstr., *31*, 3876^4 (1937).
166. Barluenga, J., Perez-Prieto, J., Asensio, G.: J. Chem. Soc., Chem. Commun., 1181 (1982).
167. Halpern, A., Jones, J. W., Gross, E. G.: J. Am. Pharm. Assoc., Sci. Ed., *37*, 333 (1948); Chem. Abstr., *43*, 2169g (1949).
168. Baker, B. R., Safir, S. R., Bernstein, S.: U.S. Patent 2,428,955 (1947); Chem. Abstr., *42*, 1317d (1948).
169. Sterzycki, R., Sobotka, W.: Symp. Pap. — IUPAC Int. Symp. Chem. Nat. Prod., 11th, *3*, 142 (1978); Chem. Abstr., *92*, 94581q (1980).
170. Szekeres, L.: Gazz. Chim. Ital., *79*, 832 (1949); Chem. Abstr., *44*, 4450i (1950).
171. Carrara, G.: Gazz. Chim. Ital., *79*, 834 (1949); Chem. Abstr., *44*, 4451b (1950).
172. Carrara, G., Mori, E.: Gazz. Chim. Ital., *73*, 113 (1943); Chem. Abstr., *38*, 4928^8 (1944).
173. Carrara, G.: Gazz. Chim. Ital., *79*, 201 (1949); Chem. Abstr., *44*, 2943f (1950).
174. Szekeres, L.: Gazz. Chim. Ital., *79*, 56 (1949); Chem. Abstr., *43*, 7921f (1949).

175. Pearson, P. E., Sigal, M. V., Jr., Krug, R. H.: J. Org. Chem., *15*, 1048 (1950).
176. Gordon, S. M., Kipnis, K.: U.S. Patent 2,343,547 (1944); Chem. Abstr., *38*, 3093[9] (1944).
177. Wiese, G. A., Jones, J. W.: J. Am. Pharm. Assoc., Sci. Ed., *39*, 286 (1950).
178. Yale, H. L.: J. Am. Chem. Soc., *79*, 4762 (1957).
179. Yale, H. L.: U.S. Patent 2,749,361 (1956); Chem. Abstr., *51*, 465h (1957).
180. Ralston, A. W., McCorkle, M. R.: U.S. Patent 2,289,590 (1943); Chem. Abstr., *37*, 498[7] (1943).
181. Ralston, A. W., McCorkle, M. R.: U.S. Patent 2,284,067 (1942); Chem. Abstr., *36*, 6302[4] (1942).
182. Ralston, A. W., McCorkle, M. R.: U.S. Patent 2,356,884 (1944); Chem. Abstr., *39*, 1246[7] (1945).
183. Floris, B., Illuminati, G.: Coord. Chem. Rev., *16*, 107 (1975).
184. Bockmühl, M., Middendorf, L., Fritzsche, P.: U.S. Patent 2,213,457 (1941?); Chem. Abstr., *35*, 583[5] (1941).
185. Robbins, E. B., Chen, K. K.: J. Am. Pharm. Assoc., Sci. Ed., *40*, 249 (1951); Chem. Abstr., *45*, 6348i (1951).
186. Whitehead, C. W.: U.S. Patent 2,698,858 (1955); Brit. Patent 733,732 (1955); Chem. Abstr., *49*, 14032g (1955).
187. Foye, W. O., Mode, R. A.: J. Am. Pharm. Assoc., Sci. Ed., *44*, 76 (1955).
188. Mingoja, Q.: Arquiv. Biol. (São Paulo), *26*, 231 (1942); Chem. Abstr., *38*, 4999[3] (1944).
189. Bodrikov, I. V., Kartashov, V. R., Koval'ova, L. I., Zefirov, N. S.: J. Organometal. Chem., *82*, C23 (1974).
190. Lam, C. H., Lie Ken Jie, M. S. F.: Chem. Phys. Lipids, *16*, 181 (1976).
191. Wendt, G., Bruce, W. F.: J. Org. Chem., *23*, 1448 (1958).
192. Wendt, G. R.: U.S. Patent 2,834,795 (1958); Chem. Abstr., *53*, 2118f (1959).
193. Molnar, N. M.: U.S. Patent 2,117,901 (1938?); Chem. Abstr., *32*, 5584[1] (1938).
194. Ehrhardt, G., Ruschig, H., Leditschke, H.: U.S. Patent 2,695,305 (1954); Chem. Abstr., *49*, 15969g (1955).
195. Oda, J., Nakagawa, T., Inoueye, Y.: Bull. Chem. Soc. Japan, *40*, 373 (1967).
196. Kohler, E. P., Addinall, C. R.: J. Am. Chem. Soc., *52*, 3728 (1930).
197. Middleton, E. B.: J. Am. Chem. Soc., *45*, 2763 (1923).
198. Ayala, A. D., Cabaleiro, M. C.: Tetrahedron Lett., 4709 (1972).
199. Corey, E. J., Tius, M. A.: Tetrahedron Lett., 2081 (1977).
200. Czernecki, S., Georgoulis, C., Provelenghiou, C.: Tetrahedron Lett., 2623 (1975).
201. Kawana, M., Emoto, S.: Bull. Chem. Soc. Japan, *40*, 618 (1967).
202. Gunstone, F. D., Inglis, R. P.: Chem. Phys. Lipids, *10*, 89 (1973).
203. Kahovcová, J., Romanuk, M.: Coll. Czech. Chem. Commun., *46*, 1413 (1981).
204. Ralston, A. W., Christensen, C. W., Josh, G.: Oil and Soap, *14*, 5 (1937).
205. Beau, J.-M., Sinay, P., Kamerling, J. P., Vliegenthart, J. F. G.: Carbohydr. Res., *67*, 65 (1978).
206. Hornstein, I.: Anal. Chem., *23*, 1329 (1951).
207. Ferrier, R. J., Prasit, P.: Carbohydr. Res., *82*, 263 (1980).
208. Waters, W. L., Traylor, T. G., Factor, A.: J. Org. Chem., *38*, 2306 (1973).
209. Collin, G., Müller-Hagen, G., Pritzkow, W.: J. Prakt. Chem., *314*, 229 (1972).
210. Manolopoulos, P. T., Mednick, M., Lichtin, N. N.: J. Am. Chem. Soc., *84*, 2203 (1962).
211. Babb, R. M., Gardner, P. D.: J. Chem. Soc., Chem. Commun., 1678 (1968).

212. Romeyn, J., Wright, G. F.: J. Am. Chem. Soc., 69, 697 (1947).
213. Rabiant, J., Wittig, G.: Bull. Soc. Chim. France, 798 (1957).
214. Hill, C. L., Whitesides, G. M.: J. Am. Chem. Soc., 96, 870 (1974).
215. Rodgman, A., Shearer, D. A., Wright, G. F.: Can. J. Chem., 35, 1377 (1957).
216. Buckles, R. E., Smith, R. J.: J. Am. Chem. Soc., 74, 3174 (1952).
217. Brook, A. G., Wright, G. F.: Can. J. Chem., 29, 308 (1951).
218. Berg, O. W., Lay, W. P., Rodgman, A., Wright, G. F.: Can. J. Chem., 36, 358 (1958).
219. Rodgman, A., Wright, G. F.: J. Org. Chem., 18, 1617 (1953).
220. Nesmeyanov, A. N., Freidlina, R. Kh.: Zh. Obshch. Khim., 7, 43 (1937).
221. Nesmeyanov, A. N., Freidlina, R. Kh.: Ber. Dt. Chem. Ges., 69, 1631 (1936).
222. Traylor, T. G.: J. Am. Chem. Soc., 86, 244 (1964).
223. Gubitz, F. W., Clarke, R. L.: J. Org. Chem., 26, 559 (1961).
224. Johnson, M. R., Rickborn, B.: J. Org. Chem., 34, 2781 (1969).
225. Johnson, M. R., Rickborn, B.: J. Chem. Soc., Chem. Commun., 1073 (1968).
226. Henbest, H. B., Nichols, B.: Proc. Chem. Soc., 61 (1957).
227. Goodman, L., Acton, E. M., Mosher, C. W., Marsh, J. P., Jr.: Methods Carbohydr. Chem., 8, 207 (1980); Chem. Abstr., 93, 239841g (1980).
228. Marsh, J. P., Jr., Mosher, C. W., Acton, E. M., Goodman, L.: J. Chem. Soc., Chem. Commun., 973 (1967).
229. Inglis, G. R., Schwartz, J. C. P., McLaren, L.: J. Chem. Soc., 1014 (1962).
230. Park, W. R. R., Wright, G. F.: Can. J. Chem., 35, 1088 (1957).
231. Sokolov, V. I., Troitskaya, L. L., Reutov, O. A.: Zh. Org. Khim., 1, 1579 (1965); J. Org. Chem. USSR, 1, 1601 (1965).
232. Robson, J. H., Wright, G. F.: Can. J. Chem., 38, 1 (1960).
233. Whistler, R. L.: Belg. Patent 887,820 (1981); Chem. Abstr., 96, 52633k (1982).
234. Jernow, J. L., Gray, D., Closson, W. D.: J. Org. Chem., 36, 3511 (1971).
235. Bach, R. D., Richter, R. F.: Tetrahedron Lett., 3915 (1971).
236. Sokolov, V. I., Troitskaya, L. L., Reutov, O. A.: Dokl. Akad. Nauk SSSR, 166, 136 (1966); Bull. Acad. Sci. USSR, Chem. Sec., 166, 45 (1966).
237. Honda, S., Izumi, K., Takiura, T.: Carbohydr. Res., 23, 427 (1972).
238. Takiura, K., Honda, S.: Carbohydr. Res., 21, 379 (1972).
239. Musker, W. K., Larson, G. L.: Tetrahedron Lett., 3481 (1968).
240. Dombrovskii, V. S., Yakushkina, N. I., Bolesov, I. G.: Zh. Org. Khim., 15, 1325 (1979); J. Org. Chem. USSR, 15, 1184 (1979).
241. Schrauth, W., Bauerschmidt, H.: Ber. Dt. Chem. Ges., 47, 2736 (1914).
242. Schaeffer, H. J., Collins, C. J.: J. Am. Chem. Soc., 78, 124 (1956).
243. Tokuyama, K.: Japan. Patent 13,469 (1967); Chem. Abstr., 68, 13322k (1968).
244. Honda, S., Kakehi, K., Takai, H., Takiura, K.: Carbohydr. Res., 29, 477 (1973).
245. Baer, H. H., Hanna, Z. S.: Can. J. Chem., 59, 889 (1981).
246. Leftin, J. H., Lichtin, N. N.: Israel J. Chem., 3, 107 (1965).
247. Belg. Patent 614,210 (1962); Chem. Abstr., 58, 1489c (1963).
248. Fritz, H., Lehmann, J., Littke, W., Schlesselmann, P.: Carbohydr. Res., 99, 82 (1982).
249. Dean, A. L., Wrenshall, R., Fujimoto, G.: J. Am. Chem. Soc., 47, 403 (1925).
250. Shabarov, Yu. S., Sychkova, L. D., Kalinichenko, A. N., Levina, R. Ya.: Zh. Obshch. Khim., 41, 1599 (1971); J. Gen. Chem. USSR, 41, 1606 (1971).

251. Tidwell, T. T., Traylor, T. G.: J. Org. Chem., *33*, 2614 (1968).
252. Rowland, R. L.: J. Am. Chem. Soc., *73*, 2381 (1951).
253. Traylor, T. G., Baker, A. W.: Tetrahedron Lett., *19*, 14 (1959).
254. Aberchrombie, M. J., Rodgman, A., Bharucha, K. R., Wright, G. F.: Can. J. Chem., *37*, 1328 (1959).
255. Tobler, E., Foster, D. J.: Helv. Chim. Acta, *48*, 366 (1965).
256. Challenger, F., Clapham, P. H.: J. Chem. Soc., 1615 (1948).
257. Müller, E.: Chem. Ber., *109*, 3793 (1976).
258. Clark, R. L., Gubitz, F. W.: U.S. Patent 2,921,068 (1960); Chem. Abstr., *54*, 8867c (1960).
259. Bach, R. D., Richter, R. F.: J. Am. Chem. Soc., *94*, 4747 (1972).
260. Seshadri, T. R., Rao, P. S.: Proc. Ind. Acad. Sci. A, *4*, 162 (1936); Chem. Zentralbl. I, 2370 (1937).
261. Rao, P. S., Sastri, V. D. N., Seshadri, T. R.: Proc. Ind. Acad. Sci. A, *9*, 22 (1939); Chem. Zentralbl. II, 638 (1939).
262. Bly, R. S., Bly, R. K., Bedenbaugh, A. O., Vail, O. R.: J. Am. Chem. Soc., *89*, 881 (1967).
263. Bach, R. D., Brummel, R. N.: J. Am. Chem. Soc., *97*, 453 (1975).
264. Gasopoulos, I.: Ber. Dt. Chem. Ges., *59*, 2184 (1924).
265. Bluthe, N., Ecoto, J., Fetizon, M., Lazare, S.: J. Chem. Soc., Perkin I, 1747 (1980).
266. Giese, B.: Chem. Ber., *108*, 2998 (1975).
267. McNeely, K. H., Rodgman, A., Wright, G. F.: J. Org. Chem., *20*, 714 (1955).
268. Cheung, T. M., Horton, D., Weckerle, W.: Carbohydr. Res., *59*, 276 (1977).
269. Cristol, S. J., Perry, J. S., Jr., Beckley, R. S.: J. Org. Chem., *41*, 1912 (1976).
270. Ortar, G., Torrini, I.: Tetrahedron, *33*, 859 (1977).
271. Bentham, S., Chamberlain, P., Whitham, G. H.: J. Chem. Soc. D, Chem. Commun., 1528 (1970).
272. Collin, G., Jahnke, U., Just, G., Lorenz, G., Pritzkow, W., Röllig, M., Winguth, L., Dietrich, P., Döring, C.-E., Hauthal, H. G., Wiedenhöft, A.: J. Prakt. Chem., *311*, 238 (1969).
273. Connor, T., Wright, G. F.: J. Am. Chem. Soc., *68*, 256 (1946).
274. Ferrier, R. J.: Adv. Carbohydr. Chem., *20*, 67 (1965).
275. Ferrier, R. J.: Adv. Carbohydr. Chem. Biochem., *24*, 199 (1969).
276. Kergomard, A., Tardivat, J. C., Vuillerme, J. P.: Bull. Soc. Chim. France, 2572 (1974).
277. Kreshkov, A. P., Balyatinskaya, L. N., Chesnokova, S. M., Kurchenko, T. V.: Zh. Obshch. Khim., *41*, 2513 (1971); J. Gen. Chem. USSR, *41*, 2541 (1971).
278. Nayak, P. L., Rout, M. K.: J. Inst. Chem., Calcutta, *43*, 32 (1971); Chem. Abstr., *75*, 48082d (1971).
279. Chaudhuri, A. K., Mallik, K. L., Das, M. N.: Tetrahedron, *19*, 1981 (1963).
280. Patnaik, A. K., Nayak, P. L., Rout, M. K.: Indian J. Chem., *8*, 722 (1970).
281. Kreshkov, A. P., Balyatinskaya, L. N.: Zh. Obshch. Khim., *37*, 2211 (1967); J. Gen. Chem. USSR, *37*, 2099 (1967).
282. Chaudhuri, A. K., Das, M. N.: Tetrahedron, *21*, 457 (1965).
283. Ambidge, I. C., Dwight, S. K., Rynard, C. M., Tidwell, T. T.: Can. J. Chem., *55*, 3086 (1977).
284. Kreshkov, A. P., Balyatinskaya, L. N., Chesnokova, S. M.: Zh. Obshch. Khim., *43*, 166 (1973); J. Gen. Chem. USSR, *43*, 164 (1973).
285. Fukuzumi, S., Kochi, J. K.: J. Am. Chem. Soc., *103*, 2783 (1981).

III. Alkoxymercuration

286. Bergmann, H. J., Collin, G., Just, G., Müller-Hagen, G., Pritzkow, W.: J. Prakt. Chem., *314*, 285 (1972).
287. Jantzen, E., Andreas, H.: Chem. Ber., *92*, 1427 (1959).
288. Müller-Hagen, G., Pritzkow, W.: J. Prakt. Chem., *311*, 874 (1969).
289. Lewis, A., Azoro, J.: Tetrahedron Lett., 3627 (1979).
290. Balyatinskaya, L. N.: Zh. Obshch. Khim., *47*, 198 (1977); J. Gen. Chem. USSR, *47*, 182 (1977).
291. Satpathy, K. K., Patnaik, A. K., Nayak, P. L., Rout, M. K.: J. Indian Chem. Soc., *48*, 847 (1971).
292. Jantzen, E., Andreas, H.: Angew. Chem., *70*, 656 (1958).
293. Raunio, E. K., Bonner, W. A.: J. Org. Chem., *31*, 396 (1966).
294. Asinger, F., Fell, B., Steffan, G., Hadik, G., Thiessen, F.: Erdöl Kohle, *18*, 178 (1965); Chem. Abstr., *63*, 474c (1965).
295. Döring, C.-E., Hauthal, H. G.: J. Prakt. Chem. (4), *22*, 59 (1963).
296. Bach, R. D., Richter, R. F.: Tetrahedron Lett., 4099 (1973).
297. Asinger, F., Fell, B., Hadik, G., Steffan, G.: Chem. Ber., *97*, 1568 (1964).
298. Wright, G. F.: Chem. in Canada, *2*, 149 (1950).
299. Kreevoy, M. M., Kowitt, F. R.: J. Am. Chem. Soc., *82*, 739 (1960).
300. Kreevoy, M. M., Gilje, J. W., Kretchmer, R. A.: J. Am. Chem. Soc., *83*, 4205 (1961).
301. Kreevoy, M. M., Ditsch, L. T.: J. Org. Chem., *25*, 134 (1960).
302. Schaleger, L. L., Watamori, N.: J. Phys. Chem., *73*, 2011 (1969).
303. Kiwan, A. M., Abdel-Hamid, A. A., Fouda, M. F. R.: J. Organometal. Chem., *23*, 19 (1970).
304. Kreevoy, M. M., Gilje, J. W., Ditsch, L. T., Batorewicz, W., Turner, M. A.: J. Org. Chem., *27*, 726 (1962).
305. Bertram, S. H.: Rec. Trav. Chim. Pays-Bas, *46*, 397 (1927).
306. Inouye, Y., Noda, M., Hirayama, O.: J. Am. Oil Chem. Soc., *32*, 132 (1955).
307. Inouye, Y., Noda, M.: Arch. Biochem. Biophys., *76*, 271 (1958).
308. Kishimoto, Y., Radin, N. S.: J. Lipid Res., *1*, 72 (1959).
309. Spengler, G.: Brit. Patent 835,779 (1960); Chem. Abstr., *55*, 3428e (1961).
310. Stearns, E. M., Jr., White, H. B., Jr., Quackenbush, F. W.: J. Am. Oil Chem. Soc., *39*, 61 (1962).
311. Timofeeva, E. A., Petryaeva, G. S., Shuikin, N. I.: Neftekhimiya, *5*, 579 (1965); Chem. Abstr., *63*, 16199d (1965).
312. White, H. B., Jr.: J. Chromatogr., *21*, 213 (1966).
313. Jantzen, E., Andreas, H., Morgenstern, K., Roth, W.: Fette, Seifen, Anstrichmittel, *63*, 685 (1961).
314. Olah, G. A., Clifford, P. R.: J. Am. Chem. Soc., *93*, 1261 (1971).
315. Napolitano, J. P., Clossen, R. D.: U.S. Patent 3,391,212 (1968); Chem. Abstr., *69*, 76600r (1968).
316. Bach, R. D., Brummel, R. N., Richter, R. F.: Tetrahedron Lett., 2879 (1971).
317. Schrauth, W., Schoeller, W., Struensee, R.: Ber. Dt. Chem. Ges., *44*, 1432 (1911).
318. Barluenga, J., Concellon, J. M., Asensio, G., Yus, M.: An. Quim., *74*, 512 (1978); Chem. Abstr., *89*, 179011s (1978).
319. Barluenga, J., Concellon, J. M., Ara, A., Asensio, G., Yus, M.: An. Quim., *74*, 785 (1978); Chem. Abstr., *91*, 57131b (1979).
320. Kurosawa, H., Okada, H., Hattori, T.: Tetrahedron Lett., 4495 (1981).
321. Wright, G. F.: Can. J. Chem., *30*, 268 (1952).
322. Horton, D., Tarelli, J. M., Wander, J. D.: Carbohydr. Res., *23*, 440 (1972).
323. Honda, S., Takiura, K.: Carbohydr. Res., *34*, 45 (1974).

324. Georgoulis, C., Paillasseur, E., Valery, J. M.: Synthesis, 67 (1979).
325. Giese, B., Heuck, K.: Tetrahedron Lett., 1829 (1980).
326. Giese, B., Heuck, K., Lüning, U.: Tetrahedron Lett., 2155 (1981).
327. Giese, B., Heuck, K.: Chem. Ber., *112*, 3759 (1979).
328. Davidson, J. M.: J. Chem. Soc., Chem. Commun., 126 (1966).
329. Hunt, D. F., Rodeheaver, G. T.: Tetrahedron Lett., 3595 (1972).
330. DeBoer, A., Thanel, C. J., Wilson, G. A.: Tetrahedron Lett., 5137 (1972).
331. Bloodworth, A. J., Savva, R. A.: J. Organometal. Chem., *152*, C29 (1978).
332. Sayre, L. M., Jensen, F. R.: J. Org. Chem., *44*, 228 (1979).
333. Watanabe, W. H., Conlon, L. E.: J. Am. Chem. Soc., *79*, 2828 (1957).
334. Weeks, G. A., Grant, W. J.: Brit. Patent 709,106 (1954); Chem. Abstr., *49*, 10360i (1955).
335. Yuki, H., Hatada, K., Nagata, K., Kajiyama, K.: Bull. Chem. Soc. Japan, *42*, 3546 (1969).
336. Burgstahler, A. W., Nordin, I. C.: J. Am. Chem. Soc., *83*, 198 (1961).
337. Büchi, G., White, J. D.: J. Am. Chem. Soc., *86*, 2884 (1964).
338. Watanabe, W. H., Conlon, L. E.: U.S. Patent 2,760,990 (1956); Chem. Abstr., *51*, 3654h (1957).
339. Adelman, R. L.: U.S. Patent 2,579,412 (1951); Chem. Abstr., *46*, 11227c (1952).
340. Adelman, R. L.: J. Am. Chem. Soc., *77*, 1669 (1955).
341. Okuyama, T., Fueno, T., Furukawa, J.: Tetrahedron, *25*, 5409 (1969).
342. Watanabe, W. H.: J. Am. Chem. Soc., *79*, 2833 (1957).
343. Adelman, R. L.: U.S. Patent 2,579,411 (1951); Chem. Abstr., *46*, 11226i (1952).
344. Adelman, R. L.: J. Am. Chem. Soc., *75*, 2678 (1953).
345. Coffman, D. D.: U.S. Patent 2,384,726 (1945); Chem. Abstr., *40*, 597^2 (1946).
346. Croxall, W. J., Glavis, F. J., Neher, H. T.: J. Am. Chem. Soc., *70*, 2805 (1948).
347. Mowry, D. T., Yanko, W. H., Ringwald, E. L.: J. Am. Chem. Soc., *69*, 2358 (1947).
348. Watanabe, W. H., Conlon, L. E., Hwa, J. C. H.: J. Org. Chem., *23*, 1666 (1958).
349. Tausz, J.: Chem. Zeitung, *42*, 349 (1918).
350. Tausz, J., Peter, W.: Petroleum, *13*, 649 (1918).
351. Martin, R. W.: Anal. Chem., *21*, 921 (1949).
352. Marquardt, R. P., Luce, E. N.: Anal. Chem., *21*, 1194 (1949).
353. Das, M. N.: Anal. Chem., *26*, 1086 (1954).
354. Budesínský, B.: Chem. Listy, *51*, 259 (1957); Chem. Abstr., *51*, 6435h (1957).
355. Budesínský, B.: Coll. Czech. Chem. Commun., *22*, 1147 (1957).
356. Johnson, J. B., Fletcher, J. P.: Anal. Chem., *31*, 1563 (1959).
357. Mallik, K. L., Das, M. N.: Chem. Ind., 162 (1959).
358. Balyatinskaya, L. N., Kreshkov, A. P., Tur'yan, Ya. I.: Zh. Anal. Khim., *19*, 1025 (1964); J. Anal. Chem. USSR, *19*, 952 (1964).
359. Triska, J., Mrnkova, A., Vodicka, L.: Sb. Vys. Sk. Chem. — Technol. Praze, Technol. Paliv, *D39*, 261 (1978); Chem. Abstr., *93*, 94853a (1980).
360. Balyatinskaya, L. N., Kreshkov, A. P.: Talanta, *27*, 1051 (1980).
361. Hofman, J., Kubesová, J.: Coll. Czech. Chem. Commun., *43*, 404 (1978).
362. Kaufmann, H. P., Pollerberg, J.: Fette, Seifen, Anstrichmittel, *59*, 815 (1957).
363. White, Jr., H. B., Quackenbush, F. W.: J. Am. Oil Chem. Soc., *39*, 511 (1962).
364. Saxena, M. P., Bhattacharyya, K. K.: Analyst (London), *102*, 878 (1977).

III. Alkoxymercuration

365. Abley, P., McQuillin, F. J., Minnikin, D. E., Kusamran, K., Maskens, K., Polgar, N.: J. Chem. Soc. D, Chem. Commun., 348 (1970).
366. Minnikin, D. E., Abley, P., McQuillin, F. J., Kusamran, K., Maskens, K., Polgar, N.: Lipids, *9*, 135 (1974).
367. Blomquist, G. J., Howard, R. W., McDaniel, C. A., Remaley, S., Dwyer, L. A., Nelson, D. R.: J. Chem. Ecol., *6*, 257 (1980).
368. Vostrowsky, O., Michaelis, K., Bestmann, H. J.: J. Liebig Ann. Chem., 1721 (1981).
369. Eldjarn, L., Jellum, E.: Acta Chem. Scand., *17*, 2610 (1963).
370. Summerbell, R. K., Lestina, G., Waite, H.: J. Am. Chem. Soc., *79*, 234 (1957).
371. Summerbell, R. K., Stephens, J. R.: J. Am. Chem. Soc., *76*, 6401 (1954).
372. Summerbell, R. K., Stephens, J. R.: J. Am. Chem. Soc., *77*, 6080 (1955).
373. Sand, J.: Ber. Dt. Chem. Ges., *34*, 1385 (1901).
374. Sand, J.: J. Liebig Ann. Chem., *329*, 135 (1903).
375. Benhamou, M.-C., Etemad-Moghadam, G., Spéziale, V., Lattes, A.: J. Heterocyclic Chem., *15*, 1313 (1978).
376. Nesmeyanov, A. N., Lutsenko, I. F.: Izv. Akad. Nauk SSSR, Otdel. Khim. Nauk, 296 (1943); Chem. Abstr., *38*, 5498^6 (1944).
377. Brown, H. C., Geoghegan, P. J., Jr., Kurek, J. T., Lynch, G. J.: Organometal. Chem. Syn., *1*, 7 (1970).
378. Werner, L. H., Scholz, C. R.: J. Am. Chem. Soc., *76*, 2701 (1954).
379. Summerbell, R. K., Kalb, G. H., Graham, E. S., Allred, A. L.: J. Org. Chem., *27*, 4461 (1962).
380. Werner, L. H., Scholz, C. R.: U.S. Patent 2,790,812 (1957); Chem. Abstr., *51*, 14833f (1957).
381. Spéziale, V., Roussel, J., Lattes, A.: J. Heterocyclic Chem., *11*, 771 (1974).
382. Benhamou, M. C., Etemad-Moghadam, G., Spéziale, V., Lattes, A.: Synthesis, 891 (1979).
383. Mihailovic, M. L., Marinkovic, D., Orbovic, N., Gojkovic, S., Konstantinovic, S.: Glas. Hem. Drus. Beograd, *45*, 497 (1980); Chem. Abstr., *95*, 61923m (1981).
384. Halpern, J., Tinker, H. B.: J. Am. Chem. Soc., *89*, 6427 (1967).
385. Scholz, C. R., Werner, L. H.: Ger. Patent 946,539 (1956); Chem. Abstr., *53*, 12208c (1959).
386. Moon, S., Waxman, B. H.: J. Org. Chem., *34*, 288 (1969).
387. Paquette, L. A., Dunkin, I. R., Freeman, J. P., Storm, P. C.: J. Am. Chem. Soc., *94*, 8124 (1972).
388. Overman, L. E.: J. Chem. Soc., Chem. Commun., 1196 (1972).
389. Overman, L. E., Campbell, C. B.: J. Org. Chem., *39*, 1474 (1974).
390. Bly, R. K., Bly, R. S.: J. Org. Chem., *28*, 3165 (1963).
391. Ganter, C., Wicker, K., Wigger, N.: Chimia, *24*, 27 (1970).
392. Ganter, C., Wicker, K.: Helv. Chim. Acta, *53*, 1693 (1970).
393. Ackermann, P., Tobler, H., Ganter, C.: Helv. Chim. Acta, *55*, 2731 (1972).
394. Barrelle, M., Apparu, M.: Tetrahedron, *33*, 1309 (1977).
395. Weinberg, N. L., Wright, G. F.: Can. J. Chem., *43*, 24 (1965).
396. Bordwell, F. G., Douglass, M. L.: J. Am. Chem. Soc., *88*, 993 (1966).
397. Ganter, C., Duthaler, R. O., Zwahlen, W.: Helv. Chim. Acta, *54*, 578 (1971).
398. Paquette, L. A., Storm, P. C.: J. Org. Chem., *35*, 3390 (1970).
399. Moon, S., Takakis, J. M., Waxman, B. H.: J. Org. Chem., *34*, 2951 (1969).
400. Sand, J., Singer, F.: J. Liebig Ann. Chem., *329*, 166 (1903).
401. Vincens, M., Dumont, C., Vidal, M.: Can. J. Chem., *57*, 2314 (1979).

402. Spéziale, V., Amat, M., Lattes, A.: J. Heterocyclic Chem., *13*, 349 (1976).
403. Amouroux, R., Chastrette, F., Chastrette, M.: J. Heterocyclic Chem., *18*, 565 (1981).
404. Amouroux, R., Chastrette, F., Chastrette, M.: Bull. Soc. Chim. France II, 293 (1981).
405. Etemad-Moghadam, G., Benhamou, M. C., Spéziale, V., Lattes, A., Bielawska, A.: Nouveau J. Chim., *4*, 727 (1980).
406. Dobrev, A., Périé, J. J., Lattes, A.: Tetrahedron Lett., 4013 (1972).
407. Brook, A. G., Rodgman, A., Wright, G. F.: J. Org. Chem., *17*, 988 (1952).
408. Sand, J., Singer, F.: Ber. Dt. Chem. Ges., *35*, 3170 (1902).
409. Wilder, P., Jr., Portis, A. R., Jr., Wright, G. W., Shepherd, J. M.: J. Org. Chem., *39*, 1636 (1974).
410. White, J. D., Avery, M. A., Carter, J. P.: J. Am. Chem. Soc., *104*, 5486 (1982).
411. Giese, B.: Chem. Ber., *108*, 2978 (1975).
412. Bartlett, P. A., Adams, J. L.: J. Am. Chem. Soc., *102*, 337 (1980).
413. Brimacombe, J. S., Miller, J. A., Zakir, U.: Carbohydr. Res., *49*, 233 (1976).
414. Lombard, R., Ambroise, G.: Bull. Soc. Chim. France, 230 (1961).
415. Coxon, J. M., Hartshorn, M. P., Mitchell, J. W., Richards, K. E.: Chem. Ind., 652 (1968).
416. Bambagiotti A., M., Vincieri, F. F., Coran, S. A.: J. Org. Chem., *39*, 680 (1974).
417. Brook, A. G., Wright, G. F.: J. Org. Chem., *22*, 1314 (1957).
418. Brieger, G., Burrows, E. P.: J. Agr. Food Chem., *20*, 1010 (1972).
419. Corey, E. J., Ponder, J. W., Ulrich, P.: Tetrahedron Lett., 137 (1980).
420. Hodjat, H., Lattes, A., Laval, J. P., Moulines, J., Périé, J. J.: J. Heterocyclic Chem., *9*, 1081 (1972).
421. Hosokawa, T., Hirata, M., Murahashi, S.-I., Sonoda, A.: Tetrahedron Lett., 1821 (1976).
422. Paquette, L. A., Thompson, G. L.: J. Am. Chem. Soc., *94*, 7118 (1972).
423. Spéziale, V., Lattes, A.: J. Heterocyclic Chem., *16*, 465 (1979).
424. Amouroux, R., Folefoc, G., Chastrette, F., Chastrette, M.: Tetrahedron Lett., 2259 (1981).
425. Cope, A. C., McKervey, M. A., Weinshenker, N. M.: J. Am. Chem. Soc., *89*, 2932 (1967).
426. Phillips, M. L., Bonjouklian, R., Jones, N. D., Hunt, A. H., Elzey, T. K.: Tetrahedron Lett., *24*, 335 (1983).
427. Mashraqui, S. H., Trivedi, G. K.: Indian J. Chem., *16B*, 849 (1978).
428. Baker, R., Evans, D. A., McDowell, P. G.: J. Chem. Soc., Chem. Commun., 111 (1977).
429. Borg-Karlson, A. K., Norin, T., Wijekoon, W. M. D., Talvitie, A.: Finn. Chem. Lett., 151 (1979); Chem. Abstr., *92*, 164097n (1980).
430. Rosen, W. E., Ziegler, J. B., Shabica, A. C.: J. Am. Chem. Soc., *77*, 762 (1955).
431. Rohela, L. C., Anand, R. C.: Indian J. Chem., *17B*, 207 (1979).
432. Beau, J.-M., Schauer, R., Haverkamp, J., Dorland, L., Vliegenthart, J. F. G., Sinay, P.: Carbohydr. Res., *82*, 125 (1980).
433. Naruta, Y., Uno, H., Maruyama, K.: J. Chem. Soc., Chem. Commun., 1277 (1981).
434. Gunstone, F. D., Inglis, R. P.: J. Chem. Soc., Chem. Commun., 12 (1972).
435. Hoye, T. R., Caruso, A. J., Dellaria, J. F., Jr., Kurth, M. J.: J. Am. Chem. Soc., *104*, 6704 (1982).

436. Cassidy, F., Moore, R. W., Wootton, G., Baggaley, K. H., Geen, G. R., Jennings, L. J. A., Tyrrell, A. W. R.: Tetrahedron Lett., 253 (1981).
437. Welzel, P., Holtmeier, W., Wessling, B.: J. Liebig Ann. Chem., 1327 (1978).
438. Pougny, J.-R., Nassr, M. A. M., Sinay, P.: J. Chem. Soc., Chem. Commun., 375 (1981).
439. Mills, L. E., Adams, R.: J. Am. Chem. Soc., *45*, 1842 (1923).
440. Spéziale, V., Dao, H. G., Lattes, A.: J. Heterocyclic Chem., *15*, 225 (1978).
441. Adams, R., Roman, F. L., Sperry, W. N.: J. Am. Chem. Soc., *44*, 1781 (1922).
442. Grundon, M. F., Stewart, D., Watts, W. E.: J. Chem. Soc., Chem. Commun., 573 (1973).
443. Nesmeyanov, A. N., Zarevich, T. S.: Ber. Dt. Chem. Ges., *68*, 1476 (1935).
444. Bartz, Q. R., Miller, R. F., Adams, R.: J. Am. Chem. Soc., *57*, 371 (1935).
445. Hosokawa, T., Miyagi, S., Murahashi, S.-I., Sonoda, A., Matsuura, Y., Tanimoto, S., Kakudo, M.: J. Org. Chem., *43*, 719 (1978).
446. Sen, A. B., Rastogi, R. P.: J. Indian Chem. Soc., *30*, 355 (1953).
447. Nageswara Sastri, V. D., Narasimhachari, N., Rajagopalan, P., Seshadri, T. R., Thiruvengadam, T. R.: Proc. Ind. Acad. Sci., *37A*, 681 (1953); Chem. Abstr., *48*, 8227*f* (1954).
448. Grundon, M. F., Steward, D., Watts, W. E.: J. Chem. Soc., Chem. Commun., 772 (1975).
449. Sand, J., Hofmann, K. A.: Ber. Dt. Chem. Ges., *33*, 1358 (1900).
450. Biilmann, E.: Ber. Dt. Chem. Ges., *33*, 1641 (1900).
451. Hofmann, K. A., Sand, J.: Ber. Dt. Chem. Ges., *33*, 2692 (1900).
452. Gunstone, F. D., Inglis, R. P.: Chem. Phys. Lipids, *10*, 105 (1973).
453. Traylor, T. G.: Acct. Chem. Res., *2*, 152 (1969).
454. Vasil'kevich, I. I., Shilov, E. A.: Ukr. Khim. Zh., *32*, 474 (1966); Chem. Abstr., *65*, 5318d (1966).
455. Iida, H., Yuasa, Y., Kibayashi, C.: Tetrahedron Lett., 3591 (1982).
456. Waters, W. L., Kiefer, E. F.: J. Am. Chem. Soc., *89*, 6261 (1967).
457. Johnson, J. R., Jobling, W. H., Bodamer, G. W.: J. Am. Chem. Soc., *63*, 131 (1941).
458. Patterson, W. I., Karabinos, J. V.: U.S. Patent 2,400,436 (1946); Chem. Abstr., *40*, 4484[4] (1946).
459. Summerbell, R. K., Lestina, G. J.: J. Am. Chem. Soc., *79*, 3878 (1957).
460. Bloodworth, A. J., Hutchings, M. G., Sotowicz, A. J.: J. Chem. Soc., Chem. Commun., 578 (1976).
461. Sharma, R. K., Shoulders, B. A., Gardner, P. D.: J. Org. Chem., *32*, 241 (1967).
462. Waters, W. L., Linn, W. S. and Caserio, M. C.: J. Am. Chem. Soc., *90*, 6741 (1968).
463. Pirkle, W. H., Boeder, C. W.: J. Org. Chem., *42*, 3697 (1977).
464. Hennion, G. F., Sheehan, J. J.: J. Am. Chem. Soc., *71*, 1964 (1949).
465. Bloodworth, A. J., Lapham, D. J., Savva, R. A.: J. Chem. Soc., Chem. Commun., 925 (1980).
466. Traylor, T. G., Baker, A. W.: J. Am. Chem. Soc., *85*, 2746 (1963).
467. Winstein, S., Shatavsky, M.: Chem. Ind., 56 (1956).
468. Takagi, E., Yokoi, Y., Yoroyuki, M.: Japan. Patent 14,714 (1961); Chem. Abstr., *56*, 12945g (1962).
469. Alexander, R. A., Baenziger, N. C., Carpenter, C., Doyle, J. R.: J. Am. Chem. Soc., *82*, 535 (1960).

470. Reppe, W., Schlichting, O., Klager, K., Toepel, T.: J. Liebig Ann. Chem., *560*, 1 (1948).
471. Lermontov, S. A., Belikova, N. A., Skornyakova, T. G., Pekhk, T. I., Lippmaa, É. T., Platé, A. F.: Zh. Org. Khim., *16*, 2322 (1980); J. Org. Chem. USSR, *16*, 1982 (1980).
472. Julia, M., Colomer, E.: An. Quim., *67*, 199 (1971).
473. Franz, H. J., Höbold, W., Höhn, R., Müller-Hagen, G., Müller, R., Pritzkow, W., Schmidt, H.: J. Prakt. Chem., *312*, 622 (1970).
474. Belikova, N. A., Lermontov, S. A., Pekhk, T. I., Lippmaa, É. T., Platé, A. F.: Zh. Org. Khim., *14*, 884 (1978); J. Org. Chem. USSR, *14*, 823 (1978).
475. Belikova, N. A., Lermontov, S. A., Skornyakova, T. G., Pekhk, T. I., Lippmaa, É. T., Platé, A. F.: Zh. Org. Khim., *15*, 492 (1979); J. Org. Chem. USSR, *15*, 436 (1979).
476. Belikova, N. A., Lermontov, S. A., Pekhk, T. I., Lippmaa, É. T., Platé, A. F.: Zh. Org. Khim., *14*, 2273 (1978); J. Org. Chem. USSR, *14*, 2101 (1978).
477. Vaidyanathaswamy, R., Devaprabhakara, D., Rao, V. V.: Tetrahedron Lett., 915 (1971).
478. Vaidyanathaswamy, R., Devaprabhakara, D.: Indian J. Chem., *13*, 287 (1975).
479. Bach, R. D.: Tetrahedron Lett., 5841 (1968).
480. Bach, R. D.: J. Am. Chem. Soc., *91*, 1771 (1969).
481. Lalithambika, M., Katiyar, S. S., Devaprabhakara, D.: J. Electroanal. Chem. Interfacial Electrochem., *31*, 219 (1971).
482. Joshi, G. C., Devaprabhakara, D.: J. Organometal. Chem., *15*, 497 (1968).
483. Löffler, H.-P., Schröder, G.: Tetrahedron Lett., 2119 (1970).
484. Schröder, G., Prange, U., Putze, B., Thio, J., Oth, J. F. M.: Chem. Ber., *104*, 3406 (1971).
485. Stille, J. K., Stinson, S. C.: Tetrahedron, *20*, 1387 (1964).
486. Poutsma, M. L., Ibarbia, P. A.: J. Am. Chem. Soc., *93*, 440 (1971).
487. Katiyar, S. S., Lalithambika, M., Devaprabhakara, D.: Electrochimica Acta, *17*, 2077 (1972).
488. Atwood, J. L., Canada, L. G., Lau, A. N. K., Ludwick, A. G., Ludwick, L. M.: J. Chem. Soc., Dalton, 1573 (1978).
489. Takagi, E., Yokoi, Y., Mangyo, M.: Japan. Patent 9,431 (1961); Chem. Abstr., *56*, 5855a (1962).
490. Sokolov, V. I., Troitskaya, L. L., Reutov, O. A.: Izv. Akad. Nauk SSSR, Ser. Khim., 1913 (1968); Bull. Acad. Sci. USSR, Div. Chem. Sci., 1826 (1968).
491. Müller, E.: Chem. Ber., *106*, 3920 (1973).
492. Zefirov, N. S., Kirin, B. N., Potekhin, K. A., Koz'min, A. S., Sadovaya, N. K., Kurkutova, E. N. and Bodrikov, I. V.: Zh. Org. Khim., *14*, 1224 (1978); J. Org. Chem. USSR, *14*, 1135 (1978).
493. Sasaki, T., Kanematsu, K., Kondo, A.: J. Chem. Soc., Perkin I, 2516 (1976).
494. Allen, J. D.: U.S. Patent 3,414,599 (1968); Chem. Abstr., *70*, 47597k (1969).
495. Yale, H. L.: U.S. Patent 2,672,472 (1954); Chem. Abstr., *49*, 3253e (1955).
496. Jantzen, E., Andreas, H.: Chem. Ber., *94*, 628 (1961).
497. Ralston, A. W., McCorkle, M. R.: U.S. Patent 2,262,430 (1942); Chem. Abstr., *36*, 1741[4] (1942).
498. Lamson, D. W., Yonetani, T.: Tetrahedron Lett., 5025 (1973).
499. Takehira, Y., Tanaka, K., Toda, F.: Chem. Lett., 1323 (1976).
500. Borisov, A. E., Vil'chevskaya, V. D., Nesmeyanov, A. N.: Dokl. Akad. Nauk SSSR, *90*, 383 (1953); Chem. Abstr., *48*, 4434f (1954).
501. Doering, W. von E., La Flamme, P. M.: Tetrahedron, *2*, 75 (1958).

502. Aranda, V. G., Beltran, F. G., Mur, J. B.: Combustibles (Zaragoza), *25*, 3 (1967); Chem. Abstr., *68*, 114722w (1968).
503. Kühnle, D., Funke, W.: Angew. Chem., *83*, 369 (1971); Angew. Chem., Int. Ed. Engl., *10*, 351 (1971).
504. Hofmann, K. A., Seiler, E.: Ber. Dt. Chem. Ges., *39*, 3187 (1906).
505. Müller, E.: Chem. Ber., *109*, 3804 (1976).
506. Cookson, R. C., Hudec, J., Marsden, J.: Chem. Ind., 21 (1961).
507. Gelin, R., Gelin, S., Albrand, M.: Bull. Soc. Chim. France, 1946 (1972).
508. Garver, L., van Eikeren, P., Byrd, J. E.: J. Org. Chem., *41*, 2773 (1976).
509. Matsuki, Y., Kodama, M., Itô, S.: Tetrahedron Lett., 2901 (1979).
510. Matsuki, Y., Kodama, M., Itô, S.: Tetrahedron Lett., 4081 (1979).
511. Audin, P., Doutheau, A., Gore, J.: Tetrahedron Lett., 4337 (1982).
512. Johnson, R. A., Lincoln, F. H., Nidy, E. G., Schneider, W. P., Thompson, J. L., Axen, U.: J. Am. Chem. Soc., *100*, 7690 (1978).
513. Hough, E., Hursthouse, M. B., Neidle, S., Rogers, D.: J. Chem. Soc., Chem. Commun., 1197 (1968).
514. Barrow, K. D., Barton, D. H. R., Chain, E. B., Ohnsorge, U. F. W., Thomas, R.: J. Chem. Soc., Chem. Commun., 1198 (1968).
515. Johnson, R. A., Nidy, E. G.: J. Org. Chem., *45*, 3802 (1980).
516. Sih, J. C., Johnson, R. A., Nidy, E. G., Graber, D. R.: Prostaglandins, *15*, 409 (1978).
517. De, B., Andersen, N. H., Ippolito, R. M., Wilson, C. H., Johnson, W. D.: Prostaglandins, *19*, 221 (1980).
518. Sih, J. C., Graber, D. R.: J. Org. Chem., *47*, 4919 (1982).
519. Toda, F., Akagi, K.: Tetrahedron, *25*, 3795 (1969).
520. Corey, E. J., Keck, G. E., Székely, I.: J. Am. Chem. Soc., *99*, 2006 (1977).
521. Plattner, R. D., Spencer, G. F., Kleiman, R.: Lipids, *11*, 222 (1976).
522. Contardi, A., Ciocca, B.: Ricerca Scient., *7*, 610 (1936); Chem. Abstr., *32*, 9039[1] (1938).
523. Reichert, J. S., Bailey, J. H., Nieuwland, J. A.: J. Am. Chem. Soc., *45*, 1552 (1923).
524. Killian, D. B., Hennion, G. F., Nieuwland, J. A.: J. Am. Chem. Soc., *56*, 1384 (1934).
525. Nieuwland, J. A., Vogt, R. R., Foohey, W. L.: J. Am. Chem. Soc., *52*, 1018 (1930).
526. Brit. Patent 231,841 (1924); Chem. Abstr., *19*, 3491 (1925).
527. Hill, H. S., Hibbert, H.: J. Am. Chem. Soc., *45*, 3108 (1923).
528. Coffman, D. D., Kalb, G. H., Ness, A. B.: J. Org. Chem., *13*, 223 (1948).
529. Coffman, D. D.: U.S. Patent 2,387,495 (1945); Chem. Abstr., *40*, 1870[6] (1946).
530. Conaway, R. F.: U.S. Patent 2,370,779 (1945); Chem. Abstr., *40*, 367[9] (1946).
531. Hennion, G. F., Murray, W. S.: J. Am. Chem. Soc., *64*, 1220 (1942).
532. Killian, D. B., Hennion, G. F., Nieuwland, J. A.: J. Am. Chem. Soc., *56*, 1786 (1934).
533. Kilian, D. B., Hennion, G. F., Nieuwland, J. A.: J. Am. Chem. Soc., *58*, 1658 (1936).
534. Chobanyan, Zh. A., Davtyan, S. Zh., Badanyan, Sh. O.: Arm. Khim. Zh., *33*, 589 (1980); Chem. Abstr., *94*, 46728n (1981).
535. Froning, J. F., Hennion, G. F.: J. Am. Chem. Soc., *62*, 653 (1940).
536. Nazarov, I. N.: Izv. Akad. Nauk SSSR, Otdel. Khim. Nauk, 203 (1940); Chem. Abstr., *36*, 744[5] (1942).

537. Nazarov, I. N.: Izv. Akad. Nauk SSSR, Otdel. Khim. Nauk, 195 (1940); Chem. Abstr., *36*, 742^9 (1942).
538. Hennion, G. F., Nieuwland, J. A.: J. Am. Chem. Soc., *57*, 2006 (1935).
539. Hennion, G. F., Lieb, D. J.: J. Am. Chem. Soc., *66*, 1289 (1944).
540. Hennion, G. F., Killian, D. B., Vaughan, T. H., Nieuwland, J. A.: J. Am. Chem. Soc., *56*, 1130 (1934).
541. Killian, D. B., Hennion, G. F., Nieuwland, J. A.: J. Am. Chem. Soc., *58*, 80 (1936).
542. Nazarov, I. N., Nagibina, T. D.: Izv. Akad. Nauk SSSR, Otdel. Khim. Nauk, 303 (1941); Chem. Abstr., *37*, 5368^9 (1943).
543. Billimoria, J. D., Maclagan, N. F.: J. Chem. Soc., 3257 (1954).
544. Riediker, M., Schwartz, J.: J. Am. Chem. Soc., *104*, 5842 (1982).
545. Hudrlik, P. F., Hudrlik, A. M.: J. Org. Chem., *38*, 4254 (1973).
546. Fabritsy, A., Kubalya, I.: Zh. Obshch. Khim., *31*, 476 (1961); J. Gen. Chem. USSR, *31*, 434 (1961).
547. Fabrycy, A., Wichert, Z.: Rocz. Chem., *42*, 35 (1968); Chem. Abstr., *69*, 35835c (1968).
548. Seher, A.: Arch. Pharm., *292*, 519 (1959); Chem. Abstr., *54*, 8623g (1960).
549. Hinton, H. D., Nieuwland, J. A.: J. Am. Chem. Soc., *52*, 2892 (1930).
550. Hill, H. S., Hibbert, H.: J. Am. Chem. Soc., *45*, 3117 (1923).
551. Hill, H. S., Hibbert, H.: J. Am. Chem. Soc., *45*, 3124 (1923).

IV. Peroxymercuration

When hydroperoxides are employed as the nucleophile in solvomercuration reactions, peroxymercuration is observed. This reaction, first reported in 1969 by Sokolov and Reutov in Russia [1] and Bloodworth and co-workers in England [2] provides a useful approach to peroxides. The work in this area is summarized in Table 4.1, which is divided into simple alkenes, alkenyl hydroperoxides and finally dienes. Our discussion will be similarly organized.

The early work on peroxymercuration was mostly carried out using mercuric acetate as the electrophile. Using this salt, however, the desired peroxymercurial is usually accompanied by 15–35% of the corresponding acetoxymercuration product [2, 13, 15]. The acetoxymercurials can be avoided by simply adding perchloric acid to the reaction [14]. This effectively converts the acetoxymercurials into the desired peroxy compounds (Eq. 1).

$$\text{(1)}$$

With α,β-unsaturated carbonyl compounds, catalytic amounts of perchloric acid are required in order to get any reaction at all [8–10, 12, 18]. Mercuric trifluoroacetate, first utilized for peroxymercuration in 1974 [6], has also been found to effectively eliminate acyloxymercuration side products. With certain olefins, such as norbornene, it is superior to the perchloric acid approach [14]. In the formation of cyclic peroxides from unsaturated peroxides [25] and dienes [23, 29, 30], mercuric nitrate monohydrate has been found to be even more effective than mercuric trifluoroacetate.

Relatively few hydroperoxides have been employed in the peroxymercuration reaction. Hydrogen peroxide has been used on several occassions, but it is not without its problems. With simple alkenes, mixtures of mono- and dimercuration products arise (Eq. 2) [3]. By appropriate control of stoichiometry, either product can be made to predominate. With ethylene, $CF_3CO_2HgCH_2CH_2O_2CCF_3$, $CF_3CO_2HgCH_2CH_2OH$ and $(CF_3CO_2HgCH_2CH_2)_2O$ are observed to accompany the usual peroxymercuration products. The latter two products obviously arise from the water present in the hydrogen peroxide.

$$H_2C{=}CHR \xrightarrow[\text{H}_2\text{O}_2]{\text{Hg(O}_2\text{CCF}_3)_2} CF_3CO_2HgCH_2CHROOH + (CF_3CO_2HgCH_2CHRO)_2 \qquad (2)$$

Table 4.1. Peroxymercuration of Alkenes

Alkene	Peroxide	Mercuric salt	Organomercurial(s) (% Yield)	Subsequent reactants	Product(s) (% Yield)	Ref.
$H_2C{=}CH_2$	$0.5\ H_2O_2$	$Hg(O_2CCF_3)_2$	$CF_3CO_2HgCH_2CH_2OOH$ (6) + $(CF_3CO_2HgCH_2CH_2O)_2$ (30) 38 : + $CF_3CO_2HgCH_2CH_2O_2CCF_3$ 43 : + $CF_3CO_2HgCH_2CH_2OH$ + $(CF_3CO_2HgCH_2CH_2)_2O$ 19	Cl^-	$(ClHgCH_2CH_2O)_2$ (52)	3
	$3\ H_2O_2$		$C^-_3CO_2HgCH_2CH_2OOH$ (6) + $(CF_3CO_2HgCH_2CH_2O)_2$ (17) 51 : + $CF_3CO_2HgCH_2CH_2O_2CCF_3$ 27 : + $CF_3CO_2HgCH_2CH_2OH$ + $(CF_3CO_2HgCH_2CH_2)_2O$ 22	—	—	3
	$t{-}BuOOH$	$Hg(OAc)_2,\ Cl^-$	$C.HgCH_2CH_2OOBu{-}t$ (89)	I_2	$ICH_2CH_2OOBu{-}t$ (100)	4
	$t{-}BuOOH$	$Hg(OAc)_2/HClO_4$	—	$NaBH_4/NaOH$ (0–5 °C)	$t{-}BuOOCH_2CH_3$ + $\underset{O}{H_2C{-}CH_2}$ >95 : <5	5
	$t{-}BuOOH$	$Hg(O_2CCF_3)_2,\ KX$	$XHgCH_2CH_2OOBu{-}t$ (~85)	—	—	6
	$t{-}BuOOH$	$Hg(O_2CCF_3)_2,\ KBr$	—	Br_2/CH_2Cl_2	$BrCH_2CH_2OOBu{-}t$ (80 from RHgBr)	7
	$HOOC(CH_3)_2C_6H_5$	$Hg(OAc)_2,\ Cl^-$	$C\ HgCH_2CH_2OOC(CH_3)_2C_6H_5$	—	—	4
$H_2C{=}CHCH_3$	H_2O_2	$Hg(O_2CCF_3)_2$	$[CF_3CO_2HgCH_2CH(CH_3)O]_2$ (49) + $CF_3CO_2HgCH_2CH(CH_3)OOH$ 87 : 13	—	—	3

Table 4.1. (continued)

Alkene	Peroxide	Mercuric salt	Organomercurial(s) (% Yield)	Subsequent reactants	Product(s) (% Yield)	Ref.
	3 H_2O_2	$Hg(O_2CCF_3)_2$	$[CF_3CO_2HgCH_2CH(CH_3)O]_2$ (68)	Br_2	$[BrCH_2CH(CH_3)O]_2$ (85 crude)	3
	n-PrOOH	$Hg(OAc)_2$, Cl^-	$ClHgCH_2CH(CH_3)OOPr$-n	—	—	4
	t-BuOOH	$Hg(OAc)_2$ (Cl^-)	$XHgCH_2CH(CH_3)OOBu$-t X = OAc, Cl (76)	I_2 (on RHgOAc)	$ICH_2CH(CH_3)OOBu$-t (47)	4
	t-BuOOH	$Hg(O_2CCF_3)_2$, KX	$XHgCH_2CH(CH_3)OOBu$-t (~85)	—	—	6
	t-BuOOH	$Hg(O_2CCF_3)_2$, KBr	—	Br_2/CH_2Cl_2	$BrCH_2CH(CH_3)OOBu$-t (81 from RHgBr)	7

$H_2C{=}C(CH_3)CHO$	t-BuOOH	$Hg(OAc)_2$ / $HClO_4$, (KBr)	$XHgCH_2C(CH_3)(OOBu$-$t)CH(OOBu$-$t)OH$
			X = OAc 75 :
			X = Br — :
			+ $XHgCH_2C(CH_3)(OOBu$-$t)CHO$
			   : 25 (92 total)
			   : — (50 total)
				—	—	8, 9

$H_2C{=}CHCOCH_3$	t-BuOOH	$Hg(OAc)_2$ / $HClO_4$, KBr	t-$BuOOCH_2CH(HgBr)COCH_3$ (50)	—	—	10
$CH_3CH{=}CHCHO$	t-BuOOH	$Hg(OAc)_2$ / $HClO_4$, (KBr)	$CH_3CH(OOBu$-$t)CH(HgX)CHO$ X = OAc (40), Br (25)	—	—	9
$H_2C{=}CHO_2CCH_3$	n-PrOOH	$Hg(OAc)_2$	—	—	CH_3CHO (100) + CH_3CH_2CHO (100)	11
	i-PrOOH	$Hg(OAc)_2$	—	—	CH_3CHO (100) + CH_3COCH_3 (100)	11
	t-BuOOH	$Hg(OAc)_2$, NaCl	$ClHgCH_2CH(OAc)OOBu$-t + $ClHgCH_2CH(OOBu$-$t)_2$ + t-$BuOCH_2CHO$ (40)	—	—	11
(1-hydroperoxytetralin)		$Hg(OAc)_2$	—	—	(1-tetralone) + CH_3CHO	11

Table 4.1. (continued)

Alkene	Peroxide	Mercuric salt	Organomercurial(s) (% Yield)	Subsequent reactants	Product(s) (% Yield)	Ref.
$H_2C=CHCO_2CH_3$	t-BuOOH	$Hg(OAc)_2$ / $HClO_4$, KBr	t-BuOOCH$_2$CH(HgBr)CO$_2$CH$_3$ (31)	NaBH$_4$/NaOH	t-BuOOCH$_2$CH$_2$CO$_2$CH$_3$ (86 crude)	10, 12
$H_2C=C(CH_3)_2$	0.5 H_2O_2	$Hg(O_2CCF_3)_2$	[CF$_3$CO$_2$HgCH$_2$C(CH$_3$)$_2$O]$_2$ (70, 80) 71 : + CF$_3$CO$_2$HgCH$_2$C(CH$_3$)$_2$OOH (33) : 29	—	—	3
	3 H_2O_2	$Hg(O_2CCF_3)_2$	CF$_3$CO$_2$HgCH$_2$C(CH$_3$)$_2$OOH (50, 62) 79 : + [CF$_3$CO$_2$HgCH$_2$C(CH$_3$)$_2$O]$_2$ (33) : 21	—	—	3
	i-PrOOH	$Hg(OAc)_2$, Cl$^-$	ClHgCH$_2$C(CH$_3$)$_2$OOPr-i	—	—	4
	t-BuOOH	$Hg(OAc)_2$,(Cl$^-$)	XHgCH$_2$C(CH$_3$)$_2$OOBu-t X = OAc, Cl (85)	Y$_2$	YCH$_2$C(CH$_3$)$_2$OOBu-t Y = Br (on RHgCl) (51), I (on RHgOAc) (71)	4
	HOOC(CH$_3$)$_2$ – CH$_2$HgO$_2$CCF$_3$	$Hg(O_2CCF_3)_2$	[CF$_3$CO$_2$HgCH$_2$C(CH$_3$)$_2$O]$_2$ (100 crude)	Br$_2$	BrCH$_2$C(CH$_3$)$_2$O]$_2$ (85 crude)	3
$H_2C=CHCH_2CH_3$	t-BuOOH	$Hg(OAc)_2$	AcOHgCH$_2$CH(OOBu-t)CH$_2$CH$_3$ + AcOHgCH$_2$CH(OAc)CH$_2$CH$_3$ (15 – 20)	KBr	BrHgCH$_2$CH(OOBu-t)CH$_2$CH$_3$ (39)	13
$CH_3CH=CHCH_3$	t-BuOOH	$Hg(O_2CCF_3)_2$, KBr	—	Br$_2$/CH$_2$Cl$_2$	CH$_3$CH(OOBu-t)CHBrCH$_3$ (64 from RHgBr)	7
cis-$CH_3CH=CHCH_3$	t-BuOOH	$Hg(OAc)_2$ / $HClO_4$, KBr	CH$_3$CH(OOBu-t)CH(HgBr)CH$_3$ (>80)	Br$_2$/CH$_2$Cl$_2$ [Br$_2$/NaBr/CH$_3$OH]	CH$_3$CH(OOBu-t)CHBrCH$_3$ (78) [93]	14
	t-BuOOH	$Hg(O_2CCF_3)_2$, (Br$^-$)	$threo$-CH$_3$CH(OOBu-t)CH(HgX)CH$_3$ X = O$_2$CCF$_3$, Br	—	—	6, 15, 16
	t-BuOOH	—	—	n-Bu$_3$SnH	CH$_3$CH(OOBu-t)CH$_2$CH$_3$ (63)	17
$trans$-$CH_3CH=CHCH_3$	t-BuOOH	$Hg(OAc)_2$ / $HClO_4$, KBr	CH$_3$CH(OOBu-t)CH(HgBr)CH$_3$ (>80)	Br$_2$/CH$_2$Cl$_2$ [Br$_2$/NaBr/CH$_3$OH]	CH$_3$CH(OOBu-t)CHBrCH$_3$ (86) [93]	14

Table 4.1. (continued)

350

Alkene	Peroxide	Mercuric salt	Organomercurial(s) (% Yield)	Subsequent reactants	Product(s) (% Yield)	Ref.
	t-BuOOH	Hg(O$_2$CCF$_3$)$_2$ (Br$^-$)	$erythro$-CH$_3$CH(OOBu-t)CH(HgX)CH$_3$ X = O$_2$CCF$_3$, Br	—	—	6, 16
H$_2$C=CHOC$_2$H$_5$	t-BuOOH	Hg(OAc)$_2$, NaCl	ClHgCH$_2$CH(OC$_2$H$_5$)OOBu-t	—	—	11
[cyclopentene]	t-BuOOH	Hg(OAc)$_2$ / HClO$_4$, KBr	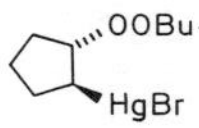 [cyclopentane: OOBu-t, HgBr] (>80)	Br$_2$/CH$_2$Cl$_2$ [Br$_2$/NaBr/CH$_3$OH]	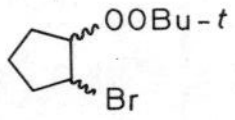 [cyclopentane: OOBu-t, Br] (80) [78]	14
	t-BuOOH	—	—	n-Bu$_3$SnH	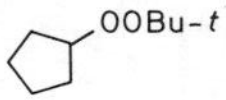 [cyclopentane: OOBu-t] (59)	17
H$_2$C=C(CH$_3$)COCH$_3$	t-BuOOH	Hg(OAc)$_2$ / HClO$_4$ (KBr)	XHgCH$_2$C(CH$_3$)(OOBu-t)COCH$_3$ X = OAc (60), Br (30)	NaBH$_4$/NaOH or Y$_2$ (on RHgBr)	YCH$_2$C(CH$_3$)(OOBu-t)COCH$_3$ + H$_2$C(-O-)C(CH$_3$)COCH$_3$ Y = H 90(85% yield) : 10 Y = Br (71 crude), I (45)	8, 10, 12, 18
$trans$-CH$_3$CH=CHCOCH$_3$	t-BuOOH	Hg(OAc)$_2$ / HClO$_4$ (KBr)	CH$_3$CH(OOBu-t)CH(HgX)COCH$_3$ X = OAc, Br	NaBH$_4$/NaOH (on RHgOAc)	E + Z - CH$_3$CH(-O-)CHCOCH$_3$	10, 19
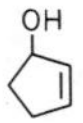 [2-cyclopenten-1-ol, OH]	(t-Bu)Me$_2$SiOOH	Hg(O$_2$CCF$_3$)$_2$	—	NaBH$_4$	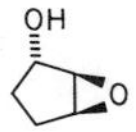 [epoxycyclopentanol: OH, O]	19
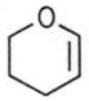 [3,4-dihydro-2H-pyran]	H$_2$O$_2$	Hg(OAc)$_2$, Cl$^-$	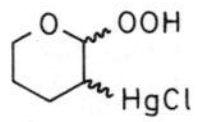 [tetrahydropyran: OOH, HgCl]	—	—	20
H$_2$C=C(CH$_3$)O$_2$CCH$_3$	t-BuOOH	Hg(OAc)$_2$	—	—	t-BuOCH$_2$COCH$_3$ (35)	11
	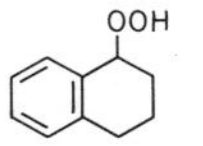 [tetralin, OOH]	Hg(OAc)$_2$	—	—	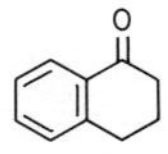 [1-tetralone] + CH$_3$COCH$_3$	11
H$_2$C=C(CH$_3$)CO$_2$CH$_3$	t-BuOOH	Hg(OAc)$_2$ / HClO$_4$, KBr	BrHgCH$_2$C(CH$_3$)(OOBu-t)CO$_2$CH$_3$ (70)	NaBH$_4$/NaOH or X$_2$	XCH$_2$C(CH$_3$)(OOBu-t)CO$_2$CH$_3$ + H$_2$C(-O-)C(CH$_3$)CO$_2$CH$_3$ X = H 70(45% yield) : 30 X = Br (79 crude), I (54)	8, 10, 12, 18

Table 4.1. (continued)

Alkene	Peroxide	Mercuric salt	Organomercurial(s) (% Yield)	Subsequent reactants	Product(s) (% Yield)	Ref.
$CH_3CH{=}CHCO_2CH_3$	t-BuOOH	$Hg(OAc)_2$ / $HClO_4$, KBr	$CH_3CH(OOBu$-$t)CH(HgBr)CO_2CH_3$	Br_2	$CH_3CH(OOBu$-$t)CHBrCO_2CH_3$ (80)	18
$trans$-$CH_3CH{=}CHCO_2CH_3$	t-BuOOH	$Hg(OAc)_2$ / $HClO_4$, KBr	$CH_3CH(OOBu$-$t)CH(HgBr)CO_2CH_3$ (58)	$NaBH_4$/NaOH	$CH_3CH(OOBu$-$t)CH_2CO_2CH_3$ (80)	10, 12
(cyclohexene ring)	H_2O_2	$Hg(OAc)_2$ (Cl⁻)	(cyclohexane: OOH, HgX) X = OAc (36), Cl (53)	—	—	1, 4, 20
	t-BuOOH	$Hg(OAc)_2$ / $HClO_4$, KBr	(cyclohexane: OOBu-t, HgBr)	Br_2/CH_2Cl_2 [Br_2/NaBr/CH_3OH]	(cyclohexane: OOBu-t, Br) (93) [82]	14
	t-BuOOH	$Hg(O_2CCF_3)_2$, KX	(cyclohexane: OOBu-t, HgX)	—	—	6
	t-BuOOH	—	—	n-Bu_3SnH	(cyclohexane: OOBu-t) (61)	17
	t-BuOOH	$Hg(O_2CCF_3)_2$, (Br⁻)	(cyclohexane: OOBu-t, HgX) X = O_2CCF_3, Br	Br_2/CH_2Cl_2 (on RHgBr)	(cyclohexane: OOBu-t, Br) (87 crude)	6, 7, 16
$(CH_3)_2C{=}CHCOCH_3$	t-BuOOH	$Hg(OAc)_2$ / $HClO_4$, KBr	$(CH_3)_2C(OOBu$-$t)CH(HgBr)COCH_3$ (95)	—	—	10
$CH_3CH{=}C(CH_3)COCH_3$	t-BuOOH	$Hg(OAc)_2$ / $HClO_4$, Cl⁻	$CH_3CH(OOBu$-$t)C(CH_3)(HgCl)COCH_3$ (25)	—	—	10
(2-cyclohexenol: OH)	$(t$-Bu$)Me_2SiOOH$	$Hg(O_2CCF_3)_2$	—	$NaBH_4$	(cyclohexane: OH, epoxide O) (66)	19
E-$CH_3CH{=}C(CH_3)CO_2CH_3$	t-BuOOH	$Hg(OAc)_2$ / $HClO_4$, KBr	$CH_3CH(OOBu$-$t)C(CH_3)(HgBr)CO_2CH_3$ (8)	—	—	10

Table 4.1. (continued)

Alkene	Peroxide	Mercuric salt	Organomercurial(s) (% Yield)	Subsequent reactants	Product(s) (% Yield)	Ref.
$(CH_3)_2C{=}CHCO_2CH_3$	$t\text{-}BuOOH$	$Hg(OAc)_2$ / $HClO_4$, KBr	$(CH_3)_2C(OOBu\text{-}t)CH(HgBr)CO_2CH_3$ (94)	$NaBH_4/NaOH$	$(CH_3)_2C(OOBu\text{-}t)CH_2CO_2CH_3$ (~79)	10, 12
$H_2C{=}C(CH_3)CH_2CH_2CH_3$	$t\text{-}BuOOH$	$Hg(OAc)_2$, Cl^-	$ClHgCH_2C(CH_3)(OOBu\text{-}t)CH_2CH_2CH_3$ (79)	Br_2	$BrCH_2C(CH_3)(OOBu\text{-}t)CH_2CH_2CH_3$ (38)	4
$H_2C{=}CH(CH_2)_3CH_3$	$t\text{-}BuOOH$	$Hg(OAc)_2$	$AcOHgCH_2CH(OOBu\text{-}t)(CH_2)_3CH_3$ (~39) + $AcOHgCH_2CH(OAc)(CH_2)_3CH_3$ (15 – 20)	$NaBH_4$ or X_2	$XCH_2CH(OOBu\text{-}t)(CH_2)_3CH_3$ X = H (50–60), Br, I	2, 13
	$t\text{-}BuOOH$	$Hg(OAc)_2$ / $HClO_4$	—	$NaBH_4/NaOH$	$CH_3CH(OOBu\text{-}t)(CH_2)_3CH_3$ (66) + $CH_3CH(OAc)(CH_2)_3CH_3$ (17.6) + $H_2\overset{O}{C{-}}CH(CH_2)_3CH_3$ (15.8)	5
$CH_3CH_2CH{=}CHCH_2CH_3$	$t\text{-}BuOOH$	$Hg(O_2CCF_3)_2$, KBr	—	Br_2/CH_2Cl_2	$CH_3CH_2CH(OOBu\text{-}t)CHBrCH_2CH_3$ (87 from RHgBr)	7
$cis\text{-}CH_3CH_2CH{=}CHCH_2CH_3$	$t\text{-}BuOOH$	$Hg(OAc)_2$ / $HClO_4$, KBr	$CH_3CH_2CH(OOBu\text{-}t)CH(HgBr)CH_2CH_3$ + $CH_3CH_2CH(OAc)CH(HgBr)CH_2CH_3$ (<2)	Br_2/CH_2Cl_2 [$Br_2/$ $NaBr/CH_3OH$] (on peroxide)	$CH_3CH_2CH(OOBu\text{-}t)CHBrCH_2CH_3$ (99) [80]	14
$trans\text{-}CH_3CH_2CH{=}CHCH_2CH_3$	$t\text{-}BuOOH$	$Hg(OAc)_2$ / $HClO_4$, KBr	$CH_3CH_2CH(OOBu\text{-}t)CH(HgBr)CH_2CH_3$ + $CH_3CH_2CH(OAc)CH(HgBr)CH_2CH_3$ (10)	Br_2/CH_2Cl_2 [$Br_2/$ $NaBr/CH_3OH$] (on peroxide)	$CH_3CH_2CH(OOBu\text{-}t)CHBrCH_2CH_3$ (67) [74]	14
	$t\text{-}BuOOH$	$Hg(O_2CCF_3)_2$, KX	$CH_3CH_2CH(OOBu\text{-}t)CH(HgX)CH_2CH_3$	—	—	6
	$t\text{-}BuOOH$	—	—	$n\text{-}Bu_3SnH$	$CH_3CH_2CH(OOBu\text{-}t)(CH_2)_2CH_3$ (24)	17
 (norbornene)	H_2O_2	$Hg(OAc)_2$	 (norbornyl) –OOH / –HgOAc (37)	—	—	20
	$t\text{-}BuOOH$	$Hg(OAc)_2$ / $HClO_4$	 (norbornyl) –OOBu-t / –HgOAc (49)	Br_2/CH_2Cl_2 [$Br_2/$ $NaBr/CH_3OH$]	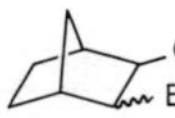 (norbornyl) –OOBu-t / –Br (87) [<100]	14

Table 4.1. (continued)

Alkene	Peroxide	Mercuric salt	Organomercurial(s) (% Yield)	Subsequent reactants	Product(s) (% Yield)	Ref.
	t-BuOOH	Hg(O$_2$CCF$_3$)$_2$, (Br$^-$)	[structure] HgX X = O$_2$CCF$_3$, Br	Br$_2$ (on RHgBr)	[structure] Br (87)	6, 7, 15, 16
H$_2$C=CHC$_6$H$_5$	H$_2$O$_2$	Hg(OAc)$_2$, (I$^-$)	XHgCH$_2$CH(OOH)C$_6$H$_5$ X = OAc (68, 75, 90, 92), I	I$_2$	ICH$_2$CH(C$_6$H$_5$)OOH (100)	1, 3, 4, 20
	0.5 H$_2$O$_2$	Hg(O$_2$CCF$_3$)$_2$	[CF$_3$CO$_2$HgCH$_2$CH(C$_6$H$_5$)O]$_2$ (66) + CF$_3$CO$_2$HgCH$_2$CH(C$_6$H$_5$)OOH (39) 63 : 37	Cl$^-$	[ClHgCH$_2$CH(C$_6$H$_5$)O]$_2$ (31)	3
	3 H$_2$O$_2$	Hg(O$_2$CCF$_3$)$_2$	CF$_3$CO$_2$HgCH$_2$CH(C$_6$H$_5$)OOH (58) + [CF$_3$CO$_2$HgCH$_2$CH(C$_6$H$_5$)O]$_2$ 76 : 24	—	—	3
	t-BuOOH	Hg(OAc)$_2$	AcOHgCH$_2$CH(OOBu-t)C$_6$H$_5$ + AcOHgCH$_2$CH(OAc)C$_6$H$_5$ (~15)	NaBH$_4$ or X$_2$	XCH$_2$CH(OOBu-t)C$_6$H$_5$ X = H (50-60), Br, I	2
	t-BuOOH	Hg(OAc)$_2$, Cl$^-$	ClHgCH$_2$CH(C$_6$H$_5$)OOBu-t (84)	X$_2$	XCH$_2$CH(C$_6$H$_5$)OOBu-t X = Br (55), I (100)	4
	t-BuOOH	Hg(OAc)$_2$, (KBr)	XHgCH$_2$CH(OOBu-t)C$_6$H$_5$ X = OAc, Br (high) + XHgCH$_2$CH(OAc)C$_6$H$_5$ X = OAc (15-20)	Y$_2$	YCH$_2$CH(OOBu-t)C$_6$H$_5$ Y = Br, I	13
	t-BuOOH	Hg(OAc)$_2$ / HClO$_4$	—	NaBH$_4$/NaOH	CH$_3$CH(OOBu-t)C$_6$H$_5$ + H$_2$C—CHC$_6$H$_5$ 45.5-86 : 14-55.5	5
	t-BuOOH	Hg(O$_2$CCF$_3$)$_2$	CF$_3$CO$_2$HgCH$_2$CH(OOBu-t)C$_6$H$_5$	—	—	21
	t-BuOOH	Hg(O$_2$CCF$_3$)$_2$, KX	XHgCH$_2$CH(C$_6$H$_5$)OOBu-t (~85)	—	—	6
	HOOCH(C$_6$H$_5$) - CH$_2$HgOAc	Hg(O$_2$CCF$_3$)$_2$, KCl	[ClHgCH$_2$CH(C$_6$H$_5$)O]$_2$ (32)	Br$_2$	[BrCH$_2$CH(C$_6$H$_5$)O]$_2$ (100 crude)	3, 6

Table 4.1. (continued)

Alkene	Peroxide	Mercuric salt	Organomercurial(s) (% Yield)	Subsequent reactants	Product(s) (% Yield)	Ref.
	$HOOC(CH_3)_2 - CH_2HgO_2CCF_3$	$Hg(O_2CCF_3)_2$	$CF_3CO_2HgCH_2CH(C_6H_5)OOC(CH_3)_2 - CH_2HgO_2CCF_3$ (94 crude)	—	—	3
	H_2O_2	$Hg(OAc)_2, Cl^-$?	(50 – 55) X = OAc, Cl ?	—	—	20
$H_2C{=}CH(CH_2)_5CH_3$	t-BuOOH	$Hg(OAc)_2$	$AcOHgCH_2CH(OOBu-t)(CH_2)_5CH_3$ (39) + $AcOHgCH_2CH(OAc)(CH_2)_5CH_3$ (15 – 20)	$NaBH_4$ or X_2	$XCH_2CH(OOBu-t)(CH_2)_5CH_3$ X = H (50 – 60), Br, I	2, 13
$C_6H_5CH{=}CHCHO$	t-BuOOH	$Hg(OAc)_2 / HClO_4$, (KBr)	$C_6H_5CH(OOBu-t)CH(HgX)CH(OOBu-t)_2$ X = OAc (33, 60), Br (50)	—	—	9
$H_2C{=}C(CH_3)C_6H_5$	H_2O_2	$Hg(O_2CCF_3)_2$	$CF_3CO_2HgCH_2C(CH_3)(C_6H_5)OOH$ (50, 64) + $[CF_3CO_2HgCH_2C(CH_3)(C_6H_5)O]_2$ (14) 90 : 10	—	—	3
	t-BuOOH	$Hg(OAc)_2, (X^-)$	$XHgCH_2C(CH_3)(C_6H_5)OOBu-t$ X = OAc, Cl, Br	Y_2	$YCH_2C(CH_3)(C_6H_5)OOBu-t$ Y = Br (on RHgCl) (100), I (on RHgOAc) (47), I (on RHgCl) (100)	4
	t-BuOOH	$Hg(OAc)_2 / HClO_4$	—	$NaBH_4 / NaOH$	$(CH_3)_2C(OOBu-t)C_6H_5$ + $H_2C{-}C(CH_3)C_6H_5$ 13.5 – 56 : 44 – 87.5	5
$C_6H_5CH{=}CHCH_3$	t-BuOOH	$Hg(O_2CCF_3)_2$, KBr	—	Br_2 / CH_2Cl_2	$C_6H_5CH(OOBu-t)CHBrCH_3$ (65 crude from RHgBr)	7
$trans$-$C_6H_5CH{=}CHCH_3$	t-BuOOH	$Hg(O_2CCF_3)_2$	$C_6H_5CH(OOBu-t)CH(HgO_2CCF_3)CH_3$ (84) + $C_6H_5CH(HgO_2CCF_3)CH(OOBu-t)CH_3$	—	—	15
$H_2C{=}C(C_6H_5)O_2CCH_3$	t-BuOOH	$Hg(OAc)_2$	—	—	t-BuOCH_2COC_6H_5 (30)	11
$H_2C{=}C(C_6H_5)CO_2CH_3$	t-BuOOH	$Hg(OAc)_2 / HClO_4$, KBr	$BrHgCH_2C(C_6H_5)(OOBu-t)CO_2CH_3$ (16.5)	—	—	8, 10
$C_6H_5CH{=}CHCO_2CH_3$	t-BuOOH	$Hg(OAc)_2 / HClO_4$, KX	$C_6H_5CH(OOBu-t)CH(HgX)CO_2CH_3$	X_2	$C_6H_5CH(OOBu-t)CHXCO_2CH_3$ X = Cl (32), Br (78)	18

Table 4.1. (continued)

Alkene	Peroxide	Mercuric salt	Organomercurial(s) (% Yield)	Subsequent reactants	Product(s) (% Yield)	Ref.
trans-$C_6H_5CH=CHCO_2CH_3$	H_2O_2	$Hg(OAc)_2$, Cl^-?	$C_6H_5CH(OOH)CH(HgX)CO_2CH_3$ $X = OAc$, Cl?	—	—	20
	t-BuOOH	$Hg(OAc)_2$ / $HClO_4$, (KBr)	$C_6H_5CH(OOBu-t)CH(HgX)CO_2CH_3$ $X = OAc$ (86), Br (70	$NaBH_4/NaOH$ ($X = Br$)	$C_6H_5CH(OOBu-t)CH_2CO_2CH_3$ (92 crude)	10, 12
$H_2C=CH(CH_2)_7CH_3$	t-BuOOH	$Hg(OAc)_2$	$AcOHgCH_2CH(OOBu-t)(CH_2)_7CH_3$ (80)	—	—	4
	t-BuOOH	$Hg(OAc)_2$	$AcOHgCH_2CH(OOBu-t)(CH_2)_7CH_3$ + $AcOHgCH_2CH(OAc)(CH_2)_7CH_3$ (15–20)	KBr	$BrHgCH_2CH(OOBu-t)(CH_2)_7CH_3$ (55)	13
	t-BuOOH	$Hg(OAc)_2$	$AcOHgCH_2CH(OOBu-t)(CH_2)_7CH_3$ + $AcOHgCH_2CH(OAc)(CH_2)_7CH_3$ (~15)	$NaBH_4$ or X_2	$XCH_2CH(OOBu-t)(CH_2)_7CH_3$ $X = H$ (50–60), Br, I	2
[structure: 1-phenylcyclohexene, C_6H_5]	H_2O_2	$Hg(OAc)_2$, Cl^-	[structure: C_6H_5, OOH, HgCl cyclohexane] (46)	—	—	20
$H_2C=C(CH_3)CH(OOBu-t)_2$	t-BuOOH	$Hg(OAc)_2$ / $HClO_4$, KBr	$BrHgCH_2C(CH_3)(OOBu-t)CH(OOBu-t)_2$ (60)	—	—	9
$H_2C=CH(CH_2)_{10}CH_3$	t-BuOOH	$Hg(OAc)_2$	—	—	$H_2C{-}CH(CH_2)_{10}CH_3$ (46) + $CH_3CH(OOBu-t)(CH_2)_{10}CH_3$ (37) + $CH_3CHOH(CH_2)_{10}CH_3$ (9)	22
$H_2C=C(C_6H_5)_2$	H_2O_2	$Hg(OAc)_2$, Cl^-?	$XHgCH_2C(C_6H_5)_2OOH$ $X = OAc$, Cl?	—	—	1, 20
	t-BuOOH	$Hg(OAc)_2$, Cl^-	$ClHgCH_2C(C_6H_5)_2OOBu-t$ (74)	X_2	$XCH_2C(C_6H_5)_2OOBu-t$ $X = Br$ (100), I (100)	4
	t-BuOOH	$Hg(OAc)_2$ / $HClO_4$	—	$NaBH_4/NaOH$	$H_2C{-}C(C_6H_5)_2$ + $CH_3C(OOBu-t)(C_6H_5)_2$ 64.5 : 35.5	5
$C_6H_5CH=CHC_6H_5$	t-BuOOH	$Hg(O_2CCF_3)_2$, KBr	—	Br_2/CH_2Cl_2	$C_6H_5CH(OOBu-t)CHBrC_6H_5$ (79 from $RHgBr$)	7

Table 4.1. (continued)

Alkene	Peroxide	Mercuric salt	Organomercurial(s) (% Yield)	Subsequent reactants	Product(s) (% Yield)	Ref.
cis-$C_6H_5CH=CHC_6H_5$	t-BuOOH	$Hg(O_2CCF_3)_2$, (Br^-)	$threo$-$C_6H_5CH(OOBu-t)CH(HgX)C_6H_5$ $X = O_2CCF_3$, Br	$Hg(O_2CCF_3)_2$ / t-BuOOH $(X = O_2CCF_3)$	$C_6H_5CH(OOBu-t)CH(O_2CCF_3)C_6H_5$ (19) + $(C_6H_5)_2CHCH(OOBu-t)_2$ (17)	15, 16
$trans$-$C_6H_5CH=CHC_6H_5$	t-BuOOH	$Hg(O_2CCF_3)_2$, (Br^-)	$erythro$-$C_6H_5CH(OOBu-t)CH(HgX)C_6H_5$	—	—	16
$C_6H_5CH=CHCOC_6H_5$	t-BuOOH	$Hg(OAc)_2$ / $HClO_4$, KX	$C_6H_5CH(OOBu-t)CH(HgX)COC_6H_5$	Y_2	$C_6H_5CH(OOBu-t)CHYCOC_6H_5$ X Y OAc Br (84) Cl Cl (55) Br Br (54) Br I (73, 84)	18
$trans$-$C_6H_5CH=CHCOC_6H_5$	t-BuOOH	$Hg(OAc)_2$ / $HClO_4$, (KBr)	$C_6H_5CH(OOBu-t)CH(HgX)COC_6H_5$ X = OAc (75), Br (60)	$NaBH_4$ / NaOH $(X = Br)$	$C_6H_5CH\overset{O}{\frown}CHCOC_6H_5$ (70) + $C_6H_5CH(OOBu-t)CH_2COC_6H_5$ (<10)	10, 12
$C_6H_5CH=CHCH(OOBu-t)_2$	t-BuOOH	$Hg(OAc)_2$ / $HClO_4$	$C_6H_5CH(OOBu-t)CH(HgOAc)CH(OOBu-t)_2$ (72)	—	—	9
cis-$CH_3(CH_2)_7CH=CH(CH_2)_7CO_2CH_3$	ROOH	$Hg(OAc)_2$	—	—	$CH_3(CH_2)_7CH\overset{O}{\frown}CH(CH_2)_7CO_2CH_3$ R = H(29), t-Bu [trans (39), cis (6)] + $CH_3(CH_2)_7CH\overset{HO\ \ H}{\diagdown\diagup}CH(CH_2)_7CO_2CH_3$ R = H(8), t-Bu (19)	22
$H_2C=CH(CH_2)_2OOH$	—	$Hg(NO_3)_2 \cdot H_2O$	(tetrahydrofuranyl peroxide)–CH_2HgNO_3	—	—	23
$H_2C=C(CH_3)(CH_2)_2OOH$	—	$Hg(NO_3)_2 \cdot H_2O$, KBr	(methyl dioxolanyl)–CH_2HgBr	$NaBH_4$ / NaOH or Br_2	(dioxolane)CH_2X + $HO(CH_2)_2C(CH_3)\overset{O}{\frown}CH_2$ X = H 90 : 10 X = Br (>74)	24, 25
$H_2C=CH(CH_2)_3OOH$	—	$Hg(NO_3)_2 \cdot H_2O$	(tetrahydropyranyl peroxide)–CH_2HgNO_3	—	—	23

Table 4.1. (continued)

Alkene	Peroxide	Mercuric salt	Organomercurial(s) (% Yield)	Subsequent reactants	Product(s) (% Yield)	Ref.
$H_2C=C(CH_3)C(OOH)(CH_3)_2$	—	$Hg(O_2CCF_3)_2$, $[Cl^-]$	$[H_2C=C(CH_2HgCl)C(OOH)(CH_3)_2$ (40)]	Br_2 [on $Hg(O_2CCF_3)_2$ product]	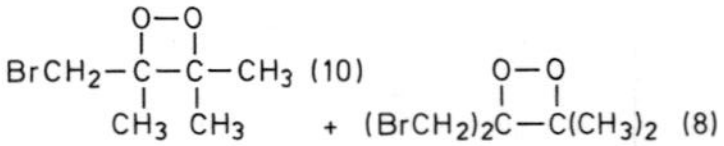 $BrCH_2-C-C-CH_3$ (10) $+$ $(BrCH_2)_2C-C(CH_3)_2$ (8)	26
$CH_3CH=CH(CH_2)_3OOH$	—	—	—	Br_2	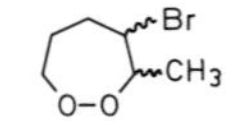 cis and trans	24
$trans-CH_3CH=CH(CH_2)_3OOH$	—	$Hg(NO_3)_2 \cdot H_2O$, KBr	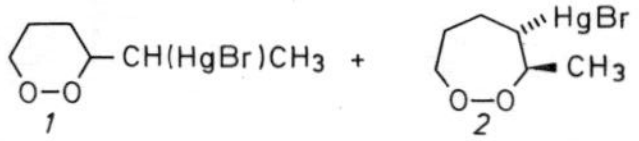 $CH(HgBr)CH_3$ (**1**) $+$ HgBr, CH_3 (**2**) (1 diastereomer) 3 : 1	$NaBH_4/NaOH$ or Br_2 (CH_2Cl_2) $[C_5H_5N]$	from **1**:  $CHXCH_3 + trans-HO(CH_2)_3CH-CHCH_3$; X = H <10 : >90; X = Br (2 diastereomers, >74) [1 stereoisomer, >74]; from **2**: 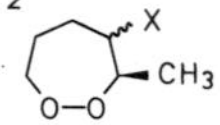 X = H, Br (2 diastereomers)	24, 25
$cis-CH_3CH_2CH=CHCH_2CH_2OOH$	—	$Hg(NO_3)_2 \cdot H_2O$, KBr	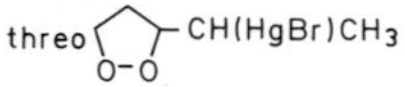 threo $CH(HgBr)CH_3$	$NaBH_4/NaOH$ or Br_2	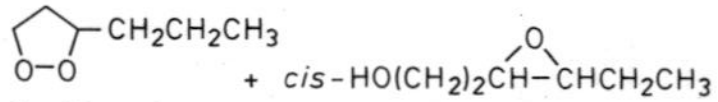 $CH_2CH_2CH_3 + cis-HO(CH_2)_2CH-CHCH_2CH_3$; X = H 3 : 1; X = Br erythro and threo (>74)	24, 25
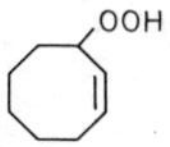 (cyclooctenyl)OOH	—	$Hg(O_2CCF_3)_2$	—	Br_2	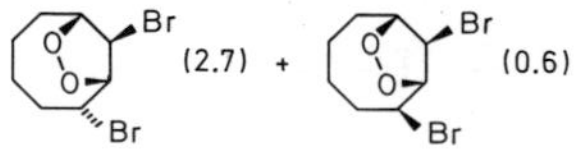 (2.7) $+$ (0.6)	27
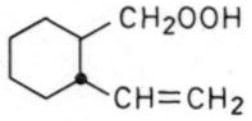 CH_2OOH, $CH=CH_2$ (cyclohexane)	—	—	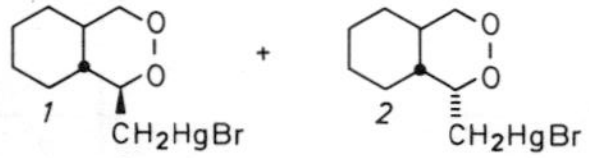 CH_2HgBr (**1**) $+$ CH_2HgBr (**2**) 3 : 1	Br_2 [$NaBH_4/NaOH$]	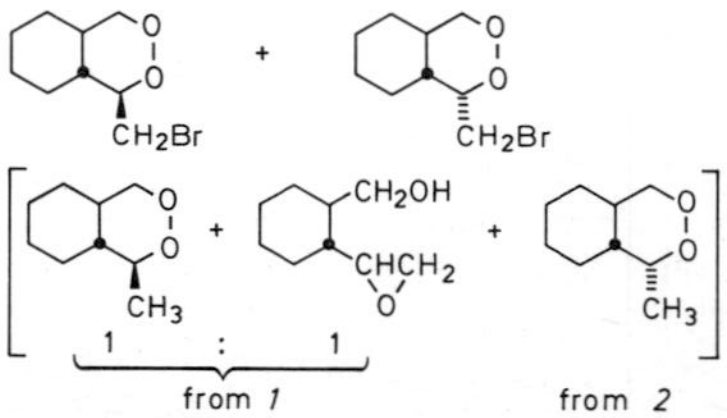 CH_2Br $+$ CH_2Br; [CH_3 : $CHCH_2$: CH_3] 1 : 1 ; from **1** ; from **2**	28

Table 4.1. (continued)

Alkene	Peroxide	Mercuric salt	Organomercurial(s) (% Yield)	Subsequent reactants	Product(s) (% Yield)	Ref.
trans-H_2C=CHCH=CHCH$_3$	H_2O_2	Hg(NO$_3$)$_2$ · H_2O, KCl	—	Br$_2$	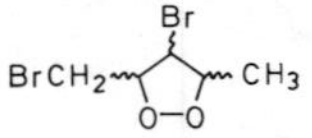 BrCH$_2$—(ring, Br, O–O)—CH$_3$ (43–73) 4 diastereomers	23
H_2C=CHCH$_2$CH=CH$_2$	H_2O_2	Hg(NO$_3$)$_2$ · H_2O, KCl	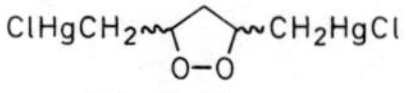 ClHgCH$_2$—(ring, O–O)—CH$_2$HgCl (91, 92) 1:1 cis/trans	NaBH$_4$/NaOH or Br$_2$	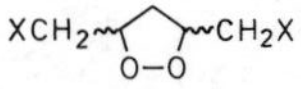 XCH$_2$—(ring, O–O)—CH$_2$X + H_2C=CHCH$_2$CHOHCH$_3$ X = H 90 : 10 (68 total) 50 : 50 cis/trans X = Br 40 : 60 cis/trans (51, 55)	23, 29, 30
(cyclohexadiene ring)	H_2O_2	Hg(O$_2$CCF$_3$)$_2$	—	Br$_2$	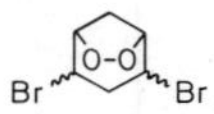 (bicyclic ring, O–O, Br, Br) (1.8)	31
H_2C=C(CH$_3$)CH$_2$CH=CH$_2$	H_2O_2	Hg(NO$_3$)$_2$ · H_2O, KCl	—	NaBH$_4$/NaOH or Br$_2$	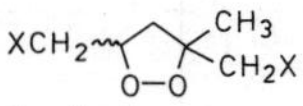 XCH$_2$—(ring, O–O, CH$_3$)—CH$_2$X X = H, Br	23
trans-H_2C=CHCH$_2$CH=CHCH$_3$	H_2O_2	Hg(NO$_3$)$_2$ · H_2O, KCl	—	Br$_2$	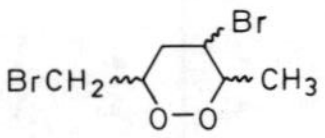 BrCH$_2$—(ring, Br, O–O)—CH$_3$	23
H_2C=CH(CH$_2$)$_2$CH=CH$_2$	H_2O_2	Hg(O$_2$CCF$_3$)$_2$, KCl	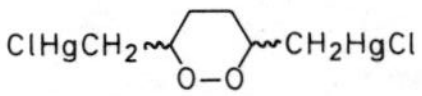 ClHgCH$_2$—(ring, O–O)—CH$_2$HgCl (26) 5 : 95 cis/trans	NaBH$_4$/NaOH [Br$_2$]	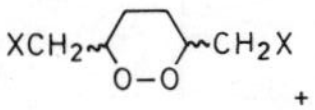 XCH$_2$—(ring, O–O)—CH$_2$X + H_2C=CH(CH$_2$)$_2$CHOHCH$_3$? X = H 90 : 10 (72 total) 20 : 80 cis/trans [X = Br 25 : 75 cis/trans (75)]	29
	H_2O_2	Hg(NO$_3$)$_2$ · H_2O, KCl	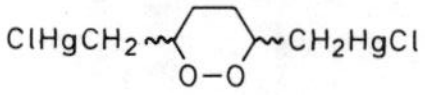 ClHgCH$_2$—(ring, O–O)—CH$_2$HgCl (72, 82 crude) 25 : 75 cis/trans	NaBH$_4$/NaOH or Br$_2$	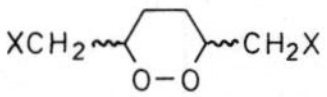 XCH$_2$—(ring, O–O)—CH$_2$X X = H (~85) 25 : 75 cis/trans X = Br (82) 25 : 75 cis/trans	23, 29 30

Table 4.1. (continued)

Alkene	Peroxide	Mercuric salt	Organomercurial(s) (% Yield)	Subsequent reactants	Product(s) (% Yield)	Ref.
trans-$H_2C{=}CHCH{=}CH(CH_2)_2OOH$	—	$Hg(NO_3)_2 \cdot H_2O$, KBr	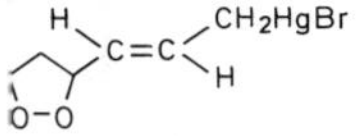 $H(CH_2HgBr)C{=}C(H)$ with O–O ring	—	—	25
$H_2C{=}C(CH_3)CH_2C(CH_3){=}CH_2$	H_2O_2	$Hg(NO_3)_2 \cdot H_2O$, KCl	—	Br_2	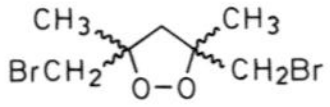 $BrCH_2$–[ring with CH_3, CH_3, O–O]–CH_2Br	23
$H_2C{=}CHC(CH_3)_2CH{=}CH_2$	H_2O_2	$Hg(NO_3)_2 \cdot H_2O$, KCl	—	$NaBH_4/NaOH$ or Br_2	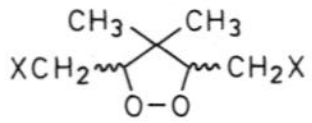 XCH_2–[ring with CH_3, CH_3, O–O]–CH_2X X = H, Br	23
 (cyclooctene ring)	H_2O_2	$Hg(O_2CCF_3)_2$	—	$NaBH_4/NaOH$	 [bicyclic O structure] (65) (no peroxides)	31
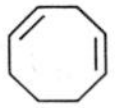 (cyclooctadiene ring)	H_2O_2	$Hg(O_2CCF_3)_2$	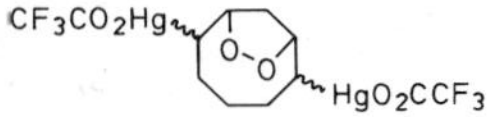 CF_3CO_2Hg–[bicyclic O–O]–HgO_2CCF_3	$NaBH_4/NaOH$ or KBr/Br_2	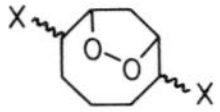 X–[bicyclic O–O]–X X = H (19, 28), Br (3 diastereomers)	31, 32
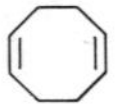 (cyclooctatriene ring)	H_2O_2	$Hg(O_2CCF_3)_2$	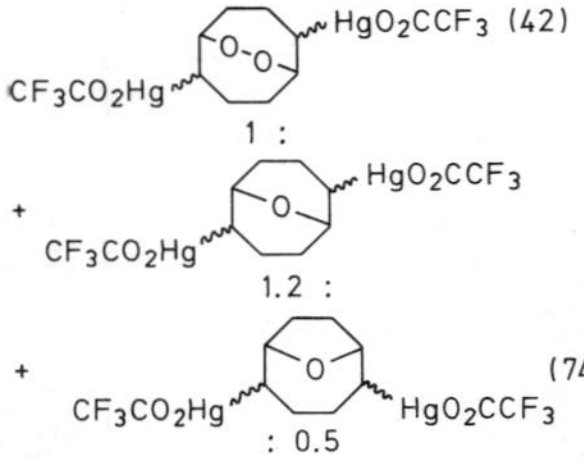 [bicyclic O–O]–HgO_2CCF_3 (42), CF_3CO_2Hg– 1 : [bicyclic O] HgO_2CCF_3, CF_3CO_2Hg– 1.2 : + CF_3CO_2Hg–[bicyclic O]–HgO_2CCF_3 : 0.5 (74 total)	$NaBH_4/NaOH$ or Br_2 (on peroxide)	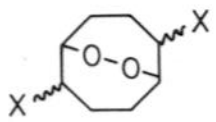 X–[bicyclic O–O]–X X = H (40) X = Br (67) 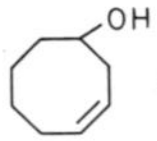 + (cyclooctenol OH) (40) 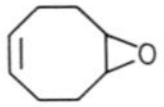 + (epoxide O) (9) (76 total)	31, 33
$H_2C{=}C(CH_3)CH_2CH_2C(CH_3){=}CH_2$	H_2O_2	$Hg(NO_3)_2 \cdot H_2O$, KCl	—	$NaBH_4/NaOH$	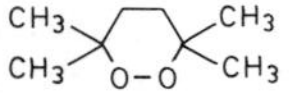 [ring with CH_3, CH_3, CH_3, CH_3, O–O]	23

Table 4.1. (continued)

Alkene	Peroxide	Mercuric salt	Organomercurial(s) (% Yield)	Subsequent reactants	Product(s) (% Yield)	Ref.
$(CH_3)_2C=CHCH=C(CH_3)_2$	H_2O_2	$Hg(NO_3)_2 \cdot H_2O$, KCl	—	$NaBH_4/H_2O_2$	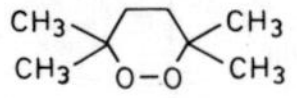 [structure]	23
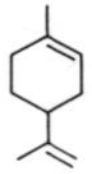 [structure]	$t\text{-BuOOH}$	$Hg(OAc)_2$, Cl^-	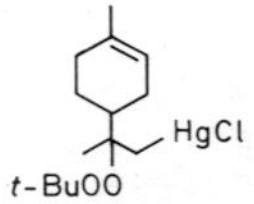 [structure] (48)	—	—	4
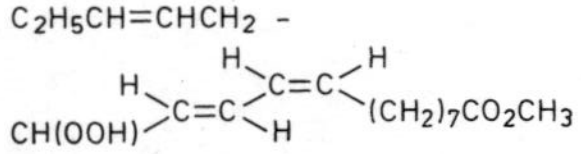 [structure] $C_2H_5CH=CHCH_2-$ ... $(CH_2)_7CO_2CH_3$	—	$Hg(O_2CCH_2Cl)_2$, NaCl	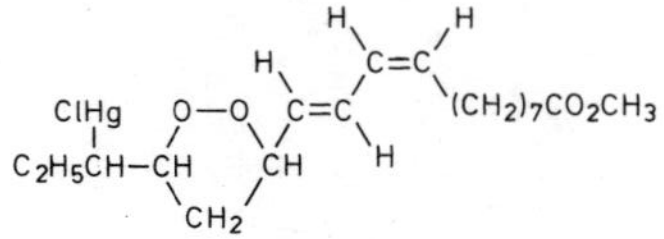 [structure]	$NaHB(OCH_3)_3$	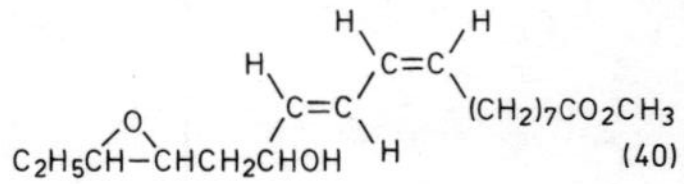 [structure] (40)	19

Alkyl hydroperoxides have been most frequently employed for peroxymercuration, since they eliminate the dimercuration and hydroxymercuration problems inherent with hydrogen peroxide. t-Butyl hydroperoxide has been widely utilized, but several other simple alkyl hydroperoxides and even mercurated alkyl hydroperoxides (Eq. 3) [3, 6] work well.

$$XHgCH_2CHROOH \ + \ H_2C{=}CHR' \ + \ HgX_2 \ \longrightarrow \ XHgCH_2CHROOCHR'CH_2HgX \qquad (3)$$

It appears that virtually all peroxymercuration reactions have been run in methylene chloride as the solvent. No studies on the effect of other solvents have been reported.

Most simple alkenes appear to react readily. Terminal olefins are more reactive than internal olefins [4], which are in turn more reactive than α,β-unsaturated carbonyl compounds, which require catalytic amounts of perchloric acid before reaction ensues. Certain aryl-substituted carbonyl systems, such as methyl α-methylcinnamate, methyl α-phenylcinnamate and 1,2-diphenylbut-1-en-3-one fail to react even under acid catalysis [10].

$$\underset{H}{\overset{C_6H_5}{\diagdown}}C{=}C\underset{CH_3}{\overset{H}{\diagup}} \quad \xrightarrow[t\text{-BuOOH}]{Hg(O_2CCF_3)_2} \quad \underset{84\,\%}{\overset{t\text{-BuOO}\ \ HgO_2CCF_3}{C_6H_5CHCHCH_3}} \ + \ \underset{\leq 16\,\%}{\overset{CF_3CO_2Hg\ \ OOBu\text{-}t}{C_6H_5CHCHCH_3}} \qquad (4)$$

The regiochemistry of peroxymercuration is the same as that of the solvomercuration reactions already discussed. Addition is exclusively Markovnikov. Only one exception has been reported, namely the peroxymercuration of $trans$-1-phenylpropene which gives a mixture of products (Eq. 4) [15]. As with alkoxymercuration, the regioselectivity of the peroxymercuration of α,β-unsaturated carbonyl systems depends on the substitution pattern of the substrate (Eqs. 5, 6) [8–10, 18].

$$\overset{O}{\overset{\|}{RCH{=}CHCR}} \quad \longrightarrow \quad \overset{ROO\ \ XHg\ \ O}{\overset{|\ \ \ \ |\ \ \ \|}{RCH-CHCR}} \qquad (5)$$

$$\overset{O}{\overset{\|}{H_2C{=}CRCR}} \quad \longrightarrow \quad \overset{ROO\ \ O}{\overset{|\ \ \ \|}{XHgCH_2CRCR}} \qquad (6)$$

The peroxymercuration reaction proceeds with high stereospecificity. Clean trans addition has been reported for cis- and $trans$-2-butene, cis- and $trans$-stilbene and cyclohexene [14, 16]. Other simple alkenes have been reported to afford a single diastereomer [7, 15], including cis- and $trans$-methyl cinnamate [18], but chalcone gives an 80:20 mixture of diastereomers from either the pure trans isomer or a mixture containing 60% cis-chalcone [18]. Norbornene gives the usual cis-exo product [16, 20].

Several minor problems have been observed during peroxymercuration. α,β-Unsaturated aldehydes readily form mercurated peroxy acetal side products (Eq. 7) [9]. As with the methoxymercuration product, the peroxymercuration product of *cis*-stilbene is unstable under the usual reaction conditions and gives solvolysis products (Eq. 8) [15]. Nevertheless, the corresponding organomercuric bromide can be easily isolated and its crystal structure has been reported [34]. This compound shows intramolecular O—Hg coordination in the solid state. The initial peroxymercuration products from enol acetates are also unstable and readily form the corresponding vinyl peroxide which thermally rearranges to various carbonyl products (Eq. 9) [11].

$$RCH{=}CHCH \longrightarrow \underset{\substack{|\quad\;\;|}}{RCHCHCH(OOBu\text{-}t)_2} \qquad (7)$$

$$\underset{\substack{|\quad\quad|}}{C_6H_5CHCHC_6H_5} \longrightarrow (C_6H_5)_2CHCH(OOBu\text{-}t)_2 \;+\; \underset{\substack{|\quad\quad|}}{C_6H_5CHCHC_6H_5} \qquad (8)$$

$$\underset{\substack{|}}{H_2C{=}CR} \longrightarrow AcOHgCH_2CR \longrightarrow H_2C{=}CR \longrightarrow R'OCH_2CR \qquad (9)$$

While the successful protodemercuration of a number of alkylperoxymercurials with alkaline sodium borohydride has been reported, complications can arise. Reduction of the *t*-butylperoxymercurials from α,β-unsaturated carbonyl compounds [12, 19], disubstituted terminal olefins [5] or internal olefins [22], and particularly aryl-substituted olefins of these types, afford significant to major amounts of the corresponding epoxides instead of the expected peroxide (Eq. 10). Cyclic peroxymercurials behave similarly (Eqs. 11, 12) [19]. The initially formed free radical presumably

$$\underset{\substack{|\quad\quad|}}{C_6H_5CHCHCOC_6H_5} \longrightarrow C_6H_5CH{-}CHCOC_6H_5 \qquad (10)$$
$$70\%$$

$$(11)$$
$$66\%$$

$$(12)$$
$$40\%$$

$$\underset{\substack{|}}{RCHCH_2} \longrightarrow RCH{-}CH_2 \;+\; t\text{-}BuO{\cdot} \qquad (13)$$

undergoes competitive intramolecular displacement (Eq. 13). These reductions are temperature dependent and best results are obtained at lower

362

temperatures [5]. Much improved yields of the *t*-butyl peroxides can be obtained by using tri-*n*-butyltin hydride instead of alkaline sodium borohydride for reduction [17].

The successful halogenation of numerous alkylperoxymercurials and even dimercurated peroxides has been achieved. Chlorination, bromination and iodination have all been reported [18]. When chlorine is used, you must use an organomercury chloride, since chlorination of RHgBr yields significant amounts of RBr. Bromination is best carried out on the organomercury halides, not RHgOAc, since the latter requires two equivalents of bromine to reach completion. Halodemercuration proceeds with extensive racemization even when polar solvents such as methanol and pyridine are employed [7, 14, 18]. While bromination in methanol (plus NaBr) versus methylene chloride is significantly faster and more stereospecific [14], halo methyl ether side-products have been observed [7, 18]. Pyridine gives still higher stereo-selectivity, but creates isolation problems [7]. Besides ether side-products, minor amounts of haloketones have also been noted (Eq. 14) [7, 13]. Many of the haloperoxides also turn out to be unstable, particularly the iodo compounds [18].

$$\underset{\displaystyle C_6H_5\overset{\displaystyle |}{\underset{\displaystyle }{C}}HCH_2HgBr}{\overset{\displaystyle t\text{-BuOO}}{}} + Br_2 \longrightarrow \underset{\displaystyle C_6H_5\overset{\displaystyle |}{\underset{\displaystyle }{C}}HCH_2Br}{\overset{\displaystyle t\text{-BuOO}}{}} + \underset{\displaystyle \underset{11\%}{C_6H_5\overset{\displaystyle O}{\overset{\displaystyle \|}{C}}CH_2Br}}{} \qquad (14)$$

Only one other reaction of these peroxymercurials has been examined. Treatment with palladium(II) salts gives high yields of the corresponding ketone presumably via intramolecular rearrangement of an intermediate alkylpalladium species (Eq. 15) [21].

$$\underset{\displaystyle C_6H_5\overset{\displaystyle |}{\underset{\displaystyle }{C}}HCH_2HgX}{\overset{\displaystyle t\text{-BuOO}}{}} \longrightarrow \quad \longrightarrow \quad C_6H_5\overset{\displaystyle O}{\overset{\displaystyle \|}{C}}CH_3 \qquad (15)$$

The mercuration of unsaturated alkyl hydroperoxides has been examined by several groups. The results reported to date are summarized in the second section of Table 4.1. Allylic hydroperoxides undergo initial allylmercurial formation followed by cyclization (Eqs. 16, 17). Subsequent bromination

$$\underset{\displaystyle \underset{OOH}{H_2C=\overset{\displaystyle CH_3}{\underset{\displaystyle }{C}}-C(CH_3)_2}}{} \longrightarrow \underset{\displaystyle \underset{OOH}{H_2C=\overset{\displaystyle XHgCH_2}{\underset{\displaystyle }{C}}-C(CH_3)_2}}{} \longrightarrow (XHgCH_2)_2\overset{\displaystyle O-O}{\underset{\displaystyle }{C}}-\overset{\displaystyle }{C}(CH_3)_2 \qquad (16)\ [26]$$

$$(17)\ [27]$$

affords low yields of the corresponding dibromoperoxides. Several alkenyl hydroperoxides with the double bond further removed from the hydroperoxide group have been effectively cyclized to form five-, six- and even seven-membered ring dialkylperoxides (Eq. 18) [23–25, 28]. These reactions are

$$(18) \ [25]$$

often highly regio- and stereospecific as shown. Bromination of these mercurials has been successful, though racemization is again prevalent [25]. Alkaline sodium borohydride reduction is again frequently accompanied by ring-opening epoxide formation, as well as some rearrangement to five- and six-membered ring ethers (Eq. 19) [25, 28]. The degree of epoxide formation appears to be heavily dependent on the stereochemistry of the mercurial.

$$(19) \ [28]$$

The peroxymercuration of dienes has also been examined as seen in the concluding section of Table 4.1. There appears to be only one example of the selective monomercuration of a diene using t-butyl hydroperoxide (Eq. 20) [4]. All other diene reactions have been effected using hydrogen

$$(20)$$

peroxide and either mercuric trifluoroacetate or mercuric nitrate monohydrate. The latter reagent appears to be more effective [29, 30]. Acyclic 1,3-, 1,4- and 1,5-dienes all afford cyclic dimercurated peroxides (Eq. 21) [23, 29, 30]. Many of these mercurials have been brominated successfully. On the other hand, alkaline sodium borohydride demercuration is sometimes accompanied by formation of unsaturated alcohols [23, 29]. On such occasions, it may be desireable to first brominate the mercurial and then reduce the resulting bromide by tri-n-butyltin hydride [23].

$$(21)$$

364

Both 1,4- and 1,5-cyclooctadiene have been reacted with hydrogen peroxide and mercuric trifluoroacetate to afford the corresponding dimercurated dioxabicyclo[5.2.1]- and -[3.3.2]decanes respectively, which have been reduced, and in the former case brominated, to the corresponding peroxides (Eqs. 22, 23).

$$(22) [32]$$

$$(23) [33]$$

Finally, one acyclic dienyl hydroperoxide has been cyclized by mercuric nitrate (Eq. 24) [25]. Note that 1,4-addition to the diene is observed.

$$(24)$$

References

1. Sokolov, V. I., Reutov, O. A.: Zh. Org. Khim., *5*, 174 (1969); J. Org. Chem. USSR, *5*, 168 (1969).
2. Ballard, D. H., Bloodworth, A. J., Bunce, R. J.: J. Chem. Soc. D, Chem. Commun., 815 (1969).
3. Bloodworth, A. J., Loveitt, M. E.: J. Chem. Soc., Perkin I, 1031 (1977).
4. Schmitz, E., Rieche, A., Brede, O.: J. Prakt. Chem., *312*, 30 (1970).
5. Bloodworth, A. J., Bylina, G. S.: J. Chem. Soc., Perkin I, 2433 (1972).
6. Bloodworth, A. J., Griffin, I. M.: J. Organometal. Chem., *66*, C1 (1974).
7. Bloodworth, A. J., Griffin, I. M.: J. Chem. Soc., Perkin I, 695 (1975).
8. Bloodworth, A. J., Bunce, R. J.: J. Chem. Soc. D, Chem. Commun., 753 (1970).
9. Bloodworth, A. J., Bunce, R. J.: J. Organometal. Chem., *60*, 11 (1973).
10. Bloodworth, A. J., Bunce, R. J.: J. Chem. Soc. C, 1453 (1971).
11. Schmitz, E., Brede, O.: J. Prakt. Chem., *312*, 43 (1970).
12. Bloodworth, A. J., Bunce, R. J.: J. Chem. Soc., Perkin I, 2787 (1972).
13. Ballard, D. H., Bloodworth, A. J.: J. Chem. Soc. C, 945 (1971).
14. Bloodworth, A. J., Courtneidge, J. L.: J. Chem. Soc., Perkin I, 3258 (1981).
15. Bloodworth, A. J., Griffin, I. M.: J. Chem. Soc., Perkin I, 195 (1975).
16. Bloodworth, A. J., Griffin, I. M.: J. Chem. Soc., Perkin II, 531 (1975).
17. Bloodworth, A. J., Courtneidge, J. L.: J. Chem. Soc., Chem. Commun., 1117 (1981).
18. Bloodworth, A. J., Griffin, I. M.: J. Chem. Soc., Perkin I, 688 (1974).
19. Corey, E. J., Schmidt, G., Shimoji, K.: Tetrahedron Lett., *24*, 3169 (1983).
20. Sokolov, V. I.: Izv. Akad. Nauk SSSR, Ser. Khim., 1089 (1972); Bull. Acad. Sci. USSR, Div. Chem. Sci., 1043 (1972).

21. Mimoun, H., Charpentier, R., Mitschler, A., Fischer, J., Weiss, R.: J. Am. Chem. Soc., *102*, 1047 (1980).
22. Gunstone, F. D., Inglis, R. P.: Chem. Phys. Lipids, *10*, 73 (1973).
23. Bloodworth, A. J., Khan, J. A.: J. Chem. Soc., Perkin I, 2450 (1980).
24. Porter, N. A., Nixon, J. R.: J. Am. Chem. Soc., *100*, 7116 (1978).
25. Nixon, J. R., Cudd, M. A., Porter, N. A.: J. Org. Chem., *43*, 4048 (1978).
26. Adam, W., Sakanishi, K.: J. Am. Chem. Soc., *100*, 3935 (1978).
27. Bloodworth, A. J., Leddy, B. P.: Tetrahedron Lett., 729 (1979).
28. Porter, N. A., Cudd, M. A., Miller, R. W., McPhail, A. T.: J. Am. Chem. Soc., *102*, 414 (1980).
29. Bloodworth, A. J., Loveitt, M. E.: J. Chem. Soc., Perkin I, 522 (1978).
30. Bloodworth, A. J., Loveitt, M. E.: J. Chem. Soc., Chem. Commun., 94 (1976).
31. Bloodworth, A. J., Khan, J. A., Loveitt, M. E.: J. Chem. Soc., Perkin I, 621 (1981).
32. Bloodworth, A. J., Khan, J. A.: Tetrahedron Lett., 3075 (1978).
33. Adam, W., Bloodworth, A. J., Eggelte, H. J., Loveitt, M. E.: Angew. Chem., *90*, 216 (1978); Angew. Chem., Int. Ed. Engl., *17*, 209 (1978).
34. Halfpenny, J., Small, R. W. H.: J. Chem. Soc., Chem. Commun., 879 (1979).

V. Acyloxymercuration

A. Alkenes

In the monograph "Organomercury Compounds in Organic Synthesis", we discussed some of the reactions which can take place between an alkene and a mercuric carboxylate salt. For example, we covered the preparation of alpha-mercurated carbonyl compounds and vinyl- or allylmercurials via mercuration of enols and enol derivatives, or certain alkenes respectively. We also discussed the Treibs reaction which involves the allylic acyloxylation of alkenes by heating alkenes with mercuric carboxylate salts. The most common reaction to occur between an alkene and a mercuric carboxylate salt is not, however, either of these processes, but simple addition of the mercury and carboxylate groups across the carbon—carbon double bond, acyloxymercuration (Eq. 1).

$$\begin{array}{c}\diagup \\ C=C \\ \diagup \end{array} \quad + \quad Hg(O_2CR)_2 \quad \longrightarrow \quad RCO_2Hg-\overset{|}{C}-\overset{|}{C}-O_2CR \tag{1}$$

This reaction has been most commonly effected using mercuric acetate in acetic acid. However, a number of other mercuric carboxylates will also undergo addition to alkenes and these reactions have been occasionally carried out in aprotic solvents including dioxane, methylene chloride, chloroform, carbon tetrachloride, benzene, tetrahydrofuran, DMF, ethyl ether, ethyl acetate, nitromethane, acetic anhydride and acetonitrile. Heating seems unnecessary, though frequently done.

The acyloxymercuration of a number of acyclic alkenes (Table 5.1) has been reported. In earlier times, this reaction was used to determine the presence of alkenes in mixtures and the degree of unsaturation [2], as well as to purify olefins [38]. Though mercuric acetate is by far the most widely employed salt for this reaction, a number of other mercuric carboxylates have been utilized, including those derived from dicarboxylic acids, such as oxalic, malonic, succinic, sebacic and phthalic acids, which are assumed to give cyclic derivatives (Eq. 2) [39].

$$RCH=CH_2 \;+\; Hg(O_2C)_2R' \quad \longrightarrow \quad \underset{\underset{RCH-CH_2Hg}{|}}{\overset{\overset{O}{\diagdown} \underset{O}{C-R'-C} \overset{O}{\diagup}}{\underset{O}{|}}} \tag{2}$$

Acyloxymercuration with mercuric acetate has been shown to exhibit second order kinetics, with 1-alkenes approximately 100 times more reactive

Table 5.1. Acyloxymercuration of Acyclic Alkenes

Alkene	Mercuric salt	Organomercurial(s) (% Yield)	Subsequent reactants	Product(s) (% Yield)	Ref.
$H_2C{=}CF_2$	$Hg(OAc)_2$	$AcOHgCH_2CO_2H$ (100)	—	—	1
$H_2C{=}CH_2$	$Hg(OAc)_2$	$AcOHgCH_2CH_2OAc$	—	—	2
	$Hg(OAc)_2$, KCl	$ClHgCH_2CH_2OAc$	—	—	3
	$Hg(OAc)_2$, NaI	$IHgCH_2CH_2OAc$	—	—	4
	$Hg(OAc)_2$	—	H_2SO_4	$(AcOCH_2CH_2Hg)_2SO_4$	5
	$Hg(OAc)_2$	$AcOHgCH_2CH_2OAc$	$C_6H_5OCH_3/H_3PO_4$	$p\text{-}CH_3OC_6H_4CH_2CH_2HgOAc$	6
	$Hg(OAc)_2$	$AcOHgCH_2CH_2OAc$	5,5-dimethyl-1,3-cyclohexanedione / $HClO_4$, NaCl	2-(2-chloromercurioethyl)-5,5-dimethyl-1,3-cyclohexanedione (18)	7
$H_2C{=}CHCF_3$	$Hg(NO_3)_2/HOAc$ (Cl⁻)	$AcOCH_2CH(HgX)CF_3$ X = NO_3 (80), Cl	—	—	8
$H_2C{=}CHCH_3$	$Hg(OAc)_2$, NaCl	$ClHgCH_2CH(OAc)CH_3$	—	—	9
$H_2C{=}CHOCH_3$	$Hg(OAc)_2$	$AcOHgCH_2CH(OAc)OCH_3$ (91.7)	—	—	10
$H_2C{=}CHCH_2CN$	$Hg(OAc)_2$, NaCl	$ClHgCH_2CH(OAc)CH_2CN$	I⁻	$IHgCH_2CH(OAc)CH_2CN$	11
$H_2C{=}CHO_2CCH_3$	$Hg(OAc)_2$	$AcOHgCH_2CH(OAc)_2$ (100)	—	—	12 – 14
	$Hg(O_2CCH_2CH_3)_2$	$CH_3CH_2CO_2HgCH_2CH(O_2CCH_2CH_3)O_2CCH_3$	—	—	15
	$Hg(O_2CC_6H_5)_2$	$C_6H_5CO_2HgCH_2CH(OAc)O_2CC_6H_5$	—	—	14

Table 5.1. (continued)

Alkene	Mercuric salt	Organomercurial(s) (% Yield)	Subsequent reactants	Product(s) (% Yield)	Ref.
$H_2C=C(CH_3)_2$	$Hg(O_2CR)_2$, X^-	$XHgCH_2C(O_2CR)(CH_3)_2$ $R = CH_3$ $X = OAc$, Cl, Br $R = C_2H_5$ $X = OAc$, O_2CEt	—	—	16
$H_2C=CHCH_2CH_3$	$Hg(OAc)_2$, X^-	$XHgCH_2CH(OAc)CH_2CH_3$ $X = OAc$, O_2CEt, Cl	—	—	16
$H_2C=CHOCH_2CH_3$	$Hg(OAc)_2$	$AcOHgCH_2CH(OAc)OCH_2CH_3$	Br_2 [KCl/H_2O]	$BrCH_2CH(OAc)OCH_2CH_3$ (54) [$ClHgCH_2CHO$ (100)]	10
$H_2C=CHCH_2HgOAc$	$Hg(OAc)_2$	$AcOHgCH_2CH(OAc)CH_2HgOAc$	—	—	17
$H_2C=CHO_2CCH_2CH_3$	$Hg(OAc)_2$	$AcOHgCH_2CH(OAc)O_2CCH_2CH_3$	—	—	15
$H_2C=CHOCH(CH_3)_2$	$Hg(OAc)_2$	$AcOHgCH_2CH(OAc)OCH(CH_3)_2$ (100)	Br_2	$BrCH_2CH(OAc)OCH(CH_3)_2$ (55.6)	10
$H_2C=CHSi(OCH_3)_3$	$Hg(OAc)_2$	$AcOCH_2CH(HgOAc)Si(OCH_3)_3$ (~100)	—	—	18
$H_2C=CHSi(CH_3)_3$	$Hg(OAc)_2$	$AcOCH_2CH(HgOAc)Si(CH_3)_3$ (100)	—	—	17, 18
$CH_3CH_2CH=CHCH_2CH_3$	$Hg(OAc)_2$, Br^-	$CH_3CH_2CH(HgBr)CH(OAc)CH_2CH_3$ erythro and threo	—	—	19
$H_2C=CHO(CH_2)_3CH_3$	$Hg(OAc)_2$	$AcOHgCH_2CH(OAc)O(CH_2)_3CH_3$ (100)	Br_2	$BrCH_2CH(OAc)O(CH_2)_3CH_3$ (43.5)	10
$H_2C=CHCH_2Si(CH_3)_3$	$Hg(OAc)_2$	$H_2C=CHCH_2HgOAc$ (21–85) + $AcOHgCH_2CH(OAc)CH_2HgOAc$ (5–36)	—	—	17, 18
cis- $CH_3CH=CHSi(CH_3)_3$	$Hg(OAc)_2$	$CH_3CH(OAc)CH(HgOAc)Si(CH_3)_3$ (84) 1:1 erythro/threo	—	—	17
$H_2C=CHO_2CC(CH_3)_3$	$Hg(OAc)_2$	$AcOHgCH_2CH(OAc)O_2CC(CH_3)_3$	—	—	15

Table 5.1. (continued)

Alkene	Mercuric salt	Organomercurial(s) (% Yield)	Subsequent reactants	Product(s) (% Yield)	Ref.
o - H_2C=$CHC_6H_4NO_2$	$Hg(OAc)_2$	o - $AcOHgCH_2CH(OAc)C_6H_4NO_2$	$NaCl/H_2O$	o - H_2C=$CHC_6H_4NO_2$	20
p - H_2C=$CHC_6H_4NO_2$	$Hg(OAc)_2$	p - $AcOHgCH_2CH(OAc)C_6H_4NO_2$	$NaCl/H_2O$	p - H_2C=$CHC_6H_4NO_2$	20
p-H_2C=$CHOC_6H_4NO_2$	$Hg(OAc)_2$	p - $AcOHgCH_2CH(OAc)OC_6H_4NO_2$ (98)	—	—	21
H_2C=CHC_6H_5	$Hg(OAc)_2$	$AcOHgCH_2CH(OAc)C_6H_5$	?	$CH_3CH(OAc)C_6H_5$ (48, 64) + $CH_3CHOHC_6H_5$ (10, 23) + H_2C=CHC_6H_5	22, 23
H_2C=$CHOC_6H_5$	$Hg(OAc)_2$	$AcOHgCH_2CH(OAc)OC_6H_5$ (93.1, 99)	—	—	10, 21
H_2C=$CHSi(O_2CCH_3)_3$	$Hg(OAc)_2$	$AcOCH_2CH(HgOAc)Si(O_2CCH_3)_3$ (~100)	—	—	18
H_2C=CHO—⬡	$Hg(OAc)_2$	$AcOHgCH_2CH(OAc)O$—⬡ (99)	—	—	21
H_2C=$CH(CH_2)_5CH_3$	$Hg(OAc)_2$	$AcOHgCH_2CH(OAc)(CH_2)_5CH_3$	$NaBH_4/NaOH$	$CH_3CH(OAc)(CH_2)_5CH_3$ (40) + $CH_3CHOH(CH_2)_5CH_3$ (31)	24
	$Hg(OAc)_2$, Cl^-	$ClHgCH_2CH(OAc)(CH_2)_5CH_3$	$NaBH_4$	$CH_3CH(OAc)(CH_2)_5CH_3$	25
H_2C=$CHSi(CH_2CH_3)_3$	$Hg(O_2CR)_2$	$RCO_2CH_2CH(HgO_2CR)Si(CH_2CH_3)_3$ (~100) R = CH_3, C_6H_5	—	—	18
H_2C=$CHO_2CC_6H_5$	$Hg(OAc)_2$	$AcOHgCH_2CH(OAc)O_2CC_6H_5$	—	—	13, 14
H_2C=$CHCH_2C_6H_5$	$Hg(OAc)_2$	$AcOHgCH_2CH(OAc)CH_2C_6H_5$ (90)	Δ	C_6H_5CH=$CHCH_2OAc$ (80–90)	26
	$Hg(OAc)_2$ (Cl^-)	$XHgCH_2CH(OAc)CH_2C_6H_5$ X = OAc, Cl	Δ	C_6H_5CH=$CHCH_2OAc$	27

Table 5.1. (continued)

Alkene	Mercuric salt	Organomercurial(s) (% Yield)	Subsequent reactants	Product(s) (% Yield)	Ref.
$CH=CH_2$ (norbornyl)	$Hg(OAc)_2$	—	$NaBH_4/NaOH$	$CH(OAc)CH_3$ (norbornyl) (48)	28
$m\text{-}H_2C=CHCH_2OC_6H_4OCH_3$	$Hg(OAc)_2$	CH_3O—ring—$OCH_2CH=CH_2$, $AcOHg$, $HgOAc$? (30)	—	—	29
carbazole-$N\text{-}CH=CH_2$	$Hg(OAc)_2$	carbazole-$N\text{-}AcOCHCH_2HgOAc$ + $AcOHgCH_2CH(OAc)_2$	—	—	30
$H_2C=CH(CH_2)_{11}CH_3$	$Hg(OAc)_2$	$AcOHgCH_2CH(OAc)(CH_2)_{11}CH_3$	—	—	2
sugar ring: $CHCH_2CH=CH_2$ / $HCOAc$ / $HCOAc$ / $AcOCH$ / CH / CH_3	$Hg(OAc)_2$ (NaCl)	sugar ring: $CHCH_2CH(OAc)CH_2HgX$ / $HCOAc$ / $HCOAc$ / $AcOCH$ / CH / CH_3 $X = OAc, Cl$	—	—	31
pyranose: $AcOCH_2$, AcO, OAc, OAc, $CH=CH_2$	$Hg(OAc)_2$ (Cl⁻)	pyranose: $AcOCH_2$, AcO, OAc, OAc, $CH(OAc)CH_2HgX$ $X = OAc, Cl$	—	—	32
$H_2C=CH(CH_2)_{13}CH_3$	$Hg(OAc)_2$ (KI)	$XHgCH_2CH(OAc)(CH_2)_{13}CH_3$ $X = OAc, I$	—	—	2
	$Hg(O_2CR)_2$	$RCO_2HgCH_2CH(O_2CR)(CH_2)_{13}CH_3$ $R = CH_3, ClCH_2, C_2H_5, n\text{-}C_{15}H_{31}, n\text{-}C_{17}H_{35}$	—	—	2

Table 5.1. (continued)

Alkene	Mercuric salt	Organomercurial(s) (% Yield)	Subsequent reactants	Product(s) (% Yield)	Ref.
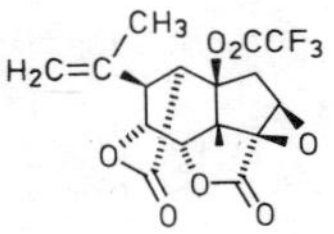 $H_2C=C$ with CH_3, O_2CCF_3 substituents (tricyclic lactone structure)	$Hg(O_2CCF_3)_2$, KCl	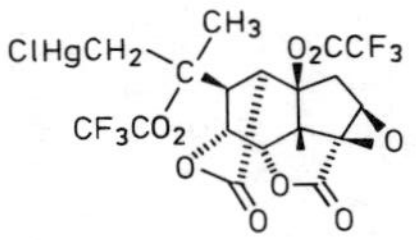 $ClHgCH_2$–$C(CH_3)$, O_2CCF_3, CF_3CO_2 (tricyclic lactone structure)	n-Bu$_3$SnH	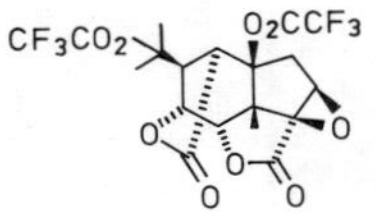 CF_3CO_2, O_2CCF_3 (tricyclic lactone structure) (35)	33
cis-$CH_3(CH_2)_7CH=CH(CH_2)_7$-$CO_2CH_3$	$Hg(OAc)_2$	$AcOHg$, OAc on $CH_3(CH_2)_7CH$–$CH(CH_2)_7CO_2CH_3$ (26–100)	$NaBH_4$	$CH_3(CH_2)_7CH(OAc)(CH_2)_8CO_2CH_3$ + $CH_3(CH_2)_8CH(OAc)(CH_2)_7CO_2CH_3$ 1 : 1 (82–93 total)	34–37

than 2-alkenes [38]. Branching in the alkyl group attached to the double bond lowers the reactivity of the alkene, while addition of a methyl group to the 2-position of a 1-alkene increases the olefin's reactivity. In this latter situation, the reaction fails to proceed to completion, but reaches an equilibrium.

Equilibria have also be observed with mercuric trifluoroacetate in THF, for which the following equilibrium constants, $K[Hg(O_2CCF_3)_2]$, have been reported: norbornene (5×10^5), 7,7-dimethylnorbornene (6.8×10^3), 1-hexene (826), 3,3-dimethyl-1-butene (91.5), cyclohexene (38), styrene (14), *trans*-2-hexene (9.2), 2-methyl-1-pentene (7.9), 1-methylcyclohexene (4.4), cyclopentene (3.4), cyclooctene (3.3), *cis*-2-hexene (2.2), α-methylstyrene (0.95) (Eq. 3) [40, 41]. The equilibria are solvent dependent as shown by the

$$\text{C=C} \;+\; Hg(O_2CCF_3)_2 \;\rightleftharpoons\; \overset{CF_3CO_2}{\underset{HgO_2CCF_3}{-C-C-}} \tag{3}$$

following equilibria constants for cyclohexene: DMF (7.5), THF (38), ethyl ether (200), ethyl acetate (413) and benzene (1330). The reaction is more complete in the less basic solvents.

Relatively little work has been carried out on the regioselectivity in acyclic systems. Methyl oleate has been acetoxymercurated numerous times, but no regioselectivity is observed [34–37]. Although most terminal olefins apparently yield only the Markovnikov product, anti-Markovnikov products are reported for 3,3,3-trifluoro-1-propene [8] and several vinylsilanes [17, 18] (Eqs. 4, 5). On the other hand, acetoxymercuration of allyltrimethylsilane affords predominantly allylmercuric acetate and its subsequent acetoxymercuration product [17].

$$H_2C-CHCF_3 \;+\; Hg(NO_3)_2 \;+\; HOAc \;\longrightarrow\; \overset{HgNO_3}{AcOCH_2CHCF_3} \tag{4}$$

$$RCH=CHSiX_3 \;+\; Hg(O_2CR')_2 \;\longrightarrow\; \overset{R'CO_2}{\underset{HgO_2CR'}{RCHCHSiX_3}} \tag{5}$$

R=H, Me   R'=Me, Ph

X=Me, Et, OMe, OAc

Relatively few monocyclic alkenes have been subjected to acyloxymercuration (Table 5.2). A simple NMR method has been developed for establishing the stereochemistry of the resulting adducts [25]. Although one exocyclic alkylidene cyclopropane has been reported to undergo a highly regiospecific acetoxymercuration (Eq. 6) [56], 1,3,3-trimethylcyclopropene

$$\overset{C_6H_5}{\underset{C_6H_5}{\triangleright=C}} \;\longrightarrow\; \overset{OAc}{\underset{AcOHg \; C_6H_5}{\triangleright-C-C_6H_5}} \tag{6}$$

Table 5.2. Acyloxymercuration of Monocyclic Alkenes

Alkene	Mercuric salt	Organomercurial(s) (% Yield)	Subsequent reactants	Product(s) (% Yield)	Ref.
(cyclobutene)	Hg(OAc)₂, Cl⁻	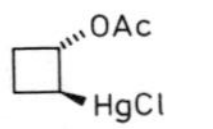 (cyclobutyl-OAc, HgCl)	NaBH₄	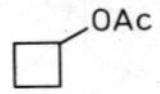 (cyclobutyl-OAc)	25
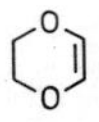 (1,4-dioxene)	Hg(OAc)₂	—	—	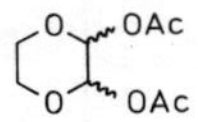 (dioxane-OAc, OAc) (low)	42
(cyclopentene)	Hg(OAc)₂	—	I₂	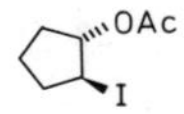 (cyclopentyl-OAc, I) (90)	43
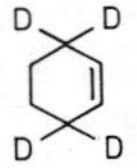 (tetradeuterio cyclohexene)	Hg(OAc)₂	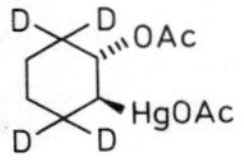 (cyclohexyl-D₄-OAc, HgOAc)	—	—	26
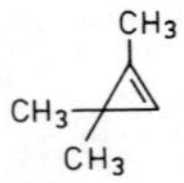 (trimethylcyclopropene)	Hg(OAc)₂	—	—	$(CH_3)_2C=C(CH_3)CH(OAc)_2$ (65)	44, 45
 (cyclohexene)	Hg(OAc)₂	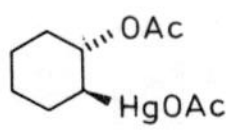 (cyclohexyl-OAc, HgOAc) (76)	—	—	19
	Hg(OAc)₂	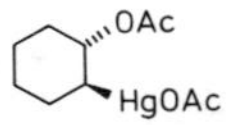 (cyclohexyl-OAc, HgOAc)	Δ	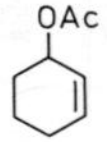 (cyclohexenyl-OAc)	27
	Hg(OAc)₂	—	I₂	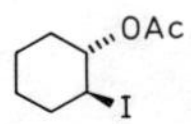 (cyclohexyl-OAc, I) (80–85)	43
	Hg(OAc)₂, (NaCl)	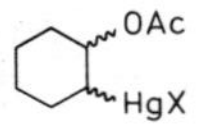 (cyclohexyl-OAc, HgX) X = OAc (91), Cl (85)	—	—	46

Table 5.2. (continued)

Alkene	Mercuric salt	Organomercurial(s) (% Yield)	Subsequent reactants	Product(s) (% Yield)	Ref.
	Hg(OAc)₂, Cl⁻	[cyclohexane with OAc and HgCl] cis or trans	?	[cyclohexyl OAc]	25
[2-cyclohexen-1-ol, OH]	Hg(OAc)₂	—	NaBH₄ / NaOH	[OH, OAc isomers] 82 : 6.8 : 4.4 : 6.8 (95 total)	47
[3-methoxycyclohexene, OCH₃]	Hg(OAc)₂	—	NaBH₄ / NaOH	[OCH₃, OAc isomers] 93.6 : 6.4 (87 total)	47
[cyclooctene oxide, epoxide]	Hg(OAc)₂	—	NaBH₄ / NaOH	[9-oxabicyclo HO isomers] 71 : 29 (65 total)	48
[cyclohexenyl acetate, OAc]	Hg(OAc)₂	—	NaBH₄ / NaOH	[OAc, OAc isomers] 92 : 8 (95 total)	47
[dimethylcyclohexene, CH₃, CH₃]	Hg(OAc)₂	—	NaBH₄	[OAc, CH₃ isomers] 87 : 13	49
[bicyclic cyclooctene]	Hg(OAc)₂	[cyclooctane with OAc and HgOAc]	ROH	[cyclooctane OR, HgOAc] R = H, CH₃	50

376

Table 5.2. (continued)

Alkene	Mercuric salt	Organomercurial(s) (% Yield)	Subsequent reactants	Product(s) (% Yield)	Ref.
	Hg(OAc)₂/NaOAc/ CH₃OH	[cyclooctane with OAc, HgOAc] 94 : [cyclooctane with OCH₃, HgOAc] 6	—	—	51
	Hg(OAc)₂, NaCl	[cyclooctane with OAc, HgCl] (70)	—	—	52
[6-nitro-2H-chromen-2-one]	Hg(OAc)₂	[substituted phenol, NO₂, OH, CHCHCO₂H, AcO, HgOAc]	—	—	53
[4,4-dimethylcyclohexene, CH₃ CH₃ CH₃]	Hg(OAc)₂	—	NaBH₄	[cyclohexane CH₃ CH₃ CH₃ OAc] 88 : [cyclohexane CH₃ CH₃ CH₃ OAc] 12	49
[glycal: AcOCH₂, O, AcO, OAc]	Hg(OAc)₂	[pyranose: AcOCH₂, O, OAc, AcO, HgOAc, OAc] (29)	NaBH₄ [H₂NCSNH₂]	[AcOCH₂, O, AcO, OAc] (41)[82] + [AcOCH₂, O, OAc, AcO, OAc] (45)	54
[3-(4-bromobenzylidene)isobenzofuranone, H, Br]	Hg(OAc)₂	[isobenzofuranone, OAc, AcOHgCH, Br] (15)	—	—	55

Table 5.2. (continued)

Alkene	Mercuric salt	Organomercurial(s) (% Yield)	Subsequent reactants	Product(s) (% Yield)	Ref.
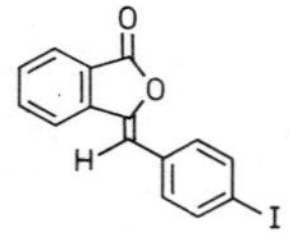 [structure: 3-(4-iodobenzylidene)phthalide]	$Hg(OAc)_2$	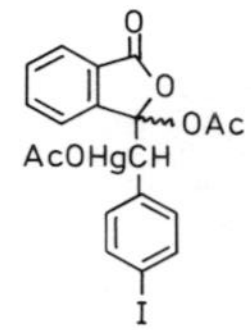 [structure: AcOHgCH(·OAc) / 4-iodophenyl phthalide] (17)	—	—	55
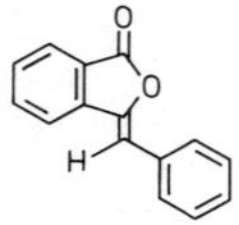 [structure: 3-benzylidenephthalide]	$Hg(OAc)_2$	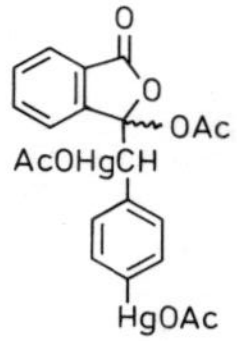 [structure: AcOHgCH(·OAc) phthalide with 4-HgOAc-phenyl] (55) 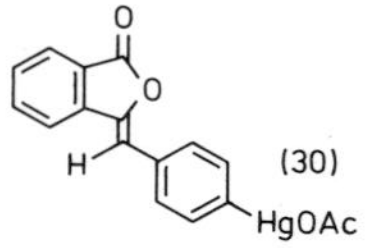 + [structure: 3-(4-HgOAc-benzylidene)phthalide] (30)	—	—	55
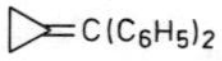 $\triangleright=C(C_6H_5)_2$	$Hg(OAc)_2$ (KBr)	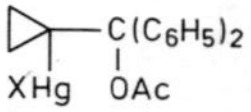 [structure: cyclopropyl–C($C_6H_5)_2$, XHg OAc] X = OAc (30), Br (80)	$NaBH_4 / NaOH$ (on $RHgOAc$)	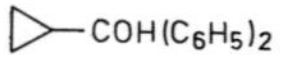 $\triangleright-COH(C_6H_5)_2$ (83)	56
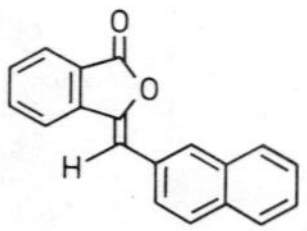 [structure: 3-(2-naphthylmethylene)phthalide]	$Hg(OAc)_2$	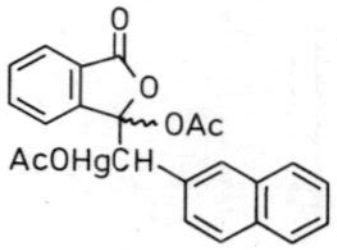 [structure: AcOHgCH(·OAc) phthalide with 2-naphthyl] (10)	—	—	55

V. Acyloxymercuration

$$\text{(7)}$$

affords only a ring-opened diacetate (Eq. 7) [44, 45]. Cyclobutene [25] and cyclopentene [43] appear to undergo anti addition upon reaction with mercuric acetate.

Cyclohexene affords trans adducts with both mercuric acetate [19, 26, 27] and mercuric trifluoroacetate [40]. With substituted cyclohexenes, the regio- and stereoselectivity of acetoxymercuration is very similar to that of the corresponding hydroxy- and methoxymercuration reactions (Eqs. 8, 9).

$$\text{(8) [49]}$$

R=H	87	:	13
R=CH$_3$	88	:	12

$$\text{(9) [47]}$$

R=H	82	:	6.8	:	4.4	:	6.8
R=CH$_3$	93.6	:	6.4	:	—	:	—
R=COCH$_3$	92	:	8	:	—	:	—

While anti addition appears to be the norm, *trans*-cyclooctene affords a syn addition compound (Eq. 10) [50]. Recall earlier that this alkene gave

$$\text{(10)}$$

substantial amounts of this mercurial even under the usual hydroxy- and methoxymercuration conditions [50–52, 57, 58]. *trans*-Cyclononene behaved similarly [58], but the stereochemistry of its acetoxymercuration product has not been established. D-Glucal triacetate appears to be the only other olefin to undergo cis addition with mercuric acetate (Eq. 11) [54]. In this case, however, it is possible that epimerization of the acetate group may have occurred.

$$\text{(11)}$$

Three other oxygenated cyclic olefins have been subjected to acetoxymercuration with interesting results. 6-Nitrocoumarin undergoes ring-opening (Eq. 12) [53], while arylidene phthalides afford the expected acetoxymercurials alongside varying amounts of arylmercurial (Eq. 13) [55]. With 2,3-dihydro-1,4-dioxin, the mercurial is apparently too unstable to even isolate (Eq. 14) [42].

$$(12)$$

$$(13)$$

$$(14)$$

The acyloxymercuration of bicyclic and polycyclic alkenes has proven quite interesting (Table 5.3). Although originally misassigned [62], the product from acetoxymercuration of norbornene is a cis-exo adduct (Eq. 15) [63]. X-ray crystallographic data [80], as well as proton [25] and carbon-13 [81] NMR data are available on various derivatives of this compound. Recall that significant acetoxymercuration accompanied the hydroxy- [68, 82, 83] and alkoxymercuration [60, 82, 84] of norbornene. Trifluoroacetoxymercuration of norbornene also affords the cis-exo adduct with a very high equilibrium constant [40, 41]. Norbornene will readily displace cyclohexene from its trifluoroacetoxymercurial [41].

$$(15)$$

The acyloxymercuration of several substituted norbornenes has also been examined. Both *exo*- and *endo*-5-methylnorbornene give mixtures of the expected exo acetates after acetoxymercuration-demercuration (Eq. 16) [64]. Even sterically hindered 7,7-dimethylnorbornene affords exclusively the cis-exo adduct (Eq. 17) [60, 82, 84]. This is another system prone to acetoxymercuration during hydroxymercuration [68, 83]. Upon trifluoroacetoxymercuration, this olefin exhibits a high equilibrium constant (6,800) and readily displaces cyclohexene from its adducts [40]. *syn*-7-Bromo- and 1,4,7,7-tetramethylnorbornene also yield cis-exo adducts exclusively [60].

Table 5.3. Acyloxymercuration of Bicyclic and Polycyclic Alkenes

Alkene	Mercuric salt	Organomercurial(s) (% Yield)	Subsequent reactants	Product(s) (% Yield)	Ref.
	$Hg(OAc)_2$, Cl^-	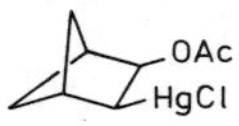 OAc, HgCl	$NaBD_4$	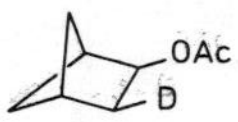 OAc, D	59
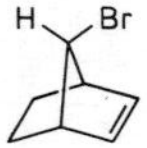 H, Br	$Hg(OAc)_2$, NaCl	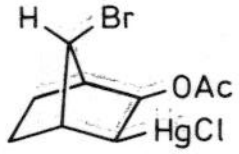 H, Br, OAc, HgCl (86.3)	—	—	60
	$Hg(O_2CCF_3)_2$	—	$NaBH_4$ (70°C, 0°C)	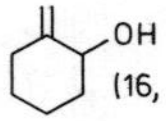 OH (16, 71) 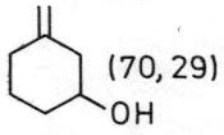 + OH (70, 29) 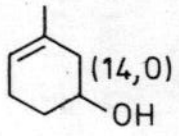 + OH (14, 0)	61
	$Hg(OAc)_2$	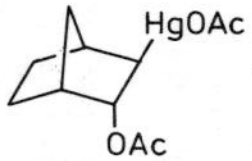 HgOAc, OAc ? (79)	—	—	62
	$Hg(OAc)_2$, NaCl	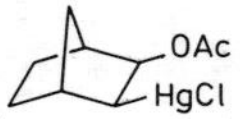 OAc, HgCl (83)	—	—	63
	$Hg(OAc)_2$, Cl^-	 OAc, HgCl	$NaBH_4$	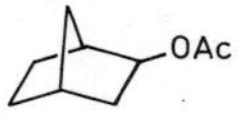 OAc	25
	$Hg(OAc)_2$, Br^-	 OAc, HgBr	—	—	19
	$HgCl_2$ / NaOAc / CH_3OH	 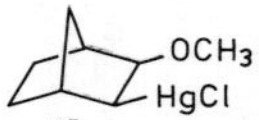 OAc, HgCl 84 : OCH3, HgCl 16	—	—	51

Table 5.3. (continued)

Alkene	Mercuric salt	Organomercurial(s) (% Yield)	Subsequent reactants	Product(s) (% Yield)	Ref.
 NC— (norbornene)	Hg(OAc)$_2$, NaCl	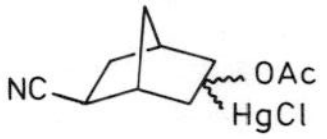 NC— ...OAc, HgCl (86)	—	—	63
 CH$_3$— 4:1 exo/endo	Hg(OAc)$_2$	—	NaBH$_4$/NaOH	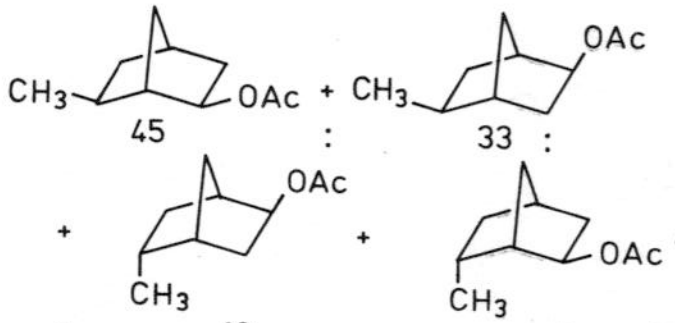 CH$_3$—...OAc 45 : CH$_3$—OAc 33 : CH$_3$—OAc 16 : CH$_3$—OAc 6 (65 total)	64
 CH$_3$— 9:1 endo/exo	Hg(OAc)$_2$	—	NaBH$_4$/NaOH	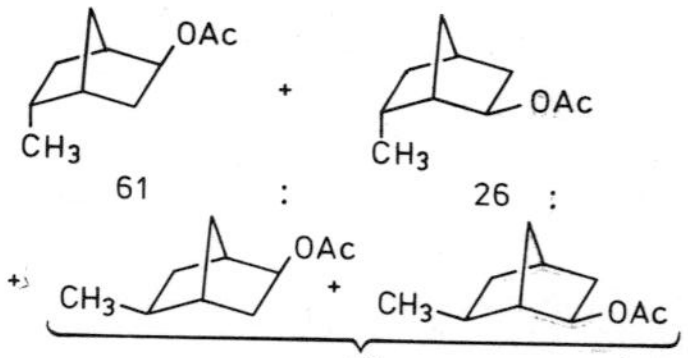 CH$_3$—OAc 61 : CH$_3$—OAc 26 : CH$_3$—OAc + CH$_3$—OAc : 13 (70 total)	64
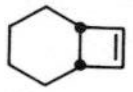 (bicyclo[4.2.0] alkene)	Hg(OAc)$_2$, Cl$^-$	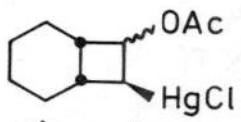 OAc, HgCl cis or trans	NaBH$_4$	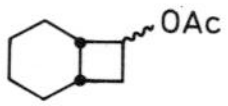 OAc	25
 (bicyclo[2.2.2]octene)	Hg(OAc)$_2$	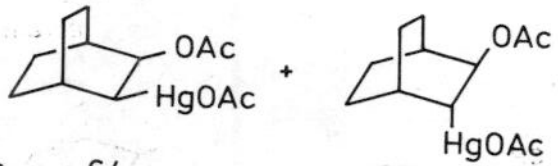 OAc, HgOAc + OAc, HgOAc; in HOAc 64 : 36; in CH$_2$Cl$_2$ or dioxane 100 : 0	—	—	65

Table 5.3. (continued)

Alkene	Mercuric salt	Organomercurial(s) (% Yield)	Subsequent reactants	Product(s) (% Yield)	Ref.
	Hg(OAc)$_2$, Cl$^-$	cis or trans	NaBH$_4$		25
	Hg(OAc)$_2$ (NaCl)	X = OAc, Cl (85)	NaOH, HCl (X = OAc)	(60 from alkene)	66
	Hg(OAc)$_2$, NaCl	(80)	—	—	63
	Hg(OAc)$_2$		NaBH$_4$/NaOH		67
	Hg(OAc)$_2$/H$_2$O	—	—	63 : 37	68
	Hg(OAc)$_2$, NaCl	(90)	LiCl / Δ	(70)	69
7:3 exo/endo	Hg(OAc)$_2$	—	NaBH$_4$/NaOH	+ 4 other compounds	64
9:1 endo/exo	Hg(OAc)$_2$	—	NaBH$_4$/NaOH	70 : 30 (53 total)	64

Table 5.3. (continued)

Alkene	Mercuric salt	Organomercurial(s) (% Yield)	Subsequent reactants	Product(s) (% Yield)	Ref.
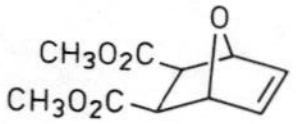 (CH₃O₂C, CH₃O₂C oxabicyclo-alkene)	Hg(OAc)₂ (NaCl)	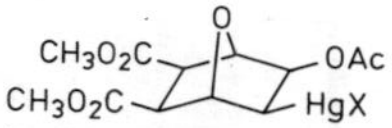 X = OAc (60), Cl (90)	X₂ (on RHgCl)	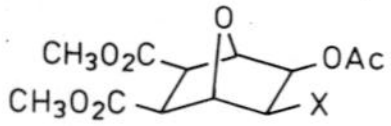 X = Cl, Br	70–72
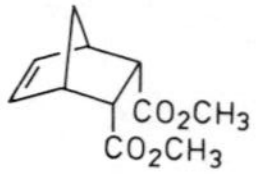 (CO₂CH₃, CO₂CH₃ bicycloalkene)	Hg(OAc)₂	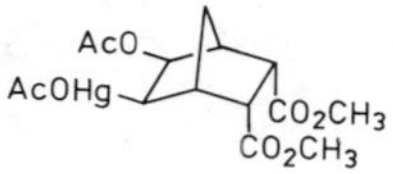 AcO, AcOHg, CO₂CH₃, CO₂CH₃	—	—	73
	Hg(OAc)₂ / HOAc, NaCl	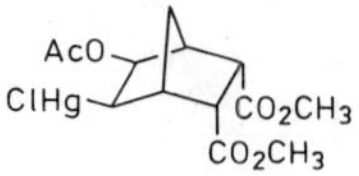 AcO, ClHg, CO₂CH₃, CO₂CH₃	—	—	74
	Hg(OAc)₂, NaCl	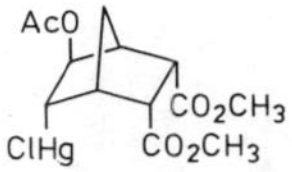 AcO, ClHg, CO₂CH₃, CO₂CH₃ ? (95)	—	—	62
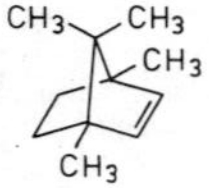 (CH₃, CH₃, CH₃ bicycloalkene)	Hg(OAc)₂, NaCl	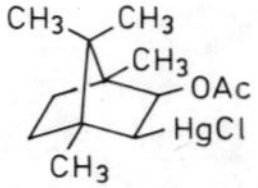 CH₃, CH₃, CH₃, OAc, HgCl, CH₃ (69)	NaBH₄ / NaOH	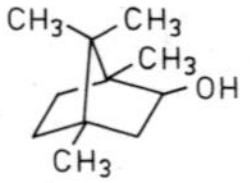 CH₃, CH₃, CH₃, OH, CH₃ (23)	60
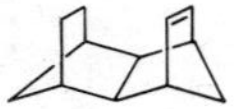	Hg(OAc)₂, NaCl	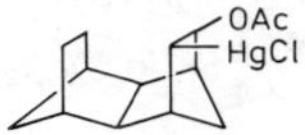 OAc, HgCl	—	—	75
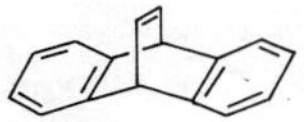	Hg(OAc)₂, NaCl	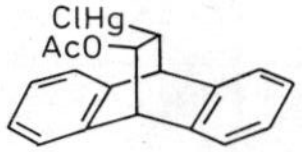 ClHg, AcO (91, 100)	NaBH₄ / NaOH	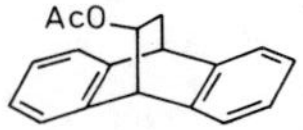 AcO (30)	76, 77

Table 5.3. (continued)

Alkene	Mercuric salt	Organomercurial(s) (% Yield)	Subsequent reactants	Product(s) (% Yield)	Ref.
	Hg(OAc)$_2$ (NaCl)	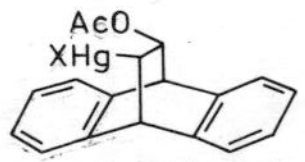 X = OAc (47), Cl (78)	NaBD$_4$ / NaOH [Na(Hg)/D$_2$O/ NaOD]	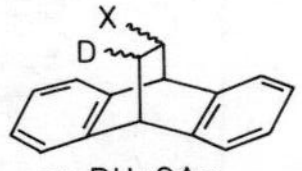 on RHgOAc X = OH (52) 25:75 cis/trans [X = OH (49) 100:0 cis/trans] on RHgCl X = OAc (88) 34:66 cis/trans [X = OH (65) 100:0 cis/trans]	78
	Hg(OAc)$_2$, NaCl	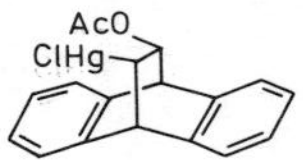 (82)	I$_2$	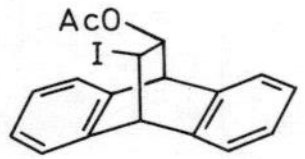	79
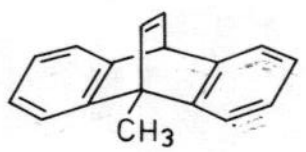	Hg(OAc)$_2$, Cl$^-$	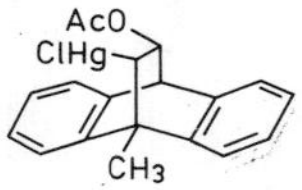 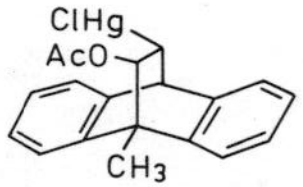 73 : 27 (~91 total)	NaBH$_4$/NaOH	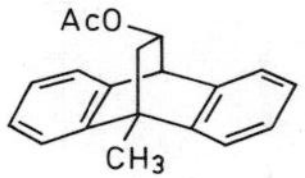 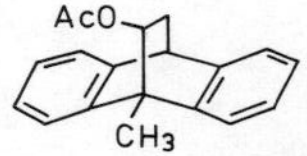 71 : 29 (86 total)	77
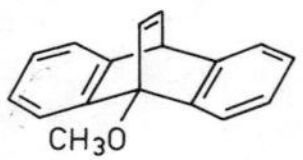	Hg(OAc)$_2$, NaCl	—	NaBH$_4$ /NaOH	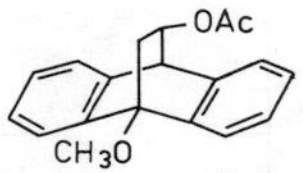 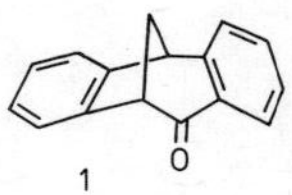 2 : 1	77

Table 5.3. (continued)

Alkene	Mercuric salt	Organomercurial(s) (% Yield)	Subsequent reactants	Product(s) (% Yield)	Ref.
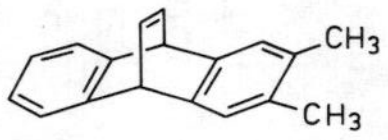	Hg(OAc)$_2$	—	NaBH$_4$	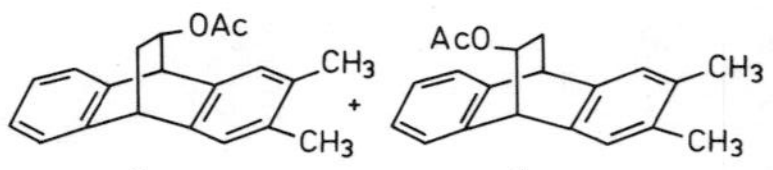 ~1 : 1	77
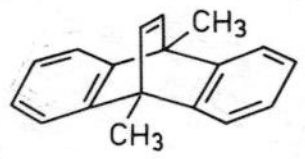	Hg(OAc)$_2$, NaCl	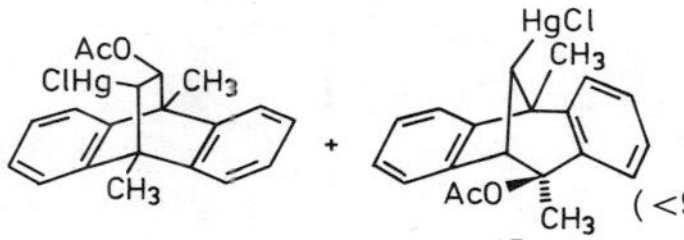 55 : 45 (<95 total)	NaBH$_4$/NaOH	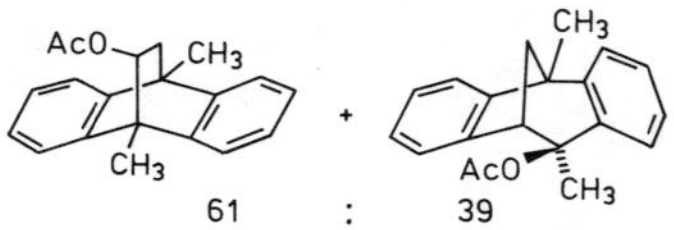 61 : 39	77

V. Acyloxymercuration

Clean regio- and stereo-control is observed in the acetoxymercuration of *endo*-5-carbomethoxy- and *endo*-5-cyanonorbornene (Eq. 18) [63]. The

$$\text{(16)}$$

$$\text{(17)}$$

$$\text{(18)} \qquad X = CO_2CH_3, \; CN$$

acetoxymercurial from *endo*-, *endo*-5,6-dicarbomethoxynorbornene was originally assigned a trans configuration [62], but appears now to be another cis-exo adduct [85]. The corresponding 7-oxa analog also appears to be a cis-exo adduct [70]. Benzonorbornadiene yields substantial quantities of the corresponding cis-exo acetoxymercuric chloride under the usual hydroxymercuration conditions [86], while bornylene affords significant amounts of two exo-acetates during hydroxymercuration-demercuration [87]. Neither olefin has been subjected to the usual acyloxymercuration conditions.

Acetoxymercuration of bicyclo[2.2.2]octene in dioxane or methylene chloride affords a pure cis-exo acetoxymercurial [65], while the same reaction in acetic acid is reported to give either the pure cis-exo adduct [66] or a 64:36 mixture of syn and anti adducts [65] (Eq. 19). The anti mercurial can be

$$\text{(19)}$$

$$
\begin{array}{lccc}
CH_2Cl_2 \;\; or \;\; C_4H_8O_2 & 100 & : & 0 \\
HOAc & 64 & : & 36
\end{array}
$$

made to predominate (54:46) in aqueous sodium acetate, but the overall yield has not been specified [65]. NMR spectral data for both isomers have been reported [25]. Both *exo*- and *endo*-5-methylbicyclo[2.2.2]octene appear to give predominantly the two anti acetates after demercuration (Eq. 20) [64]. The following tricyclic analog has not been acyloxymercurated in the usual fashion, but affords a significant amount of the exo-acetate upon hydroxymercuration-demercuration (Eq. 21) [88]. The dibenzoocta[2.2.2]triene system also undergoes exclusive cis-acetoxymercuration (Eq. 22) [25, 76–79]. Note

(20)

9 : 1   endo/exo   70   :   30
3 : 7   endo/exo   —   :   —

(plus 4 other compounds)

(21)

(22)

$R^1 = R^2 = H$	50 : 50
$R^1 = R^2 = CH_3$	50 : 50
$R^1 = CH_3$, $R^2 = H$	73 : 27
$R^1 = OCH_3$, $R^2 = H$	100 : —

that bridgehead substitution by a methyl or methoxy group favors the adduct with the more remote acetoxy group [77]. The dimethyl and methoxy derivatives also afford rearranged products. Placing two methyl groups on the two remote carbon atoms of one of the aromatic rings fails to induce any regioselectivity.

The stereochemistry of acetoxymercuration of several other polycyclic systems has also been reported or implied (Eqs. 23–25). The attempted

(23) [59]

(24) [75]

(25) [25]

trifluoroacetoxymercuration of highly strained 5-methylenebicyclo[2.1.1]-hexane results in rearranged products (Eq. 26) [61].

(26)

The relative rates of acetoxymercuration of several cyclic and bicyclic olefins has been reported: norbornene ("very large"), *exo*-2-vinylnorbornane (7), bicyclo[2.2.2]octene (2.5), cyclohexene (1) [28]. These rates are significantly different from the relative rates of hydroxymercuration [Hg(OAc)$_2$ in 1:1 H$_2$O/THF] reported for these olefins (3.7, —, 0.01 and 1 respectively) by the same authors.

A number of functional groups are readily accommodated by the acyloxymercuration reaction. A close perusal of Tables 5.1–5.3 indicates that no difficulties are encountered with most halides, ethers, esters, nitriles, cyclopropanes, epoxides, alcohols, nitro groups or vinyl silanes. Complex natural products such as carbohydrates [31, 32, 54] and a picrotin derivative [33] have presented no particular problems.

Only a few functional groups have been observed to interfere with the acyloxymercuration reaction. Certain strained olefins undergo rearrangement during the reaction as discussed earlier. Allylsilanes undergo transmetallation by mercury [17, 18]. Oxygen substituents have also presented difficulties in several cases (Eqs. 27–29).

$$\text{(27) [42]}$$

$$\text{(28) [48]}$$

$$\text{(29) [29]}$$

Vinyl halides do not undergo acyloxymercuration in the usual sense, but afford carbonyl compounds directly (Eq. 30) (Table 5.4). In Chapter II we

$$\text{(30)}$$

also noted that the hydroxymercuration of vinyl halides provides a useful carbonyl synthesis [94, 95]. Several different versions of the acyloxy approach have been reported involving Hg(OAc)$_2$/HCO$_2$H [89, 90], Hg(OAc)$_2$/HOAc/BF$_3$ · Et$_2$O [91], Hg(OAc)$_2$/CF$_3$CO$_2$H [91], and Hg(O$_2$CCF$_3$)$_2$/CH$_3$NO$_2$ [90, 92]. Both aldehydes and ketones have been prepared in this fashion and 1,1-difluoroethylene affords a mercurated carboxylic acid directly (Eq. 31) [1].

$$H_2C=CF_2 \xrightarrow[\text{HOAc}]{\text{Hg(OAc)}_2} AcOHgCH_2CO_2H \qquad \text{(31)}$$

Table 5.4. Conversion of Vinyl Halides to Carbonyl Compounds

Vinyl halide	Mercuric salt	Carbonyl compound (% Yield)	Ref.
$H_2C{=}CF_2$	$Hg(OAc)_2$ /HOAc	$AcOHgCH_2CO_2H$ (100)	1
$CH_3CCl{=}CH(CH_2)_3CH_3$	$Hg(OAc)_2$ /HCO_2H	$CH_3CO(CH_2)_4CH_3$ (80)	89
$ClCH{=}CH(CH_2)_5CH_3$	$Hg(OAc)_2$ /HCO_2H	$HCO(CH_2)_6CH_3$ (85)	89
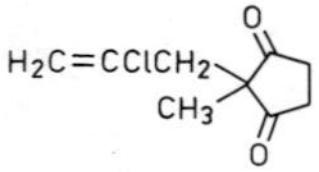 $H_2C{=}CClCH_2$— [2-methyl-2-(...)cyclopentane-1,3-dione], CH_3	$Hg(OAc)_2$ /HCO_2H	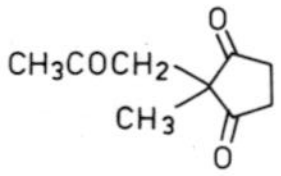 CH_3COCH_2— [2-methyl-2-(...)cyclopentane-1,3-dione], CH_3 (84)	90
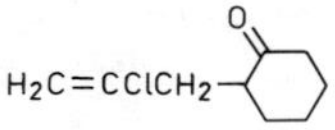 $H_2C{=}CClCH_2$—[cyclohexanone]	$Hg(OAc)_2$ /HOAc / $BF_3 \cdot Et_2O$	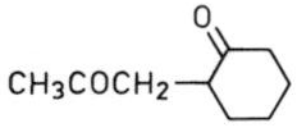 CH_3COCH_2—[cyclohexanone] (52)	91
	$Hg(O_2CCF_3)_2$ / CH_3NO_2	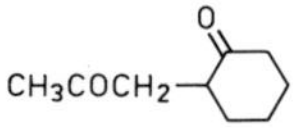 CH_3COCH_2—[cyclohexanone] (85)	90
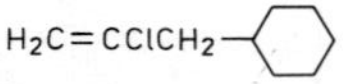 $H_2C{=}CClCH_2$—[cyclohexane]	$Hg(OAc)_2$ /CF_3CO_2H	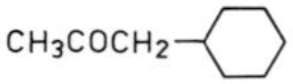 CH_3COCH_2—[cyclohexane] (41)	91
$ClCH{=}CHCH_2CH_2C_6H_5$	$Hg(OAc)_2$ /HCO_2H	$HCOCH_2CH_2CH_2C_6H_5$ (78)	89
$H_2C{=}CClCH_2CH_2C_6H_5$	$Hg(OAc)_2$ /HCO_2H	$CH_3COCH_2CH_2C_6H_5$ (88)	89
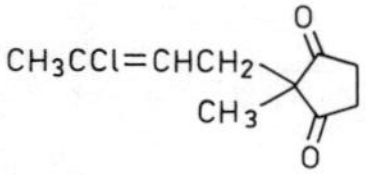 $CH_3CCl{=}CHCH_2$— [2-methyl-2-(...)cyclopentane-1,3-dione], CH_3	$Hg(O_2CCF_3)_2$ / CH_3NO_2	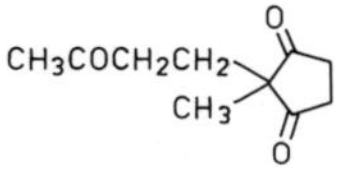 $CH_3COCH_2CH_2$— [2-methyl-2-(...)cyclopentane-1,3-dione], CH_3 (86)	90
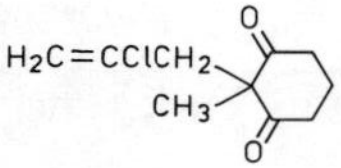 $H_2C{=}CClCH_2$— [2-methyl-2-(...)cyclohexane-1,3-dione], CH_3	$Hg(O_2CCF_3)_2$ / CH_3NO_2	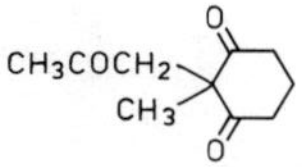 CH_3COCH_2— [2-methyl-2-(...)cyclohexane-1,3-dione], CH_3 (80)	90

Table 5.4. (continued)

Vinyl halide	Mercuric salt	Carbonyl compound (% Yield)	Ref.
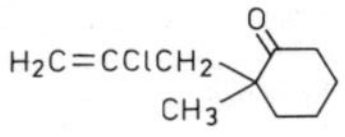 $H_2C=CClCH_2$–(cyclohexanone), CH_3	$Hg(OAc)_2 / HCO_2H$	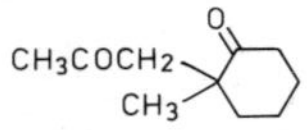 CH_3COCH_2–(cyclohexanone), CH_3 (80)	90
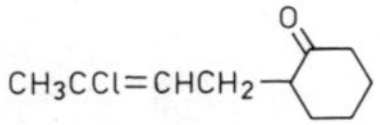 $CH_3CCl=CHCH_2$–(cyclohexanone)	$Hg(OAc)_2 / HOAc / BF_3 \cdot Et_2O$	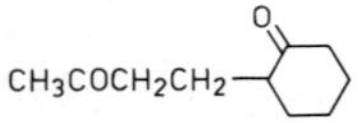 $CH_3COCH_2CH_2$–(cyclohexanone) (60)	91
	$Hg(O_2CCF_3)_2 / CH_3NO_2$	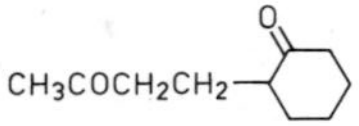 $CH_3COCH_2CH_2$–(cyclohexanone) (94)	90
$H_2C=CClCH_2CH_2COC_6H_5$	$Hg(OAc)_2 / CF_3CO_2H$	$CH_3COCH_2CH_2COC_6H_5$ (53)	91
$CH_3CCl=CH(CH_2)_2C_6H_5$	$Hg(OAc)_2 / HCO_2H$	$CH_3CO(CH_2)_3C_6H_5$ (73)	89
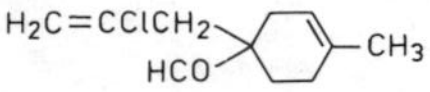 $H_2C=CClCH_2$–(cyclohexene), CH_3, HCO	$Hg(OAc)_2 / HOAc / BF_3 \cdot Et_2O$	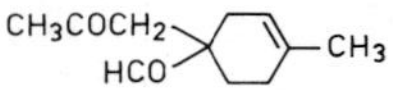 CH_3COCH_2–(cyclohexene), CH_3, HCO (71)	91
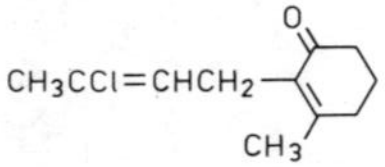 $CH_3CCl=CHCH_2$–(cyclohexenone), CH_3	$Hg(O_2CCF_3)_2 / CH_3NO_2$	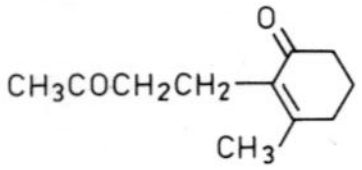 $CH_3COCH_2CH_2$–(cyclohexenone), CH_3 (95)	90
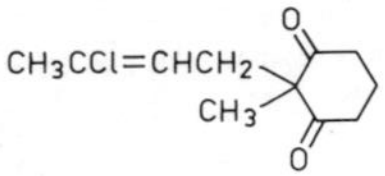 $CH_3CCl=CHCH_2$–(cyclohexanedione), CH_3	$Hg(O_2CCF_3)_2 / CH_3OH$	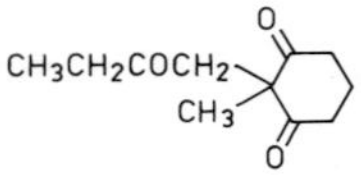 $CH_3CH_2COCH_2$–(cyclohexanedione), CH_3 (87)	90
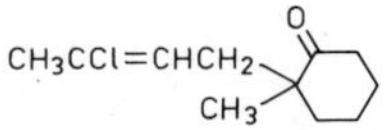 $CH_3CCl=CHCH_2$–(cyclohexanone), CH_3	$Hg(O_2CCF_3)_2 / CH_3OH$	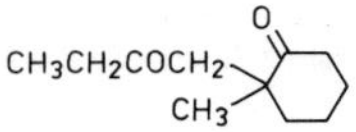 $CH_3CH_2COCH_2$–(cyclohexanone), CH_3 (83) 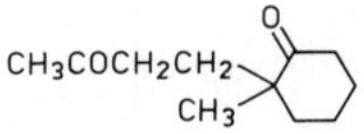 $+$ $CH_3COCH_2CH_2$–(cyclohexanone), CH_3 (6)	90
	$Hg(O_2CCF_3)_2 / CH_3NO_2$, HCl	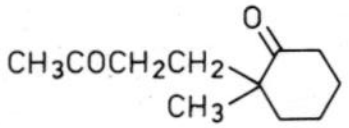 $CH_3COCH_2CH_2$–(cyclohexanone), CH_3 (97)	90

Table 5.4. (continued)

Vinyl halide	Mercuric salt	Carbonyl compound (% Yield)	Ref.
$CH_3CCl{=}CHCH_2C(CH_3)(CO_2C_2H_5)COCH_3$	$Hg(O_2CCF_3)_2$, HCl	$CH_3CO(CH_2)_2C(CH_3)(CO_2C_2H_5)COCH_3$ (60) 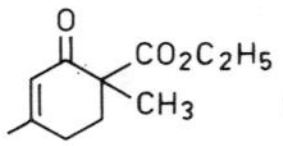 + [cyclohexenone with $CO_2C_2H_5$ and CH_3] (10) + $CH_3CH_2COCH_2C(CH_3)(CO_2C_2H_5)COCH_3$ (2.5)	92
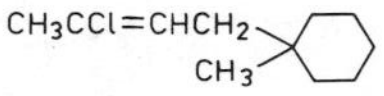 $CH_3CCl{=}CHCH_2$–[cyclohexyl with CH_3]	$Hg(O_2CCF_3)_2$ / CH_3NO_2	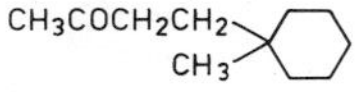 $CH_3COCH_2CH_2$–[cyclohexyl with CH_3] (60)	90
$H_2C{=}CClCH_2CH(CO_2C_2H_5)(CH_2)_4CH_3$	$Hg(OAc)_2$ / CF_3CO_2H	$CH_3COCH_2CH(CO_2C_2H_5)(CH_2)_4CH_3$ (90)	91
$CH_3CCl{=}CHCH_2CH(SO_2Ph)CH_2CH_3$	$Hg(OAc)_2$ / HCO_2H	$CH_3COCH_2CH_2CH(SO_2Ph)CH_2CH_3$ (62)	89
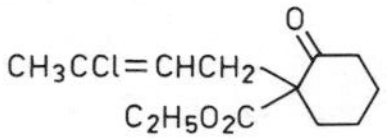 $CH_3CCl{=}CHCH_2$–[cyclohexanone with $C_2H_5O_2C$]	$Hg(OAc)_2$ / CF_3CO_2H	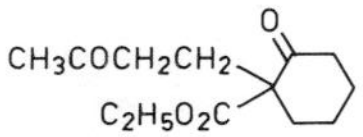 $CH_3COCH_2CH_2$–[cyclohexanone with $C_2H_5O_2C$] (55)	91
$H_2C{=}CClCH_2CH(SO_2Ph)CH{=}C(CH_3)_2$	$Hg(OAc)_2$ / HCO_2H	$CH_3COCH_2CH(SO_2C_6H_5)CH{=}C(CH_3)_2$ (19)	89
$ClCH{=}CHCH_2CH(SO_2Ph)(CH_2)_3CH_3$	$Hg(OAc)_2$ / HCO_2H	$HCOCH_2CH_2CH(SO_2Ph)(CH_2)_3CH_3$ (53)	89
$ClCH{=}CHCH_2CH(SO_2Ph)C_6H_5$	$Hg(OAc)_2$ / HCO_2H	$HCOCH_2CH_2CH(SO_2Ph)C_6H_5$ (86)	89
$H_2C{=}CClCH_2CH(SO_2Ph)C_6H_5$	$Hg(OAc)_2$ / HCO_2H	$CH_3COCH_2CH(SO_2Ph)C_6H_5$ (86)	89
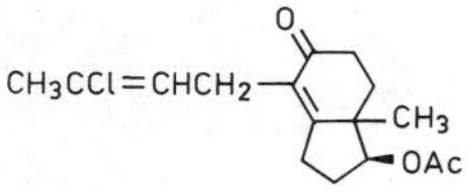 $CH_3CCl{=}CHCH_2$–[bicyclic enone with CH_3, OAc]	$Hg(O_2CCF_3)_2$ / CH_3NO_2 or $Hg(OAc)_2$ / HCO_2H	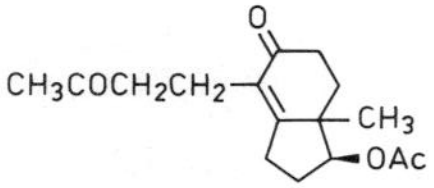 $CH_3COCH_2CH_2$–[bicyclic enone with CH_3, OAc] (91)	90

Table 5.4. (continued)

Vinyl halide	Mercuric salt	Carbonyl compound (% Yield)	Ref.
$CH_3CCl=CHCH_2CH(SO_2Ph)C_6H_5$	$Hg(OAc)_2 / HCO_2H$	$CH_3COCH_2CH_2CH(SO_2Ph)C_6H_5$ (87)	89
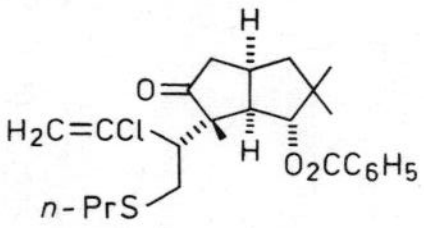 (structure: $H_2C=CCl$, n-PrS, $O_2CC_6H_5$)	$Hg(OAc)_2 / HCO_2H / NH_4O_2CH$	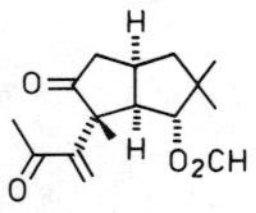 (structure) (84)	93

The reaction tolerates aldehydes, ketones, esters, sulfones and arenes, making it quite versatile synthetically. One unique aspect of this reaction is that in some substrates the regiochemistry of carbonyl formation can be reversed by simply changing the solvent from nitromethane to methanol (Eq. 32) [90]. Not all examples reported proceed as cleanly however. The mechanisms for these transformations are not immediately obvious.

$$CH_3CCl{=}CHCH_2 \quad \xrightarrow{Hg(O_2CCF_3)_2} \quad CH_3COCH_2CH_2 \quad + \quad CH_3CH_2COCH_2 \tag{32}$$

	~100%	:	0%
CH_3NO_2	~100%	:	0%
CH_3OH	0%	:	87%

One other competing reaction which has proven quite valuable synthetically is the ester exchange reaction of vinyl esters. While acyloxymercuration products have been isolated from a number of vinyl esters (Eq. 33) [12–15], upon heating in the presence of acid catalysts these mercurials revert to vinyl esters (Eq. 34). By using vinyl acetate or isopropenyl acetate, a carboxylic acid and a mercury catalyst, this sequence becomes a valuable method for preparing a wide variety of vinyl esters. The work reported to date is summarized in Table 5.5.

$$H_2C{=}CHO_2CR^1 \quad + \quad Hg(O_2CR^2)_2 \quad \longrightarrow \quad R^2CO_2HgCH_2\overset{\displaystyle O_2CR^2}{\underset{|}{C}}HO_2CR^1 \tag{33}$$

$$R^2CO_2HgCH_2\overset{\displaystyle O_2CR^2}{\underset{|}{C}}HO_2CR^1 \quad \longrightarrow \quad H_2C{=}CHO_2CR^2 \tag{34}$$

As seen in Table 5.5, a number of different mercury catalysts have been employed in this reaction. Unfortunately, little work has been reported on the relative merits of the different catalysts. One study reports catalyst efficiency as $Hg(OAc)_2/H_2SO_4 > HgSO_4 > Hg(OAc)_2/H_3PO_4$, with mercuric acetate being ineffective [98]. Another publication reports that $Hg(OAc)_2$ plus H_2SO_4 or p-TsOH is effective, but that neither $HgSO_4$, $Hg(OAc)_2$ nor H_2SO_4 alone catalyze ester exchange [12]. None of the following salts were observed to promote the exchange either: $Cd(OAc)_2$, $Zn(OAc)_2$, $Cu(OAc)_2$, $CuCl_2$, $PdCl_2$, $ZnSO_4$, $CdSO_4$, $PbSO_4$, $CuSO_4$, $BF_3 \cdot Et_2O$ [101]. The mechanistic work reported on this reaction is consistent with the scheme presented above [13–15, 108].

The demercuration of acyloxymercurials has not received a lot of attention. In most cases, sodium borohydride or alkaline sodium borohydride have been employed and either the corresponding ester or alcohol have been isolated. It has been noted that alkaline borohydride gives substantial amounts of the alcohol under conditions in which the ester is stable, suggesting that the mercury moiety in some way promotes hydrolysis of the ester group [24]. By switching from water to methylene chloride as the

Table 5.5. Mercury-Catalyzed Transesterification of Vinyl Esters

$$H_2C=C(R)OAc + R'CO_2H \longrightarrow H_2C=C(R)O_2CR'$$

R	Carboxylic acid	Mercury catalyst	% Yield of vinyl ester	Ref.
H	formic	HgO/H_2SO_4 or H_3PO_4	—	96
	propionic	$HgSO_4$	—	97
	succinic	Hg(OAc)$_2$/H_2SO_4	15	98
	crotonic	Hg(OAc)$_2$	—	99
		Hg(OAc)$_2$/Cu(OAc)$_2$/p-TsOH	52.5	100
	butyric	Hg(OAc)$_2$/HO–C$_6$H$_4$–OH	38	101
		Hg(OAc)$_2$/H_2SO_4	68	98
	2-furoic	Hg(OAc)$_2$/BF$_3$·Et$_2$O	50–78	102
	5-bromo-2-furoic	Hg(OAc)$_2$/BF$_3$·Et$_2$O	—	102
	5-nitro-2-furoic	Hg(OAc)$_2$/BF$_3$·Et$_2$O	—	102
	pivalic	$HgSO_4$	78	103
	sorbic	Hg(OAc)$_2$/Cu(OAc)$_2$/p-TsOH	50	100
	adipic	Hg(OAc)$_2$/H_2SO_4	25	98
		$HgSO_4$	16	103
	methyl adipate	Hg(OAc)$_2$/HO–C$_6$H$_4$–OH	71.6	101
H	hexanoic	Hg(OAc)$_2$/H_2SO_4	40, 70	98, 104, 105
		$HgSO_4$	40	106
	3,5,5-trimethylhexanoic	$HgSO_4$	60	103
	benzoic	$HgSO_4$	—	103
		Hg(OAc)$_2$	—	99, 107
		Hg(OAc)$_2$/BF$_3$	—	108
		Hg(OAc)$_2$/HO–C$_6$H$_4$–OH	38	101
	p-chlorobenzoic	Hg(OAc)$_2$/p-TsOH	49	12
	p-nitrobenzoic	Hg(OAc)$_2$/oleum/copper stearate	86	109
	p-dimethylaminobenzoic	Hg(OAc)$_2$/oleum/copper stearate	66.5	109
	p-hydroxybenzoic	Hg(OAc)$_2$/copper stearate	24[a]	109
	3-(2-furyl)-2-propenoic	Hg(OAc)$_2$/BF$_3$·Et$_2$O	50–78	102
	octanoic	Hg(OAc)$_2$/H_2SO_4	55, 72	98, 104, 105
		$HgSO_4$	55, 69	103, 106
	octylphthalate	$HgSO_4$	59	103

[a]Product is p-H_2C=CHO$_2$CC$_6$H$_4$OCH(OAc)CH$_3$.

Table 5.5. (continued)

$H_2C=C(R)OAc \; + \; R'CO_2H \longrightarrow H_2C=C(R)O_2CR'$

R	Carboxylic acid	Mercury catalyst	% Yield of vinyl ester	Ref.
H	nonanoic	$HgSO_4$	55	106
		$Hg(OAc)_2/H_2SO_4$	55	104, 105
	decanoic	$Hg(OAc)_2/H_2SO_4$	45, 75	98, 104, 105
		$HgSO_4$	45	106
	undecanoic	$Hg(OAc)_2/H_2SO_4$	60, 70	98, 104, 105
	dodecanoic	$Hg(OAc)_2/H_2SO_4$	72	98
		$HgSO_4$	55, 68	103, 106
	tetradecanoic	$HgSO_4$	60	104–106
	hexadecanoic	$HgSO_4$	35	106
		$Hg(OAc)_2/H_2SO_4$	35	104, 105
		$Hg(OAc)_2/Cu(OAc)_2/p\text{-TsOH}$	38.5	100
	octadecanoic	$Hg(OAc)_2/Cu(OAc)_2/p\text{-TsOH}$	37	100
		$Hg(OAc)_2/HO\text{—}\langle\text{C}_6\text{H}_4\rangle\text{—}OH$	44.9	101
		$Hg(OAc)_2/H_2SO_4$	30, 62	98, 104, 105
		$HgSO_4$	30, 55	103, 106

$H_2C=C(R)OAc \; + \; R'CO_2H \longrightarrow H_2C=C(R)O_2CR'$

R	Carboxylic acid	Mercury catalyst	% Yield of vinyl ester	Ref.
H	oleic	$HgSO_4$	32	97, 103
		$Hg(OAc)_2/Cu(OAc)_2/p\text{-TsOH}$	35	100
		$Hg(OAc)_2/HO\text{—}\langle\text{C}_6\text{H}_4\rangle\text{—}OH$	59	101
		$Hg(OAc)_2/H_2SO_4$	40, 60, ~78	98, 104, 105, 110
	linoleic	$Hg(OAc)_2/H_2SO_4$	40	98
CH_3	propionic	$Hg(OAc)_2/H_2SO_4$	—	111
	crotonic	$Hg(OAc)_2/H_2SO_4$	—	111
	butyric	$Hg(OAc)_2/H_2SO_4$	—	111
	furoic	$Hg(OAc)_2/H_2SO_4$	—	111
	hexanoic	$Hg(OAc)_2/H_2SO_4$	—	111
	benzoic	$Hg(OAc)_2/H_2SO_4$	—	111
	naphthoic (α or β ?)	$Hg(OAc)_2/H_2SO_4$	—	111
	dodecanoic	$Hg(OAc)_2/H_2SO_4$	—	111
	octadecanoic	$Hg(OAc)_2/H_2SO_4$	—	111
	oleic	$Hg(OAc)_2/H_2SO_4$	—	111

reduction solvent, the amount of alcohol can be decreased slightly [22]. Another complication observed during sodium borohydride reductions is reversion to the starting alkene. This has been observed with mercurials derived from styrene [22], D-glucal triacetate [54], and norbornene and derivatives [85]. With the latter compounds, the olefin yield increases with the solvent polarity (Eq. 35). It has been suggested that the alkene arises as follows (Eq. 36). Sodium borodeuteride has also been employed to reduce these mercurials. With the acetoxymercurial derived from bicyclo[2.1.1]-hexene, it appears that retention is observed during reduction (Eq. 37) [59].

$$\text{(35)}$$

THF	< 99	> 1
CH$_3$OH	80	20
CH$_3$NO$_2$	> 0.5	< 99.5

$$H-Hg-C-C-O_2R \longrightarrow H^+ + Hg + C=C + {}^-O_2CR \tag{36}$$

$$\text{(37)}$$

However, acetoxymercurials derived from dibenzobicyclo[2.2.2]octatriene give isomeric mixtures [78]. In this system, sodium borodeuteride in ethanol showed considerable hydrogen incorporation. As observed previously, sodium amalgam in alkaline D$_2$O gives complete retention. In one instance, tri-n-butylstannane in ethanol has been employed as a reducing agent (Eq. 38) [33]. Unfortunately, the major product was the corresponding olefin and the desired diester was obtained in only 35% yield.

$$\text{(38)}$$

Relatively few cases of acyloxymercuration and subsequent halogenation have been reported. Those that have been reported are summarized in Tables 5.1–5.3. However, acyloxymercuration and subsequent chlorination [72], bromination [1, 10, 71] and iodination [43, 79] have been accomplished on such diverse olefins as simple cyclic olefins [1], bicyclic olefins [71, 72, 79], vinyl ethers [10] and fluoro-olefins [1].

The thermolysis of acyloxymercurials is of considerable interest. While most of these compounds are easily isolated at room temperature, it has been reported that the addition of aqueous sodium chloride to the mercuric acetate adducts of o- and p-nitrostyrene results in formation of the starting olefins (Eq. 39) [20]. Facile olefin formation has also been noted in the following cases (Eqs. 40, 41). By heating mercuric acetate adducts of allylbenzene [26], cyclohexene [27], methyl 10-undecenoate [112] and methyl oleate [34, 35], allylic acetates are formed, usually in high yield. While it is obvious that these adducts form under the usual conditions for allylic acetoxylation, they are probably not intermediates in the Treibs reaction, but simply revert to olefins which are eventually converted to allylic acetates. The thermolysis

$$(39)$$

$$(40)\ [69]$$

$$(41)\ [54]$$

or photolysis of the 1-decene/mercuric acetate adduct in benzene affords a mixture of 1-phenyldecenes in low yield [113]. In toluene, thermolysis produces o- and p-tolyldecenes, but photolysis yields m- and p-tolyldecanes.

Under standard solvomercuration conditions, acetoxymercurials have also been converted into the corresponding hydroxy-, methoxy- and t-butylperoxymercurials (Eqs. 42, 43). Under low temperature super-acid conditions, mercurinium ions have been observed by NMR spectral analysis from these acyloxymercury adducts [114].

$$(42)\ [50, 52]$$

$$(43)\ [19]$$

One last reaction of acyloxymercurials which has proven interesting is their ability to form carbon—carbon bonds when reacted with arenes or β-diketones under acidic conditions (Eqs. 44, 45).

$$ArH\ +\ AcOHgCH_2CH_2OAc\ \xrightarrow{H^+}\ ArCH_2CH_2X \qquad (44)\ [6, 115, 116]$$
$$X = H,\ HgOAc,\ OAc$$

$$(45)\ [7]$$

B. Alkenoic Acids

A number of unsaturated carboxylic acids, esters and anhydrides have been subjected to intramolecular acyloxymercuration (Table 5.6). Both mercuric chloride and mercuric acetate react readily in these cyclization reactions and even protic solvents such as water or methanol can be employed without interference by hydroxy- or methoxymercuration. A number of 4-pentenoic acids [117–121, 136] and the corresponding aryl esters [117, 130] have been cyclized in this fashion (Eq. 46). The following relative rates for the phenyl

$$H_2C=CHCH_2CR^1R^2CO_2R^3 \xrightarrow{HgX_2} XHgCH_2 \qquad (46)$$

esters ($R^3=C_6H_5$) have been observed: $R^1=R^2=C_6H_5 > R^1=H, R^2=C_6H_5 > R^1=R^2=H$ [130]. In the reaction of mercuric chloride and the esters, second order kinetics are observed [117]. A number of other thermodynamic parameters have been determined for this reaction [117, 130]. The following example reported in a Japanese patent appears to be the only example of lactone formation of ring size other than five (Eq. 47) [134].

$$H_2C=CH(CH_2)_8CO_2H \longrightarrow AcOHgCH_2\overset{O-CO}{\underset{}{CH(CH_2)_8}} \qquad (47)$$

Allylic acetates and carbamates can undergo similar intramolecular cyclizations. For example, H. C. Brown reports that the hydroxymercuration of allyl acetate apparently proceeds by neighboring-group participation of the acetate to afford rearranged products (Eq. 48) [137].

$$H_2C=CHCH_2O\overset{O}{\overset{\|}{C}}CH_3 \xrightarrow{Hg(OAc)_2} AcOHgCH_2 \longrightarrow AcOHgCH_2\overset{OH}{\underset{}{CH}}CH_2O\overset{O}{\overset{\|}{C}}CH_3 + \qquad (48)$$

$$AcOHgCH_2\overset{O_2CCH_3}{\underset{}{CH}}CH_2OH$$

Overman has taken advantage of an analogous cyclization to provide a method for the contrathermodynamic allylic isomerization of allylic carbamates (Eq. 49) [126]. This approach appears limited to the conversion of carbamate esters of 2-alken-1-ols to the corresponding 1-alken-3-ols.

$$RCH=CHCH_2O_2CNMe_2 \xrightarrow{Hg(O_2CCF_3)_2} \left[\begin{array}{c} HgO_2CCF_3 \\ R \end{array} \right] \xrightarrow{PPh_3} RCH\overset{O_2CNMe_2}{\underset{}{CH}}=CH_2 \qquad (49)$$

Table 5.6. Intramolecular Acyloxymercuration

Alkene	Mercuric salt	Organomercurial(s) (% Yield)	Subsequent reactants	Product(s) (% Yield)	Ref.
$H_2C{=}CH(CH_2)_2CO_2H$	$HgCl_2$	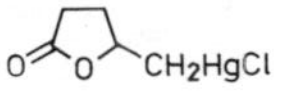 (57) CH_2HgCl	—	—	117
	$Hg(OAc)_2$, NaCl	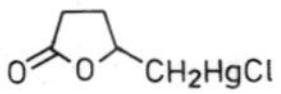 (15) CH_2HgCl	—	—	118
$H_2C{=}CHCH_2CH(C_2H_5)CO_2H$	$Hg(OAc)_2$, X^-	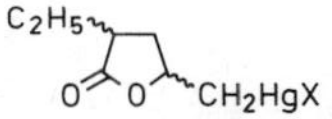 C_2H_5 CH_2HgX X = Cl (90), Br (96.02), I (95)	Br_2 (on RHgCl)	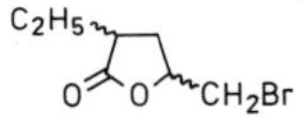 C_2H_5 CH_2Br (72.6)	119–121
$H_2C{=}CHCH_2CH(CO_2H)CH(CH_3)_2$	$Hg(OAc)_2$, KI	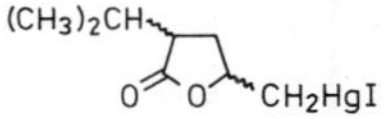 $(CH_3)_2CH$ CH_2HgI (95)	—	—	121
$H_2C{=}CHCH_2CH(CO_2H)(CH_2)_2CH_3$	$Hg(OAc)_2$, X^-	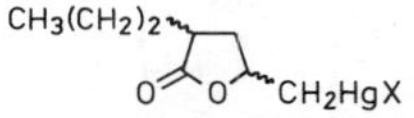 $CH_3(CH_2)_2$ CH_2HgX X = Cl (86.5), Br (86.5), I (93)	Br_2 (on RHgCl)	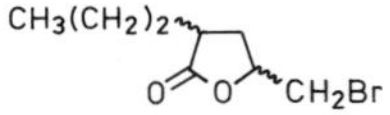 $CH_3(CH_2)_2$ CH_2Br (72.4)	119–121
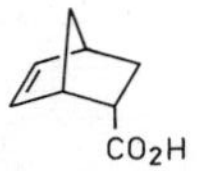 (norbornene-CO_2H)	$Hg(OAc)_2$, NaCl	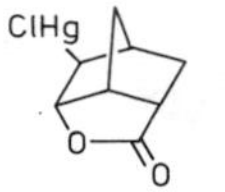 ClHg (77, 89)	$NaBH_4$	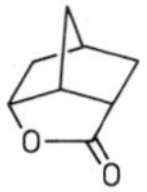 (80)	63, 122–124
	$Hg(OAc)_2$, NaCl	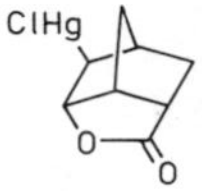 ClHg (94)	$NaHB(OCH_3)_3$	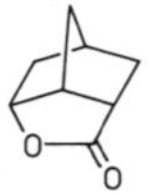	125
$H_2C{=}CHC(CH_3)_2OCON(CH_3)_2$	$Hg(O_2CCF_3)_2$	—	$P(C_6H_5)_3$	$(CH_3)_2C{=}CHCH_2OCON(CH_3)_2$ + $H_2C{=}CHC(CH_3)_2OCON(CH_3)_2$ 98.5 : 1.5 (95 total)	126

Table 5.6. (continued)

Alkene	Mercuric salt	Organomercurial(s) (% Yield)	Subsequent reactants	Product(s) (% Yield)	Ref.
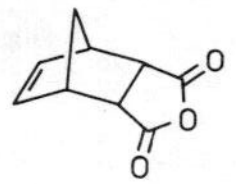	$Hg(OAc)_2 / CH_3OH$, NaCl	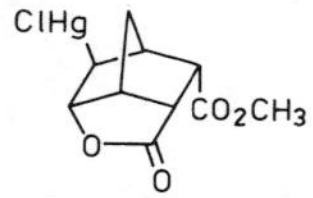 (structure) CO_2CH_3	—	—	74
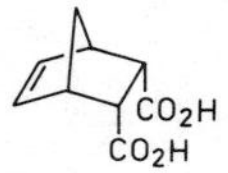	$Hg(OAc)_2$, Cl^-	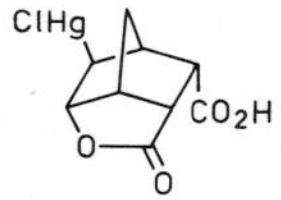 (structure) CO_2H (94)	—	—	123
	$Hg(OAc)_2$, Cl^-	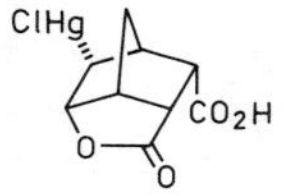 (structure) CO_2H ? (95)	—	—	62
	$Hg(OAc)_2$	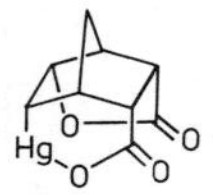 (structure) ? (98)	NaOH, HCl	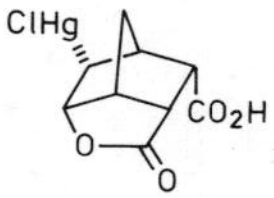 (structure) CO_2H ?	127
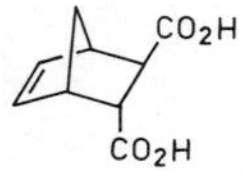	$Hg(OAc)_2$, NaCl	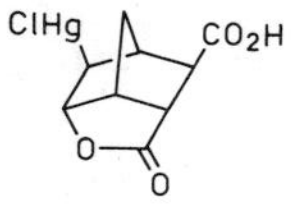 (structure) CO_2H (97)	—	—	123
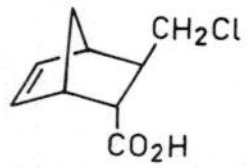	$Hg(OAc)_2$	—	$NaBH_4 / NaOH$	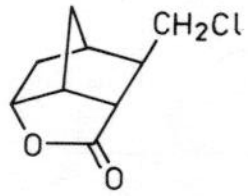 (structure) CH_2Cl (89.4)	128
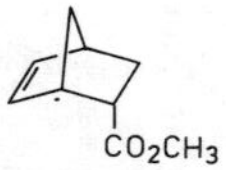	$Hg(OAc)_2 / H_2O$, Cl^-	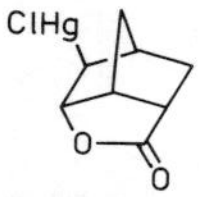 (structure) (90)	$Na(Hg) / D_2O$	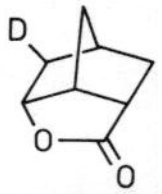 (structure)	63

Table 5.6. (continued)

Alkene	Mercuric salt	Organomercurial(s) (% Yield)	Subsequent reactants	Product(s) (% Yield)	Ref.
(structure: methylnorbornene carboxylic acid, CH₃, CO₂H)	Hg(OAc)₂, NaCl	(lactone structure, ClHg, CH₃) (88)	—	—	123
(structure: methylnorbornene carboxylic acid, CH₃, CO₂H)	Hg(OAc)₂, NaCl	(lactone structure, ClHg, CH₃) (85)	—	—	123
H₂C=CHCH₂CH(CO₂H)CH₂CH(CH₃)₂	Hg(OAc)₂, X⁻	(lactone structure, (CH₃)₂CHCH₂, CH₂HgX) X = Cl (92.2), Br (96.3), I (86)	Br₂ (on RHgCl)	(lactone structure, (CH₃)₂CHCH₂, CH₂Br) (76.2)	119–121
H₂C=CHCH₂CH(CO₂H)(CH₂)₃CH₃	Hg(OAc)₂, X⁻	(lactone structure, CH₃(CH₂)₃, CH₂HgX) X = Cl (92.4), Br (93.9), I (87)	Br₂ (on RHgCl)	(lactone structure, CH₃(CH₂)₃, CH₂Br) (71.6)	119–121
CH₃(CH₂)₂CH=CHCH₂OCON(CH₃)₂	Hg(O₂CCF₃)₂	—	P(C₆H₅)₃	CH₃(CH₂)₂CH[OCON(CH₃)₂]CH=CH₂ + CH₃(CH₂)₂CH=CHCH₂OCON(CH₃)₂ 83 : 17 (89 total)	126
(structure: norbornene dicarboxylic anhydride)	HgCl₂/H₂O	(lactone structure, ClHg, CO₂H) (85)	—	—	129
(structure: norbornene CO₂CH₃, CO₂H)	Hg(OAc)₂	(lactone structure, AcOHg, CO₂CH₃) (87)	—	—	73

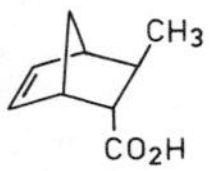

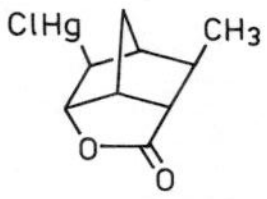

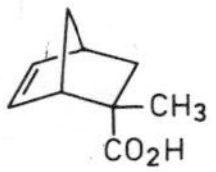

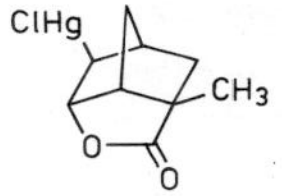

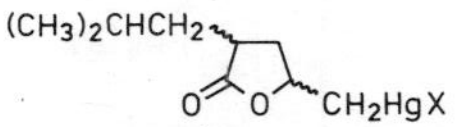

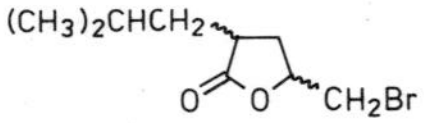

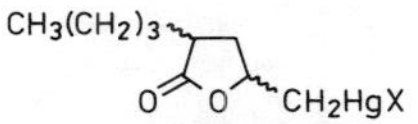

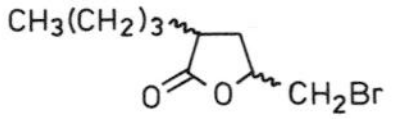

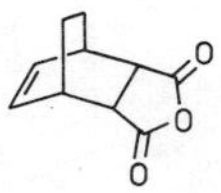

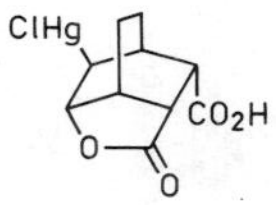

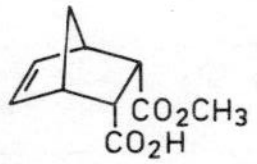

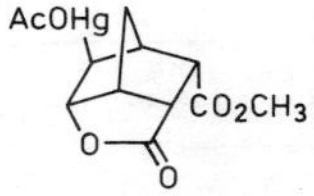

Table 5.6. (continued)

Alkene	Mercuric salt	Organomercurial(s) (% Yield)	Subsequent reactants	Product(s) (% Yield)	Ref.
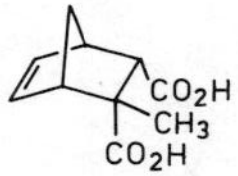 (norbornene with CO₂H, CH₃, CO₂H substituents)	—	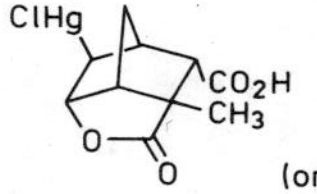 (lactone, ClHg, CO₂H, CH₃) (78) (only isomer ?)	—	—	123
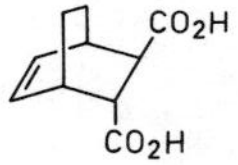 (norbornene with CO₂H, CO₂H substituents)	$HgCl_2$	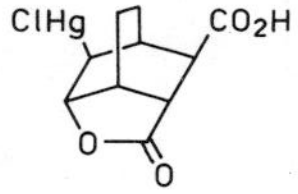 (lactone, ClHg, CO₂H) (58)	—	—	129
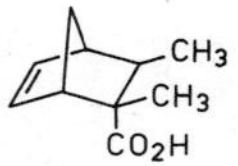 (norbornene with CH₃, CH₃, CO₂H substituents)	$Hg(OAc)_2$, NaCl	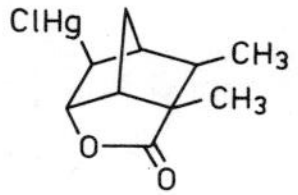 (lactone, ClHg, CH₃, CH₃) (83)	—	—	123
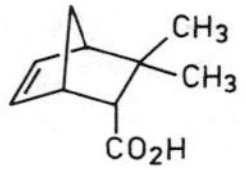 (norbornene with CH₃, CH₃, CO₂H substituents)	$Hg(OAc)_2$, NaCl	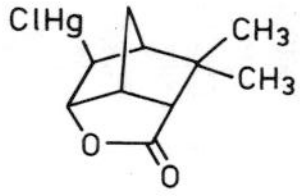 (lactone, ClHg, CH₃, CH₃) (71)	—	—	123
$H_2C=CHCH_2CH(CO_2H)CH_2-CH_2CH(CH_3)_2$	$Hg(OAc)_2$, X^-	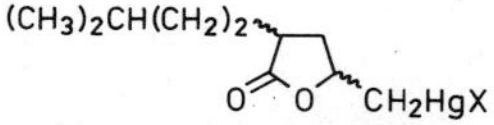 $(CH_3)_2CH(CH_2)_2$ (lactone, CH_2HgX) X = Cl (81.1), Br (65), I (92)	Br_2 (on RHgCl)	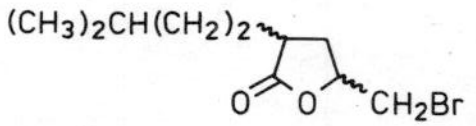 $(CH_3)_2CH(CH_2)_2$ (lactone, CH_2Br) (79.5)	119–121
$H_2C=CHCH_2CH(CO_2H)(CH_2)_4CH_3$	$Hg(OAc)_2$, KI	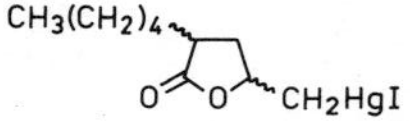 $CH_3(CH_2)_4$ (lactone, CH_2HgI) (80)	—	—	121
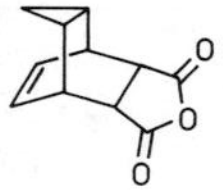 (bicyclic anhydride alkene)	$Hg(OAc)_2 / CH_3OH$, NaCl	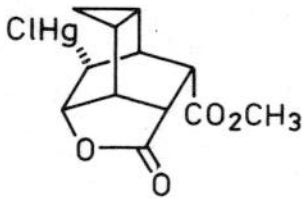 (ClHg, CO_2CH_3 lactone)	—	—	74

Table 5.6. (continued)

Alkene	Mercuric salt	Organomercurial(s) (% Yield)	Subsequent reactants	Product(s) (% Yield)	Ref.
p-$H_2C=CH(CH_2)_2CO_2C_6H_4Br$	$HgCl_2$	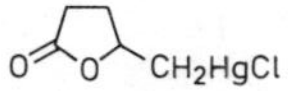 [lactone]–CH_2HgCl (~10)	—	—	117
p-$H_2C=CHCH_2CH(CO_2H)C_6H_4Cl$	$HgCl_2$	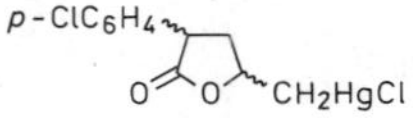 p-ClC_6H_4–[lactone]–CH_2HgCl (80)	—	—	117
p-$H_2C=CHCH_2CH(CO_2H)C_6H_4F$	$HgCl_2$	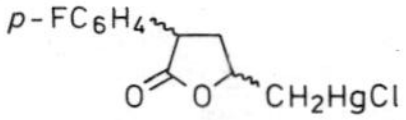 p-FC_6H_4–[lactone]–CH_2HgCl (75)	—	—	117
p-$H_2C=CH(CH_2)_2CO_2C_6H_4NO_2$	$HgCl_2$	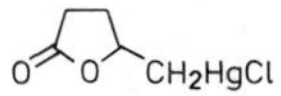 [lactone]–CH_2HgCl (~10)	—	—	117
p-$H_2C=CHCH_2CH(CO_2H)C_6H_4NO_2$	$HgCl_2$	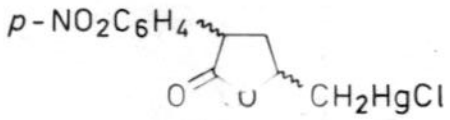 p-$NO_2C_6H_4$–[lactone]–CH_2HgCl (51)	—	—	117
$H_2C=CH(CH_2)_2CO_2C_6H_5$	$HgCl_2$	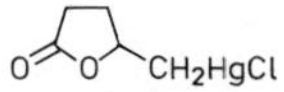 [lactone]–CH_2HgCl (10)	$NaBH_4/NaOH$	$H_2C=CH(CH_2)_2CO_2H$	117, 130
$H_2C=CHCH_2CH(C_6H_5)CO_2H$	$HgCl_2$	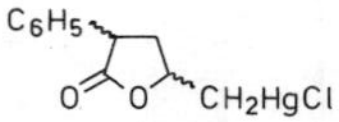 C_6H_5–[lactone]–CH_2HgCl (56)	$NaBH_4/NaOH$	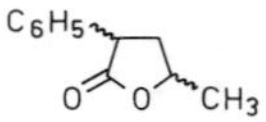 C_6H_5–[lactone]–CH_3 (85)	117
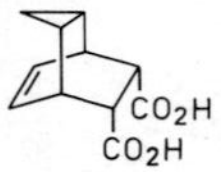 [bicyclic diene diacid, CO_2H, CO_2H]	$Hg(OAc)_2$, $NaCl$	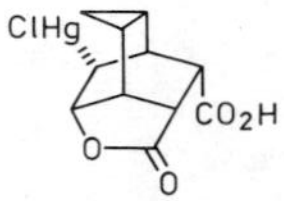 $ClHg$–[tricyclic lactone], CO_2H	—	—	74
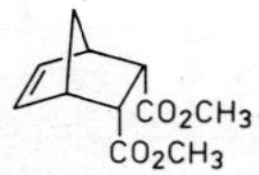 [bicyclic diene diester, CO_2CH_3, CO_2CH_3]	$Hg(OAc)_2/H_2O$, $NaCl$	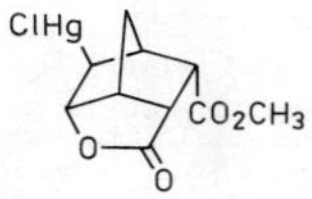 $ClHg$–[tricyclic lactone], CO_2CH_3 (93)	—	—	74, 123

Table 5.6. (continued)

Alkene	Mercuric salt	Organomercurial(s) (% Yield)	Subsequent reactants	Product(s) (% Yield)	Ref.
	Hg(OAc)$_2$, NaCl	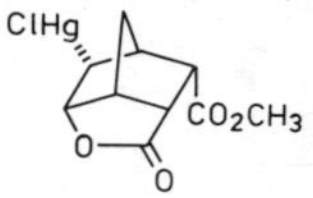 (structure: ClHg–bicyclic lactone, CO$_2$CH$_3$) ? (50)	—	—	62, 127
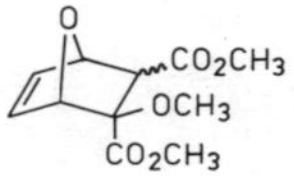 (structure: bicyclic alkene with CO$_2$CH$_3$, OCH$_3$, CO$_2$CH$_3$)	Hg(OAc)$_2$, NaCl	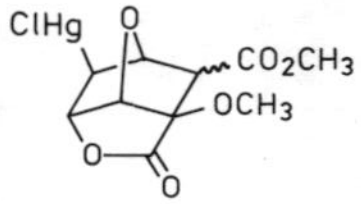 (structure: ClHg–bicyclic lactone, CO$_2$CH$_3$, OCH$_3$)	Br$_2$	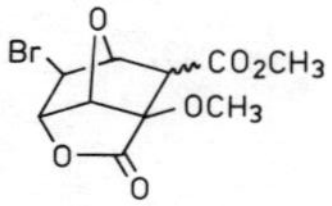 (structure: Br–bicyclic lactone, CO$_2$CH$_3$, OCH$_3$)	131–133
H$_2$C=CHCH$_2$CH(CO$_2$H)(CH$_2$)$_5$CH$_3$	Hg(OAc)$_2$, KI	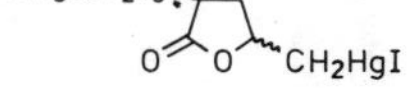 (structure: CH$_3$(CH$_2$)$_5$–lactone, CH$_2$HgI) (92)	—	—	121
H$_2$C=CH(CH$_2$)$_8$CO$_2$H	HgCl$_2$, NaOH, HOAc	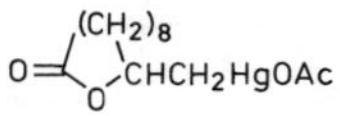 (structure: (CH$_2$)$_8$ macrolactone, CHCH$_2$HgOAc)	—	—	134
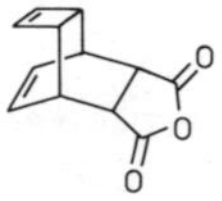 (structure: bicyclic anhydride, two C=O)	Hg(OAc)$_2$/CH$_3$OH, NaCl	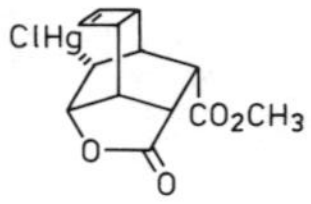 (structure: ClHg–tricyclic lactone, CO$_2$CH$_3$) (40)	—	—	74
p-H$_2$C=CH(CH$_2$)$_2$CO$_2$C$_6$H$_4$CH$_3$	HgCl$_2$	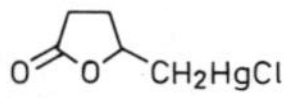 (structure: lactone, CH$_2$HgCl) (~10)	—	—	117
p-H$_2$C=CHCH$_2$CH(CO$_2$H)-C$_6$H$_4$CH$_3$	HgCl$_2$	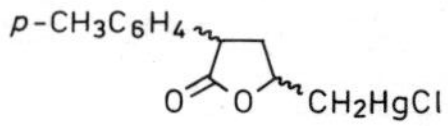 (structure: p-CH$_3$C$_6$H$_4$–lactone, CH$_2$HgCl) (77)	—	—	117
p-H$_2$C=CH(CH$_2$)$_2$CO$_2$C$_6$H$_4$OCH$_3$	HgCl$_2$	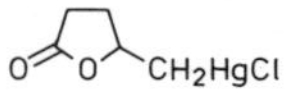 (structure: lactone, CH$_2$HgCl) (~10)	—	—	117

Table 5.6. (continued)

Alkene	Mercuric salt	Organomercurial(s) (% Yield)	Subsequent reactants	Product(s) (% Yield)	Ref.
p-$H_2C{=}CHCH_2CH(CO_2H)$-$C_6H_4OCH_3$	$HgCl_2$	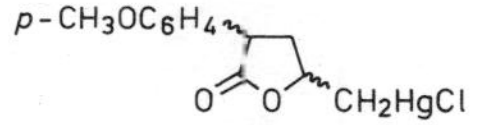 p-$CH_3OC_6H_4$ … CH_2HgCl (77)	—	—	117
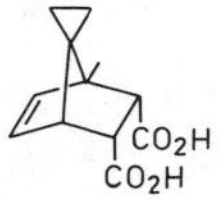 (bicyclic dicarboxylic acid structure)	$Hg(OAc)_2$	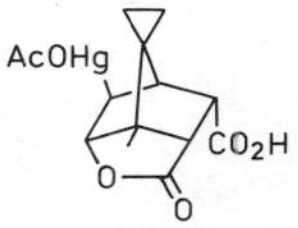 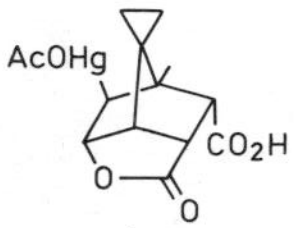 AcOHg … CO_2H + AcOHg … CO_2H 1.9 : 1	$NaBH_4$	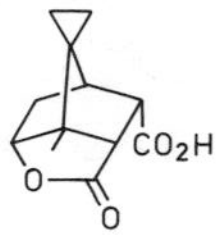 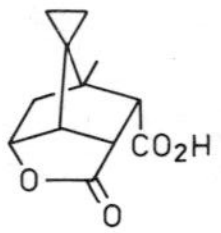 CO_2H + CO_2H (50–70 total)	135
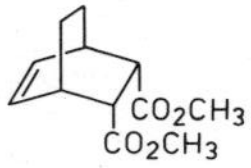 (bicyclic diester structure)	$Hg(OAc)_2$	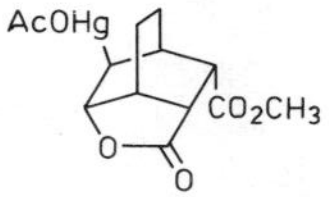 AcOHg … CO_2CH_3	—	—	129
	$Hg(OAc)_2$	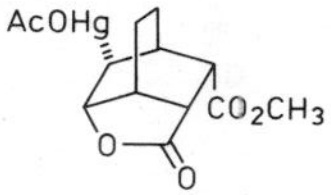 AcOHg … CO_2CH_3 ? (61 crude)	—	—	127
$H_2C{=}CHCH_2CH(CO_2H)$-$(CH_2)_6CH_3$	$Hg(OAc)_2$, KI	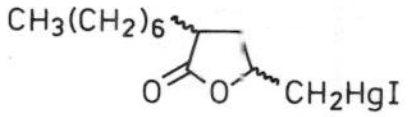 $CH_3(CH_2)_6$ … CH_2HgI (71)	—	—	121
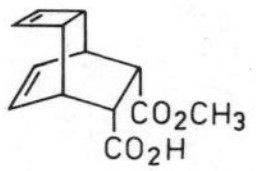 (bicyclic structure, CO_2CH_3, CO_2H)	$Hg(OAc)_2$, NaCl	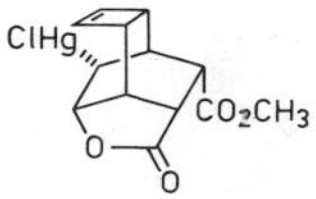 ClHg … CO_2CH_3	—	—	74
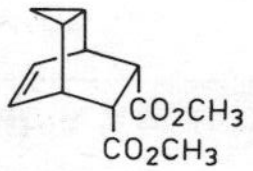 (bicyclic structure, CO_2CH_3, CO_2CH_3)	$Hg(OAc)_2$, NaCl	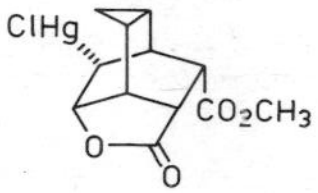 ClHg … CO_2CH_3	—	—	74

Table 5.6. (continued)

Alkene	Mercuric salt	Organomercurial(s) (% Yield)	Subsequent reactants	Product(s) (% Yield)	Ref.
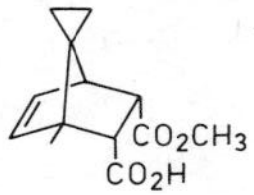 (bicyclic alkene with CO_2CH_3, CO_2H)	$Hg(OAc)_2$	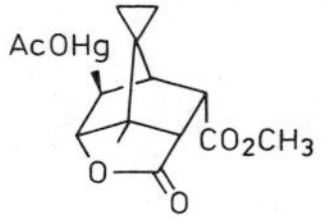 (lactone structure with AcOHg, CO_2CH_3) (83)	—	—	73
$C_6H_5CH_2CH{=}CHCH_2{-}OCON(CH_3)_2$	$Hg(O_2CCF_3)_2$	—	$P(C_6H_5)_3$	$C_6H_5CH_2CH[OCON(CH_3)_2]CH{=}CH_2$ + $C_6H_5CH_2CH{=}CHCH_2OCON(CH_3)_2$ 71 : 29 (86 total)	126
$H_2C{=}CHCH_2CH(CO_2H){-}(CH_2)_7CH_3$	$Hg(OAc)_2$, KI	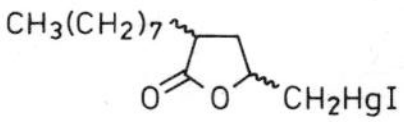 (lactone: $CH_3(CH_2)_7$, CH_2HgI) (83)	—	—	121
$H_2C{=}CHCH_2CH(C_6H_4Cl{-}p){-}CO_2C_6H_5$	$HgCl_2$	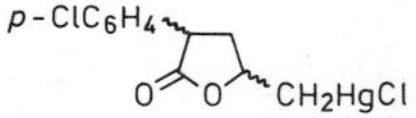 (lactone: $p{-}ClC_6H_4$, CH_2HgCl) (61)	—	—	117
$H_2C{=}CHCH_2CH(C_6H_4F{-}p){-}CO_2C_6H_5$	$HgCl_2$	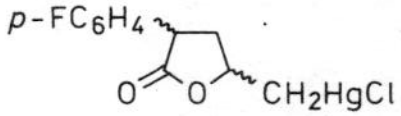 (lactone: $p{-}FC_6H_4$, CH_2HgCl) (58)	—	—	117
$H_2C{=}CHCH_2CH(C_6H_4NO_2{-}p){-}CO_2C_6H_5$	$HgCl_2$	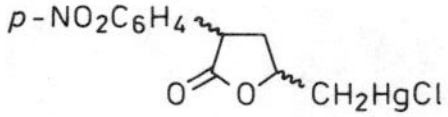 (lactone: $p{-}NO_2C_6H_4$, CH_2HgCl) (33)	—	—	117
$H_2C{=}CHCH_2CH(C_6H_5){-}CO_2C_6H_5$	$HgCl_2$	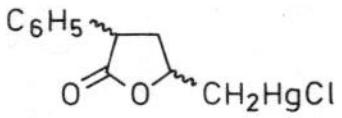 (lactone: C_6H_5, CH_2HgCl) (40)	$NaBH_4$ / $NaOH$	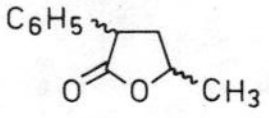 (lactone: C_6H_5, CH_3) (85)	117, 130
$H_2C{=}CHCH_2C(C_6H_5)_2CO_2H$	$HgCl_2$	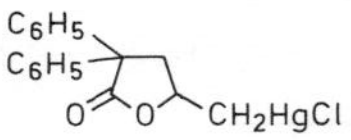 (lactone: C_6H_5, C_6H_5, CH_2HgCl) (37, 50, 82)	—	—	117, 118, 136
	$Hg(OAc)_2$, NaCl	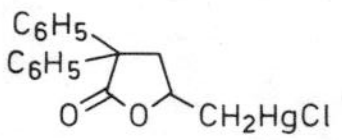 (lactone: C_6H_5, C_6H_5, CH_2HgCl) (50, 55)	Br_2	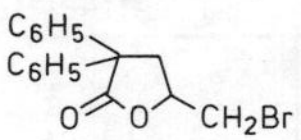 (lactone: C_6H_5, C_6H_5, CH_2Br) (80)	118

Table 5.6. (continued)

Alkene	Mercuric salt	Organomercurial(s) (% Yield)	Subsequent reactants	Product(s) (% Yield)	Ref.
$H_2C=CHCH_2CH(C_6H_4CH_3-p)-CO_2C_6H_5$	$HgCl_2$	(54)	—	—	117
$H_2C=CHCH_2CH(C_6H_4OCH_3-p)-CO_2C_6H_5$	$HgCl_2$	(51)	—	—	117
$H_2C=CHCH_2C(C_6H_5)_2CO_2C_6H_5$	$HgCl_2$		$NaBH_4/NaOH$	(80)	117, 130

V. Acyloxymercuration

A large number of bicyclic unsaturated acids have also been cyclized by mercuric salts. Numerous examples of the intramolecular acyloxymercuration of endo-norbornene carboxylic acids, the corresponding methyl esters and anhydrides have been reported (Eq. 50). Although the parent mercurial was

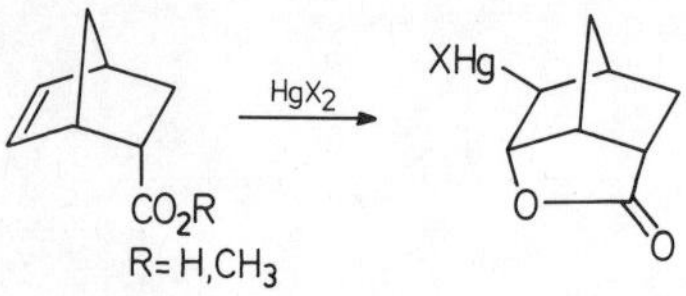

$$(50)$$

originally reported to have an endo ClHg group [124], it has since been shown to be exo [125]. Proton [63, 125] and C-13 [81] NMR data are now available for these compounds. It is interesting that the corresponding methyl ester undergoes intramolecular cyclization with mercuric acetate in aqueous acetone, but simple acetoxymercuration in acetic acid (Eq. 51) [63]. The related endo-diacid has been reported to cyclize to an endo mercurial [62, 127], but this structure seems questionable (Eq. 52). The corresponding dimethyl

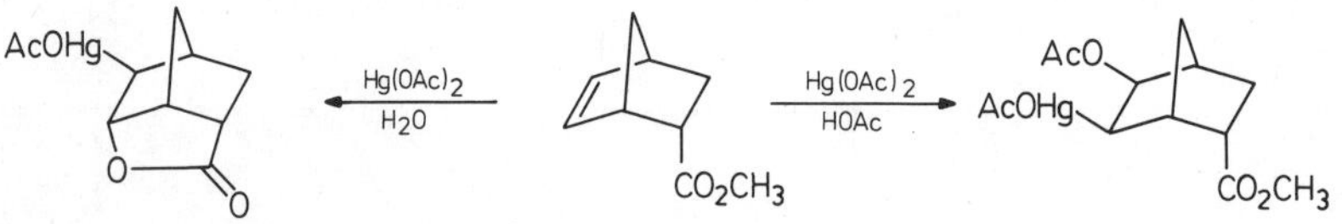

$$(51)$$

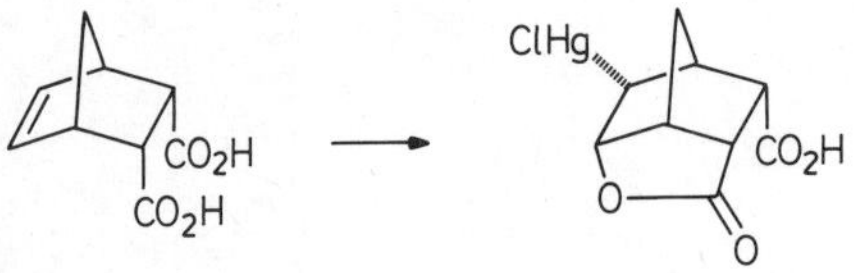

$$(52)$$

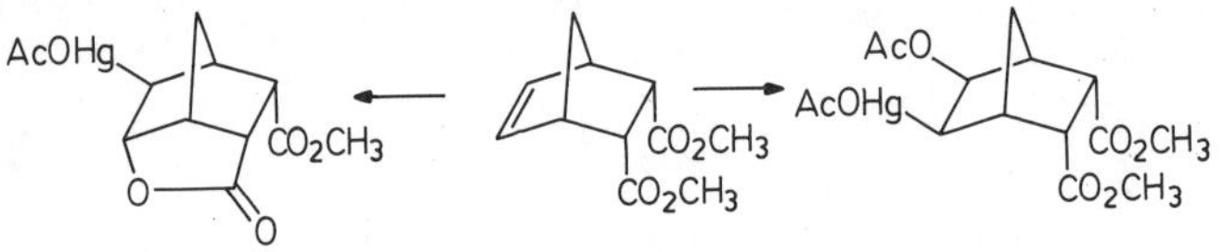

$$(53)$$

ester again gives either inter- [62, 73, 74] or intramolecular [62, 74, 127] acyloxymercuration depending on the solvent (Eq. 53). Confusion reigns as to the stereochemistry of the mercury moiety in both of these compounds, but it appears likely that the mercury is exo in both cases. The corresponding half acid-half ester [73] and cyclic anhydride [74] have also been reported to form the above mercurated lactone. Substitution of an oxygen or a cyclopropane in the seven position of these compounds still affords exo mercurials (Eqs. 54–56).

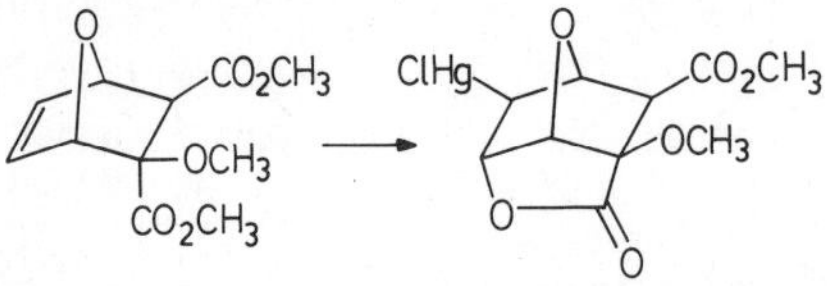

$$(54) \ [131–133]$$

$$(55) \quad [135]$$

$$(56) \quad [73]$$

Similar confusion exists with the structures of mercurials derived from the bicyclo[2.2.2]octene system. It now appears that the simple diacid, diester and anhydride afford exclusively exo mercurials (Eq. 57) [127, 129]. In tricyclic compounds in which exo attack on the double bond is hindered, acyloxymercuration proceeds by endo attack or attack elsewhere in the system (Eqs. 58, 59) [74].

$$(57)$$

$$(58)$$

$$(59)$$

Relatively few subsequent reactions of these mercurated lactones have been reported. While the successful demercuration of these compounds with sodium borohydride [117, 122, 128], sodium trimethoxyborohydride [125] and sodium amalgam in D_2O (retention) [63] has been reported, it has been noted that varying amounts of the unsaturated acids are generated with sodium borohydride [117], especially when a more polar solvent such as nitromethane is used (Eq. 60) [85]. A very limited number of the simple monocyclic [118, 119] and bicyclic [132, 133] mercurated lactones have been brominated and no other reactions of these compounds have been reported.

$$(60)$$

C. Dienes and Polyenes

A large number of dienes and polyenes have been subjected to acyloxy-mercuration (Table 5.7). Our discussion will cover acyclic, monocyclic, and bi- or polycyclic dienes or polyenes in that order.

1,5-Hexadiene can be either mono- or dimercurated depending on the stoichiometry and reaction conditions employed (Eq. 61) [139–141]. Alkaline

$$(61)$$

sodium borohydride reduction of the mono adduct provides a useful route to 5-hydroxy-1-hexene which cannot be cleanly obtained by hydroxymercuration-demercuration of the diene due to tetrahydrofuran formation [139, 140]. Under the same conditions, 6-methyl-1,5-heptadiene also affords a good yield of mono alcohol alongside carbocyclization products (Eq. 62) [140]. Other 1,5-dienes and 1,5,9-trienes have been subjected to mercuric

$$(62)$$

salts, usually under strongly acidic conditions, to afford carbomercuration products which will be discussed later in Chapter X. Methyl linoleate has also been reported to react with mercuric acetate in acetic acid to produce mono-, di- and even tri-adducts whose structures have not been established [171]. Upon heating with mercuric acetate in acetic acid, this olefin gives isomerized methyl octadienoates, acetoxy compounds and polymers [172].

The acetoxymercuration of acyclic allenes has also been examined. Optically active 2,3-pentadiene affords a cis/trans mixture of optically active vinylmercurials (Eq. 63) [138]. Vinyl allenes undergo carbomercuration

$$(63)$$

Table 5.7. Acyloxymercuration of Dienes and Polyenes

Polyene	Mercuric salt	Organomercurial(s) (% Yield)	Subsequent reactants	Product(s) (% Yield)	Ref.
(−) $CH_3CH=C=CHCH_3$	$Hg(OAc)_2$	$CH_3CH=C$ with HgOAc, CH_3, AcO, C, H 17 : 83 cis/trans	—	—	138
$H_2C=CH(CH_2)_2CH=CH_2$	$Hg(OAc)_2$ (KBr)	$XHgCH_2CH(OAc)(CH_2)_2CH=CH_2$ X = OAc, Br (100)	$NaBH_4 / NaOH$ (on $RHgOAc$)	$CH_3CHOH(CH_2)_2CH=CH_2$ (65, 80)	139, 140
	$Hg(OAc)_2$	$AcOHgCH_2CHOAc(CH_2)_2CHOAcCH_2HgOAc$	—	—	141
	$Hg(OAc)_2$, Cl^-	(norbornene OAc/HgCl organomercurial)	$NaBH_4$ ($NaBD_4$)	(OAc, H(D)) 64 : (AcO, H(D)) 36	75, 142
	$Hg(OAc)_2$, Cl^-	(AcO/HgCl organomercurial)	$NaBH_4$ ($NaBD_4$)	(OAc, H(D)) 64 : (AcO, H(D)) 36	75, 142
	$Hg(OAc)_2$, NaCl	(AcO/HgCl) 73 : (OAc/HgCl) 27 (64 total)	Cl_2	(AcO, Cl) 75 : (AcO, Cl) 25	143
	$Hg(OAc)_2$, KBr	(OAc/HgBr) (75) or (AcO/HgBr) (55)	—	—	144
(cyclooctatetraene)	$Hg(OAc)_2$	—	—	(benzocyclobutene OAc/OAc) (72)	145
(norbornadiene)	$Hg(OAc)_2$	—	$NaBH_4$	(OAc) ~85 : (OAc) ~15	28

412

Table 5.7. (continued)

Polyene	Mercuric salt	Organomercurial(s) (% Yield)	Subsequent reactants	Product(s) (% Yield)	Ref.
 (bicyclic diene)	$Hg(OAc)_2$, NaCl	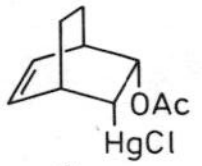 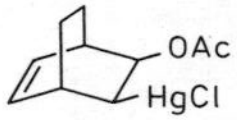 8 : 1 (73, 93 total)	Na(Hg)/NaOD	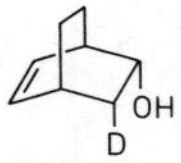 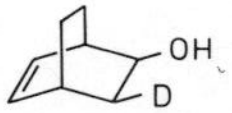 89 : 11 (68 total)	146
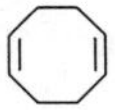 (cyclooctadiene)	$Hg(OAc)_2$	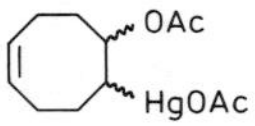 (83)	$NaBH_4$	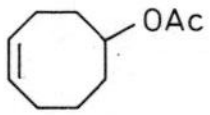 (60)	147
	$Hg(OAc)_2$	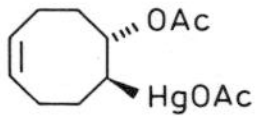 (100)	$NaBH_4$/NaOH	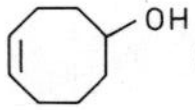 (48) 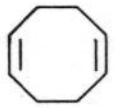 + 77 : 23	139, 148, 149
 (norbornadiene)	$Hg(OAc)_2$	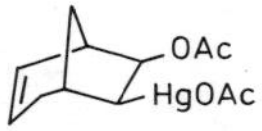 (83, 86)	$NaBH_4$ [Na(Hg)/D_2O]	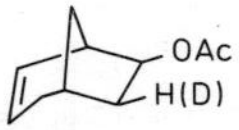	150, 151
$H_2C=CH(CH_2)_2CH=C(CH_3)_2$	$Hg(OAc)_2$	—	$NaBH_4$	$CH_3CHOH(CH_2)_2CH=C(CH_3)_2$ 83  : + : 10 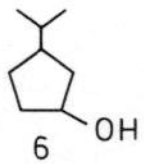 : 6 (85 total)	126
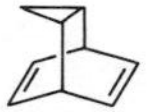 (bicyclic diene)	$Hg(OAc)_2$, NaCl	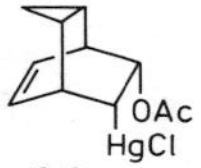 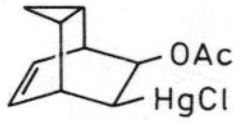 2.3 : 1 (97 total)	Na(Hg)/NaOD	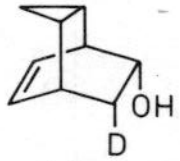 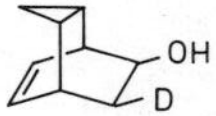 2.3 : 1 (76 total)	146
	$Hg(O_2CCF_3)_2$	—	Na(Hg)/NaOD	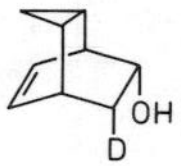 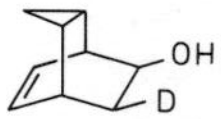 2.2 : 1 (75 total)	146

Table 5.7. (continued)

Polyene	Mercuric salt	Organomercurial(s) (% Yield)	Subsequent reactants	Product(s) (% Yield)	Ref.
(norbornadienyl OAc)	$Hg(OAc)_2$, NaCl	(OAc, OAc, HgCl) (94)	$NaBD_4$	56 : 44	152
(=CHCH$_3$ norbornene)	$Hg(OAc)_2$, NaCl	(AcO, ClHg, =CHCH$_3$) (98)	$NaBH_4$/NaOH (on RHgOAc)	(AcO, =CHCH$_3$) (71)	153, 154
$H_2C=CH$— (norbornene)	$Hg(OAc)_2$	—	$NaBH_4$/NaOH	51 : 28 : 21 (51 total)	64
(−) cyclononene	$Hg(OAc)_2$, Cl$^-$	(HgX, OAc) X = OAc, Cl	$NaBH_4$ or Br_2	(X, OAc) X = H, Br	155
cyclononene	$HgSO_4$/HCO_2H	—	—	(O_2CH)	156
CH_3O_2C, CH_3O_2C (oxabicyclic)	$Hg(OAc)_2$, NaX	(CH_3O_2C, CH_3O_2C, O, OAc, HgX) X = Cl, Br, I, CN	—	—	157
(tricyclic diene)	$Hg(OAc)_2$	(OAc, HgOAc) 3 : (HgOAc, OAc) 2 (81 total)	—	—	151

Table 5.7. (continued)

Polyene	Mercuric salt	Organomercurial(s) (% Yield)	Subsequent reactants	Product(s) (% Yield)	Ref.
 (structure)	$Hg(OAc)_2$	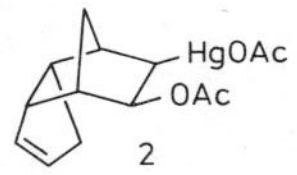 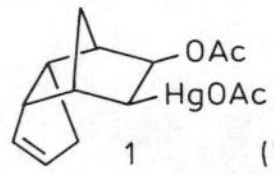 (structure) HgOAc / OAc 2 : (structure) OAc / HgOAc 1 (72 total)	—	—	151
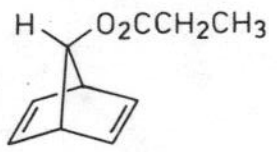 (structure) H, $O_2CCH_2CH_3$	$Hg(OAc)_2$, NaCl	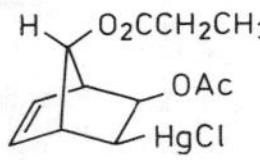 (structure) H, $O_2CCH_2CH_3$ / OAc / HgCl	$NaBH_4$	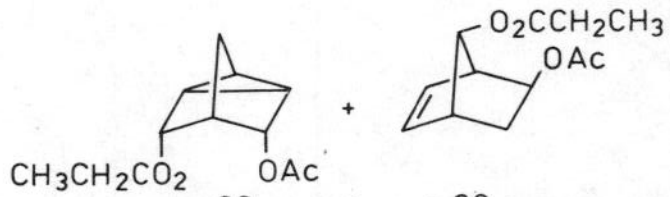 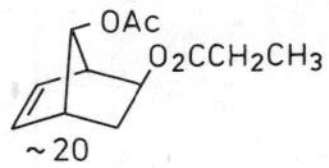 (structure) $CH_3CH_2CO_2$ / OAc 60 : (structure) $O_2CCH_2CH_3$ / OAc ~20 : (structure) OAc / $O_2CCH_2CH_3$ ~20	152
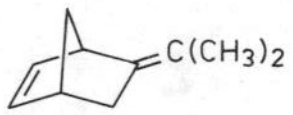 (structure) $=C(CH_3)_2$	$Hg(OAc)_2$	—	$NaBH_4$ / NaOH	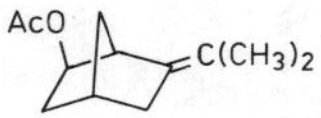 (structure) AcO / $=C(CH_3)_2$ (~65)	154
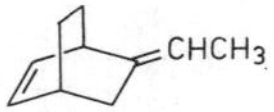 (structure) $=CHCH_3$	$Hg(OAc)_2$	—	$NaBH_4$ / NaOH	8 acetates (58 total)	28
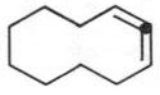 (structure)	$HgSO_4$ / HCO_2H	—	—	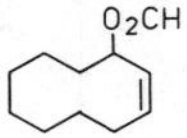 (structure) O_2CH (68)	158
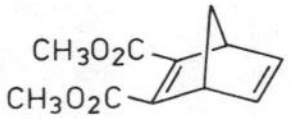 (structure) CH_3O_2C / CH_3O_2C	$Hg(OAc)_2$	 (structure) CH_3O_2C / CH_3O_2C / OAc / HgOAc	—	—	159
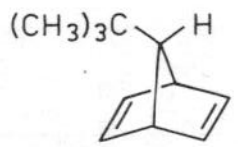 (structure) $(CH_3)_3C$, H	$Hg(O_2CCF_3)_2$	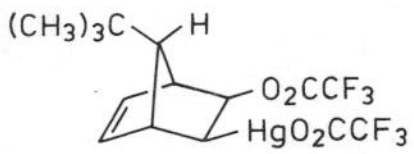 (structure) $(CH_3)_3C$, H / O_2CCF_3 / HgO_2CCF_3	—	—	160

Table 5.7. (continued)

Polyene	Mercuric salt	Organomercurial(s) (% Yield)	Subsequent reactants	Product(s) (% Yield)	Ref.
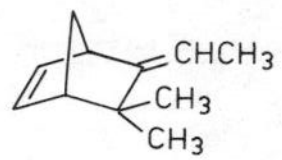 (bicyclic diene, =CHCH$_3$, CH$_3$, CH$_3$)	Hg(OAc)$_2$	—	NaBH$_4$/NaOH	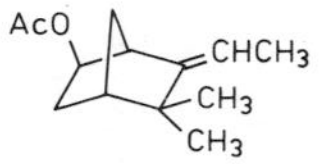 AcO– (bicyclic, =CHCH$_3$, CH$_3$, CH$_3$) (65)	154
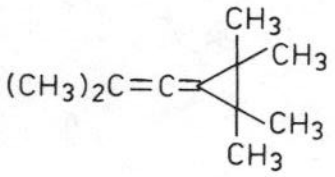 (CH$_3$)$_2$C=C=C(CH$_3$, CH$_3$)(CH$_3$, CH$_3$)	Hg(OAc)$_2$	—	NaBH$_4$/NaOH	(CH$_3$)$_2$C(OH)CCH=C(CH$_3$)$_2$, =C(CH$_3$)$_2$ 100 : + H$_2$C=C(CH$_3$)CH=C(CH$_3$)$_2$, =C(CH$_3$)$_2$ 26 : + (CH$_3$)$_2$CHC(CH$_3$)$_2$C≡CC(CH$_3$)$_2$OAc + others 15	161
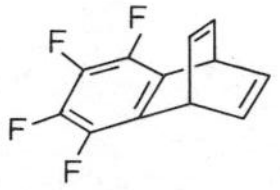 (tetrafluorobenzobicyclic diene, F, F, F, F)	Hg(OAc)$_2$, KBr	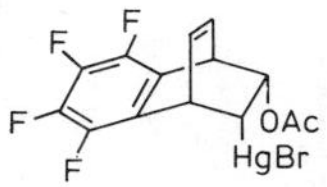 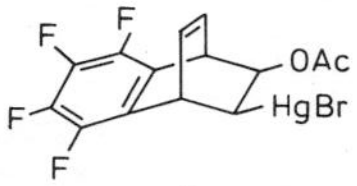 (F,F,F,F bicyclic, OAc, HgBr) 3 : (F,F,F,F bicyclic, OAc, HgBr) 2	—	—	162
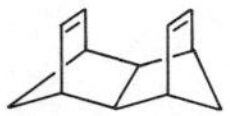 (tricyclic diene)	Hg(OAc)$_2$, NaCl	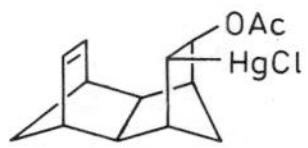 (tricyclic, OAc, HgCl)	—	—	75
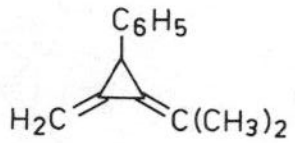 H$_2$C=(cyclopropane, C$_6$H$_5$)=C(CH$_3$)$_2$	Hg(OAc)$_2$	—	NaBH$_4$/NaOH	C$_6$H$_5$CH(OAc)CCH=C(CH$_3$)$_2$, =CH$_2$ + [C$_6$H$_5$CH(OAc)CC–]$_2$Hg, =CH$_2$, (CH$_3$)$_2$C=	163

Table 5.7. (continued)

Polyene	Mercuric salt	Organomercurial(s) (% Yield)	Subsequent reactants	Product(s) (% Yield)	Ref.
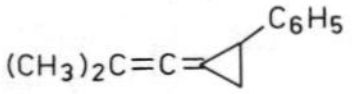 $(CH_3)_2C{=}C{=}$ (cyclopropyl)C_6H_5	$Hg(OAc)_2$	—	$NaBH_4 / NaOH$	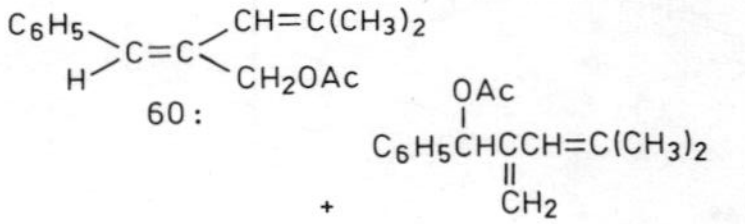 C_6H_5/H $C{=}C$ $CH{=}C(CH_3)_2$/CH_2OAc 60 : $C_6H_5CHCCH{=}C(CH_3)_2$ (OAc, $=CH_2$) : 40 + others	161
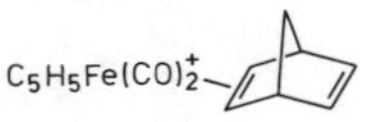 $C_5H_5Fe(CO)_2^+{-}$ (norbornadienyl) BF_4^-	$Hg(OAc)_2$	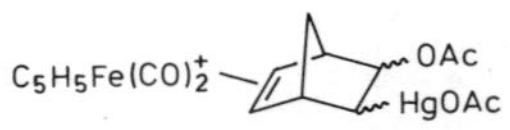 $C_5H_5Fe(CO)_2^+{-}$ (norbornene OAc, HgOAc) BF_4^-	—	—	164
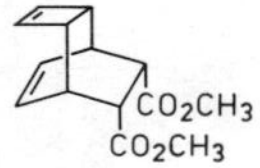 (bicyclic CO_2CH_3, CO_2CH_3)	$Hg(OAc)_2$, NaCl	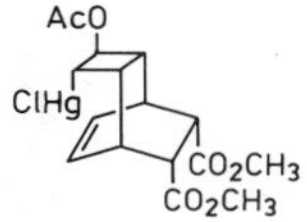 (AcO, ClHg, CO_2CH_3, CO_2CH_3)	—	—	165, 166
	$Hg(OAc)_2$, NaCl	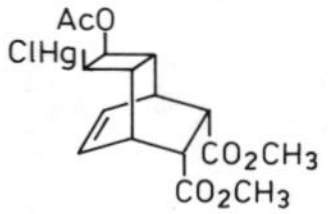 (AcO, ClHg, CO_2CH_3, CO_2CH_3) ? (89)	$NaBH_4$	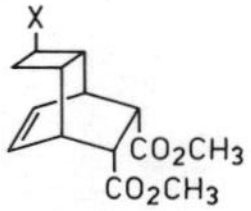 (X, CO_2CH_3, CO_2CH_3) $X = OAc, OH$	74, 167
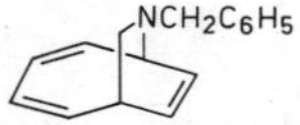 (indene $NCH_2C_6H_5$)	$Hg(OAc)_2$	—	—	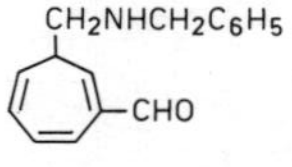 $CH_2NHCH_2C_6H_5$ (tropone) CHO (76)	168
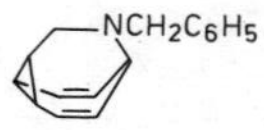 (bicyclic $NCH_2C_6H_5$)	$Hg(OAc)_2 / HOAc / Ac_2O$	—	—	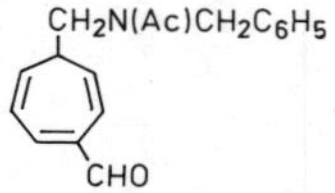 $CH_2N(Ac)CH_2C_6H_5$ (tropone) CHO	168

Table 5.7. (continued)

Polyene	Mercuric salt	Organomercurial(s) (% Yield)	Subsequent reactants	Product(s) (% Yield)	Ref.
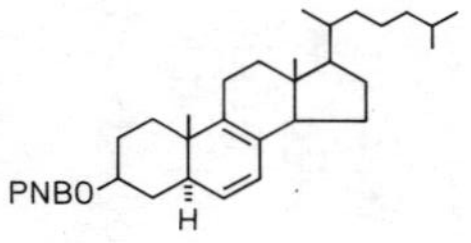 (steroid, PNBO-)	Hg(OAc)$_2$	—	NaOH	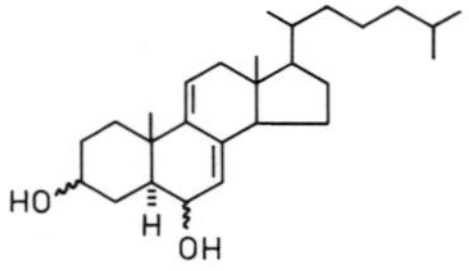 (steroid diol, HO-, OH)	169
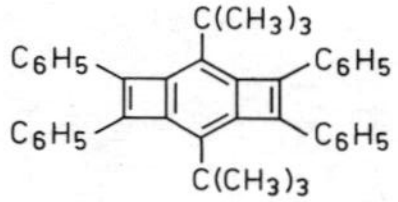 (cyclobutene/benzene derivative with C(CH$_3$)$_3$, C$_6$H$_5$ groups)	Hg(OAc)$_2$ / RCO$_2$H	—	—	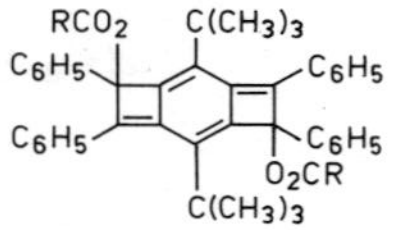 (product with RCO$_2$, O$_2$CR, C(CH$_3$)$_3$, C$_6$H$_5$ groups) R = CH$_3$ (55), C$_6$H$_5$ (17)	170

with mercuric acetate in acetic acid/perchloric acid to produce cyclopente-
nones [173, 174]. This reaction will be discussed in more detail later.

1,5-Cyclooctadiene also undergoes clean monoacetoxymercuration [139,
147–149] and sodium borohydride reduction affords the corresponding
acetate [147] (Eq. 64). Recall that hydroxymercuration of this diene afforded
primarily bicyclic ethers. Carbomercuration products are again observed

$$\text{(64)}$$

under more strongly acidic conditions [148]. Cyclooctatetraene also undergoes
carbon—carbon bond formation and oxidation upon reaction with mercuric
acetate in acetic acid (Eq. 65) [145].

$$\text{(65)}$$

Ring-containing dienes and polyenes undergo a variety of reactions with
mercuric acetate. Cyclopropane-containing dienes have been observed to
undergo ring opening (Eqs. 66–68). While optically active 1,2-cyclononadiene

$$\text{(66) [161]}$$

$$\text{(67) [161]}$$

$$\text{(68) [163]}$$

affords the expected acetoxymercuration product [155], reaction with formic
acid and catalytic amounts of mercuric sulfate affords the corresponding
allylic formate (Eq. 69) [156]. 1,2-Cyclodecadiene behaves similarly towards
formic acid and mercuric sulfate [158], but 1,2,5,8-cyclodecatetraene reacts
with mercuric sulfate in acetic acid to give cyclization products to be discussed
later [175, 176].

$$(69)$$

The acetoxymercuration and subsequent reactions of norbornadiene have received considerable attention. Upon short reaction times, norbornene affords a bicyclic adduct, while long reaction times yield predominantly a tricyclic product (Eq. 70) [75, 142]. In the tricyclic mercurial, the mercury group is estimated to be approximately 73% exo and 27% endo [143]. A

$$(70)$$

similar ratio of organic chlorides is observed upon chlorination. The demercuration of these mercurials by sodium borohydride [142, 144, 152, 177, 178], sodium borodeuteride [142, 152, 177, 179], tin hydrides and deuterides [144, 178, 180], aluminum and copper deuterides [144], and photolysis in protic solvents [178, 180, 181] have all been studied in depth. In most cases, both mercurials give the same three products in approximately the same proportions no matter what the reducing agent (Eq. 71). Reduction with sodium in naphthalene [181] or alkaline sodium/mercury analgam in D_2O [179] is

$$(71)$$

$$(72)$$

observed, however, to give a high degree of retention. The rearranged products are thought to arise by the equilibration of free radicals (Eq. 72). Similar rearrangements are observed in the acetoxymercuration-reduction of 7-acyloxynorbornadienes [152, 177]. One other approach to the norbornenyl-mercurial has been reported. Protection of one of the double bonds of norbornadiene as a cyclopentadienyliron dicarbonyl cationic complex allows clean acetoxymercuration of the remaining double bond (Eq. 73) [164].

$$(73)$$

Besides the 7-acyloxy systems mentioned above, the acyloxymercuration of a number of other norbornadienes has been studied. 7-*t*-Butylnorbornadiene gives a *cis,exo*-norbornenylmercurial with mercuric trifluoroacetate (Eq. 74) [160]. Placing electron-withdrawing ester groups on one of the double

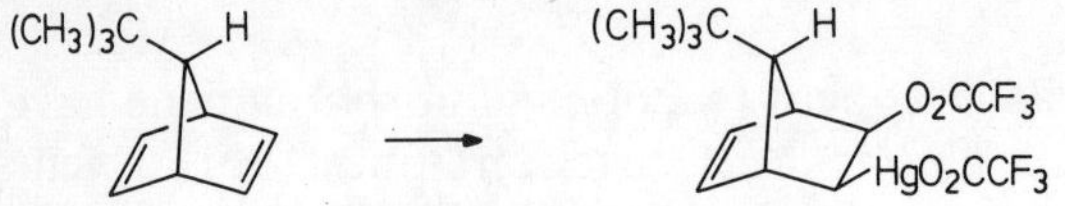

$$(74)$$

bonds of norbornadiene also provides clean bicyclic products (Eq. 75) [157, 159]. Moving one of the double bonds out of the norbornyl system reduces its reactivity sufficiently that the sole products of acetoxymercuration-demercuration are those of attack on the norbornyl double bond (Eqs. 76–78).

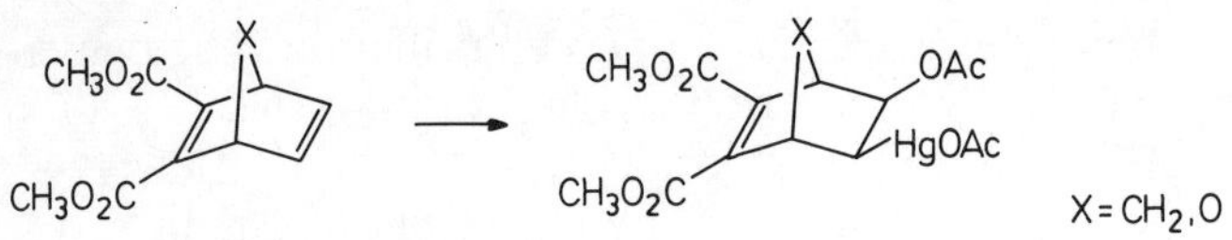

$$X = CH_2, O \qquad (75)$$

$R^1 = R^2 = H$	85	:	15
$R^1 = H$, $R^2 = CH_3$	100	:	0
$R^1 = R^2 = CH_3$	100	:	0

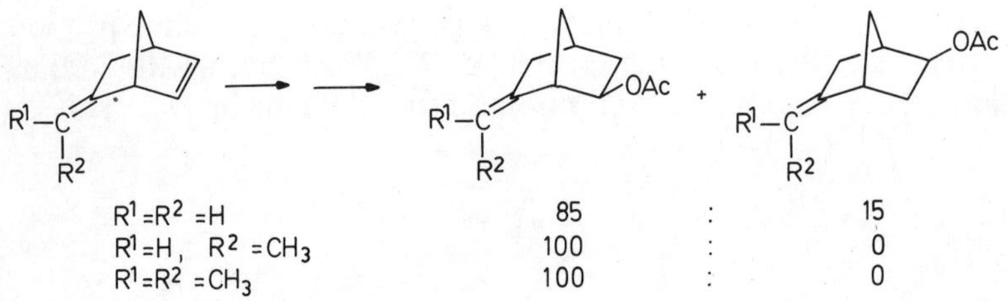

$$(76) \ [28, 153, 154]$$

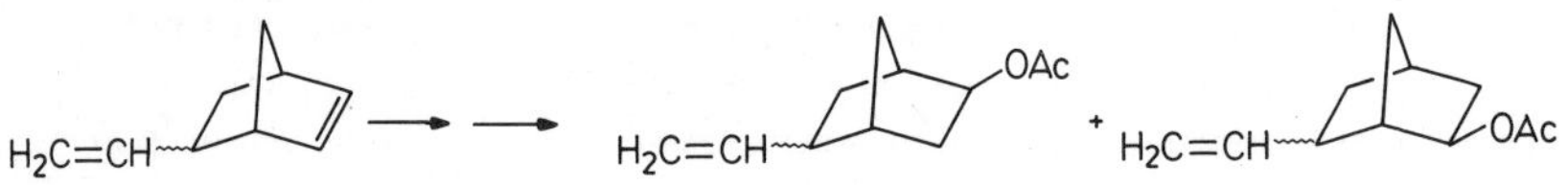

$$(77) \ [64]$$

exo	3	:	2
endo	1	:	2

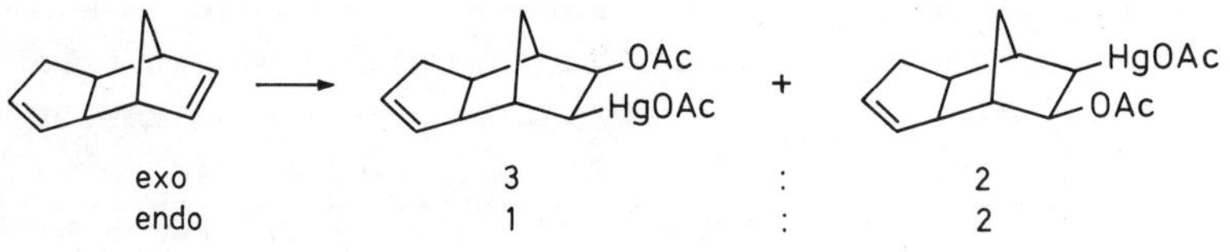

$$(78) \ [151]$$

In the bicyclo[2.2.2]octyl system, mixtures of syn and anti adducts are observed (Eqs. 79–81). The cyclopropane-containing acetoxymercurials have

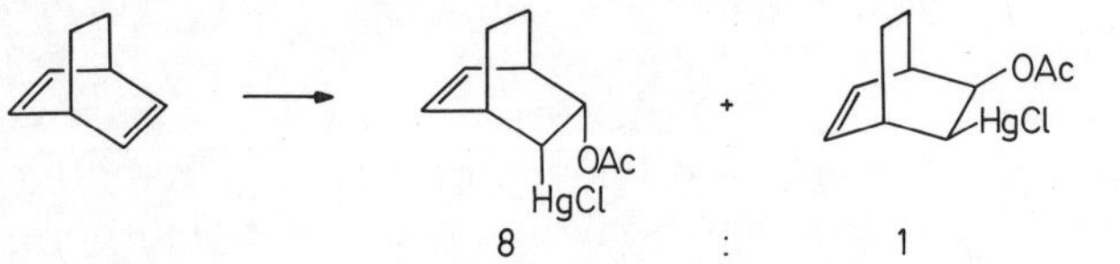

8	:	1

$$(79) \ [146]$$

420

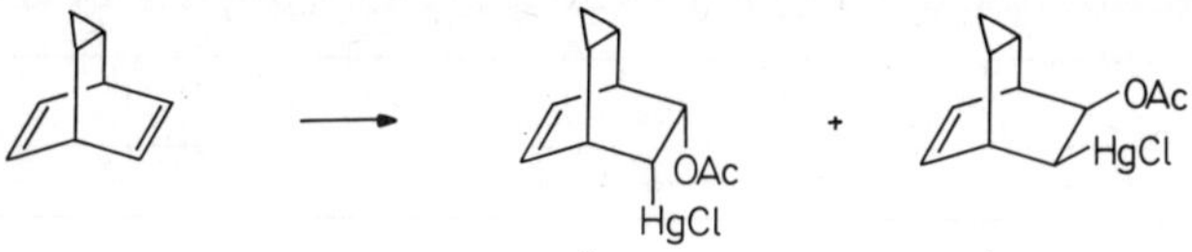

(80) [146]

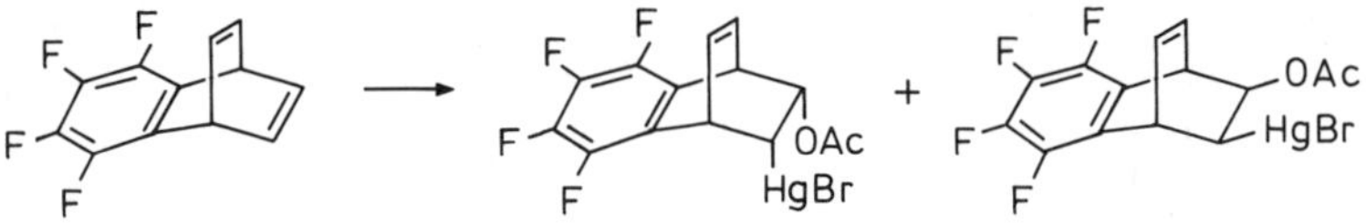

~ 2 : 1

(81) [162]

3 : 2

been cleanly reduced by Na(Hg)/NaOD with retention [146], but $NaBH_4$ in $CHCl_2$ gives a mixture of rearranged products [182].

Other polycyclic dienes have been observed to undergo clean stereo- and regiospecific monoacetoxymercuration (Eqs. 82–84). In this last example,

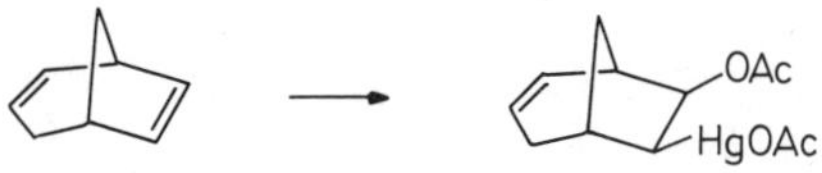

(82) [150, 151]

(83) [75]

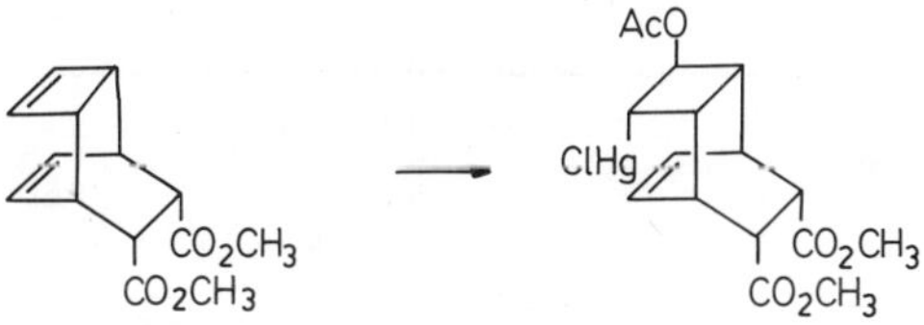

(84) [165 167]

attack of mercury on the endo side of the double bond is unusual, since most other electrophiles give exo attack.

As noted earlier, carbomercuration, rearrangement and oxidation products have also been observed during the course of some of these reactions (Eqs. 85–88).

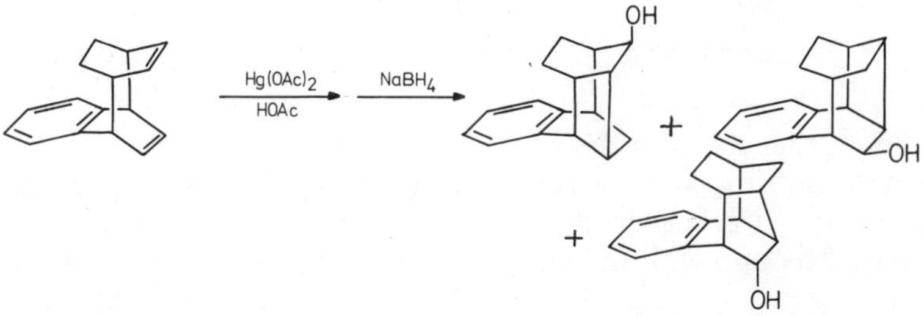

(85) [183]

421

V. Acyloxymercuration

$$C_6H_5OC\!\equiv\!CH \xrightarrow{\text{Hg(OAc)}_2} \quad (86)\ [168]$$

(86) [168]

(87) [168]

(88) [170]

D. Alkynes

The acyloxymercuration of alkynes leads to a variety of products depending on the type of acetylene employed (Table 5.8). As discussed in Chapter II, section J of the monograph "Organomercury Compounds in Organic Synthesis", terminal alkynes readily react with a variety of mercuric salts to produce dialkynylmercury compounds. Even mercuric acetate in acetic acid [200, 218], chloroform [201] or methanol [219] has been reported to form dialkynylmercurials. The reaction of a terminal acetylene and excess mercuric acetate has been reported to lead to a variety of polymercuration products, none of which seem to be very well characterized (Eqs. 89–91).

$$C_6H_5OC\!\equiv\!CH \xrightarrow{\text{Hg(OAc)}_2} C_6H_5O\overset{\overset{\displaystyle OAc}{|}}{C}\!=\!C(HgOAc)_2 \qquad (89)\ [208]$$

$$RO_2C(CH_2)_8C\!\equiv\!CH \longrightarrow RO_2C(CH_2)_8\overset{\overset{\displaystyle O}{\|}}{C}C(HgOAc)_3 \qquad (90)\ [211]$$

$$RC\!\equiv\!CH \longrightarrow R\!-\!\overset{\overset{\displaystyle OHgOAc}{|}}{C}\!=\!C(HgOAc)_2 \qquad (91)\ [204, 218]$$

Treatment of the products from the latter two reactions with aqueous HCl or with bromine affords the corresponding methyl ketones or tribromomethylketones respectively. It seems likely that both of these compounds have the same type of structure with the ketone product appearing more reasonable.

422

Table 5.8. Acyloxymercuration of Alkynes

Alkyne	Mercuric salt	Organomercurial(s) (% Yield)	Subsequent reactants	Product(s) (% Yield)	Ref.
$HC\equiv CH$	$HgSO_4/HOAc$	—	—	$H_2C=CHOAc$	184
	HgO or $Hg(OAc)_2$ or $HgSO_4/RCO_2H$	—	—	$H_2C=CHO_2CR$ $R = ClCH_2, Cl_2CH, Cl_3C, C_6H_5$ (good)	185
$HC\equiv CCH_3$	$HgO/HOAc/BF_3\cdot Et_2O/CH_3OH$	—	—	$H_2C=C(CH_3)O_2CCH_3$ (30)	186
$HC\equiv CCH_2OH$	$HgO/HOAc/BF_3\cdot Et_2O$	—	—	CH_3COCH_2OAc (30)	187
$HC\equiv CCH=CH_2$	$HgSO_4$ or $Hg_3(PO_4)_2$ or $Hg(O_3SC_6H_5)_2/RCO_2H$	—	—	$H_2C=C(O_2CR)CH=CH_2$ $R = H, Me, CH_2Cl, n-Pr$	188
$CH_3C\equiv CCH_3$	$Hg(OAc)_2$, KCl	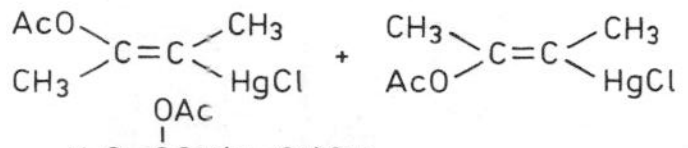 AcO,CH₃/CH₃,OAc — C=C — HgCl + CH₃,OAc — C=C — CH₃,HgCl + $H_2C=CCH(HgCl)CH_3$ (OAc)	—	—	189
	$Hg(OAc)_2$, KX	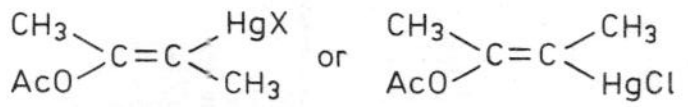 CH₃,AcO — C=C — HgX,CH₃ or CH₃,AcO — C=C — CH₃,HgCl X = Cl (70), Br (65) (40.4) or $H_2C=C(OAc)CH(HgCl)CH_3$ (17)	—	—	190
$HC\equiv CCHOHCH_3$	$HgO/HOAc/BF_3\cdot Et_2O$	—	—	$CH_3COCH(OAc)CH_3$ (41)	187
$CH_3C\equiv CCH_2OH$	$Hg(OAc)_2$	—	—	$CH_3COCH_2CH_2OAc$	191
$BrC\equiv C(CH_2)_2CO_2H$	$Hg(O_2CCF_3)_2$	—	—	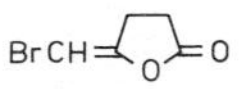 $BrCH=$⟨lactone⟩$=O$ (62) 95 : 5 Z/E	192
$ClC\equiv C(CH_2)_2CO_2H$	$Hg(O_2CCF_3)_2$	—	—	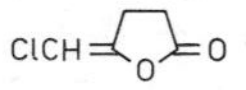 $ClCH=$⟨lactone⟩$=O$ (55)	192

Table 5.8. (continued)

Alkyne	Mercuric salt	Organomercurial(s) (% Yield)	Subsequent reactants	Product(s) (% Yield)	Ref.
$HC\equiv CC(CH_3)=CH_2$	$Hg(OAc)_2$ / RCO_2H	—	—	$H_2C=C(O_2CR)C(CH_3)=CH_2$ $R = CH_3$ (55.55), CH_2CH_3 (45.7)	193
	HgO / $trans$-$HO_2C(CH_2)_2$-$CH=CHCO_2C_2H_5$ / $BF_3\cdot Et_2O$	—	—	$trans$-$H_2C=C(CH_3)CO_2C(CH_2)_2CH=CHCO_2C_2H_5$ (with $=CH_2$ branch) (60)	194
$HC\equiv C(CH_2)_2CO_2H$	$Hg(OAc)_2$	—	—	(γ-methylene-γ-butyrolactone) (74)	195
	$Hg(O_2CCF_3)_2$	—	—	(γ-methylene-γ-butyrolactone) (81)	192
	HgO	—	—	(γ-methylene-γ-butyrolactone) (100)	196, 197
	HgO / ROH	—	—	(lactone with OR, CH_3) + (γ-methylene lactone) $R = CH_3$ 68 0 CH_2CH_3 64 0 $CH(CH_3)_2$ 0 77 $(CH_2)_3CH_3$ 42 31 $C(CH_3)_3$ 0 81	197, 198
$HC\equiv CC(CH_3)_2OH$	HgO / $HOAc$ / $BF_3\cdot Et_2O$	—	—	$CH_3COC(CH_3)_2OAc$ (49)	199
$CH_3C\equiv CCH_2O_2CCH_3$	$Hg(OAc)_2$, KCl	AcO—$C(CH_3)=C(CH_2O_2CCH_3)(HgCl)$	—	—	191
	$Hg(OAc)_2$	—	—	$CH_3COCH_2CH_2O_2CCH_3$	191
$HC\equiv C(CH_2)_3CO_2H$	$Hg(O_2CCF_3)_2$	—	—	(δ-methylene-δ-valerolactone) (80)	192
$HC\equiv C(CH_2)_3CH_3$	HgO / RCO_2H / $BF_3\cdot Et_2O$ / CH_3OH	—	—	$H_2C=C(O_2CR)(CH_2)_3CH_3$ $R = CH_3$ (31), $ClCH_2$ (68), C_6H_5 (44)	186

Table 5.8. (continued)

Alkene	Mercuric salt	Organomercurial(s) (% Yield)	Subsequent reactants	Product(s) (% Yield)	Ref.
	$Hg[C{\equiv}C(CH_2)_3CH_3]_2$ / HOAc	—	—	$H_2C{=}C(OAc)(CH_2)_3CH_3$ + $CH_3CO(CH_2)_3CH_3$ (12)	200
$CH_3CH_2C{\equiv}CCH_2CH_3$	$Hg(OAc)_2$, KCl	drawn alkene: CH_3CH_2 and AcO on one carbon, $C{=}C$, $HgCl$ and CH_2CH_3 on the other (80)	—	—	201
	$Hg(OAc)_2$ / HOAc / BF_3	—	—	$CH_3CH_2C(OAc){=}CHCH_2CH_3$ (50)	202
$HC{\equiv}C(CH_2)_4CH_3$	HgO / HOAc / Ac_2O	—	—	$H_2C{=}C(OAc)(CH_2)_4CH_3$ (34) + $CH_3CO(CH_2)_4CH_3$ (3)	203
	HgO / HOAc / $BF_3 \cdot Et_2O$ / CH_3OH	—	—	$H_2C{=}C(OAc)(CH_2)_4CH_3$ (34)	186
	$Hg(OAc)_2$ / HOAc	—	HCl or Br_2	$CX_3CO(CH_2)_4CH_3$ X = H or Br	204
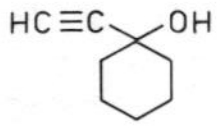 $HC{\equiv}C$–(1-hydroxycyclohexyl)	HgO / HOAc / $BF_3 \cdot Et_2O$	—	—	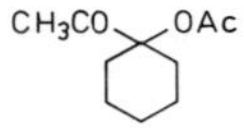 drawn cyclohexane bearing CH_3CO and OAc substituents (35)	187
$HO(CH_2)_3C{\equiv}C(CH_2)_2CO_2H$	HgO	—	—	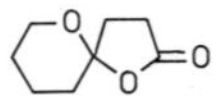 drawn spiro bicyclic lactone with ring carbonyl (=O) (73)	198
$HC{\equiv}C(CH_2)_5CH_3$	$Hg(OAc)_2$ / Ac_2O / $BF_3 \cdot Et_2O$	—	KOH	$H_2C{=}C(OAc)(CH_2)_5CH_3$ + $CH_3CO(CH_2)_5CH_3$ + $CH_3COC{\equiv}C(CH_2)_5CH_3$ 79 : 14 : 7 (89 total)	205
	$Hg(OAc)_2$ / HOAc	—	HCl or Br_2	$CX_3CO(CH_2)_5CH_3$ X = H or Br	204
$HOC(CH_3)_2C{\equiv}CC(CH_3)_2OH$	HgO / HOAc / $BF_3 \cdot Et_2O$	—	—	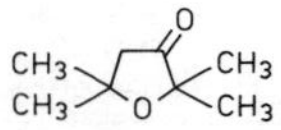 drawn tetramethyl dihydrofuranone ring (carbonyl =O; four CH_3 groups) (78)	199

Table 5.8. (continued)

Alkyne	Mercuric salt	Organomercurial(s) (% Yield)	Subsequent reactants	Product(s) (% Yield)	Ref.
$C_6H_5C\equiv CCO_2H$	$Hg(OAc)_2$ / HOAc	—	Br_2	$C_6H_5COCBr_2CO_2H$	204
$p\text{-}BrC_6H_4C\equiv CCH_3$	$Hg(OAc)_2$, KCl	$p\text{-}BrC_6H_4\!\!\diagdown\!C{=}C\!\!\diagup\!HgCl$ / AcO / CH_3 (1.95) : $p\text{-}BrC_6H_4\!\!\diagdown\!C{=}C\!\!\diagup\!OAc$ / $ClHg$ / CH_3 (1) (83 total)	—	—	206
$m\text{-}ClC_6H_4C\equiv CCH_3$	$Hg(OAc)_2$, KCl	$m\text{-}ClC_6H_4\!\!\diagdown\!C{=}C\!\!\diagup\!HgCl$ / AcO / CH_3 (1.1) : $m\text{-}ClC_6H_4\!\!\diagdown\!C{=}C\!\!\diagup\!OAc$ / $ClHg$ / CH_3 (1) (50 total)	—	—	206
$p\text{-}ClC_6H_4C\equiv CCH_3$	$Hg(OAc)_2$, KCl	$p\text{-}ClC_6H_4\!\!\diagdown\!C{=}C\!\!\diagup\!HgCl$ / AcO / CH_3 (2.9) : $p\text{-}ClC_6H_4\!\!\diagdown\!C{=}C\!\!\diagup\!OAc$ / $ClHg$ / CH_3 (1) (80 total)	—	—	206
$p\text{-}FC_6H_4C\equiv CCH_3$	$Hg(OAc)_2$, KCl	$p\text{-}FC_6H_4\!\!\diagdown\!C{=}C\!\!\diagup\!HgCl$ / AcO / CH_3 (8.7) : $p\text{-}FC_6H_4\!\!\diagdown\!C{=}C\!\!\diagup\!OAc$ / $ClHg$ / CH_3 (1) (85 total)	—	—	206
$C_6H_5C\equiv CH_3$	$Hg(OAc)_2$, KCl	$C_6H_5\!\!\diagdown\!C{=}C\!\!\diagup\!HgCl$ / AcO / CH_3 (2.85) : $C_6H_5\!\!\diagdown\!C{=}C\!\!\diagup\!OAc$ / $ClHg$ / CH_3 (1) (84 total)	—	—	206
	$Hg(OAc)_2$, KCl	$C_6H_5\!\!\diagdown\!C{=}C\!\!\diagup\!HgCl$ / AcO / CH_3 **1** (3) : $C_6H_5\!\!\diagdown\!C{=}C\!\!\diagup\!OAc$ / $ClHg$ / CH_3 (1) (96 total)	$NaBH_4$ (on **1**)	$C_6H_5\!\!\diagdown\!C{=}C\!\!\diagup\!H$ / AcO / CH_3 (3-5) : $C_6H_5\!\!\diagdown\!C{=}C\!\!\diagup\!CH_3$ / AcO / H (1)	201
	$Hg(OAc)_2$, KCl	$C_6H_5\!\!\diagdown\!C{=}C\!\!\diagup\!HgCl$ / AcO / CH_3 (3) : $C_6H_5\!\!\diagdown\!C{=}C\!\!\diagup\!OAc$ / $ClHg$ / CH_3 (1) (96 total)	HOAc, Br_2 or I_2	$C_6H_5\!\!\diagdown\!C{=}C\!\!\diagup\!X$ / AcO / CH_3 + $C_6H_5\!\!\diagdown\!C{=}C\!\!\diagup\!OAc$ / X / CH_3; X = H 3 : 1 (~100); X = Br, I — : —	207
$HC\equiv CCH_2OC_6H_5$	$Hg(OAc)_2$	$(AcOHg)_2C{=}C(OAc)CH_2OC_6H_5$?	—	—	208

Table 5.8. (continued)

Alkyne	Mercuric salt	Organomercurial(s) (% Yield)	Subsequent reactants	Product(s) (% Yield)	Ref.
$HC \equiv CCHOHC_6H_5$	$HgO / HOAc / BF_3 \cdot Et_2O$	—	—	$CH_3COCH(OAc)C_6H_5$ (50)	187
$(HC \equiv CCH_2)_2C(CO_2H)_2$	HgO	—	—	(91)	196
$HC \equiv CCH_2CH(CO_2H)CCH_3$	HgO	—	—	(98, 100)	196, 197
(cyclohexane with CO_2H and $C \equiv CH$)	HgO	—	—	(92)	196
cis, cis-$CH_3CH = CH(C \equiv C)_2CH = CHCO_2H$	$HgSO_4 / H_2SO_4$	—	—	(50) + (23) + HgCl product; R = cis-$CH_3CH = CHC \equiv C$	209
p-$CH_3C_6H_4C \equiv CCH_3$	$Hg(OAc)_2$, KCl	14 : 1	—	—	206
$C_6H_5C \equiv CCH_2CH_3$	$Hg(OAc)_2$, KCl	5 : 1 (88 total) (total 68)	—	—	201, 207
p-$CH_3OC_6H_4C \equiv CCH_3$	$Hg(OAc)_2$, KCl	(93)	—	—	206
$CH_3(CH_2)_5C \equiv CCO_2CH_3$	$Hg(OAc)_2 / HOAc$	—	Br_2	$CH_3(CH_2)_5COCBr_2CO_2CH_3$	204

428

Table 5.8. (continued)

Alkyne	Mercuric salt	Organomercurial(s) (% Yield)	Subsequent reactants	Product(s) (% Yield)	Ref.
cis-$CH_3(CH_2)_2(C{\equiv}C)_2CH{=}CHCO_2H$	$HgSO_4 / H_2SO_4$	—	—	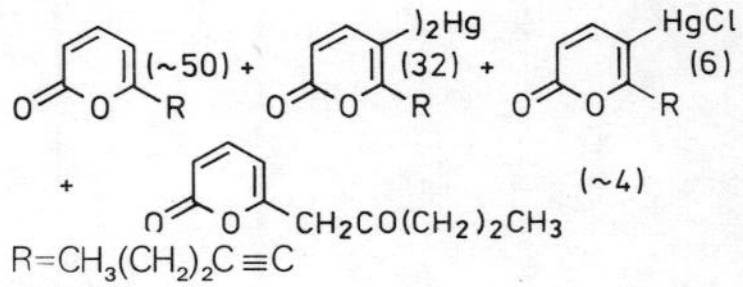 [structure] ($\sim$50) + [structure] ($)_2Hg$ (32) + [structure] HgCl (6) + [structure] $CH_2CO(CH_2)_2CH_3$ ($\sim$4), $R{=}CH_3(CH_2)_2C{\equiv}C$	209
$DC{\equiv}CCH_2CH(C_6H_5)CO_2H$	$Hg(O_2CCF_3)_2$	—	—	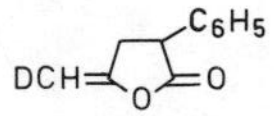 [structure] C_6H_5, $DCH{=}$ (84), 70 : 30 E/Z	192
$C_6H_5C{\equiv}CCO_2C_2H_5$	$Hg(OAc)_2 / HOAc$	—	Br_2	$C_6H_5COCBr_2CO_2C_2H_5$	204
$C_6H_5C{\equiv}C(CH_2)_2CO_2H$	HgO	—	—	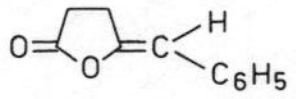 [structure] $C{=}C{<}^H_{C_6H_5}$ (83) + [structures] C_6H_5 + $C{=}C{<}^{C_6H_5}_H$ (9)	196, 197
$HC{\equiv}CCH_2CH(C_6H_5)CO_2H$	$Hg(O_2CCF_3)_2$	—	—	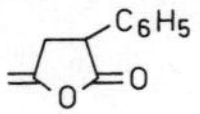 [structure] C_6H_5 (84)	192
	HgO	—	—	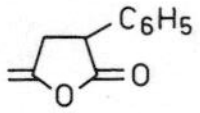 [structure] C_6H_5 (98)	196
$C_6H_5C{\equiv}CCH(CH_3)_2$	$Hg(OAc)_2$, KCl	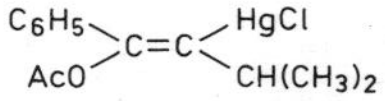 [structure] C_6H_5, AcO $C{=}C$ $HgCl$, $CH(CH_3)_2$ (52)	—	—	201
$C_6H_5C{\equiv}C(CH_2)_2CH_3$	$Hg(OAc)_2$, KCl	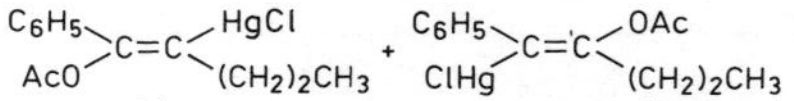 [structure] C_6H_5, AcO $C{=}C$ $HgCl$, $(CH_2)_2CH_3$ + C_6H_5, $ClHg$ $C{=}C$ OAc, $(CH_2)_2CH_3$ 11 : 1 (74 total)	—	—	201, 207

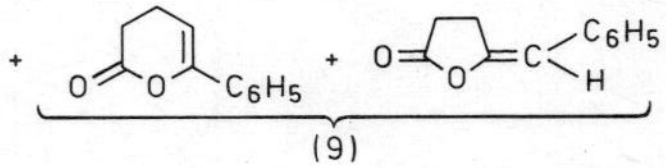

Table 5.8. (continued)

Alkyne	Mercuric salt	Organomercurial(s) (% Yield)	Subsequent reactants	Product(s) (% Yield)	Ref.
$HC{\equiv}CCH_2N(COCH_3)CH{-}$ $(CO_2H)CH_2CH(CH_3)_2$	HgO	—	—	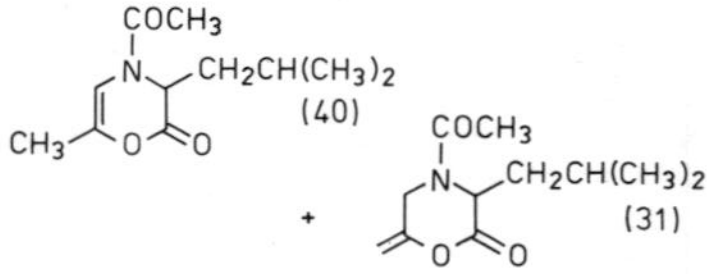 dihydrooxazinone with $COCH_3$, $CH_2CH(CH_3)_2$, CH_3 (40) + dihydrooxazinone with $COCH_3$, $CH_2CH(CH_3)_2$, exocyclic $=CH_2$ (31)	210
$HC{\equiv}C(CH_2)_8CO_2H$	$Hg(OAc)_2/HOAc$	—	HCl or Br_2	$CX_3CO(CH_2)_8CO_2H$, $X = H(100)$ or Br	204, 211
$CH_3C{\equiv}C(CH_2)_7CO_2H$	$Hg(OAc)_2/HOAc$	—	HCl	$CH_3CO(CH_2)_8CO_2H$ + $CH_3CH_2CO(CH_2)_7CO_2H$	211
$C_6H_5C{\equiv}CCH{=}C(CO_2H)_2$	HgO	—	—	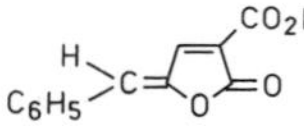 furanone: H, C_6H_5, CO_2H substituents (88)	196, 197
$HC{\equiv}CCH_2N(COC_6H_5)CH_2CO_2H$	HgO	—	—	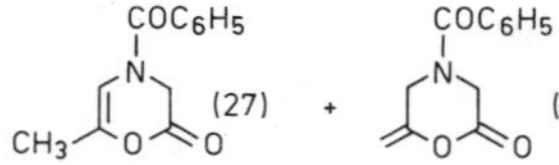 dihydrooxazinone with COC_6H_5, CH_3 (27) + oxazinone with COC_6H_5, exocyclic $=CH_2$ (36)	210
$CH_3C{\equiv}CCH_2CH(C_6H_5)CO_2H$	$Hg(O_2CCF_3)_2$	—	—	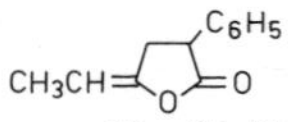 $CH_3CH{=}$ butyrolactone with C_6H_5 (86); 36 : 64 E/Z	192
$C_6H_5C{\equiv}C(CH_2)_3CH_3$	$Hg(OAc)_2$, KCl	$\underset{AcO}{\overset{C_6H_5}{>}}C{=}C\underset{(CH_2)_3CH_3}{\overset{HgCl}{<}}$ + $\underset{ClHg}{\overset{C_6H_5}{>}}C{=}C\underset{(CH_2)_3CH_3}{\overset{OAc}{<}}$ 16.5 : 11 : 1 : 1 (44 total) (74 total)	—	—	201, 207
$CH_3C{\equiv}C(CH_2)_7CO_2C_2H_5$	$HgO/HOAc$	—	HCl	$CH_3CO(CH_2)_8CO_2H$ + $CH_3CH_2CO(CH_2)_7CO_2H$	211
$C_6H_5C{\equiv}CC_6H_5$	$Hg(OAc)_2$	$\underset{AcO}{\overset{C_6H_5}{>}}C{=}C\underset{HgOAc}{\overset{C_6H_5}{<}}$	—	—	189, 212

Table 5.8. (continued)

Alkyne	Mercuric salt	Organomercurial(s) (% Yield)	Subsequent reactants	Product(s) (% Yield)	Ref.
	$Hg(OAc)_2$, Cl^-	$\underset{AcO}{\overset{C_6H_5}{>}}C{=}C\underset{HgCl}{\overset{C_6H_5}{<}}$ (80)	I_2	$\underset{AcO}{\overset{C_6H_5}{>}}C{=}C\underset{I}{\overset{C_6H_5}{<}}$	201, 213
	$Hg(OAc)_2$, NaCl	$\underset{AcO}{\overset{C_6H_5}{>}}C{=}C\underset{HgCl}{\overset{C_6H_5}{<}}$ + $\underset{AcO}{\overset{C_6H_5}{>}}C{=}C\underset{C_6H_5}{\overset{HgCl}{<}}$	I_2	$\underset{AcO}{\overset{C_6H_5}{>}}C{=}C\underset{I}{\overset{C_6H_5}{<}}$ + $\underset{AcO}{\overset{C_6H_5}{>}}C{=}C\underset{C_6H_5}{\overset{I}{<}}$	214
$HC{\equiv}CCH_2N(COCH_3)CH(CO_2H){-}CH_2C_6H_5$	HgO	—	—	[morpholinone: N-COCH3, CH2C6H5, CH3] (35) + [N-COCH3, CH2C6H5] (24)	210
$(CH_3)_3SiC{\equiv}CCH_2CH(C_6H_5){-}CO_2H$	$Hg(O_2CCF_3)_2$	—	—	$(CH_3)_3SiCH{=}$ [lactone, C6H5] (62) 45 : 55 E/Z	192
$HC{\equiv}CCH_2N(COC_6H_5)CH{-}(CO_2H)CH_2CH(CH_3)_2$	HgO	—	—	[N-COC6H5, CH2CH(CH3)2, CH3] (39) + [N-COC6H5, CH2CH(CH3)2] (35)	210
$Z{-}C_6H_5C{\equiv}CCH{=}C(C_6H_5)CO_2H$	HgO	—	—	[furanone, C6H5, C6H5, H] (72)	196
$CH_3(CH_2)_7C{\equiv}C(CH_2)_7CO_2H$	HgO / HOAc	$CH_3(CH_2)_7\overset{AcOHg}{\underset{}{C}}{=}\overset{OAc}{\underset{}{C}}(CH_2)_7CO_2H$ + $CH_3(CH_2)_7\overset{AcO}{\underset{}{C}}{=}\overset{HgOAc}{\underset{}{C}}(CH_2)_7CO_2H$	HCl	$CH_3(CH_2)_8CO(CH_2)_7CO_2H$ + $CH_3(CH_2)_7CO(CH_2)_8CO_2H$ (95 total)	215
$HC{\equiv}CCH_2N(COC_6H_5)CH{-}(CO_2H)CH_2C_6H_5$	HgO	—	—	[N-COC6H5, CH2C6H5, CH3] (32) + [N-COC6H5, CH2C6H5] (12)	210

Table 5.8. (continued)

Alkyne	Mercury reagent	Organomercurial(s) (% Yield)	Subsequent reactants	Product(s) (% Yield)	Ref.
$(C_6H_5C{\equiv}CCH_2)_2CHCO_2H$	HgO	—	—	(76) 9 : 1 +	196
(steroid, $C{\equiv}CH$, HO..., HO...)	HgO / HOAc / Ac$_2$O / BF$_3\cdot$Et$_2$O	—	—	(steroid, $C{=}CH_2$, OAc, HO..., HO...) ?	216
$CH_3(CH_2)_7C{\equiv}C(CH_2)_{11}CO_2H$	Hg(OAc)$_2$ / HOAc	—	HCl	$CH_3(CH_2)_7CO(CH_2)_{12}CO_2H$	217
(steroid, $C{\equiv}CH$, HO..., AcO...)	HgO / HOAc / Ac$_2$O / BF$_3\cdot$Et$_2$O	—	—	(steroid, $C{=}CH_2$, OAc, HO..., AcO...) ?	216

V. Acyloxymercuration

The reaction of terminal acetylenes with carboxylic acids and mercury catalysts also provides a simple approach to vinyl esters (Eq. 92). Mercury

$$RC \equiv CH + R'CO_2H \xrightarrow{Hg(II)} RC(O_2CR')=CH_2 \tag{92}$$

catalysts which have been employed in this reaction [220] are $Hg(OAc)_2$ [185, 193], $HgSO_4$ [184, 185, 188], $Hg_3(PO_4)_2$ [188], $Hg(O_3SC_6H_5)_2$ [188], $Hg(OAc)_2/Ac_2O/BF_3 \cdot Et_2O$ [205], $HgO/R'CO_2H$ [185], $Hg(OAc)_2/Ac_2O/HOAc$ [203], and $HgO/R'CO_2H/BF_3 \cdot Et_2O/CH_3OH$ [186, 194]. Besides acetic acid; formic acid, various haloacetic acids, propionic acid, butyric acid, benzoic acid and even the following carboxylic acid have been employed in this reaction (Eq. 93) [194]. Acetylene, simple alkyl acetylenes and conju-

$$H_2C=C(CH_3)C \equiv CH + HO_2C(CH_2)_2CH=CHCO_2C_2H_5 \longrightarrow H_2C=C(CH_3)-C(O_2C(CH_2)_2CH=CHCO_2C_2H_5)=CH_2 \tag{93}$$

gated enynes all react in the desired fashion. However, propargylic alcohols afford acetoxyketones rather than the anticipated enol ester (Eq. 94) [187, 191, 199, 216].

$$RCH(OH)C \equiv CH \xrightarrow[HOAc]{Hg(OAc)_2} RCH(AcO)CCH_3(O) \tag{94}$$

With many internal acetylenes, one can frequently isolate simple acyloxymercuration products (Eq. 95). 2-Butyne forms predominantly the corresponding trans adduct, but prolonged reaction times provide substantial amounts of the cis isomer [189, 190, 212]. The same holds true for 3-hexyne

$$RC \equiv CR + Hg(OAc)_2 \longrightarrow RC(AcO)=CR(HgOAc) \tag{95}$$

[201, 221]. 3-Hexyne when heated with acetic acid, mercuric acetate and boron trifluoride or perchloric acid affords 3-acetoxy-3-hexene of unknown stereochemistry [202]. On the other hand, acyloxymercuration of diphenylacetylene affords predominantly the cis isomer, although both isomers can be isolated under appropriate conditions [189, 212–214]. Iodination of these mercurials yields the corresponding vinyl iodides [213, 214]. Treatment of these acetoxymercurials with mercuric acetate and catalytic amounts of palladium acetate provides a novel approach to ene diacetates (Eq. 96) [221].

$$RC \equiv CR \longrightarrow (AcO)(R)C=C(R)(HgOAc) \xrightarrow[cat. Pd(OAc)_2]{Hg(OAc)_2} (AcO)(R)C=C(R)(OAc) \tag{96}$$

Unsymmetrical internal alkynes tend to give mixtures of the two possible regioisomers. However, 2-butynyl acetate appears to be an exception (Eq. 97) [191]. The stereochemistry of this adduct has not been established.

$$CH_3C\equiv CCH_2OAc \longrightarrow CH_3\overset{AcO}{\underset{}{C}}=\overset{HgCl}{\underset{}{C}}CH_2OAc \tag{97}$$

With aryl alkyl acetylenes a regioisomeric mixture of trans adducts is formed, the ratio being dependent on the substituents on the arene and the nature of the alkyl group (Eq. 98) [201, 206, 207]. Note that more highly

X	R	ratio 1/2
H	n-Bu	16.5
H	n-Pr	11
H	Et	5
H	Me	2.85
H	i-Pr	0
m-Cl	i-Pr	1.1
p-Br	i-Pr	1.95
p-Cl	i-Pr	2.9
p-F	i-Pr	8.7
p-Me	i-Pr	14
p-OMe	i-Pr	100

$$(98)$$

branched alkyl acetylenes such as isopropyl phenyl acetylene afford only the vinylmercurial with the mercury next to the aromatic ring. Treatment of these acetoxymercurials with acetic acid or sodium borohydride provides the corresponding enol acetates, while bromination or iodination affords the corresponding vinyl halides [201, 207]. All but the sodium borohydride reduction appear to be stereospecific.

While 9-octadecynoic acid is reported to give a regioisomeric mixture of acetoxymercurials of unknown stereochemistry which react with concentrated hydrochloric acid to afford the corresponding ketones (Eq. 99) [215], other acetylenic acids and esters have been reported to give dimercurated

$$CH_3(CH_2)_7C\equiv C(CH_2)_7CO_2H \longrightarrow \underset{AcO}{\overset{CH_3(CH_2)_7}{}}C=C\underset{(CH_2)_7CO_2H}{\overset{HgOAc}{}} + \underset{AcOHg}{\overset{CH_3(CH_2)_7}{}}C=C\underset{(CH_2)_7CO_2H}{\overset{OAc}{}}$$

$$(99)$$

species whose structures have never been firmly established (Eq. 100) [204, 211, 217]. These are most likely mercurated ketones. Upon halogenation they afford haloketones.

$$RC\equiv CR \xrightarrow[HOAc]{Hg(OAc)_2} \xrightarrow{H_2O} \underset{HgOAc}{\overset{O\ \ HgOAc}{RC-CR}} \quad or \quad \underset{HgOAc}{\overset{OHgOAc}{RC=CR}} \tag{100}$$

V. Acyloxymercuration

A number of alkynoic acids have been cyclized to unsaturated lactones using mercuric salts as catalysts. Simple 4-pentynoic acids afford γ-methylene butyrolactones in good yield using mercuric oxide [196, 197], mercuric acetate or mercuric trifluoroacetate [192, 195] (Eq. 101). Mixtures of the E

$$RC \equiv CCH_2CHR'CO_2H \longrightarrow \qquad (101)$$

and Z isomers usually result. In the presence of primary alcohols, the reaction with mercuric oxide affords the corresponding alcohol addition products (Eqs. 102, 103) [222]. Additional unsaturation in the alkynoic acid

$$HC \equiv C(CH_2)_2CO_2H \xrightarrow[ROH]{HgO} \qquad (102)$$

$$HO(CH_2)_3C \equiv C(CH_2)_2CO_2H \xrightarrow{HgO} \qquad (103)$$

$$\xrightarrow{HgO} \qquad R = CO_2H, \ C_6H_5 \qquad (104)$$

presents no problems in these lactonization reactions (Eq. 104) [196]. Cyclization to six-membered ring lactones has also been observed (Eqs. 105, 106).

$$C_6H_5C \equiv CCH_2CHRCO_2H \xrightarrow{HgO} \quad \underset{major}{} + \underset{minor}{} \qquad (105) \ [196, 197]$$

$$cis\text{-}RC \equiv CC \equiv CCH = CHCO_2H \xrightarrow{HgSO_4} \qquad (106) \ [209]$$

$$R = n\text{-}C_3H_7, \ cis\text{-}CH_3CH = CH$$

5-Hexynoic acid has also been cyclized to δ-methylene valerolactone (Eq. 107) [192]. Heterocycles have been prepared in this fashion (Eq. 108)

$$HC \equiv C(CH_2)_3CO_2H \xrightarrow{Hg(O_2CCF_3)_2} \qquad (107)$$

434

[210]. Note here that both exocyclic and endocyclic enol esters are formed depending on the substitution pattern of the starting material.

$$HC\equiv CCH_2N(COR)CHR'CO_2H \xrightarrow{HgO} \qquad\qquad\qquad\qquad \tag{108}$$

Presumably, these reactions proceed via intramolecular acyloxymercuration and subsequent protonolysis. Consistent with this is the observation noted above (Eq. 106) that vinylmercurials have actually been isolated from one reaction, and the fact that mercuric salts can be employed in only catalytic amounts.

Finally, there are two examples of diynes being subjected to the usual acyloxymercuration conditions (Eqs. 109, 110). Only carbomercuration products are observed, however.

$$\left[CH_3-\!\!\!\bigcirc\!\!\!-OCH_2C\equiv C\right]_2 \xrightarrow[\substack{HOAc\\H_2SO_4}]{Hg(OAc)_2} \qquad\qquad \tag{109}\ [223]$$

$$\tag{110}\ [224]$$

References

1. Knunyants, I. L., Pervova, E. Ya., Tyuleneva, V. V.: Izv. Akad. Nauk SSSR, Otdel. Khim. Nauk, 843 (1956); Bull. Acad. Sci. USSR, Div. Chem. Sci. 863 (1956).
2. Hugel, G., Hibou, J.: Chim. Ind., *21*, 296 (1929); Chem. Abstr., *23*, 3898 (1929).
3. Ichikawa, K., Ouchi, H., Araki, S.: J. Am. Chem. Soc., *82*, 3880 (1960).
4. Kreevoy, M. M., Bodem, G. B.: J. Org. Chem., *27*, 4539 (1962).
5. Convery, R. J.: U.S. Patent 3,031,484 (1962); Chem. Abstr., *57*, 13803f (1962).
6. Nefedov, B. K., Sergeeva, N. S., Éidus, Ya. T.: Izv. Akad. Nauk SSSR, Ser. Khim., 2497 (1972); Bull. Acad. Sci. USSR, Div. Chem. Sci., 2429 (1972).
7. Ichikawa, K., Itoh, O., Kawamura, T.: Bull. Chem. Soc. Japan, *41*, 1240 (1968).
8. Knunyants, I. L., Pervova, E. Ya., Tyuleneva, V. V.: Izv. Akad. Nauk SSSR, Otdel. Khim. Nauk, 88 (1961); Bull. Acad. Sci. USSR, Div. Chem. Sci., 77 (1961).

9. Ichikawa, K., Nishimura, K., Takayama, S.: J. Org. Chem., *30*, 1593 (1965).
10. Lutsenko, I. F., Khomutov, R. M., Eliseeva, L. V.: Izv. Akad. Nauk SSSR, Otdel. Khim. Nauk, 181 (1956); Bull. Acad. Sci. USSR, Div. Chem. Sci., 173 (1956).
11. Kreevoy, M. M., Turner, M. A.: J. Org. Chem., *29*, 1639 (1964).
12. Lüssi, H.: Helv. Chim. Acta, *49*, 1684 (1966).
13. Slinckx, G., Smets, G.: Tetrahedron, *22*, 3163 (1966).
14. Hopff, H., Osman, M. A.: Tetrahedron, *24*, 2205 (1968).
15. Hopff, H., Osman, M. A.: Tetrahedron, *24*, 3887 (1968).
16. Spengler, G., Wilderotter, M., Trommer, T.: Brennstoff-Chemie, *45*, 182 (1964).
17. Soderquist, J. A., Thompson, K. L.: J. Organometal. Chem., *159*, 237 (1978).
18. Voronkov, M. G., Chernov, N. F., Kalikhman, I. D.: Dokl. Akad. Nauk SSSR, *233*, 361 (1977); Proc. Acad. Sci. USSR, Chem. Sec., *233*, 138 (1977).
19. Bloodworth, A. J., Courtneidge, J. L.: J. Chem. Soc., Perkin I, 3258 (1981).
20. Shaborov, Yu. S., Mochalov, S. S., Oretskaya, T. S., Karpova, V. V.: J. Organometal. Chem., *150*, 7 (1978).
21. Popova, M. N., Stepanov, D. E., Yakovenko, V. S., Stepuk, E. S., Komarovskaya, I. A., Donskikh, V.: Dokl. Vses. Konf. Khim. Atsetilena, 4th, *2*, 360 (1972); Chem. Abstr., *79*, 78910x (1973).
22. Lethbridge, A., Norman, R. O. C., Thomas, C. B.: J. Chem. Soc., Perkin I, 231 (1975).
23. Wright, G. F.: J. Am. Chem. Soc., *57*, 1993 (1935).
24. Lethbridge, A., Norman, R. O. C., Thomas, C. B.: J. Chem. Soc., Perkin I, 2763 (1973).
25. Richter, R. F., Philips, C., Bach, R. D.: Tetrahedron Lett., 4327 (1972).
26. Wolfe, S., Campbell, P. G. C.: Can. J. Chem., *43*, 1184 (1965).
27. Wolfe, S., Campbell, P. G. C., Palmer, G. E.: Tetrahedron Lett., 4203 (1966).
28. Lermontov, S. A., Belikova, N. A., Skornyakova, T. G., Pekhk, T. I., Lippmaa, É. T., Platé, A. F.: Zh. Org. Khim., *16*, 2322 (1980); J. Org. Chem. USSR, *16*, 1982 (1980).
29. Lethbridge, A., Norman, R. O. C., Thomas, C. B.: J. Chem. Soc., Perkin I, 2465 (1975).
30. Sirotkina, E. E., Filimonov, V. D., Sazonova, Z. P.: Zh. Org. Khim., *9*, 381 (1973); J. Org. Chem. USSR, *9*, 385 (1973).
31. Zhdanov, Yu. A., Korol'chenko, G. A., Kubasskaya, L. A.: Dokl. Akad. Nauk SSSR, *128*, 1185 (1959); Proc. Acad. Sci. USSR, Chem. Sec., *128*, 887 (1959).
32. Zhdanov, Yu. A., Korol'chenko, G. A., Kubasskaya, L. A., Krivoruchko, R. M.: Dokl. Akad. Nauk SSSR, *129*, 1049 (1959); Proc. Acad. Sci. USSR, Chem. Sec., *129*, 1101 (1959).
33. Corey, E. J., Pearce, H. L.: Tetrahedron Lett., 1823 (1980).
34. Tubul, A., Ucciani, E., Naudet, M.: Bull. Soc. Chim. France, 464 (1967); Chem. Abstr., *67*, 2573w (1967).
35. Tubul, A., Ucciani, E., Naudet, M.: Bull. Soc. Chim. France, 2429 (1968); Chem. Abstr., *69*, 86299p (1968).
36. Gunstone, F. D., Inglis, R. P.: Chem. Phys. Lipids, *10*, 73 (1973).
37. Gunstone, F. D., Inglis, R. P.: J. Chem. Soc. D, Chem. Commun., 877 (1970).
38. Spengler, G., Schäff, R., Frömmel, H., Stein, W.: Angew. Chem., *61*, 308 (1949).
39. Spengler, G., Weber, A.: Brennstoff-Chemie, *43*, 234 (1962).
40. Brown, H. C., Rei, M.-H., Liu, K.-T.: J. Am. Chem. Soc., *92*, 1760 (1970).
41. Brown, H. C., Rei, M.-H.: J. Chem. Soc. D, Chem. Commun., 1296 (1969).

42. Summerbell, R. K., Kalb, G. H., Graham, E. S., Allred, A. L.: J. Org. Chem., *27*, 4461 (1962).
43. Georgoulis, C., Valéry, J.-M.: Bull. Soc. Chim. France, 1431 (1975).
44. Shirafuji, T., Nozaki, H.: Tetrahedron, *29*, 77 (1973).
45. Shirafuji, T., Yamamoto, Y., Nozaki, H.: Tetrahedron Lett., 4713 (1971).
46. Brook, A. G., Wright, G. F.: Can. J. Res., *28B*, 623 (1950).
47. Johnson,. M. R., Rickborn, B.: J. Org. Chem., *34*, 2781 (1969).
48. Jernow, J. L., Gray, D., Closson, W. D.: J. Org. Chem., *36*, 3511 (1971).
49. Pasto, D. J., Gontarz, J. A.: J. Am. Chem. Soc., *93*, 6909 (1971).
50. Sokolov, V. I., Troitskaya, L. L., Reutov, O. A.: J. Organometal. Chem., *17*, 323 (1969).
51. Bach, R. D., Richter, R. F.: Tetrahedron Lett., 3915 (1971).
52. Sokolov, V. I., Troitskaya, L. L., Reutov, O. A.: Dokl. Akad. Nauk SSSR, *166*, 136 (1966); Bull. Acad. Sci. USSR, Chem. Sec., *166*, 45 (1966).
53. Rao, P. S., Seshadri, T. R.: Proc. Ind. Acad. Sci. A, *4*, 630 (1936); Chem. Zentralbl. II, 3457 (1937).
54. Takiura, K., Honda, S.: Carbohydr. Res., *23*, 369 (1972).
55. Lácová, M., Volná, F., Greiplová, E., Hrnciar, P.: Chem. Zvesti, *24*, 309 (1970); Chem. Abstr., *74*, 100188p (1971).
56. Dunkelbaum, E.: Israel J. Chem., *11*, 557 (1973).
57. Sokolov, V. I., Troitskaya, L. L., Reutov, O. A.: Izv. Akad. Nauk SSSR, Ser. Khim., 2648 (1970); Bull. Acad. Sci. USSR, Div. Chem. Sci., 2501 (1970).
58. Waters, W. L., Traylor, T. G., Factor, A.: J. Org. Chem., *38*, 2306 (1973).
59. Bond, F. T.: J. Am. Chem. Soc., *90*, 5326 (1968).
60. Tidwell, T. T., Traylor, T. G.: J. Org. Chem., *33*, 2614 (1968).
61. Wiberg, K. B., Chen, W.: J. Org. Chem., *37*, 3235 (1972).
62. Aberchrombie, M. J., Rodgman, A., Bharucha, K. R., Wright, G. F.: Can. J. Chem., *37*, 1328 (1959).
63. Factor, A., Traylor, T. G.: J. Org. Chem., *33*, 2607 (1968).
64. Belikova, N. A., Lermontov, S. A., Pekhk, T. I., Lippmaa, É. T., Platé, A. F.: Zh. Org. Khim., *14*, 2273 (1978); J. Org. Chem. USSR, *14*, 2101 (1978).
65. Bach, R. D., Richter, R. F.: J. Am. Chem. Soc., *94*, 4747 (1972).
66. Traylor, T. G.: J. Am. Chem. Soc., *86*, 244 (1964).
67. Brown, H. C., Kawakami, J. H.: J. Am. Chem. Soc., *92*, 201 (1970).
68. Brown, H. C., Liu, K.-T.: J. Am. Chem. Soc., *93*, 7335 (1971).
69. Brown, H. C., Kawakami, J. H., Misumi, S.: J. Org. Chem., *35*, 1360 (1970).
70. Yur'ev, Yu. K., Zefirov, N. S., Shaiderova, L. P.: Zh. Obshch. Khim., *33*, 818 (1963); J. Gen. Chem. USSR, *33*, 803 (1963).
71. Yur'ev, Yu. K., Zefirov, N. S., Prikazchikova, D. P.: Zh. Obshch. Khim., *33*, 1793 (1963); J. Gen. Chem. USSR, *33*, 1747 (1963).
72. Zefirov, N. S., Kadzyaukas, P. P., Yur'ev, Yu. K., Bazanova, V. N.: Zh. Obshch. Khim., *35*, 1499 (1965); J. Gen. Chem. USSR, *35*, 1501 (1965).
73. Giese, B.: Chem. Ber., *108*, 2998 (1975).
74. Sasaki, T., Kanematsu, K., Kondo, A., Nishitani, Y.: J. Org. Chem., *39*, 3569 (1974).
75. Pande, K. C., Winstein, S.: Tetrahedron Lett., 3393 (1964).
76. Tanner, D. D., Van Bostelen, P. B., Lai, M.: Can. J. Chem., *54*, 2004 (1976).
77. Cristol, S. J., Perry, J. S., Jr., Beckley, R. S.: J. Org. Chem., *41*, 1912 (1976).
78. Jensen, F. R., Miller, J. J., Cristol, S. J., Beckley, R. S.: J. Org. Chem., *37*, 4341 (1972).

79. Sokolov, V. I.: Izv. Akad. Nauk SSSR, Ser. Khim., 1285 (1968); Bull. Acad. Sci. USSR, Div. Chem. Sci., 1213 (1968).
80. Halfpenny, J., Small, R. W. H., Thorpe, F. G.: Acta Cryst. B, *34*, 3077 (1978).
81. Barron, P. F., Doddrell, D., Kitching, W.: J. Organometal. Chem., *132*, 351 (1977).
82. Tobler, E., Foster, D. J.: Helv. Chim. Acta, *48*, 366 (1965).
83. Brown, H. C., Kawakami, J. H.: J. Am. Chem. Soc., *95*, 8665 (1973).
84. Kurbanov, M., Semenovsky, A. V., Smit, W. A., Shmelev, L. V., Kucherov, V. F.: Tetrahedron Lett., 2175 (1972).
85. Giese, B., Gantert, S., Schulz, A.: Tetrahedron Lett., 3583 (1974).
86. Traylor, T. G., Baker, A. W.: J. Am. Chem. Soc., *85*, 2746 (1963).
87. Coxon, J. M., Hartshorn, M. P., Lewis, A. J.: Tetrahedron, *26*, 3755 (1970).
88. Takaishi, N., Fujikura, Y., Inamoto, Y.: J. Org. Chem., *40*, 3767 (1975).
89. Julia, M., Blasioli, C.: Bull. Soc. Chim. France, 1941 (1976).
90. Yoshioka, H., Takasaki, K., Kobayashi, M., Matsumoto, T.: Tetrahedron Lett., 3489 (1979).
91. Martin, S. F., Chou, T.: Tetrahedron Lett., 1943 (1978).
92. Fráter, G.: Tetrahedron Lett., 425 (1981).
93. Schuda, P. F., Heimann, M. R.: Tetrahedron Lett., *24*, 4267 (1983).
94. Polishchuk, V. R., German, L. S., Knunyants, I. L.: Izv. Akad. Nauk SSSR, Ser. Khim., 2024 (1971); Bull. Acad. Sci. USSR, Div. Chem. Sci., 1908 (1971).
95. Arzoumanian, H., Metzger, J.: J. Organometal. Chem., *57*, C1 (1973).
96. Herrmann, W. O.: U.S. Patent 2,079,068 (1937?); Chem. Abstr., *31*, 4346[3] (1937).
97. Toussaint, W. J., McDowell, L. G., Jr.: U.S. Patent 2,299,862 (1943?); Chem. Abstr., *37*, 1722[9] (1943).
98. Asahara, T., Tomita, M.: Yushi Kagaku Kyôkaishi, *1*, 76 (1952); Chem. Abstr., *47*, 3232c (1953).
99. Herrmann, W. O., Haehnel, W.: U.S. Patent 2,245,131 (1941?); Chem. Abstr., *35*, 5908[6] (1941).
100. Morlyan, N. M., Muradyan, A. G., Kirakosyan, D. E.: Metody Poluch. Khim. Reactivov Prep. (23), 40 (1971); Chem. Abstr., *79*, 65749b (1973).
101. Freidlin, G. N., Adamov, A. A., Pershenkova, L. A.: Probl. Organ. Sinteza, Akad. Nauk SSSR, Otd. Obshch. i Tekhn. Khim., 37 (1965); Chem. Abstr., *64*, 11081a (1966).
102. Skvortsova, G. G., An, V. V., Mansurov, Yu. A., Voronov, V. K., Ratovskii, G. V., Frolov, Yu. L.: Khim. Geterotsikl. Soedin., 1443 (1972); Chem. Abstr., *78*, 43167n (1973).
103. Adelman, R. L.: J. Org. Chem., *14*, 1057 (1949).
104. Swern, D., Jordan, E. F., Jr.: Org. Syn., *30*, 106 (1950).
105. Swern, D., Jordan, E. F., Jr.: Org. Syn., Coll. Vol. 4, 977 (1963).
106. Swern, D., Jordan, E. F., Jr.: J. Am. Chem. Soc., *70*, 2334 (1948).
107. Ham, G. E., Ringwald, E. L.: J. Polym. Sci., *8*, 91 (1952).
108. Slinckx, G., Smets, G.: Tetrahedron, *23*, 1395 (1967).
109. Hopff, H., Lüssi, H.: Makromolek. Chem., *18/19*, 227 (1955).
110. Swern, D., Billen, G. N., Knight, H. B.: J. Am. Chem. Soc., *69*, 2439 (1947).
111. Dickey, J. B., Stanin, T. E.: U.S. Patent 2,646,437 (1953); Chem. Abstr., *48*, 7052d (1954).
112. Takaoka, K., Izumisawa, Y.: Kogyo Kagaku Zasshi, *67*, 1244 (1964); Chem. Abstr., *61*, 11887g (1964).

113. Takasago, M., Takaoka, K., Seno, M.: Yukagaku, *20*, 415 (1971); Chem. Abstr., *75*, 88214e (1971).
114. Olah, G. A., Clifford, P. R.: J. Am. Chem. Soc., *93*, 2320 (1971).
115. Spengler, G., Trommer, T.: Brennstoff-Chemie, *48*, 19 (1967).
116. Spengler, G., Wörle, R.: Brennstoff-Chemie, *50*, 14 (1969).
117. El Seoud, O. A., do Amaral, A. T., Moura Campos, M., do Amaral, L.: J. Org. Chem., *39*, 1915 (1974).
118. Rowland, R. L., Perry, W. L., Friedman, H. L.: J. Am. Chem. Soc., *73*, 1040 (1951).
119. Arakelyan, S. V., Dangyan, M. T., Avetisyan, A. A.: Izv. Akad. Nauk Arm. SSSR, Khim. Nauki, *15*, 435 (1962); Chem. Abstr., *59*, 5186a (1963).
120. Arakelyan, S. V., Rashidyan, L. G., Dangyan, M. T.: Izv. Akad. Nauk Arm. SSSR, Khim. Nauki, *17*, 173 (1964); Chem. Abstr., *61*, 8331b (1964).
121. Arakelyan, S. V., Gareyan, L. S., Titanyan, S. O., Dangyan, M. T.: Khim. Geterosikl. Soedin., 13 (1970); Chem. Abstr., *77*, 88614b (1972).
122. Henbest, H. B., Nicholls, B.: J. Chem. Soc., 227 (1959).
123. Ford, D. N., Kitching, W., Wells, P. R.: Austral. J. Chem., *22*, 1157 (1969).
124. Malaiyandi, M., Wright, G. F.: Can. J. Chem., *41*, 1493 (1963).
125. Jensen, F. R., Miller, J. J.: Tetrahedron Lett., 4861 (1966).
126. Overman, L. E., Campbell, C. B.: J. Org. Chem., *41*, 3338 (1976).
127. McNeely, K. H., Rodgman, A., Wright, G. F.: J. Org. Chem., *20*, 714 (1955).
128. Christol, H., Plénat, F., Revel, J.: Bull. Soc. Chim. France, 4537 (1971).
129. Chiu, D. D. K., Wright, G. F.: Can. J. Chem., *37*, 1425 (1959).
130. do Amaral, A. T., El Seoud, O. A., do Amaral, L.: J. Org. Chem., *40*, 2534 (1975).
131. Prikazchikova, L. P. quoted in Zefirov, N. S.: Russ. Chem. Rev., *34*, 527 (1965).
132. Zefirov, N. S., Prikazchikova, L. P., Bondareva, M. A., Yur'ev, Yu. K.: Zh. Obshch. Khim., *33*, 4026 (1963); J. Gen. Chem. USSR, *33*, 3967 (1963).
133. Zefirov, N. S., Prikazchikova, L. P., Yur'ev, Yu. K.: Zh. Obshch. Khim., *35*, 822 (1965); J. Gen. Chem. USSR, *35*, 825 (1965).
134. Hagihara, Y., Iritani, N.: Japan. Patent 4,222 (1955); Chem. Abstr., *51*, 16520i (1957).
135. Giese, B.: Chem. Ber., *108*, 2978 (1975).
136. Rowland, R. L., Kluchesky, E. F.: J. Am. Chem. Soc., *73*, 5490 (1951).
137. Brown, H. C., Lynch, G. J.: J. Org. Chem., *46*, 531 (1981).
138. Linn, W. S., Waters, W. L., Caserio, M. C.: J. Am. Chem. Soc., *92*, 4018 (1970).
139. Gómez-Aranda, V., Barluenga, J., Yus, M., Asensio, G.: Synthesis, 806 (1974).
140. Julia, M., Colomer Gasquez, E.: Bull. Soc. Chim. France, 1796 (1973); Chem. Abstr., *79*, 104790e (1973).
141. Gómez-Aranda, V., Barluenga-Mur, J., Yus-Astiz, M.: Rev. Acad. Cienc. Exactas, Fis.-Quim. Natur. Zaragoza, *28*, 225 (1973); Chem. Abstr., *80*, 83164g (1974).
142. Pasto, D. J., Gontarz, J. A.: J. Am. Chem. Soc., *91*, 719 (1969).
143. Vedejs, E., Salomon, M. F.: J. Org. Chem., *37*, 2075 (1972).
144. Whitesides, G. M., San Fillipo, Jr., J.: J. Am. Chem. Soc., *92*, 6611 (1970).
145. Cope, A. C., Nelson, N. A., Smith, D. S.: J. Am. Chem. Soc., *76*, 1100 (1954).
146. Müller, E.: Chem. Ber., *109*, 3793 (1976).
147. Bloodworth, A. J., Khan, J. A., Loveitt, M. E.: J. Chem. Soc., Perkin I, 621 (1981).
148. Julia, M., Colomer, E.: An. Quim., *67*, 199 (1971).
149. Franz, H. J., Höbold, W., Höhn, R., Müller-Hagen, G., Müller, R., Pritzkow, W., Schmidt, H.: J. Prakt. Chem., *312*, 622 (1970).

V. Acyloxymercuration

150. Sakai, M.: Tetrahedron Lett., 347 (1973).
151. Uemura, S., Miyoshi, H., Okano, M., Morishima, I., Inubushi, T.: J. Organometal. Chem., *171*, 131 (1979).
152. Gray, G. A., Jackson, W. R., Chambers, V. M. A.: J. Chem. Soc. C, 200 (1971).
153. Belikova, N. A., Lermontov, S. A., Pekhk, T. I., Lippmaa, É. T., Platé, A. F.: Zh. Org. Khim., *14*, 884 (1978); J. Org. Chem. USSR, *14*, 823 (1978).
154. Belikova, N. A., Lermontov, S. A., Skornyakova, T. G., Pekhk, T. I., Lippmaa, É. T., Platé, A. F.: Zh. Org. Khim., *15*, 492 (1979); J. Org. Chem. USSR, *15*, 436 (1979).
155. Bach, R. D., Mazur, U., Brummel, R. N., Lin, L.-H.: J. Am. Chem. Soc., *93*, 7120 (1971).
156. Santelli, M., Bertrand, M., Ronco, M.: Bull. Soc. Chim. France, 3273 (1964).
157. Zefirov, N. S., Prikazchikova, L. P., Yur'ev, Yu. K.: Zh. Obshch. Khim., *35*, 639 (1965); J. Gen. Chem. USSR, *35*, 642 (1965).
158. Whitham, G. H., Zaidlewicz, M.: J. Chem. Soc., Perkin I, 1509 (1972).
159. Traylor, T. G.: Acct. Chem. Res., *2*, 152 (1969).
160. Baird, W. C., Jr., Surridge, J. H.: J. Org. Chem., *37*, 304 (1972).
161. Pasto, D. J., Miles, M. F.: J. Org. Chem., *41*, 425 (1976).
162. Kartashov, V. R., Povelikina, L. N., Barkhash, V. A., Bodrikov, I. V.: Dokl. Akad. Nauk SSSR, *252*, 883 (1980); Proc. Acad. Sci. USSR, Chem. Sec., *252*, 264 (1980).
163. Pasto, D. J., Smorada, R. L., Turini, B. L., Wampfler, D. J.: J. Org. Chem., *41*, 432 (1976).
164. Nicholas, K. M.: J. Am. Chem. Soc., *97*, 3254 (1975).
165. Zefirov, N. S., Koz'min, A. S., Kirin, V. N., Sedov, B. B., Rau, V. G.: Tetrahedron Lett., 1667 (1980).
166. Sedov, B. B., Rau, V. G., Struchkov, Yu. T., Koz'min, A. S., Kirin, V. N., Nefirov, N. S.: Cryst. Struct. Commun., *9*, 995 (1980).
167. Zefirov, N. S., Kirin, B. N., Potekhin, K. A., Koz'min, A. S., Sadovaya, N. K., Kurkutova, E. N., Bodrikov, I. V.: Zh. Org. Khim., *14*, 1224 (1978); J. Org. Chem. USSR, *14*, 1135 (1978).
168. Krow, G. R., Reilly, J.: J. Org. Chem., *40*, 136 (1975).
169. Barton, D. H. R., Rosenfelder, W. J.: J. Chem. Soc., 2381 (1951).
170. Takehira, Y., Tanaka, K., Toda, F.: Chem. Lett., 1323 (1976).
171. Takaoka, K., Toyama, Y.: Kogyo Kagaku Zasshi, *70*, 1766 (1967); Chem. Abstr., *68*, 48997m (1968).
172. Toyama, Y., Takaoka, K.: Fette, Seifen, Anstrichmittel, *71*, 17 (1969); Chem. Abstr., *71*, 123466e (1969).
173. Delbecq, F., Goré, J.: Tetrahedron Lett., 3459 (1976).
174. Baudoy, R., Delbecq, F., Goré, J.: Tetrahedron, *36*, 189 (1980).
175. Thies, R. W., Hong, P.-K., Buswell, R., Boop, J. L.: J. Org. Chem., *40*, 585 (1975).
176. Thies, R. W., Hong, P. K., Buswell, R.: J. Chem. Soc., Chem. Commun., 1091 (1972).
177. Gray, G. A., Jackson, W. R.: J. Am. Chem. Soc., *91*, 6205 (1969).
178. Chambers, V. M. A., Jackson, W. R., Young, G. W.: J. Chem. Soc. D, Chem. Commun., 1275 (1970).
179. Kitching, W., Atkins, A. R., Wickham, G., Alberts, V.: J. Org. Chem., *46*, 563 (1981).
180. Chambers, V. M. A., Jackson, W. R., Young, G. W.: J. Chem. Soc. C, 2075 (1971).
181. Morrill, T. C., Vandemark, F. L.: Tetrahedron Lett., 1811 (1971).

182. Müller, E.: Chem. Ber., *109*, 3804 (1976).

183. Yang, N. C., Libman, J.: J. Am. Chem. Soc., *94*, 9228 (1972).

184. Brit. Patent 231,841 (1924); Chem. Abstr., *19*, 3491 (1925).

185. Sandler, S. R.: J. Chem. Eng. Data, *18*, 445 (1973); Chem. Abstr., *79*, 115076c (1973).

186. Hennion, G. F., Nieuwland, J. A.: J. Am. Chem. Soc., *56*, 1802 (1934).

187. Hennion, G. F., Murray, W. S.: J. Am. Chem. Soc., *64*, 1220 (1942).

188. Werntz, J. H.: U.S. Patent 1,963,108 (1934); Chem. Abstr., *28*, 4745[8] (1934).

189. Borisov, A. E., Vil'chevskaya, V. D., Nesmeyanov, A. N.: Dokl. Akad. Nauk SSSR, *90*, 383 (1953); Chem. Abstr., *48*, 4434f (1954).

190. Borisov, A. E., Vil'chevskaya, V. D., Nesmeyanov, A. N.: Izv. Akad. Nauk SSSR, Otdel. Khim. Nauk, 1008 (1954); Bull. Acad. Sci. USSR, Div. Chem. Sci., 879 (1954); Chem. Abstr., *50*, 171g (1956).

191. Matsoyan, S. G., Chukhadzhyan, G. A., Vartanyan, S. A.: Zh. Obshch. Khim., *30*, 1202 (1960); J. Gen. Chem. USSR, *30*, 1223 (1960).

192. Krafft, G. A., Katzenellenbogen, J. A.: J. Am. Chem. Soc., *103*, 5459 (1981).

193. Chobanyan, Zh. A., Davtyan, S. Zh., Badanyan, Sh. O.: Arm. Khim. Zh., *33*, 589 (1980); Chem. Abstr., *94*, 46728m (1981).

194. Shea, K. J., Wada, E.: J. Am. Chem. Soc., *104*, 5715 (1982).

195. Amos, R. A., Katzenellenbogen, J. A.: J. Org. Chem., *43*, 560 (1978).

196. Yamamoto, M.: J. Chem. Soc., Perkin I, 582 (1981).

197. Yamamoto, M.: J. Chem. Soc., Chem. Commun., 649 (1978).

198. Yamamoto, M., Yoshitake, M., Yamada, K.: J. Chem. Soc., Chem. Commun., 991 (1983).

199. Froning, J. F., Hennion, G. F.: J. Am. Chem. Soc., *62*, 653 (1940).

200. Camps, M., Montheard, J.-P.: Compt. Rend. C, *283*, 215 (1976).

201. Uemura, S., Miyoshi, H., Okano, M.: J. Chem. Soc., Perkin I, 1098 (1980).

202. Lemaire, H., Lucas, H. J.: J. Am. Chem. Soc., *77*, 939 (1955).

203. Hennion, G. F., Killian, D. B., Vaughan, T. H., Nieuwland, J. A.: J. Am. Chem. Soc., *56*, 1130 (1934).

204. Myddleton, W. W., Barrett, A. W., Seager, J. H.: J. Am. Chem. Soc., *52*, 4405 (1930).

205. Hudrlik, P. F., Hudrlik, A. M.: J. Org. Chem., *38*, 4254 (1973).

206. Spear, R. J., Jensen, W. A.: Tetrahedron Lett., 4535 (1977).

207. Uemura, S., Miyoshi, H., Sohma, K., Okano, M.: J. Chem. Soc., Chem. Commun., 548 (1975).

208. Filippova, A. Kh., Lyashenko, G. S., Frolov, Yu. L., Borisova, A. I., Ivanova, N. A., Voronkov, M. G.: Dokl. Vses. Konf. Khim. Atsetilena, 4th, *2*, 206 (1972); Chem. Abstr., *79*, 104840w (1973).

209. Hauge, K.: Acta Chem. Scand., *23*, 1059 (1969).

210. Yamamoto, M., Tanaka, S., Naruchi, K., Yamada, K.: Synthesis, 850 (1982).

211. Myddleton, W. W., Barrett, A. W.: J. Am. Chem. Soc., *49*, 2258 (1927).

212. Nesmeyanov, A. N., Borisov, A. E., Savel'eva, I. S., Osipova, M. A.: Izv. Akad. Nauk SSSR, Otdel. Khim. Nauk, 1249 (1961); Bull. Acad. Sci. USSR, Div. Chem. Sci., 1161 (1961).

213. Drefahl, G., Heublein, G., Wintzer, A.: Angew. Chem., *70*, 166 (1958).

214. Drefahl, G., Schaaf, S.: Chem. Ber., *90*, 148 (1957).

215. Seher, A.: Arch. Pharm., *292*, 519 (1959); Chem. Abstr., *54*, 8623g (1960).

216. Ruzicka, L., Meldahl, H. F.: Nature, *142*, 399 (1938).

217. Myddleton, W. W., Berchem, R. G., Barrett, A. W.: J. Am. Chem. Soc., *49*, 2264 (1927).

V. Acyloxymercuration

218. Carothers, W. H., Jacobson, R. A., Berchet, G. J.: J. Am. Chem. Soc., *55*, 4665 (1933).
219. Shostakovskii, M. F., Komarov, M. V., Yarosh, O. G.: Izv. Akad. Nauk SSSR, Ser. Khim., 908 (1968); Bull. Acad. Sci. USSR, Div. Chem. Sci., 869 (1968).
220. Kainer, F.: Kolloid-Z., *123*, 40 (1951).
221. Larock, R. C., Oertle, K., Beatty, K. M.: J. Am. Chem. Soc., *102*, 1966 (1980).
222. Yamamoto, M., Yoshitake, M., Yamada, K.: J. Chem. Soc., Chem. Commun., 991 (1983).
223. Balasubramanian, K. K., Reddy, K. V., Nagarajan, R.: Tetrahedron Lett., 5003 (1973).
224. Staab, H. A., Ipaktschi, J.: Chem. Ber., *104*, 1170 (1971).

VI. Aminomercuration

A. Alkenes

When amines are employed as the nucleophile in the solvomercuration reaction, aminomercuration takes place (Eq. 1). This reaction with simple

$$\ce{C=C} + 2\,R_2NH + HgX_2 \longrightarrow R_2N-\overset{|}{\underset{|}{C}}-\overset{|}{\underset{|}{C}}-HgX + R_2\overset{+}{N}H_2X^- \tag{1}$$

olefins, when followed by reductive demercuration, provides a valuable route to simple amines and various nitrogen heterocycles (Table 6.1). This reaction was briefly reviewed in 1972 [40].

Two studies have explored the relative reactivity of various mercury salts towards aminomercuration. In the aminomercuration of *m*-methoxy-allylbenzene, the following relative reactivity was observed: $Hg(ClO_4)_2$ $\sim Hg(NO_3)_2 \gg Hg(OAc)_2 > HgCl_2$ [38]. The reaction with mercuric nitrate required approximately 10 hours to reach completion. In the reaction of pyrrolidine and gaseous olefins, such as ethylene and propylene, the relative rates of reaction reported (as measured by olefin absorption) were reversed as far as mercuric chloride and mercuric acetate are concerned: $HgCl_2$ $> HgI_2 > Hg(OAc)_2 > CH_3HgOAc$ [3]. These reactions generally required 2 or 3 days to reach completion. In spite of the increased reactivity of the more electrophilic mercury salts, such as mercuric perchlorate and mercuric nitrate, essentially all work on the aminomercuration reaction to date has employed either mercuric chloride or mercuric acetate.

When mercuric oxide plus HBF_4 is heated in THF with alkenes and anilines, aminomercurials are not observed. Rather, vicinal diamines are isolated in 62–95% yield (Eq. 2) [17]. This approach may well be the method of choice for this transformation. If the initial reaction is carried out at

$$RCH=CH_2 + ArNHR' \xrightarrow{HgO\,\cdot\,2HBF_4} R\overset{\overset{\displaystyle ArNR'}{|}}{C}HCH_2NR'Ar \tag{2}$$

$-10\ °C$, followed by the addition of water or alcohols at $66\ °C$, hydroxy- or alkoxyamination results (Eq. 3) [13]. The addition to cyclohexene is reported to be anti. Unsymmetrical olefins tend to give mix curial in this last

$$RCH=CH_2 \xrightarrow[C_6H_5NH_2]{HgO\,\cdot\,2HBF_4} \xrightarrow{R'OH} R\overset{\overset{\displaystyle R'O}{|}}{C}HCH_2NHC_6H_5 + R\overset{\overset{\displaystyle C_6H_5NH}{|}}{C}HCH_2OR' \tag{3}$$

Table 6.1. Aminomercuration of Alkenes

Alkene	Amine	Mercuric salt	Organomercurial(s) (% Yield)	Subsequent reactants	Product(s) (% Yield)	Ref.
$H_2C=CH_2$	$(CH_3CH_2)_2NH$	$HgCl_2$	$ClHgCH_2CH_2N(CH_2CH_3)_2$	—	—	1, 2
	(pyrrolidine)	$HgCl_2$	$ClHgCH_2CH_2N$ (pyrrolidinyl) (75)	—	—	3
	(piperidine)	$HgCl_2$	$ClHgCH_2CH_2N$ (piperidinyl) (69.5)	$Na(Hg)/H_2O$	CH_3CH_2N (piperidinyl) (25)	4
	$C_6H_5NH_2$	$Hg(OAc)_2$	$C_6H_5NHCH_2CH_2HgOAc$	M, H_2O	$C_6H_5NHCH_2CH_3$ M = Li (58), Na (56), K (58)	5
	$C_6H_5NH_2$	$Hg(OAc)_2$	$C_6H_5NHCH_2CH_2HgOAc$ (78)	—	$C_6H_5NHCH_2CH_3$	6
	$C_6H_5NH_2$	$Hg(OAc)_2$, KBr	$C_6H_5NHCH_2CH_2HgBr$	PhX, Y, D_2O	$C_6H_5NHCH_2CH_2D$ X = Y = Na (76) or K (77); X = Li, Y = Na (80) or K (83)	7
	$C_6H_5NH_2$	$Hg(OAc)_2$, KBr	$C_6H_5NHCH_2CH_2HgBr$	PhLi, Li, O_2, HCl	$C_6H_5NHCH_2CH_2OH$ (79)	8
	$C_6H_5NH_2$	$Hg(OAc)_2$, KBr	$C_6H_5NHCH_2CH_2HgBr$	PhM, M, O_2, HCl	$C_6H_5NHCH_2CH_2OH$ M = Na (47), K (43)	7
	$C_6H_5NH_2$	$Hg(OAc)_2$, KBr	$C_6H_5NHCH_2CH_2HgBr$	PhLi, M, O_2, HCl	$C_6H_5NHCH_2CH_2OH$ M = Na (51), K (50)	7.
	$C_6H_5NH_2$	$Hg(OAc)_2$, KBr	$C_6H_5NHCH_2CH_2HgBr$	PhLi, Li, CO_2, EtOH/HCl	$C_6H_5NHCH_2CH_2CO_2Et$ (67)	8
	$C_6H_5NH_2$	$Hg(OAc)_2$, KBr	$C_6H_5NHCH_2CH_2HgBr$	PhLi, M, CO_2, EtOH/HCl	$C_6H_5NHCH_2CH_2CO_2Et$ M = Na (62), K (54)	7

Table 6.1. (continued)

Alkene	Amine	Mercuric salt	Organomercurial(s) (% Yield)	Subsequent reactants	Product(s) (% Yield)	Ref.
	$C_6H_5NH_2$	$Hg(OAc)_2$, KBr	$C_6H_5NHCH_2CH_2HgBr$	PhLi, Li, R^1COR^2, H_3O^+	$C_6H_5NHCH_2CH_2C(OH)R^1R^2$ R^1 R^2 H i-Pr (70) Me Ph (61) Ph Ph (66)	8
	$C_6H_5NH_2$	$Hg(OAc)_2$, KBr	$C_6H_5NHCH_2CH_2HgBr$	PhLi, Li, Me_3SiCl	$C_6H_5NHCH_2CH_2SiMe_3$ (58)	8
	$C_6H_5NH_2$	—	$C_6H_5NHCH_2CH_2HgBr$	Li, CH_3OH	$C_6H_5NHCH_2CH_3$ (78)	9
	$C_6H_5NH_2$	—	$C_6H_5NHCH_2CH_2HgBr$	Li / n-$BuNH_2$	$C_6H_5NHCH_2CH_3$ (75)	10
	$C_6H_5NH_2$	—	$C_6H_5NHCH_2CH_2HgBr$	Mg / CH_3OH	$C_6H_5NHCH_2CH_3$ (73)	11
	$C_6H_5NH_2$	—	$C_6H_5NHCH_2CH_2HgBr$	Ca / CH_3OH	$C_6H_5NHCH_2CH_3$ (82)	12
	$C_6H_5NH_2$	HgO / HBF_4	—	H_2O	$C_6H_5NHCH_2CH_2OH$ (60)	13
	$C_6H_5NHCH_3$	—	$C_6H_5N(CH_3)CH_2CH_2HgBr$	Mg / CH_3OH	$C_6H_5N(CH_3)CH_2CH_3$ (46)	11
	$C_6H_5NHCH_3$	—	$C_6H_5N(CH_3)CH_2CH_2HgBr$	Ca / CH_3OH	$C_6H_5N(CH_3)CH_2CH_3$ (59)	12
	$C_6H_5NHCH_2CH_3$	$Hg(OAc)_2$	—	M, H_2O	$C_6H_5N(CH_2CH_3)_2$ M = Li (65), Na (60), K (20)	5
	$C_6H_5NH(CH_2)_2$ - HgOAc	$Hg(OAc)_2$	—	—	$CH_3CH_2NHC_6H_5$ + $(CH_3CH_2)_2NC_6H_5$ 60 : 40	6
	$(CH_3)_3C_6H_2NH_2$	$Hg(OAc)_2$	—		$NHgCH_2CH_2NHC_6H_2(CH_3)_3$ (40) + $(\ \ NHgCH_2CH_2\ )_2NC_6H_2(CH_3)_3$	14

Table 6.1. (continued)

Alkene	Amine	Mercuric salt	Organomercurial(s) (% Yield)	Subsequent reactants	Product(s) (% Yield)	Ref.
	R_2NH	$HgCl_2$	$ClHgCH_2CH_2NR_2$ $R_2NH = C_6H_5NHCH_3$ (62), $C_6H_5NHCH_2CH_3$ (65), [cyclohexyl]$-NHCH_3$, $C_6H_5NH_2$ (70)	—	$CH_3CH_2NR_2$ $R_2NH = $ [cyclohexyl]$-NHCH_3$, $C_6H_5NH_2$	6
	R_2NH	$HgCl_2$	$ClHgCH_2CH_2NR_2$ $R_2NH = Et_2NH$ (25), [piperidine]NH (25)	$NaBH_4$	CH_3CH_2N[piperidine]	15
$H_2C{=}CHCH_3$	[pyrrolidine]	$HgCl_2$	$ClHgCH_2CH(CH_3)N$[pyrrolidine] (73)	—	—	3
	[pyrrolidine]	$HgX_2(X{=}Cl,OAc)$ $(HClO_4)$	—	$NaBH_4$	[pyrrolidine]$N{-}CH(CH_3)_2$ (25–75)	16
	$C_6H_5NH_2$	$Hg(OAc)_2$	$C_6H_5NHCH(CH_3)CH_2HgOAc$	M, H_2O	$C_6H_5NHCH(CH_3)_2$ $M = Li$ (72), Na (81), K (73)	5
	$C_6H_5NH_2$	$Hg(OAc)_2$, KBr	$C_6H_5NHCH(CH_3)CH_2HgBr$	PhLi, Li, O_2, HCl	$C_6H_5NHCH(CH_3)CH_2OH$ (83)	8
	$C_6H_5NH_2$	$Hg(OAc)_2$, KBr	$C_6H_5NHCH(CH_3)CH_2HgBr$	PhLi, Li, CO_2, EtOH/HCl	$C_6H_5NHCH(CH_3)CH_2CO_2Et$ (56)	8
	$C_6H_5NH_2$	$Hg(OAc)_2$, KBr	$C_6H_5NHCH(CH_3)CH_2HgBr$	PhLi, Li, Me_3SiCl	$C_6H_5NHCH(CH_3)CH_2SiMe_3$ (71)	8
	$C_6H_5NH_2$	HgO / HBF_4	—	—	$C_6H_5NHCH_2CH(NHC_6H_5)CH_3$ (62)	17
	$C_6H_5NH_2$	HgO / HBF_4	—	ROH	$C_6H_5NHCH_2CHORCH_3 + C_6H_5NHCH(CH_3)CH_2OR$ R = H 16 : 3 (67 total) R = CH_3 26 : 6 (54 total)	13
	$C_6H_5NH_2$	—	$C_6H_5NHCH(CH_3)CH_2HgBr$	Li, CH_3OH	$C_6H_5NHCH(CH_3)_2$ (81)	9

Table 6.1. (continued)

Alkene	Amine	Mercuric salt	Organomercurial(s) (% Yield)	Subsequent reactants	Product(s) (% Yield)	Ref.
	$C_6H_5NH_2$	—	$C_6H_5NHCH(CH_3)CH_2HgBr$	$Li/n\text{-}BuNH_2$	$C_6H_5NHCH(CH_3)_2$ (71)	10
	$C_6H_5NH_2$	—	$C_6H_5NHCH(CH_3)CH_2HgBr$	Mg/CH_3OH	$C_6H_5NHCH(CH_3)_2$ (70)	11
	$C_6H_5NH_2$	—	$C_6H_5NHCH(CH_3)CH_2HgBr$	Ca/CH_3OH	$C_6H_5NHCH(CH_3)_2$ (40)	12
	$C_6H_5NHCH_3$	—	$C_6H_5N(CH_3)CH(CH_3)CH_2HgBr$	$Mg/C_6H_5NH_2/CH_3OH$	$C_6H_5N(CH_3)CH(CH_3)_2$ (11)	11
	$C_6H_5NHCH_3$	—	$C_6H_5N(CH_3)CH(CH_3)CH_2HgBr$	Ca/CH_3OH	$C_6H_5N(CH_3)CH(CH_3)_2$ (30)	12
	R_2NH	—	—	$NaBH_4$	$CH_3CH(NR_2)CH_3$; $R_2NH = C_6H_5NH_2$ (70), pyrrolidine (83), piperidine (72)	18
$H_2C{=}CHCH_2OH$	piperidine	$HgCl_2$	$ClHgCH_2CHCH_2OH \cdot$ piperidine (74)	—	—	19
	R_2NH	$Hg(OAc)_2/\Delta$	—	—	$H_2C{=}CHCH_2NR_2$; $R_2NH =$ benzotriazole (67), carbazole (32)	20
$H_2C{=}CHO_2CCH_3$	$(CH_3CH_2)_2NH$	$Hg(OAc)_2/H_2SO_4$	—	—	$H_2C{=}CHN(CH_2CH_3)_2$ (30)	21
	pyrazole (R¹, R²)	$Hg(OAc)_2/H_2SO_4$	—	—	N-vinyl pyrazole (see table below)	22

For the row 22 product, the R^1/R^2 substituent table:

R^1	R^2	
H	H	(70)
Me	H	(78)
Me	Me	(76)
Me	Cl	(80.5)
Me	Br	(85)
Me	I	(80)
Me	Ac	(80)
Me	NO_2	(85)
Ph	Br	(86)

Table 6.1. (continued)

Alkene	Amine	Mercuric salt	Organomercurial(s) (% Yield)	Subsequent reactants	Product(s) (% Yield)	Ref.
	indole	$Hg(OAc)_2$	$AcOHgCH_2CHOAc$ (N-substituted indole) + 3-HgOAc indole	—	—	23
	carbazole	$Hg(OAc)_2$	$AcOHgCH_2CHO_2CCH_3$ (minor) + $AcOHgCH_2CH(OAc)_2$ (major)	—	—	23, 24
	carbazole	$Hg(O_2CCF_3)_2$	—	—	N-vinylcarbazole, $CH{=}CH_2$ (70)	24
	R_2NH	$Hg(OAc)_2/H_2SO_4$ or $HgSO_4/\Delta$	—	—	$H_2C{=}CHNR_2$; R_2NH = benzimidazole (58), 2-methylbenzimidazole (63), benzotriazole (72)	25
	R_2NH	$Hg(OAc)_2/H_2SO_4$	—	—	vinylamine polymer; R_2NH = Ph_2NH, indole, carbazole, tetrahydrocarbazole	21
$H_2C{=}CHCO_2CH_3$	$p\text{-}XC_6H_4NHR$	$Hg(OAc)_2$	$p\text{-}XC_6H_4NRCH_2CH(HgOAc)CO_2CH_3$ R = H, X = H (96), Cl (94), Me (94), CO_2Me (78); R = Me, X = H (90)	$NaBH_4/NaOH/$ $C_6H_5NH_2$	$p\text{-}XC_6H_4NRCH_2CH_2CO_2CH_3$ R = H, X = H (47, 56), Cl (37), Me (39) R = Me, X = H (45)	26

Table 6.1. (continued)

Alkene	Amine	Mercuric salt	Organomercurial(s) (% Yield)	Subsequent reactants	Product(s) (% Yield)	Ref.
$CH_3CH=CHCO_2H$	$C_6H_5NH_2$	$Hg(OAc)_2$	$CH_3CH(NHC_6H_5)CH(HgOAc)CO_2H$	$NaBH_4/NaOH$ (phase transfer)	$CH_3CH(NHC_6H_5)CH_2CO_2H$ (96) + $CH_3CH=CHCO_2H$ (4)	27
$H_2C=CHCH_2O_2CNH_2$	piperidine	$HgCl_2$	$ClHgCH_2CHCH_2O_2CNH_2$ (42)	—	—	19
$H_2C=C(CH_3)_2$	pyrrolidine	—	—	$NaBH_4$	N-$C(CH_3)_3$ + N-$CH(CH_3)_2$? 1 : 1	18
cis-$CH_3CH=CHCH_3$	$(CH_3)_2NH$	$HgCl_2$	threo-$CH_3CH(NMe_2)CH(HgCl)CH_3$ (24, 27.3)	$Na(Hg)/D_2O$	threo-$CH_3CH(NMe_2)CHDCH_3$ (20)	15, 28
trans-$CH_3CH=CHCH_3$	$(CH_3)_2NH$	$HgCl_2$	erythro-$CH_3CH(NMe_2)CH(HgCl)CH_3$ (4, 10)	$Na(Hg)/D_2O$	erythro-$CH_3CH(NMe_2)CHDCH_3$ (11)	15, 28
$H_2C=CHCH_2NHCONH_2$	piperidine	$HgCl_2$	$ClHgCH_2CHCH_2NHCONH_2$ (50)	—	—	19
	R_2NH	$HgCl_2$	$ClHgCH_2CH(NR_2)CH_2NHCONH_2$ R_2NH = Me_2NH, Et_2NH, $(n$-$Pr)_2NH$, NH, O NH, CH_3N NH, NH	—	—	29
	R_2NH	$HgCl_2$	$ClHgCH_2CH(NR_2)CH_2NHCONH_2$ R_2NH = NH (84), NH (68), O NH (84), CH_3N NH (64)	HCl	$ClHgCH_2CH(NR_2)CH_2NHCONH_2 \cdot HCl$ R_2NH = Me_2NH (41), Et_2NH (48), n-Pr_2NH (45), NH (69), NH (54), NH (56)	30
$H_2C=CHOCH_2CH_3$	$R^1NHCH_2CH_2NHR^1$	$Hg(O_2CR^2)_2$	—	—	R^1N NR^1 (CH_3) R^1 = H, R^2 = Ph (~41); R^1 = Ph, R^2 = Me (66)	31

Table 6.1. (continued)

Alkene	Amine	Mercuric salt	Organomercurial(s) (% Yield)	Subsequent reactants	Product(s) (% Yield)	Ref.
	$H_2N(CH_2)_3OH$	$Hg(OAc)_2$	—	—	(structure) CH_3 (60)	31
	$R^1NHC(R^2)_2CH_2OH$	$Hg(OAc)_2$	—	—	(structure) $R^1 = H,\ R^2 = Me\ (14)$; $R^1 = Ph,\ R^2 = H\ (64)$	31
$H_2C=CH(CH_2)_2OH$	(piperidine)	$HgCl_2$	$ClHgCH_2CH(CH_2)_2OH$ (53)	—	—	19
	$C_6H_5NH_2$	$Hg(OAc)_2$	—	$NaBH_4/NaOH$	$C_6H_5NHCH(CH_3)(CH_2)_2OH$ (68)	32
$CH_3CH=CHCH_2OH$	$C_6H_5NH_2$	$Hg(OAc)_2$	—	$NaBH_4/NaOH$	$C_6H_5NHCH(CH_3)(CH_2)_2OH$ (52)	32
$H_2C=C(CH_3)C\equiv CH$	$C_6H_5NHCH_3$	$Hg(OAc)_2$?/ hexane	—	—	$C_6H_5N(CH_3)CH_2CH(CH_3)CH(CH_3)N(CH_3)C_6H_5$ (31)	33
	m-$XC_6H_4NH_2$	$Hg(OAc)_2/THF$	—	$NaBH_4$ (powder)	m-$XC_6H_4NHCH_2CH(CH_3)CH(CH_3)NHC_6H_4X$-$m$ X = H, Me	33
	$XC_6H_4NH_2$	$Hg(OAc)_2/DMSO$	—	$NaBH_4/NaOH$	$XC_6H_4NHCH_2CH(CH_3)CHOHCH_3$ X = H, m-Me, p-Me, o-MeO (54-87)	33
	$XC_6H_4NH_2$	$Hg(OAc)_2/$ pentane	—	$NaBH_4/NaOH$	$XC_6H_4NHCH_2CH(CH_3)COCH_3$ X = H, m-Me, p-Me, o-MeO (57-76)	33
(cyclopentene)	$C_6H_5NH_2$	HgO/HBF_4	—	H_2O	(structure) OH, NHC_6H_5 (79)	13
$H_2C=C(CH_3)O_2CCH_3$	(benzotriazole)	$Hg(OAc)_2/H_2SO_4$	—	—	(structure) $CH_3\overset{|}{C}=CH_2$ (81)	34

Table 6.1. (continued)

Alkene	Amine	Mercuric salt	Organomercurial(s) (% Yield)	Subsequent reactants	Product(s) (% Yield)	Ref.
$H_2C{=}CHCH_2O_2CCH_3$	piperidine (NH)	$HgCl_2$	$ClHgCH_2CHCH_2O_2CCH_3 \cdot$ piperidine (NH) (74)	—	—	19
$H_2C{=}C(CH_3)CO_2CH_3$	$C_6H_5NH_2$	$Hg(OAc)_2$	$C_6H_5NHCH_2C(CH_3)(HgOAc)CO_2CH_3$? (78)	—	—	26
$H_2C{=}CHCH_2NHCOCH_3$	R_2NH	$Hg(OAc)_2$	—	$NaBH_4/NaOH$	$CH_3CH(NR_2)CH_2NHCOCH_3$ $+ H_2C{=}CHCH_2NHCOCH_3$ $R_2NH = C_6H_5NH_2$ (73) (14) $p{-}CH_3C_6H_4NH_2$ (62) (38) $C_6H_5NHCH_3$ (46) (29)	35
$H_2C{=}C(CH_3)CH_2CH_3$	$C_6H_5NH_2$	$Hg(OAc)_2$	—	$NaBH_4/NaOH$	$(CH_3)_2C(NHC_6H_5)CH_2CH_3$ (44)	36
cis- and trans- $CH_3CH{=}CHCH_2CH_3$	$C_6H_5NH_2$	$Hg(OAc)_2$	—	$NaBH_4/NaOH$	$CH_3CH(NHC_6H_5)(CH_2)_2CH_3$ $+ (CH_3CH_2)_2CHNHC_6H_5$ 2 : 1 (36 total)	36
$H_2C{=}CH(CH_2)_3OH$	piperidine (NH)	$HgCl_2$	$ClHgCH_2CH(CH_2)_3OH \cdot$ piperidine (NH) (88)	—	—	19
cis- $CH_3O_2CCH{=}CHCO_2CH_3$	$C_6H_5NH_2$	$Hg(OAc)_2$	$CH_3O_2CCH(NHC_6H_5)CH(HgOAc)CO_2CH_3$ (90)	—	—	26
cyclohexene	$C_6H_5NH_2$	$Hg(OAc)_2$	—	$NaBH_4/NaOH$	cyclohexyl-NHC_6H_5 (37)	36
	$C_6H_5NH_2$	$Hg(OAc)_2$	cyclohexyl (NHC_6H_5, $HgOAc$)	M, H_2O	cyclohexyl-NHC_6H_5 M = Li (78), Na (84), K (79)	5
	$C_6H_5NH_2$	HgO/HBF_4	—	—	cyclohexyl (NHC_6H_5, NHC_6H_5) (89)	17

Table 6.1. (continued)

Alkene	Amine	Mercuric salt	Organomercurial(s) (% Yield)	Subsequent reactants	Product(s) (% Yield)	Ref.
	$C_6H_5NH_2$	HgO/HBF_4	—	H_2O	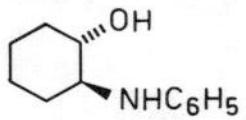 OH, NHC_6H_5 (80)	13
	$C_6H_5NH_2$	HgO/HBF_4	—	CH_3OH	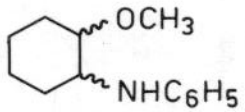 OCH_3, NHC_6H_5 (71)	13
	$C_6H_5NH_2$	—	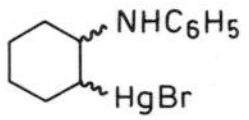 NHC_6H_5, HgBr	Li, CH_3OH	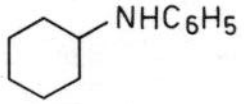 NHC_6H_5 (86)	9
	$C_6H_5NH_2$	—	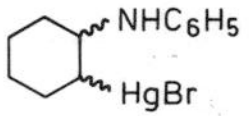 NHC_6H_5, HgBr	$Li/C_6H_5NH_2$	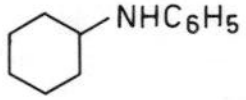 NHC_6H_5 (87)	10
	$C_6H_5NH_2$	—	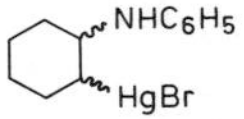 NHC_6H_5, HgBr	Mg/CH_3OH	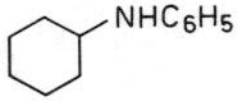 NHC_6H_5 (72)	11
	$C_6H_5NH_2$	—	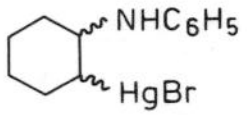 NHC_6H_5, HgBr	Ca/CH_3OH	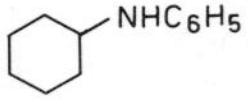 NHC_6H_5 (65)	12
	$C_6H_5NHCH_3$	$Hg(OAc)_2$	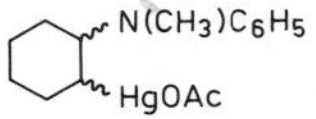 $N(CH_3)C_6H_5$, HgOAc	M, H_2O	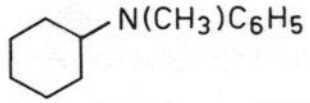 $N(CH_3)C_6H_5$ M = Li (68), Na (80), K (70)	5
	$C_6H_5NHCH_3$	$Hg(OAc)_2$	—	$Li/n\text{-}BuNH_2$	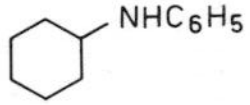 NHC_6H_5 (58)	10
	$C_6H_5NHCH_3$	$Hg(OAc)_2$	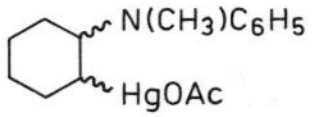 $N(CH_3)C_6H_5$, HgOAc	Ca/CH_3OH	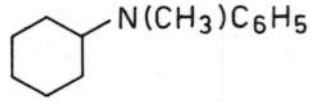 $N(CH_3)C_6H_5$ (20)	12
	$C_6H_5NHCH_3$	—	—	$Li/n\text{-}BuNH_2$	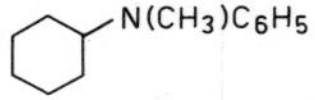 $N(CH_3)C_6H_5$ (58)	10

452

Table 6.1. (continued)

Alkene	Amine	Mercuric salt	Organomercurial(s) (% Yield)	Subsequent reactants	Product(s) (% Yield)	Ref.
	$C_6H_5NHCH_3$	—	(cyclohexyl)-$N(CH_3)C_6H_5$, HgBr	Mg/CH_3OH	(cyclohexyl)-$N(CH_3)C_6H_5$ (23)	11
$H_2C=CH(CH_2)_2CO_2CH_3$	piperidine	$HgCl_2$	$ClHgCH_2CH(CH_2)_2CO_2CH_3 \cdot$ piperidine NH (73)	—	—	19
$H_2C=CHOCH_2CH(CH_3)_2$	carbazole	$Hg(OAc)_2$	—	—	N-vinylcarbazole $CH=CH_2$	37
$H_2C=CH(CH_2)_2CHOHCH_3$	$C_6H_5NH_2$	$Hg(OAc)_2$	—	$NaBH_4/NaOH$	$C_6H_5NHCH(CH_3)(CH_2)_2CHOHCH_3$ (35)	32
$H_2C=CHCH_2NH(CH_2)_2CH_3$	$C_6H_5NH_2$	$Hg(OAc)_2$	—	$NaBH_4/NaOH$	$CH_3CH(NHC_6H_5)CH_2NH(CH_2)_2CH_3$ (22) + $H_2C=CHCH_2NH(CH_2)_2CH_3$ (46)	35
norbornene	$C_6H_5NH_2$	$Hg(OAc)_2$	—	$NaBH_4/NaOH$	norbornyl-NHC_6H_5 (32)	36
(cyclohexenyl)-CH_2OH	$C_6H_5NH_2$	$Hg(OAc)_2$	—	$NaBH_4/NaOH$	(cyclohexyl)-CH_2OH, NHC_6H_5 (57)	32
$H_2C=CH(CH_2)_4CH_3$	$C_6H_5NH_2$	$Hg(OAc)_2$	—	$NaBH_4/NaOH$	$CH_3CH(NHC_6H_5)(CH_2)_4CH_3$ (74)	36
	$C_6H_5NH_2$	$Hg(OAc)_2$	—	M, H_2O	$C_6H_5NHCH(CH_3)(CH_2)_4CH_3$ M = Li (95), Na (80), K (85)	5
	$C_6H_5NH_2$	HgO/HBF_4	—	—	$C_6H_5NHCH_2CH(NHC_6H_5)(CH_2)_4CH_3$ (70)	17

Table 6.1. (continued)

Alkene	Amine	Mercuric salt	Organomercurial(s) (% Yield)	Subsequent reactants	Product(s) (% Yield)	Ref.
	$C_6H_5NH_2$	HgO/HBF_4	—	CH_3OH	$C_6H_5NHCH_2CH(OCH_3)(CH_2)_4CH_3$ $+ CH_3OCH_2CH(NHC_6H_5)(CH_2)_4CH_3$ 16 : 7 (73 total)	13
	$C_6H_5NHCH_3$	$Hg(OAc)_2$	—	M, H_2O	$C_6H_5N(CH_3)CH(CH_3)(CH_2)_4CH_3$ M = Li (58), Na (38), K (11)	5
$H_2C=CHCH_2SC(CH_3)_3$	C_6H_5NHR	$Hg(OAc)_2$	—	$NaBH_4/NaOH$	$CH_3CH(NRC_6H_5)CH_2SC(CH_3)_3$ $+ H_2C=CHCH_2SC(CH_3)_3$ R = H (65) (12) R = CH_3 (42) (55)	35
$H_2C=CHCH_2N(CH_2CH_3)_2$	R_2NH	$Hg(OAc)_2$	—	$NaBH_4/NaOH$	$CH_3CH(NR_2)CH_2N(CH_2CH_3)_2$ $+ H_2C=CHCH_2N(CH_2CH_3)_2$ $R_2NH = C_6H_5NH_2$ (33) (37) $p-CH_3C_6H_4NH_2$ (28) (22) $C_6H_5NHCH_3$ (21) (38)	35
$H_2C=CHC_6H_5$	piperidine ($\overset{N}{\underset{H}{}}$)	$HgCl_2$	$ClHgCH_2CH(C_6H_5)N$⟨piperidino⟩ (64)	—	—	3
	$C_6H_5NH_2$	$Hg(OAc)_2$	—	$NaBH_4/NaOH$	$CH_3CH(NHC_6H_5)C_6H_5$ (51)	36
	$C_6H_5NH_2$	$Hg(OAc)_2$	$C_6H_5NHCH(C_6H_5)CH_2HgOAc$	M, H_2O	$C_6H_5NHCH(C_6H_5)CH_3$ M = Li (71), Na (78), K (20)	5
	$C_6H_5NH_2$	$Hg(OAc)_2$, KBr	$C_6H_5NHCH(C_6H_5)CH_2HgBr$	PhX, Y, D_2O	$C_6H_5NHCH(C_6H_5)CH_2D$ X = Y = Na (81) or K (83); X = Li, Y = Na (86) or K (89)	7
	$C_6H_5NH_2$	$Hg(OAc)_2$, KBr	$C_6H_5NHCH(C_6H_5)CH_2HgBr$	PhLi, Li, O_2, HCl	$C_6H_5NHCH(C_6H_5)CH_2OH$ (75)	8
	$C_6H_5NH_2$	$Hg(OAc)_2$, KBr	$C_6H_5NHCH(C_6H_5)CH_2HgBr$	PhLi, Na, O_2, HCl	$C_6H_5NHCH(C_6H_5)CH_2OH$ (58)	7

Table 6.1. (continued)

Alkene	Amine	Mercuric salt	Organomercurial(s) (% Yield)	Subsequent reactants	Product(s) (% Yield)	Ref.
	$C_6H_5NH_2$	$Hg(OAc)_2$, KBr	$C_6H_5NHCH(C_6H_5)CH_2HgBr$	PhLi, Li, CO_2, EtOH/HCl	$C_6H_5NHCH(C_6H_5)CH_2CO_2Et$ (68)	8
	$C_6H_5NH_2$	$Hg(OAc)_2$, KBr	$C_6H_5NHCH(C_6H_5)CH_2HgBr$	PhLi, Na, CO_2, EtOH/HCl	$C_6H_5NHCH(C_6H_5)CH_2CO_2Et$ (62)	7
	$C_6H_5NH_2$	$Hg(OAc)_2$, KBr	$C_6H_5NHCH(C_6H_5)CH_2HgBr$	PhLi, Li, R^1COR^2, H_3O^+	$C_6H_5NHCH(C_6H_5)CH_2C(OH)R^1R^2$ R^1 R^2 H i-Pr (81) H Ph (76) Me Me (52) Ph Ph (61)	8
	$C_6H_5NH_2$	$Hg(OAc)_2$, KBr	$C_6H_5NHCH(C_6H_5)CH_2HgBr$	PhLi, Li, Me_3SiCl	$C_6H_5NHCH(C_6H_5)CH_2SiMe_3$ (65)	8
	$C_6H_5NH_2$	HgO/HBF_4	—	H_2O	$C_6H_5NHCH_2CHOHC_6H_5$ $+ C_6H_5NHCH(C_6H_5)CH_2OH$ 26 : 4 (65 total)	13
	$C_6H_5NH_2$	—	$C_6H_5NHCH(C_6H_5)CH_2HgBr$	Li/$C_6H_5NH_2$	$C_6H_5NHCH(C_6H_5)CH_3$ (68)	10
	$C_6H_5NH_2$	—	$C_6H_5NHCH(C_6H_5)CH_2HgBr$	Mg/CH_3OH	$C_6H_5NHCH(C_6H_5)CH_3$ (99)	11
	$C_6H_5NH_2$	—	$C_6H_5NHCH(C_6H_5)CH_2HgBr$	Ca/CH_3OH	$C_6H_5NHCH(C_6H_5)CH_3$ (90)	12
	$C_6H_5NHCH_3$	$Hg(OAc)_2$	$C_6H_5N(CH_3)CH(C_6H_5)CH_2HgOAc$	M, H_2O	$C_6H_5N(CH_3)CH(C_6H_5)CH_3$ M = Li(0), Na(43), K(1)	5
	$C_6H_5NHCH_3$	—	$C_6H_5N(CH_3)CH(C_6H_5)CH_2HgBr$	Mg/CH_3OH	$C_6H_5N(CH_3)(CH_2)_2C_6H_5$ (72)	11
	$C_6H_5NHCH_3$	—	$C_6H_5N(CH_3)CH(C_6H_5)CH_2HgBr$	Ca/CH_3OH	$C_6H_5N(CH_3)CH(C_6H_5)CH_3$ (8) $+ C_6H_5N(CH_3)(CH_2)_2C_6H_5$ (8)	12

Table 6.1. (continued)

Alkene	Amine	Mercuric salt	Organomercurial(s) (% Yield)	Subsequent reactants	Product(s) (% Yield)	Ref.
	R_2NH	$HgCl_2$	$ClHgCH_2CH(NR_2)C_6H_5$	$LiAlH_4$	$C_6H_5CH(CH_3)NR_2$ + $C_6H_5CH_2CH_2NR_2$; R_2NH = piperidine 75 : 25; pyrrolidine — : —	6
	R_2NH	HgO/HBF_4	—	—	$R_2NCH_2CH(NR_2)C_6H_5$; R_2NH = $C_6H_5NH_2$ (69), o-$CH_3C_6H_4NH_2$ (95), $C_6H_5NHCH_3$ (65)	17
	R_2NH	—	—	$NaBH_4$	$CH_3CH(NR_2)C_6H_5$; R_2NH = pyrrolidine, piperidine	18
HOCH$_2$-norbornene (exo and endo)	$C_6H_5NH_2$	$Hg(OAc)_2$	—	$NaBH_4/NaOH$	norbornane-NHC$_6$H$_5$, HOCH$_2$ (30)	32
$trans$-$C_2H_5O_2CCH{=}CHCO_2C_2H_5$	$C_6H_5NH_2$	$Hg(OAc)_2$	$C_2H_5O_2CCH(NHC_6H_5)CH(HgOAc)CO_2C_2H_5$ (74)	—	—	26
cyclooctene	$C_6H_5NH_2$	HgO/HBF_4	—	—	cyclooctyl-NHC$_6$H$_5$ (80)	17
$H_2C{=}CH(CH_2)_5CH_3$	$C_6H_5NH_2$	$Hg(OAc)_2$	—	$NaBH_4/NaOH$	$CH_3CH(NHC_6H_5)(CH_2)_5CH_3$ (47)	36
$H_2C{=}CHO(CH_2)_2O(CH_2)_3CH_3$	$H_2N(CH_2)_3OH$	$Hg(O_2CC_6H_5)_2$	—	—	tetrahydrooxazine-CH$_3$ (65)	31
	$R^1R^2C(NH_2)CHR^3OH$	$Hg(O_2CC_6H_5)_2$	—	—	oxazolidine (R^1, R^2, R^3): H H H (~55); H H Me (40); Et H H (55); Me Me H (72)	31

Table 6.1. (continued)

Alkene	Amine	Mercuric salt	Organomercurial(s) (% Yield)	Subsequent reactants	Product(s) (% Yield)	Ref.
$H_2C=C(CH_3)C_6H_5$	$C_6H_5NH_2$	$Hg(OAc)_2$	—	$NaBH_4/NaOH$	$(CH_3)_2C(NHC_6H_5)C_6H_5$ (26)	36
$H_2C=CHCH_2C_6H_5$	$C_6H_5CH_2NH_2$	$Hg(NO_3)_2$	—	$NaBH_4/NaOH$	$C_6H_5CH_2NHCH(CH_3)CH_2C_6H_5$ + $C_6H_5CH_2NH(CH_2)_3C_6H_5$ (71 total); 90 : 10	38
	$C_6H_5NH_2$	HgO/HBF_4	—	—	$C_6H_5NHCH_2CH(NHC_6H_5)CH_2C_6H_5$ (88)	17
$p\text{-}H_2C=CHCH_2NHC_6H_4Cl$	$C_6H_5NH_2$	$Hg(OAc)_2$	—	$NaBH_4/NaOH$	$p\text{-}CH_3CH(NHC_6H_5)CH_2NHC_6H_4Cl$ (40) + $p\text{-}H_2C=CHCH_2NHC_6H_4Cl$ (40)	35
$H_2C=CHCH_2NHCO$-(pyridin-3-yl)	piperidine	$HgCl_2$	$ClHgCH_2CHCH_2NHCO$-(pyridin-3-yl), piperidino on CH (83)	—	—	30, 39
$H_2C=CHCH_2NHCO$-(pyridin-4-yl)	pyrrolidine	$HgCl_2$	$ClHgCH_2CHCH_2NHCO$-(pyridin-4-yl), pyrrolidino on CH (31)	—	—	30, 39
$H_2C=CHCH_2OC_6H_5$	piperidine	$HgCl_2$	$ClHgCH_2CHCH_2OC_6H_5$, piperidino on CH (46)	—	—	19
$H_2C=CHCH_2SC_6H_5$	R_2NH	$Hg(OAc)_2$	—	$NaBH_4/NaOH$	$CH_3CH(NR_2)CH_2SC_6H_5$ + $H_2C=CHCH_2SC_6H_5$; $R_2NH = C_6H_5NH_2$ (51),(49); $p\text{-}CH_3C_6H_4NH_2$ (70),(30); $C_6H_5NHCH_3$ (28),(71)	35
$H_2C=CHCH_2NHC_6H_5$	R_2NH	$Hg(OAc)_2$	—	$NaBH_4/NaOH$	$CH_3CH(NR_2)CH_2NHC_6H_5$ + $H_2C=CHCH_2NHC_6H_5$; $R_2NH = C_6H_5NH_2$ (39),(60); $p\text{-}ClC_6H_4NH_2$ (28),(71); $p\text{-}CH_3C_6H_4NH_2$ (52),(47); $C_6H_5NHCH_3$ (48),(46); $C_6H_5NHC_2H_5$ (31),(68)	35

Table 6.1. (continued)

Alkene	Amine	Mercuric salt	Organomercurial(s) (% Yield)	Subsequent reactants	Product(s) (% Yield)	Ref.
$H_2C=CH(CH_2)_2CHOHC(CH_3)_3$	$C_6H_5NH_2$	$Hg(OAc)_2$	—	$NaBH_4/NaOH$	$C_6H_5NHCH(CH_3)(CH_2)_2CHOHC(CH_3)_3$ (30)	32
m-$H_2C=CHCH_2C_6H_4CF_3$	$C_2H_5NH_2$	$Hg(ClO_4)_2$	—	$NaBH_4/NaOH$	m-$C_2H_5NHCH(CH_3)CH_2C_6H_4CF_3$ + m-$C_2H_5NH(CH_2)_3C_6H_4CF_3$ 82 : 18 (93 total)	38
	RNH_2	$Hg(NO_3)_2$	—	$NaBH_4/NaOH$	m-$RNHCH(CH_3)CH_2C_6H_4CF_3$ + m-$RNH(CH_2)_3C_6H_4CF_3$ R = $PhCH_2$ 100 : 0 (57 total) $(EtO)_2CHCH_2$ 90 : 10 (62 total)	38
$H_2C=CHCH_2O_2CC_6H_5$	R_2NH	$HgCl_2$	$ClHgCH_2CH(NR_2)CH_2O_2CC_6H_5$ (65) $R_2NH = $ NH (65), O NH (45), CH_3 NH (26), NH (27), CH_3 NH (6)	—	—	19
$C_6H_5CH=CHCO_2CH_3$	$C_6H_5NH_2$	$Hg(OAc)_2$	$C_6H_5CH(NHC_6H_5)CH(HgOAc)CO_2CH_3$ (89)	—	—	26
p-$H_2C=CHCH_2O_2CC_6H_4OH$	piperidine	$HgCl_2$	p-$ClHgCH_2CHCH_2O_2CC_6H_4OH$ (14)	—	—	19
$H_2C=CHCH_2NHCOC_6H_5$	piperidine	$HgCl_2$	$ClHgCH_2CHCH_2NHCOC_6H_5$ (53)	—	—	30, 39
m-$H_2C=CHCH_2C_6H_4OCH_3$	$C_2H_5NH_2$	$Hg(ClO_4)_2$	—	$NaBH_4/NaOH$	m-$C_2H_5NHCH(CH_3)CH_2C_6H_4OCH_3$ + m-$C_2H_5NH(CH_2)_3C_6H_4OCH_3$ 80 : 20 (87 total)	38

Table 6.1. (continued)

Alkene	Amine	Mercuric salt	Organomercurial(s) (% Yield)	Subsequent reactants	Product(s) (% Yield)	Ref.
	R_2NH	$Hg(NO_3)_2$	—	$NaBH_4/NaOH$	*m*-$CH_3CH(NR_2)CH_2C_6H_4OCH_3$ + *m*-$R_2N(CH_2)_3C_6H_4OCH_3$	38
					R_2NH =	
					$(EtO)_2CHCH_2NH_2$ 100 : 0 (87 total)	
					n-$BuNH_2$ 100 : 0 (48 total)	
					$PhCH_2NH_2$ 90 : 10 (78 total)	
					Et_2NH 60 : 40 (39 total)	
					$PhNH_2$ 100 : 0 (41 total)	
					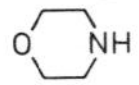 100 : 0 (68 total)	
					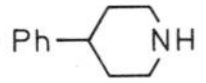 80 : 20 (86 total)	
$H_2C{=}CHCH_2SCH_2C_6H_5$	R_2NH	$Hg(OAc)_2$	—	$NaBH_4/NaOH$	$CH_3CH(NR_2)CH_2SCH_2C_6H_5$ + $H_2C{=}CHCH_2SCH_2C_6H_5$	35
					R_2NH =	
					$C_6H_5NH_2$ (71) (28)	
					p-$CH_3C_6H_4NH_2$ (35) (64)	
					$C_6H_5NHCH_3$ (67) (33)	
	R_2NH	$Hg(OAc)_2$	—	Na/NH_3	$CH_3CH(NR_2)CH_2SH$	35
					R_2NH = $C_6H_5NH_2$ (30), *p*-$CH_3C_6H_4NHCH_3$ (48), $C_6H_5NHCH_3$ (24)	
$H_2C{=}CHCH_2N(CH_3)C_6H_5$	$C_6H_5NH_2$	$Hg(OAc)_2$	—	$NaBH_4/NaOH$	$CH_3CH(NHC_6H_5)CH_2N(CH_3)C_6H_5$ (38) + $H_2C{=}CHCH_2N(CH_3)C_6H_5$ (60)	35
$H_2C{=}C(CH_3)CH_2NHC_6H_5$	R_2NH	$Hg(OAc)_2$	—	$NaBH_4/NaOH$	$(CH_3)_2C(NR_2)CH_2NHC_6H_5$ + $H_2C{=}C(CH_3)CH_2NHC_6H_5$	35
					R_2NH =	
					$C_6H_5NH_2$ (43) (56)	
					p-$CH_3C_6H_4NH_2$ (15) (80)	
					$C_6H_5NHCH_3$ (0) (100)	
$H_2C{=}CH(CH_2)_7CH_3$	$C_6H_5NH_2$	$Hg(OAc)_2$	—	$NaBH_4/NaOH$	$CH_3CH(NHC_6H_5)(CH_2)_7CH_3$ (28)	36

Table 6.1. (continued)

Alkene	Amine	Mercuric salt	Organomercurial(s) (% Yield)	Subsequent reactants	Product(s) (% Yield)	Ref.
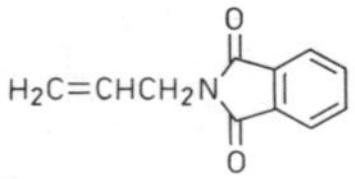 $H_2C=CHCH_2N$(phthalimide)	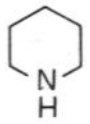 piperidine (NH)	$HgCl_2$	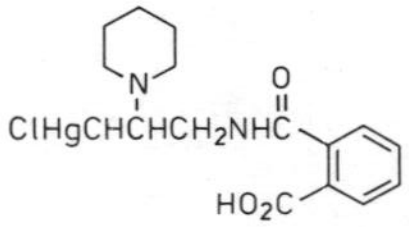 $ClHgCHCHCH_2NHC(=O)$ (with HO_2C-) piperidine	—	—	39
$o\text{-}H_2C=CHCH_2NHCOC_6H_4CO_2H$	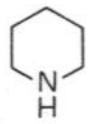 piperidine (NH)	$HgCl_2$	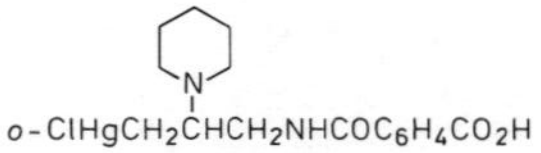 $o\text{-}ClHgCH_2CHCH_2NHCOC_6H_4CO_2H$ (51)	—	—	30
$p\text{-}H_2C=CHCH_2O_2CC_6H_4OCH_3$	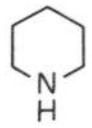 piperidine (NH)	$HgCl_2$	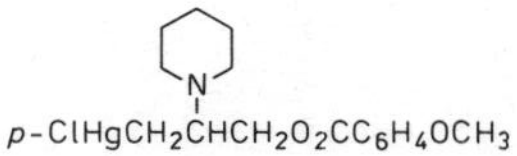 $p\text{-}ClHgCH_2CHCH_2O_2CC_6H_4OCH_3$ (40)	—	—	19
$H_2C=CHCH_2NHCOCH_2\text{-}NHCOC_6H_5$	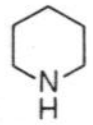 piperidine (NH)	$HgCl_2$	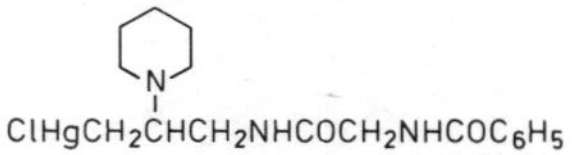 $ClHgCH_2CHCH_2NHCOCH_2NHCOC_6H_5$	—	—	39
	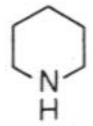 piperidine (NH)	—	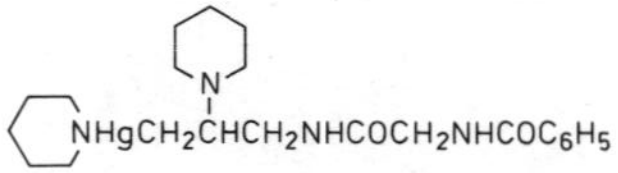 (piperidino)$NHgCH_2CHCH_2NHCOCH_2NHCOC_6H_5$? (74)	—	—	30
$H_2C=CH(CH_2)_9CH_3$	$C_6H_5NH_2$	$Hg(OAc)_2$	—	$NaBH_4/NaOH$	$CH_3CH(NHC_6H_5)(CH_2)_9CH_3$ (41)	36
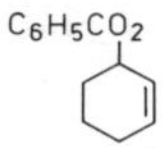 $C_6H_5CO_2$-cyclohexene	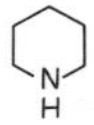 piperidine (NH)	$HgCl_2$	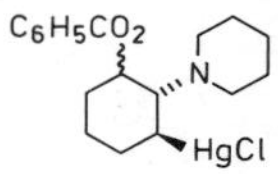 cyclohexane with $C_6H_5CO_2$, piperidino and $HgCl$ (6)	—	—	19

460

tures of regioisomers with the oxygen moiety on the more highly substituted end of the double bond. The regiochemistry cannot be reversed by simply reversing the order of addition of the oxygen and nitrogen substrates. Diamines are the sole products of such efforts.

Virtually all of the work carried out on aminomercuration has employed secondary aliphatic amines or primary or secondary anilines. It is reported that primary aliphatic amines react with mercuric acetate, mercuric chloride or mercuric bromide to afford mercury amine complexes of the type $RNH_2 \cdot HgX_2$ or $(RNH_2)_2HgX_2$ which are apparently unreactive towards at least α,β-unsaturated esters [26, 41]. However, primary aliphatic amines have been successfully employed with mercuric perchlorate or mercuric nitrate as the mercurating agent [38]. One potential problem with primary amines is the possible double mercuration of olefins (Eq. 4). While this reaction has been observed with primary anilines and ethylene [6, 14], it

$$ArNH_2 \; + \; 2H_2C\!=\!CH_2 \; + \; 2\,Hg(OAc)_2 \quad\longrightarrow\quad ArN(CH_2CH_2HgOAc)_2 \qquad (4)$$

does not appear to be a significant problem. Surprisingly, ammonia does not appear to have ever been utilized for aminomercuration-demercuration, but one can accomplish the equivalent process by amidomercuration-demercuration and subsequent amide hydrolysis as will be discussed in Chapter VII.

While excess amine has commonly been employed as the solvent for these reactions, the effect of various solvents on the rate of reaction has been studied. With gaseous olefins the rate of reaction, as measured by olefin absorption, was found to be: amine $>80\!:\!20$ THF/HMPA $>60\!:\!40$ THF/HMPA $>$ THF $>$ DME [3]. It is stated that the difficulties encountered with mercury amine complex formation can be partially overcome by the addition of water to the reaction mixture [32, 36]. Similarly, olefinic alcohols generally react smoothly with aniline and mercuric acetate in the absence of water to afford N-substituted anilines after demercuration [32]. Competitive intramolecular alkoxymercuration does not appear to be a problem with these compounds, except where the position of nucleophilic attack in the olefin is disubstituted (Eq. 5).

$$\underset{CH_3\overset{|}{C}H(CH_2)_2\overset{|}{C}HCH_3}{\overset{\underset{|}{C_6H_5NH}\qquad\overset{|}{OH}}{}} \quad\overset{R=H}{\longleftarrow}\quad R_2C\!=\!CH(CH_2)_2\overset{\overset{|}{OH}}{\underset{|}{C}}HCH_3 \quad\overset{R=CH_3}{\longrightarrow}\quad \qquad (5)$$

The effect of added perchloric acid on aminomercuration has also been examined. It is reported that 0.1 M $HClO_4$ accelerates reactions run with mercuric chloride, but the yield of alkylated amine after demercuration decreases with increasing perchloric acid concentration [3, 16]. In one of these studies, it was observed that aminomercuration can also be accomplished using β-acetoxymercurials, an amine and perchloric or acetic acids (Eq. 6) [16].

VI. Aminomercuration

$$\underset{\text{OAc}}{\text{RCHCH}_2\text{HgCl}} + \text{R}'_2\text{NH} \xrightarrow{\text{H}^+} \underset{\text{NR}'_2}{\text{RCHCH}_2\text{HgCl}} \qquad (6)$$

Relatively little has been done in the way of physically characterizing the aminomercurials. Proton NMR data [3] and one X-ray crystallographic study [1] have been reported. The latter work indicates the presence of inter-molecular mercury-nitrogen coordination, but no evidence for any intra-molecular interactions. In one study of the aminomercuration reaction, amine adducts of the expected mercurials have reportedly been isolated (Eq. 7) [19].

$$\text{RCH}=\text{CH}_2 \longrightarrow \text{RCHCH}_2\text{HgCl} \cdot \text{HN} \qquad (7)$$

Precious little has been reported on the regio- or stereoselectivity of aminomercuration. Markovnikov addition is generally observed. α,β-Unsaturated esters of various types yield solely alpha-mercurated products (Eq. 8) [26]. *cis*- and *trans*-2-Pentene yield approximately a 2:1 ratio of the 2- and 3-substituted amines, respectively [36]. However, 4-hydroxymethyl-cyclohexene [32] and 2-cyclohexenyl benzoate [19] are both reported to give

$$\text{R}^1\text{CH}=\text{CR}^2\text{CO}_2\text{R}^3 \longrightarrow \underset{\underset{\text{HgOAc}}{|}}{\overset{\overset{\text{ArNR}^4}{|}}{\text{R}^1\text{CHCR}^2\text{CO}_2\text{R}^3}} \qquad (8)$$

only one regioisomer (Eqs. 9, 10). The exact stereochemistry of these products has not been reported. However, the dimethylaminomercuration of *cis-*

$$(9)$$

$$(10)$$

and *trans*-2-butenes has been shown to yield anti adducts in $\geq 97\%$ stereospecificity [28].

Some work has been done on the aminomercuration of functionally substituted alkenes. As noted earlier, alkenols react smoothly in the absence of water, generally without interference by intramolecular alkoxymercuration [19, 32]. Allylic esters, urethanes and ureas react similarly, under conditions in which the corresponding homoallyl and higher homologs fail to react [19]. It has been suggested that a cyclic transition state of the following type is necessary for these reactions (Eq. 11). Other functional groups have also been accommodated by the aminomercuration reaction.

$$RCOCH_2CH=CH_2 \longrightarrow \longrightarrow RCOCH_2CHCH_2HgCl \tag{11}$$

With certain functional groups, however, alternative reactions occur. For example, the aminomercuration-demercuration of isopropenyl acetylene gives a variety of products involving attack on both the carbon—carbon double and triple bonds (Eq. 12) [33].

$$\tag{12}$$

The aminomercuration of vinyl ethers affords either vinyl amines or heterocycles (Eqs. 13–15). The latter two reactions are also catalyzed by

$$\tag{13}\ [37]$$

$$RHN-C_n-OH \ + \ ROCH=CH_2 \longrightarrow CH_3CH \qquad n = 2 \ or \ 3 \tag{14}\ [31]$$

$$RNHCH_2CH_2NHR \ + \ C_2H_5OCH=CH_2 \longrightarrow \tag{15}\ [31]$$

silver salts, but mercury salts proved superior.

VI. Aminomercuration

Somewhat similar results are observed with enol acetates. While indole and carbazole reportedly afford the anticipated aminomercuration products with vinyl acetate and mercuric acetate (Eq. 16) [23, 24]; by using mercuric

$$R_2NH \; + \; H_2C=CHOAc \; \xrightarrow{Hg(OAc)_2} \; AcOHgCH_2\overset{\displaystyle NR_2}{\overset{|}{C}}HOAc \qquad (16)$$

trifluoroacetate, mercuric sulfate, or mercuric acetate plus sulfuric acid; carbazole [24], benzimidazoles [25], benztriazole [25, 34], various pyrazoles and even diethylamine [21] afford the corresponding vinyl amines (Eq. 17).

$$R_2NH \; + \; H_2C=\overset{\displaystyle OAc}{\overset{|}{C}}R' \; \longrightarrow \; H_2C=\overset{\displaystyle NR_2}{\overset{|}{C}}R' \qquad (17)$$

Vinyl acetate is superior to n-butyl vinyl ether in several of these reactions [25]. Many of these same amines react with vinyl acetate and catalytic amounts of mercuric acetate plus sulfuric acid to afford predominantly the corresponding vinylamine polymers [21].

Allylation has been accomplished under conditions similar to those of the vinyl ether and ester reactions. Benztriazole and carbazole react with allyl alcohol and mercuric acetate in refluxing benzene to yield the corresponding N-allyl amines (Eq. 18) [20].

$$\text{(benzotriazole)} \; + \; H_2C=CHCH_2OH \; \longrightarrow \; \text{(N-allyl benzotriazole)} \qquad (18)$$

The demercuration of simple aminomercurials has been accomplished using a variety of reagents, but sodium borohydride is the most widely employed. In a comparison with a number of other reducing reagents, it gave higher yields of the desired amines and fewer rearrangement products [18, 42]. Two side reactions are common in these demercuration reactions, olefin formation and nitrogen migration.

The extent of olefin formation is dependent on the structure of the mercurial and the reaction conditions. For example, aminomercurials derived from methyl acrylate and anilines afford modest yields of the β-aminoester after sodium borohydride reduction, while aminomercurials derived from acrylate esters substituted about the double bond give only the starting olefin (Eq. 19) [26]. In general, the amount of olefin can be reduced by

$$RCH=CRCO_2CH_3 \; \xleftarrow{R=alkyl} \; RCH-\overset{\displaystyle R'NAr}{\overset{|}{C}}RCO_2CH_3 \; \xrightarrow{R=H} \; ArNR'CH_2CH_2CO_2CH_3 \qquad (19)$$
$$\underset{\displaystyle HgOAc}{\overset{|}{}}$$

adding 10% NaOH prior to reduction [38] and using a phase transfer catalyst [27]. This phase transfer approach appears to be the best method for demercuration of aminomercurials.

Another significant problem is nitrogen migration during reduction. The extent of rearrangement depends on the structure of the olefin and the amine employed, as seen below (Eq. 20) [38]. The reaction conditions are also

$$
\begin{array}{llll}
X=OCH_3, & R_2NH = PhNH_2 & 100 & : & 0 & (41\%\ total) \\
& = PhCH_2NH_2 & 90 & : & 10 & (78\%\ total) \\
& = Et_2NH & 60 & : & 40 & (39\%\ total) \\
X=CF_3, & R_2NH = PhCH_2NH_2 & 100 & : & 0 & (57\%\ total)
\end{array}
\tag{20}
$$

$$
\begin{array}{lll}
& NaOH & 65\% & 0\% \\
& THF/H_2O & 30\% & 30\% \\
& HOAc/H_2O & 33\% & 2\%
\end{array}
\tag{21}
$$

critical (Eq. 21) [42]. Alkaline sodium borohydride under phase transfer conditions appears superior to other reducing agents which have been employed for demercuration.

A number of other demercuration procedures have been reported however. Lithium aluminium hydride tends to give more rearrangement then sodium borohydride [6, 18, 42]. Sodium-mercury amalgam in water or D_2O reduces stereospecifically with retention, but the yields of amine are low and substantial amounts of the starting olefin are regenerated [4, 28, 42]. Lithium borohydride and hydrazine have also been used for demercuration, but offer no advantages [42]. The demercuration of aminomercurials by a number of metals in various protic solvents has also been examined: Mg in $PhNH_2$ and/or MeOH plus THF [11], Li in $PhNH_2$ or n-BuNH$_2$ plus THF [10], Ca in MeOH or $PhNH_2$ plus THF [12], Na in NH_3/THF [35]. None of these procedures appears superior to the sodium borohydride-phase transfer approach, except that the Na/NH_3/THF procedure is useful for simultaneously removing a benzyl group, thus allowing a facile synthesis of aminothiols from benzyl allyl sulfide (Eq. 22) [35].

$$
PhCH_2SCH_2CH=CH_2 \xrightarrow[Hg(OAc)_2]{R_2NH} \xrightarrow[NH_3]{Na} HSCH_2\overset{NR_2}{\underset{|}{C}}HCH_3
\tag{22}
$$

Useful transmetallation reactions have been reported for aminomercurials. While the reaction with Li, Na or K followed by an aqueous or methanolic quench affords reduction product (Eq. 23) [5, 9]; the low

465

$$RNHCH_2CH_2HgX \xrightarrow[\text{2.ROH}]{\text{1. M}} RNHCH_2CH_3 \qquad M = Li, Na, K \tag{23}$$

temperature reaction with phenyl lithium, sodium or potassium followed by the corresponding metal provides some unique beta-substituted alkali metal organometallics (Eq. 24) [7, 8]. These compounds can be reacted further with D_2O, oxygen, carbon dioxide, aldehydes or ketones, and tri-methylchlorosilane to give the expected products after work-up (Eq. 25). These reactions could prove quite useful for the preparation of beta-substituted amines.

$$RNHCH_2CH_2HgBr \xrightarrow[\text{2. M}]{\text{1. PhM}} \overset{\overset{\displaystyle M}{|}}{R}NCH_2CH_2M \tag{24}$$
$$M = Li, Na, K$$

$$\overset{\overset{\displaystyle Li}{|}}{R}NCH_2CH_2Li \begin{cases} \xrightarrow{D_2O} & RNDCH_2CH_2D \\ \xrightarrow{O_2} & RNHCH_2CH_2OH \\ \xrightarrow{CO_2} & RNHCH_2CH_2CO_2H \\ \xrightarrow{R^1COR^2} & RNHCH_2CH_2C(OH)R^1R^2 \\ \xrightarrow{Me_3SiCl} & RNHCH_2CH_2SiMe_3 \end{cases} \tag{25}$$

Relatively few other reactions of aminomercurials have been examined. The addition of HCl affords the corresponding ammonium salts which revert to starting olefin upon heating [4, 30]. Aqueous potassium cyanide, or carbon dioxide at elevated temperatures, also regenerate the starting olefin [4]. One aminomercurial has been reported to react with bromine to produce an enamine (Eq. 26) [6]. Finally, the transmetallation of an amino-mercurial by palladium chloride and subsequent lithium aluminum deuteride reduction to the hydrocarbon is reported to proceed with retention of configuration in both steps [43].

$$\langle \ \rangle \!-\! \overset{\overset{\displaystyle CH_3}{|}}{N}CH_2CH_2HgBr \xrightarrow{Br_2} \langle \ \rangle \!-\! \overset{\overset{\displaystyle CH_3}{|}}{N}\!-\!CH\!=\!CH_2 \tag{26}$$

B. Aminoalkenes

The intramolecular solvomercuration of unsaturated amines provides a valuable route to heterocyclic mercurials and amines (Table 6.2). Mercuric chloride and mercuric acetate are essentially the only mercury salts which have been employed in this reaction. Unlike the reaction of simple olefins

Table 6.2. Intramolecular Aminomercuration of Aminoalkenes

Aminoalkene	Mercuric salt	Organomercurial(s) (% Yield)	Subsequent reactant(s)	Product(s) (% Yield)	Ref.
$H_2C=CH(CH_2)_3NH_2$	HgX_2	—	$NaBH_4$ / NaOH	CH_3–pyrrolidine (NH); X = Cl (20), OAc (70)	44, 45
$H_2C=CH(CH_2)_3NHCH_3$	$HgCl_2$	—	$NaBH_4$ / NaOH	2-CH_3-1-CH_3-pyrrolidine (50)	44, 45
$H_2C=CHCH_2O(CH_2)_2NHCH_3$	$Hg(OAc)_2$	—	$NaBH_4$ / NaOH	3-CH_3-4-CH_3-morpholine (36)	46
$H_2C=CH(CH_2)_2CH-CH_2$, NCH_3	$Hg(OAc)_2$	—	$NaBH_4$ / NaOH	CH_3~pyrrolidine(CH_3)~CH_2OAc + $H_2C=CH(CH_2)_2CH-CH_2$, NCH_3 (30)	47
$H_2C=CH(CH_2)_2NH(CH_2)_2CH_3$	$HgCl_2$	—	$NaBH_4$ / NaOH	1-(CH_2)$_2CH_3$-pyrrolidine (70)	44
$H_2C=CHCH(CH_3)CH(CH_2OH)CH_2NH_2$	$Hg(OAc)_2$	—	$NaBH_4$ / NaOH	CH_3~pyrrolidine(NH)~CH_2OH (42) : CH_3~tetrahydrofuran~CH_2NH_2 (38) : 4-CH_3-piperidine-3-CH_2OH (20) (75 total)	48
$H_2C=C(CH_3)CH_2O(CH_2)_2NHCH_3$	$Hg(OAc)_2$	—	$NaBH_4$ / NaOH	3,3-diCH_3-4-CH_3-morpholine (65)	46

Table 6.2. (continued)

Aminoalkene	Mercuric salt	Organomercurial(s) (% Yield)	Subsequent reactant(s)	Product(s) (% Yield)	Ref.
$H_2C{=}CHCH_2OCH_2CH(NH_2)CH_2CH_3$	$Hg(OAc)_2$	—	$NaBH_4/NaOH$	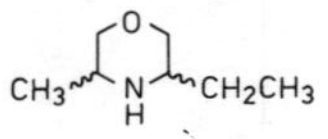 (50)	46
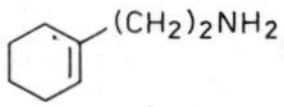 [structure: cyclohexenyl-$(CH_2)_2NH_2$]	$Hg(OAc)_2$	—	$NaBH_4/NaOH$	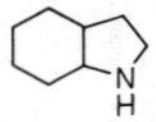 (30)	48
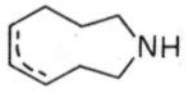 [structure: cyclic alkene NH]	$HgCl_2$	$RHgCl\cdot HCl$ (80)	$NaBH_4/NaOH$	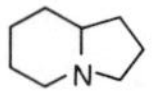 (66)	49
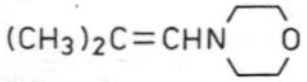 $(CH_3)_2C{=}CHN$[morpholine]	$Hg(OAc)_2$	—	$NaBH_4$	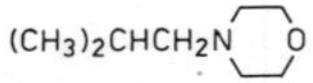 $(CH_3)_2CHCH_2N$[morpholine] (51)	50
$H_2C{=}CHCH_2O(CH_2)_2NHCH_2CH{=}CH_2$	$Hg(OAc)_2$	—	$NaBH_4/NaOH$	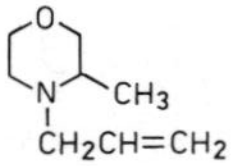 (53) CH_3 / $CH_2CH{=}CH_2$	46
$H_2C{=}CH(CH_2)_3NH(CH_2)_2CH_3$	$HgCl_2$	—	$NaBH_4/NaOH$	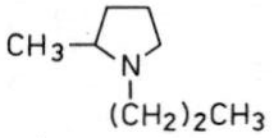 CH_3[pyrrolidine]$(CH_2)_2CH_3$ (49, 75)	44, 45
	$HgCl_2$	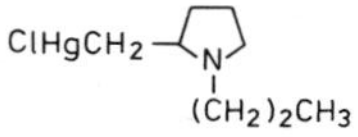 $ClHgCH_2$[pyrrolidine]$(CH_2)_2CH_3$	$NaBH_4/NaOH$ (phase transfer)	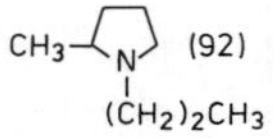 CH_3[pyrrolidine]$(CH_2)_2CH_3$ (92) $+\ H_2C{=}CH(CH_2)_3NH(CH_2)_2CH_3$ (8)	27
$p{-}H_2C{=}CHCH_2NHC_6H_4Cl$	$Hg(OAc)_2$, KBr	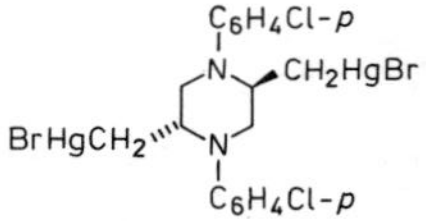 $BrHgCH_2$[piperazine with $C_6H_4Cl{-}p$, CH_2HgBr, $C_6H_4Cl{-}p$] (~87)	$NaBH_4/NaOH/$ $C_6H_5NH_2$	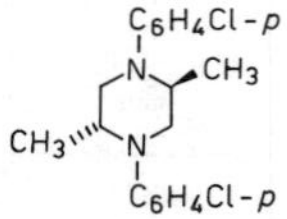 [piperazine with $C_6H_4Cl{-}p$, CH_3, CH_3, $C_6H_4Cl{-}p$] (56)	51

Table 6.2. (continued)

Aminoalkene	Mercuric salt	Organomercurial(s) (% Yield)	Subsequent reactant(s)	Product(s) (% Yield)	Ref.
$H_2C=CHCH_2NHC_6H_5$	Hg(OAc)$_2$, KBr	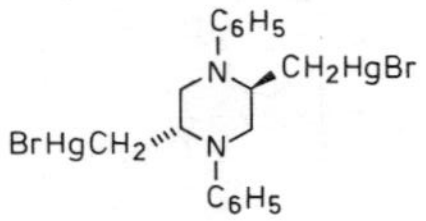 (~97)	NaBH$_4$/NaOH / C$_6$H$_5$NH$_2$	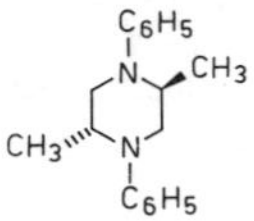 (78)	51
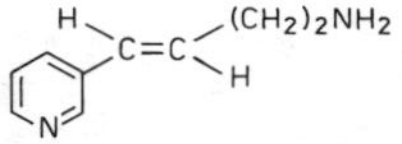 (CH$_2$)$_2$NH$_2$ (pyridyl-substituted alkene)	Hg(OAc)$_2$	—	NaBH$_4$/KOH	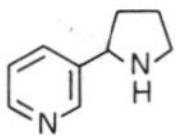 (66)	52
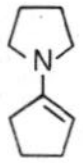 (cyclopentenyl pyrrolidine)	HgX$_2$	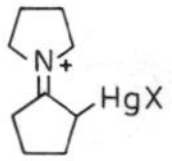 HgX$_3^-$; X = Cl (95), Br (96)	LiAlH$_4$	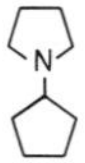 (70)	53, 54
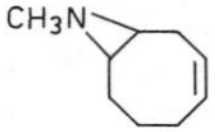 CH$_3$N (bicyclic aziridine-cyclooctene)	Hg(OAc)$_2$	—	NaBH$_4$/NaOH	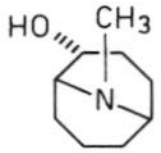 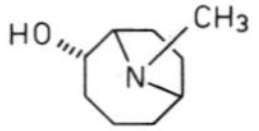 4 : 1	47
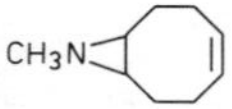 CH$_3$N (bicyclic aziridine-cyclooctene)	HgX$_2$	—	NaBH$_4$/NaOH	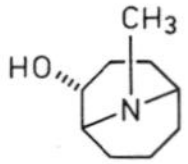 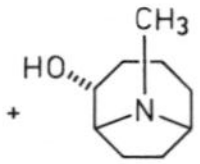 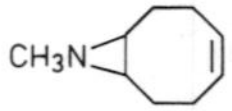 X = Cl 10–20 : 80–90; X = OAc 90 : 10 : —	47
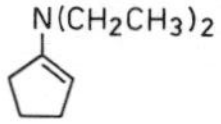 N(CH$_2$CH$_3$)$_2$ (cyclopentene)	Hg(OAc)$_2$	—	NaBH$_4$	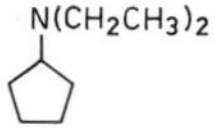 (78)	50
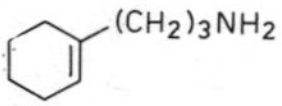 (CH$_2$)$_3$NH$_2$ (cyclohexene)	Hg(OAc)$_2$	—	NaBH$_4$/NaOH	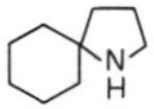 (40)	48

Table 6.2. (continued)

470

Aminoalkene	Mercuric salt	Organomercurial(s) (% Yield)	Subsequent reactant(s)	Product(s) (% Yield)	Ref.
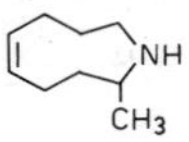 [structure: cyclooctene-fused ring with NH–CH$_3$]	HgCl$_2$	RHgCl·HCl (96)	NaBH$_4$/NaOH	 [indolizidine structures with CH$_3$] 70 : 30	49
H$_2$C=CHCH$_2$OCH(CH$_3$)CH$_2$–NHCH$_2$CH=CH$_2$	Hg(OAc)$_2$	—	NaBH$_4$/NaOH	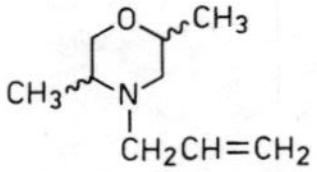 [morpholine structure with CH$_3$, CH$_3$, N–CH$_2$CH=CH$_2$] (68) 20 : 80 cis / trans	46
H$_2$C=CHCH$_2$OCH$_2$– [piperidine, HN] 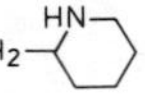	Hg(OAc)$_2$	—	NaBH$_4$/NaOH	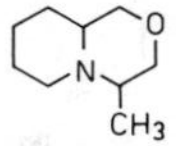 [bicyclic O,N structure with CH$_3$] (58) (mostly cis)	46
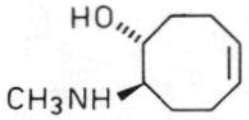 HO,,, [cyclooctene] CH$_3$NH,,,	HgX$_2$	—	NaBH$_4$/NaOH	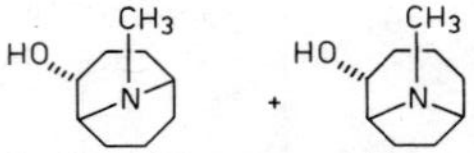 [two bicyclic structures HO,,,–N–CH$_3$] X = Cl 60 : 40; X = OAc 5 : 95	47
CH$_3$CH=C(NEt$_2$)CH$_2$CH$_3$	Hg(OAc)$_2$	—	NaBH$_4$	CH$_3$CH$_2$CH(NEt$_2$)CH$_2$CH$_3$ (71)	50
H$_2$C=CH(CH$_2$)$_3$NHC(CH$_3$)$_3$	HgCl$_2$	—	NaBH$_4$/NaOH	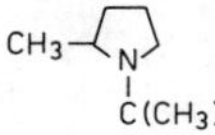 CH$_3$–[pyrrolidine, N–C(CH$_3$)$_3$] (60)	44, 45
H$_2$C=CH(CH$_2$)$_2$CH(CH$_3$)NH(CH$_2$)$_2$CH$_3$	HgCl$_2$	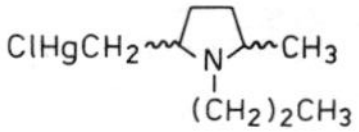 ClHgCH$_2$–[pyrrolidine N–(CH$_2$)$_2$CH$_3$]–CH$_3$ (50 ?)	NaBH$_4$/NaOH	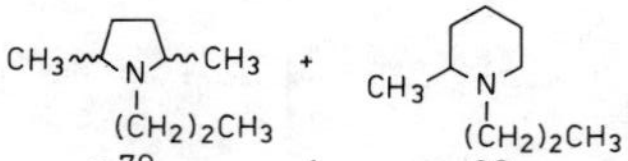 CH$_3$–[pyrrolidine N–(CH$_2$)$_2$CH$_3$]–CH$_3$ + CH$_3$–[piperidine N–(CH$_2$)$_2$CH$_3$] ~70 : ~30	45
	HgCl$_2$	—	NaBH$_4$/NaOH	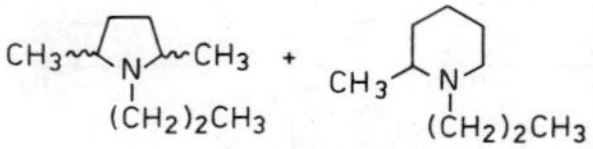 CH$_3$–[pyrrolidine N–(CH$_2$)$_2$CH$_3$]–CH$_3$ + CH$_3$–[piperidine N–(CH$_2$)$_2$CH$_3$] 2 : 1 (60 total)	44

Table 6.2. (continued)

Aminoalkene	Mercuric salt	Organomercurial(s) (% Yield)	Subsequent reactant(s)	Product(s) (% Yield)	Ref.
	HgX_2 (X = Cl, OAc)	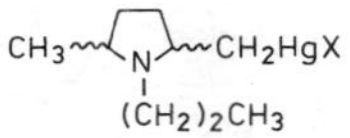 CH_3—[pyrrolidine ring, N–$(CH_2)_2CH_3$]—CH_2HgX	$NaBH_4$/NaOH (phase transfer)	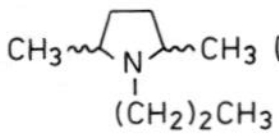 CH_3—[pyrrolidine, N–$(CH_2)_2CH_3$]—CH_3 (83) 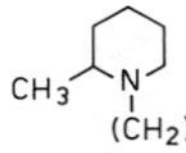 + [piperidine, CH_3, N–$(CH_2)_2CH_3$] (17)	55
$H_2C{=}CH(CH_2)_4NH(CH_2)_2CH_3$	$HgCl_2$	—	$NaBH_4$/NaOH	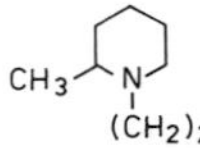 [piperidine, CH_3, N–$(CH_2)_2CH_3$] (50)	44
	$HgCl_2$	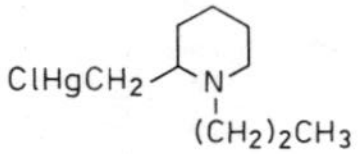 $ClHgCH_2$—[piperidine, N–$(CH_2)_2CH_3$]	$NaBH_4$/NaOH (phase transfer)	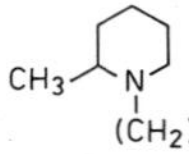 CH_3—[piperidine, N–$(CH_2)_2CH_3$] (90) + $H_2C{=}CH(CH_2)_4NH(CH_2)_2CH_3$ (10)	27
trans– $CH_3CH{=}CH(CH_2)_3NH(CH_2)_2CH_3$	$HgCl_2$	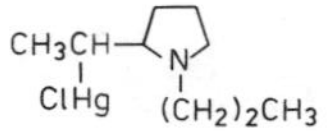 CH_3CH–[pyrrolidine, N–$(CH_2)_2CH_3$], ClHg (45 ?)	$NaBH_4$/NaOH	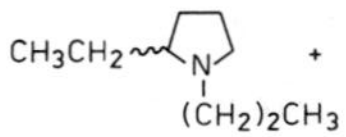 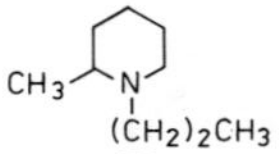 CH_3CH_2—[pyrrolidine, N–$(CH_2)_2CH_3$] + CH_3—[piperidine, N–$(CH_2)_2CH_3$] ~70 : ~30	45
	—	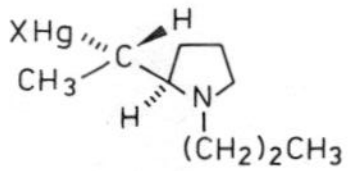 XHg···$\overset{H}{\underset{CH_3}{C}}$—[pyrrolidine, N–$(CH_2)_2CH_3$]	$NaBH_4$/NaOH (X = Cl)	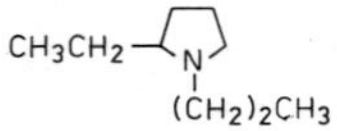 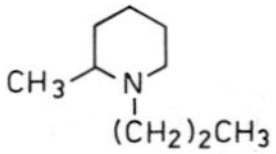 CH_3CH_2—[pyrrolidine, N–$(CH_2)_2CH_3$] + CH_3—[piperidine, N–$(CH_2)_2CH_3$] 70 : 30 (45 total)	44, 56
p–$H_2C{=}CHCH_2NHC_6H_4CH_3$	$Hg(OAc)_2$, KBr	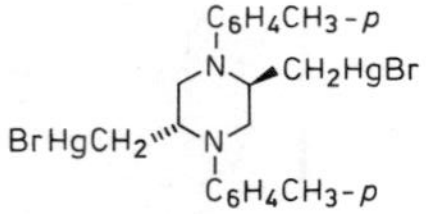 $BrHgCH_2$—[piperazine, N–$C_6H_4CH_3$-p (top), N–$C_6H_4CH_3$-p (bottom)]—CH_2HgBr (~93)	$NaBH_4$/NaOH / $C_6H_5NH_2$	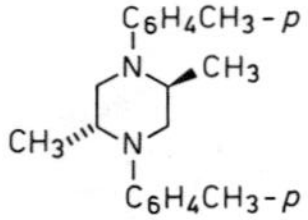 CH_3—[piperazine, N–$C_6H_4CH_3$-p (top), N–$C_6H_4CH_3$-p (bottom)]—CH_3 (76)	51
p–$H_2C{=}CHCH_2NHC_6H_4OCH_3$	$Hg(OAc)_2$, KBr	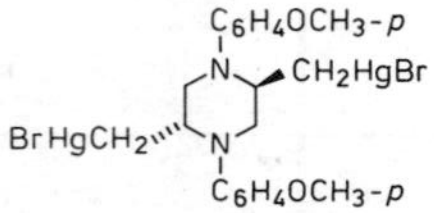 $BrHgCH_2$—[piperazine, N–$C_6H_4OCH_3$-p (top), N–$C_6H_4OCH_3$-p (bottom)]—CH_2HgBr (~98)	$NaBH_4$/NaOH / $C_6H_5NH_2$	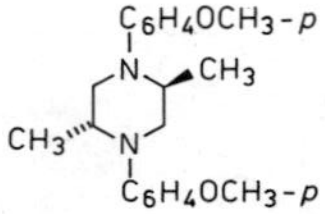 CH_3—[piperazine, N–$C_6H_4OCH_3$-p (top), N–$C_6H_4OCH_3$-p (bottom)]—CH_3 (54)	51

472

Table 6.2. (continued)

Aminoalkene	Mercuric salt	Organomercurial(s) (% Yield)	Subsequent reactant(s)	Product(s) (% Yield)	Ref.
(1-pyrrolidinyl-cyclohexene)	Hg(OAc)₂	—	NaBH₄	(cyclohexyl-pyrrolidine) (65)	50
CH₃CH₂N< (aziridinyl-cyclooctene)	HgX₂	—	NaBH₄/ NaOH	(azabicyclo products) + CH₃CH₂N< X = Cl 10—20 : 50 : 10 : 20—30 X = OAc 73 : 27 : — : — : —	47
H₂C=CH(CH₂)₂CH—CH₂ , NC(CH₃)₃	Hg(OAc)₂	—	NaBH₄/NaOH	CH₃–N(C(CH₃)₃)–CH₂OH + CH₃⋯N(C(CH₃)₃)–CH₂OH 60 : + H₂C=CH(CH₂)₂CHCH₂OH , HNC(CH₃)₃ : 40	47
N(CH₂CH₃)₂ (cyclohexene)	Hg(OAc)₂	—	NaBH₄	N(CH₂CH₃)₂ (cyclohexyl) (77)	50
(CH₂)₄NH₂ (cyclohexene)	Hg(OAc)₂	—	NaBH₄ / NaOH	(azaspiro) (48)	48
N(CH₃)₂ (cyclooctene)	Hg(OAc)₂	—	NaBH₄	N(CH₃)₂ (cyclooctyl) (65)	50

Table 6.2. (continued)

Aminoalkene	Mercuric salt	Organomercurial(s) (% Yield)	Subsequent reactant(s)	Product(s) (% Yield)	Ref.
$H_2C=CHCH_2OCH_2CH(C_2H_5)-$ $NHCH_2CH=CH_2$	$Hg(OAc)_2$	—	$NaBH_4/NaOH$	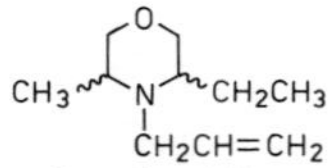 [morpholine structure] CH_3…N–CH_2CH_3, $CH_2CH=CH_2$ (46) (mostly cis)	46
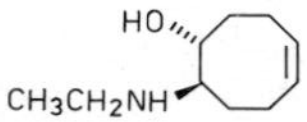 CH_3CH_2NH– [hydroxycyclooctene, HO]	HgX_2	—	$NaBH_4/NaOH$	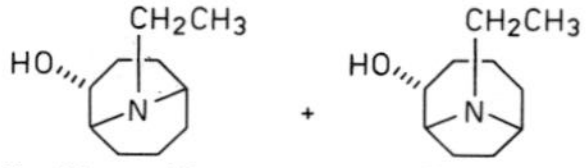 [two azabicyclic structures, HO, CH_2CH_3] $X=Cl$ 42 : 58; $X=OAc$ 18 : 82	47
$cis\text{-}C_6H_5CH=CH(CH_2)_3NH_2$	—	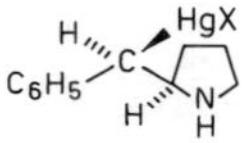 [pyrrolidine, H…C, C_6H_5, H, HgX, N–H]	$NaBH_4/NaOH$	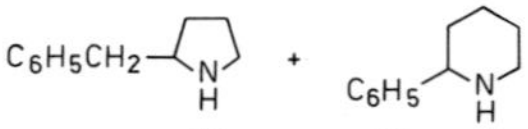 $C_6H_5CH_2$–[pyrrolidine N–H] + C_6H_5–[piperidine N–H] 70 : 30 (70 total)	44, 56
$o\text{-}CH_3CH=CHCH_2C_6H_4NHCH_3$	$Hg(OAc)_2$	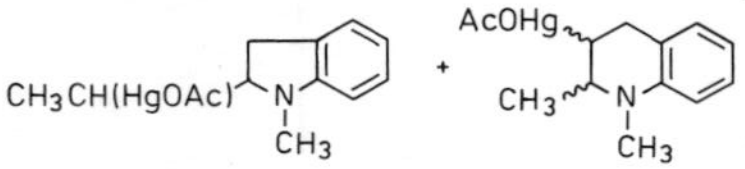 $CH_3CH(HgOAc)$–[indoline N–CH_3] + $AcOHg$…[tetrahydroquinoline CH_3, N–CH_3]	$NaBH_4/NaOH$ (phase transfer)	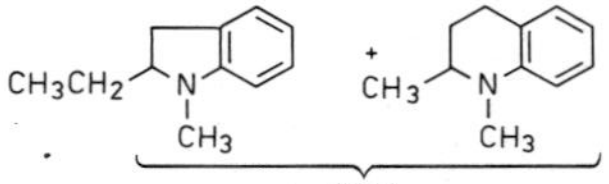 CH_3CH_2–[indoline N–CH_3] + CH_3…[tetrahydroquinoline N–CH_3] (92) + $o\text{-}CH_3CH=CHCH_2C_6H_4NHCH_3$ (8)	27
	HgX_2	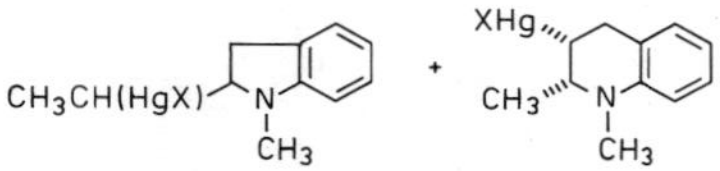 $CH_3CH(HgX)$–[indoline N–CH_3] + XHg…[tetrahydroquinoline CH_3, N–CH_3] $X=OAc$ (in pyridine) 71 : 29 (71 total); (in H_2O) 32 : 68 (58 total); $X=Cl$ (in pyridine) 14 : 86 (60 total); (in H_2O) 55 : 45 (67 total)	—	—	57
trans-o- $CH_3CH=CHCH_2C_6H_4NHCH_3$	—	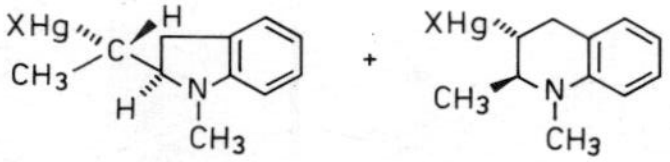 XHg…[indoline CH_3, C, H, N–CH_3] + XHg…[tetrahydroquinoline CH_3, N–CH_3]	—	—	56

Table 6.2. (continued)

Aminoalkene	Mercuric salt	Organomercurial(s) (% Yield)	Subsequent reactant(s)	Product(s) (% Yield)	Ref.
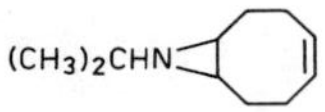 (CH$_3$)$_2$CHN— (bicyclic alkene)	HgX$_2$	—	NaBH$_4$ / NaOH	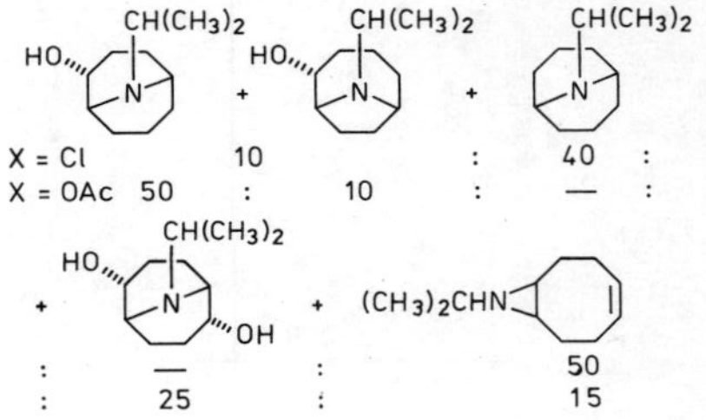 $CH(CH_3)_2$ products: X = Cl — : 10 : 40 : — ; X = OAc 50 : 10 : — : 25 : (CH$_3$)$_2$CHN— 50 / 15	47
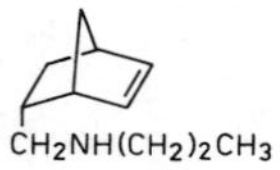 (norbornene) CH$_2$NH(CH$_2$)$_2$CH$_3$	—	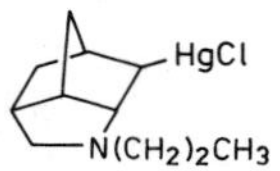 HgCl, —N(CH$_2$)$_2$CH$_3$ (bicyclic)	NaBH$_4$ / NaOH	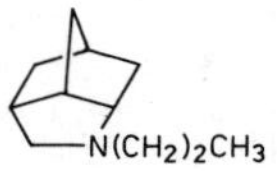 —N(CH$_2$)$_2$CH$_3$ (bicyclic)	57
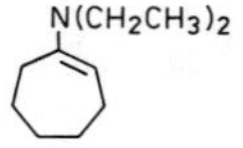 N(CH$_2$CH$_3$)$_2$ (cycloheptene)	Hg(OAc)$_2$	—	NaBH$_4$	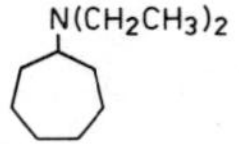 N(CH$_2$CH$_3$)$_2$ (90)	50
CH$_2$OH / H$_2$C=C(CH$_3$)CH$_2$CCH$_2$C(CH$_3$)=CH$_2$ / CH$_2$NH$_2$	Hg(OAc)$_2$	—	NaBH$_4$ / NaOH	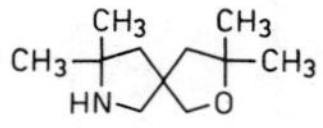 CH$_3$, CH$_3$ spiro (HN, O) (80)	48
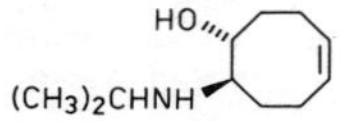 HO,,, (CH$_3$)$_2$CHNH— (cyclooctene)	HgX$_2$	—	NaBH$_4$ / NaOH	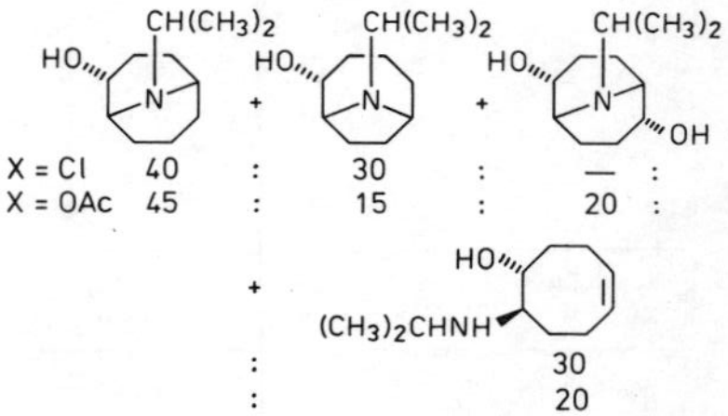 $CH(CH_3)_2$ products: X = Cl 40 : 30 : — ; X = OAc 45 : 15 : 20 ; + HO,,, (CH$_3$)$_2$CHNH— 30 / 20	47

Table 6.2. (continued)

Aminoalkene	Mercuric salt	Organomercurial(s) (% Yield)	Subsequent reactant(s)	Product(s) (% Yield)	Ref.
$(CH_3)_2C{=}CH(CH_2)_2CH(CH_3)$ – $NH(CH_2)_2CH_3$	$HgCl_2$	$RHgCl$ (50)	$NaBH_4/NaOH$	piperidine and pyrrolidine products, 1 : 1 (45 total)	44
	$Hg(OAc)_2$	$AcOHg$-piperidine (30 ?)	$NaBH_4/NaOH$	piperidine and pyrrolidine products, ~50 : ~50	45
	HgX_2 ($X = Cl, OAc$)	XHg-piperidine	$NaBH_4/NaOH$ (phase transfer)	piperidine product (100)	55
$(CH_3)_3CN$=cyclooctene	HgX_2	—	$NaBH_4/NaOH$	azabicyclic alcohols; $X = Cl$ 20 : 15 : 15 : ; $X = OAc$ 20 : 10 : 20 : 20 : ; + diol — : 10 + $(CH_3)_3CN$-cyclooctene 50 : 20	47
cyclooctene–$N(CH_2CH_3)_2$	$Hg(OAc)_2$	—	$NaBH_4$	cyclooctyl–$N(CH_2CH_3)_2$ (81)	50

Table 6.2. (continued)

Aminoalkene	Mercuric salt	Organomercurial(s) (% Yield)	Subsequent reactant(s)	Product(s) (% Yield)	Ref.
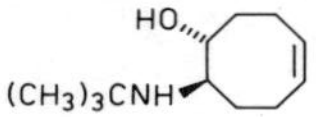 (structure)	HgX_2	—	$NaBH_4$ / NaOH	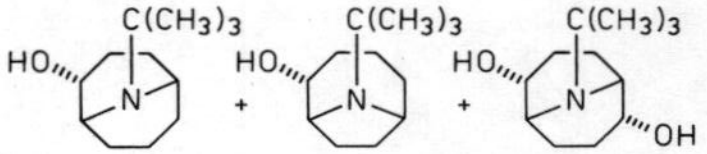 (structures) X = Cl 30 : 10 : — : ; X = OAc 40 : 20 : 25 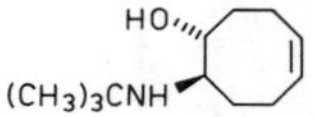 : ; + 60 : 15	47
$H_2C=CHCH(CH_3)(CH_2)_2NHCH_2C_6H_5$	—	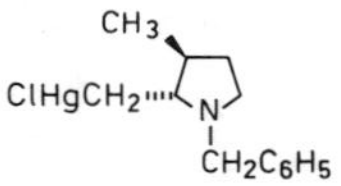 $ClHgCH_2$ (structure)	$H_2C=CClX$ / $NaHB(OCH_3)_3$	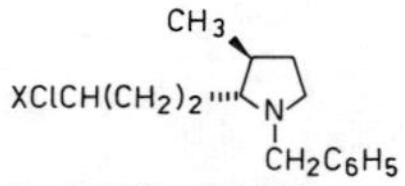 $XClCH(CH_2)_2$ (structure) X = Cl (43), CN (26)	58
p-$H_2C=CHCH(CH_3)(CH_2)_2NHC_6H_4OCH_3$	—	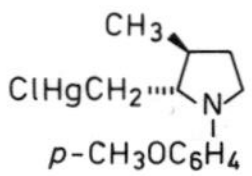 $ClHgCH_2$ (structure) p-$CH_3OC_6H_4$	$H_2C=CClX$ / $NaHB(OCH_3)_3$	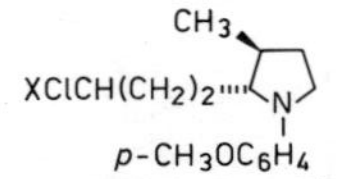 $XClCH(CH_2)_2$ (structure) p-$CH_3OC_6H_4$ X = Cl (18), CN (45)	58
$H_2C=CHCH_2CHOHCH_2CH(CH_3)NHC_6H_5$	$Hg(OAc)_2$	—	$NaBH_4$ / NaOH	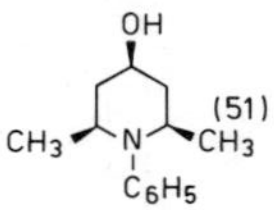 (structure) (51) 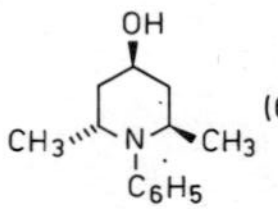 + (structure) (6)	32
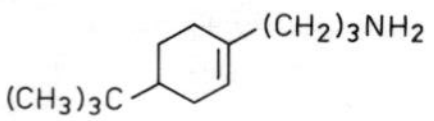 (structure) $(CH_3)_3C$ ...(CH_2)_3NH_2$	$Hg(OAc)_2$	—	$NaBH_4$ / NaOH	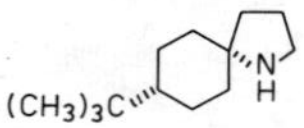 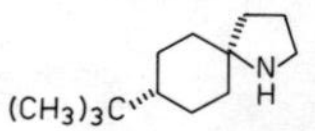 (structure) 63 : (structure) 37 (58 total)	48
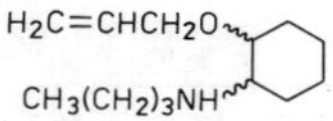 $H_2C=CHCH_2O$ (structure) $CH_3(CH_2)_3NH$	$HgCl_2$	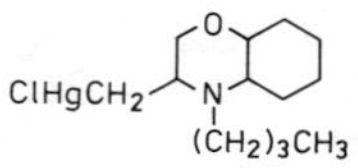 $ClHgCH_2$ (structure) $(CH_2)_3CH_3$	$NaBH_4$ / NaOH (phase transfer)	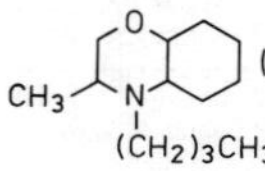 (structure) (93) 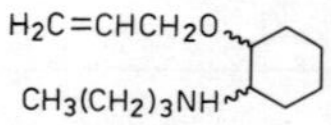 + (structure) (7)	27

Table 6.2. (continued)

Aminoalkene	Mercuric salt	Organomercurial(s) (% Yield)	Subsequent reactant(s)	Product(s) (% Yield)	Ref.
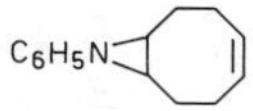 C_6H_5N (structure)	HgX_2	—	$NaBH_4$ / NaOH	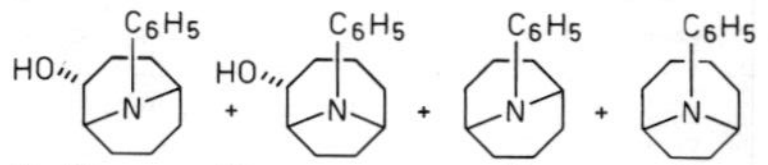 (structures) X = Cl 50 : 40 : 10; X = OAc 100 : — : — : —	47, 59
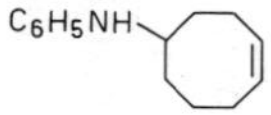 C_6H_5NH (structure)	$Hg(OAc)_2$	—	$NaBH_4$ / NaOH	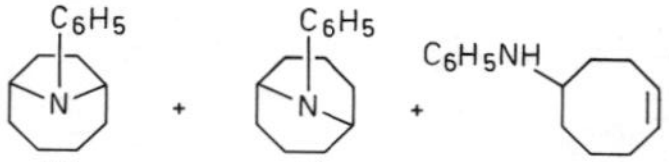 (structures) 60 : 10 : 30	59
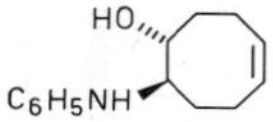 HO, C_6H_5NH (structure)	$Hg(OAc)_2$	—	$NaBH_4$ / NaOH	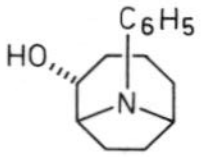 (structure)	47, 59
$CH_3CH(NH_2)(CH_2)_3CH=CH(CH_2)_9CH_3$	$Hg(OAc)_2$	—	$NaBH_4$	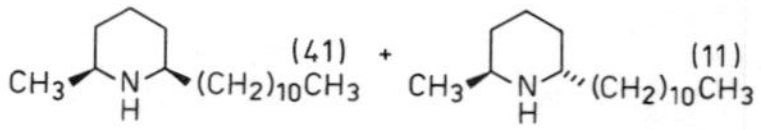 CH_3–N–$(CH_2)_{10}CH_3$ (41) + CH_3–N–$(CH_2)_{10}CH_3$ (11) + $CH_3CH(NH_2)(CH_2)_3CH=CH(CH_2)_9CH_3$ (31)	60
$CH_3CH(NH_2)(CH_2)_2CH=CH(CH_2)_{10}CH_3$	$Hg(OAc)_2$	—	$NaBH_4$	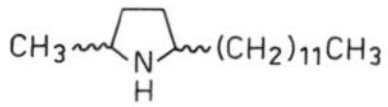 CH_3–N–$(CH_2)_{11}CH_3$	60
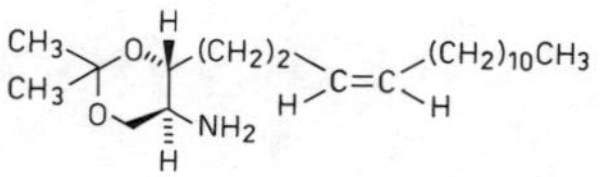 (structure with CH_3, CH_3, O, NH_2, $(CH_2)_2$C=C$(CH_2)_{10}CH_3$)	$Hg(OAc)_2$	—	$NaBH_4$	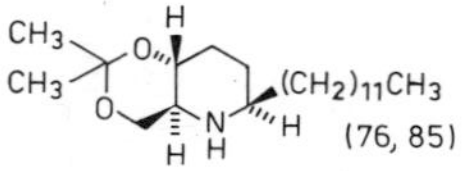 (structure) $(CH_2)_{11}CH_3$ (76, 85) 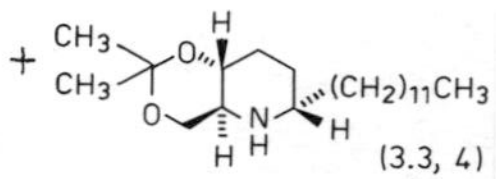 + (structure) $(CH_2)_{11}CH_3$ (3.3, 4)	61, 62

and amines which apparently requires at least two equivalents of the amine, an excess of amine is unnecessary when the reaction is carried out intramolecularly [44, 45]. The solvent appears to have relatively little influence on the yield of cyclic amine as seen in the cyclization of *n*-propyl-4-pentenylamine (Eq. 27): HOAc (10%), THF (75%), H_2O/THF (80%), H_2O (85%), CH_3OH (85%) [44, 45].

$$H_2C{=}CH(CH_2)_3NH(CH_2)_2CH_3 \longrightarrow \longrightarrow \tag{27}$$

Let's examine the effect of olefin structure on the products formed, starting with enamines. Most tertiary enamines react with mercuric halides to form ammonium salts (Eq. 28) [63]. However, the pyrrolidine enamine of

$$\tag{28}$$

cyclopentanone undergoes carbon-mercury bond formation (Eq. 29) [53, 54]. Lithium aluminium hydride reduction of this mercurial provides the corresponding saturated amine. By employing mercuric acetate, instead

$$\tag{29}$$

of mercuric halides, and reducing with sodium borohydride, a wide variety of tertiary enamines have been reduced to the corresponding saturated amines in high yield [50].

The mercuration and subsequent reduction of N-allylanilines provides an excellent stereoselective route to the piperazine ring system (Eq. 30) [51].

$$ArNHCH_2CH{=}CH_2 \longrightarrow \tag{30}$$

However, alkyl substitution on either end of the double bond or activation of the aromatic ring results in arylmercurial formation instead.

Only two examples of homoallylamine cyclization to pyrrolidines have been reported (Eqs. 31, 32). Note the anti-Markovnikov nature of this latter reaction.

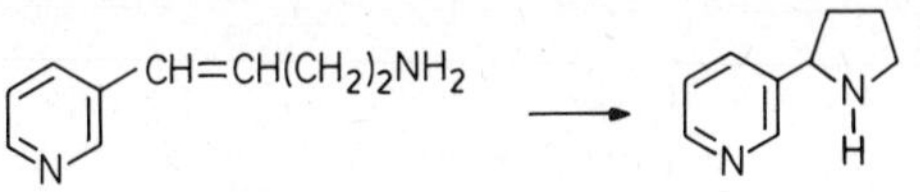

$$\text{(pyridyl)}-CH=CH(CH_2)_2NH_2 \longrightarrow \text{(pyridyl-pyrrolidine)} \qquad (31)\ [52]$$

$$\text{(cyclohexenyl)}-(CH_2)_2NH_2 \longrightarrow \text{(octahydroindole)} \qquad (32)\ [48]$$

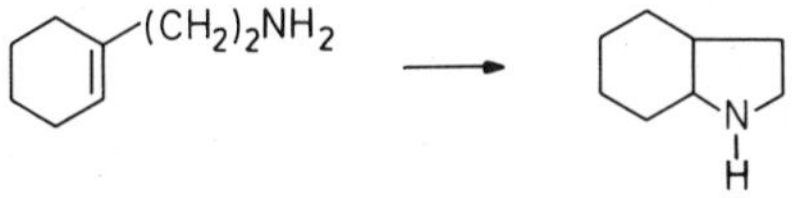

The cyclization of 4-pentenylamines has been studied much more widely. Cyclization to both 5- and 6-membered ring heterocycles has been observed depending on the substitution pattern of the unsaturated amine. Simple unsubstituted 4-pentenylamines apparently afford exclusively the pyrrolidine ring system (Eq. 33) [44, 45]. Addition of a methyl group in the one position

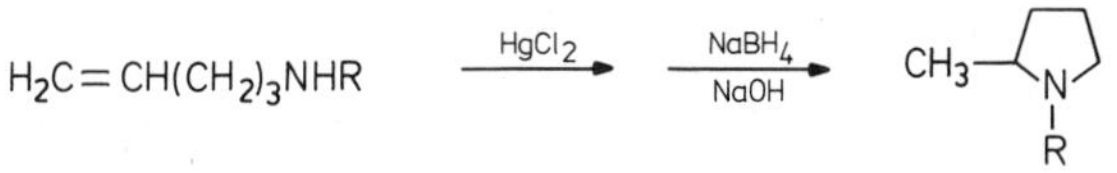

$$H_2C=CH(CH_2)_3NHR \xrightarrow{HgCl_2} \xrightarrow[NaOH]{NaBH_4} CH_3-\text{(N-R pyrrolidine)} \qquad (33)$$

affords a mercurial of unknown stereochemistry, which apparently rearranges upon sodium borohydride reduction to a mixture of 5- and 6-membered ring heterocycles (Eq. 34) [44, 45]. It is believed that nitrogen participation

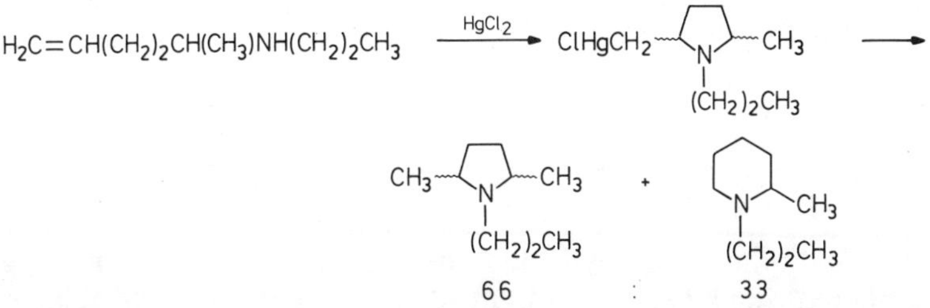

$$H_2C=CH(CH_2)_2CH(CH_3)NH(CH_2)_2CH_3 \xrightarrow{HgCl_2} ClHgCH_2\text{—(pyrrolidine)—}CH_3 \longrightarrow$$

$$CH_3\text{—(pyrrolidine)—}CH_3 \quad + \quad \text{(piperidine)—}CH_3 \qquad (34)$$

$$66 \quad : \quad 33$$

during reduction results in aziridinium ion formation which leads to the mixture of heterocycles (Eq. 35). Substitution in the 2 and 3 positions has also

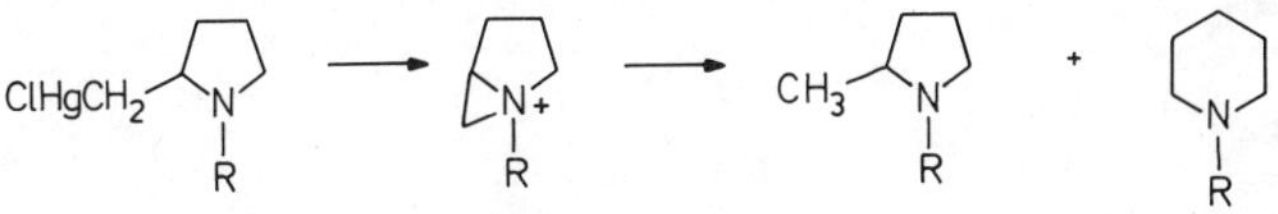

$$ClHgCH_2\text{—(N-R pyrrolidine)} \longrightarrow \text{(aziridinium N+-R)} \longrightarrow CH_3\text{—(N-R pyrrolidine)} \quad + \quad \text{(N-R piperidine)} \qquad (35)$$

resulted in mixtures of heterocycles *after reduction* (Eq. 36) [48]. Note the competing alkoxymercuration in this last example. The intermediate mer-

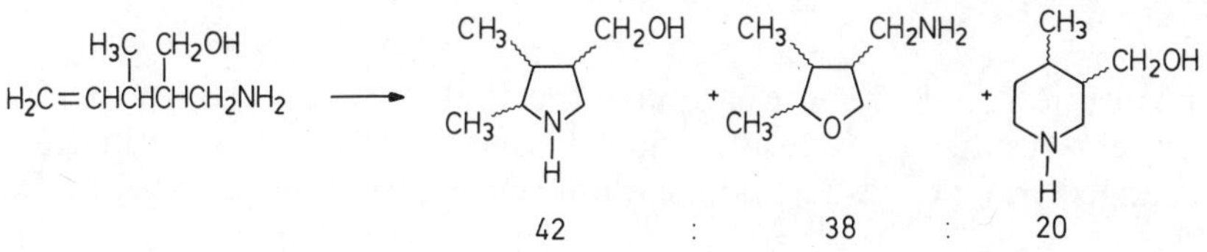

$$H_2C=CHCHCH CH_2NH_2 \longrightarrow \text{(pyrrolidine)} \quad + \quad \text{(tetrahydrofuran)} \quad + \quad \text{(piperidine)} \qquad (36)$$

$$42 \quad : \quad 38 \quad : \quad 20$$

reaction probably contains a 5-membered ring, since 3-substituted amines have been reported to stereoselectively cyclize to the *trans*-pyrrolidine mercurial (Eq. 37) [58]. Substitution in the 4-position apparently results in 5-mem-

$$H_2C=CHCH(CH_2)_2NHR \quad\longrightarrow\quad \tag{37}$$

bered ring formation as seen in the following interesting example involving simultaneous intramolecular amino- and alkoxymercuration (Eq. 38) [48].

$$\tag{38}$$

The introduction of substituents on the more remote carbon of the double bond in the 4-pentenylamine system can result in either 5- or 6-membered ring products. *cis*- or *trans*-Substitution by either one alkyl or aryl group (R^1 = Me or Ph) apparently affords exclusively the pyrrolidine mercurial (Eq. 39) [44, 45, 56, 60]. The addition across the double bond is apparently

$$R^1CH=CH(CH_2)_2CHNHR^3 \quad\longrightarrow\quad \tag{39}$$

exclusively anti [56]. Sodium borohydride reduction again affords a mixture of 5- and 6-membered ring heterocycles in most cases [44, 45, 56]. The related amine *o*-crotyl-N-methylaniline affords mixtures of aminomercurials, the ratio of which varies with the mercuric salt and the solvent employed (Eq. 40) [56, 57]. Both products arise by trans addition of mercury and nitrogen across the double bond.

$$\tag{40}$$

Disubstitution on the remote carbon of the double bond of the 4-pentenyl-amine system results in exclusive 6-membered ring mercurial formation, but rearranged products arise upon sodium borohydride reduction (Eq. 41) [45].

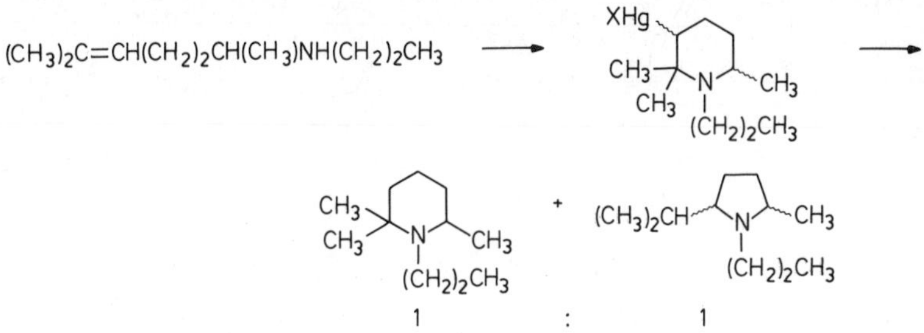

$$(CH_3)_2C{=}CH(CH_2)_2CH(CH_3)NH(CH_2)_2CH_3 \longrightarrow$$

(41)

More highly substituted cyclic and bicyclic amines have also been regioselectively cyclized (Eqs. 42–45). This last reaction is more stereoselective when the corresponding amides are cyclized (see amidomercuration, Chapter VII).

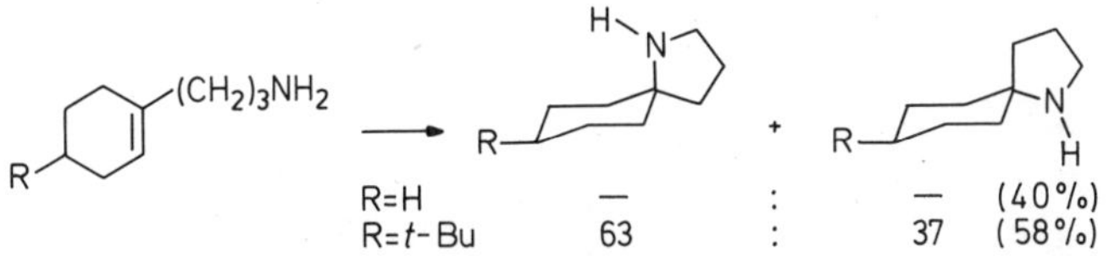

(42) [48]

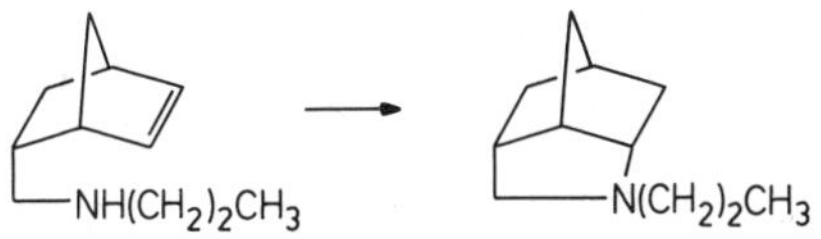

$$R{=}H \qquad — \qquad : \qquad — \quad (40\%)$$
$$R{=}t{-}Bu \qquad 63 \qquad : \qquad 37 \quad (58\%)$$

(43) [57]

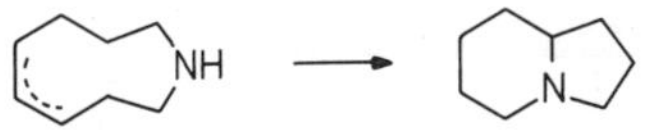

(44) [49]

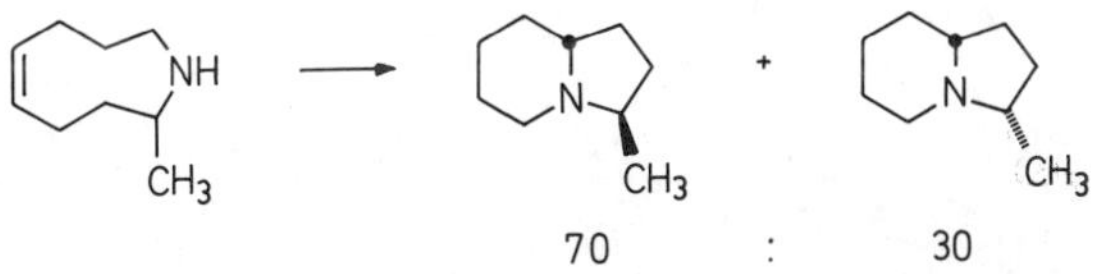

(45) [49]

$$70 \qquad : \qquad 30$$

The aminomercuration and subsequent demercuration of 5-hexenylamines affords exclusively 6-membered ring products (Eq. 46). This reaction has proved useful for the synthesis of simple piperidines (Eq. 47), naturally occurring piperidines (Eqs. 48, 49), morpholines (Eq. 50) and spirocyclic amines (Eq. 51). High stereoselectivity is frequently observed in these cyclizations.

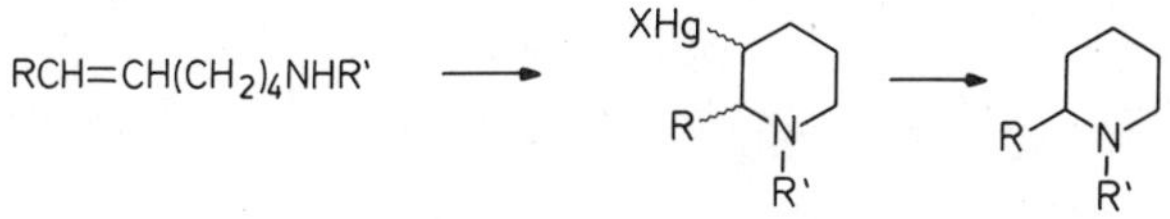

$$RCH{=}CH(CH_2)_4NHR' \longrightarrow$$

(46)

481

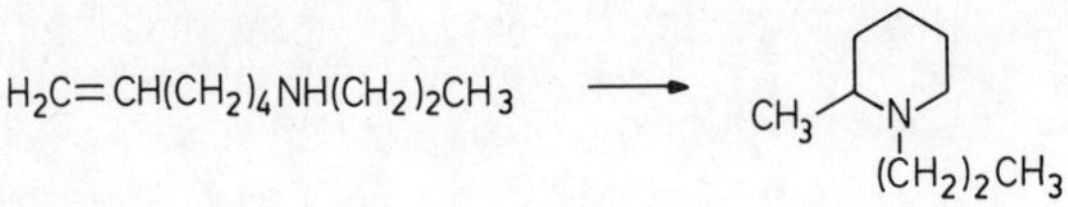

VI. Aminomercuration

$$H_2C=CH(CH_2)_4NH(CH_2)_2CH_3 \longrightarrow$$

(47) [44]

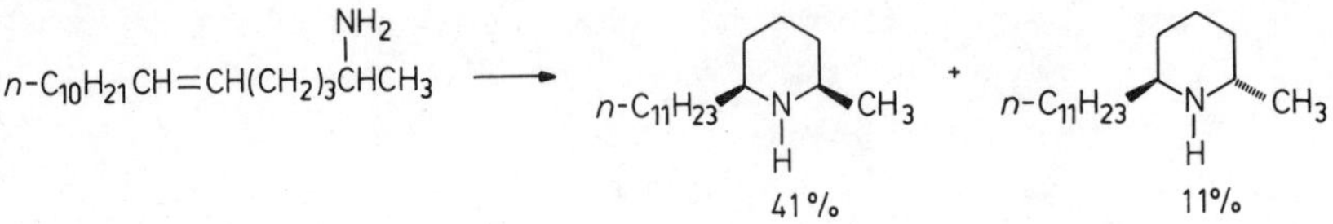

(48) [60]

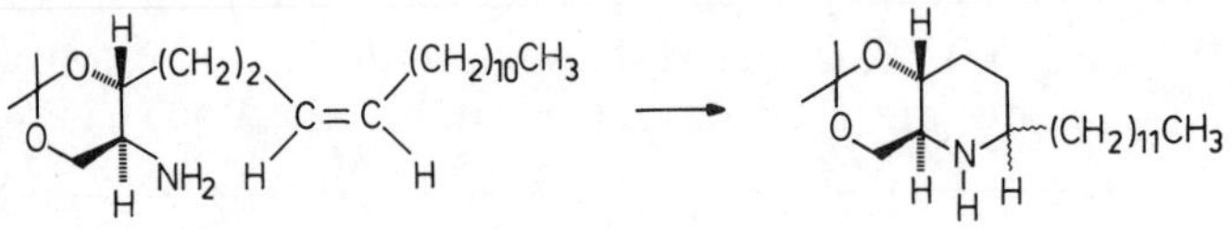

(49) [61, 62]

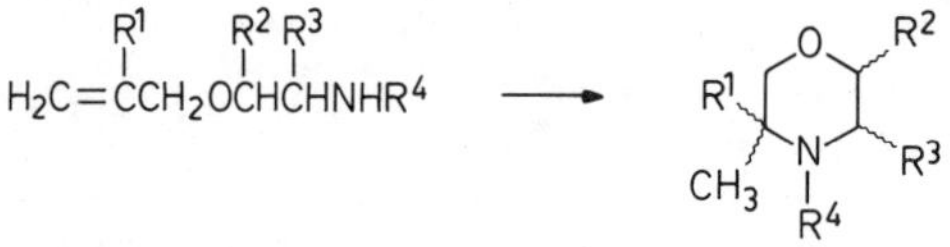

(50) [46]

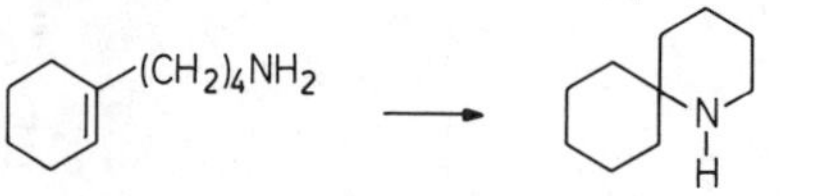

(51) [48]

The synthesis of bicyclic amines from cyclooctenyl amines has been reported (Eqs. 52) [47, 59]. The ratio of products depends on the mercuric

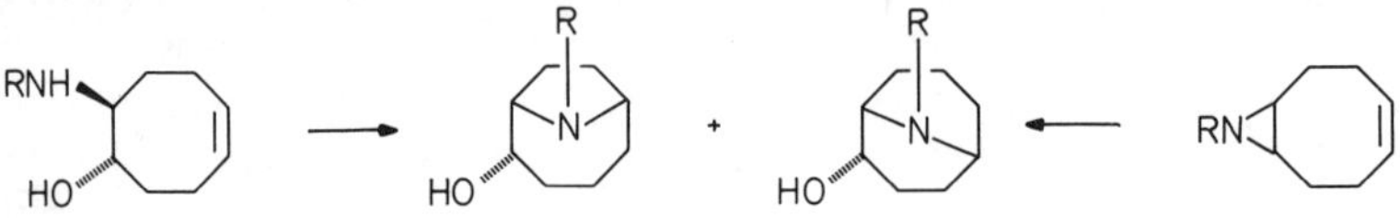

(52)

salt employed and the nature of the alkyl group on nitrogen. Several other products are also observed. Competitive intramolecular alkoxymercuration is not observed in these reactions, unless the reaction is acidified prior to reaction. The same products are observed when cyclooctenyl aziridines are submitted to the analogous reaction conditions. The tertiary amines are apparently capable of intramolecular aminomercuration with subsequent ring opening. These aziridine ring openings are not confined to cyclic systems (Eq. 53) [47].

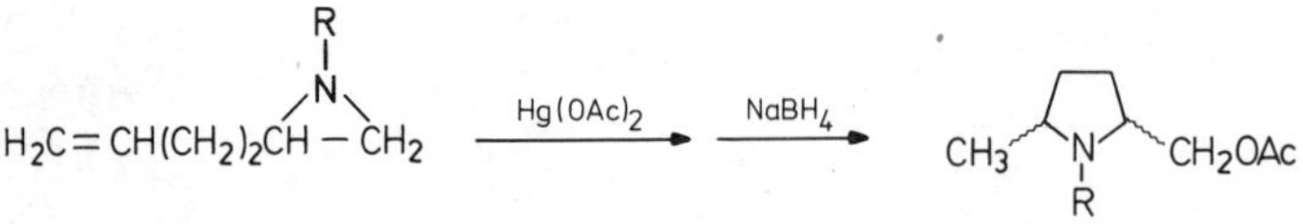

(53)

The demercuration of these cyclic aminomercurials is particularly prone to olefin formation. Excellent results are reported, however, using alkaline sodium borohydride under phase transfer conditions [21, 55]. As noted earlier, rearrangements of pyrrolidine mercurials to piperidines and vice versa have been observed even under phase transfer conditions.

Only one other reaction of these heterocyclic mercurials appears to have been studied. The pyrrolidine mercurials undergo sodium trimethoxyborohydride induced free radical additions (see Chapter VII in the monograph "Organomercury Compounds in Organic Synthesis") to electron-poor olefins, promising a valuable new route to certain alkaloids (Eq. 54) [58].

$$ \tag{54} $$

C. Dienes

The aminomercuration of dienes has been examined by several groups (Table 6.3). The nature of the products depends on the type of diene employed. No trienes or polyenes have ever been subjected to aminomercuration conditions.

The aminomercuration-demercuration of allenes has provided some unusual results. The aminomercuration of 1,2-cyclononadiene [27] and 3-methyl-1,2-butadiene [64] proceeds as anticipated, placing the mercury moiety on the central carbon of the allene and the amine on the more highly substituted terminal carbon (Eqs. 55, 56). Alkaline sodium borohydride reduction of the cyclic mercurial under phase transfer conditions proceeds

$$ \tag{55} $$

$$ (CH_3)_2C{=}C{=}CH_2 \longrightarrow (CH_3)_2C-C{=}CH_2 \longrightarrow (CH_3)_2C{=}CHCH_2NHC_6H_5 \tag{56} $$

in high yield to the expected allylic amine. However, the acyclic mercurial rearranges to the isomeric allylic amine. With N-methylaniline, this same allene also gives allylation of the aromatic ring (Eq. 57) [64]. With 1,2-pentadiene, aniline affords an 80:20 mixture of the terminal and internal

$$ (CH_3)_2C{=}C{=}CH_2 \longrightarrow (CH_3)_2C{=}CHCH_2N(CH_3)C_6H_5 \; + \tag{57} $$

Table 6.3. Aminomercuration of Dienes

Diene	Amine	Mercuric salt	Organomercurial(s) (% Yield)	Subsequent reactants	Product(s) (% Yield)	Ref.
$H_2C=C=CH_2$	$C_6H_5NH_2$	$Hg(OAc)_2$	—	$NaBH_4/NaOH$	$H_2C=CHCH_2NHC_6H_5$ (10)	64
$H_2C=CHCH=CH_2$	$p-XC_6H_4NH_2$ (X = H, Cl)	$Hg(OAc)_2$, KBr	$[p-XC_6H_4NHCH(CH_2HgBr)]_2$	$NaBH_4$	$p-XC_6H_4NHCH(CH_3)CH(CH_3)NHC_6H_4X-p$ X = H, Cl	65
$H_2C=C=C(CH_3)_2$	C_6H_5NHR	$Hg(OAc)_2$, KCl	$H_2C=C(HgCl)C(CH_3)_2NRC_6H_5$ R = H (45)	$NaBH_4/NaOH$	$(CH_3)_2C=CHCH_2NRC_6H_5$ $\quad + o-(CH_3)_2C=CHCH_2C_6H_4NHR$ R = H 100 : 0 (50 total) R = CH₃ 50 : 50 (45 total)	64
$H_2C=C(CH_3)CH=CH_2$	$C_6H_5NH_2$	$Hg(OAc)_2$	$H_2C=C(CH_3)CH(NHC_6H_5)CH_2HgOAc$	$NaBH_4/NaOH$ (phase transfer)	$H_2C=C(CH_3)CH(NHC_6H_5)CH_3$ (95) $\quad + H_2C=C(CH_3)CH=CH_2$ (5)	27
	$p-XC_6H_4NH_2$ (X = H, Cl)	$Hg(OAc)_2$, KBr	$BrHgCH_2C(CH_3)(NHC_6H_4X-p) - CH(NHC_6H_4X-p)CH_2HgBr$	$NaBH_4$	$(CH_3)_2C(NHC_6H_4X-p)CH(CH_3)NHC_6H_4X-p$	65
$H_2C=C=CHCH_2CH_3$	C_6H_5NHR	$Hg(OAc)_2$	—	$NaBH_4/NaOH$	$C_6H_5NRCH_2CH=CHCH_2CH_3$ $\quad + H_2C=CHCH(NRC_6H_5)CH_2CH_3$ R = H 80 : 20 : R = CH₃ 15(cis):25(trans) : — : $\quad + C_6H_5NRCH(CH_3)(CH_2)_2CH_3$: — (55 total) : 60 (30 total)	64
(cyclohexadiene)	$C_6H_5NH_2$	$Hg(OAc)_2$, KBr	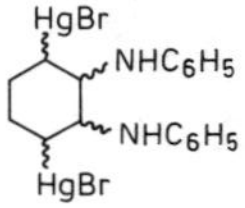 (cyclohexane with HgBr, NHC₆H₅, NHC₆H₅, HgBr substituents)	$NaBH_4$	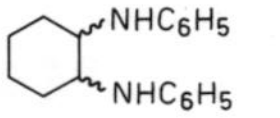 (cyclohexane with NHC₆H₅, NHC₆H₅)	65
cis- and trans- $H_2C=CHCH_2CH=CHCH_3$	$XC_6H_4NH_2$	$Hg(OAc)_2$	—	$NaBH_4/NaOH$	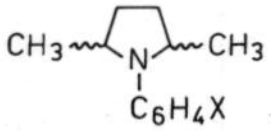 (pyrrolidine: CH_3, CH_3 on ring, N–C_6H_4X) X[cis/trans] (% yield) = p-Me[6/94](70); H[6/94](99); p-MeO[10/90](25); p-Cl[6/94](80); o-Cl[1/99](76)	66

Table 6.3. (continued)

Diene	Amine	Mercuric salt	Organomercurial(s) (% Yield)	Subsequent reactants	Product(s) (% Yield)	Ref.
$H_2C=CH(CH_2)_2CH=CH_2$	$C_6H_5NH_2$	$Hg(OAc)_2$	$AcOHgCH_2$—[N(C_6H_5) pyrrolidine]—CH_2HgOAc	M, H_2O	CH_3—[N(C_6H_5) pyrrolidine]—CH_3 + $C_6H_5NHCH(CH_3)(CH_2)_2CH=CH_2$; $M=Li$ · (31) : (47); $M=Na$ (24) : (56); $M=K$ (13) : (74)	5
	$C_6H_5NH_2$	—	$BrHgCH_2$—[N(C_6H_5) pyrrolidine]—CH_2HgBr	$Li/C_6H_5NH_2$	CH_3—[N(C_6H_5) pyrrolidine]—CH_3 (31) + $C_6H_5NHCH(CH_3)(CH_2)_2CH=CH_2$ (47)	10
	$C_6H_5NH_2$	—	$BrHgCH_2$—[N(C_6H_5) pyrrolidine]—CH_2HgBr	Mg/CH_3OH	CH_3—[N(C_6H_5) pyrrolidine]—CH_3 (26) + $C_6H_5NHCH(CH_3)(CH_2)_2CH=CH_2$ (73)	11
	$C_6H_5NH_2$	—	$BrHgCH_2$—[N(C_6H_5) pyrrolidine]—CH_2HgBr	Ca/CH_3OH	CH_3—[N(C_6H_5) pyrrolidine]—CH_3 (5) + $C_6H_5NHCH(CH_3)(CH_2)_2CH=CH_2$ (50)	12
	p-$XC_6H_4NH_2$	$Hg(OAc)_2$	$AcOHgCH_2$—[N(p-XC_6H_4) pyrrolidine]—CH_2HgOAc (X = H, Me, Cl, NO$_2$, CO$_2$CH$_3$)	$NaBH_4$ (X = H)	CH_3—[N(C_6H_5) pyrrolidine]—CH_3	67
	p-$XC_6H_4NH_2$	$Hg(OAc)_2$	—	$NaBH_4$	$ArNHCH(CH_3)CH_2CH_2CH=CH_2$; X = H (55) : ; CH_3 (27) : ; Cl (30) : ; CH_3O (35) : ; EtO_2C (85) : ; + CH_3—[N(Ar) pyrrolidine]—CH_3 + $H_2C=CH(CH_2)_2CH=CH_2$; : (9) : (22); : (12) : (53); : — : (50); : — : (40); : — : —	68

Table 6.3. (continued)

Diene	Amine	Mercuric salt	Organomercurial(s) (% Yield)	Subsequent reactants	Product(s) (% Yield)	Ref.
	$XC_6H_4NH_2$	$Hg(OAc)_2$, KBr	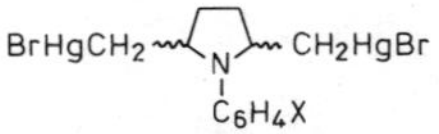 $BrHgCH_2$—[pyrrolidine, N-C_6H_4X]—CH_2HgBr; X = H (95), p-Me (89), p-MeO (88), p-Cl (90), o-Me (92)	$NaBH_4$/NaOH/ $C_6H_5NH_2$	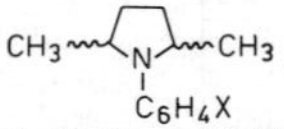 CH_3—[pyrrolidine, N-C_6H_4X]—CH_3; X = H (76), p-Me (70), p-MeO (66), p-Cl (47), o-CH_3 (44); cis/trans = 96/4, 76/24, 93/7, 89/11, 60/40	69
	$XC_6H_4NH_2$	$Hg(OAc)_2$, KY	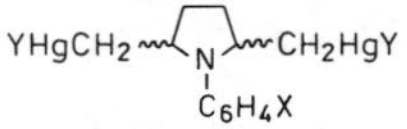 $YHgCH_2$—[pyrrolidine, N-C_6H_4X]—CH_2HgY; Y = OAc, X = p-Me (91); Y = Cl, X = p-Cl (97); Y = Br, X = H (100), p-MeO (80), o-Cl (90)	$NaBH_4$/NaOH	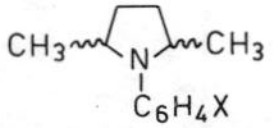 CH_3—[pyrrolidine, N-C_6H_4X]—CH_3; X[cis/trans] (% yield) = p-Me[8/92] (93); H[2/98] (99); p-MeO[32/68] (23); p-Cl[1/99] (83); o-Cl[6/94] (75)	66
	p-$XC_6H_4NH_2$	HgO/HBF_4/ DMF 80°C	—	—	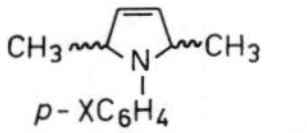 CH_3—[3-pyrroline, N-p-XC_6H_4]—CH_3; X = H (58), Me (44)	70
$(H_2C=CHCH_2)_2O$	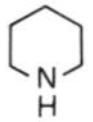 [piperidine, N-H]	$HgCl_2$	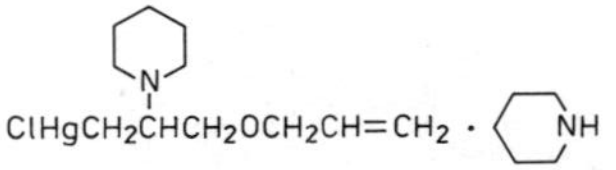 $ClHgCH_2CHCH_2OCH_2CH=CH_2 \cdot$ [piperidine, NH] (40)	—	—	19
	p-$XC_6H_4NH_2$	$Hg(OAc)_2$, KBr	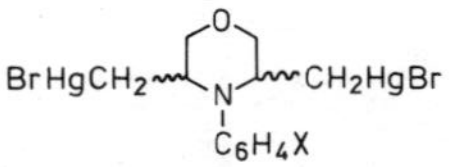 $BrHgCH_2$—[morpholine, N-C_6H_4X]—CH_2HgBr	$NaBH_4$ (neutral)	$H_2C=CHCH_2OCH_2CH(NHC_6H_4X)CH_3$; X = H (51), p-Me (43), p-MeO (48), p-NO_2 (39), p-Cl (79), o-Me (35), o-MeO (76)	71
	$XC_6H_4NH_2$	$Hg(OAc)_2$, KBr	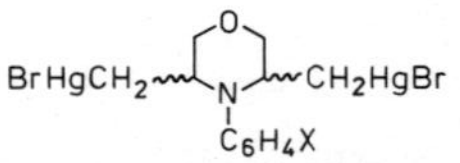 $BrHgCH_2$—[morpholine, N-C_6H_4X]—CH_2HgBr; X = H (100), p-Me (100), p-MeO (81), p-NO_2 (80), p-Cl (96), o-Me (100), o-MeO (87)	$NaBH_4$/NaOH	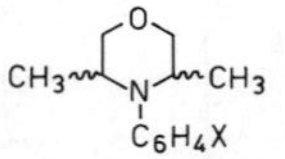 CH_3—[morpholine, N-C_6H_4X]—CH_3 [~isomer ratio]; X = H (71) [60/40], p-Me (74) [45/55], o-Me (83) [58/42], p-MeO (93) [44/56], o-MeO (55) [90/10], p-NO_2 (52) [60/40]	71
$(H_2C=CHOCH_2)_2$	$CH_3NHCH_2CH_2OH$	$Hg(OAc)_2$	—	—	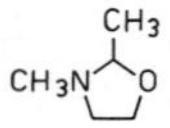 CH_3N[oxazolidine, 2-CH_3]O (68)	31

Table 6.3. (continued)

Diene	Amine	Mercuric salt	Organomercurial(s) (% Yield)	Subsequent reactants	Product(s) (% Yield)	Ref.
$(H_2C{=}CHCH_2)_2S$	$p\text{-}XC_6H_4NH_2$	$Hg(OAc)_2$, KBr	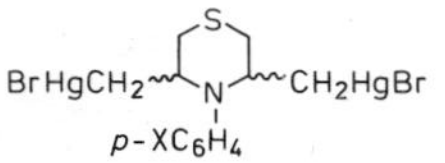 $BrHgCH_2$~...~CH_2HgBr (thiomorpholine ring, N–$p\text{-}XC_6H_4$) X = H (100), Me (85), MeO (83), NO_2 (62), Cl (98)	$NaBH_4$ (neutral)	$H_2C{=}CHCH_2SCH_2CH(NHC_6H_4X\text{-}p)CH_3$ X = H (46), Me (60), MeO (48), NO_2 (33), Cl (79)	71
$(H_2C{=}CHCH_2)_2NH$	$p\text{-}XC_6H_4NH_2$	$Hg(OAc)_2$, KBr	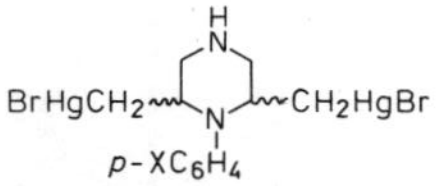 $BrHgCH_2$~...~CH_2HgBr (piperazine ring, N–H, N–$p\text{-}XC_6H_4$) X = H (100), Me (99), MeO (100), NO_2 (68), Cl (96)	$NaBH_4$ (neutral)	$H_2C{=}CHCH_2NHCH_2CH(NHC_6H_4X\text{-}p)CH_3$ X = H (49), Me (37), MeO (24), Cl (21)	71
$(H_2C{=}CHCH_2)_2CHOH$	$C_6H_5NH_2$	$Hg(OAc)_2$	—	$NaBH_4$ /NaOH	$C_6H_5NHCH(CH_3)CH_2CHOHCH_2CH{=}CH_2$ (63) 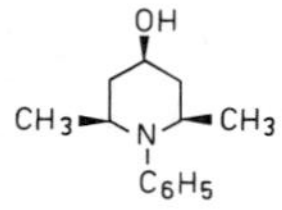 + [2,6-dimethyl-4-hydroxy-1-phenylpiperidine] (7)	32
$(H_2C{=}CHCH_2)_2NCH_3$	$p\text{-}XC_6H_4NH_2$	$Hg(OAc)_2$, KBr	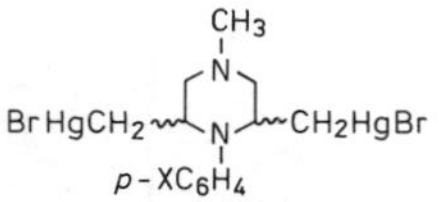 $BrHgCH_2$~...~CH_2HgBr (piperazine ring, N–CH_3, N–$p\text{-}XC_6H_4$) X = H (99), Me (97)	$NaBH_4$/NaOH	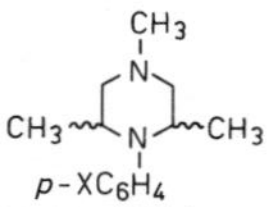 [2,6-dimethyl-1-($p\text{-}XC_6H_4$)-4-methylpiperazine] + $(H_2C{=}CHCH_2)_2NCH_3$ X = H (36) (42 : 58 cis/trans) — X = CH_3 — only product	72
 [bicyclic diene]	$p\text{-}ClC_6H_4NH_2$	$Hg(OAc)_2$	—	$NaBH_4$	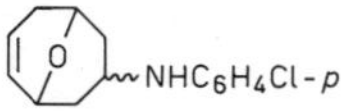 [bicyclic product]~$NHC_6H_4Cl\text{-}p$	73
	$p\text{-}NO_2C_6H_4NH_2$	$Hg(OAc)_2$	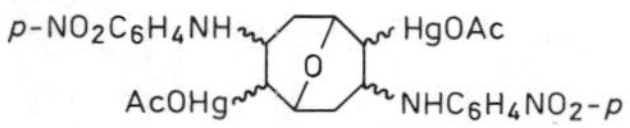 $p\text{-}NO_2C_6H_4NH$~...~HgOAc / AcOHg~...~$NHC_6H_4NO_2\text{-}p$ (bicyclic)	—	—	73

Table 6.3. (continued)

Diene	Amine	Mercuric salt	Organomercurial(s) (% Yield)	Subsequent reactants	Product(s) (% Yield)	Ref.
(structure)	$XC_6H_4NH_2$	$HgO/HBF_4/$ DMF 80 °C	—	—	(structure) NC_6H_4X; X = H (45), p-Me (62), p-Cl (56), p-Br (53), m-Me (43)	70
(structure)	$C_6H_5NH_2$	$Hg(OAc)_2$	—	$NaBH_4/NaOH$	(structure) C_6H_5 + (structure) C_6H_5	59
	$C_6H_5NH_2$	$Hg(OAc)_2$	—	$NaBH_4/NaOH/$ $C_6H_5NH_2$	(structure) C_6H_5 (>80)	74
	$C_6H_5NH_2$	$Hg(OAc)_2$	(structure) $AcOHg$ C_6H_5 $HgOAc$	M, H_2O	(structure) C_6H_5 + (structure) NHC_6H_5; M = Li (55) (26); M = Na (34) (45); M = K (3) (66)	5
	$C_6H_5NH_2$	—	(structure) $BrHg$ C_6H_5 $HgBr$	Li, CH_3OH	(structure) C_6H_5 (at −60 °C, 74) or (structure) NHC_6H_5 (at −5 °C, 68)	9
	$C_6H_5NH_2$	—	(structure) $BrHg$ C_6H_5 $HgBr$	$Li/C_6H_5NH_2$	(structure) C_6H_5 (27) + (structure) NHC_6H_5 (43)	10
	$C_6H_5NH_2$	—	(structure) $BrHg$ C_6H_5 $HgBr$	Mg/CH_3OH	(structure) C_6H_5 (5) + (structure) NHC_6H_5 (73)	11

Table 6.3. (continued)

Diene	Amine	Mercuric salt	Organomercurial(s) (% Yield)	Subsequent reactants	Product(s) (% Yield)	Ref.
	$C_6H_5NH_2$	—	[structure] Br Hg / HgBr, C_6H_5, N	Ca/CH_3OH	[structure] N, C_6H_5 (5) + [structure] NHC_6H_5 (72)	12
$(H_2C{=}CHCH_2)_2NCOCH_3$	$p{-}XC_6H_4NH_2$	$Hg(OAc)_2$, KBr	[structure] $COCH_3$, $BrHgCH_2$—N—CH_2HgBr, $p{-}XC_6H_4$; X = H (85), MeO (89)	$NaBH_4/NaOH$	[structure] $COCH_3$, CH_3—N—CH_3, $p{-}XC_6H_4$; X = H 50:50 cis/trans (65); X = MeO 6:94 (75)	72
[structure]	$C_6H_5NH_2$	$Hg(OAc)_2$, KCl	[structure] NHC_6H_5 / HgCl	$NaBH_4/NaOH$ (phase transfer)	[structure] NHC_6H_5 (93) + [structure] (7)	27
$H_2C{=}C{=}CH(CH_2)_3NH(CH_2)_2CH_3$	—	$HgCl_2$	—	$NaBH_4$ (phase transfer)	[structure] $H_2C{=}CH$—N—$(CH_2)_2CH_3$ (70)	75
$(H_2C{=}CHCH_2)_2N(CH_2)_2CH_3$	$p{-}XC_6H_4NH_2$	$Hg(OAc)_2$, KBr	[structure] $(CH_2)_2CH_3$, $BrHgCH_2$—N—CH_2HgBr, $p{-}XC_6H_4$; X = H (99), Me (88)	$NaBH_4/NaOH$	[structure] $(CH_2)_2CH_3$, CH_3—N—CH_3, $p{-}XC_6H_4$; X = H 40:60 cis/trans (52); X = Me 42:58 (68)	72
$H_2C{=}C{=}CH(CH_2)_4NH(CH_2)_2CH_3$	—	$HgCl_2$	—	$NaBH_4$ (phase transfer)	[structure] $H_2C{=}CH$—N—$(CH_2)_2CH_3$ (55)	75

Table 6.3. (continued)

Diene	Amine	Mercuric salt	Organomercurial(s) (% Yield)	Subsequent reactants	Product(s) (% Yield)	Ref.
$H_2C=CHCH_2$ (2-cyclohexenone with CH_3CH_2NH)	—	$Hg(OAc)_2$ / Δ	—	$NaBH_4$/NaOH	(indole with OH, CH_3, N-CH_2CH_3) (78)	76
$(CH_3)_2C=C=CH(CH_2)_3$ $NH(CH_2)_2CH_3$	—	$HgCl_2$	—	$NaBH_4$ (phase transfer)	$(CH_3)_2C=CH$–(pyrrolidine, N-$(CH_2)_2CH_3$) (70)	75
$H_2C=C(CH_3)CH_2CH_2C(CH_3)=CH_2$ (with CH_2OH and CH_2NH_2)	—	$Hg(OAc)_2$	—	$NaBH_4$/NaOH	(spiro with CH_3, CH_3, CH_3, CH_3, HN, O) (80)	48
p-$(H_2C=CHCH_2)_2NC_6H_4Cl$	p-$XC_6H_4NH_2$	$Hg(OAc)_2$, KBr	$BrHgCH_2$–(piperidine, N-p-ClC_6H_4, N-p-XC_6H_4)–CH_2HgBr X = H (84), Me (75), MeO (79), Cl (75)	$NaBH_4$/NaOH	(piperidine, N-p-ClC_6H_4, CH_3, CH_3, N-p-XC_6H_4) X = H 75 : 25 cis/trans (80); Me 72 : 28 (93); MeO 76 : 24 (78); Cl 78 : 22 (90)	72
$H_2C=C=CH(CH_2)_3NHC_6H_5$	—	$HgCl_2$	—	$NaBH_4$ (phase transfer)	$H_2C=CH$–(pyrrolidine, N-C_6H_5) (70)	75
$(H_2C=CHCH_2)_2NC_6H_5$	p-$XC_6H_4NH_2$	$Hg(OAc)_2$, KBr	$BrHgCH_2$–(piperidine, N-C_6H_5, N-p-XC_6H_4)–CH_2HgBr X = H (97), Cl (90)	$NaBH_4$/NaOH	(piperidine, N-C_6H_5, CH_3, CH_3, N-p-XC_6H_4) X = H 63 : 27 cis/trans (55); X = Cl 77 : 23 (61)	72

Table 6.3. (continued)

Diene	Amine	Mercuric salt	Organomercurial(s) (% Yield)	Subsequent reactants	Product(s) (% Yield)	Ref.
$H_2C=CHCH_2$- ... $CH_3(CH_2)_2NH$	—	$Hg(OAc)_2$ / Δ	—	$NaBH_4$/$NaOH$	(81)	76
$CH_3CH_2CH=C=CH(CH_2)_4$ – $NH(CH_2)_2CH_3$	—	$HgCl_2$	—	$NaBH_4$ (phase transfer)	$CH_3CH_2CH=CH$- (50)	75
$H_2C=C=CH(CH_2)_3NHCH_2C_6H_5$	—	$HgCl_2$	—	$NaBH_4$ (phase transfer)	$H_2C=CH$- (52)	75
$(H_2C=CHCH_2)_2NCH_2C_6H_5$	p-$XC_6H_4NH_2$	$Hg(OAc)_2$, KBr	$BrHgCH_2$ N CH_2HgBr p-XC_6H_4 X = H (100), Cl (84)	$NaBH_4$/$NaOH$	CH_3 N CH_3 + $(H_2C=CHCH_2)_2NCH_2C_6H_5$ p-XC_6H_4 X = H (97)(21:79 cis/trans) — X = Cl — only product	72
$H_2C=C=CH(CH_2)_4NHCH_2C_6H_5$	—	$HgCl_2$	—	$NaBH_4$ (phase transfer)	$H_2C=CH$- (54)	75
$(CH_3)_2C=C=CH(CH_2)_3$ – $NHCH_2C_6H_5$	—	$HgCl_2$	—	$NaBH_4$ (phase transfer)	$(CH_3)_2C=CH$- (52)	75

Table 6.3. (continued)

Diene	Amine	Mercuric salt	Organomercurial(s) (% Yield)	Subsequent reactants	Product(s) (% Yield)	Ref.
$CH_3CH_2CH=C=CH(CH_2)_4-$ $NHCH_2C_6H_5$	—	$HgCl_2$	—	$NaBH_4$ (phase transfer)	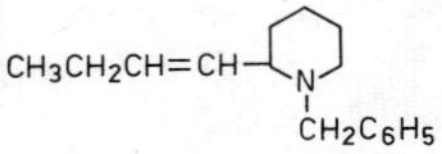 $CH_3CH_2CH=CH-$ piperidine-$CH_2C_6H_5$ (35)	75
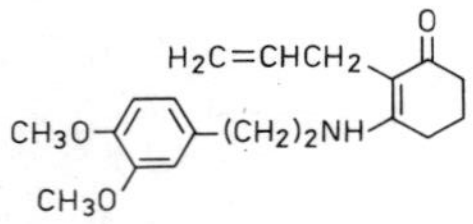 $H_2C=CHCH_2$-cyclohexenone-$(CH_2)_2NH$-dimethoxyphenyl	—	$Hg(OAc)_2/\Delta$	—	$NaBH_4/NaOH$	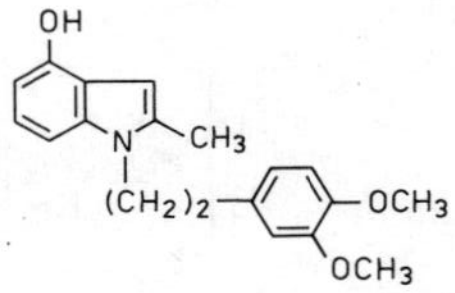 4-hydroxy-2-methylindole N-$(CH_2)_2$-dimethoxyphenyl (54)	76

allylic amines, but N-methylaniline yields only the terminal amine plus a large amount of nitrogen addition to the central carbon of the allene (Eq. 58). These reactions clearly deserve further study.

$$CH_3CH_2CH=C=CH_2 \longrightarrow CH_3(CH_2)_2\overset{\overset{\displaystyle CH_3}{|}}{C}HN(CH_3)C_6H_5 + CH_3CH_2CH=CHCH_2N(CH_3)C_6H_5$$

$$60 \quad : \quad 40$$

$$(58)$$

Allenic amines have been cyclized using mercuric chloride and subsequent demercuration with alkaline sodium borohydride under phase transfer conditions (Eq. 59) [75]. However, silver nitrate effects the same cyclization (without reduction) in higher yields.

$$R^1R^2C=CH\underset{\underset{\displaystyle R^3}{|}}{N} \xleftarrow{\; n=3 \;} R^1R^2C=C=CH(CH_2)_nNHR^3 \xrightarrow{\; n=4 \;} R^1R^2C=CH\underset{\underset{\displaystyle R^3}{|}}{N}$$

$$(59)$$

The aminomercuration of conjugated dienes has proven no less interesting. Isoprene has been regioselectively monomercurated and reduced with alkaline sodium borohydride under phase transfer conditions to afford a single allylic amine (Eq. 60) [27]. 1,3-Butadiene, isoprene and 1,3-cyclo-

$$H_2C=\overset{\overset{\displaystyle CH_3}{|}}{C}CH=CH_2 \longrightarrow H_2C=\overset{\overset{\displaystyle CH_3}{|}}{C}-\overset{\overset{\displaystyle NHC_6H_5}{|}}{C}HCH_2HgOAc \longrightarrow H_2C=\overset{\overset{\displaystyle CH_3}{|}}{C}-\overset{\overset{\displaystyle NHC_6H_5}{|}}{C}HCH_3 \qquad (60)$$

hexadiene have all been dimercurated and reduced as well, yielding vicinal diamines (Eq. 61) [65]. When mercuric oxide plus tetrafluoroboric acid is

$$H_2C=\overset{\overset{\displaystyle R}{|}}{C}CH=CH_2 \longrightarrow CH_3\overset{\overset{\displaystyle ArNH}{|}}{\underset{\underset{\displaystyle R}{|}}{C}}-\overset{\overset{\displaystyle NHAr}{|}}{C}HCH_3 \qquad (61)$$

utilized as the mercury salt, dihydropyrroles are formed instead (Eq. 62) [70]. This reaction works on 1,3-cyclooctadiene as well.

$$CH_3CH=CHCH=CHCH_3 \longrightarrow CH_3\underset{\underset{\displaystyle Ar}{|}}{N}CH_3 \qquad (62)$$

A variety of heterocycles have been prepared by aminomercuration-demercuration of non-conjugated dienes. In this manner *cis*- and *trans*-1,4-hexadiene have been cyclized to N-aryl pyrrolidines in good yield with 90 to 99% *trans* selectivity (Eq. 63) [66]. Similar results are reported for 1,5-

$$H_2C=CHCH_2CH=CHCH_3 \longrightarrow$$

(63)

hexadiene (Eq. 64) [66–68]. The stereoselectivity and the amount of un-

$$H_2C=CH(CH_2)_2CH=CH_2 \longrightarrow \quad + \quad H_2C=CH(CH_2)_2\overset{HNAr}{CHCH_3}$$

(64)

saturated amine side-product or starting diene formed on sodium borohydride reduction are dependent on the nature of the substituents on the aryl amine employed in the reaction.

The aminomercuration-demercuration of 1,6-heptadienes has provided acyclic as well as cyclic amines (Eqs. 65, 66). With 1,6-heptadiene itself

$$(H_2C=CHCH_2)_2CHOH \longrightarrow CH_3\overset{C_6H_5NH}{CHCH_2}CHOHCH_2CH=CH_2 \quad + $$

(65) [32]

63%   7%

$$(H_2C=CHCH_2)_2X \longrightarrow$$

(66)

(X=CH$_2$), the cis isomer predominates. The cis:trans ratio is once again highly dependent on the nature of the arylamine [69]. The presence of heteroatoms in the diene affords a novel route to morpholines (X = 0), tetrahydro-1,4-thiazines (X=S) and piperazines (X=NR) [71, 72]. With X=NH or low pH during the reduction, unsaturated amines are again isolated, often in reasonable yield (Eq. 67).

$$\longrightarrow H_2C=CHCH_2XCH_2\overset{ArNH}{CHCH_3}$$

(67)

Finally, the aminomercuration of several cyclic dienes has been explored. For example, the aminomercuration-demercuration of 1,5-cyclooctadiene has been examined. While only the 9-azabicyclo[3.3.1]nonane dimercurial is reported from this reaction (Eq. 68), in situ aminomercuration and subsequent alkaline sodium borohydride reduction of 1,5-cyclooctadiene is

(68)

reported to afford both the [3.3.1] and [4.2.1] bicyclic amines [59]. The addition of aniline during reduction has elsewhere been reported to give a high yield of the [3.3.1] amine [74]. The reductions of this same mercurial by Li, Na or K in H_2O [5], Li/MeOH [9], Li/RNH_2 [10], Mg/MeOH [11] and Ca/MeOH [12] have all been studied, but the cyclooctenylamine side-product is usually prominant and the alkaline sodium borohydride phase transfer approach is most likely preferable.

One bicyclic diene has also been aminomercurated (Eq. 69), but few details are available on this work [73].

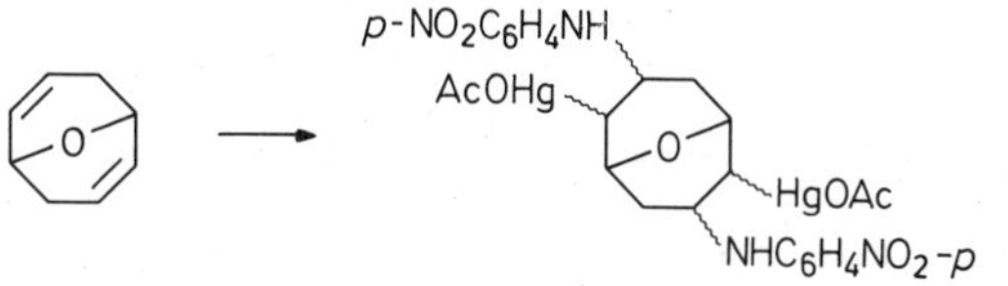

$$\text{(69)}$$

Finally, the intramolecular aminomercuration of allyl enaminoketones has been reported (Eq. 70) [76]. This reaction apparently proceeds by

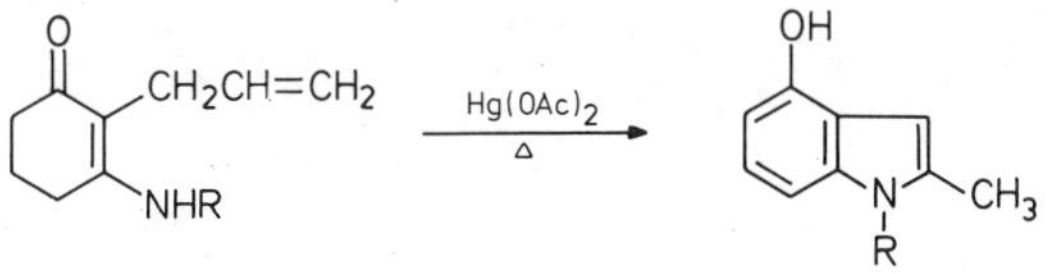

$$\text{(70)}$$

oxidation to the corresponding phenol. In fact, with certain R groups the major product is that of intramolecular phenoxymercuration of the allyl group, rather than the indole product shown above.

D. Alkynes

Relatively little work has been published on the aminomercuration of alkynes (Table 6.4). A number of older papers report that mercuric salts, particularly mercuric halides, react with aniline (or substituted anilines) and acetylene to afford "diethylideneaniline" and quinaldine (Eq. 71) [78–82, 92]. In a more recent study of the reaction of acetylene and anilines, it was observed

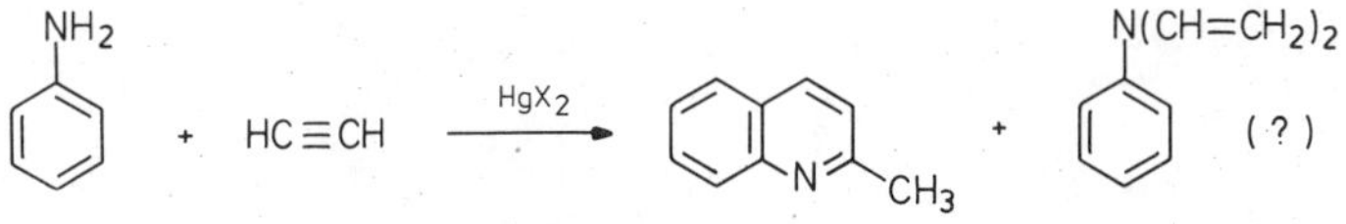

$$\text{(71)}$$

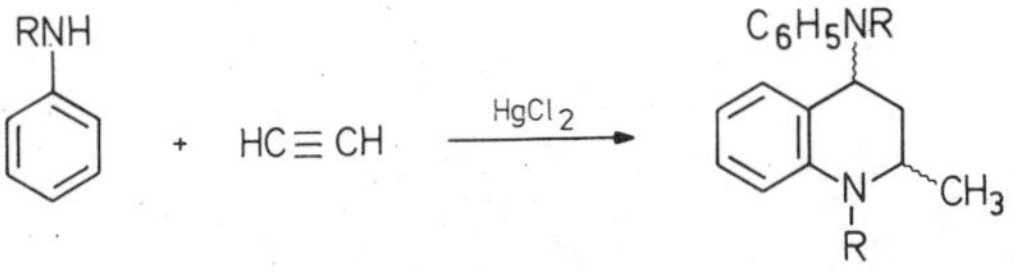

$$\text{(72)}$$

Table 6.4. Aminomercuration of Alkynes

Alkyne	Amine	Mercuric salt	Organomercurial(s) (% Yield)	Subsequent reactants	Product(s) (% Yield)	Ref.
$HC\equiv CH$	$ClNHCO_2CH_2CH_3$	$Hg(OAc)_2/H_2SO_4$	—	—	$H_2C{=}CHNClCO_2CH_2CH_3$	77
	$C_6H_5NH_2$	Hg_2Cl_2 or HgX_2 (X = Cl, Br, I)	—	—	2-methylquinoline + diethylideneaniline	78, 79
	$C_6H_5NH_2$	$HgSO_4/H_2SO_4$	—	—	2-methylquinoline	80
	o- and p-$CH_3OC_6H_4NH_2$	$HgCl_2/CuCl_2$	—	—	CH_3O-2-methylquinoline or diethylidene-anisidines	81
	$C_6H_5NHC_2H_5$	$HgCl_2$	—	—	2-methylquinoline + indole	82
	C_6H_5NHR	$HgCl_2$	—	—	4-(C_6H_5NR)-1-R-2-methyl-1,2,3,4-tetrahydroquinoline; R = H (cis and trans) (13.4); Me (trans) (18.2)	83
	C_6H_5NHR	$Hg(OAc)_2$	—	—	$CH_3CONRC_6H_5$ R = Me (81), Et (97), Ph (96)	83
	$XC_6H_4NH_2$	$Hg(OAc)_2$	—	—	$XC_6H_4N{=}C(CH_3)NHC_6H_4X$ X = H (75), 2-Me (77), 4-Me (90), 2-MeO (65), 4-MeO (78)	83
$HC\equiv CCH_2OH$	$ArNH_2$	$Hg(OAc)_2$	—	—	$ArN{=}C(NHAr)CH(CH_3)NHAr$ Ar = C_6H_5 (50), 2-$CH_3C_6H_4$ (71), 4-$CH_3C_6H_4$ (38), 4-$CH_3OC_6H_4$ (36)	84

Table 6.4. (continued)

Alkyne	Amine	Mercuric salt	Organomercurial(s) (% Yield)	Subsequent reactants	Product(s) (% Yield)	Ref.
	$ArNH_2$	$Hg(OAc)_2/K_2CO_3$	—	—	$ArN=CHC(CH_3)=NAr$ $Ar = 2\text{-}CH_3C_6H_4$ (56), $4\text{-}CH_3C_6H_4$ (32), $4\text{-}CH_3OC_6H_4$ (45), $2,6\text{-}(CH_3)_2C_6H_3$ (30)	84
$H_2C=C(CH_3)C\equiv CH$	$C_6H_5NHCH_3$	$Hg(OAc)_2$? / hexane	—	—	$C_6H_5N(CH_3)CH_2CH(CH_3)CH(CH_3)N(CH_3)C_6H_5$ (31)	33
	R_2NH	$Hg(OAc)_2$	$H_2C=C(CH_3)C(NR_2)=CHHgOAc$	$NaBH_4$	$H_2C=C(CH_3)CH(CH_3)NR_2$ (9–30) + $H_2C=C(CH_3)COCH_3$ $R_2NH = Me_2NH, Et_2NH,$ piperidine-NH, morpholine-NH	85
	R_2NH	$Hg(OAc)_2$? / hexane	—	—	$H_2C=C(CH_3)CH(CH_3)NR_2$ $R_2NH = Me_2NH,$ piperidine-NH, morpholine-NH	33
$HC\equiv C(CH_2)_2CH_3$	$C_6H_5NH_2$	HgX_2	—	—	$CH_3C(=NC_6H_5)(CH_2)_2CH_3$ $X = Cl$ (55), OAc (17)	83
$HC\equiv C(CH_2)_3CH_3$	$C_6H_5NH_2$	HgX_2	—	—	$CH_3C(=NC_6N_5)(CH_2)_3CH_3$ $X = Cl$ (67), OAc (45)	83
	C_6H_5NHR	$HgCl_2$	—	—	$CH_3C(NRC_6H_5)=CH(CH_2)_2CH_3$ $R = Me$ (51), Et (48)	83
$HC\equiv C(CH_2)_4CH_3$	$C_6H_5NH_2$	$HgO/BF_3\cdot Et_2O$	—	—	$C_6H_5N=C(CH_3)(CH_2)_4CH_3$	86
	$C_6H_5NHCH_2CH_3$	$HgO/BF_3\cdot Et_2O$	—	—	$C_6H_5NEtC(=CH_2)(CH_2)_4CH_3$	86
$HC\equiv CC_6H_5$	$XC_6H_4NH_2$	$Hg(OAc)_2$	—	$NaBH_4/NaOH$	$CH_3CC_6H_5$ with $=NC_6H_4X$ $X = H$ (100), $p\text{-}Me$ (100), $o\text{-}Me$ (89), $p\text{-}MeO$ (98), $p\text{-}Cl$ (85)	87

Table 6.4. (continued)

Alkyne	Amine	Mercuric salt	Organomercurial(s) (% Yield)	Subsequent reactants	Product(s) (% Yield)	Ref.
	C_6H_5NHR	$HgCl_2$	—	—	$H_2C=C(C_6H_5)NRC_6H_5$ R = Me (54), Et (39)	83
	C_6H_5NHR	$Hg(OAc)_2$	—	$NaBH_4/NaOH$	$H_2C=CC_6H_5$ (with RNC_6H_5) R = Me (80), Et (62), $C_6H_5CH_2$ (56)	87
$HC{\equiv}C(CH_2)_5CH_3$	aziridine	$Hg(OAc)_2$	—	$NaBH_4/NaOH$	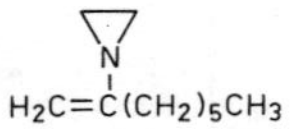 $H_2C=C(CH_2)_5CH_3$ (17.5)	88
	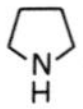 pyrrolidine	$HgCl_2$	—	$NaBH_4/NaOH$	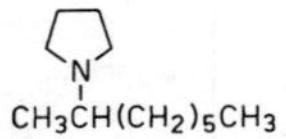 $CH_3CH(CH_2)_5CH_3$ (86)	88
	$C_6H_5NH_2$	HgX_2	—	—	$CH_3C(=NC_6H_5)(CH_2)_5CH_3$ X = Cl (69), OAc (65)	83
	C_6H_5NHR	$Hg(OAc)_2$	—	—	$CH_3C(NRC_6H_5)=CH(CH_2)_4CH_3$ X = Cl, R = Me (78) and Et (84); X = OAc, R = Me (51)	83
$CH_3(CH_2)_2C{\equiv}C(CH_2)_2CH_3$	$C_6H_5NH_2$	$HgO/BF_3 \cdot Et_2O$	—	—	$CH_3(CH_2)_2C(=NC_6H_5)(CH_2)_3CH_3$	86
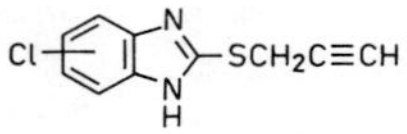 Cl-benzimidazole-$SCH_2C{\equiv}CH$	—	$Hg(OAc)_2/HOAc/ H_2SO_4$	—	—	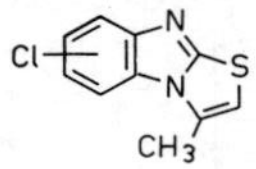 Cl-thiazolobenzimidazole-CH_3 (80)	89
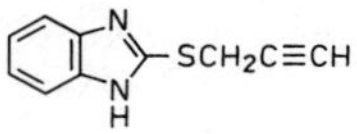 benzimidazole-$SCH_2C{\equiv}CH$	—	$Hg(OAc)_2/HOAc/ H_2SO_4$	—	—	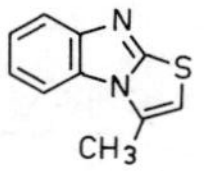 thiazolobenzimidazole-CH_3 (68)	89

Table 6.4. (continued)

Alkyne	Amine	Mercuric salt	Organomercurial(s) (% Yield)	Subsequent reactants	Product(s) (% Yield)	Ref.
5,6-dimethylbenzimidazole-2-SCH$_2$C≡CH	—	Hg(OAc)$_2$/HOAc/ H$_2$SO$_4$	—	—	(thiazolo-benzimidazole, CH$_3$, CH$_3$) (75)	89
Cl-benzimidazole-2-SCH$_2$C≡CC$_6$H$_5$	—	Hg(OAc)$_2$/HOAc/ H$_2$SO$_4$	—	—	(Cl, C$_6$H$_5$) (61)	89
benzimidazole-2-SCH$_2$C≡CC$_6$H$_5$	—	Hg(OAc)$_2$/HOAc/ H$_2$SO$_4$	—	—	(C$_6$H$_5$) (70)	89
5,6-dimethylbenzimidazole-2-SCH$_2$C≡CC$_6$H$_5$	—	Hg(OAc)$_2$/HOAc/ H$_2$SO$_4$	—	—	(CH$_3$, CH$_3$, C$_6$H$_5$) (63)	89
17-ethynyl-17-hydroxy steroid (HO, C≡CH; HO)	C$_6$H$_5$NH$_2$	HgO/BF$_3$·Et$_2$O or HgCl$_2$	—	—	(HO, C(=NC$_6$H$_5$)CH$_3$; HO)	90, 91
17-ethynyl steroid (AcO, C≡CH; AcO)	C$_6$H$_5$NH$_2$	HgCl$_2$	—	—	(AcO, C(=NC$_6$H$_5$)CH$_3$; AcO)	91

VI. Aminomercuration

that tetrahydroquinolines can be isolated from this type of reaction (Eq. 72) [83]. Totally different products are obtained, however, when mercuric acetate is used to promote the reaction (Eq. 73).

$$\underset{\text{CH}_3\overset{\displaystyle\text{NC}_6\text{H}_5}{\overset{\displaystyle\|}{\text{C}}}\text{NHC}_6\text{H}_5}{} \xleftarrow{\ R=H\ } \left[\text{C}_6\text{H}_5\text{NHR}\right] + HC\equiv CH \xrightarrow{\ R=alkyl\ } \text{CH}_3\overset{\displaystyle O}{\overset{\displaystyle\|}{\text{C}}}\text{N(R)C}_6\text{H}_5 \qquad (73)$$

Of more general interest, there are a number of examples of the reaction of terminal alkynes and secondary amines yielding enamines (Eq. 74). This

$$RC\equiv CH + R'_2NH \longrightarrow \underset{RC=CH_2}{\overset{\displaystyle NR'_2}{\overset{\displaystyle |}{}}} \qquad (74)$$

reaction has been promoted by mercuric acetate in sulfuric acid [77], mercuric oxide plus boron trifluoride etherate [86], mercuric chloride [83] and mercuric acetate followed by alkaline sodium borohydride [87, 88]. This last approach does not appear to work very well with alkyl acetylenes. It is highly dependent on the amine employed and saturated products have been obtained as well [88]. The reaction appears much better behaved when phenyl acetylene is used as the alkyne [87]. The same is true for the reaction with mercuric chloride [83]. Phenyl acetylene affords the corresponding enamines in reasonable yield, but alkyl acetylenes afford primarily isomerized enamines (Eq. 75). Intramolecular enamine formation has also been catalyzed by

$$R^1CH_2C\equiv CH + C_6H_5NHR^2 \xrightarrow{\ HgCl_2\ } \underset{R^1CH=CCH_3}{\overset{\displaystyle C_6H_5NR^2}{\overset{\displaystyle |}{}}} \qquad (75)$$

mercuric acetate (Eqs. 76, 77) [89].

$$(76)$$

$$(77)$$

The reaction of terminal and internal acetylenes with aniline provides a direct route to the corresponding phenylimines (Eq. 78). This reaction is

$$RC\equiv CR + C_6H_5NH_2 \longrightarrow \underset{RCH_2CR}{\overset{\displaystyle NC_6H_5}{\overset{\displaystyle \|}{}}} \qquad (78)$$

promoted by mercuric chloride [83, 91], mercuric oxide plus boron trifluoride etherate [86, 90], and mercuric acetate followed by alkaline sodium borohydride [87]. Quite good yields have been obtained using this latter approach on phenyl acetylene.

Other functional groups can interfere in these reactions. While isopropenyl acetylene has been aminomercurated and subsequently reduced to the corresponding allylic amine (Eq. 79) [33, 85], a variety of other products,

$$H_2C=\overset{\overset{\displaystyle CH_3}{|}}{C}C\equiv CH \quad \xrightarrow[R_2NH]{Hg(OAc)_2} \quad \xrightarrow{NaBH_4} \quad H_2C=\overset{\overset{\displaystyle CH_3}{|}}{\underset{\underset{\displaystyle NR_2}{|}}{C}}CHCH_3 \tag{79}$$

including those involving attack on the carbon—carbon double bond, have also been observed from this substrate [33], as discussed earlier in section A.

Propargyl alcohol also affords interesting products from the reaction of primary aromatic amines and mercuric acetate (Eq. 80) [84]. The diimine arises by using less aromatic amine and adding potassium carbonate to the reaction.

$$HC\equiv CCH_2OH \; + \; ArNH_2 \underset{}{\overset{\overset{\textstyle K_2CO_3}{\longrightarrow} \; CH_3\overset{\overset{\displaystyle ArN}{||}}{C}CH=NAr}{\underset{CH_3\overset{\overset{\displaystyle ArNH}{|}}{C}H-\overset{\overset{\displaystyle NAr}{||}}{C}NHAr}{\Big\langle}}} \tag{80}$$

References

1. Toman, K., Hess, G. G.: J. Organometal. Chem., *49*, 133 (1973).
2. Kochetkova, N. S., Freidlina, R. Kh., Nesmeyanov, A. N.: Izv. Akad. Nauk SSSR, Otdel. Khim. Nauk, 347 (1947); Chem. Abstr., *42*, 1572b (1948).
3. Périé, J. J., Lattes, A.: Bull. Soc. Chim. France, 583 (1970).
4. Freidlina, R. Kh., Kochetkova, N. S.: Izv. Akad. Nauk SSSR, Otdel. Khim. Nauk, 128 (1945); Chem. Abstr., *40*, 3450[7] (1946).
5. Barluenga, J., Ara, A., Asensio, G., Yus, M.: An. Quim., *74*, 455 (1978); Chem. Abstr., *89*, 163695g (1978).
6. Lattes, A., Périé, J. J.: Compt. Rend. C., *262*, 1591 (1966).
7. Barluenga, J., Fañanás, F. J., Yus, M.: J. Org. Chem., *46*, 1281 (1981).
8. Barluenga, J., Fañanás, F. J., Yus, M.: J. Org. Chem., *44*, 4798 (1979).
9. Gómez-Aranda, V., Barluenga, J., Ara, A., Asensio, G.: Synthesis, 135 (1974).
10. Barluenga, J., Ara, A., Asensio, G.: Synthesis, 116 (1975).
11. Barluenga, J., Concellón, J. M., Asensio, G.: Synthesis, 467 (1975).
12. Barluenga, J., Concellón, J. M., Asensio, G., Yus, M.: An. Quim., *74*, 512 (1978); Chem. Abstr., *89*, 179011s (1978).

VI. Aminomercuration

13. Barluenga, J., Alonso-Cires, L., Asensio, G.: Synthesis, 376 (1981).
14. Sachs, G.: J. Chem. Soc., 733 (1949).
15. DeBrule, R. F., Hess, G. G.: Synthesis, 197 (1974).
16. Périé, J. J., Lattes, A.: Tetrahedron Lett., 2289 (1969).
17. Barluenga, J., Alonso-Cires, L., Asensio, G.: Synthesis, 962 (1979).
18. Lattes, A., Périé, J. J.: Tetrahedron Lett., 5165 (1967).
19. Hall, H. K., Jr., Schaefer, J. P., Spanggord, R. J.: J. Org. Chem., *37*, 3069 (1972).
20. Hopff, H., Lüssi, H.: Helv. Chim. Acta, *46*, 1052 (1963).
21. Lopatinskii, V. P., Sirotkina, E. E., Shekhirev, Yu. P., Men'shikova, N. G., Kovaleva, L. F.: Tr. Tomskogo Gos. Univ., Ser. Khim., *170*, 35 (1964); Chem. Abstr., *63*, 1763f (1965).
22. Grandberg, I. I., Sharova, G. I.: Khim. Geterotsikl. Soedin., *4*, 1097 (1968); Chem. Abstr., *70*, 77856m (1969).
23. Sirotkina, E. E., Filimonov, V. D., Sazonova, Z. P.: Zh. Org. Khim., *9*, 381 (1973); J. Org. Chem. USSR, *9*, 385 (1973).
24. Filimonov, V. D., Sirotkina, E. E., Gaibed, I. L.: Tezisy Dokl.-Simp. Khim. Tekhnol. Geterotsikl. Soedin. Goryuch. Iskop. 2nd, 138 (1973); Chem. Abstr., *86*, 16494a (1977).
25. Hopff, H., Wyss, U., Lüssi, H.: Helv. Chim. Acta, *43*, 135 (1960).
26. Barluenga, J., Villamaña, J., Yus, M.: Synthesis, 375 (1981).
27. Etemad-Moghadam, G., Benhamou, M. C., Spéziale, V., Lattes, A., Bielawska, A.: Nouveau J. Chim., *4*, 727 (1980).
28. Bäckvall, J.-E., Åkermark, B.: J. Organometal. Chem., *78*, 177 (1974).
29. Wendt, G. R.: U.S. Patent 2,800,471 (1957); Chem. Abstr., *52*, 16381d (1958).
30. Wendt, G., Shetty, B. V., Bruce, W. F.: J. Am. Chem. Soc., *81*, 4233 (1959).
31. Watanabe, W. H.: J. Am. Chem. Soc., *79*, 2833 (1957).
32. Hodjat-Kachani, H., Périé, J. J., Lattes, A.: Chem. Lett., 409 (1976).
33. Chobanyan, Zh. A., Davtyan, Zh. S., Badanyan, Sh. O.: Arm. Khim. Zh., *34*, 854 (1981); Chem. Abstr., *96*, 51893h (1982).
34. Hopff, H., Wyss, U., Lüssi, H.: Helv. Chim. Acta, *43*, 1967 (1960).
35. Barluenga, J., Jimenez, C., Nájera, C., Yus, M.: Synthesis, 201 (1981).
36. Gasc, M. B., Périé, J., Lattes, A.: Tetrahedron, *34*, 1943 (1978).
37. Wolf, R. E., Gorman, S. B.: U.S. Patent 3,729,483 (1973); Chem. Abstr., *79*, 5862y (1973).
38. Griffith, R. C., Gentile, R. J., Davidson, T. A., Scott, F. L.: J. Org. Chem., *44*, 3580 (1979).
39. Wendt, G. R.: U.S. Patent 2,922,789 (1960); Chem. Abstr., *54*, 9973a (1960).
40. Lattes, A.: Afinidad, *29*, 153 (1972); Chem. Abstr., *77*, 48544x (1972).
41. Barluenga, J., Nájera, C., Yus, M.: An. Quim., *75*, 341 (1979); Chem. Abstr., *91*, 192741r (1979).
42. Périé, J. J., Lattes, A.: Bull. Soc. Chim. France, 1378 (1971).
43. Bäckvall, J.-E., Åkermark, B.: J. Chem. Soc., Chem. Commun., 82 (1975).
44. Périé, J. J., Laval, J. P., Roussel, J., Lattes, A.: Tetrahedron, *28*, 675 (1972).
45. Périé, J., Laval, J. P., Lattes, A.: Compt. Rend. C, *272*, 1141 (1972).
46. Dobrev, A., Périé, J. J., Lattes, A.: Tetrahedron Lett., 4013 (1972).
47. Barrelle, M., Apparu, M.: Tetrahedron, *33*, 1309 (1977).
48. Hodjat, H., Lattes, A., Laval, J. P., Moulines, J., Périé, J. J.: J. Heterocyclic Chem., *9*, 1081 (1972).
49. Wilson, S. R., Sawicki, R. A.: J. Org. Chem., *44*, 330 (1979).
50. Bach, R. D., Mitra, D. K.: J. Chem. Soc. D, Chem. Commun., 1433 (1971).
51. Barluenga, J., Nájera, C., Yus, M.: J. Heterocyclic Chem., *16*, 1017 (1979).

52. Frank, W. C., Kim, Y. C., Heck, R. F.: J. Org. Chem., *43*, 2947 (1978).
53. Brodersen, K., Optiz, G., Breitinger, D.: Chem. Ber., *97*, 2046 (1964).
54. Brodersen, K.: Angew. Chem., *76*, 151 (1964); Angew. Chem., Int. Ed. Engl., *3*, 70 (1964).
55. Benhamou, M. C., Etemad-Moghadam, G., Spéziale, V., Lattes, A.: Synthesis, 891 (1979).
56. Roussel, J., Périé, J. J., Laval, J. P., Lattes, A.: Tetrahedron, *28*, 701 (1972).
57. Périé, J. J., Laval, J. P., Roussel, J., Lattes, A.: Tetrahedron Lett., 4399 (1971).
58. Kozikowski, A. P., Scripko. J.: Tetrahedron Lett., *24*, 2051 (1983).
59. Barrelle, M., Apparu, M.: Tetrahedron Lett., 2611 (1976).
60. Moriyama, Y., Doan-Huynh, D., Monneret, C., Khuong-Huu, Q.: Tetrahedron Lett., 825 (1977).
61. Saitoh, Y., Moriyama, Y., Takahashi, T., Khuong-Huu, Q.: Tetrahedron Lett., 75 (1980).
62. Saitoh, Y., Moriyama, Y., Hirota, H., Takahashi, T., Khuong-Huu, Q.: Bull. Chem. Soc. Japan, *54*, 488 (1981).
63. Brodersen, K., Opitz, G., Breitinger, D., Menzel, D.: Chem. Ber., *97*, 1155 (1964).
64. Hodjat-Kachani, H., Périé, J. J., Lattes, A.: Chem. Lett., 405 (1976).
65. Gómez-Aranda, V., Barluenga-Mur, J., Yus-Astiz, M., Aznar, F.: Rev. Acad. Cienc. Exactas, Fis.-Quim. Natur. Zaragoza, *29*, 321 (1974); Chem. Abstr., *85*, 20716w (1976).
66. Barluenga, J., Nájera, C., Yus, M.: J. Heterocyclic Chem., *18*, 1297 (1981).
67. Gómez-Aranda, V., Barluenga-Mur, J., Yus-Astiz, M.: An. Quim., *68*, 221 (1972).
68. Gómez-Aranda, V., Barluenga, J., Yus, M., Asensio, G.: Synthesis, 806 (1974).
69. Barluenga, J., Nájera, C., Yus, M.: Synthesis, 896 (1979).
70. Barluenga, J., Perez-Prieto, J., Asensio, G.: J. Chem. Soc., Chem. Commun., 1181 (1982).
71. Barluenga, J., Nájera, C., Yus, M.: Synthesis, 911 (1978).
72. Barluenga, J., Nájera, C., Yus, M.: J. Heterocyclic Chem., *17*, 917 (1980).
73. Gómez-Aranda, V., Barluenga-Mur, J., Asensio-Aguilar, G.: Rev. Acad. Cienc. Exactas, Fis.-Quim. Natur. Zaragoza, *28*, 215 (1973); Chem. Abstr., *80*, 83174k (1974).
74. Gómez-Aranda, V., Barluenga-Mur, J., Asensio, G., Yus, M.: Tetrahedron Lett., 3621 (1972).
75. Arseniyadis, S., Gore, J.: Tetrahedron Lett., *24*, 3997 (1983).
76. Iida, H., Yuasa, Y., Kibayashi, C.: Tetrahedron Lett., 3591 (1982).
77. Staudinger, J. J. P., Tuerck, K. H. W.: Brit. Patent 573,752 (1945); Chem. Abstr., *43*, 3031i (1949).
78. Kozlov, N. S., Dinaburskaya, B., Rubina, T.: Zh. Obshch. Khim., *6*, 1349 (1936); Chem. Abstr., *31*, 1374[6] (1937).
79. Kozlov, N. S., Pachankova, R.: Zh. Obshch. Khim., *6*, 1352 (1936); Chem. Abstr., *31*, 1374[7] (1937).
80. Vogt, R. R., Nieuwland, J. A.: J. Am. Chem. Soc., *43*, 2071 (1921).
81. Kozlov, N. S., Bogdanovskaya, R.: Zh. Obshch. Khim., *6*, 1346 (1937); Chem. Abstr., *31*, 1374[5] (1937).
82. Kryuk, I. F.: Zh. Obshch. Khim., *10*, 1507 (1940); Chem. Abstr., *35*, 2518[5] (1941).
83. Barluenga, J., Aznar, F., Liz, R., Rodes, R.: J. Chem. Soc., Perkin I, 2732 (1980).
84. Barluenga, J., Aznar, F., Liz, R.: J. Chem. Soc., Chem. Commun., 1181 (1981).

VI. Aminomercuration

85. Chobanyan, Zh. A., Davtyan, S. Zh., Badanyan, Sh. O.: Arm. Khim. Zh., *33*, 1003 (1980); Chem. Abstr., *95*, 6409y (1981).
86. Loritsch, J. A., Vogt, R. R.: J. Am. Chem. Soc., *61*, 1462 (1939).
87. Barluenga, J., Aznar, F.: Synthesis, 704 (1975).
88. Hudrlik, P. F., Hudrlik, A. M.: J. Org. Chem., *38*, 4254 (1973).
89. Balasubramanian, K. K., Nagarajan, R.: Synthesis, 189 (1976).
90. Stavely, H. E.: J. Am. Chem. Soc., *62*, 489 (1940).
91. Goldberg, M. W., Aeschbacher, R.: Helv. Chim. Acta, *22*, 1188 (1939).
92. Kozlov, N. S., Mitskevich, D.: Zh. Obshch. Khim., *7*, 1082 (1937); Chem. Abstr., *31*, 6209[4] (1937).

VII. Amidomercuration

The reaction of alkenes, mercuric salts and either nitriles or amides results in the formation of β-amidomercurials (Eq. 1). The majority of the early

$$\ce{>C=C<} \quad \xrightarrow[\text{RCN or RCONH}_2]{\text{HgX}_2} \quad \xrightarrow{\text{H}_2\text{O}} \quad \ce{XHg-C-C-NHCR} \overset{\text{O}}{\underset{}{}} \tag{1}$$

work was carried out with nitriles and only recently has the direct use of amides been explored. Our discussion will cover nitriles first and amides later.

Virtually all work on the amidomercuration with nitriles has been effected using mercuric nitrate, either anhydrous or hydrated, in the appropriate nitrile as the solvent (Table 7.1). Mercuric acetate, mercuric trifluoroacetate and mercuric sulfate have proved unsatisfactory for this reaction [1, 3]. It has also been reported that amidomercuration can be effected by electrolysis of cyclohexene, acetonitrile, nitric acid and lithium perchlorate at a mercury anode [6].

Although the majority of work has been carried out using acetonitrile, a variety of other nitriles can be employed. It is reported, however, that trichloroacetonitrile and pentaacetyl-α-gluconitrile fail to react [7]. Alkylthiocyanates (RSCN) and dialkylcyanamides (R_2NCN) also proved unreactive [1].

The amidomercuration of a number of simple olefins has been effected. Simple cyclic and bicyclic alkenes react well, as does 1,5-cyclooctadiene which gives an uncharacterized bis adduct [1]. There appears to be only one example of intramolecular amidomercuration of a nitrile (Eq. 2), but the product was not well characterized [9]. All additions apparently proceed by

$$\tag{2}$$

anti addition in a Markovnikov manner.

While the amidomercuration of a number of olefins has been successful, there are some important limitations. Olefins of the type $R_2C{=}CH_2$ and $R_2C{=}CHR$ fail to react in the anticipated fashion [1, 3]. With α- and

Table 7.1. Amidomercuration of Alkenes Using Nitriles

Alkene	Nitrile	Mercuric salt	Organomercurial(s) (% Yield)	Subsequent reactants	Product(s) (% Yield)	Ref.
$H_2C=CH_2$	CH_3CN	$Hg(NO_3)_2 \cdot 2\ H_2O$, NaCl	$ClHgCH_2CH_2NHCOCH_3$ (52)	—	—	1
$H_2C=CHCH_2Cl$	CH_3CN	$Hg(NO_3)_2 \cdot 2\ H_2O$, NaCl	$ClHgCH_2CH(NHCOCH_3)CH_2Cl$ (31)	—	—	1
$H_2C=CHCH_3$	RCN	$Hg(NO_3)_2 \cdot 2\ H_2O$, NaCl	$ClHgCH_2CH(CH_3)NHCOR$ R = Me (30), Ph (22)	$LiAlH_4$ (R = Me)	$(CH_3)_2CHNHCH_2CH_3$ (> 95) + $CH_3(CH_2)_2NHCH_2CH_3$ (~3–4)	1
cis‑$CH_3CH=CHCH_3$	CH_3CN	$Hg(NO_3)_2$, NaCl	threo‑$CH_3CH(HgCl)CH(NHCOCH_3)CH_3$	$H_2C=CHX$ / $NaHB(OCH_3)_3$	$X(CH_2)_2CH(CH_3)CH(CH_3)NHCOCH_3$ X = CN (74) 68 : 32 threo/erythro X = CO_2CH_3 (26)	2
	CH_3CN	$Hg(NO_3)_2 \cdot 2\ H_2O$, NaCl	$CH_3CH(HgCl)CH(NHCOCH_3)CH_3$ (36)	—	—	1
cyclopentene	CH_3CN	$Hg(NO_3)_2$	—	$NaBH_4$ / NaOH	(70)	3
	CH_3CN	$Hg(NO_3)_2 \cdot H_2O$, NaCl	(32)	Δ	(31) + (60)	4
cyclohexene	CH_3CN	$Hg(NO_3)_2$	—	$NaBH_4$ / NaOH	(95)	3
	CH_3CN	$Hg(NO_3)_2$, NaCl		$H_2C=CHX$ / $NaHB(OCH_3)_3$	X = CN (78) 40 : 60 cis/trans X = CO_2CH_3 (15)	2
	CH_3CN	$Hg(NO_3)_2$ / HNO_3, NaCl	(94)	Na(Hg) / H_2O	(66)	5

Table 7.1. (continued)

Alkene	Nitrile	Mercuric salt	Organomercurial(s) (% Yield)	Subsequent reactants	Product(s) (% Yield)	Ref.
	CH_3CN	electrolysis at a mercury anode, NaCl	cyclohexyl-NHCOCH$_3$, HgCl (65)	—	—	6
	RCN	$Hg(NO_3)_2$, NaCl	cyclohexyl-NHCOR, HgCl; R = Ph (74), PhCH$_2$ (64), EtO$_2$CCH$_2$ (78)	—	—	7
	RCN	$Hg(NO_3)_2 \cdot 2H_2O$ NaCl	cyclohexyl-NHCOR, HgCl	LiAlH$_4$, HCl	cyclohexyl-$\overset{+}{N}H_2CH_2R$ Cl$^-$; R = Me (60), Et (24)	1
	RCN	$Hg(NO_3)_2 \cdot H_2O$, NaCl	cyclohexyl-NHCOR, HgCl; R = Me (93), Ph (50)	Δ	bicyclic O–C(=N)CR + cyclohexene (60); R = Me (25), Ph (0)	4
	RCN	$Hg(NO_3)_2 \cdot 2 H_2O$, NaCl	cyclohexyl-NHCOR, HgCl; R = Me (81), Ph (39), PhCH$_2$ (27)	NaBH$_4$/NaOH [or Na(Hg)/H$_2$O] (R = Me)	cyclohexyl-NHCOCH$_3$ (64) [46]	1
	RCN	$Hg(NO_3)_2 \cdot 2 H_2O$, NaCl	cyclohexyl-NHCOR, HgCl	X$_2$	cyclohexyl-NHCOR, X; X = Cl, R = Me (55); X = Br, R = Me (41), Ph (69), PhCH$_2$ (69)	1
$H_2C{=}CHC(CH_3)_3$	CH_3CN	$Hg(NO_3)_2$	—	NaBH$_4$/NaOH	$CH_3CH(NHCOCH_3)C(CH_3)_3$ (90)	3
$H_2C{=}CH(CH_2)_3CH_3$	CH_3CN	$Hg(NO_3)_2$	—	NaBH$_4$/NaOH	$CH_3CH(NHCOCH_3)(CH_2)_3CH_3$ (92)	3
	RCN	$Hg(NO_3)_2 \cdot 2 H_2O$, NaCl	$ClHgCH_2CH(NHCOR)(CH_2)_3CH_3$; R = Me (80), Ph (18)	LiAlH$_4$, HCl	$CH_3CH(\overset{+}{N}H_2CH_2CH_3)(CH_2)_3CH_3$ Cl$^-$ (57)	1

Table 7.1. (continued)

Alkene	Nitrile	Mercuric salt	Organomercurial(s) (% Yield)	Subsequent reactants	Product(s) (% Yield)	Ref.
(norbornene)	CH_3CN	$Hg(NO_3)_2 \cdot 2\,H_2O$, NaCl	(norbornyl) NHCOCH$_3$ / HgCl (83)	—	—	1
(1-methylcyclohexene)	CH_3CN	$Hg(NO_3)_2$	—	$NaBH_4/NaOH$	(product with CH$_3$, NHCOCH$_3$) (80)	8
(cycloheptene)	CH_3CN	$Hg(NO_3)_2 \cdot H_2O$, NaCl	(cycloheptyl) NHCOCH$_3$ / HgCl (3.2)	Δ	(bicyclic oxazoline, CCH$_3$) (11)	4
	RCN	$Hg(NO_3)_2 \cdot 2\,H_2O$, NaCl	(cycloheptyl) NHCOR / HgCl; R = Me (35), Et (55), Ph (38)	—	—	1
$H_2C=CHC_6H_5$	CH_3CN	$Hg(NO_3)_2$	—	$NaBH_4/NaOH$	$CH_3CH(C_6H_5)NHCOCH_3$ (50)	3
	CH_3CN	$Hg(NO_3)_2$, NaCl	$ClHgCH_2CH(NHCOCH_3)C_6H_5$	$H_2C=CHX$ / $NaHB(OCH_3)_3$	$X(CH_2)_3CH(NHCOCH_3)C_6H_5$; X = CN (67), CO_2CH_3 (22)	2
	CH_3CN	$Hg(NO_3)_2 \cdot 2\,H_2O$, NaCl	$ClHgCH_2CH(NHCOCH_3)C_6H_5$ (14)	—	—	1
(5-cyanonorbornene)	—	$Hg(NO_3)_2/H_2O$, NaCl	(lactam, HgCl) ? (14) + (cyanonorbornanol, OH, HgCl)	—	—	9
(cyclooctadiene)	CH_3CN	$Hg(NO_3)_2 \cdot 2\,H_2O$, NaCl	CH_3CONH—(cyclooctane)—NHCOCH$_3$ / ClHg, HgCl (52)	—	—	1

Table 7.1. (continued)

Alkene	Nitrile	Mercuric salt	Organomercurial(s) (% Yield)	Subsequent reactants	Product(s) (% Yield)	Ref.
(bicyclic alkene structure)	CH_3CN	$Hg(NO_3)_2 \cdot 2\,H_2O$, NaCl	(bicyclic structure with NHCOCH$_3$, HgCl) (63)	—	—	1
$H_2C=CH-$(cyclohexyl)	CH_3CN	$Hg(NO_3)_2 \cdot H_2O$, NaCl	$ClHgCH_2CH(NHCOCH_3)-$(cyclohexyl) (22)	Δ	(4-cyclohexyl-2-methyloxazoline) (21)	4
$H_2C=CH(CH_2)_5CH_3$	CH_3CN	$Hg(NO_3)_2$, NaCl	$ClHgCH_2CH(NHCOCH_3)(CH_2)_5CH_3$	$H_2C=CHX$ / $NaHB(OCH_3)_3$	$X(CH_2)_3CH(NHCOCH_3)(CH_2)_5CH_3$ $X = CN\,(67)$, $CO_2CH_3\,(40)$	2
cis-$CH_3(CH_2)_2CH=CH-$$(CH_2)_2CH_3$	CH_3CN	$Hg(NO_3)_2 \cdot 2\,H_2O$, NaCl	$CH_3(CH_2)_2CH(HgCl)CH(NHCOCH_3)(CH_2)_2CH_3$ (96)	—	—	1
$trans$-$CH_3(CH_2)_2CH=CH-$$(CH_2)_2CH_3$	CH_3CN	$Hg(NO_3)_2 \cdot H_2O$, NaCl	$erythro$-$CH_3(CH_2)_2CH(HgCl)CH-$$(NHCOCH_3)(CH_2)_2CH_3$ (65)	Δ	(4,5-dipropyl-2-methyloxazoline) (16)	4
	CH_3CN	$Hg(NO_3)_2 \cdot 2\,H_2O$, NaCl	$CH_3(CH_2)_2CH(HgCl)CH(NHCOCH_3)(CH_2)_2CH_3$ (72)	—	—	1
(1-methyl-6-(1-hydroxyethyl)cyclohexene)	CH_3CN	$Hg(NO_3)_2$	—	$NaBH_4$ / NaOH	(hydroxyethyl-methyl-cyclohexyl-NHCOCH$_3$) (52) + (methyl-cyclohexenyl-CHCH$_3$-NHCOCH$_3$) (37)	8
$H_2C=CH(CH_2)_7CH_3$	CH_3CN	$Hg(NO_3)_2$	—	$NaBH_4$ / NaOH	$CH_3CH(NHCOCH_3)(CH_2)_7CH_3$ (86)	3
	CH_3CN	$Hg(NO_3)_2 \cdot 2\,H_2O$, NaCl	$ClHgCH_2CH(NHCOCH_3)(CH_2)_7CH_3$ (93)	$NaBH_4$ / NaOH [or Na(Hg) / H_2O]	$CH_3CH(NHCOCH_3)(CH_2)_7CH_3$ (60) [58]	1
	CH_3CN	$Hg(NO_3)_2 \cdot 2\,H_2O$, NaCl	$ClHgCH_2CH(NHCOCH_3)(CH_2)_7CH_3$ (93)	$LiAlH_4$, HCl	$CH_3CH(\overset{+}{N}H_2CH_2CH_3)(CH_2)_7CH_3\ Cl^-$ (59)	1

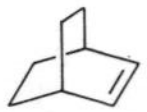

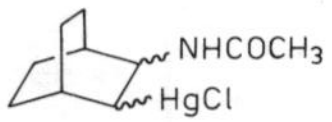

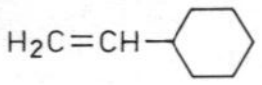

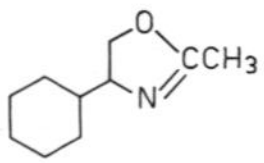

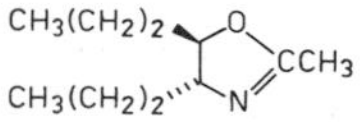

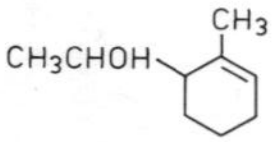

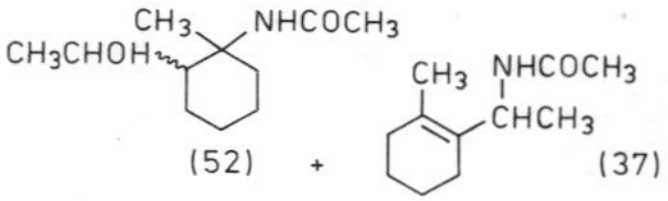

Table 7.1. (continued)

Alkene	Nitrile	Mercuric salt	Organomercurial(s) (% Yield)	Subsequent reactants	Product(s) (% Yield)	Ref.
(β-pinene, exo-methylene)	(indol-3-yl)CH₂CN structure	Hg(NO₃)₂	—	NaBH₄/NaOH	(17)	10
(α-pinene)	CH₃CN	Hg(NO₃)₂	—	NaBH₄/NaOH, HCl	[structure] Cl⁻ (51)	11
(indol-3-yl)CH₂CN structure	Hg(NO₃)₂	—	NaBH₄/NaOH	(11)	10
C₆H₅-cyclohexene	CH₃CN	Hg(NO₃)₂	—	NaBH₄/NaOH	C₆H₅, NHCOCH₃ cyclohexene (70)	8
CH₃CHOH–C₆H₅-cyclohexene	CH₃CN	Hg(NO₃)₂	—	NaBH₄/NaOH	C₆H₅, NHCOCH₃ cyclohexene (90)	8
trans- CH₃(CH₂)₅CH=CH(CH₂)₅CH₃	CH₃CN	Hg(NO₃)₂ · H₂O, NaCl	erythro-CH₃(CH₂)₅CH(HgCl)CH-(NHCOCH₃)(CH₂)₅CH₃ (22)	—	—	4
H₂C=CH(CH₂)₁₂CH₃	CH₃CN	Hg(NO₃)₂ · 2 H₂O, NaCl	ClHgCH₂CH(NHCOCH₃)(CH₂)₁₂CH₃ (91)	Br₂	BrCH₂CH(NHCOCH₃)(CH₂)₁₂CH₃ (15)	1
H₂C=CH(CH₂)₁₅CH₃	CH₃CN	Hg(NO₃)₂ · 2 H₂O, NaCl	ClHgCH₂CH(NHCOCH₃)(CH₂)₁₅CH₃ (84)	NaBH₄/NaOH	CH₃CH(NHCOCH₃)(CH₂)₁₅CH₃ (93)	1
cis- CH₃(CH₂)₇CH=CH-(CH₂)₇CO₂CH₃	CH₃CN	Hg(NO₃)₂	—	NaBH₄/NaOH	CH₃(CH₂)₇CH(H)(NHCOCH₃)CH(CH₂)₇CO₂CH₃ (100)	12

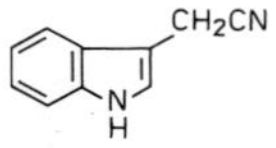

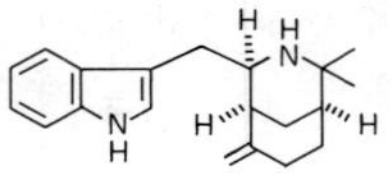

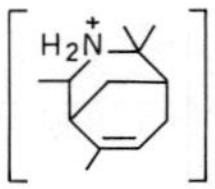

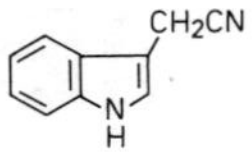

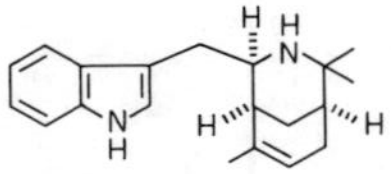

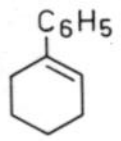

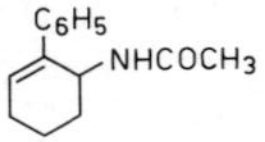

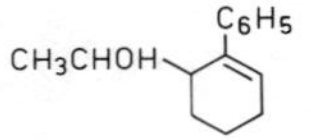

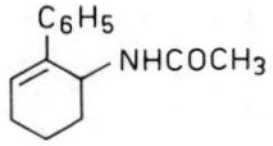

β-pinene, amidomercuration proceeds with rearrangement (Eq. 3) [8]. This reaction has been beautifully exploited in the efficient synthesis of the alkaloids (+)-makomakine (Eq. 4) and racemic hobartine (Eq. 5) [10].

$$(3)$$

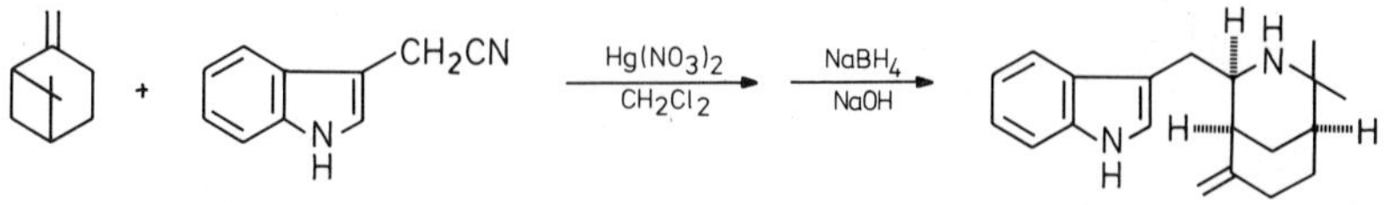

$$(4)$$

$$(5)$$

Allylic amidation has also been observed with certain trisubstituted olefins (Eq. 6) [8]. Carbon—carbon bond cleavage and other rearrangements have also been observed as side reactions (Eqs. 7, 8). Several distinct mechanistic pathways appear to be involved here (Scheme 7.1). While the

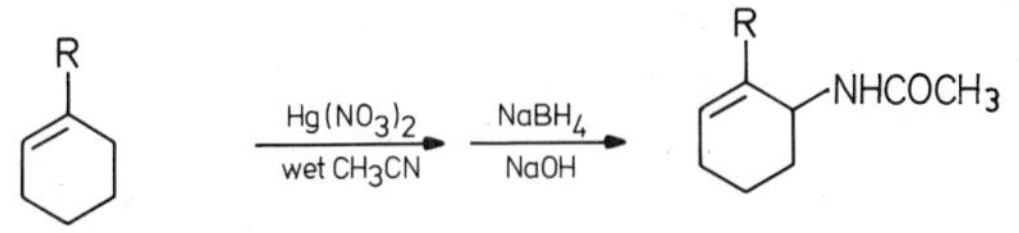

$$(6)$$

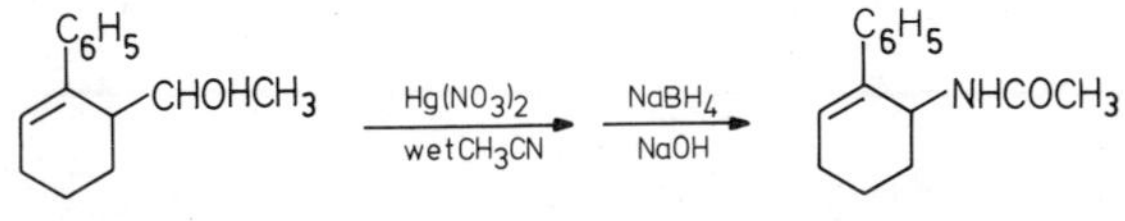

$$(7)$$

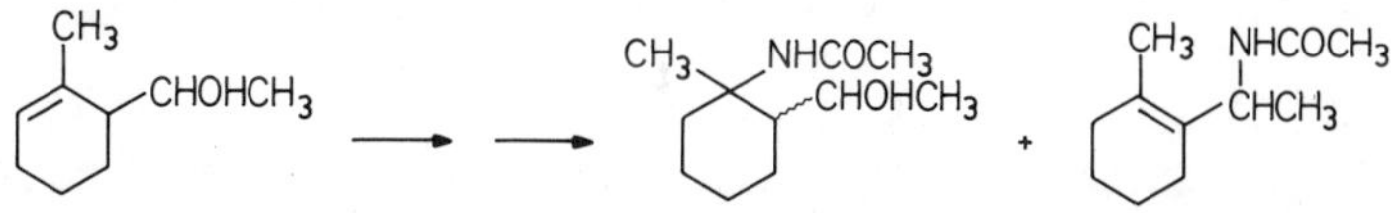

$$(8)$$

Scheme 7.1

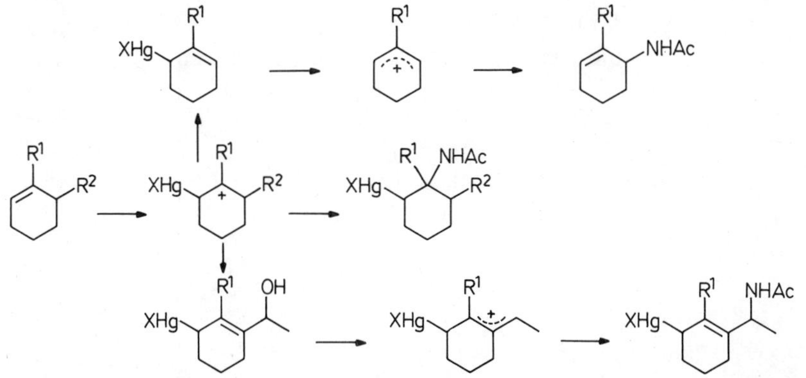

direct observation of a mercurinium ion in the amidomercuration reaction has been claimed [13], this report is suspect. It is noteworthy that 3,3-dimethyl-1-butene fails to rearrange under the usual amidomercuration conditions [3].

Several different demercuration procedures have been reported for the β-amidomercurials. The most general appears to be alkaline sodium borohydride [3]. Sodium-mercury amalgam works well [1, 5], but appears to give slightly lower yields than sodium borohydride. Lithium aluminum hydride reduction affords the corresponding alkylamine, thus providing an alternative to the aminomercuration-demercuration procedure (Eq. 9) [1]. The

$$XHg - \overset{|}{\underset{|}{C}} - \overset{|}{\underset{|}{C}} - NH\overset{O}{\overset{||}{C}}R \xrightarrow{\text{LiAlH}_4} H - \overset{|}{\underset{|}{C}} - \overset{|}{\underset{|}{C}} - NHCH_2R \qquad (9)$$

acetonitrile amidomercuration-LiAlH$_4$ demercuration of propene is reported to afford $>95\%$ of isopropyl ethyl amine and ~ 3–4% of n-propyl ethyl amine. It is not clear if the latter product arises via anti-Markovnikov amidomercuration or rearrangement during reduction.

Several other useful reactions of amidomercurials have been reported. For example, halogenation affords β-haloamides (Eq. 10) [1]. Thermolysis

$$ClHg - \overset{|}{\underset{|}{C}} - \overset{|}{\underset{|}{C}} - NH\overset{O}{\overset{||}{C}}R \xrightarrow[X=Cl,Br]{X_2} X - \overset{|}{\underset{|}{C}} - \overset{|}{\underset{|}{C}} - NH\overset{O}{\overset{||}{C}}R \qquad (10)$$

results in oxazoline formation (Eq. 11) [4]. The yields in this reaction are

$$\xrightarrow{180-240°C} \qquad (11)$$

low, however, and substantial amounts of olefin are formed. The sodium trimethoxyborohydride-induced addition to electron-deficient alkenes (see Chapter VII in the monograph "Organomercury Compounds in Organic Synthesis") generates compounds of potential use in alkaloid synthesis (Eq. 12) [2]. Since this latter reaction proceeds via free radicals, there is some loss of stereochemistry during alkylation.

$$R\overset{O}{\overset{||}{C}}NH - \overset{|}{\underset{|}{C}} - \overset{|}{\underset{|}{C}} - HgCl + H_2C=CHX \xrightarrow{\text{NaHB(OCH}_3)_3} R\overset{O}{\overset{||}{C}}NH - \overset{|}{\underset{|}{C}} - \overset{|}{\underset{|}{C}} - \overset{|}{\underset{|}{C}} - \overset{|}{\underset{|}{C}} - X \qquad (12)$$
$$X = CN, CO_2CH_3$$

Amides can be used instead of nitriles for amidomercuration (Table 7.2). In 1981 a simple procedure for intermolecular amidomercuration-demercuration using amides was introduced [17]. Equivalent amounts of olefin and anhydrous mercuric nitrate plus 10 equivalents of amide are refluxed 6–24 hours in methylene chloride (Eq. 13). Alkaline sodium borohydride in the presence of n-butylamine was used for demercuration. Modest to ex-

$$\diagdown C = C \diagup \quad + \quad RCNH_2 \quad \xrightarrow[CH_2Cl_2]{Hg(NO_3)_2} \quad \xrightarrow[\substack{NaOH \\ n\text{-}BuNH_2}]{NaBH_4} \quad H - \overset{|}{\underset{|}{C}} - \overset{|}{\underset{|}{C}} - NH\overset{O}{\overset{||}{C}}R \qquad (13)$$

cellent yields of amides are obtained. Urea and urethane also produced good yields of the corresponding amides, but N-propyl acetamide and succinimide cannot be employed in this reaction. From the examples reported, it appears that this reaction may be subject to limitations in olefin structure similar to those of the nitrile procedure. However, α-methylstyrene does provide a high yield of the corresponding amide. Mercuric acetate fails in this reaction, but the addition of n-butylamine during reduction apparently increases the amide yield.

The amidomercuration of vinyl acetate and allyl alcohol was studied some time ago. With mercuric acetate and saccharin, the corresponding N-vinyl [16] and N-allyl [14] derivatives are formed in fair yields (Eq. 14). Certain heterocyclic amines undergo analogous reactions as reported in the

$$(14)$$

preceding chapter. These reactions are not very general, however, since phthalimide and several other amines fail to react. 2-Pyridone and O-trimethylsilyl lactims undergo an analogous reaction with vinyl acetate, mercuric acetate and sulfuric acid, affording N-vinylpyridones and N-vinyl pyrimidinones (Eqs. 15, 16) [15].

$$(15)$$

$$(16)$$

513

Table 7.2. Amidomercuration of Alkenes Using Amides

Alkene	Amide	Mercuric salt	Organomercurial(s) (% Yield)	Subsequent reactants	Product(s) (% Yield)	Ref.
$H_2C{=}CHCH_2OH$	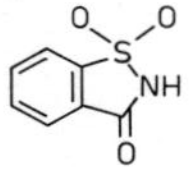 (saccharin, benzisothiazolone dioxide, NH)	$Hg(OAc)_2$	—	—	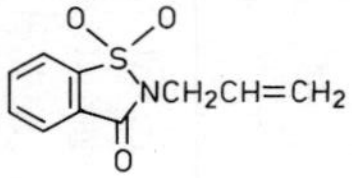 (N-allyl saccharin, $NCH_2CH{=}CH_2$) (50)	14
$H_2C{=}CHO_2CCH_3$	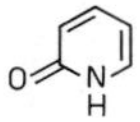 (2-pyridone, NH)	$Hg(OAc)_2 / H_2SO_4$	—	—	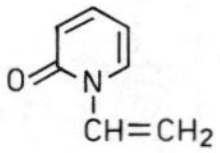 (N-vinyl 2-pyridone, $CH{=}CH_2$) (14.3)	15
	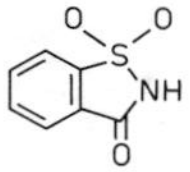 (saccharin, NH)	$Hg(OAc)_2 / H_2SO_4$	—	—	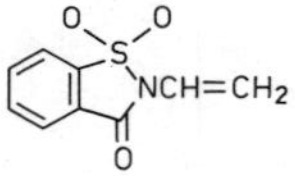 (N-vinyl saccharin, $NCH{=}CH_2$) (47)	16
	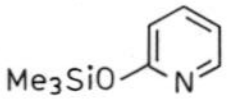 (Me_3SiO-pyridine)	$Hg(OAc)_2 / H_2SO_4$	—	—	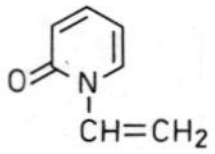 (N-vinyl 2-pyridone, $CH{=}CH_2$) (75)	15
	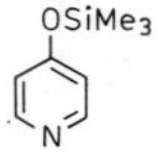 (4-$OSiMe_3$-pyridine)	$Hg(OAc)_2 / H_2SO_4$	—	—	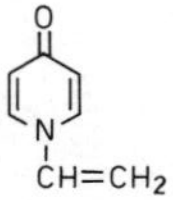 (N-vinyl 4-pyridone, $CH{=}CH_2$) (37.6)	15
	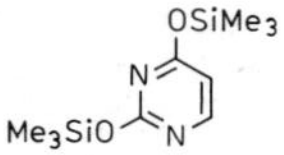 (2,4-bis-$OSiMe_3$-pyrimidine)	$Hg(OAc)_2 / H_2SO_4$	—	—	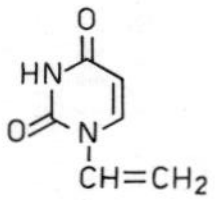 (N-vinyl uracil, $CH{=}CH_2$) (30)	15
	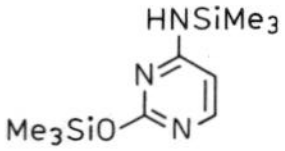 (4-$HNSiMe_3$-2-Me_3SiO-pyrimidine)	$Hg(OAc)_2 / H_2SO_4$	—	—	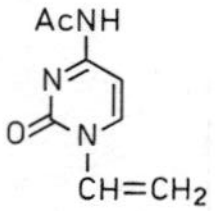 (AcNH N-vinyl cytosine, $CH{=}CH_2$) (13.8)	15

Table 7.2. (continued)

Alkene	Amide	Mercuric salt	Organomercurial(s) (% Yield)	Subsequent reactants	Product(s) (% Yield)	Ref.
	AcNSiMe₃ / Me₃SiO — (pyrimidine)	Hg(OAc)₂/H₂SO₄	—	—	AcNH-pyrimidinone, CH=CH₂ (21.1)	15
(cyclopentene)	CH_3CONH_2	$Hg(NO_3)_2$	—	$NaBH_4/NaOH$ / n-BuNH₂	cyclopentyl-NHCOCH₃ (42)	17
	p-$CH_3C_6H_4SO_2NH_2$	$Hg(NO_3)_2$	—	$NaBH_4/NaOH$ / n-BuNH₂	cyclopentyl-NHSO₂C₆H₄CH₃-p (52)	18
$H_2C=CHCH_2CH=CHCH_3$	p-$CH_3C_6H_4SO_2NH_2$	$Hg(NO_3)_2$	—	$NaBH_4/NaOH$ / n-BuNH₂	2,5-dimethylpyrrolidine (N-p-CH₃C₆H₄SO₂) (80)	18
$H_2C=CH(CH_2)_2CH=CH_2$	p-$CH_3C_6H_4SO_2NH_2$	$Hg(NO_3)_2$	—	$NaBH_4/NaOH$ / n-BuNH₂	2,6-dimethylpiperidine (N-p-CH₃C₆H₄SO₂) (63)	18
(cyclohexene)	CH_3CONH_2	$Hg(NO_3)_2$	—	$NaBH_4/NaOH$ / n-BuNH₂	cyclohexyl-NHCOCH₃ (92)	17
	p-$CH_3C_6H_4SO_2NH_2$	$Hg(NO_3)_2$	—	$NaBH_4/NaOH$ / n-BuNH₂	cyclohexyl-NHSO₂C₆H₄CH₃-p (66)	18
$H_2C=CHCH(CH_3)CH_2CONH_2$	—	—	CH₃-pyrrolidinone-CH₂HgCl	$H_2C=CClX$ / $NaHB(OCH_3)_3$	$H_2C=CHCH(CH_3)CH_2CONH_2$	2
$H_2C=CH(CH_2)_3CH_3$	p-$CH_3C_6H_4SO_2NH_2$	$Hg(NO_3)_2$	—	$NaBH_4/NaOH$ / n-BuNH₂	$CH_3CH(NHSO_2C_6H_4CH_3$-$p)(CH_2)_3CH_3$ (52)	18

Table 7.2. (continued)

516

Alkene	Amide	Mercuric salt	Organomercurial(s) (% Yield)	Subsequent reactants	Product(s) (% Yield)	Ref.
$H_2C=CH(CH_2)_2$ 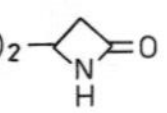 [azetidinone structure]	—	$Hg(OAc)_2$	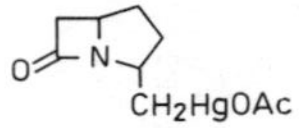 [pyrrolidinone-CH_2HgOAc] ($\sim$100)	$NaBH_4$	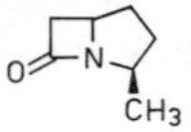 [pyrrolidinone-CH_3] (75)	19
$H_2C=CH(CH_2)_4CH_3$	CH_3CONH_2	$Hg(NO_3)_2$	—	$NaBH_4/NaOH/$ n-$BuNH_2$	$CH_3CH(NHCOCH_3)(CH_2)_4CH_3$ (81)	17
	p-$CH_3C_6H_4SO_2NH_2$	$Hg(NO_3)_2$	—	$NaBH_4/NaOH/$ n-$BuNH_2$	$CH_3CH(NHSO_2C_6H_4CH_3$-$p)(CH_2)_4CH_3$ (46)	18
$H_2C=CHC_6H_5$	p-$CH_3C_6H_4SO_2NH_2$	$Hg(NO_3)_2$	—	$NaBH_4/NaOH/$ n-$BuNH_2$	$CH_3CH(C_6H_5)NHSO_2C_6H_4CH_3$-$p$ (31)	18
	$RCONH_2$	$Hg(NO_3)_2$	—	$NaBH_4/NaOH/$ n-$BuNH_2$	$CH_3CH(NHCOR)C_6H_5$ $R = Me (84)$, $Ph(53)$, $NH_2(75)$, $OEt(99)$	17
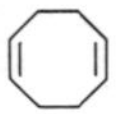 [cyclooctadiene]	p-$CH_3C_6H_4SO_2NH_2$	$Hg(NO_3)_2$	—	$NaBH_4/NaOH/$ n-$BuNH_2$	[bicyclic p-$CH_3C_6H_4SO_2$ amine] 56 + [bicyclic p-$CH_3C_6H_4SO_2$ amine] 44 (73 total) 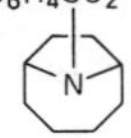 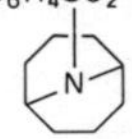	18
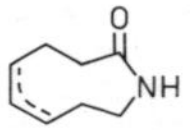 [cyclononenone lactam]	—	$Hg(OAc)_2$	—	$NaBH_4/NaOH$	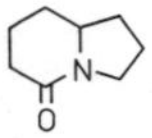 [indolizidinone] (85)	20
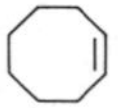 [cyclooctene]	CH_3CONH_2	$Hg(NO_3)_2$	—	$NaBH_4/NaOH/$ n-$BuNH_2$	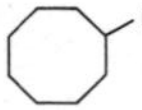 [cyclooctyl-$NHCOCH_3$] (44)	17
$H_2C=CH(CH_2)_2CH(CH_3)$ - $NHCOCH_3$	—	$Hg(OAc)_2$	—	$NaBH_4/NaOH$	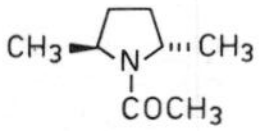 CH_3-[pyrrolidine-N-$COCH_3$]-CH_3 (98)	21

Table 7.2. (continued)

Alkene	Amide	Mercuric salt	Organomercurial(s) (% Yield)	Subsequent reactants	Product(s) (% Yield)	Ref.
$H_2C{=}CH(CH_2)_2CH(CH_3)-NHCO_2CH_3$	—	$Hg(OAc)_2$	—	$NaBH_4/NaOH$	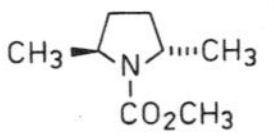 (pyrrolidine) CH_3–N(CO_2CH_3)–CH_3 (90)	21
$H_2C{=}CH(CH_2)_5CH_3$	CH_3CONH_2	$Hg(NO_3)_2$	—	$NaBH_4/NaOH/ n\text{-}BuNH_2$	$CH_3CH(NHCOCH_3)(CH_2)_5CH_3$ (97)	17
	$p\text{-}CH_3C_6H_4SO_2NH_2$	$Hg(NO_3)_2$	—	$NaBH_4/NaOH/ n\text{-}BuNH_2$	$CH_3CH(NHSO_2C_6H_4CH_3\text{-}p)(CH_2)_5CH_3$ (74)	18
$H_2C{=}C(CH_3)C_6H_5$	CH_3CONH_2	$Hg(NO_3)_2$	—	$NaBH_4/NaOH/ n\text{-}BuNH_2$	$(CH_3)_2C(NHCOCH_3)C_6H_5$ (80)	17
$H_2C{=}CHCH_2C_6H_5$	CH_3CONH_2	$Hg(NO_3)_2$	—	$NaBH_4/NaOH/ n\text{-}BuNH_2$	$CH_3CH(NHCOCH_3)CH_2C_6H_5$ (17)	17
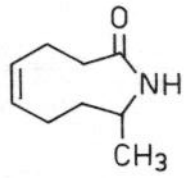 (2-methyl azacyclonon-en-one)	—	$Hg(OAc)_2$	—	$NaBH_4/NaOH$	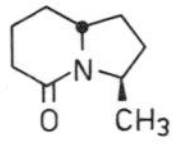 (indolizidine) CH_3 (83)	20
$o\text{-}H_2C{=}CHCH_2C_6H_4-NHCOCH{=}CH_2$	—	$Hg(OAc)_2$	—	$NaBH_4$	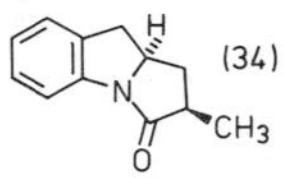 (pyrrolo-indolinone) CH_3 (34) 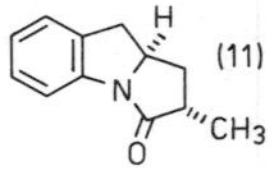 + (11)	22
$o\text{-}H_2C{=}CHCH_2C_6H_4-NHCOC(CH_3){=}CH_2$	—	$Hg(OAc)_2$	—	$NaBH_4$	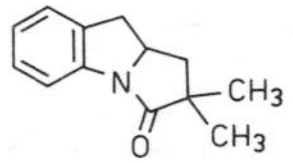 (dimethyl pyrrolo-indolinone) CH_3, CH_3	22
$H_2C{=}CH(CH_2)_3NHCO_2CH_2C_6H_5$	—	$Hg(OAc)_2$	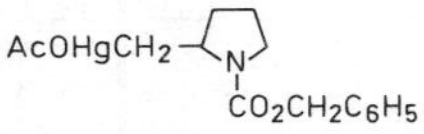 $AcOHgCH_2$–(pyrrolidine)–N–$CO_2CH_2C_6H_5$	$H_2C{=}CHCO_2CH_3/ NaHB(OCH_3)_3$	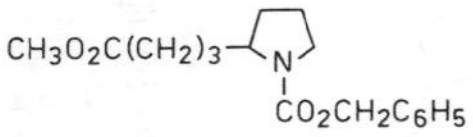 $CH_3O_2C(CH_2)_3$–(pyrrolidine)–N–$CO_2CH_2C_6H_5$	23

Table 7.2. (continued)

Alkene	Amide	Mercuric salt	Organomercurial(s) (% Yield)	Subsequent reactants	Product(s) (% Yield)	Ref.
$H_2C{=}CH(CH_2)_2CH(CH_3)$ – $NHCO_2CH_2C_6H_5$	—	$Hg(OAc)_2$	CH_3-pyrrolidine-CH_2HgOAc, $CO_2CH_2C_6H_5$	$H_2C{=}CHCO_2CH_3/$ $NaBH_4$	CH_3-pyrrolidine-$(CH_2)_3CO_2CH_3$, $CO_2CH_2C_6H_5$ (41)	23
	—	$Hg(OAc)_2$	—	$NaBH_4/NaOH$	CH_3-pyrrolidine-CH_3, $CO_2CH_2C_6H_5$ (95)	21
o- $H_2C{=}CHCH_2C_6H_4$ – $NHCO_2CH_2C_6H_5$	—	$Hg(OAc)_2$	—	$H_2C{=}CHCN/$ NaH_3BCN	indoline-$(CH_2)_3CN$, $CO_2CH_2C_6H_5$ (79)	23

The intramolecular amidomercuration of unsaturated amides has been known since 1979. All reactions reported to date have used mercuric acetate. Subsequent demercuration with alkaline sodium borohydride has provided a valuable route to cyclic amides (Eqs. 17–21). The yields are usually

$$(17)\ [2]$$

$$(18)\ [21, 23]$$

$$R = CH_3,\ OCH_3,\ OCH_2C_6H_5$$

$$(19)\ [19]$$

$$(20)\ [20]$$

$$(21)\ [20]$$

high and the reaction frequently exhibits high regio- and stereoselectivity as shown in the preceding examples. Intramolecular amidomercuration exhibits higher stereoselectivity than the corresponding aminomercuration reactions.

Recently the borohydride-induced free radical coupling of these cyclic amidomercurials and olefins has received attention. Although the following example afforded primarily the ring-opened product (Eq. 22) [2], closely

$$(22)$$

related mercurials obtained by ureidomercuration have been reported to couple in good yield (Eqs. 23, 24) [23]. Intramolecular coupling has also been

$$R = H,\ CH_3$$

$$(23)$$

519

VII. Amidomercuration

achieved (Eq. 25) [22]. These reactions look particularly promising for alkaloid synthesis.

$$\text{(24)}$$

$$\text{(25)}$$

$$R = H, CH_3$$

Finally, using a procedure essentially identical to that used for intermolecular amidomercuration-demercuration using amides, one can readily effect sulphonamidomercuration (Eq. 26) [18]. 1,4- and 1,5-Hexadiene undergo cyclization to afford exclusively the *cis*-pyrrolidine derivative (Eq. 27), while 1,5-cyclooctadiene yields a mixture of bicyclic amides (Eq. 28). The tosyl group can be removed by Na/NH_3 providing another alternative to aminomercuration-demercuration.

$$\text{(26)}$$

$$H_2C=CHCH_2CH=CHCH_3 \longrightarrow \qquad \longleftarrow H_2C=CH(CH_2)_2CH=CH_2 \qquad \text{(27)}$$

$$\text{(28)}$$

References

1. Beger, J., Vogel, D.: J. Prakt. Chem. *311*, 737 (1969).
2. Kozikowski, A. P., Scripko, J.: Tetrahedron Lett., *24*, 2051 (1983).
3. Brown, H. C., Kurek, J. T.: J. Am. Chem. Soc. *91*, 5647 (1969).
4. Kretchmer, R. A., Daly, P. J.: J. Org. Chem., *41*, 192 (1976).
5. Chow, D., Robson, J. H., Wright, G. F.: Can. J. Chem., *43*, 312 (1965).
6. Weinberg, N. L.: Tetrahedron Lett., 4823 (1970).
7. Sokolov, V. I., Reutov, O. A.: Izv. Akad. Nauk SSSR, Ser. Khim., 222 (1968); Bull. Acad. Sci. USSR, Div. Chem. Sci., 225 (1968).
8. Fry, A. J., Simon, J. A.: J. Org. Chem., *47*, 5032 (1982).

9. Factor, A., Traylor, T. G.: J. Org. Chem. *33*, 2607 (1968).
10. Stevens, R. V., Kenny, P. M.: J. Chem. Soc., Chem. Commun., 384 (1983).
11. Delpech, B., Khuong-Huu, Q.: J. Org. Chem., *43*, 4898 (1978).
12. Gunstone, F. D., Inglis, R. P.: Chem. Phys. Lipids, *10*, 73 (1973).
13. Sokolov, V. I., Yustynyuk, Yu. A., Reutov, O. A.: Dokl. Akad. Nauk SSSR, *173*, 1103 (1967); Proc. Acad. Sci. USSR, Chem. Sec., *173*, 356 (1967).
14. Hopff, H., Lüssi, H.: Helv. Chim. Acta, *46*, 1052 (1963).
15. Kaye, H., Chang, S.-H.: Tetrahedron, *26*, 1369 (1970).
16. Hopff, H., Wyss, U., Lüssi, H.: Helv. Chim. Acta, *43*, 135 (1960).
17. Barluenga, J., Jiménez, C., Nájera, C., Yus, M.: J. Chem. Soc., Chem. Commun., 670 (1981).
18. Barluenga, J., Jiménez, C., Nájera, C., Yus, M.: J. Chem. Soc., Chem. Commun., 1178 (1981).
19. Aida, T., Legault, R., Dugat, D., Durst, T.: Tetrahedron Lett., 4993 (1979).
20. Wilson, S. R., Sawicki, R. A.: J. Org. Chem., *44*, 330 (1979).
21. Harding, K. E., Burks, S. R.: J. Org. Chem., *46*, 3920 (1981).
22. Danishefsky, S., Taniyama, E.: Tetrahedron Lett., *24*, 15 (1983).
23. Danishefsky, S., Taniyama, E., Webb, R. R., II: Tetrahedron Lett., *24*, 11 (1983).

VIII. Azidomercuration

Other solvomercuration methods for the introduction of nitrogen functionality into organic substrates have been developed. One of these involves the reaction of olefins and sodium azide in the presence of mercury salts which results in azidomercuration (Table 8.1) (Eq. 1).

$$\text{C=C} \xrightarrow[\text{NaN}_3]{\text{HgX}_2} \quad XHg-\overset{|}{C}-\overset{|}{C}-N_3 \tag{1}$$

Several different mercury salts and solvent combinations have been used in this reaction including $Hg(OAc)_2/H_2O/THF$, $Hg(OAc)_2/CH_3OH$, $Hg(O_2CCF_3)_2/H_2O/THF$ and $Hg(NO_3)_2/DMF$. Elevated temperatures and lengthy reaction times are frequently employed in these reactions.

An interesting variety of olefins have been subjected to azidomercuration. Simple terminal mono- and disubstituted alkenes generally react well, but styrene gives only a very low yield of azide after alkaline sodium borohydride reduction [2, 3]. 3,3-Dimethyl-1-butene reacts well to give the anticipated pinacolyl azide without rearrangement in 61% yield [3]. More highly substituted alkenes, such as cyclohexene and 1-methylcyclohexene, react very poorly [3], but strained olefins such as norbornene [3] and cyclopropenes [1] proceed nicely (Eqs. 2. 3). Note the regio- and stereochemistry

$$\tag{2}$$

$$\tag{3}$$

in these reactions. Azidomercuration appears promising for the synthesis of azide-containing carbohydrates and by further reduction the corresponding aminosugars (Eqs. 4, 5). Note the stereochemistry of these reactions.

$$\tag{4}\ [5\text{--}7]$$

Table 8.1. Azidomercuration of Alkenes

Alkene	Mercuric salt	Organomercurial(s) (% Yield)	Subsequent reactants	Product(s) (% Yield)	Ref.
(cyclopropane alkene)	$Hg(OAc)_2$ / NaN_3, NaBr	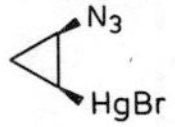 (cyclopropane, N_3, HgBr) (96)	—	—	1
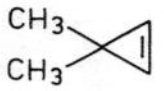 (dimethylcyclopropane alkene)	$Hg(OAc)_2$ / NaN_3, NaBr	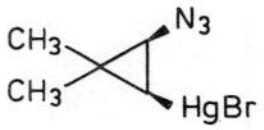 (CH_3, CH_3, N_3, HgBr cyclopropane) (81)	Na(Hg) / D_2O	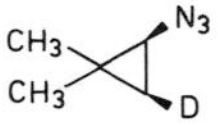 (CH_3, CH_3, N_3, D cyclopropane) (5)	1
	$Hg(OAc)_2$ / NaN_3	—	$NaBH_4$ / KOH	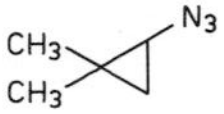 (CH_3, CH_3, N_3 cyclopropane) (5)	1
 (trimethylcyclopropane alkene)	$Hg(OAc)_2$ / NaN_3, NaBr	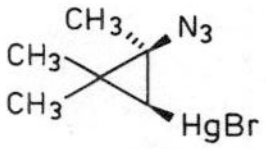 (CH_3, CH_3, CH_3, N_3, HgBr cyclopropane) (100)	Na(Hg) / D_2O	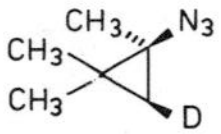 (CH_3, CH_3, CH_3, N_3, D cyclopropane) (10)	1
	$Hg(OAc)_2$ / NaN_3	—	$NaBH_4$ / KOH	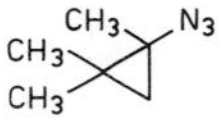 (CH_3, CH_3, CH_3, N_3 cyclopropane) (26)	1
(cyclohexene)	$Hg(NO_3)_2$ / NaN_3, NaCl	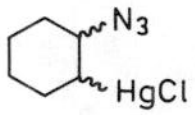 (cyclohexane, N_3, HgCl) (10 – 25)	—	—	2
	$Hg(OAc)_2$ / NaN_3	—	$NaBH_4$ / KOH	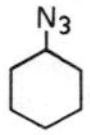 (cyclohexane, N_3) (4)	3
$H_2C=CHC(CH_3)_3$	$Hg(OAc)_2$ / NaN_3	—	$NaBH_4$ / KOH	$CH_3CHN_3C(CH_3)_3$ (61)	3
 (norbornene)	$Hg(OAc)_2$ / NaN_3	—	$NaBH_4$ / KOH	 (norbornane, N_3) (75)	3

Table 8.1. (continued)

Alkene	Mercuric salt	Organomercurial(s) (% Yield)	Subsequent reactants	Product(s) (% Yield)	Ref.
 (methylenecyclohexane)	Hg(OAc)$_2$ / NaN$_3$	—	NaBH$_4$ / KOH	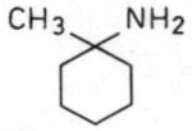 (1-amino-1-methylcyclohexane) CH$_3$, NH$_2$ (60)	3
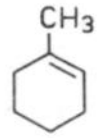 (1-methylcyclohexene)	Hg(OAc)$_2$ / NaN$_3$	—	NaBH$_4$ / KOH	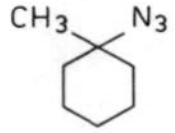 (1-azido-1-methylcyclohexane) CH$_3$, N$_3$ (<1)	3
H$_2$C=CH(CH$_2$)$_4$CH$_3$	Hg(OAc)$_2$ / NaN$_3$	—	NaBH$_4$ / KOH	CH$_3$CHN$_3$(CH$_2$)$_4$CH$_3$	3
H$_2$C=CHC$_6$H$_5$	Hg(NO$_3$)$_2$ / NaN$_3$, NaCl	ClHgCH$_2$CHN$_3$C$_6$H$_5$ (10 – 25)	—	—	2
	Hg(OAc)$_2$ / NaN$_3$	—	NaBH$_4$ / KOH	CH$_3$CHN$_3$C$_6$H$_5$ (<1)	3
H$_2$C=C(CH$_3$)(CH$_2$)$_4$CH$_3$	Hg(OAc)$_2$ / NaN$_3$	—	NaBH$_4$ / KOH	(CH$_3$)$_2$CN$_3$(CH$_2$)$_4$CH$_3$ (50)	3
H$_2$C=CH(CH$_2$)$_5$CH$_3$	Hg(OAc)$_2$ / NaN$_3$	—	NaBH$_4$ / KOH	CH$_3$CHN$_3$(CH$_2$)$_5$CH$_3$ (55)	3
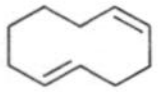 (cyclodecadiene)	Hg(N$_3$)$_2$	—	NaBH$_4$ / KOH	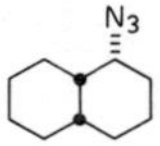 (decahydronaphthalenyl azide) N$_3$ (11) 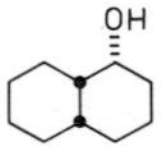 + (decahydronaphthalenol) OH (39)	4
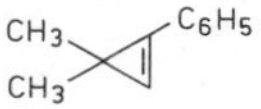 (1,2-dimethyl-3-phenylcyclopropene) CH$_3$, CH$_3$, C$_6$H$_5$	Hg(OAc)$_2$ / NaN$_3$, NaBr	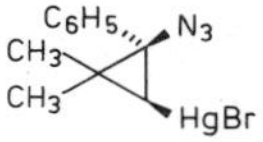 (cyclopropane) C$_6$H$_5$, N$_3$, CH$_3$, CH$_3$, HgBr (87)	Na(Hg) / D$_2$O	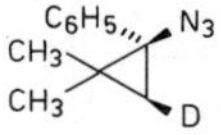 (cyclopropane) C$_6$H$_5$, N$_3$, CH$_3$, CH$_3$, D (3)	1
	Hg(OAc)$_2$ / NaN$_3$	—	NaBH$_4$ / KOH	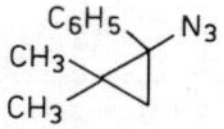 (cyclopropane) C$_6$H$_5$, N$_3$, CH$_3$, CH$_3$ (25)	1

Table 8.1. (continued)

Alkene	Mercuric salt	Organomercurial(s) (% Yield)	Subsequent reactants	Product(s) (% Yield)	Ref.
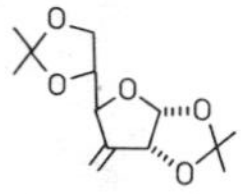	$Hg(OAc)_2$ / NaN_3	—	$NaBH_4$ / $NaOH$	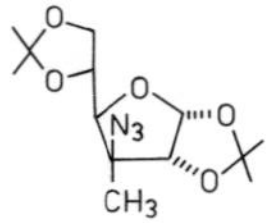 (43)	5–7
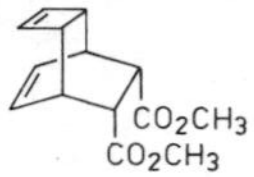	$Hg(OAc)_2$ / NaN_3, NaX	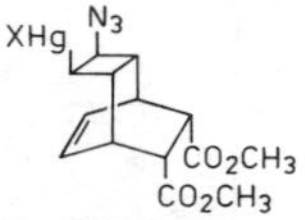 X = Cl (95), I	I_3^- (on RHgCl)	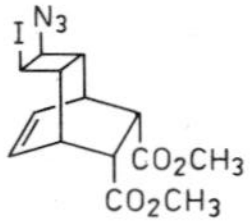	8, 9
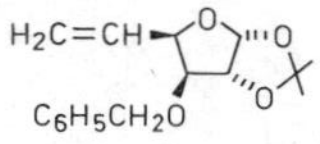	$Hg(O_2CCF_3)_2$ / NaN_3	—	KBH_4 / KOH	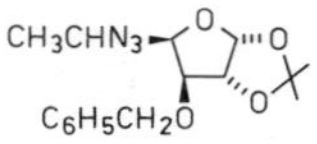 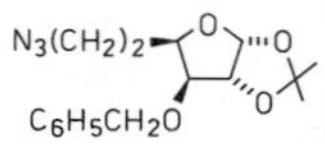 80 : ·20 (90 total)	10

VIII. Azidomercuration

$$(5) \ [10]$$

The latter reaction appears to be the only example in which anti-Markovnikov azidomercuration has been observed. This substrate is also reported to generate substantial amounts of the anti-Markovnikov alcohol upon hydroxymercuration-demercuration [11].

Two dienes have been subjected to azidomercuration (Eqs. 6, 7).

$$(6) \ [8, 9]$$

$$(7) \ [4]$$

In the tricyclic diene, cis addition to the less hindered double bond is observed, while the reaction of *cis, trans*-1,5-cyclodecadiene proceeds with carbon—carbon bond formation (see carbomercuration, Chapter X).

Of the subsequent reactions of these β-azidomercurials, protodemercuration is certainly the most important. Once again alkaline sodium borohydride seems to be the reagent of choice. For stereospecific deuterium incorporation, sodium-mercury amalgam in D_2O has been employed, but the yields are lower than with alkaline sodium borohydride and amines have been observed as side products [1]. One example of iodination has been reported (Eq. 8) [8]. Attempts to induce free radical coupling of electron-deficient olefins and *trans*-β-azidocyclohexylmercuric chloride failed, the only product being cyclohexene [12].

$$(8)$$

526

References

1. Galle, J. E., Hassner, A.: J. Am. Chem. Soc., *94*, 3930 (1972).
2. Sokolov, V. I., Reutov, O. A.: Izv. Akad. Nauk SSSR, Ser. Khim., 1632 (1967); Bull. Acad. Sci. USSR, Div. Chem. Sci., 1581 (1967).
3. Heathcock, C. H.: Angew. Chem., *81*, 148 (1969); Angew. Chem., Int. Ed. Engl., *8*, 134 (1969).
4. Traynham, J. G., Hsieh, H. H.: J. Org. Chem., *38*, 868 (1973).
5. Brimacombe, J. S., Miller, J. A., Zakir, U.: Carbohydr. Res., *41*, C3 (1975).
6. Brimacombe, J. S., Miller, J. A., Zakir, U.: Carbohydr. Res., *44*, C9 (1975).
7. Brimacombe, J. S., Miller, J. A., Zakir, U.: Carbohydr. Res., *49*, 233 (1976).
8. Mehta, G., Pandey, P. N.: J. Org. Chem., *40*, 3631 (1975).
9. Sasaki, T., Kanematsu, K., Kondo, A.: Tetrahedron, *31*, 2215 (1975).
10. Czernecki, S., Georgoulis, C., Provelenghiou, C.: Tetrahedron Lett., 4841 (1979).
11. Christol, H., Plénat, F., Revel, J.: Bull. Soc. Chim. France, 4537 (1971).
12. Kozikowski, A. P., Scripko, J.: Tetrahedron Lett., *24*, 2051 (1983).

IX. Nitromercuration

The reaction of olefins, mercuric salts and alkali metal nitrites in water provides a convenient method for nitromercuration (Table 9.1) (Eq. 1). This reaction, first reported in 1967 [1], is best carried out using mercuric

$$\text{C=C} + HgX_2 + NaNO_2 \longrightarrow XHg - \overset{|}{C} - \overset{|}{C} - NO_2 \qquad (1)$$

chloride or mercuric nitrate. The latter salt reacts five times faster than the former. Mercuric acetate apparently undergoes preferential hydroxymercuration, which is best avoided by running the reaction in aqueous HOAc at pH 4.3–4.5. Mercuric perchlorate also undergoes initial hydroxymercuration, but sodium nitrite is observed to react with the hydroxymercurial under the usual reaction conditions to eventually afford the nitromercurial (Eq. 2) [1, 2]. The mechanism of the latter step of this reaction has been studied in detail.

$$CH_3CH=CH_2 \quad \xrightarrow[\substack{NaNO_2 \\ H_2O}]{Hg(ClO_4)_2} \quad \left[CH_3\overset{OH}{\overset{|}{C}}HCH_2HgClO_4 \right] \longrightarrow CH_3\overset{NO_2}{\overset{|}{C}}HCH_2HgClO_4 \qquad (2)$$

For simple olefins, nitromercuration is best run in water as the solvent, using one equivalent of the mercury salt and two equivalents of sodium (or potassium) nitrite per olefin [1]. A phosphate buffer of pH 5.5–6.5 has also been added [4]. For water-insoluble, unsaturated polymers, nitromercuration is best carried out using aqueous methylene chloride in the presence of a phase transfer catalyst such as tetra-*n*-butylammonium bisulfate [5–7]. This latter procedure does present some problems with competitive nitritomercuration, not observed using other procedures (Eq. 3).

$$\text{C=C} \longrightarrow XHg - \overset{|}{C} - \overset{|}{C} - O - NO \qquad (3)$$

A variety of olefins undergo facile nitromercuration, including mono- and disubstituted olefins, as well as α,β-unsaturated ketones and esters [1]. However, trisubstituted olefins give only low yields and tetrasubstituted olefins don't react. Electron-deficient olefins such as vinyl chloride, 1,2-dichloroethylene and diethyl maleate also fail to react. *cis*-4-Methyl-2-pentene reacts, while *trans*-4-methyl-2-pentene does not. As noted earlier, even un-

528

Table 9.1. Nitromercuration of Alkenes

Alkene	Mercuric salt	Organomercurial(s) (% Yield)	Subsequent reactants	Product(s) (% Yield)	Ref.
$H_2C{=}CH_2$	$HgCl_2/NaNO_2$	$ClHgCH_2CH_2NO_2$ (69)	—	—	1
$H_2C{=}CHCH_3$	$HgCl_2/NaNO_2$	$ClHgCH_2CHNO_2CH_3$ (80)	—	—	1
	$Hg(ClO_4)_2/NaNO_2$	$ClO_4HgCH_2CHNO_2CH_3$	—	—	2, 3
$H_2C{=}CHCO_2CH_3$	$HgCl_2/NaNO_2$	$NO_2CH_2CH(HgCl)CO_2CH_3$? (69)	—	—	1
cis-$CH_3CH{=}CHCH_3$	$Hg(ClO_4)_2/NaNO_2$	$CH_3CH(HgClO_4)CHNO_2CH_3$	—	—	3
trans-$CH_3CH{=}CHCH_3$	$Hg(ClO_4)_2/NaNO_2$	$CH_3CH(HgClO_4)CHNO_2CH_3$	—	—	3
(cyclopentene)	$HgCl_2/NaNO_2$	—	NaOH	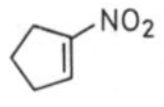 (1-nitrocyclopentene) (61)	4
$H_2C{=}C(CH_3)CH_2CH_3$	$HgCl_2/NaNO_2$	$ClHgCH_2C(NO_2)(CH_3)CH_2CH_3$ (75)	—	—	1
(cyclohexene)	$HgCl_2/NaNO_2$	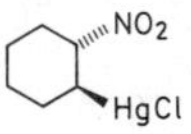 (trans-1-nitro-2-chloromercuricyclohexane) (80)	NaOH	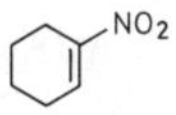 (1-nitrocyclohexene) (>98)	4
(cyclohexene)	$HgCl_2/NaNO_2$	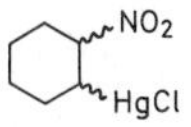 (1-nitro-2-chloromercuricyclohexane)	Br_2	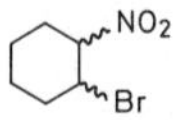 (1-nitro-2-bromocyclohexane)	1
$(CH_3)_2C{=}CHCOCH_3$	$HgCl_2/NaNO_2$	$(CH_3)_2C(NO_2)CH(HgCl)COCH_3$? (25)	—	—	1
cis-$CH_3CH{=}CHCH(CH_3)_2$	$HgCl_2/NaNO_2$	$CH_3CH(ClHg)(NO_2)CHCH(CH_3)_2$ (12)	—	—	1
cis- or trans-$CH_3CH_2CH{=}CHCH_2CH_3$	$HgCl_2/NaNO_2$	—	NaOH	$CH_3CH_2(H)C{=}C(CH_2CH_3)(NO_2)$ (65)	4
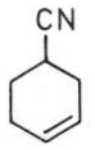 (4-cyanocyclohexene)	$HgCl_2/NaNO_2$	—	NaOH	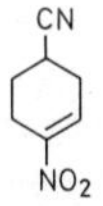 (4-cyano-1-nitrocyclohexene) (75)	4
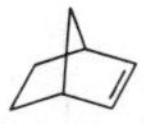 (norbornene)	$HgCl_2/NaNO_2$	—	DBU	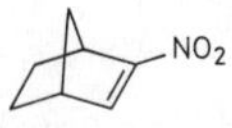 (2-nitronorbornene) (77)	4
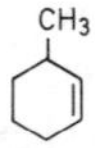 (1-methylcyclohexene)	$HgCl_2/NaNO_2$	—	NaOH	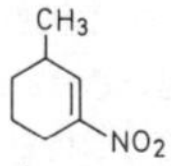 (1-methyl-2-nitrocyclohexene) (80)	4
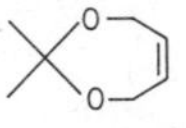 (2,2-dimethyl-4,7-dihydro-1,3-dioxepine)	$HgCl_2/NaNO_2$	—	NaOH	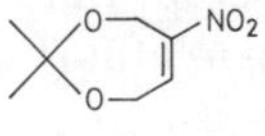 (nitro dioxepine) (67)	4
$H_2C{=}CH(CH_2)_3CH_3$	$HgCl_2/NaNO_2$	—	Et_3N	$H_2C{=}C(NO_2)(CH_2)_3CH_3$ (71)	4

Table 9.1. (continued)

Alkene	Mercuric salt	Organomercurial(s) (% Yield)	Subsequent reactants	Product(s) (% Yield)	Ref.
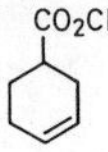 CO_2CH_3	$HgCl_2/NaNO_2$	—	NaOH	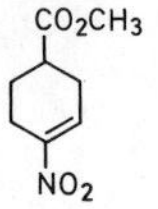 CO_2CH_3, NO_2 (78)	4
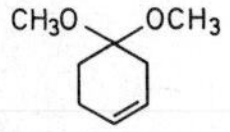 CH_3O OCH_3	$HgCl_2/NaNO_2$	—	NaOH	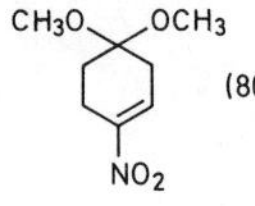 CH_3O OCH_3, NO_2 (80)	4

saturated polymers can be nitromercurated [5–7]. The following relative reactivities have been reported: cyclohexene > propene > ethylene [1].

The reaction is highly regioselective for cyclohexenes, as indicated by the position of nitro incorporation noted below [4].

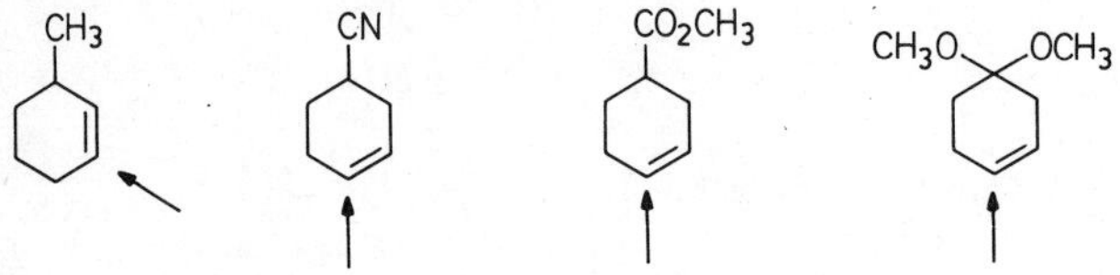

The β-nitromercurials undergo two synthetically useful reactions. They react cleanly with sodium hydroxide, triethylamine or DBU to generate nitroolefins in high yield (Eq. 4) [4]. As noted above, the nitromercuration

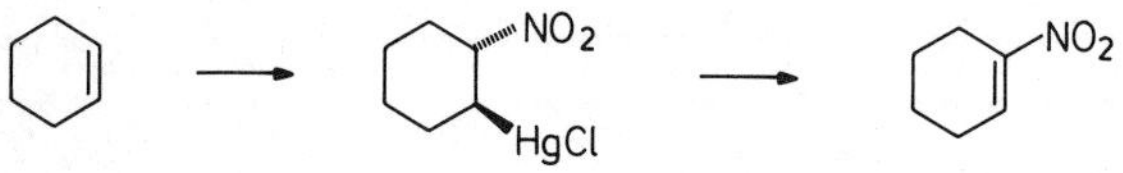

$$(4)$$

of substituted cyclohexenes is regioselective. Subsequent base-induced olefin formation provides a useful regiospecific route to the corresponding nitro-olefins, which can be useful in a variety of subsequent synthetic transformations. The reaction is also highly stereoselective. *cis*- and *trans*-3-Hexene yield different diastereomeric nitromercurials, but elimination affords solely the E-nitroolefin (Eq. 5).

$$CH_3CH_2CH = CHCH_2CH_3 \longrightarrow \longrightarrow \quad \underset{H}{\overset{CH_3CH_2}{\diagdown}} C = C \underset{NO_2}{\overset{CH_2CH_3}{\diagup}} \tag{5}$$

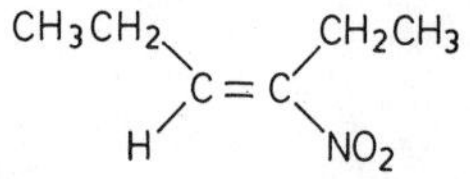

The bromination of one nitromercurial has been successful (Eq. 6), but

$$\text{(6)}$$

the chlorination of another nitromercurial was reported to give a dozen products [1].

The nitromercurials are decomposed by 5% HCl, HOAc, DMF, pyridine, aqueous ammonia, borane, Raney nickel and ethyl phosphite, but no products from these reactions have been reported [1]. An attempt to induce coupling with electron-deficient olefins using sodium trimethoxyborohydride failed [8]. No successful protodemercuration procedures for these nitromercurials have yet been reported.

References

1. Bachman, G. B., Whitehouse, M. L.: J. Org. Chem., *32*, 2303 (1967).
2. Shinoda, S., Saito, Y.: J. Organometal. Chem., *90*, 1 (1975).
3. Matsuo, M., Saito, Y.: J. Organometal. Chem., *27*, C41 (1971).
4. Corey, E. J., Estreicher, H.: J. Am. Chem. Soc., *100*, 6294 (1978).
5. Chien, J. C. W., Lillya, C. P.: Gov. Rep. Announce. Index (U.S.), *80*, 2873 (1980); Chem. Abstr., *93*, 169390h (1980).
6. Chien, J. C. W., Lillya, C. P.: Gov. Rep. Announce. Index (U.S.), *81*, 3622 (1981); Chem. Abstr., *95*, 170750y (1981).
7. Chien, J. C. W., Kohara, T., Lillya, C. P., Sarubbi, T., Su, B.-H.: J. Polym. Sci., Polym. Chem. Ed., *18*, 2723 (1980).
8. Kozikowski, A. P., Scripko, J.: Tetrahedron Lett., *24*, 2051 (1983).

X. Carbomercuration

A. Alkenes

The addition of carbon and mercury across the double bond of an alkene, carbomercuration, is a much less general reaction than those reported in the earlier sections of this chapter (Eq. 1). Nevertheless, there are a number of reactions here worth discussing.

$$\text{C}=\text{C} \longrightarrow -\overset{|}{\text{C}}-\overset{|}{\text{C}}-\overset{|}{\text{C}}-\text{HgX} \tag{1}$$

Very few organomercurials have been observed to add directly across the carbon—carbon double bond of an alkene (Table 10.1). The most extensively studied organomercurial of this type is the salt prepared from mercuric oxide and trinitromethane. It adds readily to a variety of alkenes (Eq. 2). With

$$\text{C}=\text{C} + \text{Hg}[\text{C}(\text{NO}_2)_3]_2 \longrightarrow (\text{NO}_2)_3\text{CHg}-\overset{|}{\text{C}}-\overset{|}{\text{C}}-\text{C}(\text{NO}_2)_3 \tag{2}$$

ethylene either mono or bis adducts can be obtained depending on the solvent employed (Eq. 3) [3]. Water, benzene and carbon tetrachloride give the

$$(\text{NO}_2)_3\text{CHg}(\text{CH}_2)_2\text{C}(\text{NO}_2)_3 \longleftarrow \text{H}_2\text{C}=\text{CH}_2 \longrightarrow \text{Hg}[(\text{CH}_2)_2\text{C}(\text{NO}_2)_3]_2 \tag{3}$$

single insertion product, while alcohols, nitromethane, nitrobenzene, DMF and methylene chloride favor the bis adduct. The former product has been found to react readily with ethylene to give the bis adduct [1, 2]. Single insertion products of a number of other olefins have also been reported (see Table 10.1). Only cyclopentene appears to give a mixture of the mono and bis adducts [2].

Simple mono- and disubstituted olefins react well with $\text{Hg}[\text{C}(\text{NO}_2)_3]_2$, but no carbomercuration products have been reported for any tri- or tetra-substituted olefins. The reactions of cyclohexene [3], 3,3,6,6-tetradeutero-cyclohexene [10], cycloheptene [12] and *cis*-cyclooctene [12] are all reported to proceed by trans addition, while norbornene forms the cis adduct [10].

Table 10.1. Simple Carbomercuration of Alkenes

Alkene	Mercury reagent	Organomercurial(s) (% Yield)	Subsequent reactants	Product(s) (% Yield)	Ref.
$H_2C{=}CH_2$	$ClHgC(NO_2)_3$	$ClHgCH_2CH_2C(NO_2)_3$ (92)	—	—	1, 2
	$HgCl_2 / HC(NO_2)_3$	$ClHgCH_2CH_2C(NO_2)_3$ (80)	—	—	2
	$Hg[C(NO_2)_3]_2$	$(NO_2)_3CHgCH_2CH_2C(NO_2)_3$ (93)	HCl	$ClHgCH_2CH_2C(NO_2)_3$	2
	$Hg[C(NO_2)_3]_2$	$(NO_2)_3CHgCH_2CH_2C(NO_2)_3$	Br_2	$BrCH_2CH_2C(NO_2)_3$ (81)	2
	$Hg[C(NO_2)_3]_2$	$(NO_2)_3CHgCH_2CH_2C(NO_2)_3$ (93)	$H_2C{=}CH_2$	$Hg[CH_2CH_2C(NO_2)_3]_2$ (65.5, 84)	1, 2
	$Hg[C(NO_2)_3]_2$	$(NO_2)_3C(CH_2)_2HgC(NO_2)_3$ or $[(NO_2)_3C(CH_2)_2]_2Hg$	—	—	3
	$C_6H_5HgC(NO_2)_3$	$Hg[CH_2CH_2C(NO_2)_3]_2$ (80.5) + $(C_6H_5)_2Hg$	—	—	2
	$Hg[C(NO_2)_2C(NO_2)_2CH_3]_2$	$Hg[(CH_2)_2C(NO_2)_2C(NO_2)_2CH_3]_2$ (84)	HCl	$ClHg(CH_2)_2C(NO_2)_2C(NO_2)_2CH_3$ (95)	4
	$HgO / HC(NO_2)_2CH_2C(NO_2)_2CH_3$	$Hg[(CH_2)_2C(NO_2)_2CH_2C(NO_2)_2CH_3]_2$ (84)	—	—	5
	$Hg(OAc)_2 / HC(NO_2)_2CH_2C(NO_2)_2CH_3$	$Hg[(CH_2)_2C(NO_2)_2CH_2C(NO_2)_2CH_3]_2$ (38)	—	—	5
$H_2C{=}CHCH_3$	$Hg[C(NO_2)_3]_2$	$(NO_2)_3CHgCH_2CH(CH_3)C(NO_2)_3$ (97)	Br_2	$BrHgCH_2CH(CH_3)C(NO_2)_3$	1, 2
	$Hg[C(NO_2)_3]_2$	$[(NO_2)_3CCH(CH_3)CH_2]_2Hg$ (79)	F_2	$(NO_2)_3CCH(CH_3)CH_2F$ + $[(NO_2)_3CCH(CH_3)CH_2]_2$	2, 6
$H_2C{=}CHCH_2OH$	$Hg[C(NO_2)_3]_2$	$(NO_2)_3CHgCH_2CH(CH_2OH)C(NO_2)_3$ (54)	—	—	7
	$Hg[C(NO_2)_3]_2$	$Hg[CH_2CH(CH_2OH)C(NO_2)_3]_2$ (70, 80)	—	—	1, 7
$H_2C{=}CHCO_2CH_3$	$Hg[C(NO_2)_3]_2$	$(NO_2)_3CCH_2CH(CO_2CH_3)HgC(NO_2)_3$ (82)	—	—	1

Table 10.1. (continued)

Alkene	Mercury reagent	Organomercurial(s) (% Yield)	Subsequent reactants	Product(s) (% Yield)	Ref.
$H_2C=C(CH_3)_2$	$Hg[C(NO_2)_3]_2$	$(NO_2)_3CC(CH_3)_2CH_2HgC(NO_2)_3$ (80) or $[(NO_2)_3CC(CH_3)_2CH_2]_2Hg$	Br_2 (on R_2Hg)	$(NO_2)_3CC(CH_3)_2CH_2Br$	3
$H_2C=CHCH_2CH_3$	$Hg[C(NO_2)_3]_2$	$[(NO_2)_3CCH(C_2H_5)CH_2]_2Hg$	F_2	$(NO_2)_3CCH(C_2H_5)CH_2F$	6
cis-$CH_3CH=CHCH_3$	$Hg[C(NO_2)_3]_2$	$[(NO_2)_3CCH(CH_3)CH(CH_3)]_2Hg$	F_2	$(NO_2)_3CCH(CH_3)CHFCH_3$	6
trans-$CH_3CH=CHCH_3$	$Hg[C(NO_2)_3]_2$	$(NO_2)_3CCH(CH_3)CH(CH_3)HgC(NO_2)_3$	—	—	6
$H_2C=CH(CH_2)_2OH$	$Hg[C(NO_2)_3]_2$	$(NO_2)_3CHgCH_2CH(CH_2CH_2OH)C(NO_2)_3$ (75)	$H_2C=CH(CH_2)_2OH$	$Hg[CH_2CH(CH_2CH_2OH)C(NO_2)_3]_2$	7
3,5-dinitrocyclopentene	$Hg[C(NO_2)_3]_2$	3,5-dinitro-cyclopentane bearing $C(NO_2)_3$ and $HgC(NO_2)_3$ substituents (100)	—	—	8
cyclopentene	$Hg[C(NO_2)_3]_2$	cyclopentane with $C(NO_2)_3$ and $HgC(NO_2)_3$ substituents + cyclopentane with $C(NO_2)_3$ and $)_2Hg$ substituents	—	—	2
$H_2C=CH(CH_2)_3OH$	$Hg[C(NO_2)_3]_2$	$Hg[CH_2CH(CH_2CH_2CH_2OH)C(NO_2)_3]_2$ (86)	—	—	7
$H_2C=CHSi(CH_3)_3$	$Hg[C(NO_2)_3]_2$	$[(NO_2)_3CCH_2CH(Si(CH_3)_3)-]_2Hg$ (90)	Br_2	$(NO_2)_3CCH_2CHBrSi(CH_3)_3$ (80)	9
3,3,6,6-tetradeuteriocyclohexene	$Hg[C(NO_2)_3]_2$, Cl^-	tetradeuteriocyclohexane bearing $C(NO_2)_3$ and $HgCl$ substituents	—	—	10
$(CH_3)_2C=C(CN)_2$	$Hg[C(CH_3)_3]_2$	$(CH_3)_3CC(CH_3)_2C(CN)_2HgC(CH_3)_3$ (40)	—	—	11

Table 10.1. (continued)

Alkene	Mercury reagent	Organomercurial(s) (% Yield)	Subsequent reactants	Product(s) (% Yield)	Ref.
(cyclohexene)	$Hg[C(NO_2)_3]_2$	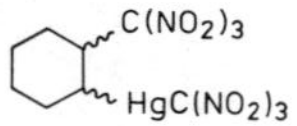 (cyclohexane with $C(NO_2)_3$ and $HgC(NO_2)_3$) (55)	—	—	1, 2
	$Hg[C(NO_2)_3]_2$, HCl	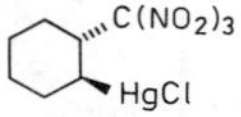 (cyclohexane with $C(NO_2)_3$ and $HgCl$) (65 – 75)	—	—	3
$H_2C=CHCH_2Si(CH_3)_3$	$Hg[C(NO_2)_3]_2$	$[(NO_2)_3CHgCH_2]_2C[C(NO_2)_3]_2$? (62)	Br_2	$(BrCH_2)_2C[C(NO_2)_3]_2$? (50)	9
 (norbornene)	$Hg[C(NO_2)_3]_2$	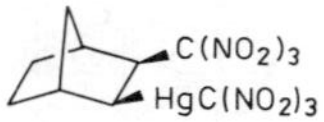 (norbornane with $C(NO_2)_3$ and $HgC(NO_2)_3$)	—	—	10
 (cycloheptene)	$Hg[C(NO_2)_3]_2$, $Hg(OAc)_2$, NaCl	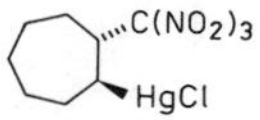 (cycloheptane with $C(NO_2)_3$ and $HgCl$)	—	—	12
$H_2C=CH(CH_2)_3O_2CCH_3$	$Hg[C(NO_2)_3]_2$	$(NO_2)_3CHgCH_2CH(CH_2CH_2CH_2O_2CCH_3)C(NO_2)_3$ (100)	—	—	7
$H_2C=CH(CH_2)_2Si(CH_3)_3$	$Hg[C(NO_2)_3]_2$	$Hg\left[CH_2CH(CH_2)_2Si(CH_3)_3\big|_{C(NO_2)_3}\right]_2$ (94)	Br_2	$BrCH_2CH(CH_2)_2Si(CH_3)_3\big|^{C(NO_2)_3}$ (74)	9
$H_2C=CHC_6H_5$	$HgCl_2 / HC(NO_2)_3$	$ClHgCH_2CH(C_6H_5)C(NO_2)_3$ (60)	—	—	2
	$Hg[C(NO_2)_3]_2$	$(NO_2)_3CHgCH_2CH(C_6H_5)C(NO_2)_3$ (80, 97)	—	—	1, 2
$(CH_3)_2C=C(CN)CO_2C_2H_5$	$Hg[C(CH_3)_3]_2$	$(CH_3)_3CC(CH_3)_2C(CN)(CO_2C_2H_5)HgC(CH_3)_3$ (35 – 40)	—	—	11
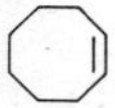 (cyclooctene)	$Hg[C(NO_2)_3]_2$, $Hg(OAc)_2$, NaCl	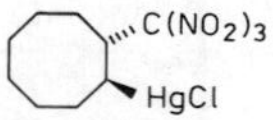 (cyclooctane with $C(NO_2)_3$ and $HgCl$)	—	—	12

Table 10.1. (continued)

Alkene	Mercury reagent	Organomercurial(s) (% Yield)	Subsequent reactants	Product(s) (% Yield)	Ref.
p-ClC$_6$H$_4$CH=C(CN)$_2$	Hg[C(CH$_3$)$_3$]$_2$	(CH$_3$)$_3$CCH(C$_6$H$_4$Cl-p)C(CN)$_2$HgC(CH$_3$)$_3$ (65)	—	—	11
C$_6$H$_5$CH=C(CN)$_2$	Hg[C(CH$_3$)$_3$]$_2$	(CH$_3$)$_3$CCH(C$_6$H$_5$)C(CN)$_2$HgC(CH$_3$)$_3$ (75)	—	—	11
p-CH$_3$C$_6$H$_4$CH=C(CN)$_2$	Hg[C(CH$_3$)$_3$]$_2$	(CH$_3$)$_3$CCH(C$_6$H$_4$CH$_3$-p)C(CN)$_2$HgC(CH$_3$)$_3$ (45)	—	—	11
C$_6$H$_5$CH=C(CN)COC$_6$H$_5$	Hg[C(CH$_3$)$_3$]$_2$	—	—	(CH$_3$)$_3$CCH(C$_6$H$_5$)C(COC$_6$H$_5$)=C=NC(CH$_3$)$_3$	11

Addition occurs in a Markovnikov fashion except where certain functional groups direct otherwise (Eqs. 4, 5). A number of functional groups are

$$H_2C=CHCO_2CH_3 \longrightarrow (NO_2)_3CCH_2\overset{\overset{\displaystyle HgC(NO_2)_3}{|}}{C}HCO_2CH_3 \qquad (4)\ [1]$$

$$H_2C=CHSi(CH_3)_3 \longrightarrow \left[(NO_2)_3CCH_2\overset{\overset{\displaystyle (CH_3)_3Si}{|}}{C}H\right]_2 Hg \qquad (5)\ [9]$$

apparently readily accommodated. Both mono and bis adducts from allyl alcohol and 4-hydroxy-1-butene have been prepared in good yield (Eq. 6) [1, 7]. 5-Hydroxy-1-pentene affords either carbomercuration or intramole-

$$(NO_2)_3CHgCH_2\overset{\overset{\displaystyle C(NO_2)_3}{|}}{C}HCH_2OH \longleftarrow H_2C=CHCH_2OH \longrightarrow Hg\left[CH_2\overset{\overset{\displaystyle C(NO_2)_3}{|}}{C}HCH_2OH\right]_2 \qquad (6)$$

cular alkoxymercuration products depending on the solvent used in the reaction (Eq. 7) [7]. While 3-butenyltrimethylsilane affords the normal Markovnikov bis adduct in high yield, allyltrimethylsilane undergoes silyl cleavage [9].

$$Hg\left[CH_2\overset{\overset{\displaystyle C(NO_2)_3}{|}}{C}H(CH_2)_3OH\right]_2 \xleftarrow[CH_3NO_2]{} H_2C=CH(CH_2)_3OH \xrightarrow[H_2O]{} (NO_2)_3CHgCH_2\text{-}\langle O\rangle \qquad (7)$$

Several variations of this general carbomercuration procedure have been reported. For example, to prepare the corresponding organomercuric chlorides one can either react the initial mono adducts with HCl [2, 3], or react the olefin directly with trinitromethylmercuric chloride or mercuric chloride plus trinitromethane (Eq. 8) [1, 2]. Phenyltrinitromethylmercury also inserts

$$H_2C=CH_2 \xrightarrow[HgCl_2/HC(NO_2)_3]{ClHgC(NO_2)_3\ or} ClHgCH_2CH_2C(NO_2)_3 \qquad (8)$$

ethylene, but undergoes disproportionation (Eq. 9) [2]. The dialkylmercurial

$$2\,H_2C=CH_2 + 2\,C_6H_5HgC(NO_2)_3 \longrightarrow (C_6H_5)_2Hg + Hg\left[(CH_2)_2C(NO_2)_3\right]_2 \qquad (9)$$

prepared from 1,1,3,3-tetranitrobutane and mercuric oxide also smoothly reacts with ethylene to give a bis adduct (Eq. 10) [4, 5].

$$CH_3\overset{\overset{\displaystyle NO_2}{|}}{\underset{\underset{\displaystyle NO_2}{|}}{C}}CH_2\overset{\overset{\displaystyle NO_2}{|}}{\underset{\underset{\displaystyle NO_2}{|}}{C}}H \xrightarrow{HgO} \xrightarrow{H_2C=CH_2} \left[CH_3\overset{\overset{\displaystyle NO_2}{|}}{\underset{\underset{\displaystyle NO_2}{|}}{C}}CH_2\overset{\overset{\displaystyle NO_2}{|}}{\underset{\underset{\displaystyle NO_2}{|}}{C}}(CH_2)_2\right]_2 Hg \qquad (10)$$

X. Carbomercuration

Both the protonolysis and halogenation of these nitro-containing organomercurials have been studied. As noted above, the mono adducts react with HCl to give organomercuric chlorides (Eq. 11) [2, 3]. The resulting

$$\text{(11)}$$

mercurials are very unreactive towards further protonolysis. The bis ethylene adduct derived from the 1,1,3,3-tetranitrobutane mercurial also reacts with HCl to form the corresponding organomercuric chloride (Eq. 12) [4].

$$\left[\begin{array}{c}\text{NO}_2 \quad \text{NO}_2 \\ | \qquad | \\ \text{CH}_3\text{CCH}_2\text{C(CH}_2)_2 \\ | \qquad | \\ \text{NO}_2 \quad \text{NO}_2\end{array}\right]_2 \text{Hg} \xrightarrow{\text{HCl}} \begin{array}{c}\text{NO}_2 \quad \text{NO}_2 \\ | \qquad | \\ \text{CH}_3\text{CCH}_2\text{C(CH}_2)_2\text{HgCl} \\ | \qquad | \\ \text{NO}_2 \quad \text{NO}_2\end{array} \qquad \text{(12)}$$

Some work on the halogenation of these organomercurials has appeared. It is reported that $(NO_2)_3C(CH_2)_2HgC(NO_2)_3$ reacts with bromine to give the corresponding organomercuric bromide and $BrC(NO_2)_3$ [2]. The corresponding organomercuric chloride is very unreactive towards bromine and chlorine. Dialkylmercurials derived from double addition of $Hg[C(NO_2)_3]_2$ to isobutene [3] and trimethylsilyl-substituted olefins [9] do, however, react with bromine to afford the corresponding organic bromides. The successful fluorination of similar dialkylmercurials derived from ethylene, propene and 1- and 2-butene has also been achieved [6, 13]. Only one alkyl group appears to be cleaved by fluorine. 3,3,3-Trinitropropylmercuric bromide also undergoes fluorination, however, but a significant amount of 1,1,1,6,6,6-hexanitrohexane is also observed. This product appears to arise by a free radical coupling process.

There appears to be only one other example of the addition of a simple organomercurial across the double bond of an olefin. It is reported that di-*t*-butylmercury reacts with electron-deficient olefins in just such a fashion (Table 10.1) (Eq. 13) [11]. No 1,4-addition is evident. Additions to alkynes

$$RCH{=}C(CN)_2 \; + \; Hg\big[C(CH_3)_3\big]_2 \longrightarrow \begin{array}{c}(CH_3)_3C \quad HgC(CH_3)_3 \\ | \qquad\quad | \\ R\dot{C}H\dot{C}(CN)_2\end{array} \qquad \text{(13)}$$

(see Chapter II, section J.3 of the monograph "Organomercury Compounds in Organic Synthesis") and diaza compounds have also been observed (Eq. 14) [11, 14]. Note the disproportionation in this last example. Neither di-

$$EtO_2CN{=}NCO_2Et \; + \; Hg\big[C(CH_3)_3\big]_2 \longrightarrow \begin{array}{c}EtO_2C \quad\;\; CO_2Et \\ \diagdown\; N{-}N \;\diagup \\ \diagup \qquad\quad \diagdown \\ (CH_3)_3C \qquad C(CH_3)_3\end{array} + \begin{array}{c}EtO_2C \quad\;\; CO_2Et \\ \diagdown\; N{-}N \;\diagup \\ \diagup \qquad\quad \diagdown \\ (CH_3)_3CHg \qquad HgC(CH_3)_3\end{array}$$

$$\text{(14)}$$

phenylmercury nor diethylmercury react with the electron-deficient olefins, but the latter mercurial apparently reacts with the diaza compound to give unspecified products [11].

The carbomercuration of olefins and subsequent protonolysis or solvolysis provides a useful method by which two carbon groups, a carbon and a proton, or a carbon and an oxygen can be added across the double bond of an olefin. A number of these reactions were discussed in Chapter IV, section B of the monograph "Organomercury Compounds in Organic Synthesis" and will be only briefly covered here. Table 10.2 summarizes much of this chemistry.

The reactions of simple olefins, mercuric salts (usually mercuric acetate) and relatively electron-rich arenes actually leads to β-phenethylmercuric salts (Eq. 15) [18–20, 29]. This reaction is usually catalyzed by the addition

$$\text{ArH} + \text{RCH=CH}_2 + \text{HgX}_2 \longrightarrow \overset{\overset{\displaystyle\text{Ar}}{|}}{\text{RCHCH}_2\text{HgX}} \tag{15}$$

of strong acids such as H_3PO_4 [19] or $HClO_4$ [18]. The reaction may first involve addition of the mercury salt across the double bond and then subsequent aryl substitution (Eq. 16). The latter transformation, aryl substitution,

$$\text{RCH=CH}_2 \xrightarrow{\text{Hg(OAc)}_2} \overset{\overset{\displaystyle\text{AcO}}{|}}{\text{RCHCH}_2\text{HgOAc}} \xrightarrow{\text{ArH}} \overset{\overset{\displaystyle\text{Ar}}{|}}{\text{RCHCH}_2\text{HgOAc}} \tag{16}$$

has been studied in some detail. The kinetics of the reaction are consistent with an electrophilic aromatic substitution reaction which has been suggested to proceed through an intermediate mercurinium ion [16, 17]. The same reaction is also observed when one starts with the appropriate aromatic mercurial and an olefin (Eq. 17) [20].

$$\text{(17)}$$

As indicated above, the reaction as commonly run results in the formation of β-phenethyl esters (Eq. 18) [19, 20, 23]. Ethylene, propylene and 2-butene

$$\text{ArH} + \text{RCH=CH}_2 + \text{R'CO}_2\text{H} \longrightarrow \overset{\overset{\displaystyle\text{O}_2\text{CR'}}{|}}{\text{ArCH}_2\text{CHR}} \tag{18}$$

appear to be the only olefins so far employed in this particular reaction. Mercuric acetate is usually used, but mercuric sulfate can also be employed. Besides anisole [19, 23] and various dimethoxybenzenes [20], a number of other arenes appear to react satisfactorily [19]. Several carboxylic acids other than acetic acid have also been successfully utilized [19]. The reaction appears to involve solvolysis of an intermediate β-phenethylmercurial to

Table 10.2. Additional Carbomercuration Reactions of Alkenes

Alkene	Carbon nucleophile	Mercuric salt and other reagents	Organomercurial(s) (% Yield)	Subsequent reactants	Product(s) (% Yield)	Ref.
$H_2C{=}CH_2$	C_6H_6	$HgSO_4 / H_2SO_4$	—	—	$C_6H_5CH_2CH_2C_6H_5$	15
	$C_6H_5OCH_3$	$Hg(OAc)_2 / HClO_4 /$ HOAc	$p\text{-}CH_3OC_6H_4(CH_2)_2HgOAc$	—	—	16–18
	$C_6H_5OCH_3$	$Hg(OAc)_2 / H_3PO_4 /$ HOAc, KI	$p\text{-}CH_3OC_6H_4(CH_2)_2HgI$	—	—	19
	$C_6H_5OCH_3$	$- / RCO_2H$	—	—	$p\text{-}CH_3OC_6H_4(CH_2)_2O_2CR$ $R = H(46),\ Et(42),\ Am(27)$	19
	$o\text{-}C_6H_4(OCH_3)_2$	$Hg(OAc)_2 / HBF_4$	$3,4\text{-}(CH_3O)_2C_6H_3CH_2CH_2HgOAc$	—	$3,4\text{-}(CH_3O)_2C_6H_3CH_2CH_2OAc$ (60)	20
	$m\text{-}C_6H_4(OCH_3)_2$	$Hg(OAc)_2$	—	—	$2,4\text{-}(CH_3O)_2C_6H_3CH_2CH_2OAc$	20
	$p\text{-}C_6H_4(OCH_3)_2$	$Hg(OAc)_2$	—	—	$2,5\text{-}(CH_3O)_2C_6H_3CH_2CH_2OAc$	20
	ArH	$Hg(OAc)_2$	—	—	$ArCH_2CH_2OAc$ $Ar = C_6H_5,\ p\text{-}CH_3C_6H_4,\ 3,4\text{-}(CH_3)_2C_6H_3,$ ClC_6H_4 (isomer ?), α-naphthyl, $p\text{-}CH_3OC_6H_4$ (70)	19
	$CH_3COCH_2COCH_3$	$Hg(OAc)_2 / HClO_4,$ NaCl	$(CH_3CO)_2CH(CH_2)_2HgCl$ (87)	KOH	$(CH_3CO)_2C{<}^{CH_2}_{CH_2}$ (1,1-diacetylcyclopropane)	21
	$CH_3COCH_2CO_2Et$	$Hg(OAc)_2 / BF_3 /$ HOAc [NaCl]	$[CH_3COCH(CO_2Et)(CH_2)_2HgCl]$	—	$CH_3COCH(CO_2Et)(CH_2)_2OAc$ (37)	22

Table 10.2. (continued)

Alkene	Carbon nucleophile	Mercuric salt and other reagents	Organomercurial(s) (% Yield)	Subsequent reactants	Product(s) (% Yield)	Ref.
	$C_6H_5COCH_2COCH_3$	$Hg(OAc)_2/HClO_4$, NaCl	$C_6H_5COCH(COCH_3)(CH_2)_2HgCl$ (95)	KOH	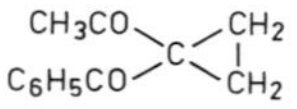 CH_3CO, C_6H_5CO C CH_2 CH_2	21
	$(C_6H_5CO)_2CH_2$	$Hg(OAc)_2/HClO_4$, NaCl	$(C_6H_5CO)_2CH(CH_2)_2HgCl$ (19) + $C_6H_5C(OAc){=}C(COC_6H_5)(CH_2)_2HgCl$? (11)	—	—	21
$H_2C{=}CHCH_3$	$C_6H_5OCH_3$	$Hg(OAc)_2/HClO_4/$ HOAc	$p\text{-}CH_3OC_6H_4CH(CH_3)CH_2HgOAc$	—	—	17
	$C_6H_5OCH_3$	$Hg(OAc)_2/BF_3/HOAc$ —		—	$p\text{-}CH_3OC_6H_4CHCH(OAc)CH_3$ (54 – 67)	23
	—	$p\text{-}AcOHgC_6H_4OCH_3/$ $HBF_4/HOAc$ —		—	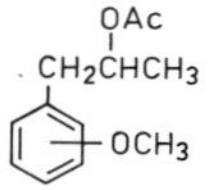 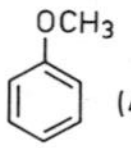 CH_2CHCH_3 (OAc) —OCH_3 (45 – 55) + OCH_3 (47) (o- and p-)	20
	$o\text{-}C_6H_4(OCH_3)_2$	$Hg(OAc)_2$	—	—	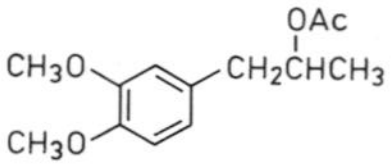 CH_3O— CH_2CHCH_3 (OAc), CH_3O (65)	20
	$m\text{-}C_6H_4(OCH_3)_2$	$Hg(OAc)_2$	—	—	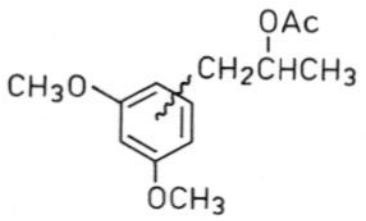 CH_3O— CH_2CHCH_3 (OAc), OCH_3	20
	$p\text{-}C_6H_4(OCH_3)_2$	$Hg(OAc)_2$	—	—	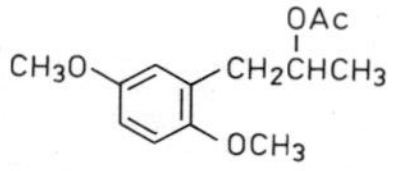 CH_3O— CH_2CHCH_3 (OAc), OCH_3	20

Table 10.2. (continued)

542

Alkene	Carbon nucleophile	Mercuric salt and other reagents	Organomercurial(s) (% Yield)	Subsequent reactants	Product(s) (% Yield)	Ref.
	$CH_3COCH_2COCH_3$	$Hg(OAc)_2/HClO_4$, NaCl	$(CH_3CO)_2CHCH(CH_3)CH_2HgCl$ (75)	KOH	$(CH_3CO)_2C{<}^{CHCH_3}_{CH_2}$	21
	$CH_3COCH_2CO_2Et$	$Hg(OAc)_2/BF_3/HOAc$	—	—	$CH_3COCH(CO_2Et)CH(CH_3)CH_2OAc$ (26)	22
$H_2C{=}CHO_2CCH_3$	$CH_3COCH_2CO_2Et$	$Hg(OAc)_2/H_2SO_4$	—	—	$CH_3\overset{O}{\underset{\underset{HCCH_3}{\parallel}}{C}}CCO_2Et$ (48)	24, 25
	$CH_2(CO_2Et)_2$	$Hg(OAc)_2/H_2SO_4$	—	—	$CH_3CH{=}C(CO_2Et)_2$ + $H_2C{=}CHCH(CO_2Et)_2$	25
$CH_3CH{=}CHCH_3$	$C_6H_5OCH_3$	$Hg(OAc)_2/BF_3/HOAc$	—	—	$p{-}CH_3OC_6H_4CH(CH_3)CH(OAc)CH_3$ (10–40)	23
$H_2C{=}CHC_6H_5$	$C_6H_5OCH_3$	$Hg(OAc)_2/HClO_4/HOAc$	$p{-}CH_3OC_6H_4CH(C_6H_5)CH_2HgOAc$	—	—	17
	$C_6H_5OCH_3$	$Hg(OAc)_2/HClO_4/HOAc$	—	—	$p{-}CH_3OC_6H_4CH(C_6H_5)CH_2C_6H_4OCH_3{-}p$ (63)	23, 26
	$C_6H_5OCH_3$	$Hg(OAc)_2/HClO_4/HOAc$	—	I_2	$p{-}CH_3OC_6H_4CH(C_6H_5)CH_2I$ + $p{-}CH_3OC_6H_4CH_2CHIC_6H_5$ ~2 : ~1	23
	$CH_3COCH_2COCH_3$	$Hg(OAc)_2/HClO_4$, NaCl	$(CH_3CO)_2CHCH(C_6H_5)CH_2HgCl$ (75)	—	—	21
	$CH_3COCH_2CO_2Et$	$Hg(OAc)_2$	—	—	$\underset{CH_3}{\overset{EtO_2C}{\diagdown}}\!\!\diagup\!\!O\!\!\diagup C_6H_5$ or $\underset{CH_3}{\overset{EtO_2C}{\diagdown}}\!\!\diagup\!\!O\!\!\diagup C_6H_5$?	22
	$C_6H_5COCH_2COCH_3$	$Hg(OAc)_2/HClO_4$, NaCl	$C_6H_5COCH(COCH_3)CH(C_6H_5)CH_2HgCl$ (98)	—	—	21
	$C_6H_5COCH_2CO_2Et$	$Hg(OAc)_2/HClO_4$, NaCl	$C_6H_5COCH(CO_2Et)CH(C_6H_5)CH_2HgCl$ (48)	—	—	21

Table 10.2. (continued)

Alkene	Carbon nucleophile	Mercuric salt and other reagents	Organomercurial(s) (% Yield)	Subsequent reactants	Product(s) (% Yield)	Ref.
$H_2C{=}CH(CH_2)_2C_6H_5$	—	$Hg(OAc)_2 / HClO_4$	—	—	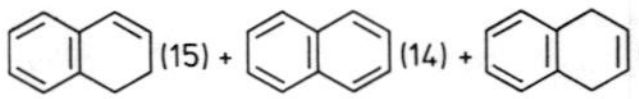 (15) + (14) + (6)	27
	—	$Hg(ClO_4)_2 / HOAc$	—	—	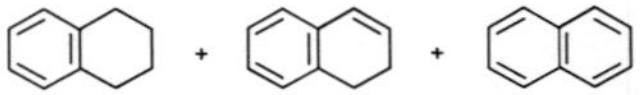 + + (35 total)	28
$H_2C{=}CH(CH_2)_3C_6H_5$	—	$Hg(OAc)_2 / HClO_4$ (HOAc)	—	—	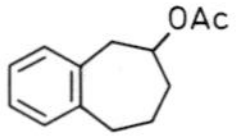 OAc (57, 60)	27, 28
$H_2C{=}CHCH_2C(CH_3)_2C_6H_5$	—	$Hg(OAc)_2 / HClO_4 /$ HOAc	—	—	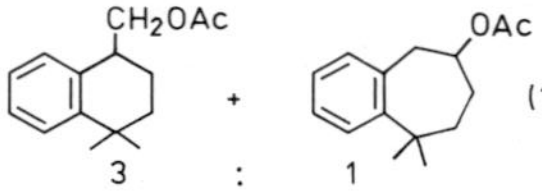 CH_2OAc 3 + OAc 1 (10–25)	28
$H_2C{=}C(CH_3)(CH_2)_3C_6H_5$	—	$Hg(OAc)_2 / HClO_4 /$ HOAc	—	—	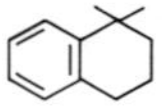 (60)	28
$(CH_3)_2C{=}CH(CH_2)_2C_6H_5$	—	$Hg(ClO_4)_2 / HOAc$	—	—	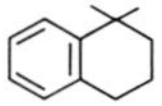 (60)	28
$m\text{-}H_2C{=}CH(CH_2)_3C_6H_4OCH_3$	—	$Hg(OAc)_2 / HClO_4 /$ HOAc	—	—	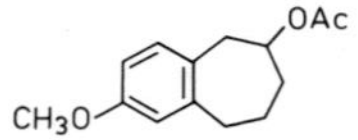 CH_3O OAc (80)	28
$m\text{-}H_2C{=}C(CH_3)(CH_2)_3 - C_6H_4OCH_3$	—	$Hg(BF_4)_2 / H_2O$	—	—	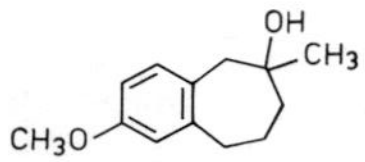 CH_3O OH CH_3 (60)	28

Table 10.2. (continued)

544

Alkene	Carbon nucleophile	Mercuric salt and other reagents	Organomercurial(s) (% Yield)	Subsequent reactants	Product(s) (% Yield)	Ref.
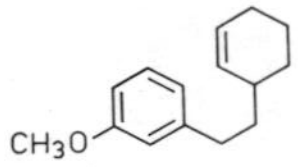 (structure)	—	Hg(OAc)₂/HClO₄/ HOAc	—	—	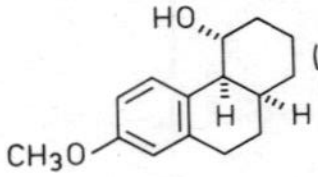 (structures, 40 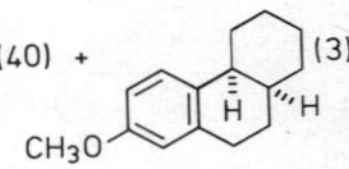 + 3 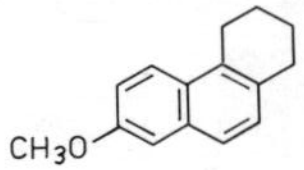 + 3)	28
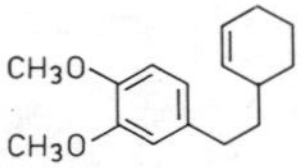 (structure)	—	Hg(OAc)₂/HClO₄/ HOAc	—	—	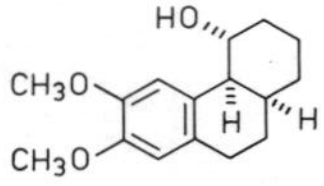 (structure, 40)	28

a bridged cation which is then trapped by the carboxylic acid present. β-Phenethylmercurials of this type have been shown to solvolyze in this fashion [20, 29]. Note that the resulting ester group ends up on the more highly substituted end of the original double bond, although the intermediate mercurial contains the aryl group on that carbon.

When these solvolytic reactions are run in the presence of an excess of olefin, diarylation can be achieved (Eq. 19) [15, 19, 23, 26]. To date,

$$\text{ArH} + \text{RCH}{=}\text{CH}_2 \longrightarrow \overset{\overset{\displaystyle \text{Ar}}{|}}{\text{RCHCH}_2\text{Ar}} \tag{19}$$

only ethylene and styrene have been employed as olefins in this reaction. Anisole and benzene are the sole arenes utilized and either $\text{Hg(OAc)}_2/\text{HClO}_4/\text{HOAc}$ or $\text{HgSO}_4/\text{H}_2\text{SO}_4$ apparently work well.

Under similar reaction conditions using certain aryl-substituted olefins, products of intramolecular carbomercuration and subsequent solvolysis or protonolysis have also been observed (Table 10.2). The following examples seem to be the more preparatively useful of these reactions (Eqs. 20–22).

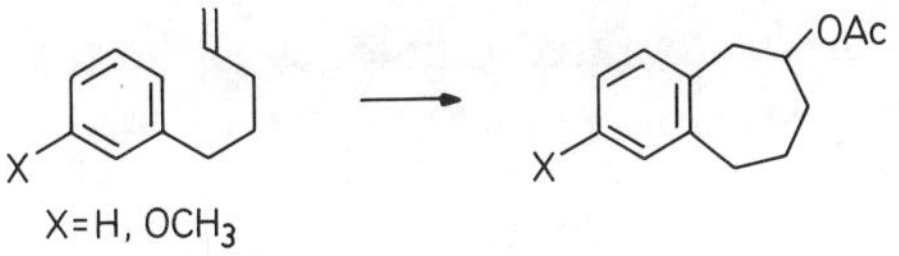

$$\tag{20}\ [27,\ 28]$$

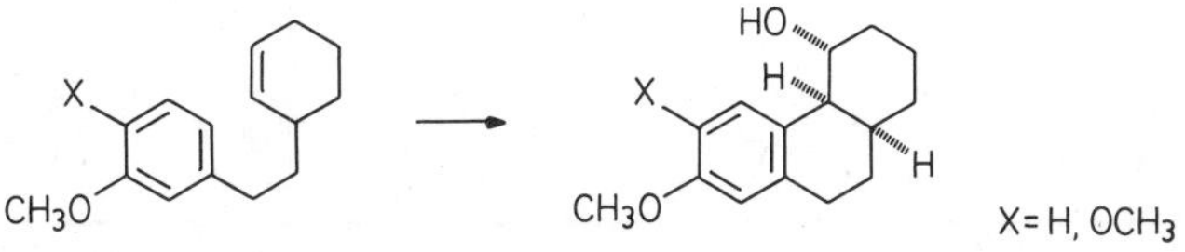

X = H, OCH$_3$

$$\tag{21}\ [28]$$

X = H, OCH$_3$

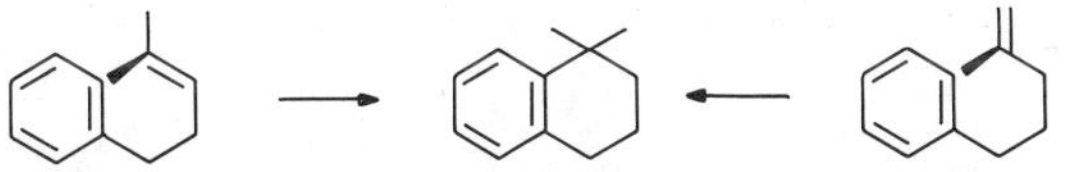

$$\tag{22}\ [28]$$

Many dienes and polyenes undergo similar cyclization reactions to be discussed in the next section.

These carbomercuration reactions are not limited to the introduction of aryl groups. Olefins also react with beta dicarbonyl compounds in the present of mercuric acetate and acetic acid, plus acids such as BF_3 and HClO_4, to afford carbomercuration products (Table 10.2) (Eq. 23) [21, 22]. Diketones

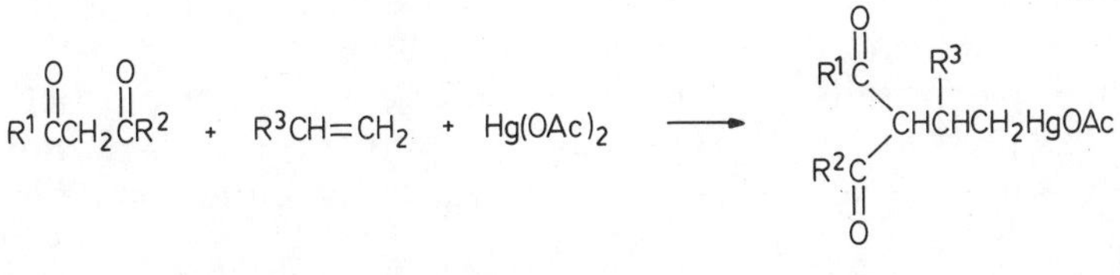

$$\tag{23}$$

and ketoesters work well, but simple ketones or diesters are not suitable for this reaction. Only ethylene and monosubstituted alkenes afford stable mercurials. The mercurials themselves can be isolated in high yield [21], or solvolysis to the corresponding acetate esters can be effected, albeit in low yield (Eq. 24) [22]. Note the substitution pattern in this reaction. When

$$CH_3CCH_2COEt \; + \; CH_3CH{=}CH_2 \; \xrightarrow[\substack{HOAc \\ BF_3}]{Hg(OAc)_2} \; CH_3CCHCOEt \tag{24}$$

vinyl acetate is employed as the olefin in this type of reaction, unsaturated esters are obtained instead (Eq. 25) [24, 25].

$$RCCH_2COEt \; + \; H_2C{=}CHOAc \; \longrightarrow \; RCCCOEt \tag{25}$$
$$R = CH_3, OEt$$

One particularly useful reaction of the mercurated diketones is the formation of cyclopropanes upon treatment with 10% aqueous KOH (Eq. 26) [21].

$$\xrightarrow{KOH} \tag{26}$$

B. Dienes and Polyenes

The reaction of dienes or polyenes with mercury salts has in many cases resulted in cyclic mercurial formation via intramolecular carbomercuration. Many of these reactions have been observed while attempting simple hydroxy- or alkoxymercuration of the diene or polyene in question. These and many more examples are summarized in Table 10.3. Only a few of the more interesting examples will be singled out for discussion, but many of these reactions provide a very simple approach to the ring system in question.

Let us start our discussion with the reactions of allenes since they are quite different from the majority of reactions we will be discussing in this section. A German patent indicates that allene (plus 10% propyne) reacts with phenol in the presence of $HgSO_4$ and H_2SO_4 to form a bis-phenol (Eq. 27) [30]. The following allenic alcohol reacts with mercuric acetate in

Table 10.3. Carbomercuration of Dienes and Polyenes

Olefin	Mercuric salt / solvent	Organomercurial(s) (% Yield)	Subsequent reactants	Product(s) (% Yield)	Ref.
10:1 $H_2C=C=CH_2$ / $CH_3C\equiv CH$	$HgSO_4$ / C_6H_5OH / H_2SO_4	—	—	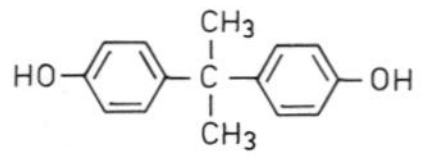 [structure: HO—C$_6H_4$—C(CH$_3$)$_2$—C$_6H_4$—OH]	30
$H_2C=C=C(CH_3)CH=CH_2$	$Hg(OAc)_2$ / HOAc, $HClO_4$	—	—	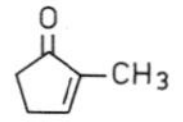 [2-methylcyclopent-2-enone] CH_3 (31) 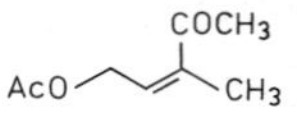 + [$AcOCH_2$ $COCH_3$ CH_3] (7)	31
$H_2C=CH(CH_2)_2CH=CH_2$	$Hg(OAc)_2$ / $HClO_4$ / BF_3 / HOAc / $NaClO_4$	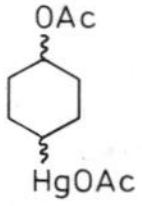 [cyclohexenyl OAc, HgOAc] (80–85)	—	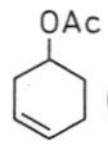 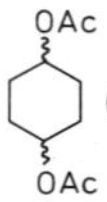 [4-acetoxycyclohexene] (50–60) + [1,4-diacetoxycyclohexane] (7–18)	27, 32
$CH_3CH=C=C(CH_3)CH=CH_2$	$Hg(OAc)_2$ / HOAc, $HClO_4$	—	—	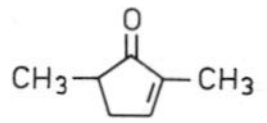 [2,3-dimethylcyclopent-2-enone] (20)	31
 [cyclooctatetraene]	$Hg(OAc)_2$ / CH_3OH	—	—	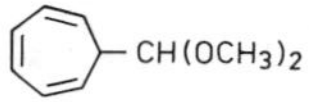 [cycloheptatrienyl—$CH(OCH_3)_2$] (88)	33
	$Hg(OAc)_2$ / HOAc	—	—	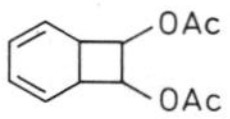 [bicyclic diacetate, OAc/OAc] (72)	33
	$HgSO_4$ / H_2O [or $Hg(OAc)_2$ / H_2O]	—	—	$C_6H_5CH_2CHO$ (54) [19]	33
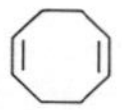 [cyclooctadiene]	$Hg(OAc)_2$ / $HClO_4$	—	—	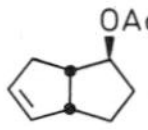 [bicyclopentane OAc] (14) 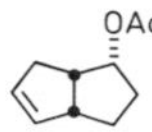 + [bicyclopentane OAc] (23) + 3 diacetates	34

548

Table 10.3. (continued)

Olefin	Mercuric salt / solvent	Organomercurial(s) (% Yield)	Subsequent reactants	Product(s) (% Yield)	Ref.
$H_2C=C(CH_3)(CH_2)_2C(CH_3)=CH_2$	$Hg(OAc)_2 / HClO_4$	—	—	(54) + (2)	32
$H_2C=CH(CH_2)_2CH=C(CH_3)_2$	$Hg(OAc)_2 / HOAc$	—	$NaBH_4$	$CH_3CHOH(CH_2)_2CH=C(CH_3)_2$ + + 83 : 10 : 6	32
	$Hg(OAc)_2 / HClO_4$	—	—	(30)	27
	$Hg(O_2CCF_3)_2$	—	$NaBH_4 / NaOH$	(75)	35
	$Hg(O_2CCF_3)_2$	—	$NaBH_4 / NaOH$	(62) 80:20 E/Z	35
	$Hg(OAc)_2 / HOAc, HClO_4$	—	—	(75)	36
	$Hg(O_2CCF_3)_2$	—	$NaBH_4 / NaOH$	(50) 80:20 E/Z	35
	$Hg(OAc)_2 / H_2O / THF$	—	$NaBH_4 / NaOH$	+ +	37

Table 10.3. (continued)

Olefin	Mercuric salt / solvent	Organomercurial(s) (% Yield)	Subsequent reactants	Product(s) (% Yield)	Ref.
	HgSO$_4$/HOAc	—	LiAlH$_4$	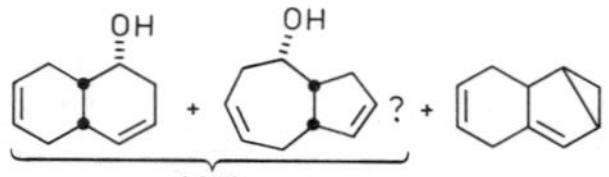 (23) ? + (28)	38, 39
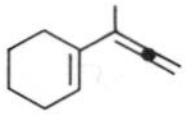	Hg(OAc)$_2$ / HOAc, HClO$_4$	—	—	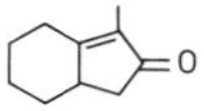 (75)	36
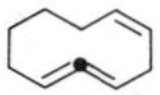	Hg(OAc)$_2$ / HOAc [HgCl$_2$]	—	—	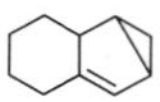 (55) [72]	38
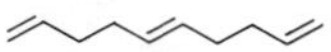	Hg(OAc)$_2$ / HBF$_4$ / HOAc	—	—	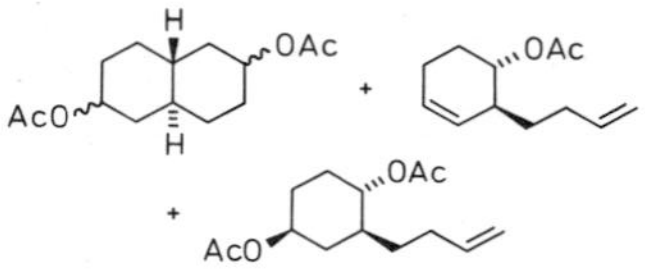	40
	Hg(OAc)$_2$ / HBF$_4$ / HOAc	—	NaBH$_4$	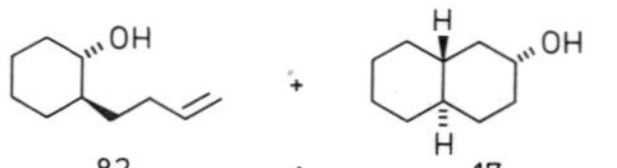 82 : 17	40
H$_2$C=C=C(n-C$_5$H$_{11}$)CH=CH$_2$	Hg(OAc)$_2$ / HOAc, HClO$_4$	—	—	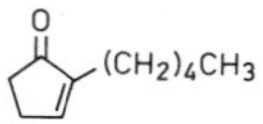 (70)	31, 36
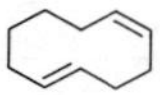	Hg(OAc)$_2$ / H$_2$O / THF	—	NaBH$_4$/NaOH	 (60) 83 : 17	41

Table 10.3. (continued)

Olefin	Mercuric salt / solvent	Organomercurial(s) (% Yield)	Subsequent reactants	Product(s) (% Yield)	Ref.
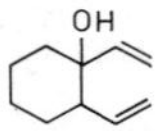 (structure)	Hg(O₂CCF₃)₂	—	NaBH₄ / NaOH	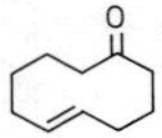 (structure) (35)	35
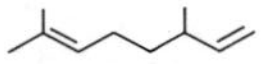 (structure)	Hg(OAc)₂ / H₂O / THF	—	NaBH₄ / NaOH	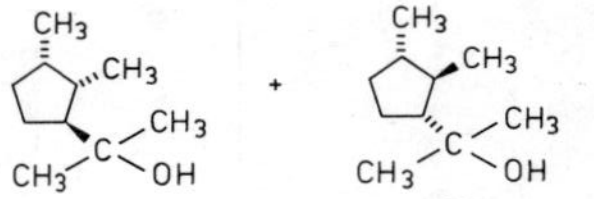 (structure) 70 : (structure) 30	42
	Hg(OAc)₂ / ROH	—	NaBH₄ / NaOH	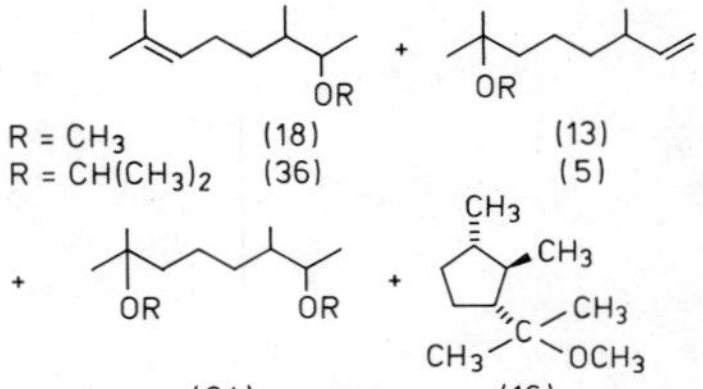 R = CH₃ (18), (13); R = CH(CH₃)₂ (36), (5); (24), (4); (16), (18)	42
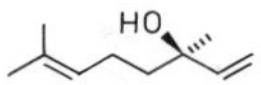 (structure)	Hg(OAc)₂ / H₂O / THF [NaCl]	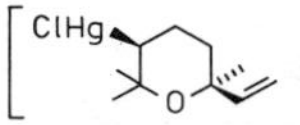 [(structure) (25)]	NaBH₄ / NaOH	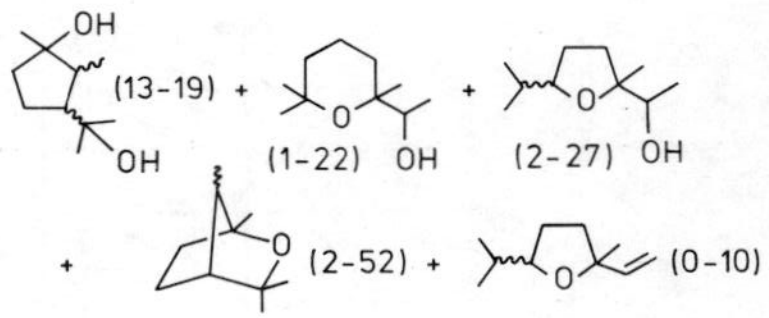 (13-19) + (1-22) + (2-27) + (2-52) + (0-10)	43, 44
(structure)	Hg(OAc)₂ / H₂O	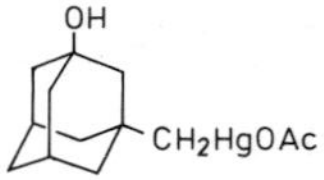 CH₂HgOAc (75)	I₂	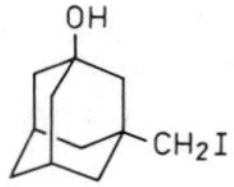 CH₂I (88)	45

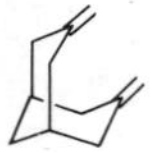

Table 10.3. (continued)

Olefin	Mercuric salt / solvent	Organomercurial(s) (% Yield)	Subsequent reactants	Product(s) (% Yield)	Ref.
$CH_3CH=C=C(n\text{-}C_5H_{11})CH=CH_2$	$Hg(OAc)_2$ / HOAc, $HClO_4$	—	—	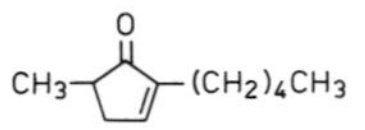 (50)	31, 36
$H_2C=C=C(n\text{-}C_5H_{11})C(CH_3)=CH_2$	$Hg(OAc)_2$ / HOAc, $HClO_4$	—	—	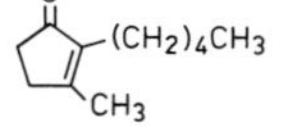 (79)	36
$cis\text{-}H_2C=C=C(n\text{-}C_5H_{11})\text{-}CH=CHCH_3$	$Hg(OAc)_2$ / HOAc, $HClO_4$	—	—	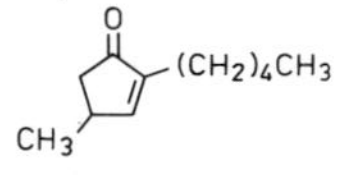 (50)	36
$trans\text{-}H_2C=C=C(n\text{-}C_5H_{11})\text{-}CH=CHCH_3$	$Hg(OAc)_2$ / HOAc, $HClO_4$	—	—	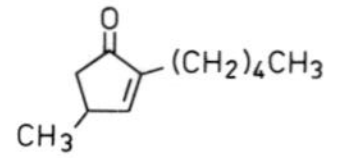 (78)	36
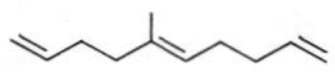	$Hg(OAc)_2$ / HCO_2H [NaCl]	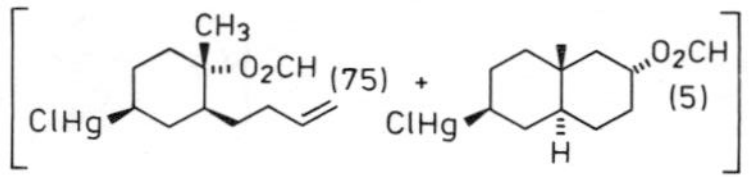 (75) + (5)	$NaBH_4$	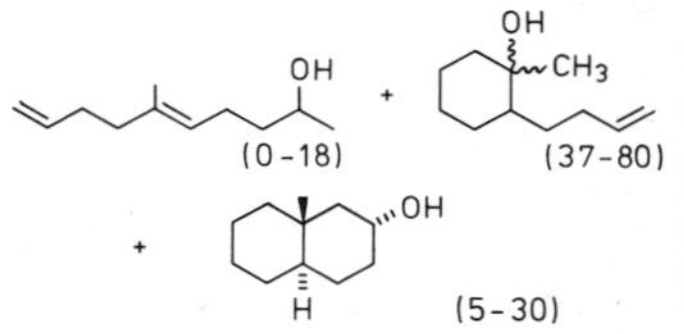 (0–18) + (37–80) + (5–30)	46, 47
	$Hg(OAc)_2$ / HBF_4 / HCO_2H	—	—	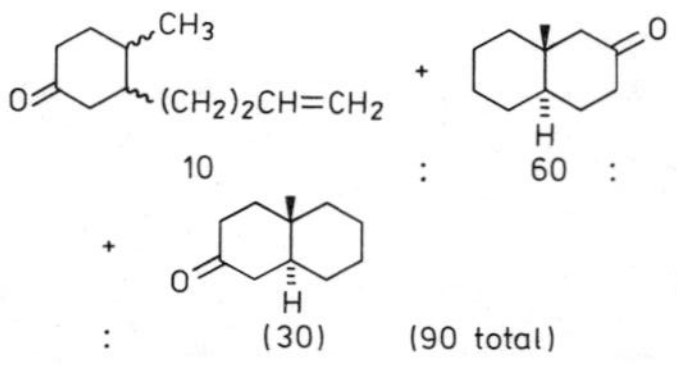 10 : 60 : (30) (90 total)	46

Table 10.3. (continued)

Olefin	Mercuric salt / solvent	Organomercurial(s) (% Yield)	Subsequent reactants	Product(s) (% Yield)	Ref.
	$Hg(OAc)_2$ / HBF_4 / HOAc	—	—	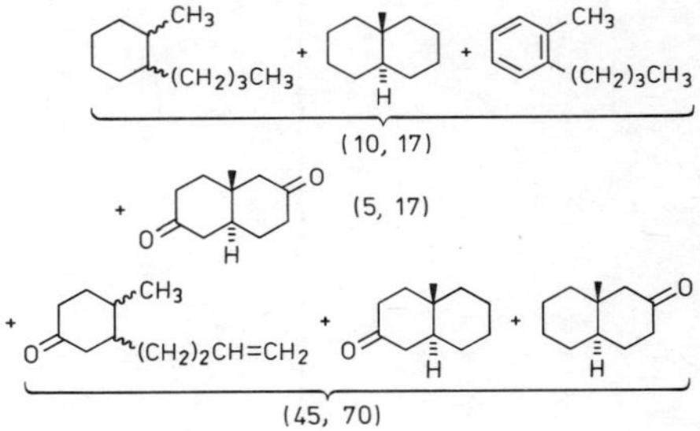 (10, 17); (5, 17); (45, 70)	46
	$Hg(OAc)_2$ / HOAc, HBF_4	—	$NaBH_4$ / NaOH	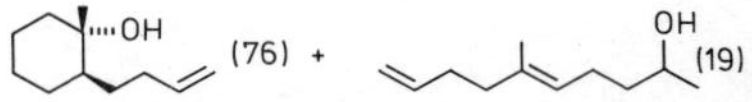 (76) + (19)	46
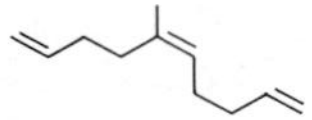	$Hg(OAc)_2$ / HCO_2H, NaCl	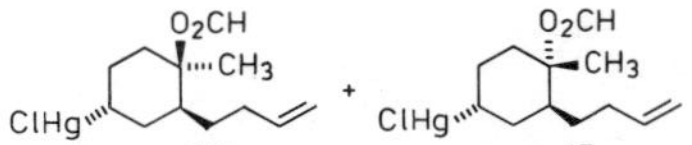 85 : 15	—	—	48
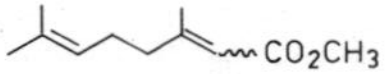 CO_2CH_3 (E and Z)	$Hg(O_2CCF_3)_2$ / CH_3NO_2	—	$NaBH_4$ / NaOH	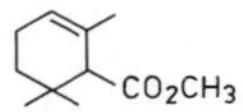 (30)	49
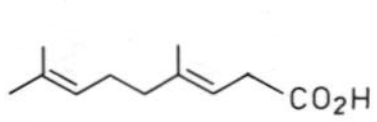 CO_2H	$Hg(O_2CCF_3)_2$ / CH_3NO_2, X^-	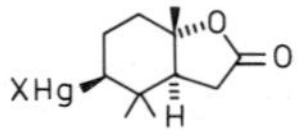 $X = Br, I$	Y_3^- [or Y_2 / $h\nu$]	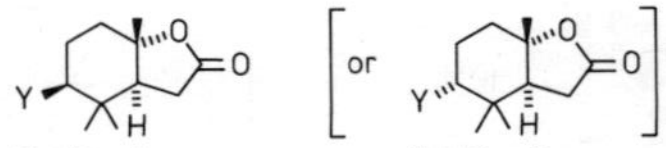 $Y = Br, I$ [or $Y = Br, I$]	50, 51
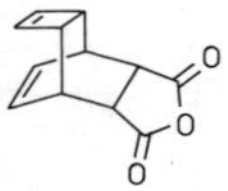	$Hg(OAc)_2$ / CH_3OH, NaCl	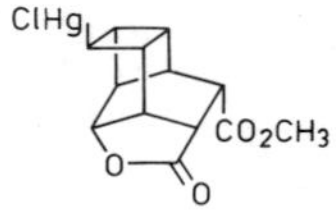 ?	—	—	52

Table 10.3. (continued)

Olefin	Mercuric salt / solvent	Organomercurial(s) (% Yield)	Subsequent reactants	Product(s) (% Yield)	Ref.
	$Hg(OAc)_2$ / HOAc, $HClO_4$	—	—	(49)	36
	$Hg(OAc)_2$ / HCO_2H, NaCl	(90)	$LiAlH_4$ or $NaBH_4$/OH^-	(95)	53
$CH_3CH=C=C(n-C_5H_{11})C(CH_3)=CH_2$	$Hg(OAc)_2$ / HOAc, $HClO_4$	—	—	(79)	36
	$Hg(O_2CCF_3)_2$ / CH_3NO_2, KBr	—	Br_2/KBr/O_2	(25)	51
	$Hg(O_2CCF_3)_2$ / CH_3NO_2	—	$NaBH_4$/NaOH	(60)	54
	$Hg(O_2CCF_3)_2$ / CH_3NO_2	—	$NaBH_4$/NaOH	(60)	54
(E and Z)	$Hg(O_2CCF_3)_2$ / CH_3NO_2	—	$NaBH_4$	plus *cis*-chromene	55

Table 10.3. (continued)

Olefin	Mercuric salt / solvent	Organomercurial(s) (% Yield)	Subsequent reactants	Product(s) (% Yield)	Ref.
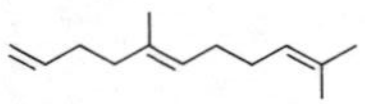	Hg(OAc)$_2$/HCO$_2$H (10 min), NaCl	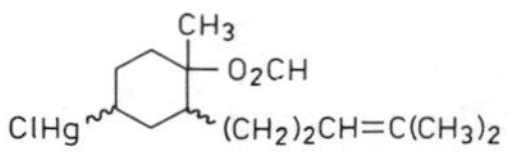 CH$_3$, O$_2$CH, ClHg~(CH$_2$)$_2$CH=C(CH$_3$)$_2$ (60)	NaBH$_4$/NaOH	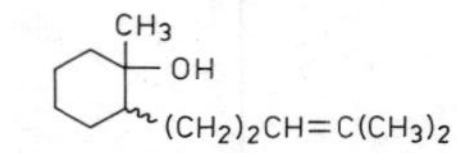 CH$_3$, OH, ~(CH$_2$)$_2$CH=C(CH$_3$)$_2$ (40) (1:1 *cis/trans*)	56
	Hg(OAc)$_2$/HCO$_2$H (6 h), NaCl	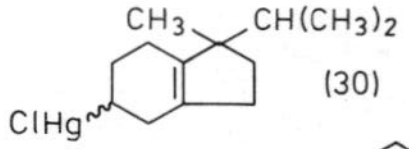 CH$_3$, CH(CH$_3$)$_2$, ClHg~ (30) 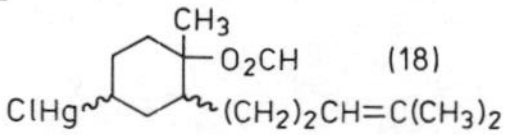 + CH$_3$, O$_2$CH, ClHg~(CH$_2$)$_2$CH=C(CH$_3$)$_2$ (18)	NaBH$_4$/NaOH	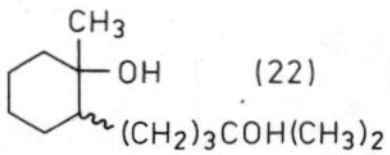 CH$_3$, OH, ~(CH$_2$)$_3$COH(CH$_3$)$_2$ (22) plus other products	56
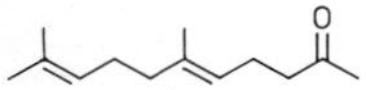	Hg(O$_2$CCF$_3$)$_2$/CH$_3$NO$_2$ [NaCl]	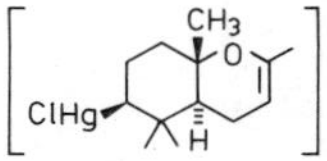 CH$_3$, O, ClHg~ H	NaBH$_4$/NaOH	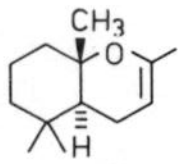 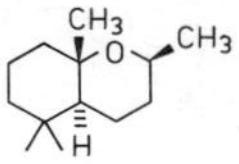 CH$_3$, O 62 : CH$_3$, O, CH$_3$ 38	57
	Hg(O$_2$CCF$_3$)$_2$/CH$_3$NO$_2$ —		NaBH$_4$ or NaBD$_4$/NaOH	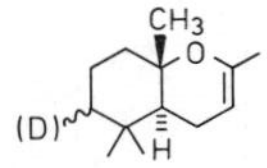 CH$_3$, O, (D)~ H (62)	49, 58
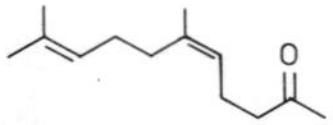	Hg(O$_2$CCF$_3$)$_2$/CH$_3$NO$_2$ [NaCl]	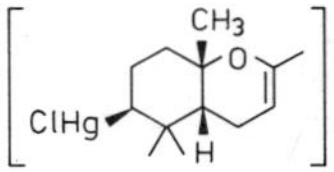 CH$_3$, O, ClHg~ H	NaBH$_4$/NaOH	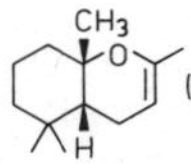 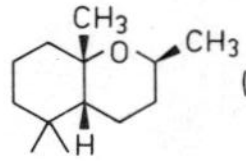 CH$_3$, O (58.5) + CH$_3$, O, CH$_3$ (37.5)	57
	Hg(O$_2$CCF$_3$)$_2$/CH$_3$NO$_2$ —		NaBH$_4$ or NaBD$_4$/NaOH	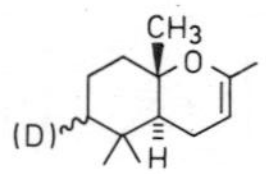 CH$_3$, O, (D)~ H (62)	49, 58
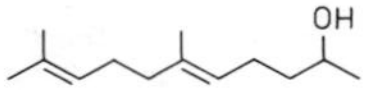	Hg(O$_2$CCF$_3$)$_2$/CH$_3$NO$_2$, KBr —		Br$_2$/KBr/O$_2$ [or Br$_2$/hν]	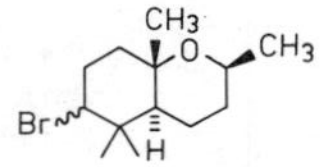 CH$_3$, O, CH$_3$, Br~ H β−Br(16) [α−Br(12)] 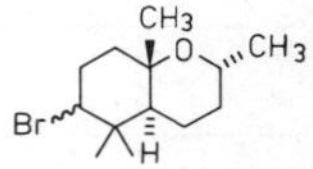 + CH$_3$, O, CH$_3$, Br~ H β−Br(12) [α−Br(10)]	51

Table 10.3. (continued)

Olefin	Mercuric salt / solvent	Organomercurial(s) (% Yield)	Subsequent reactants	Product(s) (% Yield)	Ref.
	Hg(OAc)$_2$/H$_2$O/THF	—	NaBH$_4$/NaOH	(61)	59
	Hg(O$_2$CCF$_3$)$_2$/CH$_3$NO$_2$, KBr	—	Br$_2$/KBr/O$_2$	(11) + (11)	51
	Hg(O$_2$CCF$_3$)$_2$/CH$_3$NO$_2$, KBr	—	Br$_2$/KBr/O$_2$ [or Br$_2$/hν]	β – Br (26) [α – Br (21)]	51
	Hg(OAc)$_2$/H$_2$O/THF	—	NaBH$_4$/NaHCO$_3$	(major) + +	60
	Hg(OAc)$_2$/H$_2$O/THF	—	NaBH$_4$/NaHCO$_3$	+ +	60

Table 10.3. (continued)

Olefin	Mercuric salt / solvent	Organomercurial(s) (% Yield)	Subsequent reactants	Product(s) (% Yield)	Ref.
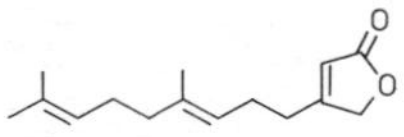	$Hg(O_2CCF_3)_2$, NaCl	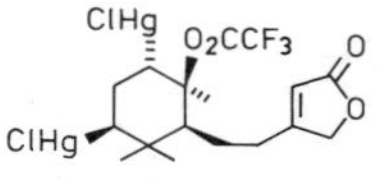 (61)	—	—	61
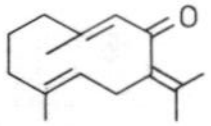	$Hg(OAc)_2/H_2O/THF$	—	$NaBH_4/NaOH$	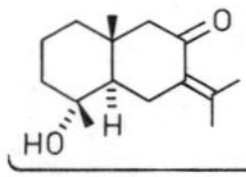 + 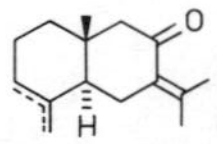 (65) + 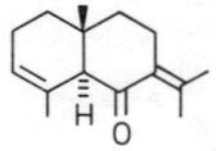 (4)	62
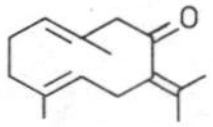	$Hg(OAc)_2/H_2O/THF$	—	$NaBH_4/NaOH$	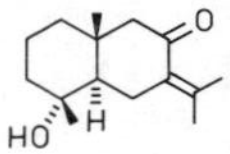 (70) + 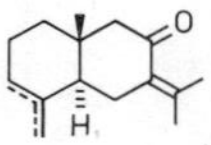 (15)	62
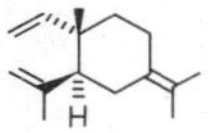	$Hg(OAc)_2/H_2O/THF$	—	$NaBH_4/NaOH$	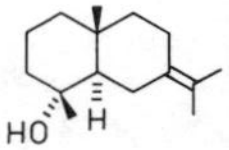 (14)	63
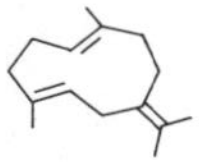	$Hg(OAc)_2/H_2O$	—	reduction	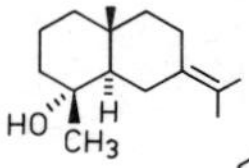 (16) + 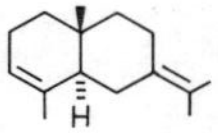 (9.5) + 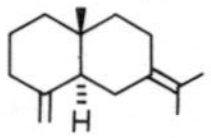 (9.5)	64
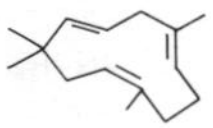	$Hg(OAc)_2/H_2O/THF$	—	$NaBH_4$	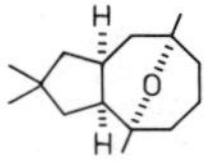 (45) + 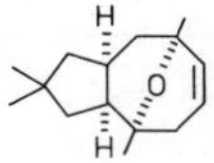 (15)	65

Table 10.3. (continued)

Olefin	Mercuric salt / solvent	Organomercurial(s) (% Yield)	Subsequent reactants	Product(s) (% Yield)	Ref.
	Hg(OAc)$_2$	—	NaBD$_4$	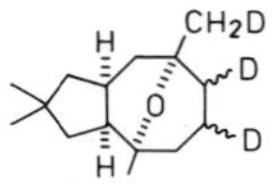 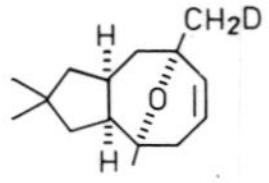 CH$_2$D products	66
	Hg(NO$_3$)$_2$	—	NaBD$_4$	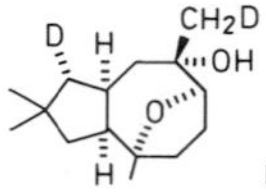 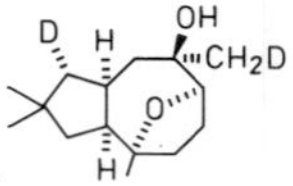 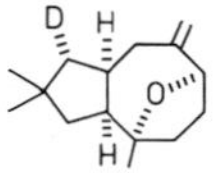 CH$_2$D / OH products	66
	Hg(NO$_3$)$_2$ / H$_2$O / HOAc	—	NaBH$_4$	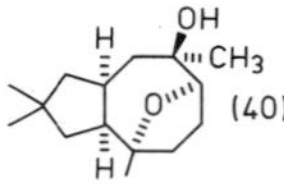 (40) 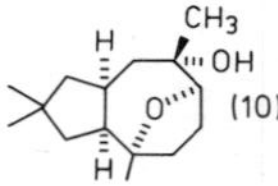 + (10) 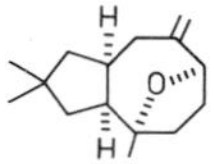 + (5)	67
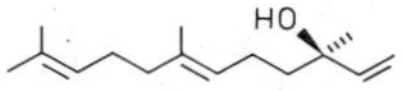	HgCl$_2$ / H$_2$O / HOAc or CdCO$_3$	—	NaBH$_4$ / NaOH	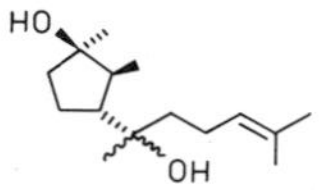 (16 – 22) 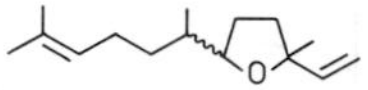 + (19)	43
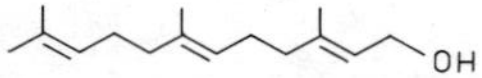	Hg(O$_3$SCF$_3$)$_2$ · C$_6$H$_5$–N(CH$_3$)$_2$, NaCl	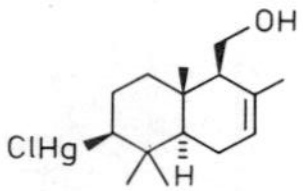 (51)	—	—	68

Table 10.3. (continued)

Olefin	Mercuric salt / solvent	Organomercurial(s) (% Yield)	Subsequent reactants	Product(s) (% Yield)	Ref.
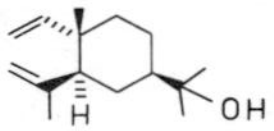 (structure)	Hg(OAc)$_2$/H$_2$O/THF	—	NaBH$_4$/NaOH	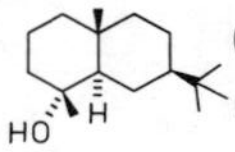 (structure) (69) 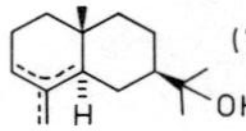 + (structure) (7)	63
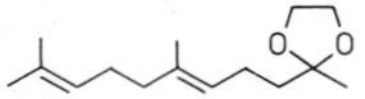 (structure)	Hg(O$_2$CCF$_3$)$_2$/CH$_3$CN	—	NaBH$_4$/NaOH	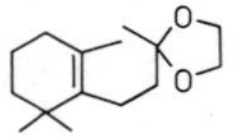 (structure) (65)	49
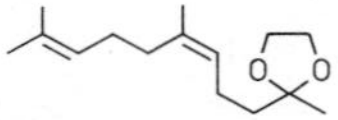 (structure)	Hg(O$_2$CCF$_3$)$_2$/CH$_3$CN	—	NaBH$_4$/NaOH	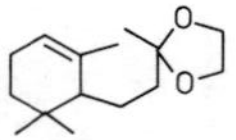 (structure) (60)	49
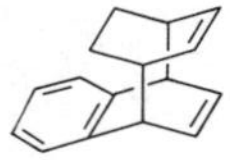 (structure)	Hg(OAc)$_2$/HOAc	—	NaBH$_4$	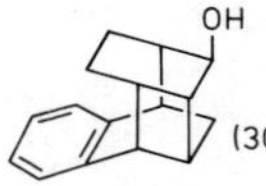 (structure) (30) 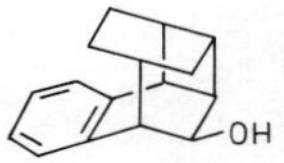 + (structure) (12) 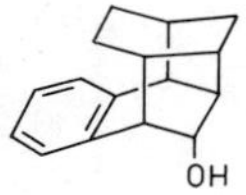 + (structure) (6)	69
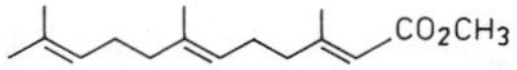 (structure) CO$_2$CH$_3$	Hg(O$_3$SCF$_3$)$_2$ · C$_6$H$_5$ – N(CH$_3$)$_2$	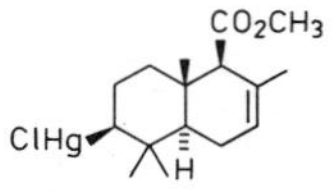 (structure) CO$_2$CH$_3$, ClHg– (56)	—	—	68
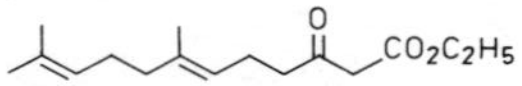 (structure) CO$_2$C$_2$H$_5$	Hg(O$_3$SCF$_3$)$_2$ · C$_6$H$_5$ – N(CH$_3$)$_2$, NaCl	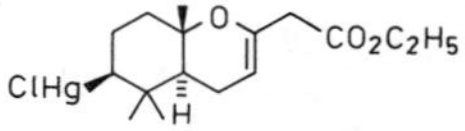 (structure) CO$_2$C$_2$H$_5$, ClHg– (74)	—	—	68
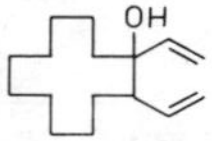 (structure) OH	Hg(O$_2$CCF$_3$)$_2$	—	NaBH$_4$/NaOH	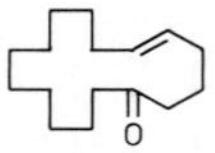 (structure) (70) 70 : 30 *E/Z*	35

Table 10.3. (continued)

Olefin	Mercuric salt / solvent	Organomercurial(s) (% Yield)	Subsequent reactants	Product(s) (% Yield)	Ref.
(triene) OH, CO_2CH_3	$Hg(O_2CCF_3)_2$ / CH_3NO_2, KBr	—	Br_2 / KBr / O_2	CH_3, O, $CH_2CO_2CH_3$ (>22) + $CH_2CO_2CH_3$, O, CH_3 (>22)	51
(triene) CO_2R (R = Me or Et ?)	$Hg(O_2CCF_3)_2$ / CH_3NO_2	—	$NaBH_4$ / NaOH	CH_3, CO_2R (60)	70
(triene) CO_2R (R = Me or Et ?)	$Hg(O_2CCF_3)_2$ / CH_3NO_2	—	$NaBH_4$ / NaOH	CH_3, CO_2R (≤25) + CO_2R (20–25) + CH_3, CO_2R ? (≤50)	70
(triene) O_2CCCl_3	$Hg(O_3SCF_3)_2 \cdot C_6H_5-N(CH_3)_2$, NaCl	O_2CCCl_3, ClHg (58)	—	—	68
(triene) O_2CCH_3	$Hg(O_3SCF_3)_2 \cdot R_3N$, NaCl	O_2CCH_3, OH, ClHg + O_2CCH_3, ClHg; $R_3N = C_6H_5N(CH_3)_2$ (44) (23); $R_3N = (CH_3)_3C$—pyridine—$C(CH_3)_3$ (30) (33)	—	—	68
(triene) $CO_2C_2H_5$	$Hg(O_2CCF_3)_2$ / CH_3NO_2	—	$NaBH_4$ / NaOH	$CO_2C_2H_5$ (60)	49

Table 10.3. (continued)

560

Olefin	Mercuric salt / solvent	Organomercurial(s) (% Yield)	Subsequent reactants	Product(s) (% Yield)	Ref.
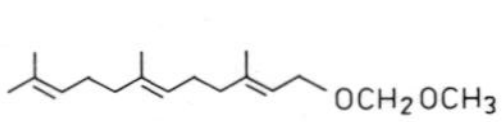 OCH_2OCH_3	$Hg(O_3SCF_3)_2 \cdot C_6H_5-N(CH_3)_2$, NaCl	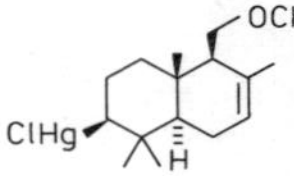 (40) 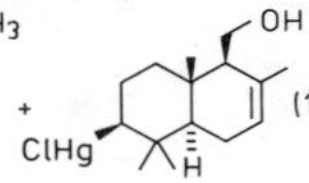 + (10)	—	—	68
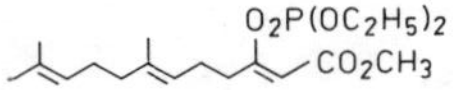 $O_2P(OC_2H_5)_2$ CO_2CH_3	$Hg(O_2CCF_3)_2$ / CH_3NO_2, NaCl	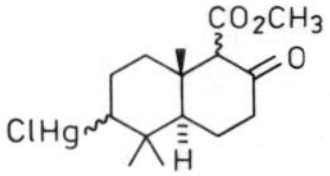 CO_2CH_3	I_2, ICl or KI_3	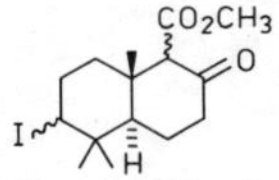 $3\alpha : 3\beta - I = 1:1$ to $5:1$	71
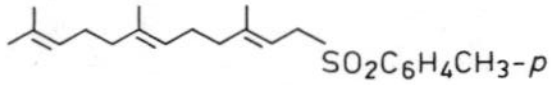 $SO_2C_6H_4CH_3-p$	$Hg(O_3SCF_3)_2 \cdot C_6H_5-N(CH_3)_2$, NaCl	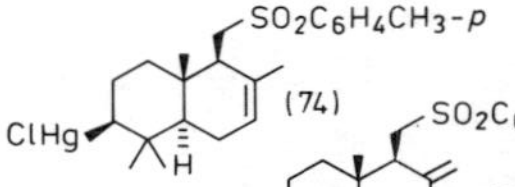 $SO_2C_6H_4CH_3-p$ (74) 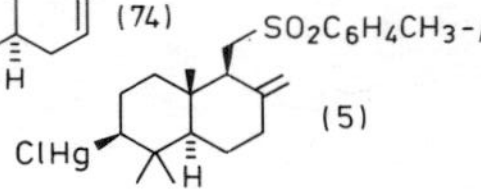 + $SO_2C_6H_4CH_3-p$ (5)	—	—	68
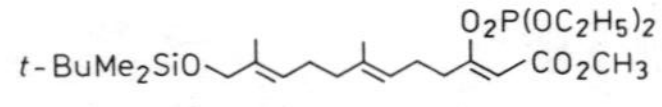 t-BuMe$_2$SiO $O_2P(OC_2H_5)_2$ CO_2CH_3	$Hg(O_2CCF_3)_2$, NaCl	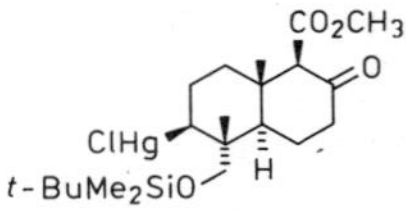 (60)	$NaBH_4 / O_2$ (on ketal)	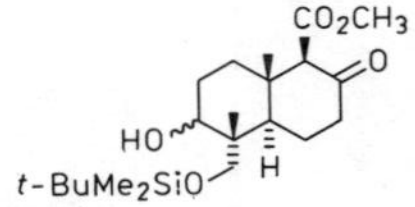 (both epimers)	72
$(C_6H_5)_2C=C=C(Br)C(C_6H_5)_2OH$	$Hg(OAc)_2$	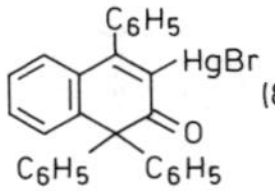 (8) 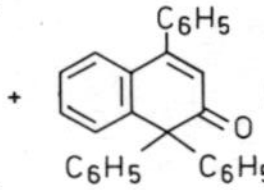 + (27)	$NaBH_4$	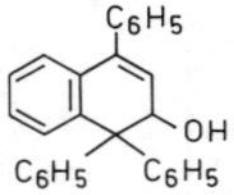 (90)	73
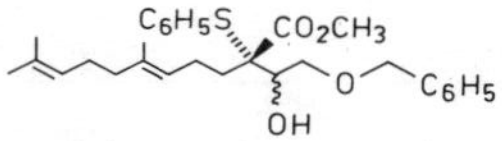 C_6H_5S CO_2CH_3 OH C_6H_5 (both epimers)	$Hg(O_2CCF_3)_2$, KBr	—	$Br_2 / LiBr / O_2$	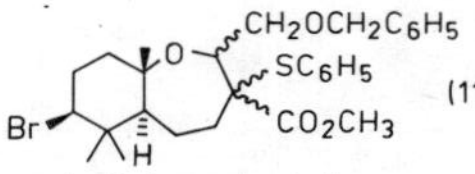 (11–15 each) (all 4 products)	74

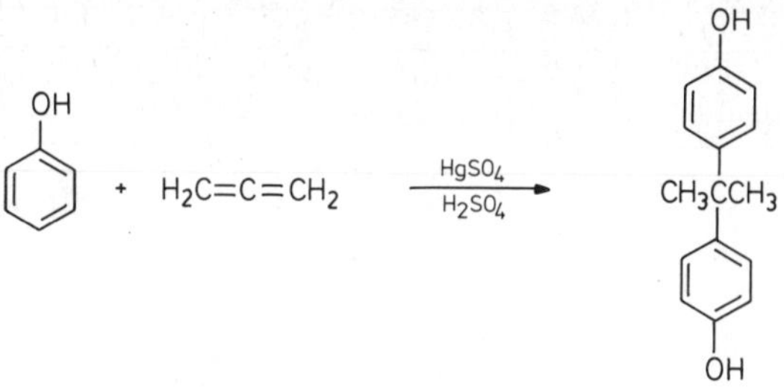

$$(27)$$

acetone to also yield products involving formation of a new carbon—carbon bond (Eq. 28) [73]. A variety of other products are observed when the solvent is changed.

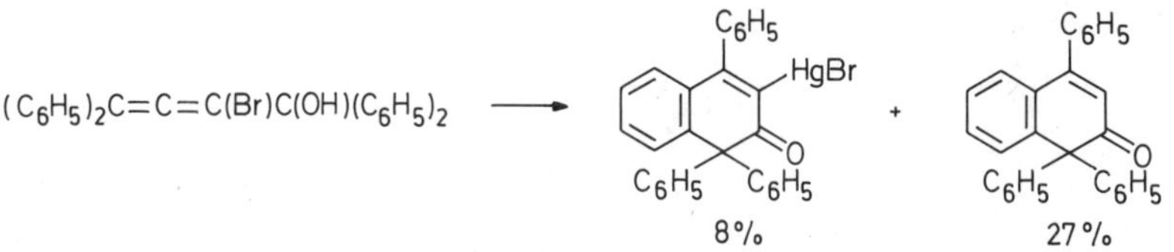

$$(C_6H_5)_2C=C=C(Br)C(OH)(C_6H_5)_2 \longrightarrow$$

$$(28)$$

Vinylallenes undergo a very interesting reaction when treated with mercuric acetate in acetic acid, followed by the addition of one equivalent of perchloric acid (Eq. 29) [31, 36]. The yields of cyclopentenone are

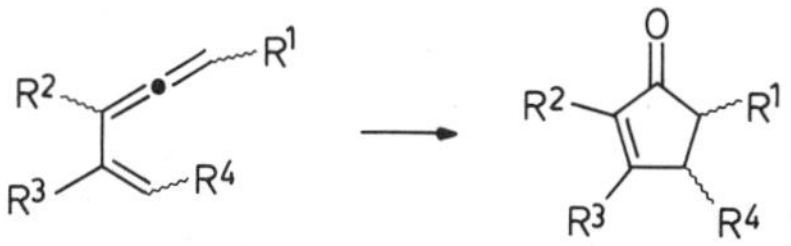

$$(29)$$

generally good and the reaction is applicable to the synthesis of bicyclic ketones as well (Eq. 30). Two possible mechanisms for this transformation have been suggested (Scheme 10.1).

$$(30)$$

Scheme 10.1

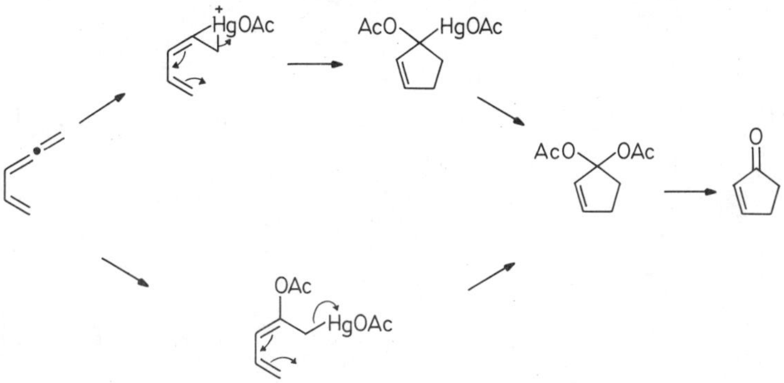

X. Carbomercuration

Two cyclic unsaturated allenes have been subjected to mercury-promoted cyclization (Eqs. 31, 32) [38, 39]. The products from the latter allene have been explained as follows (Scheme 10.2).

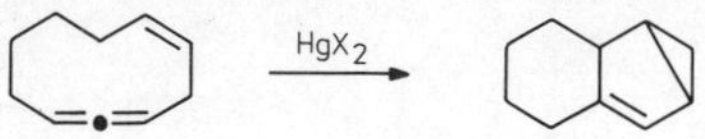

$$(31)$$

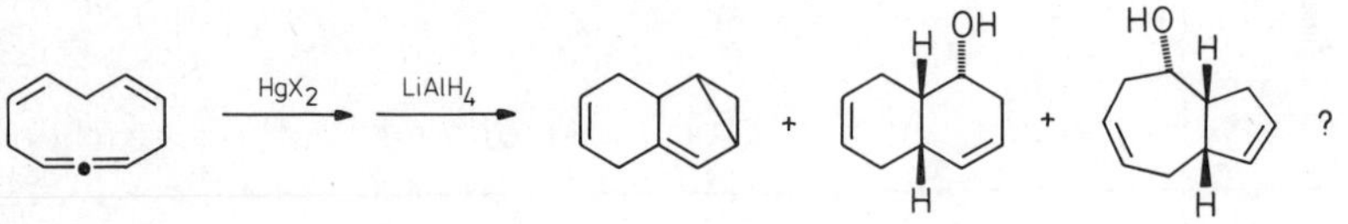

$$(32)$$

Scheme 10.2

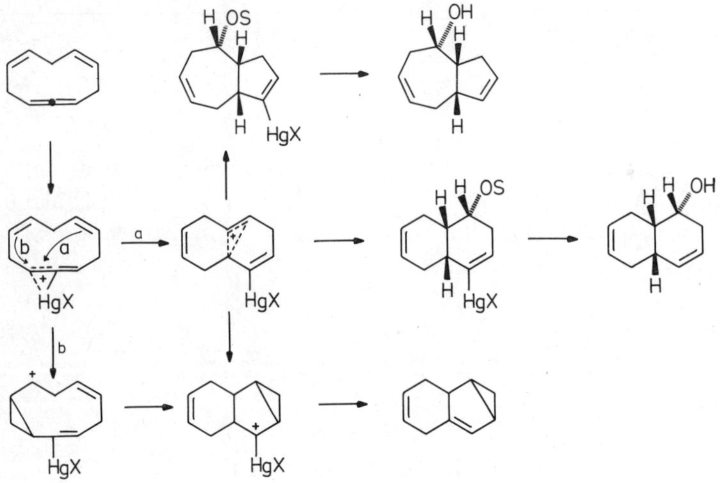

The mercuration of non-conjugated, acyclic dienes, particularly those of terpene origin, can provide a useful route to cyclic products. For example, the carbomercuration-solvolysis approach provides a simple route to the following cyclic esters (Eqs. 33, 34) [27, 32]. Even simple hydroxymercuration-demercuration conditions are sufficient to cyclize certain dienes (Eqs. 35, 36).

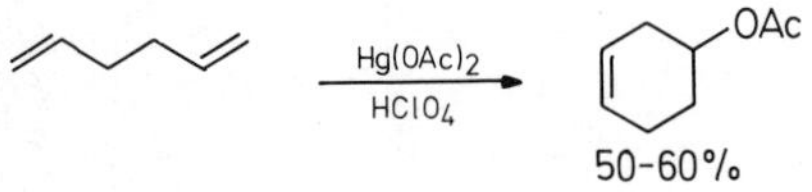

$$(33)$$

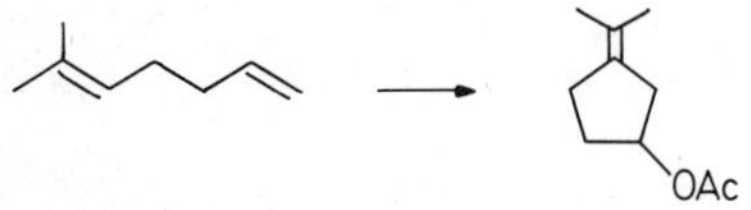

$$(34)$$

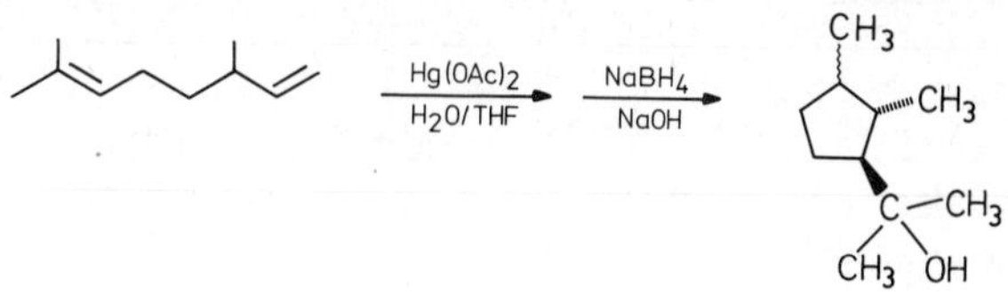

$$(35)\ [42]$$

$$(36)\ [63]$$

Recently, mercuric trifluoroacetate has been shown to promote the oxy-Cope rearrangement of tertiary 1,5-hexadien-3-ols to the corresponding unsaturated ketones (Eq. 37) [35]. The yields are decent and the reaction can be used for ring expansions (Eq. 38).

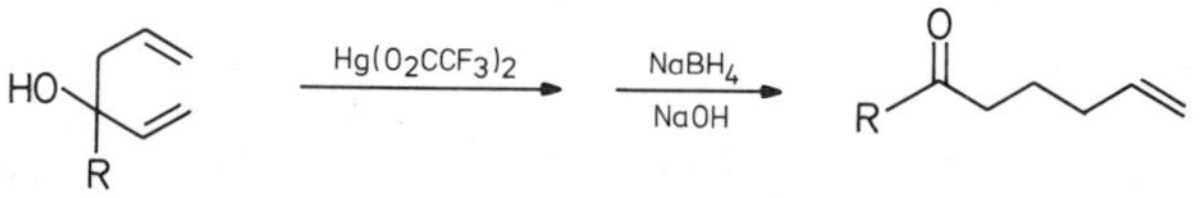

$$(37)$$

$$(38)$$

With dienes of terpene origin, a number of useful cyclization reactions have been reported. Internal nucleophiles often trap the cationic intermediate quite nicely, providing polycyclic products in high yields. The following example nicely illustrates this point (Eq. 39) [74]. Note the

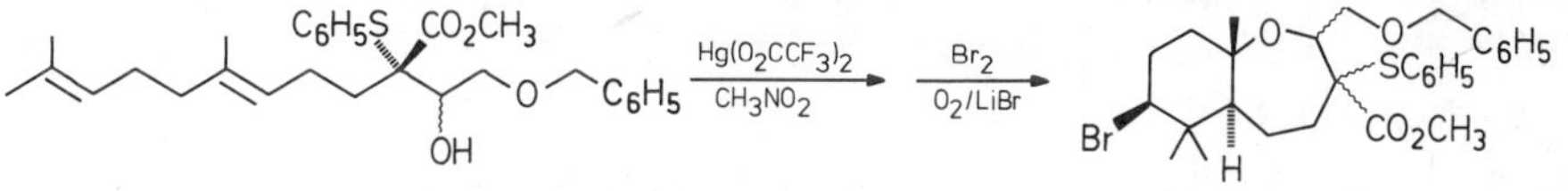

$$(39)$$

stereoselective formation of the trans bicyclic system and the fact that bromination of the mercurial proceeds with retention. A number of other dienols have been similarly cyclized and brominated (Eq. 40) [51]. Pure E-olefins

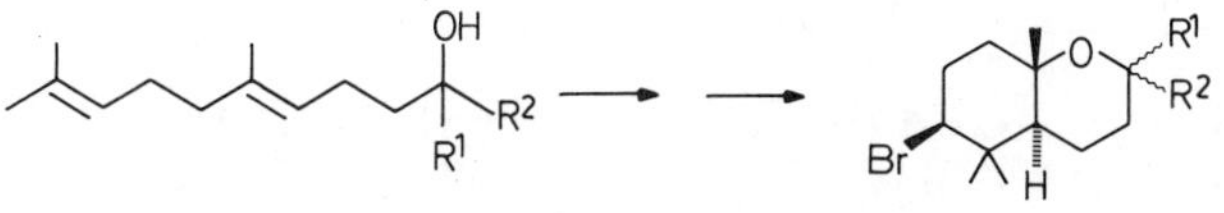

$$(40)$$

yield only the trans-fused bicyclic system. With enantiomeric substrates ($R^1 \neq R^2$) nearly equal amounts of the two possible diastereomers are observed, implying that approach of the mercury electrophile is not influenced by the nature of the stereocenter at the remote end of the diene system. This reaction sequence has been used to synthesize the marine natural product *d,1*-3β-bromo-8-epicaparrapi oxide. While products of carbon—carbon bond formation are also observed in the hydroxymercuration-demercuration of the terpene linalool, the predominant reaction appears to be intramolecular alkoxymercuration (Eq. 41) [43, 44]. The observed multitude of cyclization products appear to arise during the reduction step. Similar results are observed with the related terpene nerolidol.

Ketones can also serve as intramolecular traps (Eqs. 42–44). Unlike

$$(41)$$

$$(42) \ [68]$$

$$(43) \ [49, 54, 55, 57, 58]$$

$$(44) \ [49, 54, 57, 58]$$

the corresponding acid-promoted cyclization, the geranylacetone reactions are highly stereospecific. The corresponding *cis*- and *trans*-ketals give only monocyclic, unsaturated products, but the position of the double bond is quite specific (Eqs. 45, 46) [49].

$$(45)$$

$$(46)$$

Intramolecular trapping by a carboxylic acid group is evident in the cyclization of homogeranic acid (Eq. 47) [51]. Bromination of this inter-

$$(47)$$

mediate proceeds with clean retention when carried out using Br_2, KBr and O_2 in pyridine. Clean inversion is observed upon photolysis with Br_2 in pyridine. These intermediates have recently proved valuable in the synthesis of $(\pm)$-loliolide [50]. The related ester of geranic acid has also been cyclized, but the ester group does not participate in the reaction (Eq. 48) [49].

$$(48)$$

A number of terpenoid trienes have also been cyclized by mercury reagents. The carbomercuration of 1,5,9-decatriene and derivatives affords a variety of products. For example, the parent compound give a mixture of monocyclic and bicyclic products (Eq. 49) [40]. The E- and z-5-methyl

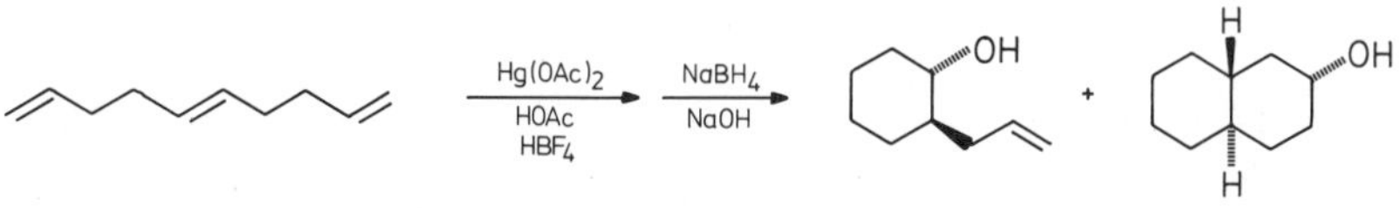

$$(49)$$

homologs have been observed to cyclize stereospecifically (Eqs. 50, 51).

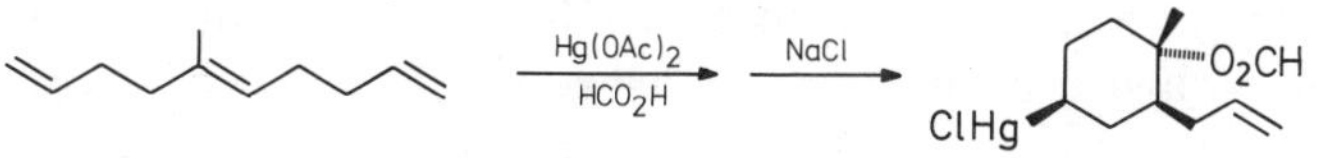

$$(50)\ [47]$$

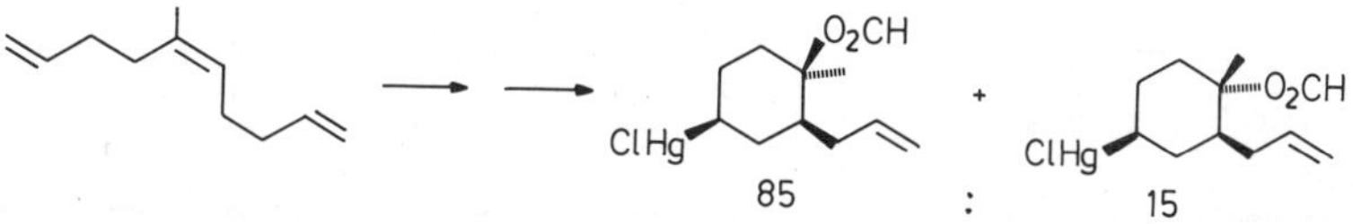

$$(51)\ [48]$$

85 : 15

Under solvolytic conditions, the former triene gives both monocyclic and bicyclic products (Eq. 52) [46]. The following related triene cyclizations have also been reported (Eqs. 53, 54).

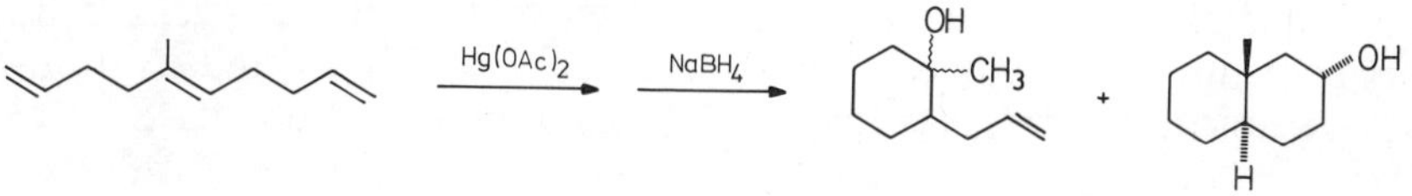

$$(52)$$

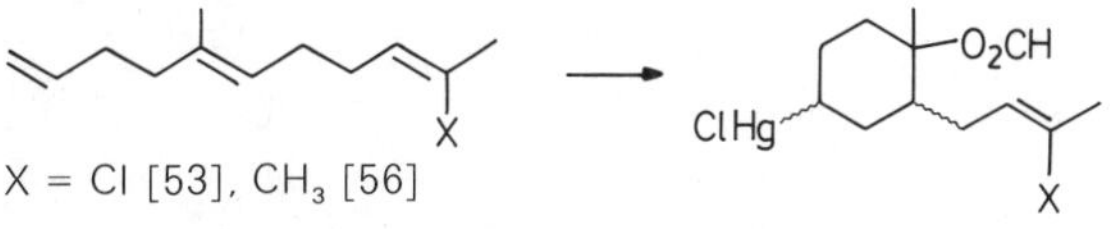

$$(53)$$

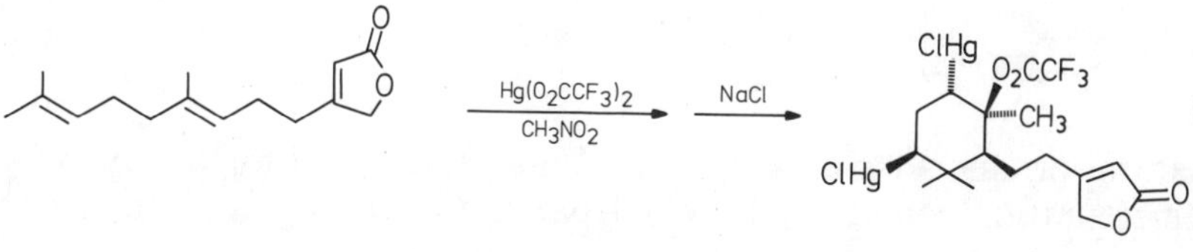

X = Cl [53], CH_3 [56]

$$(54)\ [61]$$

X. Carbomercuration

Many farnesol derivatives have been cyclized to the corresponding bicyclic products in high yields (Eq. 55). A new reagent, mercuric trifluoro-

$$(55)$$

R=CO$_2$R [49, 68, 70], CH$_2$SO$_2$C$_6$H$_5$CH$_3$ [68], CH$_2$OCH$_2$OCH$_3$ [68]
and CH$_2$O$_2$CCCl$_3$ [68]

methanesulfonate plus N,N-dimethylaniline or 2,6-di-*tert*-butylpyridine, has recently been developed for effective cyclization of these compounds [68]. It is interesting that farnesyl acetate affords an alcohol product as the major product under these same conditions (Eq. 56) [68]. The acetate group must be

$$(56)$$

undergoing neighboring group participation by oxygen, since the corresponding trichloroacetate ester gives only the expected olefin. While geranylacetoacetic ester undergoes exclusive cyclization on oxygen (Eq. 42). the corresponding enol phosphate and a closely related derivative have been shown to undergo facile cyclization on carbon (Eq. 57). These mercurials have been iodinated [71] and oxidized (O$_2$/NaBH$_4$) [72] to mixtures of the corresponding epimeric iodides and alcohols.

$$(57)$$

R=CH$_3$ [71], CH$_2$OSiMe$_2$(*t*-Bu) [72]

A number of cyclic dienes and polyenes have been further cyclized by mercury reagents. For example, 1,5-cyclooctadiene (Eq. 58) [34] and *cis,trans*-1,5-cyclodecadiene (Eq. 59) [41] give good yields of bicyclic products. Several

$$(58)$$

$$(59)$$

related natural products have been cyclized under the usual hydroxymercuration-demercuration conditions (Eqs. 60–62). On the other hand, the mercuration of humulene is quite complex as shown by the following labelling results (Eqs. 63, 64) [65–67]. A complex mechanism has been advanced to explain these results [66].

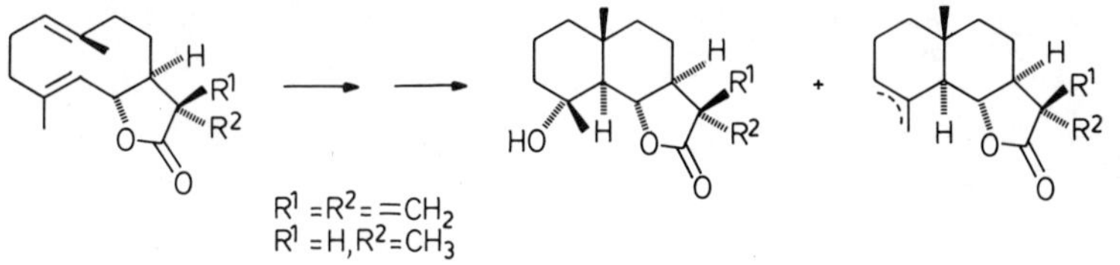
H
R1
R2
O
HO
R1 = R2 = =CH2
R1 = H, R2 = CH3

(60) [60]

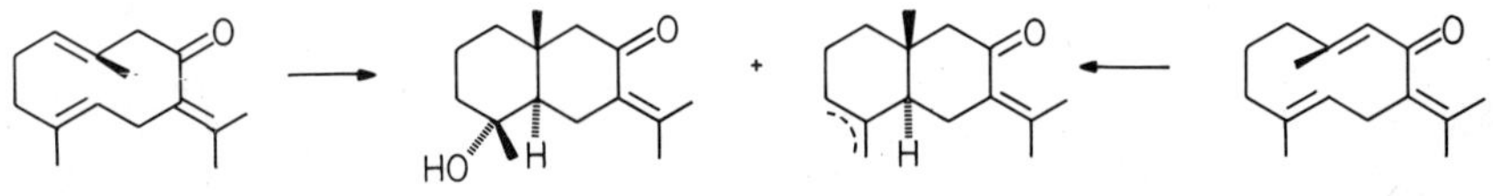
O
HO
H

(61) [62]

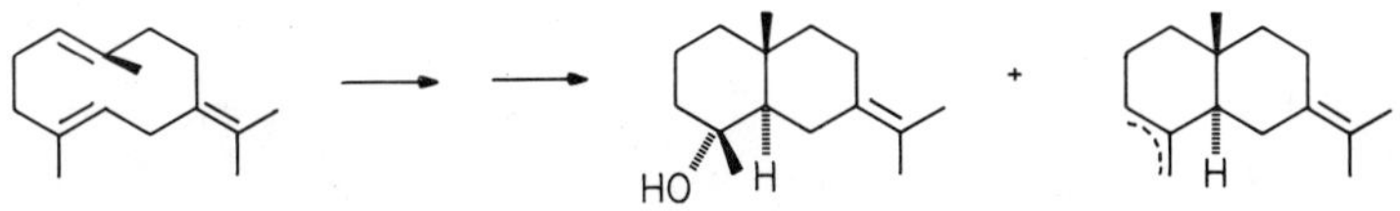
HO
H

(62) [64]

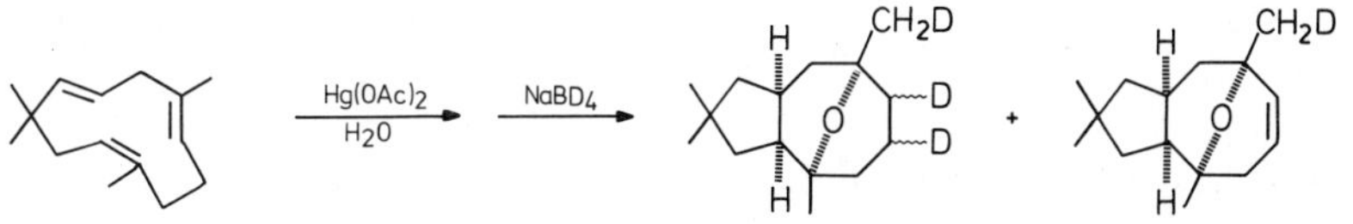
Hg(OAc)2
H2O
NaBD4
CH2D
D
O
H

(63)

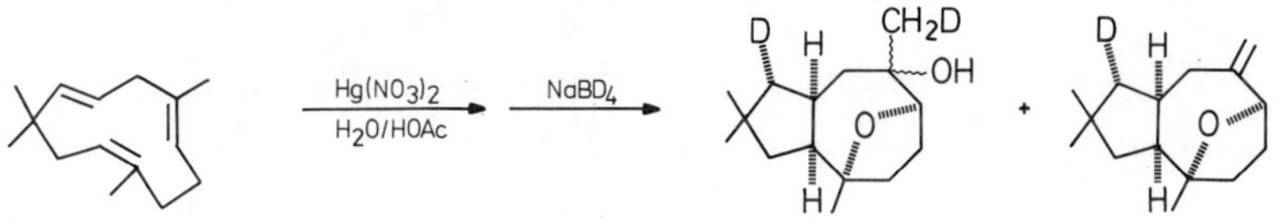
Hg(NO3)2
H2O/HOAc
NaBD4
CH2D
OH
D
H
O

(64)

Finally, the mercuration of cyclooctatetraene has been examined. Depending on the reaction conditions, a variety of ring-contracted products are observed (Eq. 65) [33].

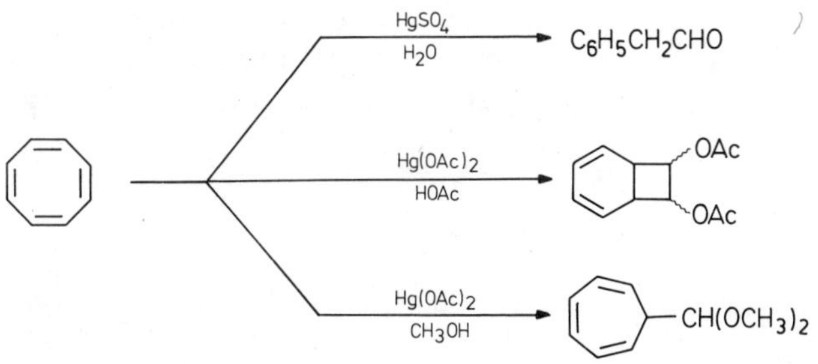
HgSO4
H2O
C6H5CH2CHO
Hg(OAc)2
HOAc
OAc
OAc
Hg(OAc)2
CH3OH
CH(OCH3)2

(65)

Bicyclic and polycyclic dienes in which the double bonds are held in close proximity to each other, often afford products involving carbon—carbon bond formation (Eqs. 66–68). Bullvalene also provides rearranged products,

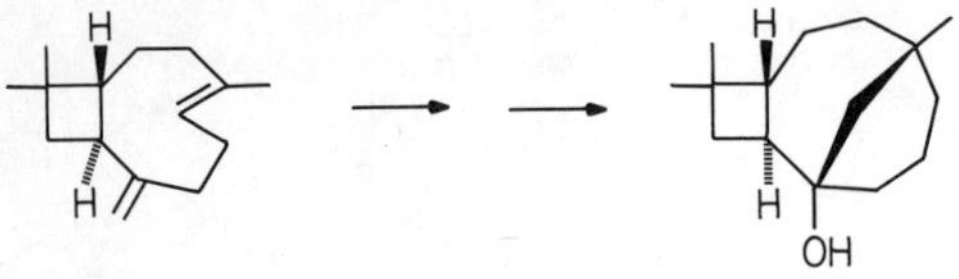

(66) [59]

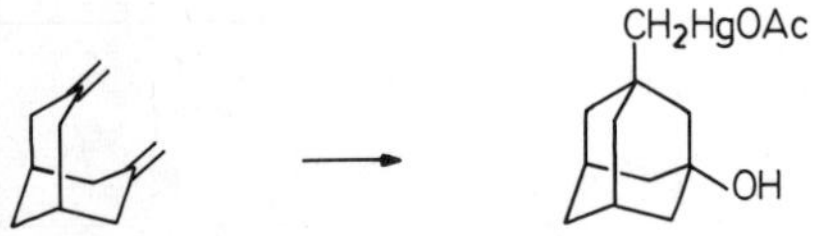

CH2HgOAc
OH

(67) [45]

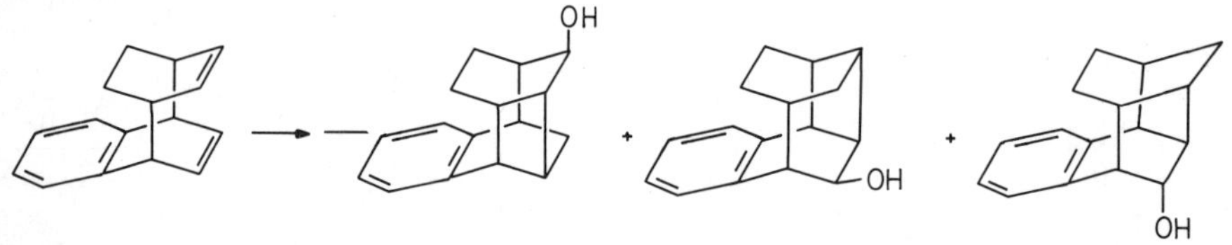

OH
OH
OH

(68) [69]

but the products may well be formed via free radical rearrangement during the demercuration step (Eq. 69) [37].

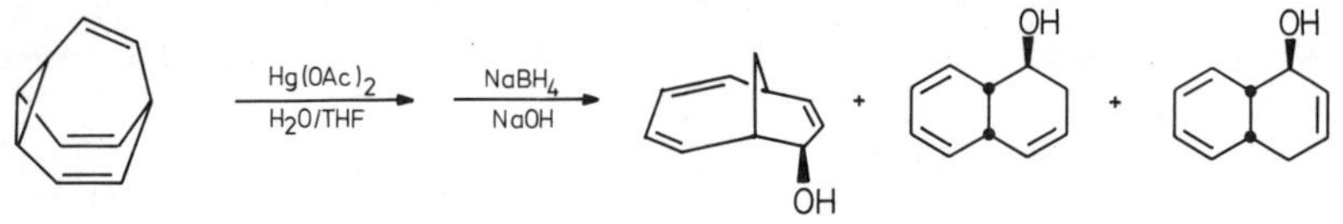

Hg(OAc)2
H2O/THF
NaBH4
NaOH
OH
OH
OH

(69)

C. Alkynes

Relatively few reactions of alkynes and mercury reagents lead to carbon—carbon bond formation (Table 10.4). The most straightforward reaction is that of di-*t*-butylmercury and electron-deficient alkynes (Eq. 70) [11, 14].

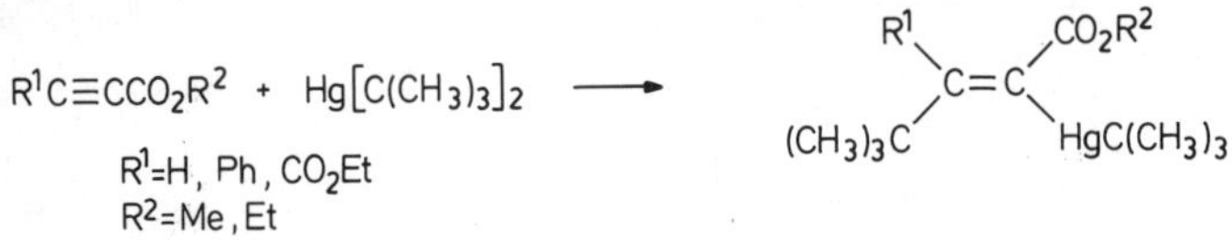

R1C≡CCO2R2 + Hg[C(CH3)3]2
R1=H, Ph, CO2Et
R2=Me, Et
R1
CO2R2
C=C
(CH3)3C
HgC(CH3)3

(70)

Table 10.4. Carbomercuration of Alkynes

Alkyne	Mercury reagent	Organomercurial(s) (% Yield)	Subsequent reactants	Product(s) (% Yield)	Ref.
HC≡CH	HgSO$_4$ / CH$_3$COCH$_2$CO$_2$Et	—	—	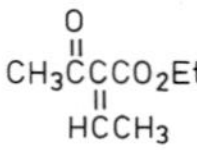 CH$_3$C(=O)CCO$_2$Et (=HCCH$_3$) (20–50)	24
HC≡CCO$_2$CH$_3$	Hg[C(CH$_3$)$_3$]$_2$	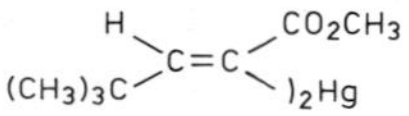 (CH$_3$)$_3$C / H >C=C< CO$_2$CH$_3$ /)$_2$Hg	HCl or H$_2$O	trans-(CH$_3$)$_3$CCH=CHCO$_2$CH$_3$	11, 14
C$_2$H$_5$O$_2$CC≡CCO$_2$C$_2$H$_5$	Hg[C(CH$_3$)$_3$]$_2$	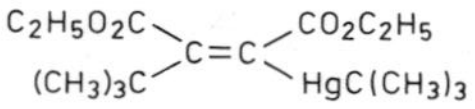 C$_2$H$_5$O$_2$C / (CH$_3$)$_3$C >C=C< CO$_2$C$_2$H$_5$ / HgC(CH$_3$)$_3$	HCl or H$_2$O	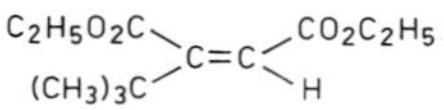 C$_2$H$_5$O$_2$C / (CH$_3$)$_3$C >C=C< CO$_2$C$_2$H$_5$ / H	11, 14
o-HC≡CCH$_2$OC$_6$H$_4$Cl	Hg(O$_2$CCF$_3$)$_2$	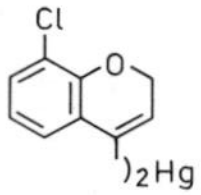 8-chloro-2H-chromen-4-yl mercurial, ()$_2$Hg (54)	—	—	76
HC≡CCH$_2$OC$_6$H$_5$	Hg(O$_2$CCF$_3$)$_2$	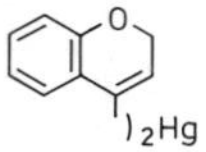 2H-chromen-4-yl mercurial, ()$_2$Hg (75)	—	—	76
p-HC≡CCH$_2$OC$_6$H$_4$CH$_3$	Hg(O$_2$CCF$_3$)$_2$	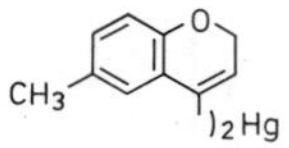 6-methyl-2H-chromen-4-yl mercurial, ()$_2$Hg (83)	—	—	76
C$_6$H$_5$OCH$_2$C≡CCH$_3$	Hg(O$_2$CCF$_3$)$_2$, NaCl	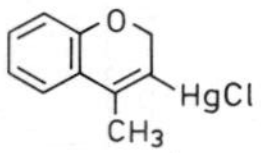 4-methyl-2H-chromen-3-yl HgCl	—	—	77
p-HC≡CCH$_2$OC$_6$H$_4$OCH$_3$	Hg(O$_2$CCF$_3$)$_2$	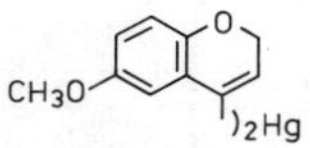 6-methoxy-2H-chromen-4-yl mercurial, ()$_2$Hg (97)	—	—	76

Table 10.4. (continued)

Alkyne	Mercury reagent	Organomercurial(s) (% Yield)	Subsequent reactants	Product(s) (% Yield)	Ref.
$C_6H_5C{\equiv}CCO_2C_2H_5$	$Hg[C(CH_3)_3]_2$	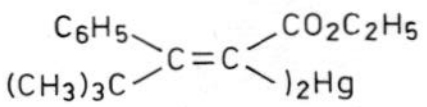	—	—	11
m-$CH_3OC_6H_4(CH_2)_2C{\equiv}C(CH_2)_2CH_3$	$Hg(O_2CCF_3)_2$, NaCl	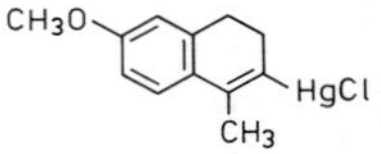 (30)	—	—	77
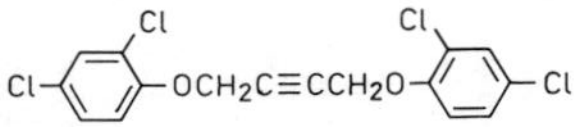	HgO/H_2SO_4/HOAc	—	—	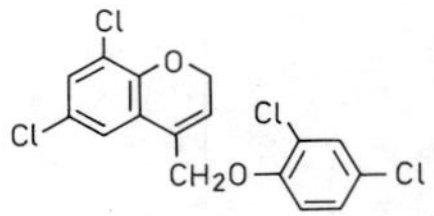 (30)	78
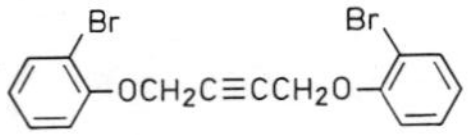	HgO/H_2SO_4/HOAc	—	—	 (24)	78
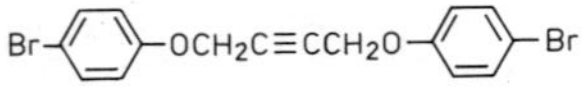	HgO/H_2SO_4/HOAc	—	—	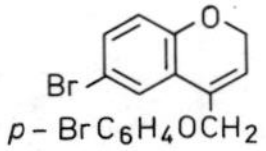 (30)	78
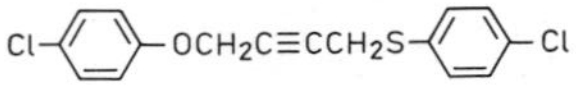	HgO/H_2SO_4/HOAc	—	—	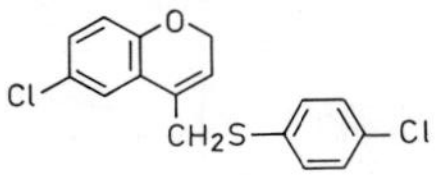 (24)	79
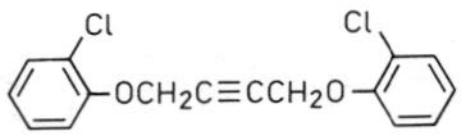	HgO/H_2SO_4/HOAc	—	—	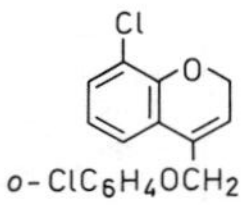 (31)	78

Table 10.4. (continued)

Alkyne	Mercury reagent	Organomercurial(s) (% Yield)	Subsequent reactants	Product(s) (% Yield)	Ref.
Cl—$OCH_2C{\equiv}CCH_2O$—Cl	HgO/H_2SO_4 / HOAc	—	—	(chromene structure; p-$ClC_6H_4OCH_2$) (26)	78
p-$C_6H_5OCH_2C{\equiv}CCH_2SC_6H_4Cl$	HgO/H_2SO_4 / HOAc	—	—	(chromanone structure; HC=S—C$_6$H$_4$—Cl) (37)	79
$C_6H_5OCH_2C{\equiv}CCH_2OC_6H_5$	$HgSO_4$ / HOAc	—	—	(chromene structure; $C_6H_5OCH_2$) (30)	80
CH_3—$OCH_2C{\equiv}CCH_2S$—Cl	HgO/H_2SO_4 / HOAc	—	—	(CH_3-chromanone structure; HC=S—C$_6$H$_4$—Cl) (27)	79
Cl—$OCH_2C{\equiv}CCH_2O$—Cl (dimethyl)	HgO/H_2SO_4 / HOAc	—	—	(chromene structure; CH_3, Cl, CH_2O—C$_6$H$_3$(CH$_3$)Cl) (20)	78
$OCH_2C{\equiv}CCH_2S$ (ditolyl)	HgO/H_2SO_4 / HOAc	—	—	(CH_3-chromanone structure; HC=S—C$_6$H$_4$—CH$_3$) (44)	79
CH_3—$OCH_2C{\equiv}CCH_2S$—(o-tolyl)	HgO/H_2SO_4 / HOAc	—	—	(CH_3-chromanone structure; HC=S—C$_6$H$_4$—CH$_3$) (37)	79

Table 10.4. (continued)

Alkyne	· Mercury reagent	Organomercurial(s) (% Yield)	Subsequent reactants	Product(s) (% Yield)	Ref.
(2-CH₃ phenyl OCH₂C≡CCH₂O 2-CH₃ phenyl)	HgO / H₂SO₄ / HOAc	—	—	chromene, o-CH₃C₆H₄OCH₂ (58) or benzofuran (CH₂)₂OC₆H₄CH₃-o (15)	78
(4-CH₃ phenyl OCH₂C≡CCH₂O 4-CH₃ phenyl)	HgO / H₂SO₄ / HOAc	—	—	chromene, p-CH₃C₆H₄OCH₂ (58) or benzofuran (CH₂)₂OC₆H₄CH₃-p (11)	78
(2-OCH₃ phenyl OCH₂C≡CCH₂O 2-OCH₃ phenyl)	HgO / H₂SO₄ / HOAc	—	—	chromene, o-CH₃OC₆H₄OCH₂ (60)	78
[p-CH₃C₆H₄OCH₂C≡C—]₂	HgO / H₂SO₄ / HOAc	—	—	chromene (61) /₂	81
(2,5-diCH₃ phenyl OCH₂C≡CCH₂O 2,5-diCH₃ phenyl)	HgO / H₂SO₄ / HOAc	—	—	chromene, CH₂O (60) or benzofuran (CH₂)₂O (20)	78

Table 10.4. (continued)

Alkyne	Mercury reagent	Organomercurial(s) (% Yield)	Subsequent reactants	Product(s) (% Yield)	Ref.
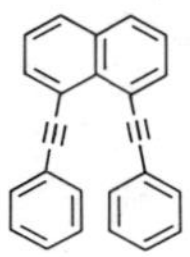 (structure)	$Hg(OAc)_2$ / HOAc	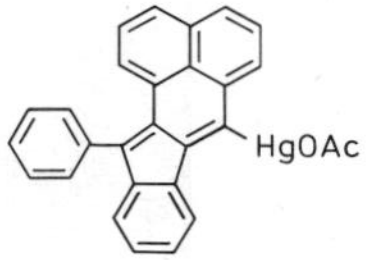 (structure) HgOAc (56)	—	—	82
	$Hg(OAc)_2$ / H_2SO_4 / HOAc	—	—	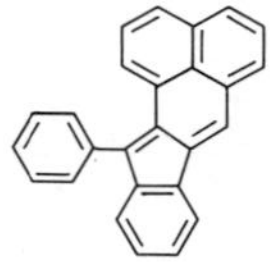 (structure) (57)	82
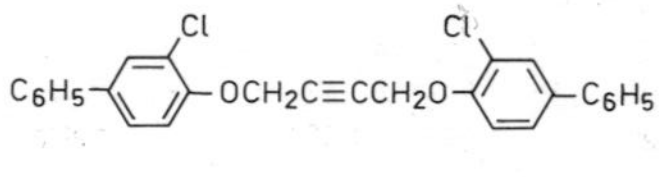 C_6H_5—(structure)—$OCH_2C{\equiv}CCH_2O$—(structure)—C_6H_5	HgO / H_2SO_4 / HOAc	—	—	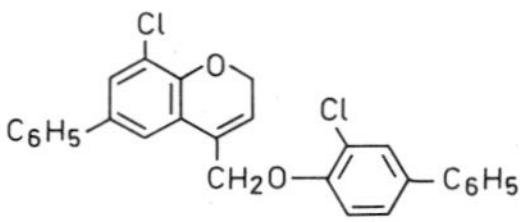 (structure) CH_2O—(structure)—C_6H_5 (26)	78
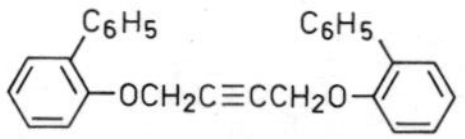 C_6H_5—(structure)—$OCH_2C{\equiv}CCH_2O$—(structure)—C_6H_5	HgO / H_2SO_4 / HOAc	—	—	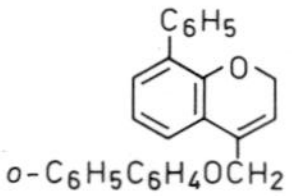 (structure) o-$C_6H_5C_6H_4OCH_2$ (26)	78
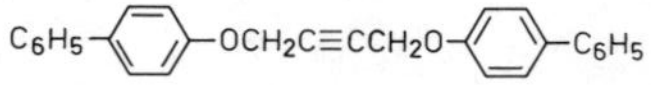 C_6H_5—(structure)—$OCH_2C{\equiv}CCH_2O$—(structure)—C_6H_5	HgO / H_2SO_4 / HOAc	—	—	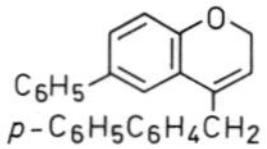 (structure) p-$C_6H_5C_6H_4CH_2$ (41)	78
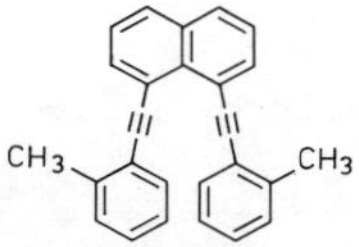 (structure) CH_3 / CH_3	$Hg(OAc)_2$ / H_2SO_4 / HOAc	—	—	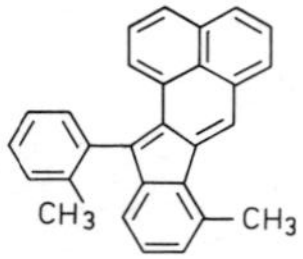 (structure) (60)	82

Table 10.4. (continued)

Alkyne	Mercury reagent	Organomercurial(s) (% Yield)	Subsequent reactants	Product(s) (% Yield)	Ref.
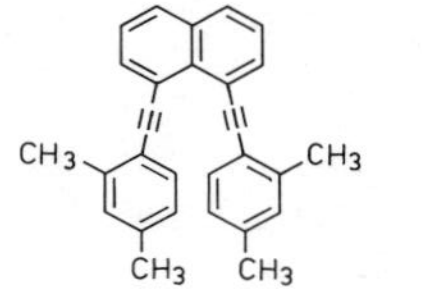 (structure)	Hg(OAc)$_2$ / H$_2$SO$_4$ / HOAc	—	—	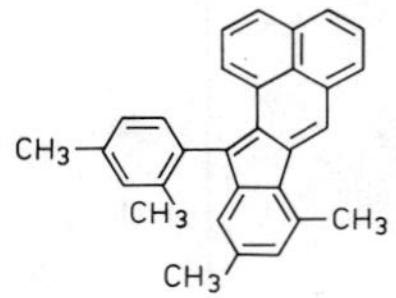 (structure) (60)	82
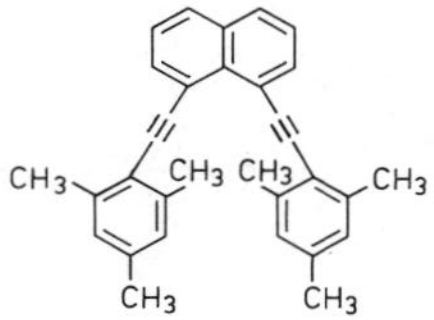 (structure)	Hg(OAc)$_2$ / CH$_3$OH	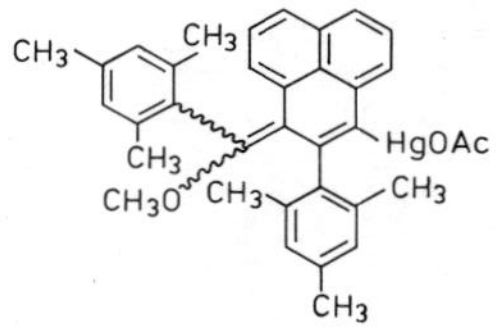 (structure)	—	—	82
	Hg(OAc)$_2$ / H$_2$SO$_4$ / HOAc	—	—	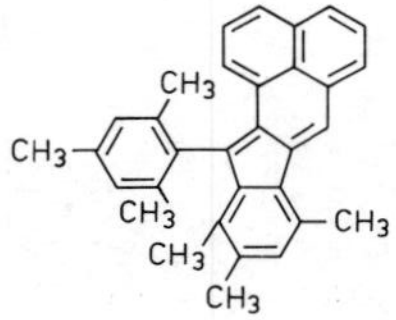 (structure) (25)	82

The reaction proceeds at room temperature to give clean syn-addition and the mercury moiety can be stereospecifically replaced by a proton when the mercurial is reacted with HCl. Unfortunately, no such addition to alkynes is observed with diethyl- or diphenylmercury.

The mercuration of certain aryl acetylenes results in cyclic carbon—carbon bond formation (Eqs. 71–73). The mechanism of this first example (Eq. 71)

$$X-C_6H_4-OCH_2C{\equiv}CH \xrightarrow{Hg(O_2CCF_3)_2} \text{(chromene, }\tfrac{1}{2}Hg) \tag{71) [76]}$$

$$C_6H_5-OCH_2C{\equiv}CCH_3 \xrightarrow[\ ]{Hg(O_2CCF_3)_2} \xrightarrow{NaCl} \text{(product, HgCl, CH}_3) \tag{72) [77]}$$

$$CH_3O-C_6H_4-(CH_2)_2C{\equiv}C(CH_2)_2CH_3 \xrightarrow[\ ]{Hg(O_2CCF_3)_2} \xrightarrow{NaCl} \text{(product, CH}_3\text{O, HgCl, CH}_3(CH_2)_2) \tag{73) [77]}$$

is unclear, but the latter reactions presumably involve intramolecular electrophilic aromatic substitution by a mercury-alkyne complex. Several related cyclizations have been reported in which use of a strong acid prevented isolation of the intermediate mercurials (Eqs. 74–76). In equation 74, the

$$XC_6H_4OCH_2C{\equiv}CCH_2OC_6H_4X \xrightarrow[\substack{H_2SO_4 \\ HOAc}]{HgO} X-\text{(CH}_2OC_6H_4X) \ + \ X-\text{((CH}_2)_2OC_6H_4X) \tag{74) [78, 80]}$$

$$\left[CH_3-C_6H_4-OCH_2C{\equiv}C \right]_2 \longrightarrow CH_3-\text{(product)}_{\tfrac{1}{2}} \tag{75) [81]}$$

$$XC_6H_4OCH_2C{\equiv}CCH_2SC_6H_4Y \longrightarrow X-\text{(YC}_6H_4SCH_2) \quad or \quad X-\text{(HCSC}_6H_4Y) \tag{76) [79]}$$

chromene is usually the major product, while the sulfur analog in equation 76 usually affords the vinyl sulfide as the major product. Analogous reactions on the corresponding disulfides or disulfones yield primarily products of hydration of the carbon—carbon triple bond [79]. Cyclizations across triple

X. Carbomercuration

bonds have also been observed in the following case (Eq. 77) [82]. In the presence of sulfuric acid, this latter reaction proceeds to the simple hydrocarbon minus the mercury moiety.

$$\text{(77)}$$

Arenes are not the only groups which can be added to acetylenes. There is one example of the carbomercuration of acetylene using ethyl acetoacetate (Eq. 78) [24].

$$\text{(78)}$$

References

1. Novikov, S. S., Godovikova, T. I., Tartakovskii, V. A.: Dokl. Akad. Nauk SSSR, *124*, 834 (1959); Proc. Acad. Sci. USSR, Chem. Sec., *124*, 89 (1959).
2. Tartakovskii, V. A., Novikov, S. S., Godovikova, T. I.: Izv. Akad. Nauk SSSR, Otdel. Khim. Nauk, 1042 (1961); Bull. Acad. Sci. USSR, Div. Chem. Sci., 963 (1961).
3. Tartakovskii, V. A., Nikonova, L. A., Novikov, S. S.: Izv. Akad. Nauk SSSR, Ser. Khim., 706 (1967); Bull. Acad. Sci. USSR, Div. Chem. Sci., 690 (1967).
4. Tartakovskii, V. A., Savost'yanova, I. A., Novikov, S. S.: Izv. Akad. Nauk SSSR, Ser. Khim., 1204 (1963); Bull. Acad. Sci. USSR, Div. Chem. Sci., 1206 (1963).
5. Tartakovskii, V. A., Savost'yanova, I. A., Novikov, S. S.: Izv. Akad. Nauk SSSR, Ser. Khim., 1330 (1963); Bull. Acad. Sci. USSR, Div. Chem. Sci., 1206 (1963).
6. Eremenko, L. T., Nesterenko, G. N.: Izv. Akad. Nauk SSSR, Ser. Khim., 1601 (1969); Bull. Acad. Sci. USSR, Div. Chem. Sci., 1483 (1969).
7. Novikov, S. S., Tartakovskii, V. A., Godovikova, T. I., Gribov, B. G.: Izv. Akad. Nauk SSSR, Otdel. Khim. Nauk, 272 (1962); Bull. Acad. Sci. USSR, Div. Chem. Sci., 249 (1962).
8. Khmel'nitskii, L. I., Godovikova, T. I., Igritskaya, A. V., Novikov, S. S.: Izv. Akad. Nauk SSSR, Ser. Khim., 2108 (1970); Bull. Acad. Sci. USSR, Div. Chem. Sci., 1981 (1970).
9. Shvekhgeimer, G. A., Sobtsova, N. I., Baranski, A.: Rocz. Chem., *47*, 1243 (1973); Chem. Abstr., *80*, 27321u (1974).

10. Tartakovskii, V. A., Zefirov, N. S., Chekulaeva, V. N.: Izv. Akad. Nauk SSSR, Ser. Khim., 1638 (1968); Bull. Acad. Sci. USSR, Div. Chem. Sci., 1549 (1968).

11. Blaukat, U., Neumann, W. P.: J. Organometal. Chem., *49*, 323 (1973).

12. Sokolov, V. I., Troitskaya, L. L., Reutov, O. A.: Zh. Org. Khim., *1*, 1579 (1965); J. Org. Chem. USSR, *1*, 1601 (1965).

13. Eremenko, L. T., Natsibullin, F. Ya., Nesterenko, G. N.: Izv. Akad. Nauk SSSR, Ser. Khim., 1360 (1968); Bull. Acad. Sci. USSR, Div. Chem. Sci., 1280 (1968).

14. Neumann, W. P., Blaukat, U.: Angew. Chem., *81*, 625 (1969); Angew. Chem., Int. Ed. Engl., *8*, 611 (1969).

15. Ichikawa, K., Tozaki, H., Ueki, I., Shingu, H.: J. Chem. Soc. Japan, Pure Chem. Sec., *72*, 267 (1951); Chem. Abstr., *46*, 3025a (1952).

16. Ichokawa, K., Fujita, K., Ouchi, H.: J. Am. Chem. Soc., *81*, 5316 (1959).

17. Ichikawa, K., Fujita, K., Itoh, O.: J. Am. Chem. Soc., *84*, 2632 (1962).

18. Ichikawa, K., Fukushima, T., Ouchi, H.: Japan. Patent 13,015 (1960); Chem. Abstr., *55*, 1529e (1961).

19. Ichikawa, K., Fukushima, S., Ouchi, H., Tsuchida, M.: J. Am. Chem. Soc., *80*, 6005 (1958).

20. Julia, M., Colomer Gasquez, E., Labia, R.: Bull. Soc. Chim. France, 4145 (1972); Chem. Abstr., *78*, 71598x (1973).

21. Ichikawa, K., Itoh, O., Kawamura, T., Fujiwara, M., Ueno, T.: J. Org. Chem., *31*, 447 (1966).

22. Ichikawa. K., Ouchi, H., Fukushima, S.: J. Org. Chem., *24*, 1129 (1959).

23. Ichikawa, K., Fukushima, S., Ouchi, H., Tsuchida, M.: J. Am. Chem. Soc., *81*, 3401 (1959).

24. Adelman, R. L.: J. Org. Chem., *14*, 1057 (1949).

25. Adelman, R. L.: U.S. Patent 2,550,439 (1951); Chem. Abstr., *45*, 8035e (1951).

26. Ichikawa, K., Fukushima, T., Ouchi, H., Tsuchida, M.: Japan. Patent 21,529 (1961); Chem. Abstr., *57*, 16493b (1962).

27. Julia, M., Colomer, E., Julia, S.: Bull. Soc. Chim. France, 2397 (1966); Chem. Abstr., *65*, 18437a (1966).

28. Julia, M., Labia, R.: Bull. Soc. Chim. France, 4151 (1972); Chem. Abstr., *78*, 124337v (1973).

29. Julia, M., Labia, R.: Compt. Rend. C, *268*, 104 (1969); Chem. Abstr., *70*, 78112c (1969).

30. Schuster, L., Raff, P.: Ger. Patent 1,297,616 (1969); Chem. Abstr., *71*, 49548q (1969).

31. Delbecq, F., Goré, J.: Tetrahedron Lett., 3459 (1976).

32. Julia, M., Colomer Gasquez, E.: Bull. Soc. Chim. France, 1796 (1973); Chem. Abstr., *79*, 104790e (1973).

33. Cope, A. C., Nelson, N. A., Smith, D. S.: J. Am. Chem. Soc., *76*, 1100 (1954).

34. Julia, M., Colomer, E.: An. Quim., *67*, 199 (1971).

35. Bluthe, N., Malacria, M., Gore, J.: Tetrahedron Lett., 4263 (1982).

36. Baudouy, R., Delbecq, F., Gore, J.: Tetrahedron, *36*, 189 (1980).

37. Bergter, L., Seidl, P. R.: J. Org. Chem., *47*, 73 (1982).

38. Thies, R. W., Hong, P.-K., Buswell, R., Boop, J. L.: J. Org. Chem., *40*, 585 (1975).

39. Thies, R. W., Hong, P. K., Buswell, R.: J. Chem. Soc., Chem. Commun., 1091 (1972).

40. Julia, M., Colomer Gasquez, E.: Bull. Soc. Chim. France, 4148 (1972); Chem. Abstr., *78*, 83888e (1973).

X. Carbomercuration

41. Traynham, J. G., Franzen, G. R., Knesel, G. A., Northington, D. J., Jr.: J. Org. Chem., *32*, 3285 (1967).
42. McQuillin, F. J., Parker, D. G.: J. Chem. Soc., Perkin I, 809 (1974).
43. Matsuki, Y., Kodama, M., Itô, S.: Tetrahedron Lett., 2901 (1979).
44. Matsuki, Y., Kodama, M., Itô, S.: Tetrahedron Lett., 4081 (1979).
45. Stetter, H., Gärtner, J.: Chem. Ber., *99*, 925 (1966).
46. Julia, M., Fourneron, J.-D.: Bull. Soc. Chim. France, 770 (1975).
47. Julia, M.: Fr. Demande 2,224,477 (1974); Chem. Abstr., *83*, 10407r (1975).
48. Julia, M., Fourneron, J.-D.: J. Chem. Research (S), 466 (1978).
49. Kurbanov, M., Semenovsky, A. V., Smit, W. A., Shmelev, L. V., Kucherov, V. F.: Tetrahedron Lett., 2175 (1972).
50. Rouessac, F., Zamarlik, H., Gnonlongoun, N.: Tetrahedron Lett., *24*, 2247 (1983).
51. Hoye, T. R., Kurth, M. J.: J. Org. Chem., *44*, 3461 (1979).
52. Cookson, R. C., Hudec, J., Marsden, J.: Chem. Ind., 21 (1961).
53. Julia, M., Fourneron, J. D.: Tetrahedron, *32*, 1113 (1976).
54. Dyadchenko, A. I., Semenovskii, A. V., Smit, V. A.: Izv. Akad. Nauk SSSR, Ser. Khim., 2276 (1976); Bull. Acad. Sci. USSR, Div. Chem. Sci., 2123 (1976).
55. Mustafaeva, M. T., Smit, V. A., Semenovskii, A. V., Kucherov, V. F.: Izv. Akad. Nauk SSSR, Ser. Khim., 1151 (1973); Bull. Acad. Sci. USSR, Div. Chem. Sci., 1111 (1973).
56. Julia, M., Fourneron, J.-D., Thal, C.: An. Quim., *70*, 888 (1974).
57. Dyadchenko, A. I., Semenovskii, A. V., Smit, V. A.: Izv. Akad. Nauk SSSR, Ser. Khim., 2281 (1976); Bull. Acad. Sci. USSR, Div. Chem. Sci., 2127 (1976).
58. Semenovskii, A. V., Smit, V. A., Kucherov, V. F.: Izv. Akad. Nauk SSSR, Ser. Khim., 2155 (1970); Bull. Acad. Sci. USSR, Div. Chem. Sci., 2039 (1970).
59. Misra, S. C., Chandra, G.: Indian J. Chem., *11*, 613 (1973).
60. Govindan, S. V., Bhattacharyya, S. C.: Indian J. Chem., *16B*, 1 (1978).
61. Tius, M. A., Takaki, K. S.: J. Org. Chem., *47*, 3166 (1982).
62. Tsankova, E., Ognyanov, I., Norin, T.: Tetrahedron, *36*, 669 (1980).
63. Renold, W., Ohloff, G., Norin, T.: Helv. Chim. Acta, *62*, 985 (1979).
64. Brown, E. D., Sam, T. W., Sutherland, J. K., Torre, A.: J. Chem. Soc., Perkin I, 2326 (1975).
65. Misumi, S., Ohfune, Y., Furusaki, A., Shirahama, H., Matsumoto, T.: Tetrahedron Lett., 2865 (1976).
66. Misumi, S., Ohtsuka, T., Hashimoto, H., Ohfune, Y., Shirahama, H., Matsumoto, T.: Tetrahedron Lett., 35 (1979).
67. Misumi, S., Ohtsuka, T., Ohfune, Y., Sugita, K., Shirahama, H., Matsumoto, T.: Tetrahedron Lett., 31 (1979).
68. Nishizawa, M., Takenaka, H., Nishide, H., Hayashi, Y.: Tetrahedron Lett., *24*, 2581 (1983).
69. Yang, N. C., Libman, J.: J. Am. Chem. Soc., *94*, 9228 (1972).
70. Dyadchenko, A. I., Semenovskii, A. V., Smit, V. A., Kurbanov, M.: Izv. Akad. Nauk SSSR, Ser. Khim., 1345 (1977); Bull. Acad. Sci. USSR, Div. Chem. Sci., 1243 (1977).
71. Corey, E. J., Tius, M. A., Das, J.: J. Am. Chem. Soc., *102*, 7612 (1980).
72. Corey, E. J., Tius, M. A., Das, J.: J. Am. Chem. Soc., *102*, 1742 (1980).
73. Toda, F., Akagi, K.: Tetrahedron, *25*, 3795 (1969).
74. Hoye, T. R., Caruso, A. J., Dellaria, J. F., Jr., Kurth, M. J.: J. Am. Chem. Soc., *104*, 6704 (1982).
75. Julia, M., Fournenon, J.-D.: Tetrahedron Lett., 3429 (1973).
76. Bates, D. K., Jones, M. C.: J. Org. Chem., *43*, 3775 (1978).

77. Larock, R. C., Harrison, L. W.: J. Am. Chem. Soc., *106*, 4218 (1984).
78. Majumdar, K. C., Thyagarajan, B. S.: J. Heterocyclic Chem., *9*, 489 (1972).
79. Thyagarajan, B. S., Majumdar, K. C., Bates, D. K.: J. Heterocyclic Chem., *12*, 59 (1975).
80. Thyagarajan, B. S., Balasubramanian, K. K., Rao, R. B.: Tetrahedron, *23*, 1893 (1967).
81. Balasubramanian, K. K., Reddy, K. V., Nagarajan, R.: Tetrahedron Lett., 5003 (1973).
82. Staab, H. A., Ipaktschi, J.: Chem. Ber., *104*, 1170 (1971).

XI. Halomercuration

A. Alkenes

The addition of fluorine and mercury across a carbon—carbon double bond is well documented, but there are few examples of chlorine-mercury additions and none of bromine or iodine and mercury (Table 11.1).

The reaction of 1,1-dimethoxy-2-methyl-1-propene and mercuric chloride in the presence of zinc chloride results in demethylation and formation of the corresponding mercurated ester (Eq. 1) [17]. While coumarin and derivatives

$$(CH_3)_2C = C(OCH_3)_2 \quad + \quad HgCl_2 \quad \xrightarrow{\text{ZnCl}_2} \quad ClHgC(CH_3)_2CO_2CH_3 \qquad \text{(1)}$$
$$99\%$$

have been reported to undergo mercuric chloride addition to the double bond, the products have not been well characterized [15, 18]. The addition of mercuric chloride to ketene appears to be the only well documented case of chloromercuration of a carbon—carbon double bond (Eq. 2) [4].

$$H_2C=C=O \quad + \quad HgCl_2 \quad \longrightarrow \quad ClHgCH_2COCl \qquad \text{(2)}$$

There are numerous examples of the fluoromercuration of alkenes. The reagents which have been employed are HgF_2 [5, 11, 13, 14], HgF_2/HF [6–8, 10], HgF_2/KF [11, 13], HgF_2/AsF_3 [1, 2], $HgF_2/HgCl_2$ [7], $HgF_2/HgCl_2/HF$ [8], $HgCl_2/KF$ [11, 12] and $Hg(O_2CCF_3)_2/KF$ [9]. Organomercuric halides [11] and trifluoroacetates [9] plus fluoride salts undergo analogous addition reactions to afford the corresponding diorganomercury compounds. The reagents containing mercuric chloride give rise to either organomercuric chlorides or diorganomercurials, while the straight fluoride reagents afford only dialkylmercury compounds with only two exceptions [5]. The reactions are usually carried out by heating in an excess of the reagent or solvents such as DMF, THF, DME or diethyl ether.

Only polyhalogenated alkenes apparently undergo fluoromercuration. The majority of work has been carried out on perfluorinated alkenes in which case the fluorine always adds to the carbon bearing the most fluorines. With 1,1-difluoroethylene or trifluoroethylene, the fluorine again adds to the most highly fluorinated carbon [1, 2]. The fluoromercuration of bis(trifluoromethyl)-ketene places the fluorine on the carbonyl carbon [13]. Mixed results are reported when organomercuric salts are used for fluoromercuration. Phenyl-

580

Table 11.1. Halomercuration of Alkenes

Alkene	Mercury reagent	Organomercurial(s) (% Yield)	Subsequent reactants	Product(s) (% Yield)	Ref.
$F_2C=CFH$	HgF_2/AsF_3	$(CF_3CHF)_2Hg$ (66)	—	—	1, 2
	HgF_2	—	NOCl	$CF_3CFClNO$ + CF_3CFCl_2 + CF_3CHFCl	3
$F_2C=CH_2$	HgF_2/AsF_3	$(CF_3CH_2)_2Hg$ (66)	—	—	1, 2
$H_2C=C=O$	$HgCl_2$	$ClHgCH_2COCl$	Na_2CO_3/H_2O	$ClHgCH_2CO_2H$ (100)	4
$F_2C=CFCl$	HgF_2	$CF_3CFClHgF$ (24)	X_2 or Na_2SnO_2 (X = Cl, I)	CF_3CFClX X = Cl (84), I (81), H (89)	5
	HgF_2/AsF_3	$(CF_3CFCl)_2Hg$ (31)	—	—	1, 2
	HgF_2	—	NOCl	$CF_3CFClNO$	3
$F_2C=CCl_2$	HgF_2/HF	$(CF_3CCl_2)_2Hg$ (69)	—	—	6
	HgF_2	—	NOCl	CF_3CCl_2NO	3
$F_2C=CF_2$	HgF_2	CF_3CF_2HgF ? (2)	—	—	5
	HgF_2/AsF_3	$(CF_3CF_2)_2Hg$ (56)	S / Δ	$(CF_3CF_2S)_2$	1, 2
$F_2C=CHCF_3$	HgF_2/HF	$[(CF_3)_2CH]_2Hg$ (80)	Br_2	$(CF_3)_2CHBr$	7, 8
$F_2C=CBrCF_3$	$CF_3HgO_2CCF_3/$ CsF	$CF_3CFBrCF_2HgCF_3$? (61)	—	—	9
$F_2C=CFCF_2Cl$	HgF_2/HF	$[ClF_2CCF(CF_3)]_2Hg$	—	—	10

Table 11.1. (continued)

Alkene	Mercury reagent	Organomercurial(s) (% Yield)		Subsequent reactants	Product(s) (% Yield)	Ref.
$F_2C=CFCF_3$	HgF_2/HF	$[(CF_3)_2CF]_2Hg$	(60–80)	X_2	$(CF_3)_2CFX$ $X = Br$ (68), I (74, 79)	6–8
	HgF_2	—		NOCl	$(CF_3)_2CFNO$	3
	$HgF_2/HgCl_2/HF$	$(CF_3)_2CFHgCl$	(54)	—	—	7, 8
	$HgCl_2/KF$	$(CF_3)_2CFHgCl$	(25.8)	—	—	11, 12
	$HgCl_2/KF$	$[(CF_3)_2CF]_2Hg$	(65)	—	—	11, 12
	$(CF_3)_3CHgCl/KF$	$[(CF_3)_2CF]_2Hg$	(67)	—	—	11
$F_2C=CFCF_2CHF_2$	HgF_2/HF	$[F_2CHCF_2CF(CF_3)]_2Hg$	(73)	—	—	6
$H_2C=C(CF_3)_2$	$HgCl_2/KF$	$[(CF_3)_3C]_2Hg$		—	—	12
$F_2C=CFCF_2CF_2Cl$	HgF_2/HF	$[ClCF_2CF_2CF(CF_3)]_2Hg$	(5)	—	—	6
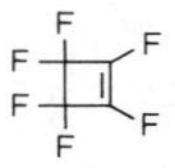 (cyclobutene diagram)	$HgCl_2/KF$	$F_7\square$—HgCl	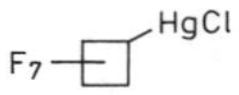 (46.5)	—	—	11
	HgF_2/KF	$(F_7\square$—$)_2Hg$	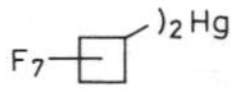 (33)	—	—	11
$(CF_3)_2C=C=O$	HgF_2/KF	$[(CF_3)_2C(COF)]_2Hg$	(70)	Na_2CO_3	$[F_2C=C(CF_3)]_2Hg$ (55) $+ (CF_3)_2CHHgC(CF_3)=CF_2$ (34.7) $+ [(CF_3)_2CH]_2Hg$ (7.6)	13

Table 11.1. (continued)

Alkene	Mercury reagent	Organomercurial(s) (% Yield)	Subsequent reactants	Product(s) (% Yield)	Ref.
$F_2C=C(CF_3)_2$	HgF_2	$[(CF_3)_3C]_2Hg$ (66.5)	—	—	11
	HgF_2 / HF	$[(CF_3)_3C]_2Hg$ (33)	X_2	$(CF_3)_3CX$ $X = Br$ (68), I	6
	$HgCl_2 / KF$	$(CF_3)_3CHgCl$ (42)	—	—	11
	$Hg(O_2CCF_3)_2 / KF$	$[(CF_3)_3C]_2Hg$ (80)	—	—	9
	C_6H_5HgF	$C_6H_5HgC(CF_3)_3$ (61)	—	—	11
	$CF_3HgO_2CCF_3 / KF$	$(CF_3)_2CFCF_2HgCF_3$?	—	—	9
$F_2C=CF(CF_2)_5CF_3$	HgF_2	—	S / Δ	$CF_3CS(CF_2)_5CF_3$	14
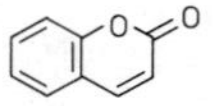 (coumarin)	$HgCl_2$	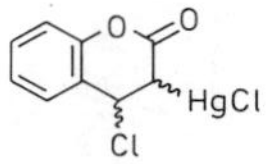 (chroman-2-one, 3-HgCl, 4-Cl) ?	—	—	15
$F_2C=CF(CF_2)_6CF_3$	—	—	S / Δ	$CF_3CS(CF_2)_6CF_3$	14
$F_2C=CF(CF_2)_7CF_2H$	HgF_2 / HF	$[HCF_2(CF_2)_7CF(CF_3)]_2Hg$ (low)	—	—	6
$(CH_3O-C_6H_4-)_2C=C=C=C=C=C(-C_6H_4-OCH_3)_2$	$HgCl_2$	$(CH_3O-C_6H_4-)_2C=C=C=C-\overset{+}{C}(-C_6H_4-OCH_3)$, $\overset{|}{HgCl}$ $HgCl_3^-$ (70)	—	—	16

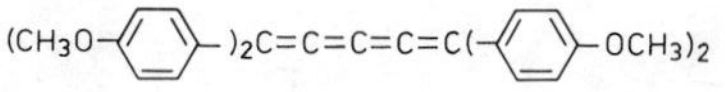

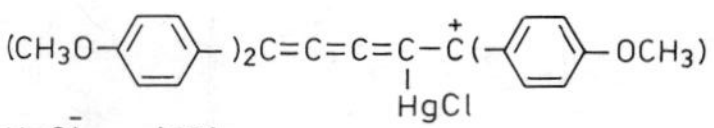

mercuric fluoride reacts with perfluoroisobutylene to afford the *t*-butyl derivative (Eq. 3) [11]. Perfluoro-*t*-butylmercuric chloride plus potassium

$$(CF_3)_2C{=}CF_2 \ + \ FHgC_6H_5 \ \longrightarrow \ (CF_3)_3CHgC_6H_5 \tag{3}$$

fluoride reacts with perfluoropropene with similar regiochemical results, but trifluoromethylmercuric trifluoroacetate in the presence of fluoride salts reportedly adds in the opposite direction to polyhalogenated olefins (Eq. 4) [9].

$$CF_3CX{=}CF_2 \ + \ CF_3CO_2HgCF_3 \ \xrightarrow{F^-} \ CF_3CXFCF_2HgCF_3 \tag{4}$$
$$X = Br, \ CF_3$$

Only a few reactions have been carried out on the halomercuration products. While the perhaloalkylmercurials are stable to most strong acids and bases, demercuration to the corresponding hydrocarbons can be effected by a number of reagents [5, 6]. Successful chlorination, bromination and iodination of these highly halogenated organomercurials have also been reported [5, 7, 8]. Reaction with NOCl provides the corresponding nitroso-alkenes (Eq. 5) [3]. Upon heating with elemental sulfur, primary per-

$$[(CF_3)_2CF]_2Hg \ \xrightarrow{NOCl} \ (CF_3)_2CFNO \tag{5}$$

fluoroalkylmercurials have been reported to afford disulfides, while the secondary mercurials yield thioketones (Eqs. 6, 7) [2, 14].

$$(CF_3CF_2)_2Hg \ \xrightarrow[\Delta]{S} \ (CF_3CF_2S)_2 \tag{6}$$

$$[(CF_3)_2CF]_2Hg \ \longrightarrow \ CF_3\overset{\overset{\textstyle S}{\|}}{C}CF_3 \tag{7}$$

There appears to be only one example of a polyene reacting with a mercuric halide (Eq. 8) [16].

$$(CH_3O{-}C_6H_4{-})_2C{=}C{=}C{=}C{=}C({-}C_6H_4{-}OCH_3)_2 \ + \ HgCl_2 \ \longrightarrow$$
$$[(CH_3O{-}C_6H_4{-})_2\overset{+}{C}{-}\underset{\underset{\textstyle HgCl}{|}}{C}{=}C{=}C{=}C({-}C_6H_4{-}OCH_3)_2]HgCl_3^- \tag{8}$$

B. Alkynes

The reaction of mercuric halides and alkynes appears to be much more general. This reaction was discussed in some detail in Chapter II, section J, of the monograph "Organomercury Compounds in Organic Synthesis" and all results to date are summarized in Table 11.2. A few brief comments seem in order.

Only one example of the fluoromercuration of an alkyne has been reported (Eq. 9) [9]. All other examples of the halomercuration of alkynes involve mercuric chloride.

$$CF_3C\equiv CCF_3 \xrightarrow[\text{CsF}]{Hg(O_2CCF_3)_2} \quad \underset{\underset{66\%}{}}{F\diagdown \underset{CF_3}{}C=C\diagup \overset{CF_3}{\underset{)_2Hg}{}}} \tag{9}$$

Mercuric chloride readily adds to acetylene to give the trans adduct [19–22]. Thermal [23] and free radical [40] procedures do exist for preparation of the cis isomer. Propyne forms a much less stable trans adduct with mercuric chloride [26]. Other than the cis adduct obtained from cyclo-octyne [38], there appear to be no other examples of simple, unsubstituted alkynes undergoing chloromercuration.

On the other hand, many functionally substituted alkynes react readily with mercuric chloride to give stable adducts. These include vinyl acetylene which gives a trans adduct (Eq. 10) [26, 31]. Earlier reports of the formation of 1-chloro-4-chloromercurio-1,3-butadiene appear to be in error [32, 33]. Certain propargylic alcohols also form stable trans addition compounds (Eq. 11) [28–30]. α,β-Unsaturated ketones [39], carboxylic acids [24, 25, 35],

$$HC\equiv CCH=CH_2 \ + \ HgCl_2 \ \longrightarrow \ \underset{H}{ClHg\diagdown}C=C\underset{Cl}{\diagup CH=CH_2} \tag{10}$$

$$\underset{HC\equiv C\overset{OH}{\overset{|}{C}}R_2}{} \ + \ HgCl_2 \ \longrightarrow \ \underset{H}{ClHg\diagdown}C=C\underset{Cl}{\diagup \overset{OH}{\overset{|}{C}R_2}} \tag{11}$$

and esters [24], as well as 1,4-dichloro-2-butyne [28], undergo analogous Markovnikov additions to produce stable adducts. Where the stereochemistry of these addition compounds has been established, only trans adducts are reported. The kinetics of the ester reactions has been examined [34]. Our work on the mercuration of 4-hydroxy-2-alkyn-1-ones has provided evidence for initial cis chloromercuration products which are readily dehydrated to furylmercurials (Eq. 12) [36]. Acetylenic ethers also undergo

$$\underset{CH_3\overset{OH}{\overset{|}{C}}HC\equiv C\overset{O}{\overset{\|}{C}}CH_3}{} \xrightarrow{HgCl_2} \underset{CH_3CH\diagup \underset{OH}{}}{Cl\diagdown}C=C\underset{\underset{O}{}}{\diagup \overset{HgCl}{}\diagdown CCH_3} \longrightarrow \underset{CH_3\ \ O\ \ CH_3}{Cl\diagdown \diagup HgCl} \tag{12}$$

Table 11.2. Halomercuration of Alkynes

Alkyne	Mercuric salt	Organomercurial(s) (% Yield)	Subsequent reactants	Product(s) (% Yield)	Ref.
HC≡CH	HgCl$_2$/HCl	Cl,H / C=C / H,HgCl	I$_2$	ClCH=CHI	19–22
	HgCl$_2$/90–100°C	Cl,H / C=C / HgCl,H	—	—	23
HC≡CCO$_2$H	HgCl$_2$	Cl,H / C=C / CO$_2$H,HgCl (23, 91)	—	—	24, 25
HC≡CCH$_3$	HgCl$_2$	ClHg,H / C=C / CH$_3$,Cl (35)	I$_2$ or CuCl$_2$	XCH=C(Cl)CH$_3$ X = I, Cl	26
HC≡COCH$_3$	HgCl$_2$	ClHgCH=CCl(OCH$_3$) (92)	—	—	27
HC≡CCH$_2$OH	HgCl$_2$	H,ClHg / C=C / Cl,CH$_2$OH (54, 55)	CO/Li$_2$PdCl$_4$	butenolide (Cl-substituted furan-2(5H)-one) (96)	28–30
CF$_3$C≡CCF$_3$	Hg(O$_2$CCF$_3$)$_2$/CsF	F,CF$_3$ / C=C / CF$_3$,)$_2$Hg (66)	—	—	9
HC≡CCH=CH$_2$	HgCl$_2$	ClHg,H / C=C / CH=CH$_2$,Cl (45, 50)	—	—	26, 31
	HgCl$_2$	ClHgCH=CHCH=CHCl ? (95)	—	—	32, 33
ClCH$_2$C≡CCH$_2$Cl	HgCl$_2$	Cl,ClCH$_2$ / C=C / CH$_2$Cl,HgCl (77)	—	—	28
HC≡CCO$_2$CH$_3$	HgCl$_2$	ClHC=C(HgCl)CO$_2$CH$_3$	—	—	24, 34

Table 11.2. (continued)

Alkyne	Mercuric salt	Organomercurial(s) (% Yield)	Subsequent reactants	Product(s) (% Yield)	Ref.
$CH_3C\equiv CCO_2H$	$HgCl_2$	$CH_3CCl=C(HgCl)CO_2H$ (54)	—	—	24, 35
$HC\equiv COCH_2CH_3$	$HgCl_2$	$ClHgCH=CCl(OCH_2CH_3)$	—	—	27
$HOCH_2C\equiv CCH_2OH$	$HgCl_2$	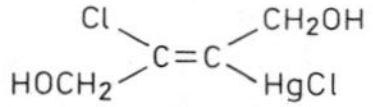 (45, 87)	X_2	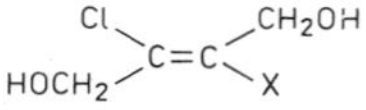 $X = Br$ (10.7), I (45)	28, 30
$HOCH_2C\equiv CCOCH_3$	$HgCl_2$	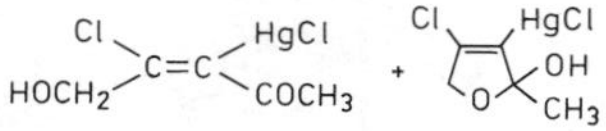 (75 total)	—	—	36
$HOCH_2C\equiv CCO_2CH_3$	$HgCl_2$	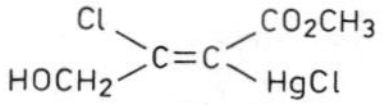 (100)	—	—	36
$CH_3CH_2C\equiv COCH_3$	$HgCl_2$	$CH_3CH_2C(HgCl)=CCl(OCH_3)$	—	—	27
$HC\equiv CC(CH_3)_2OH$	$HgCl_2$	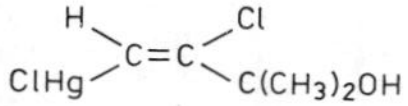 (31, 38)	CO/Li_2PdCl_4	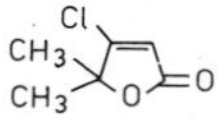 (100)	28–30
$CH_3O_2CC\equiv CCO_2CH_3$	$HgCl_2$	$CH_3O_2CCCl=C(HgCl)CO_2CH_3$ (73)	—	—	24, 34
$CH_3C\equiv CCO_2C_2H_5$	$HgCl_2$	$CH_3CCl=C(HgCl)CO_2C_2H_5$ (55)	—	—	24
$CH_3CHOHC\equiv CCOCH_3$	$HgCl_2$	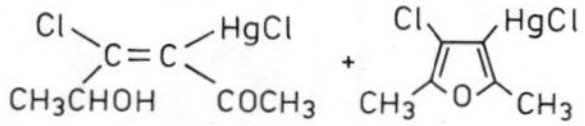 (90 total)	—	—	36
$HC\equiv CO(CH_2)_3CH_3$	$HgCl_2$	$ClHgCH=C(Cl)O(CH_2)_3CH_3$	—	—	27
$(CH_3)_2CHC\equiv COCH_3$	$HgCl_2$	$(CH_3)_2CHC(HgCl)=CCl(OCH_3)$ (95)	—	—	27

Table 11.2. (continued)

Alkyne	Mercuric salt	Organomercurial(s) (% Yield)	Subsequent reactants	Product(s) (% Yield)	Ref.
$HC{\equiv}CCOH(CH_3)CH_2CH_3$	$HgCl_2$	(26)	CO/Li_2PdCl_4	(98)	29, 30
$HC{\equiv}C$—OH (cyclopentyl)	$HgCl_2$	(37)	CO/Li_2PdCl_4	(99)	29, 30
$(CH_3)_2COHC{\equiv}CCOCH_3$	$HgCl_2$	(100)	—	—	36
$HC{\equiv}COC_6H_5$	$HgCl_2/HCl$	$(ClHg)_2C{=}CClCH_2OC_6H_5$	—	—	37
(cyclooctyne)	$HgCl_2$	(52)	—	—	38
$HC{\equiv}C$—OH (cyclohexyl)	$HgCl_2$	(31)	CO/Li_2PdCl_4	(95)	29, 30
$CH_3CHOHC{\equiv}CCO(CH_2)_3OH$	$HgCl_2$	(17)	—	—	36
$HOC(CH_3)_2C{\equiv}CC(CH_3)_2OH$	$HgCl_2$	(85, 94.5)	CO/Li_2PdCl_4	(94)	28–30

Table 11.2. (continued)

Alkyne	Mercuric salt	Organomercurial(s) (% Yield)	Subsequent reactants	Product(s) (% Yield)	Ref.
$HC\equiv C{-}OH$ (cycloheptane)	$HgCl_2$	$ClHg$–$CH{=}C(Cl)$–(1-hydroxycycloheptyl) (17)	CO/Li_2PdCl_4	(spiro lactone, Cl) (81)	30
$C_6H_5C\equiv CCOCH_3$	$HgCl_2$	$(Cl)(C_6H_5)C{=}C(COCH_3)(HgCl)$ (97)	HBr	$(Cl)(C_6H_5)C{=}C(COCH_3)(H)$	39
$C_6H_5C\equiv CCO_2CH_3$	$HgCl_2$	$C_6H_5CCl{=}C(HgCl)CO_2CH_3$ (74)	—	—	24, 34

mercuric chloride addition, but the stereochemistry has not been established due to the limited thermal stability of the products (Eq. 13) [27]. Phenoxy acetylene, on the other hand, is reported to undergo dimercuration (Eq. 14) [37]. Our own work on the mercuration of propargylic amines has provided some interesting anti-Markovnikov, trans adducts (Eq. 15) [41].

$$RC\equiv COR' \; + \; HgCl_2 \; \longrightarrow \; \underset{R}{\overset{ClHg}{\diagdown}}C=C\underset{Cl}{\overset{OR'}{\diagup}} \tag{13}$$

$$HC\equiv COC_6H_5 \; + \; HgCl_2 \; \xrightarrow{HCl} \; (ClHg)_2C=C(Cl)CH_2OC_6H_5 \tag{14}$$

$$HC\equiv CCH_2NR_2 \; \xrightarrow[HCl]{HgCl_2} \; \left[\underset{H}{\overset{Cl}{\diagdown}}C=C\underset{HgCl}{\overset{CH_2\overset{+}{N}HR_2}{\diagup}} \right]_2 Hg_nCl_{2n+2}^{2-} \tag{15}$$

$$n = 1 \text{ or } 2$$

Relatively few reactions have been reported for the vinylmercurials resulting from the halomercuration of alkynes. Protonolysis of the acetylenic ketone products by HBr and acetyl chloride has been reported [39]. The propargylic alcohol adducts undergo protonolysis by $SnCl_2$ [28]. Bromination and iodination occur readily using Br_2 and I_2 and chlorination can be effected by $CuCl_2$ [21, 26, 28]. Palladium-promoted carbonylation of the halomercuration products has also been reported [26, 36]. This reaction provides a novel route from propargylic alcohols to butenolides (Eq. 16) [29, 30].

$$\underset{R_2\overset{|}{C}C\equiv CH}{\overset{OH}{}} \; \xrightarrow{HgCl_2} \; \underset{R_2C}{\overset{Cl}{\diagdown}}C=C\underset{HgCl}{\overset{H}{\diagup}} \; \xrightarrow[Li_2PdCl_4]{CO} \; \tag{16}$$

The reaction of acetylenes, anilines and mercuric chloride has proven to be a useful procedure for the preparation of substituted quinolines (Eqs. 17, 18) [32, 42–44]. It is not clear if the alkyne chloromercuration products are actual intermediates in these reactions or not. Mercuric chloride also promotes the reaction of acetylenes and imines (Eq. 19) [45].

$$C_6H_5NH_2 \; + \; HC\equiv CH \; \xrightarrow{HgCl_2} \; \tag{17}$$

$$4\text{-}RC_6H_4NH_2 \; + \; HC\equiv CCH=CH_2 \; \xrightarrow{HgCl_2} \; \tag{18}$$

$$C_6H_5C\equiv CH \; + \; C_6H_5CH=NR \; \xrightarrow{HgCl_2} \; C_6H_5COCH_2CH(C_6H_5)NHR \tag{19}$$

References

1. Krespan, C. G.: J. Org. Chem., 25, 105 (1960).
2. Krespan, C. G.: U.S. Patent 2,844,614 (1958); Chem. Abstr., 53, 3061e (1959).
3. Tarrant, P., O'Connor, D. E.: J. Org. Chem., 29, 2012 (1964).
4. Kazankova, M. A., Trostyanskaya, I. G., Satina, T. Ya., Lutsenko, I. F.: Zh. Obshch. Khim., 46, 1412 (1976); J. Gen. Chem. USSR, 46, 1387 (1976).
5. Goldwhite, H., Haszeldine, R. N., Mukherjee, R. N.: J. Chem. Soc., 3825 (1961).
6. Aldrich, P. E., Howard, E. G., Linn, W. J., Middleton, W. J., Sharkey, W. H.: J. Org. Chem., 28, 184 (1963).
7. Miller, Jr., W. T., Freedman, M. B., Fried, J. H., Koch, H. F.: J. Am. Chem. Soc., 83, 4105 (1961).
8. Miller, W. T., Jr., Freedman, M. B.: J. Am. Chem. Soc., 85, 180 (1963).
9. Martynov, B. I., Sterlin, S. R., Dyatkin, B. L.: Izv. Akad. Nauk SSSR, Ser. Khim., 1642 (1974); Bull. Acad. Sci. USSR, Div. Chem. Sci., 1564 (1974).
10. Yakubovich, A. Ya., Rozenshtein, S. M., Gitel, P. O.: Zh. Obshch. Khim., 37, 278 (1967); J. Gen. Chem. USSR, 37, 261 (1967).
11. Dyatkin, B. L., Sterlin, S. R., Martynov, B. I., Mysov, E. I., Knunyants, I. L.: Tetrahedron, 27, 2843 (1971).
12. Dyatkin, B. L., Sterlin, S. R., Martynov, B. I., Knunyants, I. L.: Tetrahedron Lett., 1387 (1970).
13. Dyatkin, B. L., Zhuravkova, L. G., Martynov, B. I., Mysov, E. I., Sterlin, S. R., Knunyants, I. L.: J. Organometal. Chem., 31, C15 (1971).
14. Howard, E. G., Jr., Middleton, W. J.: U.S. Patent 2,970,173 (1961); Chem. Abstr., 55, 14311i (1961).
15. Seshadri, T. R., Rao, P. S.: Proc. Ind. Acad. Sci. A, 4, 162 (1936); Chem. Zentralbl. I, 2370 (1937).
16. Fischer, H., Fischer, H.: Chem. Ber., 97, 2959 (1964).
17. McElvain, S. M., Aldridge, C. L.: J. Am. Chem. Soc., 75, 3987 (1953).
18. Seshadri, T. R., Rao, P. S.: Proc. Ind. Acad. Sci. A, 4, 157 (1936); Chem. Zentralbl. I, 4621 (1937).
19. Nesmeyanov, A. N., Freidlina, R. Kh., Borisov, A. E.: Izv. Akad. Nauk SSSR, Otdel. Khim. Nauk, 150 (1945); Chem. Abstr., 40, 3451[7] (1946).
20. Nesmeyanov, A. N., Freidlina, R. Kh., Borisov, A. E.: Izv. Akad. Nauk SSSR, Otdel. Khim. Nauk, 137 (1945); Chem. Abstr., 40, 3451[3] (1946).
21. Freidlina, R. Kh., Nesmeyanov, A. N.: Dokl. Akad. Nauk SSSR, 26, 60 (1940); Chem. Abstr., 34, 6567[4] (1940).
22. Chapman, D. L., Jenkins, W. J.: J. Chem. Soc., 115, 847 (1919).
23. Freidlina, R. Kh., Nogina, O. V.: Izv. Akad. Nauk SSSR, Otdel. Khim. Nauk, 105 (1947); Chem. Abstr., 42, 4149d (1948).
24. Nesmeyanov, A. N., Kochetkov, N. K., Dashunin, V. M.: Izv. Akad. Nauk SSSR, Otdel. Khim. Nauk, 77 (1950); Chem. Abstr., 44, 7225g (1950).
25. Kurtz, A. N., Billups, W. E., Greenlee, R. B., Hamil, H. F., Pace, W. T.: J. Org. Chem., 30, 3141 (1965).
26. Shestakova, V. S., Brailoskii, S. M., Temkin, O. N., Azbel', B. I.: Zh. Org. Khim. 14, 2039 (1978); J. Org. Chem. USSR, 14, 1891 (1978).
27. Kazankova, M. A., Satina, T. Ya., Lutsenko, I. F.: Zh. Obshch. Khim., 45, 712 (1975); J. Gen. Chem. USSR, 45, 701 (1975).
28. Nesmeyanov, A. N., Kochetkov, N. K.: Izv. Akad. Nauk SSSR, Otdel. Khim. Nauk, 76 (1949); Chem. Abstr., 43, 7412h (1949).
29. Larock, R. C., Riefling, B.: Tetrahedron Lett., 4661 (1976).

XI. Halomercuration

30. Larock, R. C., Riefling, B., Fellows, C. A.: J. Org. Chem., *43*, 131 (1978).
31. Mikhailov, B. M., Vasil'ev, L. S., Veselovskii, V. V., Kiselev, V. G.: Izv. Akad. Nauk SSSR, Ser. Khim., 725 (1975); Bull. Acad. Sci. USSR, Div. Chem. Sci., 655 (1975).
32. Kozlov, N. S., Korotyshova, G. P.: Dokl. Akad. Nauk SSSR, *222*, 111 (1975); Proc. Acad. Sci. USSR, Chem. Sec., *222*, 290 (1975).
33. Kozlov, N. S., Korotyshova, G. P.: Vestsi Akad. Navuk B. SSR, Ser. Khim. Navuk, 52 (1974); Chem. Abstr., *82*, 43558y (1975).
34. Dvorko, G. F., Shilov, E. A.: Ukr. Khim. Zh., *28*, 833 (1962); Chem. Abstr., *59*, 1449a (1963).
35. Babayan, A. T., Grigoryan, A. A.: Zh. Obshch. Khim., *26*, 1945 (1956); J. Gen. Chem. USSR, *26*, 2167 (1956).
36. Larock, R. C., Liu, C.-L.: J. Org. Chem., *48*, 2151 (1983).
37. Filippova, A. Kh., Lyashenko, G. S., Frolov, Yu. L., Borisova, A. I., Ivanova, N. A., Voronkov, M. G.: Dokl. Vses. Konf. Khim. Atsetilena, 4th, *2*, 206 (1972); Chem. Abstr., *79*, 104840w (1973).
38. Wittig, G., Fischer, S.: Chem. Ber., *105*, 3542 (1972).
39. Nesmeyanov, A. N., Kochetkov, N. K.: Izv. Akad. Nauk SSSR, Otdel. Khim. Nauk, 305 (1949); Chem. Abstr., *43*, 7413d (1949).
40. Nesmeyanov, A. N., Borisov, A. E., Novikova, N. V.: Izv. Akad. Nauk SSSR, Ser. Khim., 857 (1970); Bull. Acad. Sci. USSR, Div. Chem. Sci., 804 (1970).
41. Larock, R. C., Burns, L. D., Varaprath, S., Russell, C. E.: unpublished work.
42. Kozlov, N. S., Korotyshova, G. P.: Dokl. Akad. Nauk Beloruss. SSR, *16*, 335 (1972); Chem. Abstr., *77*, 100972b (1972).
43. Kozlov, N. S., Korotyshova, G. P.: Vestsi Akad. Navuk B. SSR, Ser. Khim. Navuk, 130 (1972); Chem. Abstr., *78*, 3495m (1973).
44. Kozlov, N. S., Korotyshova, G. P.: USSR Patent, 432,142 (1972); Chem. Abstr., *81*, 91373k (1974).
45. Kozlov, N. S., Korotyshova, G. P.: Vestsi Akad. Navuk B. SSR, Ser. Khim. Navuk, 106 (1975); Chem. Abstr., *84*, 4589f (1976).

XII. Miscellaneous Mercuration Reactions

There are a few mercuration reactions which do not fit neatly into any of the preceding chapters, but, nevertheless, seem important enough to mention. These are organized according to the element added to the olefin or alkyne alongside mercury, starting with oxygen and proceeding to sulfur, nitrogen, phosphorus and certain metals.

Several oxygen nucleophiles other than those already covered will add to alkenes or alkynes. For example, sulfuric acid in the presence of catalytic amounts of mercuric sulfate adds to acetylene (Eq. 1) [1]. Mercuric nitrate addition to dienes has also been observed (Eqs. 2, 3). The intramolecular cyclization of an allenic phosphonic acid has been shown to proceed predominantly by trans addition (Eq. 4) [4]. Oximes have also been

$$HC\equiv CH \;+\; H_2SO_4 \xrightarrow{\text{cat. } HgSO_4} H_2C=CHOSO_3H \tag{1}$$

$$\tag{2) [2}$$

$$\tag{3) [3}$$

$$\tag{4}$$

77 %

reported to undergo intramolecular mercuration, although the products have not been well characterized (Eqs. 5, 6) [5, 6].

$$\tag{5}$$

XII. Miscellaneous Mercuration Reactions

$$(CH_3)_2C=CH(CH_2)_2\overset{\overset{\displaystyle NOH}{\|}}{C}CH_3 \longrightarrow \left[\text{...}\right]_n \cdot HgI_2 \qquad (6)$$

$$n = 1, 2$$

$$\qquad \xrightarrow[NaCl]{Hg(OAc)_2} \qquad (7)$$

$$RC\equiv CR \xrightarrow[-SCN]{HgCl_2} \qquad (8)$$

There appear to be only two examples of thiomercuration. We have observed the intramolecular formation of mercurated benzothiophenes upon reaction of mercuric acetate and 2-(1-pentynyl)thioanisole (Eq. 7) [7]. The thiocyanate group has also been added to alkynes (Eq. 8) [8]. This reaction is accompanied by nitrogen attack as well.

There appears to be one other nitrogen addition reaction not covered previously. The rearrangement of allylic trichloroacetimidates is markedly catalyzed by mercuric trifluoroacetate or nitrate (Eq. 9) [9, 10]. This 1,3-transposition of hydroxyl and amino functions presumably proceeds via cyclic iminomercuration as shown. This reaction is only synthetically useful, however, for imidates derived from 2-alken-1-ols.

Mercury and phosphorus are also capable of adding across double or triple bonds. Ketene reacts with triisopropylphosphite and mercuric chloride or bromide to give organomercurials which appear to arise by initial mercury-phosphorus addition to the carbon—carbon double bond followed by a second ketene insertion (Eq. 10) [11, 12]. The same mercury-phosphite complexes have been observed to add to alkynyl ethers to give vinylmercurials (Eq. 11) [12, 13]. Two geometric isomers are evident in this reaction.

$$\xrightarrow{HgX_2} \qquad \xrightarrow{HX} \qquad (9)$$

$$(i\text{-PrO})_3P \cdot HgX_2 \; + \; H_2C=C=O \longrightarrow \left[(i\text{-PrO})_2\overset{\overset{\displaystyle O}{\|}}{P}-\overset{\overset{\displaystyle O}{\|}}{C}-CH_2HgX\right] \longrightarrow \qquad (10)$$

$$X = Cl, Br$$

$$(i\text{-PrO})_2\overset{\overset{\displaystyle O}{\|}}{P}-\overset{\overset{\displaystyle O_2CCH_2HgX}{|}}{C}=CH_2$$

$$(i\text{-PrO})_3\text{P}\cdot\text{HgCl}_2 \ +\ \text{RC}\equiv\text{COR'} \ \longrightarrow \ \underset{R}{\overset{\text{ClHg}}{}}\text{C}=\text{C}\overset{\text{OR'}}{\underset{\overset{\|}{\text{O}}}{\text{P}(\text{O}-i\text{-Pr})_2}} \tag{11}$$

Metallomercuration of double and triple bonds is a known reaction, but few examples presently exist. *Bis*(trimethylsilyl)mercury apparently adds to highly halogenated alkenes to give 1:1 adducts which have not been well characterized due to their rapid decomposition (Eq. 12) [14]. Certain platinum-alkyne complexes react with mercuric halides to afford products of platinum-mercury addition across the acetylene triple bond (Eq. 13) [15].

$$\text{F}_2\text{C}=\text{CFX} \ +\ \text{Hg}\big[\text{Si}(\text{CH}_3)_3\big]_2 \ \longrightarrow \ (\text{CH}_3)_3\text{SiCF}_2\text{CFXHgSi}(\text{CH}_3)_3 \tag{12}$$
$$X = \text{Cl}, \text{CF}_3 \qquad\qquad X=\text{Cl}, \text{CF}_3$$
$$+\ (\text{CH}_3)_3\text{SiHgCF}_2\text{CFXSi}(\text{CH}_3)_3$$
$$X=\text{Cl}$$

$$\text{Pt}(\text{F}_3\text{CC}\equiv\text{CCF}_3)\text{L}_2 \ +\ \text{HgX}_2 \ \longrightarrow \ \text{PtX}\big[\text{C}(\text{CF}_3)=\text{C}(\text{HgX})\text{CF}_3\big]\text{L}_2 \tag{13}$$
$$\text{L}=\text{PPh}_3 \ \text{or} \ \text{MePPh}_2 \quad X=\text{Cl}, \text{Br}$$

One can fairly safely assume that a number of other variations of the solvomercuration reaction will be reported in the years just ahead due to the broad synthetic utility of the resulting organomercurials.

References

1. Plauson, H.: U.S. Patent 1,436,288 (1921?); Chem. Abstr., *17*, 564 (1923).
2. Bach, R. D., Mazur, U., Brummel, R. N., Lin, L.-H.: J. Am. Chem. Soc., *93*, 7120 (1971).
3. Sokolov, V. I., Reutov, O. A.: Izv. Akad. Nauk SSSR, Ser. Khim., 222 (1968); Bull. Acad. Sci. USSR, Div. Chem. Sci., 225 (1968).
4. Macomber, R. S.: J. Am. Chem. Soc., *99*, 3072 (1977).
5. Sand, J.: J. Liebig Ann. Chem., *329*, 135 (1903).
6. Sand, J., Singer, F.: J. Liebig Ann. Chem., *329*, 166 (1903).
7. Larock, R. C., Harrison, L. W.: J. Am. Chem. Soc., *106*, 4218 (1984).
8. Blaukat, U., Neumann, W. P.: J. Organometal. Chem., *49*, 323 (1973).
9. Overman, L. E.: J. Am. Chem. Soc., *96*, 597 (1974).
10. Overman, L. E.: J. Am. Chem. Soc., *98*, 2901 (1976).
11. Kazankova, M. A., Trostyanskaya, I. G., Satina, T. Ya., Lutsenko, I. F.: Zh. Obshch. Khim., *48*, 301 (1978); J. Gen. Chem. USSR, *48*, 267 (1978).
12. Kazankova, M. A., Satina, T. Ya., Lutsenko, I. F.: Zh. Obshch. Khim., *49*, 2414 (1979); J. Gen. Chem. USSR, *49*, 2131 (1979).

13. Kazankova, M. A., Satina, T. Ya., Lutsenko, I. F.: Zh. Obshch. Khim., *45*, 1195 (1975); J. Gen. Chem. USSR, *45*, 1178 (1975).
14. Fields, R., Haszeldine, R. N., Hubbard, A. F.: J. Chem. Soc. D, Chem. Commun., 647 (1970).
15. Kemmitt, R. D. W., Kimura, B. Y., Littlecott,, G. W.: J. Chem. Soc., Dalton, 636 (1973).

XIII. Subject Index